Catastrophe Theory for Scientists and Engineers

Catastrophe Theory for Scientists and Engineers

ROBERT GILMORE

Drexel University
Philadelphia, Pennsylvania

Dover Publications, Inc.
New York

Published in Canada by General Publishing Company, Ltd., 30 Lesmill Road, Don Mills, Toronto, Ontario.
Published in the United Kingdom by Constable and Company, Ltd., 3 The Lanchesters, 162–164 Fulham Palace Road, London W6 9ER.

This Dover edition, first published in 1993, is an unabridged, unaltered republication of the work first published by John Wiley & Sons ("A Wiley-Interscience Publication"), New York, 1981.

Manufactured in the United States of America
Dover Publications, Inc., 31 East 2nd Street, Mineola, N.Y. 11501

Library of Congress Cataloging-in-Publication Data

Gilmore, Robert, 1941–
Catastrophe theory for scientists and engineers / Robert Gilmore.
—Dover ed.
p. cm.
Originally published: New York : Wiley, c1981.
Includes bibliographical references and index.
ISBN 0-486-67539-4 (pbk.)
1. Catastrophes (Mathematics) 2. Mathematical physics. 3. Science—Mathematics. 4. Engineering mathematics. I. Title.
QC20.C36G55 1993
514′.74—dc20 92-43547
CIP

Prologue

The traditional approach for determining the properties of equations has been to exhibit a complete set of solutions, and then to study the properties of these solutions. Thus spoke Legendre, Laguerre, Bessel, Hermite, Gegenbauer, Jacobi, and Chebyshev, for the study of linear second-order ordinary differential equations is not excessively difficult. But what of more complicated equations—nonlinear or of degree exceeding 2? And is a complete set of eigenfunctions really that useful for determining the qualitative properties of an equation which models some system? The properties of sines, cosines, and exponentials are well known, but the studies of Fourier series and transforms is somewhat delicate.

The American astronomer G. W. Hill was among the first to realize the limitations and inadequacies of the traditional approach, and that the classical special functions of mathematical physics had come from a mined out lode. The brilliant French savant H. Poincaré further realized that in many instances it is only information of a qualitative nature, or only limited quantitative information, which is the ultimate goal of the study of some system of equations. In such cases a full spectrum of solutions to an equation, obtained by much hard work if at all, may be more a hindrance than a help in understanding the qualitative properties of the equation or system of equations.

Just one century ago, in 1880, Poincaré began laying the groundwork to the modern approach for determining the qualitative properties of the solutions of ordinary differential equations. The ingredients in this formulation involve three concepts: structural stability, dynamical stability, and critical sets. Poincaré was particularly interested in studying how the qualitative properties of a system change as the parameters describing the system change.

This work was apparently far ahead of its time. Poincaré himself was astounded by the pathologies he had created. Among his contemporaries, only A. M. Lyapunov took up the program, which Poincaré had outlined, with his study of critical solutions.

After Lyapunov, work in bifurcation theory lapsed, although topology took on a life of its own and flourished, independent of Poincaré's dynamical systems program. It was not until the 1930s that the Russian mathematicians A. A. Andronov and L. S. Pontryagin again took up Poincaré's challenge by developing the concept of structural stability.

Work in this area accelerated during the 1950s and 1960s. Differential topology and qualitative dynamics were synthesized into a topological theory of dynamical systems by S. Smale in 1967. This work represents a formulation, in modern mathematical language, of the program initiated by Poincaré. As such it is a watershed in, but not a culmination of, the development of qualitative dynamical systems theory.

Along other mathematical lines, M. Morse had established (around 1930) the structure for canonical forms of a function near an isolated critical point, and H. Whitney described canonical forms for mappings at singular points. In the 1950s R. Thom introduced the concept of transversality as a mechanism for discussing structural stability, and then used this tool to describe canonical forms for certain singularities of mappings $f: \mathbb{R}^n \to \mathbb{R}^1$ (functions) which he called "catastrophes."

Thom's Catastrophe Theory lies at the intersection of the line of thought initiated by Poincaré and the sequence (in differential calculus) of canonical forms for functions and mappings, including the Implicit Function Theorem and the results of Morse and Whitney. Specifically, Thom's Elementary Catastrophe Theory is Poincaré's program applied to the equilibria of dynamical systems which are derivable from a potential function.

I came into this field quite by accident. Several years ago I was studying a laser physics problem which exhibited a bifurcation. It took me three weeks to recognize this fact. As a physicist I had seen a wide variety of linear problems, but had had virtually no exposure to nonlinear problems, their methods, or their techniques. In the process of acquiring these tools, I ran across some cryptic references to a new branch of mathematics called "Catastrophe Theory." Unfortunately the writings on this subject proved more cryptic than the references. Through a chain of friendly associations I was led finally to Tim Poston, who removed the lids from my eyes with explanations of elegant simplicity. I am indebted to him for showing me that Thom's elementary catastrophes belong to an accessible branch of mathematics rather than an inaccessible branch of philosophy.

This book is divided into four parts. Parts 1 and 4 describe the mathematics of Catastrophe Theory. Part 1 is written at an elementary level, requiring nothing more than a knowledge of differential calculus. Part 4 goes into enough mathematical detail that Thom's Theorem can at least be stated. No proof is given here. (For a proof, see Zeeman's collection.) It was at the level of Part 1 that I came to terms with Catastrophe Theory. This part has been "impedance-matched" to the needs of a typical scientist or engineer.

The last two chapters of Part 1 lay the groundwork for applications of Elementary Catastrophe Theory. Seven representative applications are drawn from the "hard" sciences. They by no means represent a complete spectrum of applications in mathematics, physics, chemistry, and the engineering disciplines. A selection of applications in the "soft" sciences can be found in Zeeman's book.

Part 3 deals with the ways in which Elementary Catastrophe Theory can be extended. This includes a discussion of catastrophe functions more complicated than those discussed by Thom, as well as a description of the qualitative properties of dynamical systems.

I am indebted to the many persons who have helped make this book possible. These include Tim Poston, who taught me the essentials and made his work available to me prior to publication. Ian Stewart and N. Bazley provided useful information about Catastrophe Theory and nonlinear mathematics.

This book grew out of a chapter that I had written for a book on mathematical physics. I am indebted to S. R. Deans, my coauthor in this venture, for suggesting that this material should stand alone. He also helped with some of the numerical computations, as did E. Ratigan, P. Draper, N. A. Pope, and J. A. Ross. L. Miller and his staff have done a superb job in the preparation of the figures. Finally, I thank my wife, Claire, for suffering through the preparation of this work.

ROBERT GILMORE

Arlington, Virginia
March 1981

Contents

PART 1

Catastrophe Theory for Scientists and Engineers

PART 1

Catastrophe Theory for Pedestrians

In which we present Elementary Catastrophe Theory from a very elementary point of view.

CHAPTER 1

The Program of Catastrophe Theory

Catastrophe Theory is a mathematical program in much the same way that Felix Klein's Erlangen Program is a mathematical program. The Erlangen Program attempts to classify geometries by classifying the transformation group which leaves the theorems of the geometry invariant. Catastrophe Theory attempts to study how the qualitative nature of the solutions of equations depends on the parameters that appear in the equations.

To make concrete the program which is Catastrophe Theory, we look for solutions $\psi_1(t, x; c_\alpha)$, $\psi_2(t, x; c_\alpha)$, ... for a system of n equations defined over a space $\mathbb{R}^N$ whose coordinates are $x = (x_1, x_2, \ldots, x_N)$:

$$F_i\left(\psi_j; c_\alpha; t, \frac{d\psi_j}{dt}, \frac{d^2\psi_j}{dt^2}, \ldots; x_l; \frac{\partial\psi_j}{\partial x_l}, \frac{\partial^2\psi_j}{\partial x_l\,\partial x_m}, \ldots; \int dx_l, \ldots\right) = 0 \quad (1.1)$$

where

$$1 \le i, \quad \le n$$
$$1 \le l, m \le N$$
$$1 \le \alpha \quad \le k$$

The variables x_l and t may conveniently be regarded as space and time coordinates. Since the solutions ψ_j describe the state of some system, they will be called **state variables.** The equations $F_i = 0$ are assumed to depend on k parameters c_α (Reynold's number, fine structure constant, magnetic field, and so on). Since the parameters c_α may control the qualitative properties of the solutions ψ_j, the c_α are called **control parameters.**

The problem of determining the solutions of (1.1), let alone studying how these solutions change as the control parameters c_α change, is formidable. This problem can be made increasingly more tractable by a sequence of simplifying assumptions.

1. We assume that (1.1), which may in general be an integro-differential equation (or much worse), in fact contains no integrals. In fact, we assume that (1.1) is simply a set of (nonlinear) partial differential equations. The program of Catastrophe Theory is still too difficult to carry out for such a system of equations.

2. We simplify further by assuming that the resulting system of equations involves no space derivatives of any order:

$$F_i = F_i\left(\psi_j; c_\alpha; t, \frac{d\psi_j}{dt}, \frac{d^2\psi_j}{dt^2}, \ldots; x_l; \underline{\hspace{4em}}; \underline{\hspace{2em}}\right) \tag{1.2}$$

This system of equations is still too difficult for anything to be said.

3. We simplify yet again by assuming that the resulting system of equations is independent of the space coordinates x_l entirely:

$$F_i = F_i\left(\psi_j; c_\alpha; t, \frac{d\psi_j}{dt}, \frac{d^2\psi_j}{dt^2}, \ldots; \underline{\hspace{2em}}; \underline{\hspace{4em}}; \underline{\hspace{2em}}\right) \tag{1.3}$$

Still, we cannot say much in general about the solutions of such equations.

4. We further simplify by assuming that no higher time derivatives than the first appear in (1.3). Moreover, we assume that the time derivatives occur in a special ("canonical") way in the simplified function:

$$F_i = \frac{d\psi_i}{dt} - f_i(\psi_j; c_\alpha; t) \tag{1.4}$$

Systems of equations of this type ($F_i = 0$) are called **dynamical systems**. Again, they are very difficult to study.

5. We continue to simplify by assuming that the functions f_i in (1.4) are independent of time. The resulting system of equations

$$F_i = \frac{d\psi_i}{dt} - f_i(\psi_j; c_\alpha; \underline{\hspace{2em}}) = 0 \tag{1.5}$$

is called an **autonomous dynamical system**. A few useful and powerful statements can be made about autonomous dynamical systems which depend on a small number of control parameters ($k \leq 4$).

6. We complete this long series of simplifying assumptions by observing that the functions f_i look in many ways like the components of a force. Marvelous simplifications result in classical mechanics when the force can be derived from a potential. When all the functions f_i can be derived as the negative gradient (with respect to the ψ_i) of some potential function $V(\psi_j; c_\alpha)$,

$$f_i = -\frac{\partial V(\psi_j; c_\alpha)}{\partial \psi_i}$$

$$F_i = \frac{d\psi_i}{dt} + \frac{\partial V(\psi_j; c_\alpha)}{\partial \psi_i} = 0 \tag{1.6}$$

the resulting system is called a **gradient system** ($\dot{\boldsymbol{\psi}} = -\boldsymbol{\nabla}_\psi V$). A great deal can be said about such systems.

7. Of particular interest are the equilibria $d\psi_i/dt = 0$ of dynamical and gradient systems. The equilibria $\psi_j(c_\alpha)$ of a gradient system are defined by the equation

$$\frac{\partial V(\psi_j; c_\alpha)}{\partial \psi_i} = 0 \tag{1.7}$$

This equation may have no solutions $[V(\psi) = \psi]$, one solution $[V(\psi) = \psi^2]$, or more than one solution $[V(\psi; c) = \psi^4 + c\psi^2$, one solution for $c > 0$, three for $c < 0]$. A great many useful and powerful statements can be made about the equilibria of gradient systems, and how these equilibria depend on the control parameters c_α.

Elementary Catastrophe Theory is the study of how the equilibria $\psi_j(c_\alpha)$ *of* $V(\psi_j; c_\alpha)$ *change as the control parameters* c_α *change.*

The foregoing considerations are summarized in Table 1.1.

Table 1.1 Where Elementary Catastrophe Theory Lives

Simplification Assumption	Structural Form of Equations $F_i = 0$	Remark
0	$F_i\left(\psi_j; c_\alpha; t, \frac{d\psi_j}{dt}, \frac{d^2\psi_j}{dt^2}, \ldots; x_l; \frac{\partial\psi_j}{\partial x_l}, \frac{\partial^2\psi_j}{\partial x_l\,\partial x_m}, \ldots; \int dx_l, \ldots\right)$	
1	$F_i\left(\psi_j; c_\alpha; t, \frac{d\psi_j}{dt}, \frac{d^2\psi_j}{dt^2}, \ldots; x_l; \frac{\partial\psi_j}{\partial x_l}, \frac{\partial^2\psi_j}{\partial x_l\,\partial x_m}, \ldots; \text{———}\right)$	
2	$F_i\left(\psi_j; c_\alpha; t, \frac{d\psi_j}{dt}, \frac{d^2\psi_j}{dt^2}, \ldots; x_l; \text{————}; \text{———}\right)$	
3	$F_i\left(\psi_j; c_\alpha; t, \frac{d\psi_j}{dt}, \frac{d^2\psi_j}{dt^2}, \ldots; -; \text{————}; \text{———}\right)$	
4	$F_i = \frac{d\psi_i}{dt} - f_i(\psi_j; c_\alpha; t)$	Dynamical system
5	$F_i = \frac{d\psi_i}{dt} - f_i(\psi_j; c_\alpha; -)$	Autonomous dynamical system
6	$F_i = \frac{d\psi_i}{dt} + \frac{\partial V(\psi_j; c_\alpha)}{\partial \psi_i}$	Gradient system
7	$0 = 0 + \frac{\partial V(\psi_j; c_\alpha)}{\partial \psi_i}$	Equilibria of gradient systems

CHAPTER 2

The Local Character of Potentials

The state variables ψ_j which appear in the potential $V(\psi_j; c_\alpha)$ of (1.7) are in reality generalized coordinates for the system under consideration. In keeping with normal practice we shall make a change of notation, replacing ψ_j by x_j with the understanding that x_j generally has nothing to do with the space coordinates introduced in (1.1). In the case of a structureless, spinless point particle at rest in the presence of a potential $V(x; c)$, $x \in \mathbb{R}^n$, $c \in \mathbb{R}^k$, the particle state is completely described by the particle position, so that in this particular case the solutions $x_j(c_\alpha)$ of (1.7) really are the coordinates of the particle equilibria. More generally, however, the variables x_j may be things like Fourier coefficients or physical order parameters.

In this chapter we consider the local properties both of potentials $V(x)$ and of families of potentials $V(x; c)$, which may be considered as mappings $\mathbb{R}^n \to \mathbb{R}^1$ and $\mathbb{R}^n \otimes \mathbb{R}^k \to \mathbb{R}^1$, respectively. The local properties, both of potentials and of families of potentials, are determined by a sequence of theorems of functional analysis which involve an increasing intimacy with topology. These theorems, the Implicit Function Theorem of advanced calculus, the Morse Lemma, and the latest, the Thom Theorem, are summarized in Table 2.1.

1. The Implicit Function Form

The generalized force acting on a system which is described by a potential is the negative gradient of the potential. If the gradient is nonzero at a particular

Table 2.1 Three Important Results of Functional Analysis

Conditions	$V \doteq$	Equation	Theorem/Lemma	Reference and Date
1. $\nabla V \neq 0$	y_1	(2.1)	Implicit Function Theorem	
2. $\nabla V = 0$ but $\det V_{ij} \neq 0$	$\sum_{i=1}^{n} \lambda_i y_i^2$	(2.2a)	Morse Lemma	2, 1931
	$M_i^n(\tilde{y})$	(2.2b)		
3. $\nabla V = 0$ and $\det V_{ij} = 0$	$f_{NM}(y_1, \ldots, y_l) + M_i^{n-l}$	(2.3a)	Splitting Lemma	3, 1969
V "general"	$CG(l) + M_i^{n-l}$	(2.3b)	Thom Theorem	4, early 1960s
$k \leq 5$	$\mathrm{Cat}(l, k) + M_i^{n-l}$	(2.4)	Thom Theorem	4, early 1960s

[a] Mathematicians did not become overly concerned with rigorous proofs of theorems until the late 19th century. Knowledge of this theorem predates the era of rigor. In fact, it even predates the calculus, as it was certainly known to Galileo.[1]

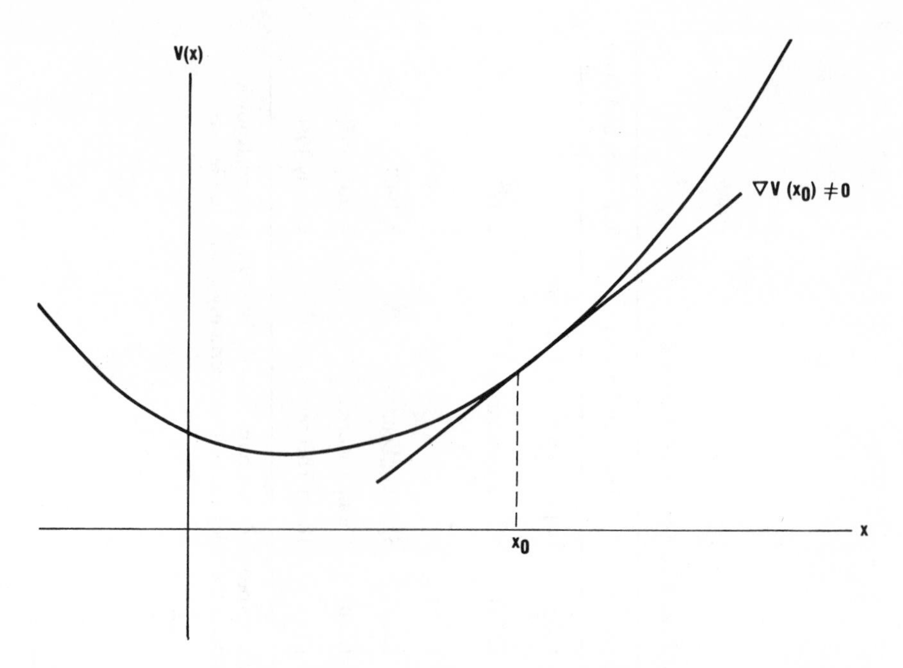

Figure 2.1 At a point x_0 where $\nabla V \neq 0$, a smoothe change of coordinates can be found which transforms V to a local linear function: $V \to a + (y - y_0)b$.

point in state space, then the corresponding force at this point is nonzero: $F = -\nabla V \neq 0$ (Fig. 2.1). It is then possible to choose a new coordinate system in the neighborhood of this point so that the force has only one nonvanishing component—say in the 1 direction. This is made precise by the Implicit Function Theorem,[1] which guarantees the existence of a smooth (derivatives of arbitrarily high order exist) change of variables:

$$\begin{aligned} y_1 &= y_1(x_1, x_2, \ldots, x_n) \\ y_2 &= y_2(x_1, x_2, \ldots, x_n) \\ &\vdots \\ y_n &= y_n(x_1, x_2, \ldots, x_n) \end{aligned} \tag{2.0}$$

so that, in the new coordinate system,

$$V \doteq y_1 \quad (+ \text{ a constant}) \tag{2.1}$$

Here $\doteq$ means "... is equal, after a smooth change of variables, to ... ." The constant term is not important when dealing with the local properties of a potential, and will usually be dropped from consideration. The constant may also be removed by shifting the origin in $\mathbb{R}^1$.

2. The Morse Forms

If the physical system is in equilibrium (stable or unstable) at a particular point in the state space, $\nabla V = 0$. The stability properties of the equilibrium may be determined from the eigenvalues of the stability matrix or the Hessian matrix $V_{ij} = \partial^2 V/\partial x_i \, \partial x_j$. The conditions required for the validity of the Implicit Function Theorem ($\nabla V \neq 0$) are no longer satisfied and the canonical form (2.1) is no longer valid. However, if det $V_{ij} \neq 0$, a theorem of Morse[2] guarantees the existence of a smooth change of variables so that the potential can be written locally as a quadratic form:

$$V \doteq \sum_{i=1}^{n} \lambda_i y_i^2 \tag{2.2a}$$

Here λ_i are the eigenvalues of the stability matrix V_{ij}, evaluated at the equilibrium. By absorbing a "length scale" into the new coordinates according to $\tilde{y}_i = |\lambda_i|^{1/2} y_i$, the quadratic form (2.2a) can be written in the Morse canonical form:

$$V \doteq -\tilde{y}_1^2 - \cdots - \tilde{y}_i^2 + \tilde{y}_{i+1}^2 + \cdots + \tilde{y}_n^2 = M_i^n(\tilde{y}) \tag{2.2b}$$

The function $M_i^n(\tilde{y})$ is called a **Morse *i*-saddle.** Only Morse 0-saddles have a local minimum at the equilibrium, so that only the 0-saddles are locally stable. The three Morse functions on $\mathbb{R}^2$ are illustrated in Fig. 2.2.

Definition. The **equilibrium points,** or **critical points,** of a smooth function $V(x_1, \ldots, x_n)$ are the points at which $\nabla V = 0$. The critical points at which det $V_{ij} \neq 0$ are called **isolated, nondegenerate,** or **Morse critical points.**

3. The Thom Forms

Definition. The critical points of $V(x_1, \ldots, x_n)$ at which det $V_{ij} = 0$ are called **nonisolated, degenerate,** or **non-Morse critical points.**

If the potential depends on one or more control parameters $c_1, \ldots, c_k$, the stability matrix V_{ij} and its eigenvalues $\lambda_1, \lambda_2, \ldots, \lambda_n$ also depend on these control parameters. Then one or more of the eigenvalues $\lambda_i(c)$ may assume the value zero for certain values of the control parameters c. When this happens, det $V_{ij} = 0$, the conditions required for the validity of the Morse Lemma ($\nabla V = 0$, det $V_{ij} \neq 0$) are no longer satisfied, and the canonical form (2.2) for a potential at an equilibrium is no longer valid. However, it is still possible to find a canonical form for the potential at a non-Morse critical point. If l eigenvalues $\lambda_1(c), \ldots, \lambda_l(c)$ vanish at $c = c^0$, then the Thom Splitting Lemma[3] may be used to split the potential into a non-Morse part and a Morse part:

$$V(x; c) \doteq f_{NM}(y_1(x; c), \ldots, y_l(x; c); c) + \sum_{j=l+1}^{n} \lambda_j(c)(y_j(x))^2 \tag{2.3a}$$

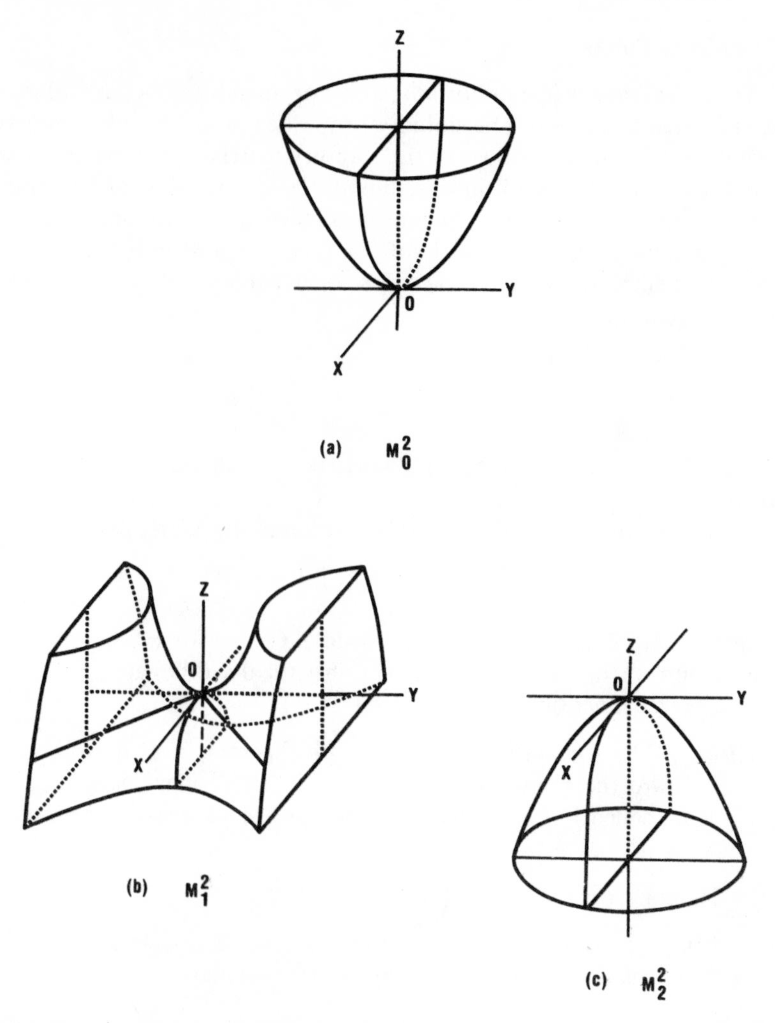

Figure 2.2 The three Morse saddles on $\mathbb{R}^2$ are: (*a*) M_0^2, which has a local minimum; (*b*) M_1^2, which has a saddle shape; (*c*) M_2^2, which has a local maximum.

The l "bad" coordinates $y_1(x; c), \ldots, y_l(x; c)$, associated with the l vanishing eigenvalues $\lambda_1(c), \ldots, \lambda_l(c)$, are smooth functions of both the n state variables $x_1, \ldots, x_n$ and the k control parameters $c_1, \ldots, c_k$. The "good" coordinates $y_{l+1}(x), \ldots, y_n(x)$, associated with the nonvanishing eigenvalues $\lambda_{l+1}(c), \ldots,$ $\lambda_n(c)$, are smooth functions only of the original state variables x. At $(x^0; c^0)$ the stability matrix $\partial^2 f_{NM}/\partial y_i\, \partial y_j$ $(1 \leq i, j \leq l)$ vanishes (all matrix elements are zero) while the $(n - l) \times (n - l)$ stability matrix of the Morse function in (2.3a) is nonsingular. Under suitable conditions [$k \leq 5$ and no special or symmetry conditions are present in the family of potentials $V(x_1, \ldots, x_n; c_1, \ldots, c_k)$] a

Table 2.2 Elementary Catastrophes of Thom

Name	k	Germ	Perturbation
A_2	1	x^3	a_1x
$A_{\pm 3}$	2	$\pm x^4$	$a_1x + a_2x^2$
A_4	3	x^5	$a_1x + a_2x^2 + a_3x^3$
$A_{\pm 5}$	4	$\pm x^6$	$a_1x + a_2x^2 + a_3x^3 + a_4x^4$
A_6	5	x^7	$a_1x + a_2x^2 + a_3x^3 + a_4x^4 + a_5x^5$
D_{-4}	3	$x^2y - y^3$	$a_1x + a_2y + a_3y^2$
D_{+4}	3	$x^2y + y^3$	$a_1x + a_2y + a_3y^2$
D_5	4	$x^2y + y^4$	$a_1x + a_2y + a_3x^2 + a_4y^2$
D_{-6}	5	$x^2y - y^5$	$a_1x + a_2y + a_3x^2 + a_4y^2 + a_5y^3$
D_{+6}	5	$x^2y + y^5$	$a_1x + a_2y + a_3x^2 + a_4y^2 + a_5y^3$
$E_{\pm 6}$	5	$x^3 \pm y^4$	$a_1x + a_2y + a_3xy + a_4y^2 + a_5xy^2$

theorem of Thom[4] guarantees the existence of a smooth change of variables so that the potential can be written in the canonical form

$$V \doteq CG(l) + \sum_{j=l+1}^{n} \lambda_j y_j^2 \tag{2.3b}$$

The function $CG(l)$ is called a **catastrophe germ**. In Table 2.2 we list all the canonical catastrophe germs for $k \leq 5$. These correspond to only one ($l = 1$) or two ($l = 2$) vanishing eigenvalues.

4. Neighborhoods of Validity

The canonical forms (2.3a) and (2.3b) appear analogous. In fact, they are not.

The Thom Splitting Lemma, which is used in the construction of the canonical form (2.3a), is based on nothing more complicated than advanced calculus, and in particular the Implicit Function Theorem. The decomposition (2.3a) is valid for V in an open neighborhood of $(x^0; c^0)$ in $\mathbb{R}^n \otimes \mathbb{R}^k$, where x^0 is a non-Morse critical point for value $c = c^0$ of the control parameters. However, the Splitting Lemma says nothing about the function $f_{NM}(y_1, \ldots, y_l)$ except that it has no first- or second-degree terms in $(y_i - y_i^0)$ in its Taylor series expansion at the degenerate critical point when $c = c^0$.

The Thom Theorem, which is used in the construction of the canonical form (2.3b), is very deep and has required and stimulated the development of new branches of mathematics. The decomposition in (2.3b) is valid for V in an open neighborhood of x^0 in $\mathbb{R}^n$ when c has the fixed value c^0. The Thom result (2.3b) does list specific canonical forms for the ways in which the bad non-Morse variables $y_1, \ldots, y_l$ may occur at a degenerate critical point, after a suitable change of variables.

Thus the results (2.3a) and (2.3b) are not analogous. The former decomposition is valid in a neighborhood of $(x^0; c^0)$ in $\mathbb{R}^n \otimes \mathbb{R}^k$, but does not give specific forms for f_{NM}; the latter is only valid in a neighborhood of x^0 in $\mathbb{R}^n$, but gives specific forms, called catastrophe germs, for f_{NM}.

What is desirable is a canonical decomposition for a potential with the strengths of both decompositions (2.3a) and (2.3b) and the weaknesses of neither. This result has also been provided by Thom. If x^0 is a non-Morse critical point of $V(x; c)$ for $c = c^0$, then in an open neighborhood of $(x^0; c^0)$ in $\mathbb{R}^n \otimes \mathbb{R}^k$

$$V \doteq \text{Cat}(l, k) + \sum_{j=l+1}^{n} \lambda_j(c) y_j^2 \tag{2.4}$$

The function $\text{Cat}(l, k)$ is called a **catastrophe function**, or simply a **catastrophe**. Here l is the dimension of the null space of V_{ij} at a non-Morse critical point, and k is the number of control parameters. These functions are described in the following section.

5. Catastrophe Functions

Table 2.2 lists canonical forms for 1- and 2-(state) variable catastrophes. This table of elementary catastrophes, first described by Thom, actually contains two functions under each entry, the catastrophe germ $CG(l)$ and the perturbation $\text{Pert}(l, k)$. The catastrophe function $\text{Cat}(l, k)$ is the sum of the two functions under each entry on this list.

Definition.

$$\text{Cat}(l, k) = CG(l) + \text{Pert}(l, k) \tag{2.5}$$

The catastrophe germ $CG(l)$ is the l-variable non-Morse function which appears in the canonical form (2.3b) in the neighborhood of a non-Morse critical point x^0 (or y^0) *at* the fixed value $c = c^0$. The catastrophe function $\text{Cat}(l, k)$ is the l-(state) variable, k-(control) parameter function which appears in the canonical form (2.4) in the neighborhood of a non-Morse critical point x^0 *around* the value c^0. The catastrophe function $\text{Cat}(l, k)$ reduces to the catastrophe germ $CG(l)$ when the physical control parameters c_α equal c_α^0 in $\mathbb{R}^k$ or the mathematical control parameters $a_\alpha = 0$ in $\mathbb{R}^k$. In general, the mathematical control parameters $a_1, \ldots, a_k$ are related to the physical control parameters $c_1, \ldots, c_k$ through a smooth transformation with nonsingular Jacobian

$$\det \left. \frac{\partial a_\alpha(c)}{\partial c_\beta} \right|_{c^0} \neq 0 \qquad 1 \leq \alpha, \beta \leq k \tag{2.6}$$

Example 1. The catastrophe function labeled D_{-4} in Table 2.2 is

$$D_{-4}\colon\ f(x, y; a_1, a_2, a_3) = (x^2 y - y^3) + (a_1 x + a_2 y + a_3 y^2) \tag{2.7}$$

where for convenience we have taken $x_1 = x$, $x_2 = y$.

Example 2. Suppose a potential $V(x; c)$ depends on 10 state variables $x_1, \ldots, x_{10}$ and three control parameters c_1, c_2, c_3. Suppose also that a non-Morse critical point occurs at $(x^0; c^0)$ and that at this point the 10×10 matrix V_{ij} has nullity 2 (i.e., 2 eigenvalues are zero), with three negative and five positive eigenvalues. What is the local character of this potential at $(x^0; c^0)$? According to (2.4),

$$V \doteq \mathrm{Cat}(l, k) + f_M(y_3, \ldots, y_{10}) \tag{2.8}$$

Since $l = 2$, $k = 3$, we determine immediately from Table 2.2 that Cat(2, 3) $= D_{\pm 4}$, so that

$$V \doteq (y_1^2 y_2 \pm y_2^3) + (a_1 y_1 + a_2 y_2 + a_3 y_2^2) + \sum_{j=3}^{10} \lambda_j(c) y_j^2 \tag{2.9}$$

This canonical form is valid in an open neighborhood of $(x^0; c^0)$ in $\mathbb{R}^n \otimes \mathbb{R}^k$. In this form, $\lambda_j(c^0) = -1, j = 3, 4, 5$, and $\lambda_j(c^0) = +1, j = 6, 7, 8, 9, 10$.

6. Summary

The local character of a potential is described by a sequence of theorems of functional analysis requiring an increasing intimacy with topology.

When $\nabla V \neq 0$, the Implicit Function Theorem can be used to transform V into the canonical form (2.1).

When $\nabla V = 0$, the Implicit Function Theorem is not valid. But when det $V_{ij} \neq 0$, the Morse Lemma can be used to cast V into the canonical form (2.2).

When $\nabla V = 0$ and det $V_{ij} = 0$, the Morse Lemma is not valid. But if $V(x; c)$, $x \in \mathbb{R}^n$ and $c \in \mathbb{R}^k$, is a typical k-parameter family of potentials, a theorem of Thom can be used to cast V into the canonical form (2.3). For $k \leq 5$, Thom's theorem can be used to put V into the stronger canonical form (2.4).

The terms "critical point" and "equilibrium" or "stationary point" have different meanings in the literature of mathematics and some of the sciences, as indicated in Table 2.3. To avoid confusion, we shall adopt the mathematics terminology except where specifically stated otherwise.

Table 2.3 Notational Conventions

Property	Mathematics	Physics
$\nabla V \neq 0$	Noncritical point	Nonequilibrium point
$\nabla V = 0$, det $V_{ij} \neq 0$	Critical point (also Morse, or normal critical point)	Equilibrium point
$\nabla V = 0$, det $V_{ij} = 0$	Non-Morse critical point, degenerate critical point	Critical point

References

1. Galileo Galilei, *Dialogues Concerning Two New World Sciences* (transl. H. Crew and A. De Salvio), Evanston: Northwestern Univ. Press, 1950.
2. M. Morse, The Critical Points of a Function of n Variables, *Trans. Am. Math. Soc.* **33**, 72–91 (1931).
3. D. Gromoll and W. Meyer, On Differentiable Functions with Isolated Critical Points, *Topology* **8**, 361–370 (1969).
4. R. Thom, *Stabilité Structurelle et Morphogénèse*, New York: Benjamin, 1972; transl. *Structural Stability and Morphogenesis*, Reading: Benjamin, 1975.

CHAPTER 3

Change of Variables. 1 Canonical Forms

The canonical forms for potentials given in (2.1)–(2.4) are obtained after a smooth change of variables. In this chapter we will attempt to see, in a rather intuitive way, just why the canonical forms for potentials should have the forms (2.1) at noncritical points, (2.2) for Morse critical points, and the form (2.4) for non-Morse critical points in families of functions depending on fewer than six control parameters.

In brief, the qualitative properties of a potential at a point are governed by the earliest (lowest degree) terms of its Taylor series expansion about that point. When the function depends on control parameters, so also do its Taylor series coefficients. Certain choices of the control parameters may cause these leading terms to vanish. Then the qualitative properties of the function also change. A coordinate transformation can then be introduced to transform away the later (higher degree) terms in the Taylor series expansion. The terms which remain determine the qualitative properties of the function at a point.

These terms are called the **germ of the function**. The germ resides between the early terms which are killed off by the control parameters and the later terms which are killed off by a coordinate transformation. When the germ is linear, the Implicit Function Theorem is applicable (Sec. 2). When the germ is a non-degenerate quadratic form, the Morse Lemma is applicable (Sec. 3). When the germ contains a degenerate quadratic form, the Splitting Lemma is applicable (Sec. 4). Germs for non-Morse functions in $1, 2, \ldots, l$ "bad" state variables are discussed in the remaining sections (Secs. 5–9).

1. Change of Variables

In order to describe physical processes in $\mathbb{R}^n$, it is useful to set up a coordinate system $(x_1, x_2, \ldots, x_n)$. Any coordinate system will do. If another coordinate system $(x'_1, x'_2, \ldots, x'_n)$ is established, it is possible to transform back and forth between them:

$$\begin{aligned} x'_1 &= x'_1(x_1, x_2, \ldots, x_n) \\ x'_2 &= x'_2(x_1, x_2, \ldots, x_n) \\ &\vdots \\ x'_n &= x'_n(x_1, x_2, \ldots, x_n) \end{aligned} \tag{3.1}$$

at any point $p \in \mathbb{R}^n$ at which the Jacobian $\det|\partial x'_i/\partial x_j|_{p \in \mathbb{R}^n}$ is nonzero. At such a point the transformation (3.1) may in principle be inverted: $x_j = x_j(x'_i)$. This inverse transformation exists by the Inverse Function Theorem.

For many purposes it is useful to express the coordinates $x'_i = x'_i(x_j)$ in a Taylor series expansion:

$$x'_i = A_i + A_{i,j}x_j + A_{i,jk}x_jx_k + A_{i,jkl}x_jx_kx_l + \cdots \tag{3.2}$$

The transformation (3.2) is invertible at $(x_1, x_2, \ldots, x_n) = (0, 0, \ldots, 0)$ if $\det|A_{i,j}| \neq 0$. The factorial coefficients which are sometimes written explicitly in expansions of type (3.2) have been omitted for simplicity.

For computational purposes in this chapter and the next, it will often be convenient to perform the general nonlinear transformation (3.2) in a sequence of simpler steps:

$$x_i \to x'_i = x_i + A_i \tag{3.3ih}$$

$$x'_i \to x''_i = A_{i,j}x'_j \tag{3.3hl}$$

$$x''_i \to x'''_i = \delta_{ij}x''_j + A_{i,jk}x''_jx''_k + A_{i,jkl}x''_jx''_kx''_l + \cdots \tag{3.3nl}$$

The transformation (3.3ih) is an inhomogeneous linear transformation which amounts to a simple displacement of the origin and a rigid displacement of the coordinate lines. The transformation (3.3hl) is a homogeneous linear transformation. Such a transformation rotates and stetches the coordinate axes. It is nonsingular if $\det|A_{i,j}| \neq 0$. The transformation (3.3nl) is an "axis-preserving" nonlinear transformation. It is clearly nonlinear, but it preserves the coordinate axes at the origin because $\partial x'''_i/\partial x''_j = \delta_{ij}$. The inhomogeneous and homogeneous linear transformations are easily invertible, but the nonlinear transformation is not.

The general nonlinear transformation (3.2) can be obtained by successively performing the three transformations (3.3ih), (3.3hl), and (3.3nl). In this chapter we are particularly concerned with the qualitative properties of a potential at a point (which we take as the origin throughout), so we will have no need for the origin-displacing inhomogeneous transformations (3.3ih). However, these will be of use in the following chapter.

Since the primes can easily get out of hand, we will usually arrange things so that the initial coordinate system in a coordinate transformation is unprimed, and the new coordinate system is denoted by a single prime.

Once one or more coordinate systems have been established in $\mathbb{R}^n$, it is possible to represent a physical process or concept (e.g., a potential V) in functional form:

$$\begin{array}{ccc} & V & \\ \text{in } (x) \swarrow & & \searrow \text{in } (x') \\ f(x) & = & f'(x') \end{array} \tag{3.4}$$

The functional form depends on the coordinate system chosen. The equality $f'(x') = f(x)$ in (3.4) means

$$f'(x'(x)) = f(x) \tag{3.4'}$$

One physical process is represented by two different functions in two different coordinate systems.

Example. The potential on $\mathbb{R}^2$ shown as a saddle embedded in $\mathbb{R}^2 \otimes \mathbb{R}^1$ in Fig. 3.1 has the functional expressions

$$\begin{array}{ccc} & V \text{ (saddle)} & \\ \text{in } (x, y) \swarrow & & \searrow \text{in } (x', y') \\ -x^2 + y & = & 2x'y' \end{array}$$

The coordinate systems (x, y) and (x', y') are here related by a linear transformation:

$$\begin{aligned} x' &= x + y/\sqrt{2} \\ y' &= -x + y/\sqrt{2} \end{aligned}$$

Two *different* physical processes are represented by different functions in the *same* coordinate system (x, y):

$$\begin{array}{ccc} V & & V' \\ \downarrow \text{in } (x) & & \downarrow \text{in } (x) \\ f(x, y) & \neq & f'(x, y) \end{array}$$

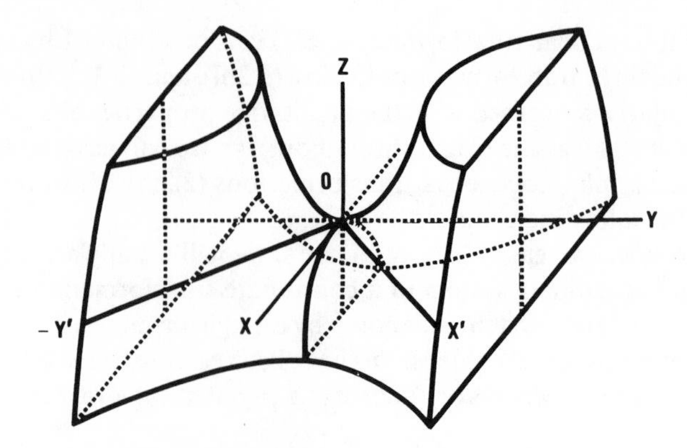

Figure 3.1 The representation of a saddle in $\mathbb{R}^2 \otimes \mathbb{R}^1$ depends on the coordinate system chosen.

For example, the two different saddles in Fig. 3.2 are represented by two different functions in the coordinate system (x, y):

$$V_1 = -x^2 + y^2$$
$$V_2 = 2xy$$

However, these two saddles are qualitatively the same because one can be "changed" into the other by rotating a coordinate system.

More generally, two different physical processes V, V' have different functional forms in one coordinate system: $f(x) \neq f'(x)$. But the two processes are *qualitatively* similar if we can find a smooth change of coordinates (3.1) so that the functional form for V', *expressed in terms of the new coordinates*, is equal to the functional form of V in the original coordinate system:

$$f'(x') = f(x) \tag{3.5}$$

In particular, this means

$$f'[x'(x)] = f(x) \tag{3.6a}$$
$$f'(x') = f[x(x')] \tag{3.6b}$$

If f and f' are qualitatively similar, then all qualitative properties of one are the same as the qualitative properties of the other.

These ideas are subtle and profound* and will be illustrated by applications and examples in the following sections.

* The difference between (3.4) and (3.5) is the same as the difference between the "passive" and the "active" interpretations of the application of Group Theory in Quantum Mechanics.

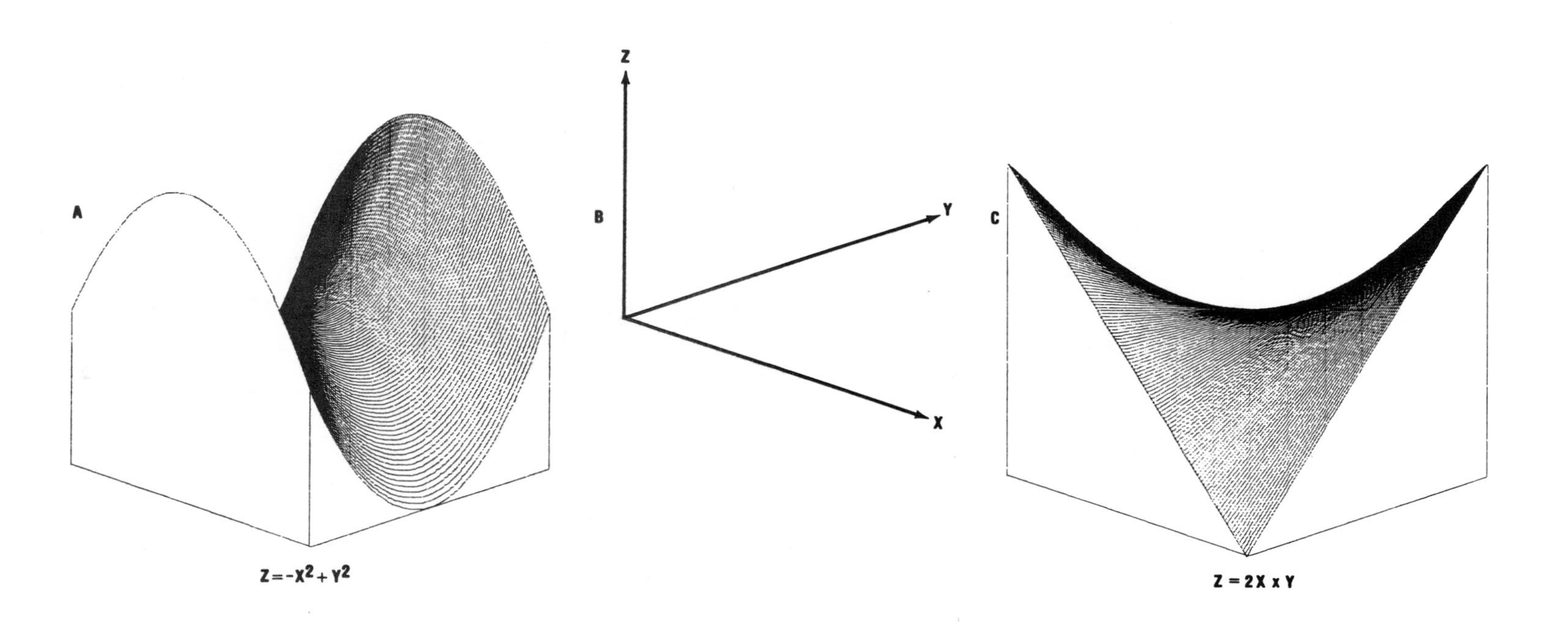

Figure 3.2 In a particular coordinate system different functions represent different surfaces. Here the two saddles $f(x, y) = -x^2 + y^2$ and $f'(x, y) = 2xy$ are shown in the same coordinate system. Both saddles are shown on the pedestal $(x, y) \in [-1, +1] \times [-1, +1]$ with $z \in [-2, +2]$.

2. Application 1: The Implicit Function Theorem

As a first example of the meaning of (3.5), we consider two physical processes which result in nonzero forces at $(x_1, x_2, \ldots, x_n) = (0, 0, \ldots, 0) \in \mathbb{R}^n$:

$$\begin{aligned} V &= f(x_1, x_2, \ldots, x_n), \qquad & \mathbf{F} &= -\nabla V \\ V' &= x_1 \qquad , & \mathbf{F}' &= -\hat{\mathbf{e}}_1 \end{aligned} \tag{3.7}$$

where $\hat{\mathbf{e}}_1$ is the unit vector in the "1" direction. Since nonzero force produces nonzero motion, any two potentials which produce nonzero forces at a point must be qualitatively the same at that point. To show that V and V' are qualitatively similar, we must find a smooth transformation so that (3.5) holds. One such transformation is given by

$$\begin{aligned} x_1' &= f(x_1, x_2, \ldots, x_n) \\ x_2' &= a_{21}x_1 + a_{22}x_2 + \cdots + a_{2n}x_n \\ &\vdots \\ x_n' &= a_{n1}x_1 + a_{n2}x_2 + \cdots + a_{nn}x_n \end{aligned} \tag{3.8}$$

The Jacobian of this transformation is nonsingular, provided

$$\det \left| \frac{\partial x_i'}{\partial x_j} \right|_{x=0} = \det \begin{bmatrix} \dfrac{\partial f}{\partial x_1} & \dfrac{\partial f}{\partial x_2} & \cdots & \dfrac{\partial f}{\partial x_n} \\ a_{21} & a_{22} & \cdots & a_{2n} \\ \vdots & \vdots & & \vdots \\ a_{n1} & a_{n2} & \cdots & a_{nn} \end{bmatrix} \neq 0 \tag{3.9}$$

If the top row is not identically zero, the real numbers a_{ij} can be chosen so that the Jacobian is nonzero. The transformation (3.8) is then invertible, and

$$V'(x') = x_1' = f(x_1, x_2, \ldots, x_n) = V(x) \tag{3.10}$$

Since $V'[x'(x)] = V(x)$ and $V'(x') = V[x(x')]$, we find

$$V[x(x')] = V'(x') = x_1' \tag{3.11}$$

Expression (3.11) is the same as expression (2.1).

3. Application 2: The Morse Lemma

When $\nabla f = 0$, the first row in the determinant (3.9) is zero, the transformation defined in (3.8) does not define a new coordinate system and is not invertible, the Implicit Function Theorem is not applicable, and $f(x) \neq x_1'$.

In this case we must look at the quadratic and higher degree terms in the Taylor series expansion of $f(x)$. If $\det f_{ij} \neq 0$, the Morse Lemma is applicable and the canonical form (2.2a) is valid. To see why, we look initially at a function of two variables with no constant or linear terms:

$$f(x, y) = ax^2 + bxy + cy^2 + \text{third-degree terms} + \cdots \tag{3.12}$$

By a linear transformation

$$\begin{pmatrix} \tilde{x} \\ \tilde{y} \end{pmatrix} = \begin{pmatrix} \alpha & \beta \\ \gamma & \delta \end{pmatrix} \begin{pmatrix} x \\ y \end{pmatrix} \tag{3.13hl}$$

it is a simple matter to transform the quadratic terms to diagonal form:

$$f(\tilde{x}, \tilde{y}) = \lambda_1 \tilde{x}^2 + \lambda_2 \tilde{y}^2 + (d\tilde{x}^3 + e\tilde{x}^2\tilde{y} + f\tilde{x}\tilde{y}^2 + g\tilde{y}^3) + (h\tilde{x}^4 + \cdots) \tag{3.14}$$

where $2\lambda_1 = (a + c) + \sqrt{(a - c)^2 + b^2}$ and $2\lambda_2 = (a + c) - \sqrt{(a - c)^2 + b^2}$. For typographic convenience we will suppress the $\tilde{}$ which appear in (3.14).

Next we define a nonlinear axis-preserving transformation:

$$\begin{aligned} x' &= x + \Delta_1 = x + (A_{20}x^2 + A_{11}xy + A_{02}y^2) + (A_{30}x^3 + \cdots) \\ y' &= y + \Delta_2 = y + (B_{20}x^2 + B_{11}xy + B_{02}y^2) + (B_{30}x^3 + \cdots) \end{aligned} \tag{3.13nl}$$

The idea is to choose the disposable coefficients A_{pq}, $B_{pq}(p + q \geq 2)$ in (3.13nl) in such a way as to transform (3.14) to Morse canonical form. By putting $f'(x', y') = \lambda_1 x'^2 + \lambda_2 y'^2$ and equating $f'(x', y')$ with $f(x, y)$ in (3.14), we place constraints on the disposable coefficients A_{pq}, B_{pq}:

$$\begin{aligned} &\lambda_1 x'^2 + \lambda_2 y'^2 = \lambda_1 x^2 + \lambda_2 y^2 + (dx^3 + ex^2y + fxy^2 + gy^3) + (hx^4 + \cdots) \\ &\qquad \| \\ &\lambda_1\Big(x + \sum_{p+q\geq 2} A_{pq}x^p y^q\Big)^2 + \lambda_2\Big(y + \sum_{p+q\geq 2} B_{pq}x^p y^q\Big)^2 \end{aligned} \tag{3.15}$$

These constraints may be obtained by expanding the squares in the second line of (3.15) and equating coefficients of the corresponding monomials $x^p y^q$ on each side of the equation. For the third degree terms

$$\begin{aligned} x^3&: \quad \lambda_1 2A_{20} &&= d \\ x^2y&: \quad \lambda_1 2A_{11} + \lambda_2 2B_{20} &&= e \\ xy^2&: \quad \lambda_1 2A_{02} + \lambda_2 2B_{11} &&= f \\ y^3&: \quad \lambda_2 2B_{02} &&= g \end{aligned} \tag{3.16}$$

The coefficients A, B in (3.14) are not uniquely determined because there are six of them and only four cubic coefficients.

If we go out to the quartic terms in (3.15), then the details get messier $[h = \lambda_1(2A_{30} + A_{20}^2)]$, but the conclusions do not change. The quartic terms bring in five new coefficients (of x^4, x^3y, x^2y^2, xy^3, y^4) on the right-hand side of (3.15). On the left-hand side we must keep terms up to cubic in the coordinate transformation (3.13nl). This brings in eight new coefficients ($A_{30}, A_{21}, A_{12}, A_{03}; B_{30}, B_{21}, B_{12}, B_{03}$). And so it goes. As we consider higher degree terms in (3.14), we must consider higher degree terms in (3.13nl). Since the number of disposable coefficients A_{pq}, B_{pq} increases faster than the number of coefficients which appear in $f(x, y)$ as we proceed to terms of higher degree, we can always find smooth transformations (3.13nl) so that

$$f'(x', y') = \lambda_1 x'^2 + \lambda_2 y'^2 = f(x, y) \tag{3.17}$$

Since the transformation (3.13) is invertible, we have, by (3.6b),

$$f[x(x', y'), y(x', y')] = f'(x', y') = \lambda_1 x'^2 + \lambda_2 y'^2 \tag{3.18}$$

The argument follows along similar lines in the case of a function of n variables. By a linear transformation (3.3hl) we can always bring the Taylor series expansion of f into the form

$$f(x) = \sum_{i=1}^{n} \lambda_i x_i^2 + \text{third-degree terms} + \text{fourth-degree terms} + \cdots \tag{3.19}$$

The number of terms of degree exactly $k(>2)$ is $(n + k - 1)!/k!(n - 1)!$. A nonlinear transformation (3.3nl) has the form

$$x_i' = x_i + \Delta_i = x_i + \sum_{p+q+\cdots+r\geq 2} A_{i;\,pq\cdots r} x_1^p x_2^q \cdots x_n^r \tag{3.20}$$

For any i the number of terms of degree exactly k in (3.20) is also

$$\frac{(n + k - 1)!}{k!(n - 1)!}, \qquad k > 1.$$

The disposable coefficients $A_{i;\,pq\cdots r}$ in (3.20) are constrained by the known coefficients in (3.19) and the relation between $f'(x')$ and $f(x)$:

$$f'(x') = f(x) \tag{3.21}$$

where

$$f'(x') = \sum_{i=1}^{n} \lambda_i x_i'^2 = \sum \lambda_i x_i^2 + \sum 2\lambda_i x_i \Delta_i + \sum \Delta_i^2 \tag{3.22}$$

The number of coefficients of terms of degree $3, 4, \ldots, k$ in (3.19) is

$$N_{\text{Taylor series}} = N_{TS}(n, k) = \binom{n+k}{k} - \binom{n+2}{2}$$

where $\binom{p}{q}$ is the binomial coefficient $p!/q!(p - q)!$.

The terms of degree $3, 4, \ldots, k$ in (3.22) come from the terms of degree $2, \ldots, k - 1$ in (3.20). For each x_i' there are

$$\binom{n+k-1}{k-1} - \binom{n+1}{1}$$

such terms, and there are n coordinates, so the number of disposable coefficients in the change of variables (3.20) contributing to terms of degree $\leq k$ in the expansion (3.22) is

$$N_{\text{change variable}} = N_{CV}(n, k) = n\binom{n+k-1}{k-1} - n\binom{n+1}{1}$$

It is a simple matter to verify that

$$N_{CV}(n, k) \geq N_{TS}(n, k), \qquad n \geq 1, \quad k \geq 3 \tag{3.23}$$

As a result, there is always more than sufficient "room" to choose the disposable coefficients in the smooth change of variables so that the Morse canonical form is obtained. Since the transformation (3.20) is invertible, we have, by (3.6b),

$$\begin{aligned} f(x) &= f[x(x')] = f'(x') \\ &= \sum_{i=1}^{n} \lambda_i x_i'^2 \end{aligned} \tag{3.24}$$

Expression (3.24) is the same as expression (2.2a).

4. Application 3: The Splitting Lemma

When $\det f_{ij} = 0$, one or more eigenvalues in (3.22) are zero, the Morse Lemma is not applicable, and $f(x) \neq \sum \lambda_i x_i'^2$.

To see why, we can set $\lambda_1 = 0$ in (3.15) in the 2-variable case. Then we can see from (3.16) that the disposable coefficients B_{20}, B_{11}, B_{02} in the y' transformation can be uniquely chosen to eliminate the terms x^2y, xy^2, y^3 in the Taylor series expansion of $f(x, y)$. But it is not possible to eliminate the term x^3. Were we to write down a system of equations analogous to (3.16), but for the fourth-degree terms in the Taylor series expansion of $f(x, y)$, we would find that the four disposable coefficients B_{30}, B_{21}, B_{12}, B_{03} (multiplied by λ_2) enter linearly into expressions for the coefficients of x^3y, x^2y^2, xy^3, y^4. They can therefore be chosen uniquely to eliminate these monomials, but it is not possible to eliminate the term x^4. In the rth-degree terms in the Taylor series expansion of $f(x, y)$, the r disposable coefficients $B_{r-1,0}, \ldots, B_{0,r-1}$ (multiplied by λ_2) enter linearly into the expressions for the coefficients of the r monomials $x^{r-1}y, \ldots, y^r$ so that these terms can all be transformed away. As a result, the change of variables $y \to y'$ given in (3.13nl) can be chosen to eliminate all monomials of the form x^py^q, $p + q > 2$, $q \geq 1$. Therefore, when $\lambda_1 = 0$,

$$\begin{aligned} f(x, y) &\doteq px'^3 + qx'^4 + \cdots + \lambda_2 y'^2 \\ &= f_{NM}(x') + \lambda_2 y'^2 \end{aligned} \tag{3.25}$$

In transforming $f(x, y)$ to the form (3.25) we have made no use at all of our freedom to choose the nonlinear coordinate transformation $x \to x'$ to tighten up the canonical form of the non-Morse function.

We can proceed in a similar fashion to search for canonical forms for a function of n state variables at a non-Morse critical point involving one or even l "bad" variables. In the following paragraph we discuss the general coordinate transformation which produces a canonical form. In the succeeding paragraphs we discuss the cases $l = 1$ and $l > 1$, where l is the number of eigenvalues that vanish at the non-Morse critical point.

In the Taylor series expansion of a function of n state variables which has a non-Morse critical point at $x_1 = x_2 = \cdots = x_n = 0$, all first-degree terms are absent ($\nabla f = 0$). The constant term is unimportant in determining the qualitative properties of a function at a point, so the Taylor series expansion essentially begins with the quadratic terms. By means of a linear transformation (3.3hl) we can put the quadratic terms into canonical diagonal form: $\lambda_1 x_1^2 + \lambda_2 x_2^2 + \cdots + \lambda_n x_n^2$. Next we can make a nonlinear transformation $x_i \to x_i'$ (3.3nl) to see how close we can come to a canonical form in this new coordinate system. We do this by comparing coefficients of corresponding terms on both sides of the equation:

$$\sum_{i=1}^{n} \lambda_i x_i'^2 \qquad = f(x_1, x_2, \ldots, x_n)$$

$$\downarrow \qquad\qquad\qquad \downarrow$$

$$\sum_{i=1}^{n} \lambda_i x_i^2 + 2\sum_{i=1}^{n} \lambda_i x_i \Delta_i + \sum_{i=1}^{n} \lambda_i \Delta_i^2 = \sum_{i=1}^{n} \lambda_i x_i^2 + \frac{1}{3!}\sum_{ijk} f_{ijk} x_i x_j x_k + \cdots \tag{3.26}$$

When all eigenvalues are nonzero, the nonlinear parts of the transformation (the Δ_i) can be chosen to transform away all terms of degree 3 and higher. This was shown in Sec. 3.

When one eigenvalue is zero (say λ_1), then the term x_1^2 does not appear on either side of (3.26). The cross term $\lambda_i x_i \Delta_i$ ($i \neq 1$) gives rise to terms of the form $x_1^{p_1} \cdots x_i^{p_i} \cdots x_n^{p_n}$, where $p_i \geq 1$ and $\sum_{j=1}^{n} p_j \geq 3$ (since Δ_i begins with quadratic terms). As a result, all terms of the form $x_1^{p_1} \cdots x_n^{p_n}$ except those with $p_2 = p_3 = \cdots = p_n = 0$ can be transformed away by nonlinear change of coordinates, and

$$f(x_1, x_2, \ldots, x_n) \doteq (t_3 x_1'^3 + t_4 x_1'^4 + \cdots) + \sum_{i=2}^{n} \lambda_i x_i'^2$$

$$= f_{NM}(x_1') + \sum_{i=2}^{n} \lambda_i x_i'^2 \tag{3.27i}$$

The argument proceeds in similar fashion when $l > 1$ eigenvalues vanish. If $\lambda_1 = \lambda_2 = \cdots = \lambda_l = 0$ in (3.26), then nonlinear changes of variables can be chosen for $x_i \to x_i' = x_i + \Delta_i$, $i = l + 1, \ldots, n$, which transform away all monomials of the form $x_1^{p_1} x_2^{p_2} \cdots x_n^{p_n}$ in (3.26), except those for which $p_{l+1} = \cdots = p_n = 0$. As a consequence,

$$f(x_1, x_2, \ldots, x_n) \doteq \left(\sum_{p_1 + \cdots + p_l \geq 3} t_{p_1 \cdots p_l} x_1'^{p_1} \cdots x_l'^{p_l} \right) + \sum_{i=l+1}^{n} \lambda_i x_i'^2$$

$$= f_{NM}(x_1', \ldots, x_l') + \sum_{i=l+1}^{n} \lambda_i x_i'^2 \tag{3.27l}$$

Expression (3.27l) is the same as expression (2.3a).

Remark. In obtaining the Splitting Lemma (3.27l) we have exploited the nonlinear transformation in the "good" state variables to transform away all terms in the Taylor series expansion which are "mixed," that is, products involving both "good" and "bad" variables. We have also transformed away all terms of degree > 2 in the "good" variables. We have *not* exploited the possibility of nonlinear transformation among the "bad" variables. In fact, the result (3.27l) can be obtained choosing $\Delta_1 = \Delta_2 = \cdots = \Delta_l = 0$, so that $x_1' = x_1, \ldots, x_l' = x_l$. In the following five sections we will exploit this possibility to search for canonical forms for the non-Morse functions $f_{NM}(x_1, \ldots, x_l)$. Before doing so, we recall that non-Morse critical points may generally be encountered in families of functions, or functions $f = f(x; c)$ depending on control parameters $c \in \mathbb{R}^k$, because at a critical point the eigenvalues $\lambda_i(c)$ of the stability matrix $f_{ij} = \partial^2 f(x; c)/\partial x_i \, \partial x_j$ are functions of the control parameters. It is important to know how many eigenvalues can simultaneously vanish in a typical k-parameter family of functions.

By the Splitting Lemma (2.3a), in the neighborhood of a non-Morse critical point at which l eigenvalues vanish simultaneously, it is possible to choose l "bad" variables and $n - l$ "good" variables. In this coordinate system, the stability matrix has a block-diagonal structure:

$$V_{ij} \rightarrow \left[\begin{array}{c|c} \dfrac{\partial^2 f_{NM}}{\partial x_i \, \partial x_j} & \bigcirc \\ \hline \bigcirc & \dfrac{\partial^2 f_M}{\partial x_r \, \partial x_s} \end{array} \right] \begin{array}{l} 1 \leq i, j \leq l \\ \\ l + 1 \leq r, s \leq n \end{array} \tag{3.28}$$

The $l \times l$ real symmetric matrix $\partial^2 f_{NM}/\partial x_i \, \partial x_j$ has $l(l + 1)/2$ independent matrix elements. These are functions of the k control parameters. We would therefore expect that not all the $l(l + 1)/2$ matrix elements $\partial^2 f_{NM}/\partial x_i \, \partial x_j$ can be made to vanish simultaneously unless there are at least as many control parameters. Therefore we could expect to find l simultaneously vanishing eigenvalues in a k-parameter family of functions, provided l and k are related by

$$k \geq l(l + 1)/2 \tag{3.29}$$

These intuitive considerations are correct and will be made more precise in Chapter 22.

5. Application 4: $l = 1$ (One Vanishing Eigenvalue)

In a k-parameter family of functions ($k \geq 1$) it is in general possible to encounter non-Morse critical points at which there is only one "bad" variable, which we call x. Near such a point the non-Morse function in (2.3a) has the Taylor expansion

$$f_{NM}(x; c) = t_0 + t_1 x + t_2 x^2 + t_3 x^3 + t_4 x^4 + \cdots \tag{3.30}$$

around the degenerate critical point at $x = 0$. The constant term is unimportant and the linear term is zero since $x = 0$ is a critical point. (i.e., $t_0 = t_1 = 0$). The remaining Taylor series coefficients $t_2(c) = \lambda(c)$, $t_3(c), \ldots, t_n(c), \ldots$ are functions of the k control parameters c_α. We could expect to choose the control parameters to make exactly k coefficients $t_j(c)$ vanish. If we choose $c \in \mathbb{R}^k$ so that $t_2 = t_3 = \cdots = t_{k+1} = 0$ at $c = c^0$, then at c^0 the first nonzero term in (3.30) is x^{k+2}:

$$f_{NM}(x; c^0) = t_{k+2} x^{k+2} + \text{higher terms} \tag{3.31}$$

We now show that all higher terms can be removed by a smooth change of variables, and that f_{NM} has a canonical form given by $\pm x^{k+2}$. We set

$$f(x) = t_n x^n + t_{n+1} x^{n+1} + \cdots \tag{3.32}$$

$$x' = A_1 x + A_2 x^2 + A_3 x^3 + \cdots \tag{3.33}$$

Since we want to show that $f(x)$ is qualitatively the same as $\pm x^n$, we set $f'(x') = \pm(x')^n$ and set $f'(x') = f(x)$:

$$\pm(A_1 x + A_2 x^2 + A_3 x^3 + \cdots)^n = t_n x^n + t_{n+1} x^{n+1} + \cdots \tag{3.34}$$

Proceeding as before [cf. (3.16)], we find

$$x^n: \qquad \pm A_1^n = t_n \tag{3.35n}$$

$$x^{n+1}: \qquad \pm nA_1^{n-1} A_2 = t_{n+1} \tag{3.35 $n+1$}$$

$$x^{n+2}: \qquad \pm\left(nA_1^{n-2} A_3 + \binom{n}{2} A_1^{n-2} A_2^2\right) = t_{n+2} \tag{3.35 $n+2$}$$

$$\vdots$$

If n is even, we must choose the $\pm$ sign on the left-hand side of (3.35) according to whether t_n is positive or negative. If n is odd, we can always choose this sign to be positive. Once A_1 is determined from (3.35n), A_2 can be uniquely determined from (3.35 $n + 1$), which is even linear in this unknown. In fact, once $A_1, A_2, \ldots, A_j$ have been determined from (3.35 $n \to n + j - 1$), A_{j+1} is determined from (3.35 $n + j$), in which the single unknown A_{j+1} occurs linearly.

In this way the transformation $x' = x'(x)$ (3.33) may be computed. Then from (3.6b) again,

$$f_{NM}[x(x')] = f'_{NM}(x') = \pm(x')^n \tag{3.36}$$

As a result, in the case of k control parameters but only one vanishing eigenvalue, the leading term in the Taylor series expansion of $f_{NM}(x; c)$ is typically no worse that $\pm x^{k+2}$ and the non-Morse function can be put into the canonical form $\pm x^{k+2}$ at the degenerate critical point.

Remark. It is sometimes useful to choose a canonical form $\pm x^n/n$. This is easily done by rescaling x' by the factor $n^{1/n}$. We will often do this below without specific mention.

Important Observation. The control variables can be used to get rid of the early terms in the Taylor series expansion of a function at a critical point. A smooth change of variables can be used to get rid of the later terms of the expansion. The no-man's-land in the Taylor series expansion at which these operations collide is called the germ of the function at a non-Morse critical point. The single state variable germs in k-parameter families are $\pm x^{j+2}$ ($j \leq k$) and are listed in Table 2.2.

6. Application 5: $l = 2$ (Two Vanishing Eigenvalues)

In a k (≥ 3) parameter family of functions it is in general possible to encounter non-Morse critical points with two "bad" variables which we call x, y. The leading part of the expansion of a function about the critical point at $(x, y) = (0, 0)$ is

$$\begin{aligned} f(x, y) = {} & t_{20}x^2 + t_{11}xy + t_{02}y^2 \\ & + t_{30}x^3 + t_{21}x^2y + t_{12}xy^2 + t_{03}y^3 \\ & + t_{40}x^4 + \cdots \end{aligned} \tag{3.37}$$

Three control parameters can be used to annihilate the quadratic terms (t_{20}, t_{11}, t_{02}). A smooth change of variables can then be introduced in an attempt to put the cubic terms in (3.37) into some sort of canonical form, and to transform away the remaining (quartic and higher degree) terms in the Taylor series expansion. To do this it is convenient to guess a suitable canonical form for the cubic terms. Useful guesses are

$$f'(x', y') = x'^3 + ay'^3 \tag{3.38a}$$

$$f'(x', y') = x'^2y' + by'^3 \tag{3.38b}$$

We shall work for now with the assumed canonical form (3.38b). We proceed exactly as in the Morse case for two variables (Sec. 3). We first make a linear transformation (3.13hl) and compare $\tilde{x}^2\tilde{y} + b\tilde{y}^3$ with the cubic terms in (3.37).

The resulting system of four equations in five unknowns

$$\begin{aligned}
x^3:&\quad \alpha^2\gamma \qquad\qquad + \ b\gamma^3 = t_{30}\\
x^2y:&\quad 2\alpha\beta\gamma + \alpha^2\delta + 3b\gamma^2\delta = t_{21}\\
xy^2:&\quad \beta^2\gamma + 2\alpha\beta\delta + 3b\gamma\delta^2 = t_{12}\\
y^3:&\quad \qquad\qquad \beta^2\delta + b\delta^3 = t_{03}
\end{aligned} \tag{3.39}$$

can always be solved for $\alpha, \beta, \lambda, \delta$. The expansion (3.37) then takes the form at the degenerate critical point:

$$\tilde{f}(\tilde{x}, \tilde{y}) = \tilde{x}^2\tilde{y} + b\tilde{y}^3 + (t_{40}\tilde{x}^4 + t_{31}\tilde{x}^3\tilde{y} + \cdots) + \cdots \tag{3.40}$$

It is an easy matter now to show that terms of degree 4 and higher can be removed by a smooth change of variables if $b \neq 0$. Again suppressing the $\tilde{}$ for typographic convenience, inserting (3.13nl) into the canonical expression $f'(x', y') = x'^2y' + by'^3$, expanding, and making term by term comparisons, we generate a series of simple equations for the disposable coefficients in (3.13nl):

$$\begin{aligned}
x^4:&\quad B_{20} = t_{40}\\
x^3y:&\quad 2A_{20} + B_{11} = t_{31}\\
x^2y^2:&\quad 2A_{11} + B_{02} + 3bB_{20} = t_{22}\\
xy^3:&\quad 2A_{02} \qquad + 3bB_{11} = t_{13}\\
y^4:&\quad \qquad\qquad + 3bB_{02} = t_{04}
\end{aligned} \tag{3.41}$$

Since there are six disposable coefficients and five Taylor series coefficients, all quartic terms can easily be removed. This argument extends to terms of degree 5, 6, ... using the counting arguments of Sec. 3.

Finally, if $b \neq 0$, then the variables x', y' can be rescaled to give canonical germs

$$\begin{aligned}
D_{+4}:&\quad x'^2y' + y'^3, \qquad b > 0\\
D_{-4}:&\quad x'^2y' - y'^3, \qquad b < 0
\end{aligned} \tag{3.42}$$

In the case of a 4-parameter family of functions, three degrees of freedom may be used to kill off the three quadratic coefficients, and the remaining degree of freedom can be used to annihilate the parameter b in (3.38b). In this case the smooth change of variables (3.13nl) cannot be used to eliminate all terms with degree > 3. This is clear from the last line in (3.41), which indicates that the terms x^4, x^3y, x^2y^2, and xy^3 can be transformed away, but y^4 cannot. So in the case of a 4-parameter family we have functions of the form $x'^2y' + b'y'^4$, with $b' \neq 0$ in general, because all four degrees of freedom in the control parameter space have been used. For $b' \neq 0$, x', y' can always be rescaled to give the canonical germ

$$\pm D_5: \quad \pm(x'^2y' + y'^4) \tag{3.43}$$

In a 5-parameter family there is one additional degree of freedom that can be used to annihilate b' also. Then by now familiar arguments, the term $b''y'^5$

cannot be eliminated, and we have the canonical germs

$$\begin{aligned} D_{+6}&: \ x'^2y' + y'^5, \qquad b'' > 0 \\ D_{-6}&: \ x'^2y' - y'^5, \qquad b'' < 0 \end{aligned} \tag{3.44}$$

Continuing in this vein, in k-parameter families we can encounter canonical germs

$$D_{\pm(k+1)}: \ x'^2y' \pm y'^k \tag{3.45}$$

Exactly similar arguments can be applied to the assumed canonical form (3.38a). In a 3-parameter family, in general $a \neq 0$. The coordinate y' can be rescaled to give the canonical form

$$D_{+4}: \ x'^3 + y'^3 \tag{3.46}$$

The functions (3.38a) for $a \neq 0$ are equivalent to the functions (3.38b) for $b > 0$. Because of the relation between a and b ($a^2 \sim b$), it generally takes two additional control parameters to annihilate a (only one additional for b), so that the corresponding canonical germ in a 5-parameter family of functions is

$$E_{\pm 6}: \ x'^3 \pm y'^4 \tag{3.47}$$

7. Application 6: $l = 2, k \geq 7$

In the case that the control parameter space has dimension exceeding 6, seven of these degrees can be used to eliminate all quadratic *and* cubic ($3 + 4 = 7$) terms in the Taylor series expansion (3.37). In this case we can proceed as before:

1. Carry out a linear transformation (3.13hl) to put the quartic terms into some sort of canonical form.
2. Carry out a nonlinear transformation (3.13nl) to get rid of the higher terms.

At this point we encounter a new difficulty. The linear transformation (3.13hl) has four parameters (α, β, γ, δ), while the quartic term is determined by five coefficients ($t_{40}, t_{31}, t_{22}, t_{13}, t_{04}$). There is no way to cast the quartic terms into a canonical form in which *all* terms have canonical coefficients ± 1 or 0. The best that we can do is a canonical form depending on one ($1 = 5 - 4$) parameter:

$$\pm x'^4 + ax'^2y'^2 \pm y'^4, \qquad a^2 \neq 4 \tag{3.48}$$

Remark. The canonical form for the cubic coefficients does not depend on any parameters (b could be fixed at ± 1) because we had just enough freedom in the linear transformation (four degrees) to fix the four cubic coefficients uniquely.

8. Application 7: $l = 3, k \geq 6$

In a k-parameter family of functions it is in general possible to encounter non-Morse critical points with three "bad" variables which we call x, y, z when $k \geq 6$. A Taylor series expansion of $f_{NM}(x, y, z; c)$ about the critical point $(x, y, z) = (0, 0, 0)$ has six quadratic terms, ten cubic terms, ... Then six of the k degrees of freedom can be used to kill off the six quadratic coefficients.

We can then look for a nonlinear transformation to bring the remainder of the Taylor series expansion to standard form. This is easily done by exact analogy with the 2-variable case:

1. Find a linear transformation $(x, y, z) \to (\tilde{x}, \tilde{y}, \tilde{z})$ to bring the cubic terms into a standard form.
2. Find a nonlinear axis-presevering transformation to remove quartic and higher terms.

Here we encounter the same problem encountered in the previous section. The linear transformation has nine degrees of freedom, since it is a real 3×3 matrix, but 10 coefficients parameterize the cubic terms. Therefore any canonical forms for the cubic terms must depend on one real parameter. One such canonical form is

$$T_{3,3,3}: \quad x'^3 + y'^3 + z'^3 + ax'y'z', \qquad a^3 \neq -27 \tag{3.49}$$

9. Application 8: $l \geq 4$

In a k-parameter family of functions we can expect to encounter non-Morse critical points with l "bad" variables when $k \geq l(l + 1)/2$ [cf. (3.29)]. In a Taylor series expansion of such a function about the non-Morse critical point there are $l(l + 1)/2$ quadratic terms, $l(l + 1)(l + 2)/6$ cubic terms, and so on. When the quadratic terms have been killed off by the control parameters, we can try to find a canonical form for the cubic terms. The linear part of the nonlinear transformation has dimension l^2 since it is an $l \times l$ matrix. This leaves a canonical form depending on

$$\frac{l(l + 1)(l + 2)}{6} - l^2 = \frac{l(l - 1)(l - 2)}{6} \tag{3.50}$$

real parameters. These parameters are called **moduli**. A germ depending on t moduli is called ***t*-modal**. A germ that does not depend on moduli is called **zero-modal** or **simple**.[1,2]

Once we know how many moduli a germ depends on, it is sometimes simple to guess its canonical form. For example, it is not generally possible to have a non-Morse critical point with four vanishing eigenvalues in a k-parameter family of functions unless $k \geq 4(4 + 1)/2 = 10$. In a 10-parameter family all 10 quadratic coefficients can simultaneously vanish. If we search for a canonical form for the cubic terms, it must depend on four moduli [see (3.50)]. This immediately suggests the following canonical form for the cubic terms:

$$f_{NM}(x, y, z, w) = x^3 + y^3 + z^3 + w^3 \\ + a_1 yzw + a_2 xzw + a_3 xyw + a_4 xyz \tag{3.51}$$

It may then be verified that a smooth change of variables of the form (3.2) with $A_i = 0$ ($i = 1, 2, 3, 4$) can transform away all quartic and higher degree terms unless the a_i (in (3.51)) accidentally obey a very stringent relation.

In the more general case of l "bad" state variables in a $k \geq l(l + 1)/2$ parameter family, we can search for a canonical form for the cubic terms. This depends on $\binom{l}{3}$ moduli. The binomial nature of this numerical dependence, together with the forms (3.49) for $l = 3$ and (3.51) for $l = 4$, suggests the general canonical form for $l \geq 3$:

$$f_{NM}(x_1, x_2, \ldots, x_l) = \sum_{i=1}^{l} x_i^3 + \sum_{1 \leq i < j < k \leq l} a_{ijk} x_i x_j x_k \tag{3.52}$$

There are precisely $\binom{l}{3}$ different coefficients a_{ijk}. Once this form has been guessed, it is a simple matter of algebra to verify that a smooth change of variables of the form (3.3nl) can be used to transform away all quartic and higher degree terms, unless the coefficients a_{ijk} obey a very stringent relation.

Moduli occur in the canonical forms for cubic terms when the number of "bad" state variables is too large [$l > 2$ from (3.50)]. They can also occur for $l = 2$ when the numbers of control parameters becomes too large [$k \geq 7$ from (3.48)]. They do not occur in the canonical forms for $l = 1$, no matter what k is.

Simple germs (i.e., germs that do not depend on moduli) occur for $l = 1$, k arbitrary, and $l = 2$, $k < 7$. They do not occur for $l \geq 3$. But the case $l = 3$ can occur when $k = 6$. So for $k \leq 5$, all germs that are encountered are simple. This is why Thom's list of elementary (=simple) catastrophes cuts off after $k = 5$.[3] For $k \geq 6$ there are stable simple germs and also nonsimple germs (Table 3.1).

Table 3.1[a] Simple and Nonsimple Catastrophe Germs

l \ k	1	2	3	4	5	6	7	8	9	10	11
1	S	S	S	S	S	S	S	S	S	S	S
2			S	S	S	S	S, M	S, M	S, M	S, M	S, M
3						M	M	M	M	M	M
4										M	M

[a] In a k-parameter family of functions with a non-Morse critical point having l "bad" directions, the canonical germs may be simple (S) or may depend on moduli (M), as shown. Only simple germs occur for $k \leq 5$.

Simple germs occur only in the cases $l = 1$ and $l = 2$. These germs are

$$\begin{array}{lll} A_{\pm k}: & \pm x^{k+1}, & k = 2, 3, \ldots \\ D_{\pm k}: & x^2y \pm y^{k-1}, & k = 4, 5, \ldots \\ E_{\pm 6}: & x^3 \pm y^4 & \\ E_{\pm 7}: & \pm(x^3y + y^4) & \\ E_8: & x^3 + y^5 & \end{array} \tag{3.53}$$

10. Summary

The local character of a potential or of a function is determined by the early terms in its Taylor series expansion. Further, two functions are qualitatively alike if the two are related by a smooth change of variables (3.6). Such a coordinate transformation can often be used to transform away the later terms in a Taylor series expansion. Then a function has qualitatively the same properties as the leading terms in its Taylor series expansion, that is, a simple polynomial.

In this chapter we saw how control parameters could be used to "annihilate" the early terms in the Taylor series expansion of a function, thereby changing its qualitative properties. We saw also how a coordinate transformation could be used to cut off the Taylor tail. The piece remaining, which resides between the terms removed by the control parameters and those removed by the coordinate transformation, could also be put into some standard form. This is the "canonical form" referred to in the title of this chapter.

For a single function depending on no control parameters, we generally encounter only noncritical points and isolated critical points, whose canonical forms are given in (2.1) and (2.2) and derived in Secs. 2 and 3.

Non-Morse critical points may be encountered in families of functions. For such families there is the general canonical form (2.3a), derived in Sec. 4. The canonical germs encountered in k-parameter families of functions with one and two "bad" variables are constructed in Secs. 5 and 6. In Secs. 7, 8, and 9 we discussed the problems that can arise with germs when the number of control parameters becomes too large ($l = 2$, $k \geq 7$) or when the number of "bad" variables becomes too large ($l \geq 3, k \geq l(l + 1)/2$): the germs depend on moduli. For $k < 6$ only simple germs are encountered. This is why Thom's list of elementary catastrophes terminates at $k = 5$. Canonical forms for all germs with $k < 6$ are given in Secs. 5 and 6. The list of all simple germs is given in (3.53).

References

1. V. I. Arnol'd, Critical Points of Smooth Functions, *Proc. Int. Cong. Math.*, Vancouver, 1974, pp. 19–39. This beautiful paper exists in expanded form in Ref. 2.

2. V. I. Arnol'd, "Critical Points of Smooth Functions and Their Normal Forms," *Russian Math. Surveys* **30**, 1–75 (1975).

3. T. Poston and I. N. Stewart, *Catastrophe Theory and Its Applications*, London: Pitman, 1978.

CHAPTER 4

Change of Variables. 2 Perturbations

The non-Morse functions which appear in the canonical form (2.3b), and which are listed in the left-hand column of Table 2.2, occur in conjunction with perturbation functions listed in the right-hand column of Table 2.2. For the Implicit Function canonical form (2.1) and the Morse canonical form (2.2) no such additional baggage was required. Why is it required for the Thom canonical form? The answer is that perturbations do not affect the qualitative properties of a function at a noncritical point or at a Morse critical point, but they do affect the qualitative properties of a function at a degenerate critical point. In essence, the list of perturbations is required to describe all the qualitative changes that a perturbation can produce in the neighborhood of a degenerate critical point.

It is often useful to know about the qualitative properties of a function or family of functions in the neighborhood of some point. If $f(x; c)$ is a family of functions with a noncritical point, a Morse critical point, or a degenerate critical point at $x^0 \in \mathbb{R}^n$ when $c = c^0 \in \mathbb{R}^k$, then the difference

$$p(x_1, \ldots, x_n; c_1, \ldots, c_k) = f(x; c) - f(x; c^0) \tag{4.1}$$

can be regarded as a perturbation of $f(x; c^0)$ in the neighborhood of (x^0, c^0). In this chapter we shall study what happens when we perturb a function at a noncritical point, an isolated critical point, or a degenerate critical point. In particular, we shall search for (and find) canonical forms for the perturbations of a function, just as we searched for, and found, canonical forms for the function itself in the previous chapter.

To study the effects of perturbations of functions in the neighborhood of x^0, it is sufficient to study a perturbation of the canonical form for $f(x; c^0)$ in the neighborhood of x^0. For convenience we will always move the point $x^0 \in \mathbb{R}^n$ to the origin in $\mathbb{R}^n$. It is also convenient to choose $c^0 = 0$ in $\mathbb{R}^k$.

The analysis proceeds in much the same way as that of Chapter 3 as far as technical details are concerned. The viewpoint, however, is inverted. In Chapter 3 we tried to see how many of the leading terms in the Taylor series expansion could become zero as the control parameters were varied. We then introduced a smooth change of variables in an attempt to eliminate the higher degree terms in the "Taylor tail" and to leave the remaining terms in some canonical form.

In this chapter we begin with the canonical form which is already missing more or fewer of the leading terms of its Taylor series expansion and all of its Taylor tail. Then we make an arbitrary perturbation and see which of the perturbing terms can be transformed away by a smooth and continuous change of variables. As a general rule of thumb we find that if some term could be transformed away to get a canonical form in Chapter 3, the corresponding perturbation term can be transformed away in this chapter. It is only the earliest terms in the Taylor series expansion of the perturbation which cannot be transformed away. Thus the germ is "incomplete" without these earliest terms. With them it is equivalent, under a continuous change of variables, to an arbitrary perturbation. Thus the early perturbation terms, which cannot be transformed away, become the canonical form for the most general perturbation of a catastrophe germ, and the coefficients of the monomials in these terms become the canonical mathematical control parameters.

1. Perturbations

The effects of a perturbing function $p(x; c)$ (4.1) on the function $f(x; c^0 = 0)$ are most easily studied by first constructing the Taylor series expansion of $p(x; c)$ at $x^0 = 0 \in \mathbb{R}^n$:

$$p(x; c) = p(0; c) + p_i x_i + p_{ij} x_i x_j + \cdots \tag{4.2}$$

Since the qualitative properties of a function at a point are independent of the value of the function at that point, we will often set the constant term $p(0; c)$ in (4.2) equal to zero. All derivatives are evaluated at $x = 0$. We have not indicated the factorial coefficients explicitly, again for convenience. All doubled indices are summed over. If $p(x; c)$ depends on k control parameters, the Taylor series coefficients $p_i, p_{ij}, \ldots$ also depend on these k control parameters. This dependence will be understood but will not be shown explicitly throughout this chapter.

2. Application 1: The Implicit Function Form

As a first example of how an arbitrary perturbation affects a function, we consider a perturbation at a noncritical point. In the neighborhood of such a point, $f(x)$ has the canonical form $f(x) = x_1$, so that the perturbed function is

$$F = f + p = x_1 + p \tag{4.3}$$

In general $\nabla F \neq 0$ at $x = 0$, so that the Implicit Function Theorem is applicable to $F(x)$ as well as to $f(x)$. Therefore the perturbed function $F(x)$ can be put into the canonical form $F(x) \doteq x_1'$ following the procedure described in Chapter 3, Sec. 2.

The effect of a perturbation on a function f at a point where $\nabla f \neq 0$ is twofold (Fig. 4.1):

1. The value of the perturbed function is changed slightly, to $\mathcal{O}(p(0; c))$.
2. The gradient $\nabla(f + p)$ is changed slightly, to $\mathcal{O}(p_i)$.

Perturbation of $f(x)$ at a noncritical point does not produce a qualitative change in the nature of $f(x)$ in the neighborhood of the noncritical point.

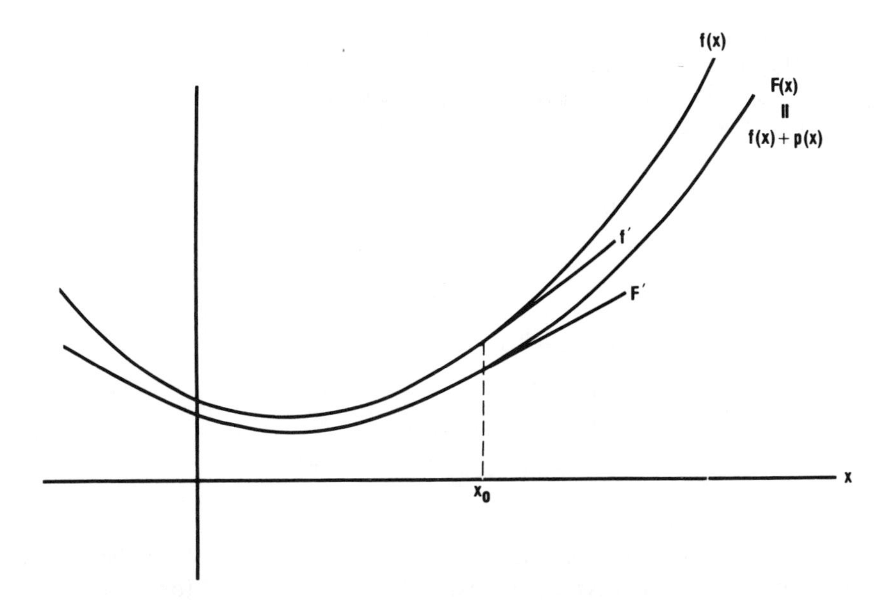

Figure 4.1 Perturbing a function at a noncritical point changes slightly the value of the function and its gradiant, but produces no local qualitative change.

3. Application 2: The Morse Forms

As a second example we consider what happens to a function when it is perturbed in the neighborhood of a Morse critical point. We choose (2.2a) as the canonical form for f in the neighborhood of the isolated critical point at $x = 0$. Then the perturbed function $F = f + p$ has Taylor series expansion

$$F = p(0; c) + p_i x_i + (\lambda_i \delta_{ij} + p_{ij}) x_i x_j + \text{cubics} + \cdots \tag{4.4}$$

Then in general at $x = 0$, $\nabla F = \nabla(f + p) = \nabla p \neq 0$, so it would appear that the perturbation has changed the qualitative character of the function $f(x)$, namely, from equilibrium to nonequilibrium at $x = 0$.

However, it is more useful to view this change from a different perspective. To motivate this alternative viewpoint, we consider first the case of a single state variable ($n = 1$). Then the Taylor series expansion (4.4) becomes ($\lambda \neq 0$)

$$F(x) = p(0; c) + p_1 x + (\lambda + p_2) x^2 + p_3 x^3 + \cdots \tag{4.5}$$

where the perturbation coefficients p_j are all considered small. As long as $p_1 \neq 0$ we can define a new independent variable y as in Chapter 3, Sec. 2:

$$y = p_1 x + (\lambda + p_2) x^2 + p_3 x^3 + \cdots \tag{4.6}$$

The inverse transformation exists and can be determined by making the ansatz

$$x = A_1 y + A_2 y^2 + A_3 y^3 + \cdots \tag{4.7}$$

The coefficients $A_1, A_2, \ldots$ can be determined by plugging (4.6) into (4.7) and demanding that the identity $x = x$ be satisfied. Equating coefficients of the powers of x on both sides of the resulting equation leads to a sequence of equations involving increasing numbers of coefficients A_j:

$$\begin{aligned} x^1&: \quad 1 = A_1 p_1 \\ x^2&: \quad 0 = A_1(\lambda + p_2) + A_2 p_1^2 \\ x^3&: \quad 0 = A_1 p_3 + 2A_2 p_1(\lambda + p_2) + A_3 p_1^3 \\ &\quad \vdots \end{aligned} \tag{4.8}$$

The coefficients are given explicitly by

$$\begin{aligned} A_1 &= 1/p_1 \\ A_2 &= -(\lambda + p_2)/p_1^3 \\ A_3 &= 2(\lambda + p_2)^2 p_1^{-5} - p_3 p_1^{-4} \\ &\vdots \end{aligned} \tag{4.9}$$

This is very bad. It clearly becomes increasingly difficult to construct the inverse transformation $x(y)$ (4.7) as the perturbation becomes smaller. More specifically, the inverse transformation (4.7) is discontinuous in the neighborhood of $p_1 = 0$.

This suggests another approach. Since the difficulty occurs because the coefficient p_1 appears in the denominator of (4.9) when we try to transform away the quadratic and higher degree terms, we might be luckier and find a transformation with a continuous inverse as $p \to 0$ if we settle for a transformation which removes cubic and higher degree terms. This is because the coefficient of the quadratic term does not go to zero as $p \to 0$.

To see whether this is the case, we look for a transformation of the form (3.3nl):

$$x = y + B_2 y^2 + B_3 y^3 + \cdots \tag{4.10}$$

which transforms away terms of degree >2 in the expression (4.5) for $F(x)$. By substituting (4.10) into (4.5) and equating the coefficients of $y^3, y^4, \ldots$ to zero, we generate a sequence of equations:

$$\begin{aligned}
y^1:&\quad \text{arbitrary} = p_1 \\
y^2:&\quad \text{arbitrary} = p_1 B_2 + (\lambda + p_2) \\
y^3:&\quad \quad 0 \quad\; = p_1 B_3 + (\lambda + p_2)\cdot 2B_2 + p_3 \\
y^4:&\quad \quad 0 \quad\; = p_1 B_4 + (\lambda + p_2)\cdot(2B_3 + B_2^2) + p_3 \cdot 3B_2 + p_4
\end{aligned} \tag{4.11}$$

If $p_1 = 0$, the system of equations determined by the coefficients of $y^3, y^4, \ldots$ is linear and can be solved easily:

$$\begin{aligned}
B_2 &= -\frac{p_3}{2(\lambda + p_2)} \\
B_3 &= \frac{5p_3^2}{8(\lambda + p_2)^2} - \frac{p_4}{2(\lambda + p_2)} \\
&\vdots
\end{aligned} \tag{4.12}$$

All coefficients $B_2, B_3, \ldots$ are well defined because the only factor which occurs in the denominators is the far-from-zero term $(\lambda + p_2)$. In case $p_1 \neq 0$ the system of equations can still be solved, and the coefficients $B_2(p_1), B_3(p_1), \ldots$ depend on p_1 in a continuous way (ascending power series).

The important point is that if $\lambda \neq 0$ in (4.5), a smooth invertible transformation (4.10) can always be found which transforms away all terms of degree greater than 2.

Remark (*Continuity Argument*). This argument extends to more general cases. If the first nonvanishing term in the Taylor series expansion of $f(x)$ is x^n, then a perturbation of x^n will have the form

$$F(x) = f + p = p_1 x + \cdots + p_{n-1}x^{n-1} + (1 + p_n)x^n + p_{n+1}x^{n+1} + \cdots \tag{4.13}$$

A smooth change of variables (4.10) can always be found to transform away all terms of degree greater than n for reasons best seen in the denominators of (4.9) and (4.12). The calculation to verify this proceeds along lines outlined above and in Chapter 3, Sec. 5. This calculation will not be carried out in this work because it is somewhat more messy algebraically than that of (3.35) because of contributions from terms of degree less than n, as seen in (4.11). This argument extends to the n-state-variable case also. Any high-degree terms which can be transformed away by a smooth change of variables to put a function $f(x_1, \dots, x_n)$ into some canonical form can also be transformed away in the perturbed function $F = f + p$. The transformation is smooth, invertible, and a continuous function of the Taylor series coefficients given in (4.1). We shall call this line of reasoning the Continuity Argument.

We now return to the perturbed function $F(x) = \lambda x^2 + p(x)$, given in (4.5). In a new coordinate system y related to x by (4.10), $F(y)$ is given by

$$\begin{aligned} F[x(y)] &= p(0; c) + p_1 y + \{(\lambda + p_2) + p_1 B_2\} y^2 \\ &= p(0; c) + p_1 y + \lambda' y^2 \end{aligned} \tag{4.14}$$

The coordinate y may be rescaled if desired ($y \to \tilde{y} = |\lambda'|^{1/2} y$) to give the quadratic term $\tilde{y}^2$ a canonical coefficient of ± 1. The origin may also be displaced ($y' = y + p_1/2\lambda'$) to give a canonical form:

$$F(y') = \left\{ p(0;c) - \frac{p_1^2}{4\lambda'} \right\} + \lambda' y'^2 \tag{4.15}$$

In short, we have shown that perturbing a function of a single variable at a Morse critical point does not change the qualitative nature of the function. The critical point is moved slightly $\mathcal{O}(p_1)$ and the value of the function at the critical point is changed slightly $\mathcal{O}(p(0; c))$, but the type of the critical point is unchanged.

We can now discuss the qualitative effect of a perturbation of a function with a Morse critical point at $0 \in \mathbb{R}^n$. The Taylor series expansion of such a function is given in (4.4). We will discuss this problem from two points of view:

1. *Scientist's or Engineer's Viewpoint.* If we want to determine the qualitative properties of the function (4.4) in the neighborhood of the origin, then we should recall that x_i are all small.

(a) Therefore the series can be truncated beyond terms of degree 2, so that

$$F \simeq p(0; c) + p_i x_i + (\lambda_i \delta_{ij} + p_{ij}) x_i x_j \tag{4.16a}$$

(b) The quadratic terms can then be diagonalized, leading to

$$F \simeq p(0; c) + p_i' x_i' + \sum_{i=1}^{n} \lambda_i' x_i'^2 \tag{4.16b}$$

(c) The origin can finally be shifted ($x_i' \to x_i'' = x_i' + (p_i/2\lambda_i')$), leading to

$$F \simeq \left\{ p(0; c) - \sum_{i=1}^{n} \frac{p_i'^2}{4\lambda_i'} \right\} + \sum_{i=1}^{n} \lambda_i' x_i''^2 \tag{4.16c}$$

2. *Mathematician's Viewpoint.*

(a) The quadratic terms in (4.4) can be diagonalized by a homogeneous linear transformation of the type (3.3hl), leading to

$$F' = p(0; c) + p_i' x_i' + \sum_{i=1}^{n} \lambda_i'' x_i'^2 + \text{third-degree terms} + \cdots \tag{4.17a}$$

(b) Terms of degree 3 and higher may be removed by a nonlinear transformation of type (3.3nl) along the lines described in Chapter 3, Sec. 3, and suggested in the Continuity Argument above:

$$F'' = p(0; c) + p_i' x_i'' + \sum_{i=1}^{n} \lambda_i''' x_i''^2 \tag{4.17b}$$

(c) The linear terms may be removed by a displacement of the origin according to $x_i'' \to x_i''' = x_1'' + p_i'/2\lambda_i'''$. The result is the canonical form

$$F''' = \left\{ p(0; c) - \sum_{i=1}^{n} \frac{p_i'^2}{4\lambda_i'''} \right\} + \sum_{i=1}^{n} \lambda_i''' x_i'''^2 \tag{4.17c}$$

Since the eigenvalues λ_i''' depend continuously on the Taylor series coefficients $p_i, p_{ij}, \ldots$ appearing in (4.1), the perturbed function $F = f + p$ has a Morse i-saddle near $0 \in \mathbb{R}^n$ if the original function has a Morse i-saddle at $0 \in \mathbb{R}^n$.

Remark. A careful reader will notice that the intuitive approach described by (4.16) and the rigorous approach described by (4.17) lead eventually to the same canonical form and the same conclusions. The question may be raised: why bother with a (somewhat tedious) rigorous approach if the intuitive one will suffice? Aside from questions of rigor, the answer is as follows. The rigorous approach leads directly, by the series of transformations given in (4.17a), (4.17b), and (4.17c), to canonical forms for perturbations of all catastrophe germs constructed in Chapter 3. The intuitive approach has not done so.

Remark. It may also be noticed that the sequence in which the coordinate transformations (3.3) are performed is different in the two viewpoints described by (4.16) and (4.17), as follows:

1. *Intuitive Approach.* (3.3nl), followed by (3.3hl), followed by (3.3ih).
2. *Rigorous Approach.* (3.3hl), followed by (3.3nl), followed by (3.3ih).

That the canonical forms (4.16c) and (4.17c) obtained are identical suggests that it really does not matter in what order the transformations are carried out, except for computational purposes. To be more precise, since the Taylor series coefficients $p_i, p_{ij}, \ldots$ appearing in (4.1) are all small, the appropriate transformations (3.3) are all near the identity. This means that it does not matter in what order they are carried out. Composition of these transformations in any order produces the same overall transformation to lowest degree in the perturbation

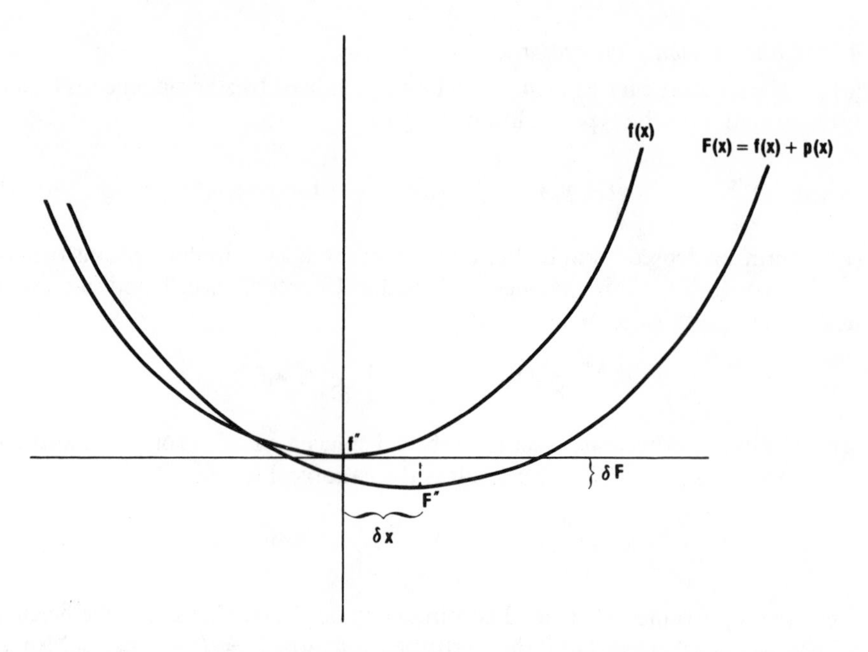

Figure 4.2 Perturbing a function at a Morse critical point changes slightly the location of the critical point, the critical value, and the eigenvalues, but produces no local qualitative change.

coefficients $p_i, p_{ij}, \ldots$. If $\lambda_1, \ldots, \lambda_n$ are the eigenvalues of the canonical form for $f(x_1, \ldots, x_n)$, then

$$\begin{aligned} \lambda_i' &= \lambda_i + p_{ii} + \mathcal{O}(p^2) \quad \text{in (4.16c)} \\ \lambda_i''' &= \lambda_i + p_{ii} + \mathcal{O}(p^2) \quad \text{in (4.17c)} \\ \lambda_i''' - \lambda_i' &= \mathcal{O}(p^2) \end{aligned} \tag{4.18}$$

This result can already be seen from (4.14). The commutativity of the transformations (3.3) to lowest order in the small perturbation coefficients p_i, p_{ij} will be made more precise in Chapter 23, Sec. 1.

The effect of a perturbation on a function f at a point where $\nabla f = 0$ but $\det f_{ij} \neq 0$ is threefold (Fig. 4.2):

1. The location of the equilibrium is changed slightly, to $\mathcal{O}(p_i)$.
2. The value of the perturbed function at the new equilibrium is changed slightly, to $\mathcal{O}(p_i^2)$ if $p(0; c) = 0$.
3. The eigenvalues are changed slightly, to $\mathcal{O}(p_{ii})$, but the type of the Morse critical point is unchanged.

Perturbation of $f(x)$ at a Morse critical point does not produce a qualitative change in the nature of $f(x)$ in the neighborhood of the Morse critical point.

4. Application 3: The Splitting Lemma Form

As a third example we consider what happens to a function in the neighborhood of a non-Morse critical point when it is perturbed. From the Splitting Lemma (2.3a) and the discussion of it in Chapter 3, Sec. 4, the unperturbed function has the form

$$f(x_1, \ldots, x_n) = f_{NM}(x_1, \ldots, x_l) + \sum_{i=l+1}^{n} \lambda_i x_i^2 \tag{4.19}$$

where $\lambda_i \neq 0$ $(l + 1 \leq i \leq n)$ and the function f_{NM} of l "bad" variables has no linear or quadratic terms in its Taylor series expansion around the degenerate critical point at $x = 0 \in \mathbb{R}^n$. The most general perturbation of f is given by (4.1), so the perturbed function $F = f + p$ has Taylor series expansion [setting $p(0; c) = 0$]

$$\begin{aligned} F &= f + p = (2.3\text{a}) + (4.1) \\ &= p_i x_i + (\lambda_i \delta_{ij} + p_{ij}) x_i x_j + f_{NM}(x_1, \ldots, x_l) \\ &\quad + \text{terms of degree 3 and higher} \end{aligned} \tag{4.20}$$

where $\lambda_1 = \lambda_2 = \cdots = \lambda_l = 0$. We search for a canonical form for (4.20).

To find such a canonical form, it is most convenient first to introduce a homogeneous linear transformation to diagonalize the quadratic terms in (4.20). In the new coordinate system

$$F' = p_i' x_i' + \sum_{i=1}^{n} \lambda_i' x_i'^2 + f_{NM}' + \text{terms of degree 3 and higher} \tag{4.21}$$

Here $\lambda_i' - \lambda_i$ are all small, so that $\lambda_1', \ldots, \lambda_l'$ are small and $\lambda_{l+1}', \ldots, \lambda_n'$ are not. A homogeneous linear transformation of type (3.3hl) maps cubics into cubics and, more generally, nth-degree terms into nth-degree terms by a simple homogeneity argument.

Next we look for an axis-preserving nonlinear transformation of type (3.3nl) to remove as many higher degree terms as possible. By the Continuity Argument of Sec. 3 we cannot remove all terms higher than linear because all linear coefficients are small. Similarly, we cannot look for continuous transformations $x_i' \to x_i''$ $(i = 1, 2, \ldots, l)$ to remove terms of degree > 2 because the coefficients of the $x_i'^2$ are small. However, the coefficients of $x_j'^2$ $(j = l + 1, \ldots, n)$ are not small, so we can look for continuous nonlinear transformations of type (3.3nl) on these $n - l$ variables. In particular, we can try to see which terms of degree > 2 can be removed by such transformations and which terms cannot.

This is a calculation which we have already done in Chapter 3, Sec. 4 [cf. (3.26) and (3.27)] when the linear terms are absent. Once again [cf. Sec. 3, (4.11)] the linear terms make the algebra somewhat messy. Once again we omit computational details. Once again the results have been previewed in Chapter 3.

We found there that all terms of the form $x_1'^{r_1} \cdots x_n'^{r_n}$ with $r_1 + r_2 + \cdots + r_n \geq 3$ can be removed by a nonlinear transformation on the "good" variables except those for which $r_{l+1} = \cdots = r_n = 0$. This means that in the Taylor series expansion (4.21) all terms of degree 3 or higher involving cross products between the l "bad" variables $x_1', \ldots, x_l'$ and the $n - l$ "good" variables $x_{l+1}', \ldots, x_n'$ can be removed by a continuous nonlinear transformation of type (3.2). All terms of degree > 2 in the "good" variables can also be removed by such a transformation. As a result of this sequence of transformations the perturbed function F (4.20) splits into a sum of two functions:

$$F \doteq F_{NM} + F_M \tag{4.22}$$

$$F_{NM} = f_{NM}(x_1'', \ldots, x_l'') + p(x_1'', \ldots, x_l'')$$

$$p = p_i'' x_i'' + p_{ij}'' x_i'' x_j'' + \cdots, \qquad 1 \leq i, j, \ldots \leq l \tag{4.22NM}$$

$$F_M = \sum_{i=l+1}^{n} p_i'' x_i'' + \sum_{i=l+1}^{n} \lambda_i'' x_i''^2 \tag{4.22M}$$

The canonical form for the perturbed Morse function F_M has been discussed in the previous section (4.17c). Canonical forms for perturbed non-Morse functions F_{NM} will be discussed in the following sections.

5. Application 4: Perturbation of $\pm x^n$

The implications of the principal result (4.22) of the previous section cannot be overestimated. If a function $f(x; c^0)$ with a non-Morse critical point at x^0 occurs in a family of functions $f(x; c)$, then (4.22) says that for any other function $f(x; c)(c \neq c^0)$ near $f(x; c^0)$, a coordinate system can be found so that the perturbed function $f(x; c) = f(x; c^0) + p(x)$ is obtained by perturbing separately the Morse part $\sum_{i=l+1}^{n} \lambda_i x_i^2$ of f and the non-Morse part f_{NM} of f. Perturbation of the former produces no qualitative change, perturbation of the latter does.

We begin our search for the canonical forms for perturbations of non-Morse functions by investigating the perturbations of catastrophe germs for a single variable: $\pm x^n$. From (4.22) and (4.1) the most general perturbation of $\pm x^n$ is

$$F(x) = p(0; c) + p_1 x + \cdots + p_{n-1} x^{n-1} + (\pm 1 + p_n) x^n + p_{n+1} x^{n+1} + \cdots \tag{4.23}$$

From the Continuity Argument of Sec. 3 it is not possible to find a continuous transformation which removes all terms of degree $> j$ for $j = 1, 2, \ldots, n - 1$ because the corresponding coefficients of x^j are small, but it is possible to transform away terms of degree $> n$ because the coefficient of x^n is not small. The nonlinear transformation of type (3.3nl) which does this can be constructed exactly as in (4.10)–(4.12). A homogeneous linear transformation of type

(3.3hl) can be used to rescale x and bring the coefficient of x^n once again into canonical form, so that after (3.3nl) and (3.3hl) we have

$$F''(x'') = p''_1 x'' + \cdots + p''_{n-1} x''^{n-1} \pm x''^n \tag{4.23''}$$

Finally an inhomogeneous linear transformation $x'' \to x''' = x'' - s$ (3.3ih) can be introduced. This leads to

$$\begin{aligned} F'''(x''') = p''_1(s + x''') \\ + p''_2(s^2 + 2sx''' + x'''^2) + \cdots \\ + p''_{n-1}(s^{n-1} + \cdots + x'''^{n-1}) \\ \pm (s^n + \cdots + nsx'''^{n-1} + x'''^n) \end{aligned} \tag{4.23'''}$$

The coefficients of the various powers of x''' can then be collected together and the shift s chosen in an effort to eliminate one of the powers of x. The simplest power to eliminate is the $(n - 1)$st because its coefficient

$$p''_{n-1} \pm ns$$

is the simplest. This coefficient can always be set equal to zero by setting $s = \mp p''_{n-1}/n$. The resulting canonical form for the perturbed function is

$$F'''(x''') = p'''_1 x''' + \cdots + p'''_{n-2} x'''^{n-2} \pm x'''^n \tag{4.24}$$

By now dropping primes and equating the several times transformed Taylor series coefficients p'''_j in (4.24) with the canonical perturbation parameters a_j, we have

$$\pm x^n + \text{arbitrary perturbation} \doteq \pm x^n + \sum_{\alpha=1}^{n-2} a_\alpha x^\alpha \tag{4.25}$$

In short, in a general k-parameter family of functions with one "bad" variable, the canonical form for the worst non-Morse critical point likely to be encountered is $\pm x^{k+2}$ and the canonical form for any function near that function is $\pm x^{k+2} + \sum_{\alpha=1}^{k} a_\alpha x^\alpha$.

Remark. The k canonical coefficients a_α are functions of all the Taylor series coefficients $p_i, p_{ij}, \ldots$ of the perturbing function. These, in turn, are coefficients of the physical control parameters $c_1, \ldots, c_k$. We can use either the canonical mathematical coefficients a_α or the physical control parameters c_β as long as there is a nonsingular transformation between them. In general there is, and (2.6) is satisfied.

Example. In a 1-parameter family of functions it is in general possible to encounter only one "bad" variable at a non-Morse critical point [Chapter 3, Sec. 4, (3.28)]. In a suitable coordinate system the "good" and "bad" variables can be split apart [(2.3a) and Chapter 3, Sec. 4, (3.27)]. The non-Morse function of the "bad" variable x has canonical form x^3 [Chapter 3, Sec. 5]:

$$f_{NM}(x) = x^3 \tag{4.26}$$

A perturbation of the original function, such as changing the value of the control parameter, will perturb separately the Morse part and the non-Morse part of the original function in a suitable coordinate system [Sec. 4, (4.22)]. The perturbation of the non-Morse function x^3 (4.26) can be truncated above terms of degree 3 because a nonlinear transformation can transform them away. In fact, the perturbation can be truncated beyond terms of degree 2 because the perturbation of the cubic term can be scaled away by renormalizing x. Thus,

$$f_{NM}(x) + \text{perturbation} = p_0 + p_1 x + p_2 x^2 + x^3 \tag{4.27}$$

A shift in the origin ($x = x' + s$) gives

$$\begin{aligned} f_{NM}(x') + p(x') &= a_0 + a_1 x' + a_2 x'^2 + x'^3 \\ a_0 &= p_0 + p_1 s + p_2 s^2 + s^3 \\ a_1 &= p_1 + 2p_2 s + 3s^2 \\ a_2 &= p_2 + 3s \end{aligned} \tag{4.28}$$

The constant term is unimportant and can be dropped. The coefficient a_2 can always be set equal to zero, but a_1 cannot be set equal to zero by appropriate choice of s if $p_2^2 - 3p_1 < 0$. Since we want the same canonical form to exist for all perturbations, we have

$$x^3 \xrightarrow[\text{perturbation}]{\text{canonical}} x^3 + a_1 x \tag{4.29}$$

The properties of this 1-parameter family of functions $F(x; a_1)$ are easy to study (Fig. 4.3). For $a_1 = 0$, $F(x; a_1 = 0)$ has a degenerate critical point at $x = 0$. For $a_1 < 0$, $F(x; a_1)$ has two isolated Morse critical points. As a_1 increases toward zero, the two critical points approach each other, and become degenerate at $a_1 = 0$. For $a_1 > 0$, $F(x; a_1)$ has no critical points.

In general, perturbation of a catastrophe germ, which has a degenerate critical point, will split the degenerate critical point up into a number of nondegenerate critical points ("**morsification**"). The maximum number of isolated critical points that can be obtained by perturbing a catastrophe germ is indicated by a subscript on the germ. For the single-state-variable catastrophes A_k: x^{k+1} this is simple to see:

$$\frac{d}{dx}\left(x^{k+1} + \sum_{j=1}^{k-1} a_j x^j\right) = 0 \tag{4.30}$$

is a polynomial equation of degree k, so it has at most k real roots.

The effect of a perturbation on a function f at a point where $\nabla f = 0$ and $\det f_{ij} = 0$ is fourfold (Fig. 4.3):

1. The number of equilibria is changed. A k-fold degenerate critical point is split into at most k isolated critical points, possibly fewer.
2. The locations of the critical points are changed.
3. The values of the function at the critical points (critical values) are changed.
4. The Morse saddle types of the isolated equilibria are related to each other in a way which is characteristic of the catastrophe.

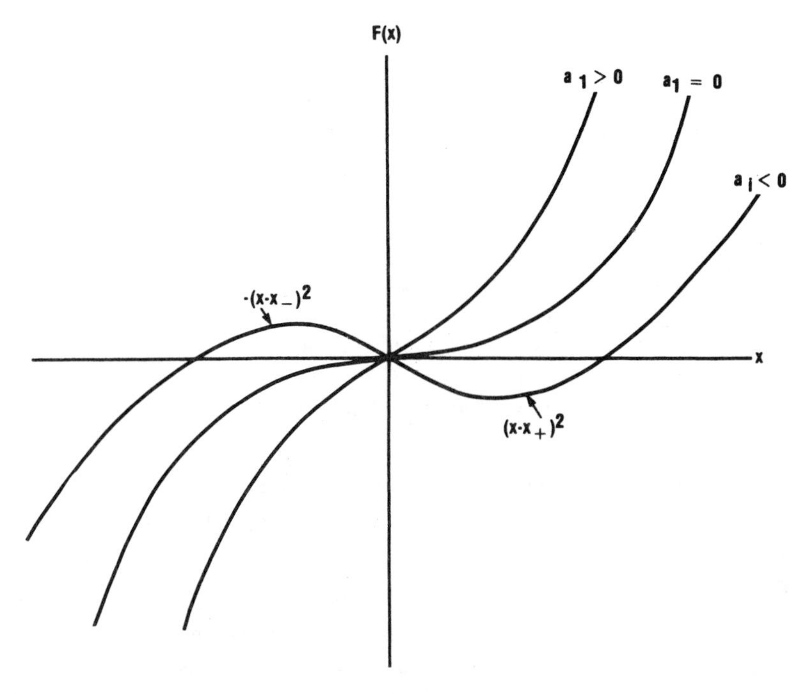

Figure 4.3 The 1-parameter family of functions $F(x; a_1) = x^3 + a_1 x$ contains members with two isolated critical points and with no critical points. Such functions are separated by the function $F(x; a_1 = 0)$ with a doubly degenerate critical point.

Perturbation of $f(x)$ at a non-Morse critical point produces a qualitative change in the nature of $f(x)$ in the neighborhood of the critical point. The number of critical points, their locations, and their critical values depend on the perturbation coefficients $p_i, p_{ij}, \ldots$ in a more complicated way than the corresponding dependences for the Implicit Function form (Sec. 2) or the Morse form (Sec. 3). The dependence on the parameters a_α appearing in the canonical perturbation (Table 2.2) is canonical, so it need be studied only once. We do this in detail for two simple catastrophes in Chapter 6.

6. Application 5: Perturbation of $x^2y \pm y^k$

We next consider perturbations of the catastrophe germs labeled $D_{\pm(k+1)}$ in Table 2.2. The most general perturbation has the form (neglecting the constant term)

$$p(x, y) = \sum_{r+s\geq 1} p_{rs} x^r y^s \tag{4.31}$$

For simplicity we study first the germ

$$D_4(x, y) = x^2y + y^3 \tag{4.32}$$

The most general perturbation of $D_4(x, y)$ is then

$$\begin{aligned} F &= (4.32) + (4.31) \\ &= \text{linear terms} + \text{quadratic terms} \\ &\quad + p_{30}x^3 + (1 + p_{21})x^2y + p_{12}xy^2 + (1 + p_{03})y^3 \\ &\quad + \text{fourth-degree terms} + \cdots \end{aligned} \tag{4.33}$$

By the Continuity Argument of Sec. 3 we should use the two terms of D_4 itself as the "pivotal" terms beyond which we try to remove the higher degree terms by a continuous homogeneous nonlinear transformation of type (3.2). This is because the only coefficients in (4.33) which are large are those of x^2y and y^3.

We have seen already (Chapter 3, Sec. 6) that a homogeneous linear transformation can be used to put the cubic terms back into the canonical form of D_{+4} so that

$$\begin{aligned} F' &= (\text{linear terms})' + (\text{quadratic terms})' \\ &\quad + (x^2y + y^3)' + (\text{fourth-degree terms})' + \cdots \end{aligned} \tag{4.34}$$

We saw there also that a nonlinear transformation of type (3.3nl) could be used to transform away all terms of degree 4 and higher. Here the details of the calculation are somewhat more involved because of contributions of the linear and quadratic terms. However, a homogeneous nonlinear transformation of type (3.2) can be applied to either (4.33) or (4.34) which transforms away all terms of degree ≥ 4, and which leaves the cubic terms in the canonical form (4.32). The details of the calculation are omitted. Dropping primes, we have

$$F \doteq \text{linear terms} + \text{quadratic terms} + (x^2y + y^3) \tag{4.35}$$

Finally it is possible to shift the origin through an inhomogeneous transformation of type (3.3ih) with $x' = x + s_1$, $y' = y + s_2$. If the quadratic terms are

$$p_{20}x^2 + p_{11}xy + p_{02}y^2 \tag{4.36}$$

then the choice $(s_1, s_2) = (p_{11}/2, p_{20})$ eliminates the coefficients of x^2 and xy, while the choice $(p_{11}/2, p_{02}/3)$ eliminates xy and y^2. It is convenient to eliminate the terms x^2 and xy so that the most general perturbation of $D_4(x, y)$ is

$$D_4(x, y) + \text{arbitrary perturbation} \doteq x^2y + y^3 + a_1x + a_2y + a_3y^2 \tag{4.37}$$

The most general perturbation of the catastrophe germ $D_{k+1}(x, y) = x^2y + y^k$ is

$$\begin{aligned} F &= \text{linear terms} + \text{quadratic terms} \\ &\quad + (1 + p_{21})x^2y + \cdots + (1 + p_{0k})y^k + \cdots \end{aligned} \tag{4.38}$$

All terms not written in explicitly are small. By a linear transformation the cubic terms can be put in the canonical form

$$x^2y + by^3 \tag{4.39}$$

where b is small for a small perturbation because $b = 0$ in the absence of a perturbation ($k > 3$). By the Continuity Argument of Sec. 3 we can hope to find a continuous transformation which transforms away only those terms which can be obtained from the monomials in (4.38) with large coefficients: x^2y and y^k. Following the standard algorithm, and carrying out a calculation similar to that of Chapter 3, Sec. 6, we find

$$F \doteq \text{linear terms} + \text{quadratic terms}$$
$$+ x^2y + p_{03}y^3 + \cdots + p_{0,k-1}y^{k-1} + y^k \quad (4.40)$$

If the quadratic terms are as given as in (4.36), the shift of origin $(s_1, s_2) = (p_{11}/2, p_{20})$ can be used to eliminate the terms x^2 and xy, giving for the canonical form for a function near $D_{k+1}(x, y)$

$$F \doteq (x^2y + y^k) + a_1x + \sum_{j=2}^{k} a_jy^{j-1} \quad (4.41)$$

The canonical perturbations of the germs $D_{-(k+1)}$ are the same as the perturbations of $D_{+(k+1)}$.

7. Application 6: Perturbation of $x^3 \pm y^4$

By now the arguments we have been using have become a bit tedious. Rather than repeat them ad nauseum, we shall quicken the pace and cut a few corners. It should be understood that rigorous arguments underlie our corner-cutting methods. Thus these "intuitive" methods are really quite rigorous.

The general perturbation of a 2-state-variable germ can be represented schematically as in Fig. 4.4. Associated with each monomial (e.g., xy^2) is a corresponding Taylor series coefficient (p_{12}). Since all the Taylor series coefficients p_{ij} are small, we can ask which of these coefficients can be made zero by a transformation of type (3.2) which is near the identity transformation:

$$\begin{aligned} x \to x' &= x + \Delta_1 \\ y \to y' &= y + \Delta_2 \end{aligned} \quad (4.42)$$

Here the Δ_i are obtained by replacing the finite coefficients A which appear in (3.2) by infinitesimal coefficients δA. Then this transformation, applied to the germ $x^3 + y^4$, gives

$$(x')^3 + (y')^4 \qquad = x^3 + y^4 + (4.1)$$
$$\downarrow$$
$$\begin{aligned} &x^3 + 3x^2\Delta_1 + 3x\Delta_1^2 + \Delta_1^3 \\ &+ y^4 + 4y^3\Delta_2 + 6y^2\Delta_2^2 + 4y\Delta_2^3 + \Delta_2^4 \end{aligned} \quad (4.43)$$

Since Δ_1 and Δ_2 are "infinitesimals," we can "linearize" (4.43). Only those terms that can be written in the form $3x^2\Delta_1 + 4y^3\Delta_2$ could possibly be transformed

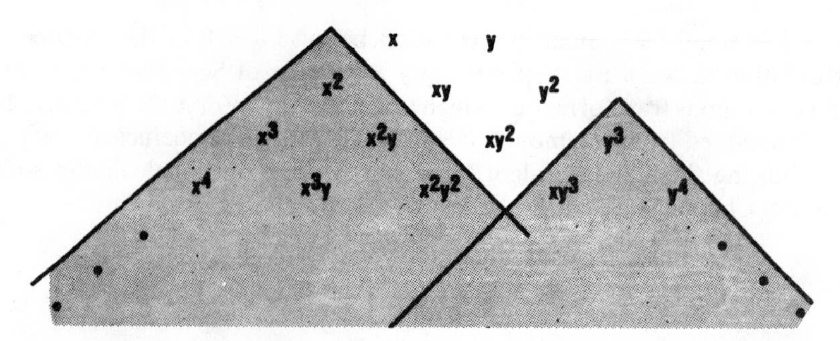

Figure 4.4 The general perturbation of a function of two variables is represented by a vector in a linear vector space whose basis vectors $x^p y^q$ may conveniently be represented in a Pascal triangle arrangement.

away by the change of variables (4.42). These terms are shown in the shadow regions in Fig. 4.4. In fact, *all* terms that occur in these shadow regions *can* be transformed away by using the counting argument of Chapter 3, Sec. 3. The only terms that cannot be transformed away from the perturbation $p(x, y)$ in (4.43) are therefore x, y, xy, y^2, xy^2. Therefore the canonical form of any perturbation of $x^3 + y^4$ is

$$(x^3 + y^4) + (a_1 x + a_2 y + a_3 xy + a_4 y^2 + a_5 xy^2) \tag{4.44}$$

The canonical perturbations of the germ $E_{-6} = x^3 - y^4$ are the same as the perturbations of $E_{+6} = x^3 + y^4$.

This infinitesimal method is extremely powerful and will be described more fully in Chapter 23.

8. Summary

A perturbation does not affect the qualitative properties of a function in the neighborhood of a noncritical point or a Morse critical point. In the former case the magnitude and direction of the gradient ∇f are changed slightly. In the latter case the critical point is moved slightly and the critical values are changed, but the Morse saddle type is unchanged. The absence of qualitative changes brought about by perturbations in neighborhoods of such points is the reason that the Implicit Function Theorem and the Morse Lemma do not come equipped with lists of perturbations.

When a family of functions contains a member with a non-Morse critical point, we showed in Chapter 3, Sec. 4, that it was possible to choose a coordinate system which splits this function into a "bad" non-Morse part and a "good" Morse function. We showed in this chapter, Sec. 4, that the same splitting result holds for all other functions in the family near the non-Morse function. That is, it is always possible to find a coordinate system in which the perturbed function is split into two parts. The two parts may then be studied separately. The perturbed Morse function has no qualitatively new properties, but the perturbed non-Morse function does.

The perturbed catastrophe germs were studied by the methods presented in Chapter 3. A most general perturbation (4.1) was introduced, and we tried to see how much of the Taylor tail could be transformed away by a smooth change of variables. The Continuity Argument of Sec. 3 precluded the possibility of transforming away the very earliest terms in the Taylor series expansion of the perturbed function. For continuity we had to study those terms with finite coefficients. We found, not at all surprisingly, that those terms which could be transformed away in Chapter 3 to find the canonical germ, could also be transformed away in the perturbed germ. It was only the earliest coefficients which could not be removed by a continuous change of variables. Some of the early coefficients could be removed by an inhomogeneous transformation. In studying families of functions it is not important where we choose the origin, so we were able to allow inhomogeneous transformations in this chapter. In Chapter 3 we wanted to consider the qualitative properties of a function *at a fixed point*, so only homogeneous transformations were considered there.

Perturbation of a function at a non-Morse critical point has a significant qualitative effect on the properties of the function in the neighborhood of the point. In general a k-fold degenerate critical point will split up into at most k

Table 4.1 Elementary Catastrophe Functions (Zero Modal)

Name	Catastrophe Germ[a]	Perturbation
$A_{\pm k}$[b]	$\pm x^{k+1}$	$\sum_{j=1}^{k-1} a_j x^j$
$D_{\pm k}$	$x^2y \pm y^{k-1}$ k even $\pm(x^2y + y^{k-1})$ k odd	$\sum_{j=1}^{k-3} a_j y^j + \sum_{j=k-2}^{k-1} a_j x^{j-(k-3)}$[c]
$E_{\pm 6}$	$\pm(x^3 + y^4)$	$\sum_{j=1}^{2} a_j y^j + \sum_{j=3}^{5} a_j x y^{j-3}$
E_7	$x^3 + xy^3$	$\sum_{j=1}^{4} a_j y^j + \sum_{j=5}^{6} a_j x y^{j-5}$[c]
E_8	$x^3 + y^5$	$\sum_{j=1}^{3} a_j y^j + \sum_{j=4}^{7} a_j x y^{j-4}$

[a] For typographical convenience we use $y_1 = x$, $y_2 = y$.
[b] $A_{+k} \doteq A_{-k}$ if k is even.
[c] The expression of the perturbation in terms of the monomials $x^p y^q$ is not unique.

isolated critical points, possibly fewer. Fortunately the most general form that a perturbation can have can be reduced, by a smooth change of variables, to a canonical form. It was these canonical forms which were constructed in Secs. 5–7.

The existence of significant qualitative effects of perturbations on catastrophe germs, together with the reduction of an arbitrary perturbation of a germ to canonical form, are the reasons that Thom's list of elementary catastrophes in fact includes one list of simple catastrophe germs and a collateral list of canonical perturbations. We summarize the list of canonical perturbations for all simple catastrophe germs (3.53) in Table 4.1.

The methods described in this chapter can be used to compute canonical forms for the nonsimple germs discussed in Chapter 3, Secs. 7–9. However, simpler methods based on the "corner-cutting" approach of Sec. 7 are available and will be discussed in Chapter 23, so we defer such a discussion until then.

CHAPTER 5

The "Crowbar Principle"

A crowbar is a powerful tool for prying things apart. In normal use it is inserted into a crack or a hole (measure zero) and "wiggled" until the adjacent pieces are broken apart.

In Chapter 2 we have described some theorems that can be used in the same spirit when we consider either a single isolated function or else a family of functions. For a single isolated function most points $x \in \mathbb{R}^n$ are noncritical, yet it is the isolated critical points that completely organize the global qualitative properties of the function (Sec. 1). For a family of functions most points $c \in \mathbb{R}^k$ parameterize Morse functions, yet it is the measure zero set in $\mathbb{R}^k$ that parameterizes functions with degenerate critical points which completely organizes the global qualitative properties of the family of functions (Sec. 2). This separatrix is called the bifurcation set and denoted by $\mathscr{S}_B$.

The set of points $c \in \mathbb{R}^k$ which parameterize non-Morse functions also divides the control parameter space $\mathbb{R}^k$ into open regions, each of which parameterizes functions that are qualitatively similar. We determine these separatrices, and the qualitative type of function associated with each open region, for each of the elementary catastrophes listed in Table 2.2 whose control parameter space can easily be visualized (i.e., $k \leq 3$). The elementary catastrophes A_2, A_3, A_4, D_{+4}, and D_{-4} are studied in Secs. 3, 4, 5, 6, and 7, respectively.

In Sec. 8 we introduce another kind of separatrix which parameterizes functions with degenerate critical values at isolated critical points. This separatrix is called the Maxwell set and denoted by $\mathcal{S}_M$. The Maxwell set is determined by the Clausius–Clapeyron equations (Sec. 9).

1. The "Crowbar Principle" for a Potential

Suppose we begin by pulling functions of n (state) variables, or more specifically potentials $V(x_1, x_2, \ldots, x_n)$, out of a hat. Then most of the functions we encounter which have any critical points at all, have only isolated or nondegenerate critical points. (Functions with only isolated critical points are called **Morse functions.**) In fact, almost all of the functions we encounter in this way will be Morse functions. The term "almost all" will be made more precise in Chapter 21, Sec. 1. As a consequence, Thom's Theorem is not needed when considering typical potentials.

Now we consider a typical potential $V(x_1, \ldots, x_n)$. If x^0 is a random, or typical, point in $\mathbb{R}^n$, then $\nabla V \neq 0$ at x^0 because a typical potential has only isolated critical points. Since $\nabla V \neq 0$ and the Implicit Function Theorem is valid almost everywhere in $\mathbb{R}^n$, why bother with the Morse Lemma for the occasional point at which $\nabla V = 0$?

The answer is that the critical points have an importance that the noncritical points do not have. Basically they organize the entire qualitative nature of the potential $V(x)$.

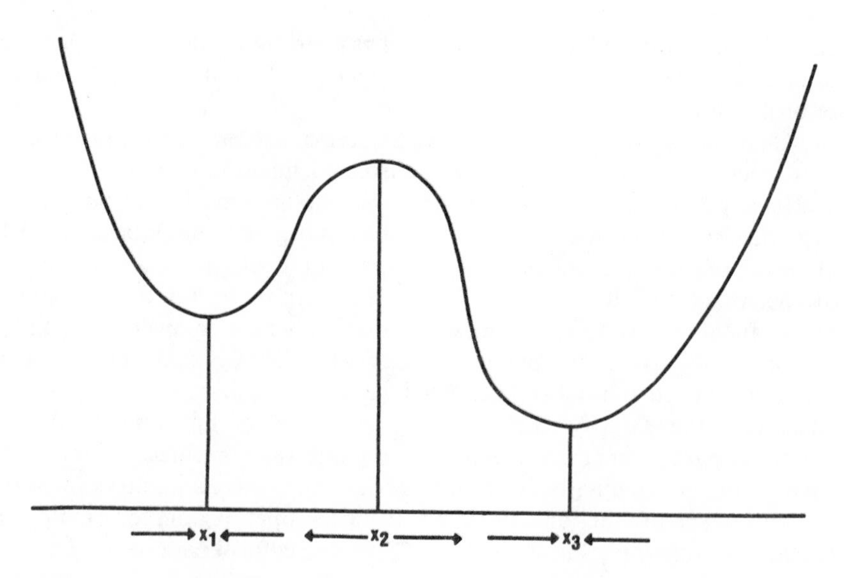

Figure 5.1 In one dimension the locations of the isolated critical points and their types determine completely the qualitative nature of the potential.

This is easiest to see in one dimension (Fig. 5.1). If $V(x)$ has critical points at x_1, x_2, x_3 as shown, then at any point to the left of x_1, $F = -dV/dx$ points toward the right, at any point between x_1 and x_2, F points to the left, and so on. The direction of the force F changes upon passing through a nondegenerate critical point. Thus if $V(x)$ has only isolated critical points and if we know the locations of all critical points, then all qualitative properties of $V(x)$ are known if we know the direction of F in any one intermediate open region between critical points, or if we know the Morse type of any one critical point.

Similar considerations hold in two or more dimensions. For example, if $V(x_1, x_2)$ has two isolated minima, then there must be some path connecting these 0-saddles which passes through a 1-saddle. By plotting the location and saddle type of each isolated critical point in $\mathbb{R}^2$, the qualitative features of $V(x_1, x_2)$ may readily be inferred.

The qualitative nature of $V(x_1, x_2)$ may be depicted in one of two generally convenient ways (Fig. 5.2). In one representation (Fig. 5.2*a*) the contours, or curves $V(x_1, x_2) =$ constant, are shown and the corresponding values of the constant are given. In the neighborhood of a local maximum or minimum, the contours are roughly elliptic, while in the neighborhood of a saddle they are roughly hyperbolic. Mountain tops and saddles are connected by ridges, lake bottoms and saddles by valleys. This topographic representation is particularly convenient for hikers and mountain climbers because it contains a great deal more information than simply the locations of mountains, lakes, and saddles.

A representation which is particularly convenient for our purposes is shown in Fig. 5.2*b*. Here each critical point is represented by a dot. The type of critical point is indicated by arrows. These arrows are determined from the stability matrix V_{ij} at the critical point. Diagonalization of V_{ij} at a critical point gives the principal axes (directions of maximum/minimum ascent/descent). The eigenvalues associated with the principal axes indicate the magnitude and direction of the force along the principal axes. We take as standard convention that an arrow pointing toward a critical point indicates a force in that direction—the longer the arrow, the stronger the force. The local maxima, minima, and the saddles of the contour map in Fig. 5.2*a* are shown in Fig. 5.2*b*. It is clear that the principal axes in Fig. 5.2*b* can be connected to give the valleys and ridges which are apparent in Fig. 5.2*a*.

If a representation of $V(x_1, x_2)$ is given in terms of the locations, principal axes, and eigenvalues of critical points as in Fig. 5.2*b*, then the associated contour plot may be roughly sketched out, at least in the neighborhood of each critical point. How this is done is indicated in Fig. 5.2*c*.

If we pour water onto the contour map Fig. 5.2*a*, it will collect in lakes at the bottom of each valley. The minima around which the water gravitates are called **attractors**. Each attractor sits at the bottom of some watershed or basin, called a **basin of attraction**. There is one attractor (local minimum) within each basin of attraction. Attractors are separated by saddles, ridges, and maxima, which constitute the boundary between different basins of attraction. The basins of

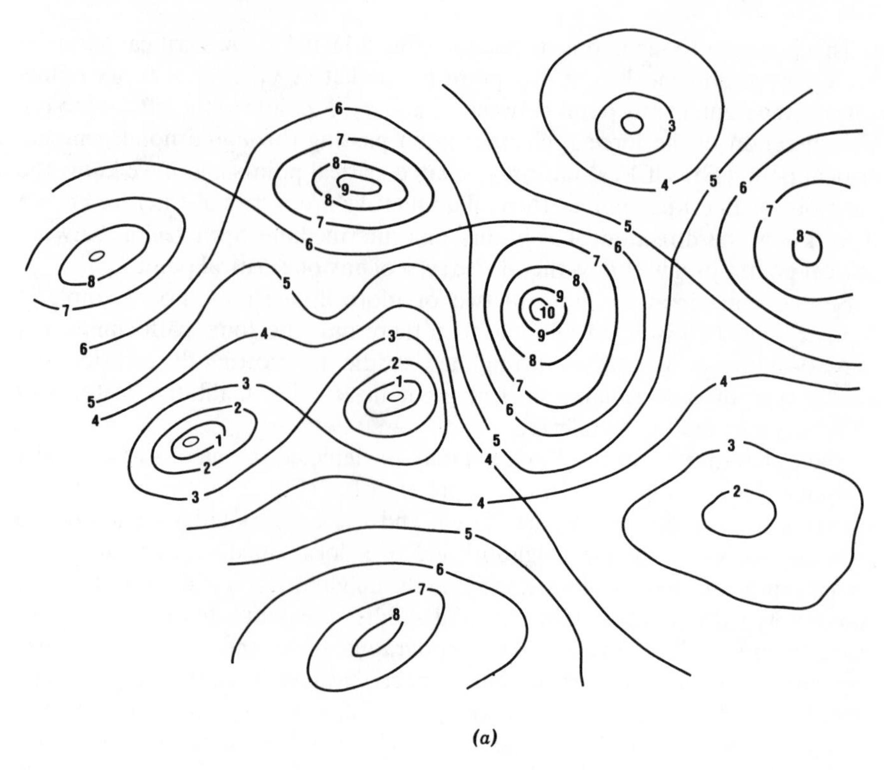

(a)

Figure 5.2 (*a*) A contour representation for a potential function of two variables.

attraction for the contour map of Fig. 5.2*a* can easily be located, at least qualitatively, from Fig. 5.2*b*.

The representation of $V(x)$ given in Fig. 5.2*b* can more easily be extended to the case of n (>2) state variables than the contour representation shown in Fig. 5.2*a*. There is one basin of attraction for each Morse 0-saddle. The outlines of the basin may be determined by connecting i-saddles with adjacent $(i+1)$-saddles $(i > 0)$. This process was illustrated in Fig. 5.2*b* for $\mathbb{R}^2$. In Fig. 5.1 for $\mathbb{R}^1$, the only non-0-saddle occurs at x_2, which divides $\mathbb{R}^1$ into the left-hand basin with attractor at x_1 and a right-hand basin with attractor at x_3. The principal axis is the x axis, and the corresponding force directions and eigenvalues of d^2V/dx^2 are shown below the critical points.

2. The "Crowbar Principle" for a Family of Potentials

We consider next a typical family of potentials $V(x_1, \ldots, x_n; c_1, \ldots, c_k)$. If c^0 is a random, or typical, point in $\mathbb{R}^k$, then the potential which this point parameterizes is a Morse function with only isolated critical points. Since $V(x; c^0)$ is

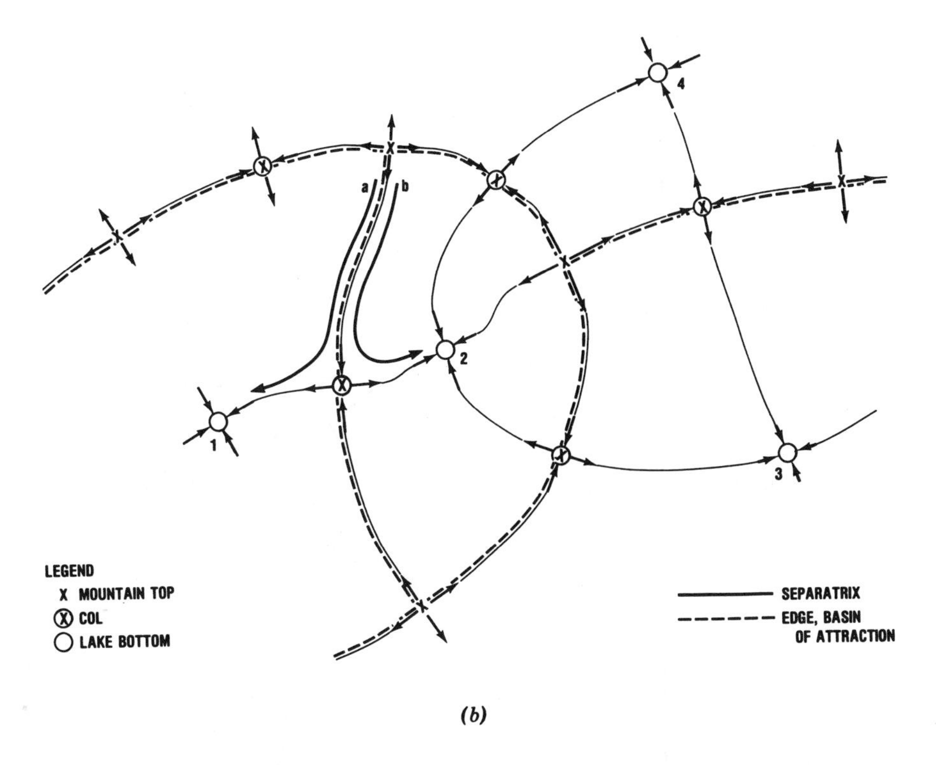

Figure 5.2 (*b*) A dynamical system representation of the same function of two variables.

a Morse function and either the Implicit Function Theorem or the Morse Lemma is valid at every point $x \in \mathbb{R}^n$ for almost every point $c^0 \in \mathbb{R}^k$, why bother with the Thom Theorem for the occasional point (measure zero, but generally not isolated) $c^0 \in \mathbb{R}^k$ at which V is not a Morse function?

The answer is that the non-Morse functions $V(x; c^0)$ have an importance that the Morse functions do not have. Basically they organize the entire qualitative nature of the *family* of potentials $V(x; c)$.

To see how this comes about, we choose a typical family of potentials $V(x; c)$, and we let $c^0 \in \mathbb{R}^k$ be a point at which $V(x; c^0)$ is a Morse function. Then $V(x; c^0)$ has only isolated critical points. We want to investigate how the positions of these critical points change as the control parameters c^0 change. Because the isolated critical points determine the attractors and basins of attraction of V, we are in reality studying how the "surface topography" of the potential $V(x; c^0)$ changes with changing c^0. Assume that x^0 is some critical point when $c = c^0$. When c^0 changes to $c^0 + \delta c^0$, the position of the critical point moves from x^0 to $x^0 + \delta x^0$. We wish to compute the changes δx_j^0 in terms of the δc_α^0 for any critical point.

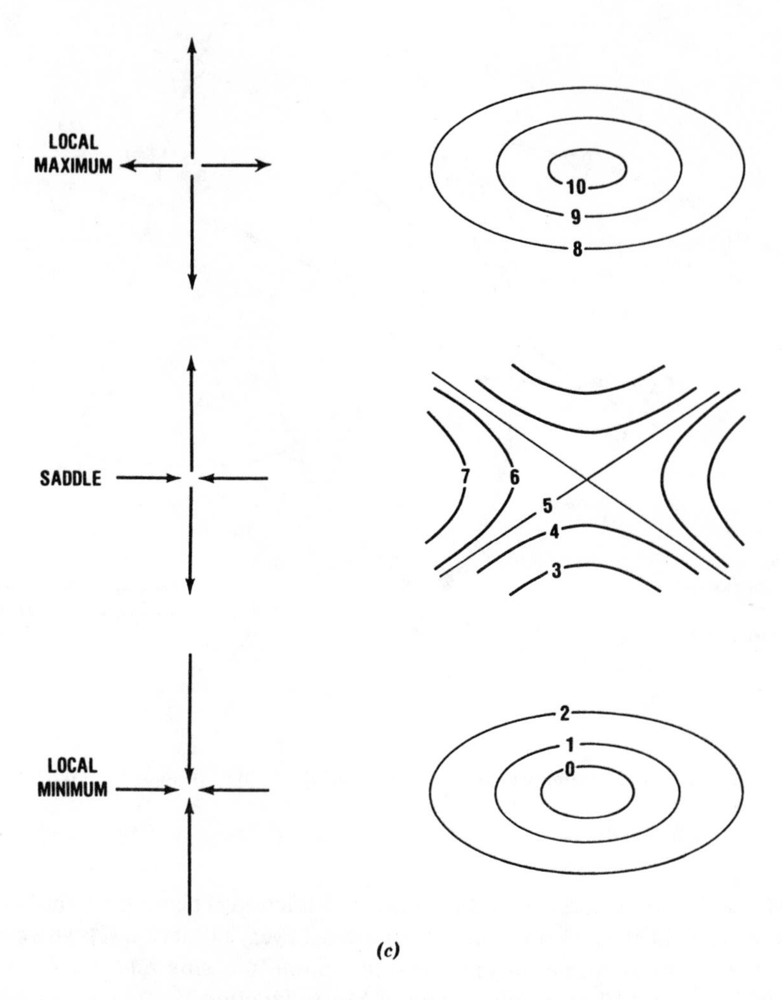

Figure 5.2 (*c*) The relation between representations (*a*) and (*b*) is particularly easy to carry out in the neighborhood of isolated critical points.

This is done most simply by making a Taylor series expansion of V about the point $(x^0; c^0)$ in $\mathbb{R}^n \otimes \mathbb{R}^k$:

$$\begin{aligned} V(x; c) = \; & V(x^0; c^0) \\ & + (x - x^0)_i V_i + (c - c^0)_\alpha V_\alpha \\ & \frac{1}{2!}(x - x^0)_i (x - x^0)_j V_{ij} + (x - x^0)_i (c - c^0)_\alpha V_{i\alpha} + \frac{1}{2!}(c - c^0)_\alpha \\ & \times (c - c^0)_\beta V_{\alpha\beta} + \mathcal{O}(3) \end{aligned} \quad (5.1)$$

All derivatives are evaluated at $(x^0; c^0)$. Since x^0 is a critical point, $V_i = 0$. Now we let $c = c^0 + \delta c^0$ and search for the equilibrium of $V(x; c^0 + \delta c^0)$ which is near x^0. This is done by setting the first derivatives $\partial V/\partial x_i = 0$:

$$\frac{\partial V}{\partial x_i} = V_{ij}\delta x_j^0 + V_{i\alpha}\delta c_\alpha^0 + \mathcal{O}(2) = 0$$

Neglecting the second-order terms, we can solve for δx_j^0 in terms of the changes δc_α^0 whenever V_{ij} is nonsingular at $(x^0; c^0)$:

$$\delta x_j^0 = -(V^{-1})_{ji} V_{i\alpha} \delta c_\alpha^0$$

or

$$\frac{\partial x_j^0}{\partial c_\alpha^0} = -(V^{-1})_{ji} V_{i\alpha} \tag{5.2}$$

As the control parameters change, the location of the critical point changes and the value of the function at the critical point also changes. The change in the critical value due to a change δc in the control parameters is determined by substituting (5.2) into (5.1), with the result

$$V(x^0 + \delta x; c^0 + \delta c) = V(x^0; c^0) + \delta^{(1)}V + \delta^{(2)}V + \cdots \tag{5.3}$$

$$\delta^{(1)}V = V_\alpha \delta c^\alpha \tag{5.3;1}$$

$$\begin{aligned}\delta^{(2)}V &= \tfrac{1}{2}\delta c^\alpha[V_{\alpha\beta} - V_{\alpha i}(V^{-1})^{ij}V_{j\beta}]\delta c^\beta \\ &= \tfrac{1}{2}g_{\alpha\beta}\delta c^\alpha \delta c^\beta\end{aligned} \tag{5.3;2}$$

Equations (5.2) give the "linear response" of the locations of isolated critical points to small changes in the control parameters. Equation (5.3;1) gives the linear response of the critical value to small changes in the control parameters. It is convenient to represent the quadratic response in terms of the "metric tensor" $g_{\alpha\beta}$, defined in (5.3;2).

The important thing we learn from (5.2) is that small changes in the control parameters produce only small changes in the positions of isolated critical points. Further, since the eigenvalues and eigenvectors of V_{ij} are smooth functions of c at isolated critical points, small changes in the control parameters produce only small changes in the eigenvalues and the eigenvectors. From the discussion of Sec. 1 this means that the attractors and the basins of attraction of $V(x; c^0 + \delta c^0)$ differ only slightly from those of $V(x; c^0)$, so that these two functions are qualitatively (topographically) similar.

However, if $V(x; c^0)$ is not a Morse function, then this potential has at least one degenerate critical point (x^0), and (5.2) is not valid at that critical point because the matrix V_{ij} is not invertible there. As $c \to c^0$, two or more isolated critical points converge toward $x = x^0$. As they do, their principal axes become mutually parallel, and the eigenvalues associated with one or more principal directions tend toward zero.

The occurrence of degenerate critical points corresponds to a qualitative change in the topography of the family of potentials. This should not be too surprising. The topography is determined by the distribution and type of isolated critical points. When two or more critical points become degenerate, the topography must change.

A function is said to be **structurally stable** if its qualitative properties (number and types of critical points, basins of attraction, etc.) are not changed by a sufficiently small perturbation. It is clear from (5.2) that the Morse functions in a family of functions are structurally stable. Further, the set of $c \in \mathbb{R}^k$ at which $V(x; c)$ is not a Morse function forms a separatrix dividing $\mathbb{R}^k$ into open regions parameterizing structurally stable functions of qualitatively different types. In the following five sections we determine the separatrices parameterizing structurally unstable functions for those catastrophes listed in Table 2.2 for which the space of control parameters can easily be visualized ($k \leq 3$).

3. Example 1: Control Space of A_2

Suppose $V(x_1, x_2, \ldots, x_n; c)$ is a general 1-parameter family of potentials. Then in this family we can expect to find individual functions which have a non-Morse critical point. For such a function the decomposition (2.3b) is valid in the neighborhood of the critical point (Chapter 3, Sec. 4). In fact, for any function near the non-Morse function the decomposition (2.4) is valid (Chapter 4, Sec. 4). The Morse part of this decomposition is not qualitatively affected by perturbations, so it is only necessary to study how the qualitative properties of the catastrophe function are changed as the control parameters are changed. These remarks are valid independent of the catastrophe, and will not be repeated in the introductions to the following four sections. In those sections we already assume that the catastrophe has been pulled out of the family of functions, and we get right down to the business of studying it.

The catastrophe A_2 is given by

$$A_2\colon \quad F(x; a) = \tfrac{1}{3}x^3 + ax \tag{5.4;0}$$

We remark here the coefficients in simple catastrophe germs can be chosen to have canonical values like $\pm 1, 0$. Sometimes, as when taking derivatives, other canonical values are more convenient. Such values are simply generated by a scale change. Such factors of convenience may also be introduced into the perturbation.

The critical points of $F(x; a)$ are determined by setting the gradient of $F(x; a)$ equal to zero. The doubly degenerate critical point of $F(x; a)$ is determined by setting $d^2F/dx^2 = 0$ also:

1. Critical points: $\qquad x^2 + a = 0 \qquad$ (5.4;1)
2. Two-fold degenerate: $\qquad 2x \quad = 0 \qquad$ (5.4;2)

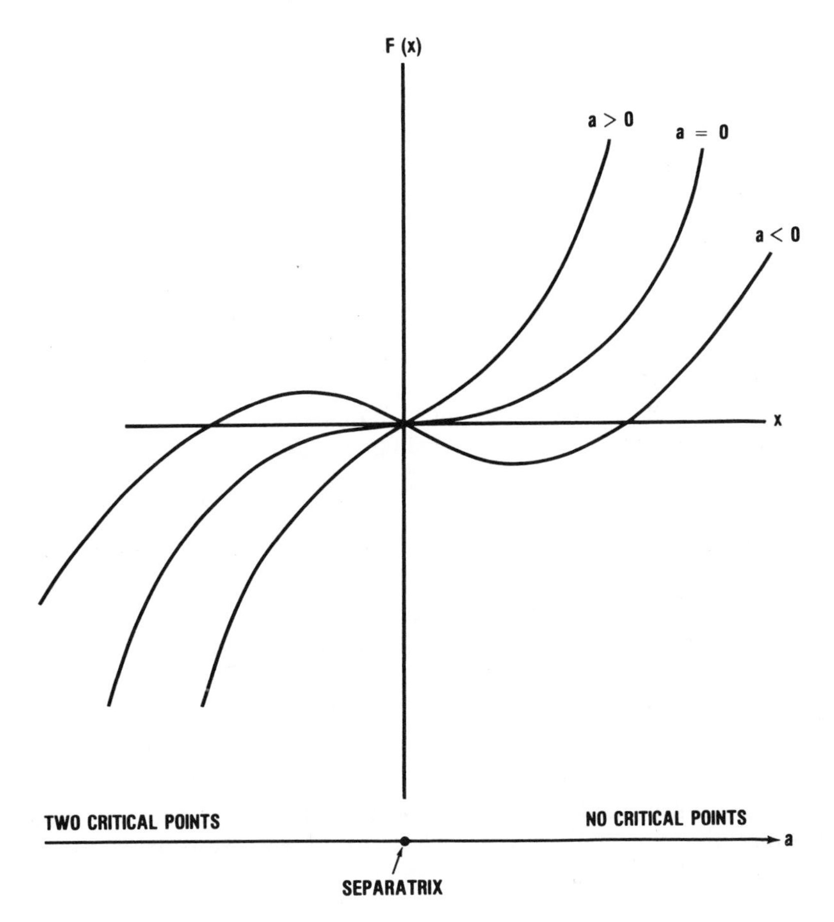

Figure 5.3 All functions $F(x; a) = \frac{1}{3}x^3 + ax$ with $a > 0$ are of qualitatively the same type, having no critical points. All functions with $a < 0$ are also qualitatively similar. The two qualitatively distinct types have a common boundary, the separatrix with $a = 0$.

The doubly degenerate critical point occurs at $x = 0$ from (5.4;2), and from (5.4;1) for $a = 0$. Thus the isolated point $a = 0$ is the separatrix in control parameter space between functions of two qualitatively different types. To determine these types, we have only to choose some $a > 0$ and determine the qualitative properties of $F(x; a)$. Then from (5.4) all functions $F(x; a)$ with $a > 0$ are qualitatively of the same type (no critical points). The same argument applies to functions $F(x; a)$ with $a < 0$ (Fig. 5.3).

4. Example 2: Control Space of A_3

The critical points, the doubly degenerate critical points, and the triply degenerate critical point of the A_3 catastrophe are determined by setting the first, second, and third derivatives of $F(x; a, b)$ equal to zero:

$$A_3: \quad F(x; a, b) = \tfrac{1}{4}x^4 + \tfrac{1}{2}ax^2 + bx \tag{5.5;0}$$

1. Critical points: $x^3 + ax + b = 0$ (5.5;1)
2. Two-fold degenerate: $3x^2 + a = 0$ (5.5;2)
3. Three-fold degenerate: $6x = 0$ (5.5;3)

At critical points (5.5;1) is valid. At doubly degenerate critical points both (5.5;1) and (5.5;2) are valid. At a triply degenerate critical point (5.5;1), (5.5;2), and (5.5;3) are all valid.

The location of the point in control parameter space which describes a function with a triply degenerate critical point is determined as follows:

$$(5.5;3) \Rightarrow x = 0 \xrightarrow{(5.5;2)} a = 0 \xrightarrow{(5.5;1)} b = 0 \tag{5.6}$$

The corresponding function $F(x; a = 0, b = 0) = x^4/4$ has a triply degenerate critical point at the origin.

The points in control parameter space which parameterize functions with a doubly degenerate critical point are determined from (5.5;2) and (5.5;1):

$$(5.5;2) \Rightarrow a = -3x^2 \xrightarrow{(5.5;1)} b = 2x^3 \tag{5.7}$$

If the location of the doubly degenerate critical point is x_c, then (5.7) defines the values of the control parameters a, b which describe a function with a doubly degenerate critical point at x_c. Equation (5.7) provides a parametric representation of the relation between a and b. A more direct relation between a and b can be obtained by eliminating x from (5.7):

$$\left(-\frac{a}{3}\right)^{1/2} = x = \left(\frac{b}{2}\right)^{1/3}$$

$$\left(-\frac{a}{3}\right)^{3} = \left(\frac{b}{2}\right)^{2}$$

$$\left(\frac{a}{3}\right)^{3} + \left(\frac{b}{2}\right)^{2} = 0 \tag{5.8}$$

Curve (5.8) is shown in Fig. 5.4.

The separatrix in control parameter space consists of the point $(a, b) = (0, 0)$ (5.6) and the "fold curve" (5.8). To find the qualitative properties of functions $F(x; a, b)$ parameterized by points in Region I, it is sufficient to choose *any* point in this region and study the qualitative properties of the corresponding function. We pick an easy point $(a, b) = (+1, 0)$. Then (5.5;1) has a single real solution at $x = 0$. This solution is a minimum. So by (5.3), all functions parameterized by controls in Region I have a single minimum. The qualitative

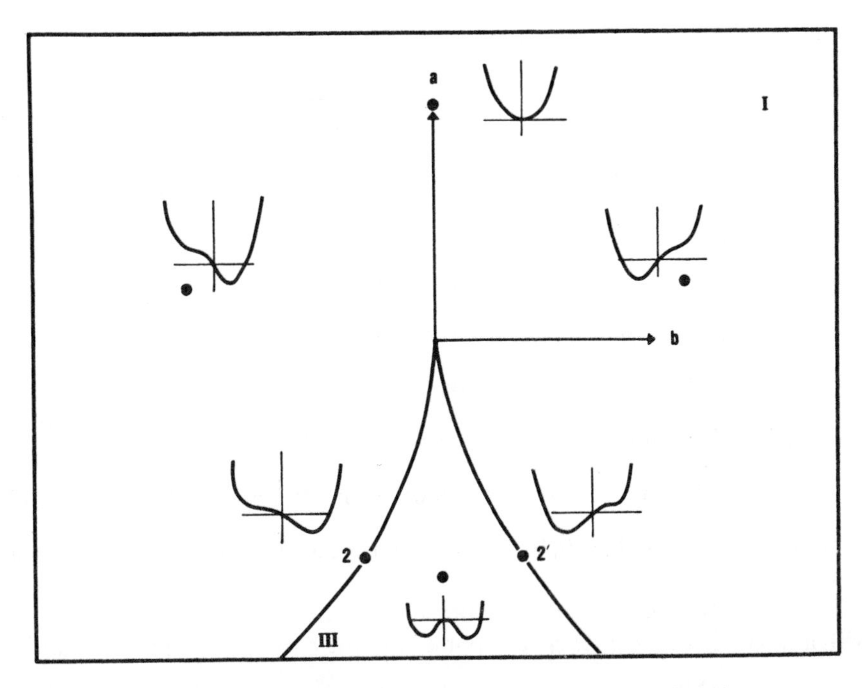

Figure 5.4 The separatrix of the cusp catastrophe, defined by $dF/dx = 0$, $d^2F/dx^2 = 0$, divides the control parameter space $\mathbb{R}^2$ into two open regions representing functions with one critical point and functions with three critical points. Points on the fold lines (5.8) represent non-Morse functions, while the point at the origin in $\mathbb{R}^2$ represents the function $x^4/4$ with a three-fold degenerate critical point at the origin in $\mathbb{R}^1$. The separatrix has components of dimension 1 (fold lines, with $ab \neq 0$) and of dimension 0. The potentials associated with some points in the control plane $\mathbb{R}^2$ have shapes as shown.

properties of functions in Region III are determined similarly. A convenient point in this region is $(a, b) = (-1, 0)$. At this point (5.5;1) has three solutions, a local maximum at $x = 0$ and two local minima at $x = +1$ and $x = -1$. Therefore all functions parameterized by points in Region III must have two local minima and one local maximum.

What about functions parameterized by points on the cusp-shaped fold lines? At the point 2 in Fig. 5.4, both a and b assume negative values. From (5.7), the corresponding doubly degenerate critical point x_c must also have a negative value. Since the sum of the three roots of (5.5;1) is zero (because the coefficient of the quadratic term is zero), and the two degenerate roots are negative, the third must be positive. Therefore, along the left-hand fold line the local maximum and the left-hand minimum become degenerate. Similarly, at 2′ the local maximum and right-hand minimum become degenerate (Fig. 5.4). At the intersection of the two fold lines at $(a, b) = (0, 0)$, the local maximum and the two minima of the corresponding function are degenerate.

5. Example 3: Control Space of A_4

The critical points of degeneracy j (isolated critical points have $j = 1$) are obtained by setting the first j derivatives of $F(x; a, b, c)$ equal to zero:

$$A_4: \quad F(x; a, b, c) = \tfrac{1}{5}x^5 + \tfrac{1}{3}ax^3 + \tfrac{1}{2}bx^2 + cx \tag{5.9;0}$$

1. Critical point: $x^4 + ax^2 + bx + c = 0$ (5.9;1)
2. Twofold degenerate: $4x^3 + 2ax + b = 0$ (5.9;2)
3. Threefold degenerate: $12x^2 + 2a = 0$ (5.9;3)
4. Fourfold degenerate: $24x = 0$ (5.9;4)

The fourfold degenerate critical point is determined as in (5.6):

$$(5.9;4) \Rightarrow x = 0 \xrightarrow{(5.9;\,3)} a = 0 \xrightarrow{(5.9;\,2)} b = 0 \xrightarrow{(5.9;\,1)} c = 0 \tag{5.10}$$

That is, the function $F(x; 0, 0, 0)$ has a fourfold degenerate critical point at $x = 0$.

The curves of points in control parameter space $\mathbb{R}^3$ which describe functions with a threefold degenerate critical point have the parametric representation

$$(5.9;3) \Rightarrow a = -6x^2 \xrightarrow{(5.9;\,2)} b = 8x^3 \xrightarrow{(5.9;\,1)} c = -3x^4 \tag{5.11}$$

These lines are shown in Fig. 5.6, whose construction we now discuss.

The surfaces of points in control parameter space which describe functions with a twofold degenerate critical point have the parametric representation

$$(5.9;2) \Rightarrow b = -4x^3 - 2ax \xrightarrow{(5.9;\,1)} c = 3x^4 + ax^2 \tag{5.12}$$

Now both the location of the doubly degenerate critical point x and the value of the control parameter a enter into the parametric representation of the 2-dimensional surface (5.12). To study the surface (5.12), we make the following observation. If we scale x by a factor λ and a by a factor λ^2, then b scales like λ^3 and c like λ^4:

$$\text{If} \quad \begin{matrix} x \to \lambda x \\ a \to \lambda^2 a \end{matrix} \quad \text{then} \quad \begin{matrix} b \to \lambda^3 b \\ c \to \lambda^4 c \end{matrix} \tag{5.13}$$

Therefore we need to determine the b–c cross sections of the surface (5.12) in three planes, say $a = +1$, $a = 0$, $a = -1$. Then we can scale these cross sections using (5.13) to obtain the entire surface. These cross sections are shown in Fig. 5.5. They are computed by fixing the value of a and letting x vary. The cross sections are pieced together by scaling to give the surface shown in Fig. 5.6.

This surface divides $\mathbb{R}^3$ into three open regions. The qualitative properties of functions parameterized by points within any one region are the same. These properties change as we pass through this surface. Once again, to determine the

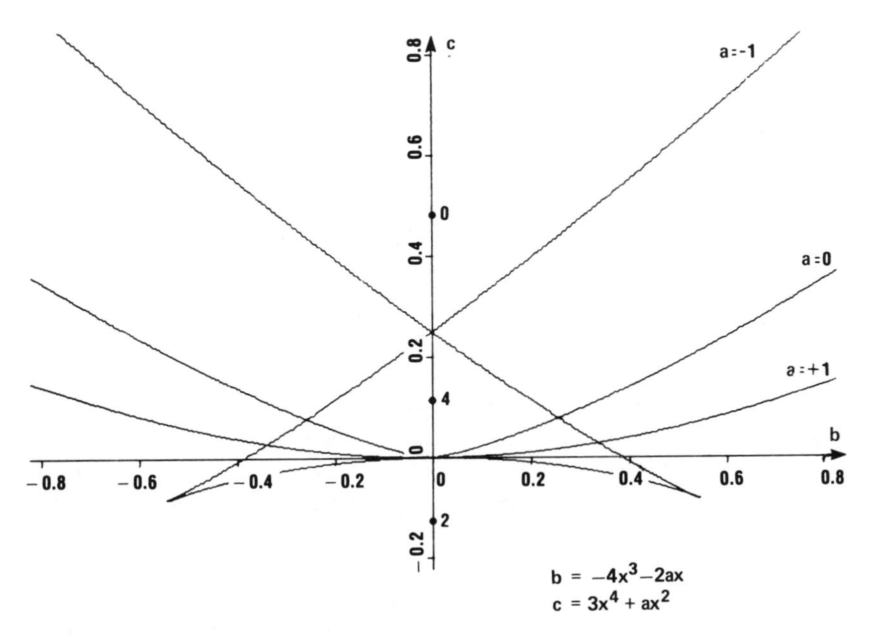

Figure 5.5 The three cross sections with $a = +1$, $a = 0$, $a = -1$ of the separatrix of A_4 in $\mathbb{R}^3$ are computed by fixing a and letting x vary, giving parametric representations (5.12) for b and c.

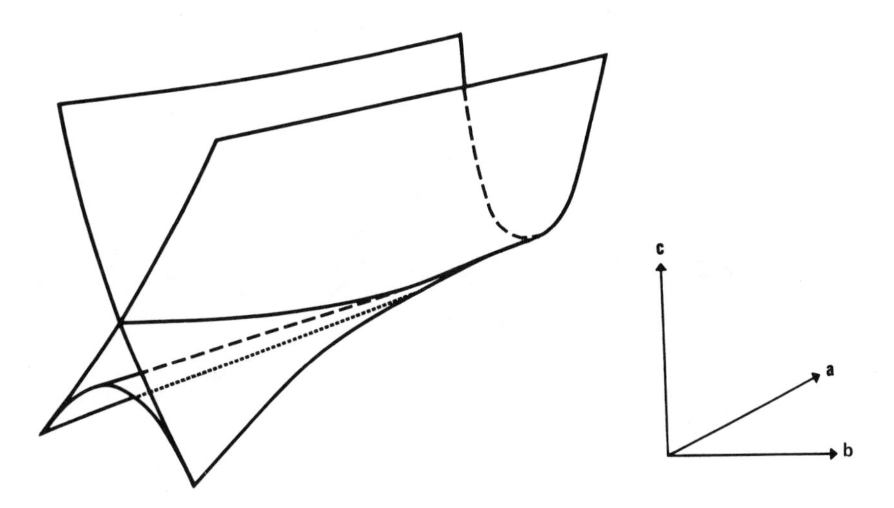

Figure 5.6 The three cross sections shown in Fig. 5.5 may be joined together by the scaling relations (5.13) to contruct the separatrix for A_4 in $\mathbb{R}^3$.

qualitative types of any region it is totally adequate to look at convenient points. Three such points are shown as 0, 2, and 4 in Fig. 5.5. The equation

$$(x^2)^2 + ax^2 + c = 0$$

has as solutions

$$x = \pm\left[-\frac{a}{2} \pm \sqrt{\left(\frac{a}{2}\right)^2 - c}\right]^{1/2}$$

For $a > 0$, there are no real solutions for $c > 0$ and two real solutions for $c < 0$. When $a < 0$, there are no real solutions for $c > (a/2)^2$, two real solutions for $c < 0$, and four real solutions for $0 < c < (a/2)^2$.

We can also ask what kind of functions correspond to points in the surface (5.12), on the edges (5.11), or on the self-intersection line. At point 2 in Fig. 5.7 the doubly degenerate critical point occurs at $x = 0$ from (5.12), so that isolated critical points occur on either side of it. At 1 the threefold degenerate critical point occurs for $x < 0$ from (5.11), and at 3 for $x > 0$. The remainder of Fig. 5.7 can be filled in by commonsense or, if this fails, by continuity arguments.

On the line S_l in Fig. 5.7 the two left-hand critical points are degenerate, while on S_r the two right-hand critical points are degenerate. The line of intersection of the left- and right-hand sheets (Fig. 5.6) represents functions with

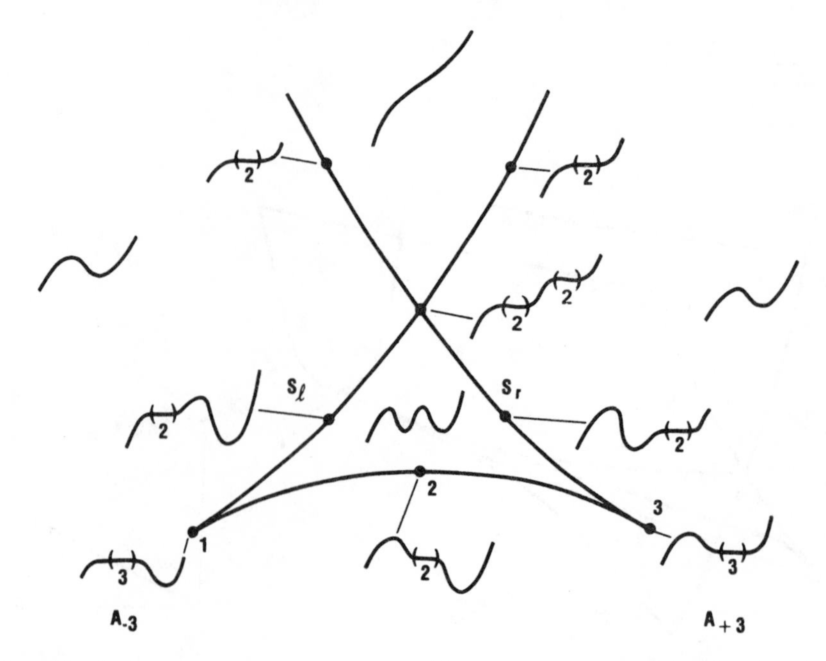

Figure 5.7 Each point on the separatrix shown in Fig. 5.6 represents a non-Morse function. Some non-Morse functions associated with different components of the separatrix are shown here.

two doubly degenerate critical points. On this line $b = 0$ and $c = (a/2)^2$ so that (5.9;1) becomes

$$x^4 + ax^2 + \left(\frac{a}{2}\right)^2 = \left(x^2 + \frac{a}{2}\right)^2 = 0$$

This has double roots at $x = -\sqrt{-a/2}$ and $x = +\sqrt{-a/2}$ for $a < 0$. The values of the potential (5.9;0) at these two roots are shown in Fig. 5.8.

In short, the set of points at which $F(x; a, b, c)$ has a non-Morse critical point divides the control parameter space $\mathbb{R}^3$ into three open regions. Any point in $\mathbb{R}^3$ can be approximated arbitrarily closely by a sequence of points lying completely within one of these regions. That is, even the non-Morse functions can be approximated arbitrarily closely by Morse functions. The separatrix consists of a point, describing a function with a fourfold degenerate critical point,

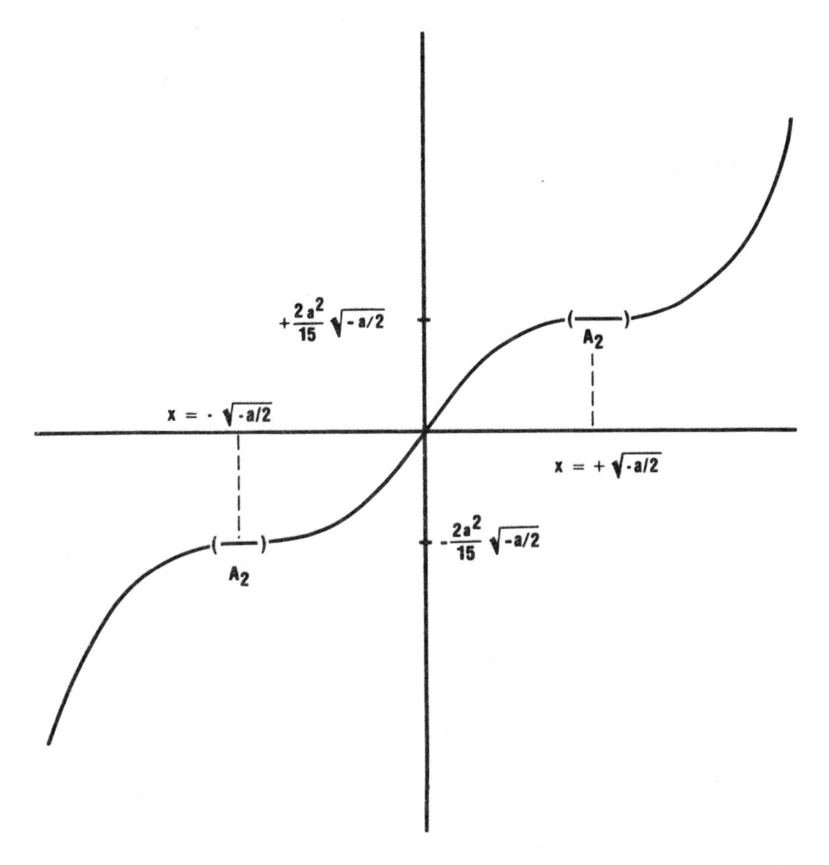

Figure 5.8 Points along the self-intersection of the two "fold" wings of the separatrix shown in Fig. 5.6 represent functions with a pair of doubly degenerate critical points.

two curves describing functions with threefold degenerate critical points, a curve describing functions with two twofold degenerate critical points, and three surfaces describing functions with doubly degenerate critical points.

6. Example 4: Control Space of D_{+4}

For catastrophes involving two state variables with three or more control parameters, we have seen already (Chapter 4, Sec. 6) that a perturbation can be chosen which eliminates two of the three quadratic monomials. More generally, a perturbation can be chosen which removes two of three independent linear combinations of degree 2 monomials, leaving only one such combination to be kept, along with the linear terms, as the most general perturbation. In this and the following section we choose the degree 2 terms in the perturbation to minimize our computational efforts.

The catastrophe function D_{+4} which we study is

$$D_{+4}\colon \quad F(x, y; a, b, c) = x^2y + \tfrac{1}{3}y^3 + a(y^2 - x^2) + bx + cy \qquad (5.14;0)$$

We observe now that this function has a number of symmetries, of which a useful one is $F \to +F$ when $(x, y) \to (-x, +y)$ and $(a, b, c) \to (+a, -b, +c)$. The critical points of F are determined by

$$\text{1a.} \qquad \frac{\partial F}{\partial x} = 2xy - 2ax + b = 0 \qquad (5.14;1a)$$

$$\text{1b.} \qquad \frac{\partial F}{\partial y} = x^2 + y^2 + 2ay + c = 0 \qquad (5.14;1b)$$

and the stability matrix is

$$F_{ij} = 2\begin{bmatrix} y - a & x \\ x & y + a \end{bmatrix} \qquad (5.14;2)$$

The fourfold degenerate critical point occurs when the stability matrix vanishes identically.

$$F_{ij} = 0 \xRightarrow{(5.14;2)} x = y = a = 0 \begin{array}{l} \xRightarrow{(5.14;1a)} b = 0 \\ \xRightarrow{(5.14;1b)} c = 0 \end{array} \qquad (5.15)$$

As a result the point $(a, b, c) = (0, 0, 0) \in \mathbb{R}^3$ in control parameter space parameterizes the function

$$F(x, y; 0, 0, 0) = x^2y + \tfrac{1}{3}y^3 \qquad (5.16)$$

which has a fourfold degenerate critical point at $(x, y) = (0, 0) \in \mathbb{R}^2$ in state variable space. The function (5.16) is shown in Fig. 5.9.

Z=X³+3XY²

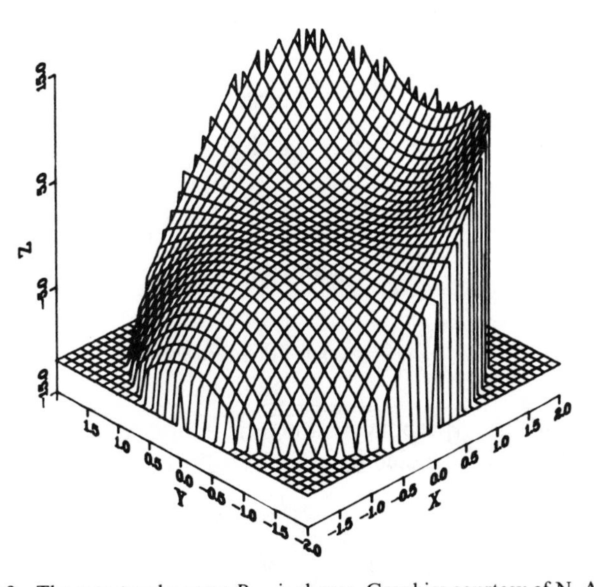

Figure 5.9 The catastrophe germ D_{+4} is shown. Graphics courtesy of N. A. Pope.

The twofold and threefold degenerate critical points are found by requiring one of the eigenvalues of the stability matrix to vanish. When this occurs, the determinant of (5.14;2) is zero:

$$\det F_{ij} = 4(y^2 - a^2 - x^2) = 0 \tag{5.17}$$

So the "bad set" is the hyperbola

$$y^2 - x^2 = a^2 \tag{5.18}$$

This means that whenever a critical point lies on this hyperbola, it is doubly or triply degenerate.

Now we can use the two variables x and y to give a parametric representation for the separatrix in control parameter space. The parametric representation for $a(x, y)$ is given in (5.18), and the parametric representations for b, c can be determined from (5.14;1a, b):

$$\begin{aligned} -b &= 2xy - 2ax \\ -c &= x^2 + y^2 + 2ay \end{aligned} \tag{5.19}$$

However, there is an easier way to approach this problem. We note that if x and y both scale like λ, then so also does a ($x \to \lambda x$, $y \to \lambda y \Rightarrow a \to \lambda a$). Then b and c both scale like λ^2:

$$\begin{matrix} x \to \lambda x \\ y \to \lambda y \end{matrix} \Rightarrow \begin{matrix} a \to \lambda a \\ b \to \lambda^2 b \\ c \to \lambda^2 c \end{matrix} \quad V \to \lambda^3 V \tag{5.20}$$

Therefore it is sufficient to determine the b–c cross section of the separatrix in planes $a = +1$, $a = 0$, $a = -1$, and then reconstruct the entire separatrix through the scaling relation (5.20). For $a > 0$, we determine how the upper half of the hyperbola parameterizes the b–c separatrix by making the substitutions

$$\begin{aligned} y &= a \cosh \theta \\ x &= a \sinh \theta \end{aligned} \qquad \text{(5.21 upper)}$$

Then

$$\begin{aligned} -\frac{b}{a^2} &= \sinh 2\theta - 2 \sinh \theta \\ -\frac{c}{a^2} &= \cosh 2\theta + 2 \cosh \theta \end{aligned} \qquad \text{(5.22 upper)}$$

The lower half of the hyperbola parameterizes another part of this separatrix, which can be easily constructed from the substitutions

$$\begin{aligned} y &= -a \cosh \theta \\ x &= -a \sinh \theta \end{aligned} \qquad \text{(5.21 lower)}$$

Then

$$\begin{aligned} -\frac{b}{a^2} &= \sinh 2\theta + 2 \sinh \theta \\ -\frac{c}{a^2} &= \cosh 2\theta - 2 \cosh \theta \end{aligned} \qquad \text{(5.22 lower)}$$

The symmetry $\theta \to -\theta \Rightarrow (x, y) \to (-x, +y)$ and $(a, b, c) \to (+a, -b, +c)$ has already been noted. The curves (5.22) are easily plotted and are shown in Fig. 5.10.

The slice $a = 0$ may be obtained easily from (5.18) and (5.19), and is shown in Fig. 5.10. Finally, the $a = -1$ slice may be obtained from (5.21) and (5.22) and it, too, is shown in Fig. 5.10.

The three cross sections shown in Fig. 5.10 may be strung together using the scaling relations (5.20) to give the separatrix, or bifurcation set, of the catastrophe D_{+4}, which is shown in Fig. 5.11. The separatrix can be viewed as two surfaces which intersect each other in the $a = 0$ plane. In front of the plane $a = 0$ one surface has a crease while behind the plane it has none. The other surface is a mirror image. These surfaces divide control parameter space $\mathbb{R}^3$ into four open regions which describe functions of four qualitatively different types.

To determine the qualitative properties of functions parameterized by points in each of these open regions, we investigate the properties of functions

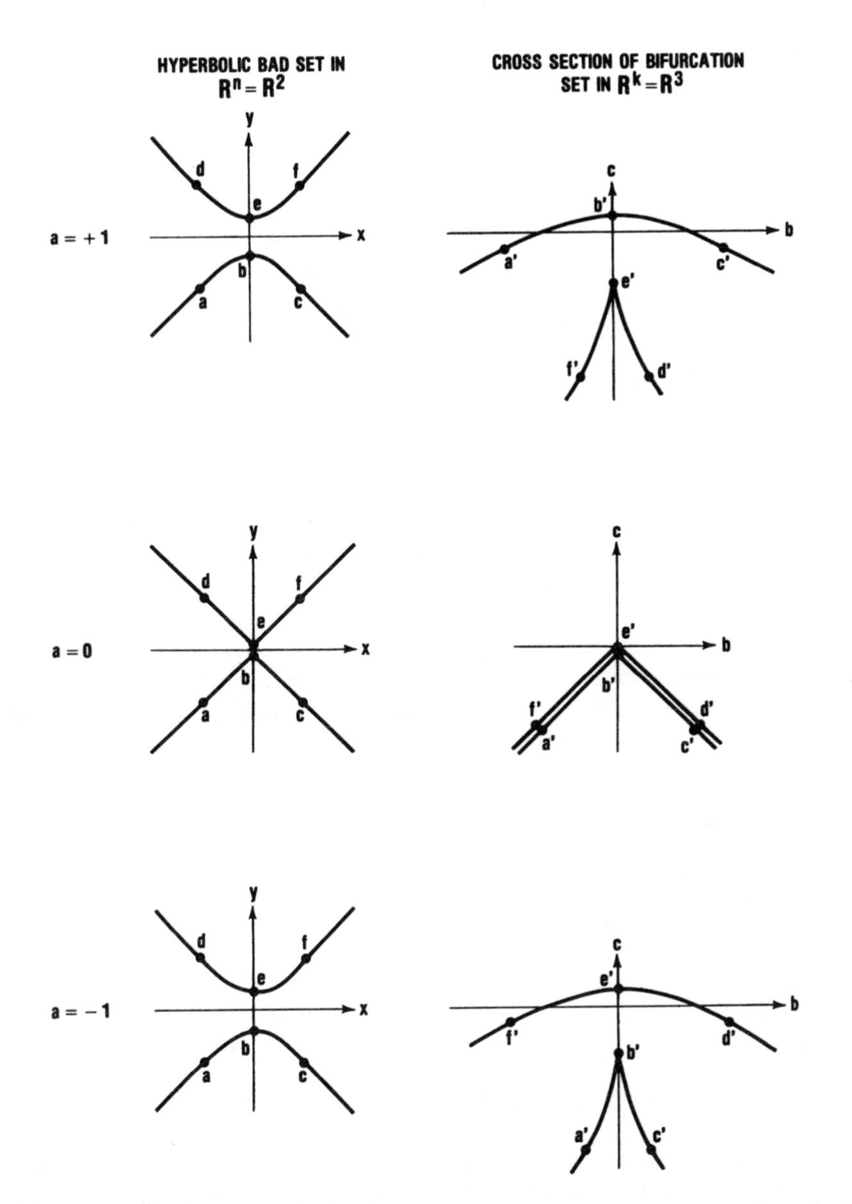

Figure 5.10 The relationship between the hyperbolic "bad set" and the separatrix of D_{+4} in $\mathbb{R}^3$ is shown for $a = +1$; $a = 0$; and $a = -1$.

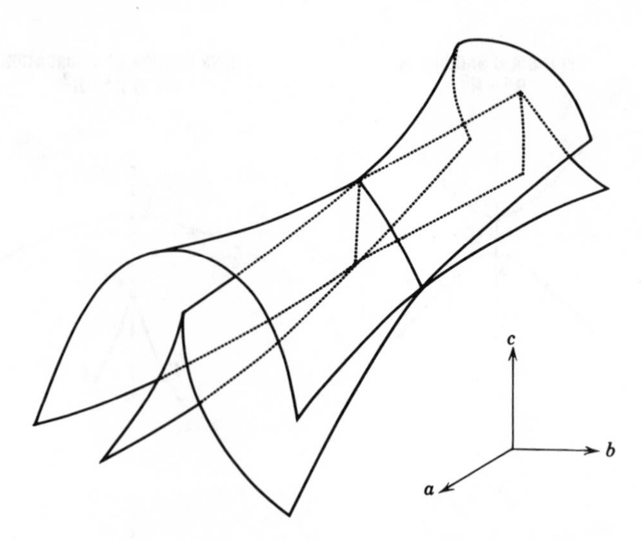

Figure 5.11 The separatrix for D_{+4} in $\mathbb{R}^3$ may be constructed from the three cross sections shown in Fig. 5.10 by using the scaling relations (5.20).

parameterized by points along the line $a = +1, b = 0$ in $\mathbb{R}^3$ (Fig. 5.12). Along this line the equations (5.14), which determine the critical points, are explicitly

$$2xy - 2x = 0 \tag{5.23a}$$

$$x^2 + y^2 + 2y + c = 0 \tag{5.23b}$$

From (5.23a) the critical points must have as coordinates either $x = 0$ or $y = +1$. If $x = 0$, we have, from (5.23b),

$$x = 0 \xRightarrow{(5.23b)} y = -1 \pm \sqrt{1 - c} \tag{5.24}$$

There is a real pair of critical points for $c < +1$. If $y = +1$, we have, from (5.23b),

$$y = 1 \xRightarrow{(5.23b)} x = \pm\sqrt{-(3 + c)} \tag{5.25}$$

This real pair of critical points exists for $c < -3$. The locations and range of c values for which the critical points (5.24) and (5.25) exist is summarized in Fig. 5.12.

Once the locations of the critical points are known, their stability properties (Morse saddle type) can be determined from the stability matrix (5.14;2). At the critical points (5.24), the stability matrix is

$$F_{ij} = 2\begin{bmatrix} y - 1 & 0 \\ 0 & y + 1 \end{bmatrix}, \qquad x = 0, \quad y = -1 \pm \sqrt{1 - c} \tag{5.26}$$

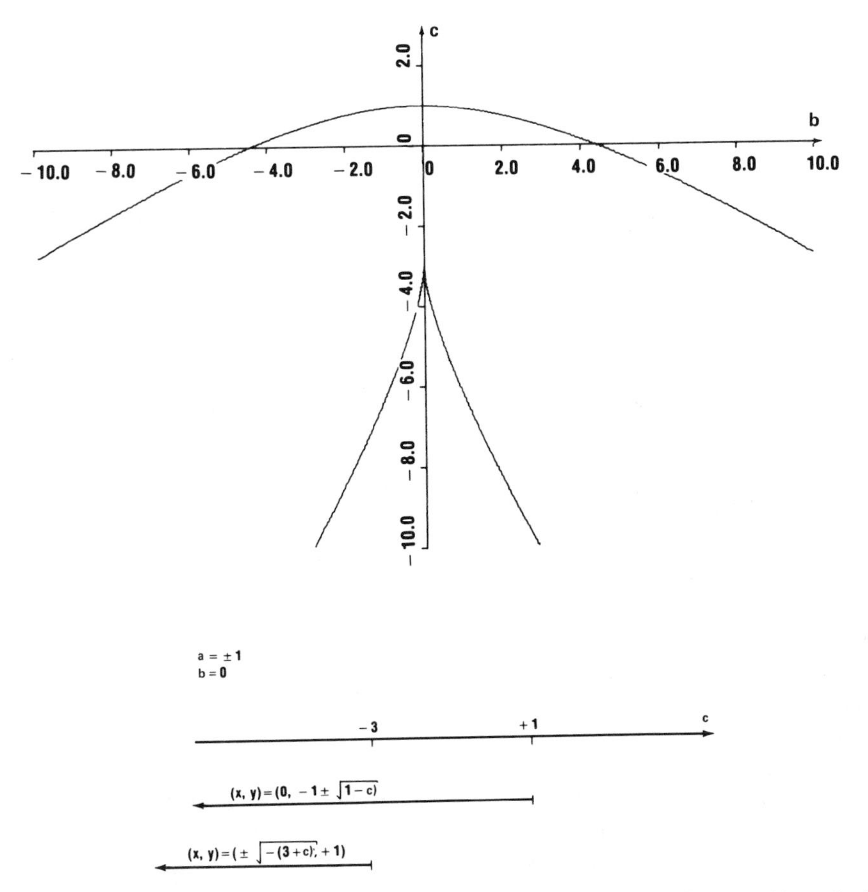

Figure 5.12 In the plane $a = +1$ the structurally stable Morse functions parameterized by each of the three open regions may be determined by setting $b = 0$ and varying c.

Thus the x and y axes are principal directions, with the eigenvalues in the x and y principal directions equal to $2(y - 1)$ and $2(y + 1)$, respectively. The x eigenvalue of the upper critical point changes sign when y goes through $+1$, while the y eigenvalue is zero when $y = -1$. These are the points where the critical points (5.24) cross the "bad set" (5.18), where we expect interesting things to happen. The locations and stability properties of these critical points are shown in Fig. 5.13.

The stability properties and principal axes of the critical points (5.25) are determined similarly. From the stability matrix we find

$$F_{ij} \xrightarrow{(5.25)} 2\begin{bmatrix} 0 & \pm\sqrt{-(3+c)} \\ \pm\sqrt{-(3+c)} & 2 \end{bmatrix} \Rightarrow \begin{array}{l} \lambda_1 = 2(1 + \sqrt{1 - (3+c)}) \\ \lambda_2 = 2(1 - \sqrt{1 - (3+c)}) \end{array} \tag{5.27}$$

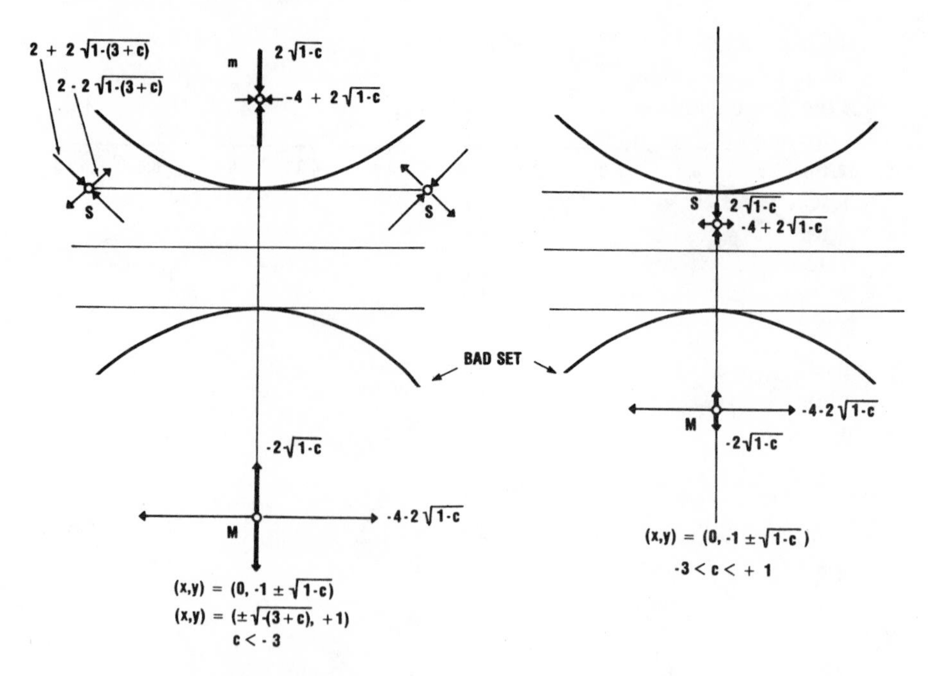

Figure 5.13 The critical points of the catastrophe D_{+4} are shown in the x–y plane as a function of the control parameter c for $a = +1, b = 0$.

The principal directions are no longer along the x and y axes, so they must be determined by a simple matrix diagonalization. The results are shown in Fig. 5.13. These two roots intersect the "bad set" (5.18) at the point $(x, y) = (0, 1)$ when $c = -3$. At this point the eigenvalue $\lambda_2 = 0$, and we expect more interesting things to happen.

Before discussing the interesting things that occur when $c = -3$ or $c = +1$, we should not lose sight of the fact that we have answered the question we originally set out to answer. For $c > 1$ there are no critical points, for $-3 < c < +1$ there is one maximum and one saddle, and for $c < -3$ there is one maximum, one minimum, and two saddles. This determines completely the qualitative properties of all functions parameterized by three of the four open regions shown in Fig. 5.11. In the fourth region (the "middle" region for $a < 0$) there is one minimum and one saddle.

As c approaches $+1$ from below, the two critical points at

$$(x, y) = (0, -1 \pm \sqrt{1 - c})$$

approach each other, become degenerate at $(0, -1)$ when $c = +1$, and disappear for $c > 1$. The eigenvalues in the x direction approach -4, while the eigenvalues in the y direction become increasingly small, and vanish in the limit $c = +1$. This is a catastrophe, because the number of isolated critical points has changed as c passes through $+1$ where det $F_{ij} = 0$. Since two critical points are involved, we expect this to be a "fold" or A_2 catastrophe. Since the x eigenvalue is nonzero and the y eigenvalue becomes zero, we expect the x direction to be the "good" direction and the y direction to be the "bad" direction.

These notions can be made precise by determining the canonical form for the non-Morse function $F(x, y; a, b, c)$ for $(a, b, c) = (1, 0, 1)$ at the doubly degenerate critical point $(x, y) = (0, -1)$

$$F(x, y; 1, 0, 1) = x^2y + \tfrac{1}{3}y^3 + (y^2 - x^2) + y \tag{5.28}$$

It is useful to expand this function about the degenerate critical point by making an inhomogeneous linear transformation $x = X$, $y = (y + 1) - 1 = Y - 1$. Then

$$F(X, Y) = X^2Y + \tfrac{1}{3}Y^3 - 2X^2 - \tfrac{1}{3} \tag{5.29}$$

The constant term is unimportant and can be dropped. The stability matrix for $F(X, Y)$ at the doubly degenerate critical point $(X, Y) = (0, 0)$ is

$$F_{ij}|_{(0,0)} = \begin{bmatrix} -4 & 0 \\ 0 & 0 \end{bmatrix} \tag{5.30}$$

Therefore X is a "good" direction and Y the "bad" direction.

We now attempt to put the function (5.29) into the canonical form (2.3b) by making a smooth change of variables:

$$\begin{aligned} x' &= X + \Delta_1 = X + A_{20}X^2 + A_{11}XY + A_{02}Y^2 + \cdots \\ y' &= Y + \Delta_2 = Y + B_{20}X^2 + B_{11}XY + B_{02}Y^2 + \cdots \end{aligned} \tag{5.31}$$

Since we suspect that the catastrophe germ is of type A_2 and the coefficient of x'^2 is -2, we can see if (5.31) can be chosen to satisfy the equality:

$$\begin{aligned} -2X^2 + X^2Y + \tfrac{1}{3}Y^3 &\overset{?}{=} -2x'^2 + ky'^3 \\ &= -2(\underset{2}{X^2} + \underset{3}{2X\Delta_1} + \underset{4}{\Delta_1^2}) \\ &\quad + k(\underset{3}{Y^3} + \underset{4}{3Y^2\Delta_2} + \underset{5}{3Y\Delta_2^2} + \underset{6}{\Delta_2^3}) \end{aligned} \tag{5.32}$$

The small number under each term indicates the monomials of lowest degree in that particular product. The equality is satisfied for terms of degree 2. For the terms of degree 3 we find

$$X^2Y + \tfrac{1}{3}Y^3 \overset{?}{=} -4X(A_{20}X^2 + A_{11}XY + A_{02}Y^2) + kY^3 \tag{5.33}$$

This equation is satisfied by choosing

$$A_{20} = 0, \qquad A_{11} = -\tfrac{1}{4}, \qquad A_{02} = 0, \qquad k = \tfrac{1}{3} \tag{5.34}$$

For the terms of degree 4 and higher, all coefficients on the left-hand side of (5.32) are zero. The corresponding terms on the right-hand side depend on the disposable coefficients A_{ij}, B_{ij} $(i + j \geq 2)$ which appear in (5.31). If, for a given degree, there are more disposable coefficients on the right-hand side of (5.32) than fixed coefficients on the left-hand side, then $F(X, Y)$ can be put into the canonical form $-2x'^2 + \frac{1}{3}y'^3$. For example, for terms of degree 5, there are six fixed coefficients (all zero) on the left-hand side of (5.32). Terms of degree up to degree 4 enter from Δ_1 (those of degree exactly 4 enter linearly), and those of degree up to 3 enter from Δ_2 (those of degree 3 enter linearly). The number of new (not present from degree 4 considerations) disposable coefficients is $5 + 4 = 9$, so there is enough "room" to choose these coefficients so that all monomials of degree 5 have vanishing coefficients. This counting argument is summarized for the lowest degrees in Table 5.1. From this table it is clear that $F(X, Y)$ can in fact be put into the canonical form we have assumed. Further, by a simple scaling transformation, we have the canonical form

$$F(X, Y) \doteq -x''^2 + y''^3 \tag{5.35}$$

where $x'' = (2)^{1/2}x'$ and $y'' = (3)^{-1/3}y'$.

We now turn our attention to the other interesting point along the line $a = 1$, $b = 0$, namely, $c = -3$. As c approaches -3 from below, the three critical points at $(x, y) = (\pm\sqrt{-(3 + c)}, 1)$ and $(0, -1 + \sqrt{1 - c})$ approach each other, form a triply degenerate critical point when $c = -3$ at $(x, y) = (0, 1)$, and continue as a single critical point as c increases above -3. The principal axes of these three critical points rotate to the x and y directions as c approaches

Table 5.1 Summary of the Counting Argument Used to Transform $F(x, y; a, b, c)$ to the Canonical Form (2.3b) at $(a, b, c) = (1, 0, 1)$

Left-Hand Side		Right-Hand Side			
		Δ_1		Δ_2	
Degree	Number of New Fixed Coefficients	Degree	Number of New Disposable Coefficients	Degree	Number of New Disposable Coefficients
4	5	3	4	2	3
5	6	4	5	3	4
6	7	5	6	4	5
Increase by 1 ↓		Increase by 1 ↓		Increase by 1 ↓	

-3 from below, and remain along these directions for c greater than -3. The eigenvalues in the y direction approach $+4$ near $c = -3$, while the eigenvalues in the x direction vanish when $c = -3$. The x eigenvalue of the critical point $(x, y) = (0, -1 + \sqrt{1 - c})$ changes sign as c passes through -3. This suggests strongly a catastrophe of type $A_{\pm 3}$, with y as "good" direction and x as "bad" direction.

These ideas can be put into concrete form by proceeding as described above. First, we expand $F(x, y; a, b, c)$ for $(a, b, c) = (1, 0, -3)$ about the threefold degenerate critical point $(x, y) = (0, 1)$ using $x = X, y = (y - 1) + 1 = Y + 1$:

$$F(X, Y) = X^2 Y + \tfrac{1}{3}Y^3 + 2Y^2 - \tfrac{5}{3} \tag{5.36}$$

The constant term is unimportant and will be dropped. The stability matrix for $F(X, Y)$ at $(0, 0)$ is

$$F_{ij}|_{(0,0)} = \begin{bmatrix} 0 & 0 \\ 0 & 4 \end{bmatrix} \tag{5.37}$$

Clearly X is the "bad" variable and Y the "good" variable.

Next we try to put (5.36) into a canonical form by introducing the axis-preserving nonlinear transformation (5.31). The canonical form we guess is $kx'^4 + 2y'^2$. Proceeding as before,

$$\begin{aligned} 2Y^2 + X^2 Y + \tfrac{1}{3}Y^3 &\stackrel{?}{=} 2y'^2 + kx'^4 \\ &= 2(\underset{2}{Y^2} + \underset{3}{2Y\Delta_2} + \underset{4}{\Delta_2^2}) \\ &\quad + k(\underset{4}{X^4} + \underset{5}{4X^3\Delta_1} + \underset{6}{6X^2\Delta_1^2} + \underset{7}{4X\Delta_1^3} + \underset{8}{\Delta_1^4}) \end{aligned} \tag{5.38}$$

The equality is satisfied for terms of degree 2. For terms of degree 3 we have

$$X^2 Y + \tfrac{1}{3}Y^3 \stackrel{?}{=} 4Y(B_{20}X^2 + B_{11}XY + B_{02}Y^2) \tag{5.39}$$

This is satisfied for $B_{20} = \frac{1}{4}, B_{11} = 0, B_{02} = \frac{1}{12}$. All terms of degree 4 are given by

$$\begin{aligned} 0 \stackrel{?}{=} &\ 4Y(B_{30}X^3 + B_{21}X^2Y + B_{12}XY^2 + B_{03}Y^3) \\ &+ 2(\tfrac{1}{4}X^2 + \tfrac{1}{12}Y^2)^2 + kX^4 \end{aligned} \tag{5.40}$$

This is satisfied provided

$$\begin{aligned} \tfrac{1}{8} + k &= 0 & \\ 4B_{21} + \tfrac{1}{12} &= 0 \qquad & B_{30} = 0 \\ 4B_{03} + \tfrac{1}{72} &= 0 & B_{12} = 0 \end{aligned} \tag{5.41}$$

That (5.38) can be satisfied for terms of all degrees greater than 4 can be shown using the type of counting arguments summarized in Table 5.1. For the sake of maintaining interest, we vary the argument somewhat by taking account of the symmetry $F(X, Y) = F(-X, Y)$ of (5.36). Under this symmetry, coefficients of monomials X^pY^q must necessarily be zero if p is odd. We demand that the canonical form for $F(X, Y)$ have the same type of symmetry. This means that $(X, Y) \to (-X, Y) \Rightarrow (x', y') \to (-x', y')$. Then not all coefficients in Δ_1, Δ_2 are disposable; some are constrained to be zero. We can now construct a table, analogous to Table 5.1, which shows the number of coefficients on the left- and right-hand sides of (5.38) which are not required by symmetry considerations to be zero. It is clear from Table 5.2 and (5.41) that

$$F(X, Y) \doteq 2y'^2 - \tfrac{1}{8}x'^4 \tag{5.42}$$

The coordinates can be rescaled by $x'' = (8)^{-1/4}x'$ and $y'' = (2)^{1/2}y'$ to give the canonical form

$$F(X, Y) \doteq -x''^4 + y''^2 \tag{5.43}$$

Remark. In view of the similarity between (5.29) and (5.36) one might be tempted to guess a canonical form $2y'^2 + kx'^3$ in (5.38). However, with this guess the equality in (5.38) cannot be satisfied. This is easy to see because a term kX^3 would then appear on the right-hand side of (5.39). This term could not be removed by any choice of disposable coefficients.

The transformation (5.31) from Cartesian coordinates (X, Y) to curvilinear coordinates (x', y') is illustrated in Fig. 5.14. A cross section of $F(x', y')$ for various values of $y' =$ constant shows clearly the occurrence of the "dual cusp" catastrophe $-x^4 = A_{-3}$.

This analysis has been carried out for $a = +1$. Further, we have simplified the calculations by doing them in the symmetry plane $b = 0$. When $b \neq 0$, the symmetry apparent in Fig. 5.13 is destroyed. Then for fixed $a = 1$ and $b \neq 0$, a line with c increasing will strike either one side or the other of the folded sheet in Fig. 5.11. These sheets represent fold catastrophes, where the minimum in Fig. 5.13 collides with and annihilates either the left-hand saddle or the right-hand saddle as they meet on the upper part of the hyperbolic "bad set" (5.18). The remaining of the upper saddles and the maximum collide and annihilate when they meet on the lower branch of the hyperbolic "bad set" as c reaches the upper curved surface in Fig. 5.11. When $b < 0$, the minimum collides with the right-hand saddle, and the maximum with the left-hand saddle.

This entire analysis can be repeated in the plane $a = -1$. This is unnecessary, however. The two saddles lie between the upper and lower branches of the hyperbolic "bad set." The minimum lies above the upper branch, and the maximum lies below the lower branch. As c increases, the maximum collides with and annihilates the left-hand saddle (if $b < 0$) on the lower hyperbolic branch, and for larger values of c the minimum and right-hand saddle annihilate each other on the upper branch.

In the plane $a = 0$, the self-intersection of the two surfaces describes the occurrence of two fold catastrophes. That is, a minimum and saddle annihilate each other at one point in the x–y plane, while a maximum and saddle simultaneously annihilate each other at some other point on the "bad set."

Table 5.2 Summary of the Counting Argument, Including Symmetry Considerations, Used to Transform $F(x, y; a, b, c)$ to the Canonical Form (2.3b) at $(a, b, c) = (1, 0, -3)$

Left-Hand Side		Right-Hand Side				
		Δ_1		Δ_2		
Degree	Number of New Fixed Coefficients	Degree	Number of New Disposable Coefficients	Degree	Number of New Disposable Coefficients	Total Number of New Disposable Coefficients
5	3	2	1	4	3	4
6	4	3	2	5	3	5
7	4	4	2	6	4	6
8	5	5	3	7	4	7
9	5	6	3	8	5	8
	Increase by "$\frac{1}{2}$" ↓					Increase by 1 ↓

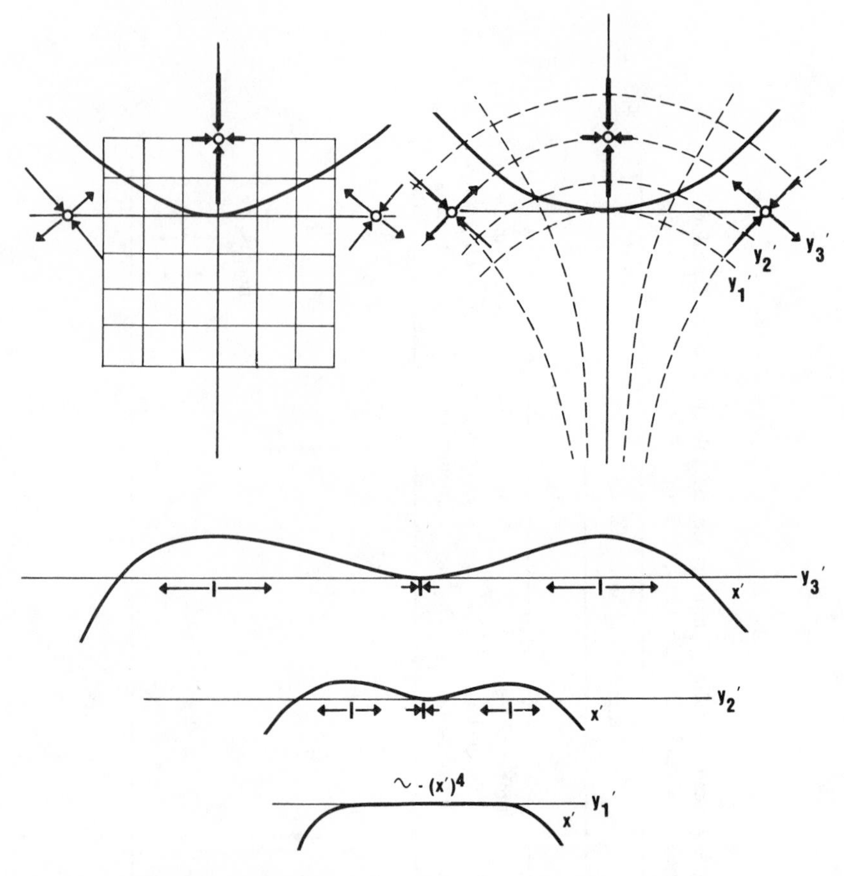

Figure 5.14 A smooth transformation from a Cartesian to a curvilinear coordinate system shows clearly how the dual cusp catastrophe is associated with the D_{+4} catastrophe.

7. Example 5: Control Space of D_{-4}

The catastrophe function D_{-4} which we study is

$$D_{-4}: \quad F(x, y; a, b, c) = x^2y - \tfrac{1}{3}y^3 + a(y^2 + x^2) + bx + cy \qquad (5.44;0)$$

We observe now that this function has a number of symmetries, of which one useful one is $F \to +F$ when $(x, y) \to (-x, +y)$ and $(a, b, c) \to (+a, -b, +c)$. The critical points of F are determined by

1a. $$\frac{\partial F}{\partial x} = 2xy + 2ax + b = 0 \qquad (5.44;1a)$$

1b. $$\frac{\partial F}{\partial y} = x^2 - y^2 + 2ay + c = 0 \qquad (5.44;1b)$$

and the stability matrix is

$$F_{ij} = 2\begin{bmatrix} y + a & x \\ x & -y + a \end{bmatrix} \tag{5.44;2}$$

The fourfold degenerate critical point occurs when the stability matrix vanishes identically:

$$F_{ij} = 0 \xrightarrow{(5.44;\,2)} x = y = a = 0 \xrightarrow{(5.44;\,1a)} b = 0$$
$$\xrightarrow{(5.44;\,1b)} c = 0 \tag{5.45}$$

As a result the point $(a, b, c) = (0, 0, 0) \in \mathbb{R}^3$ in control parameter space parameterizes the function

$$F(x, y; 0, 0, 0) = x^2y - \tfrac{1}{3}y^3 \tag{5.46}$$

which has a fourfold degenerate critical point at $(x, y) = (0, 0) \in \mathbb{R}^2$ in state variable space. The function (5.46) is shown in Fig. 5.15.

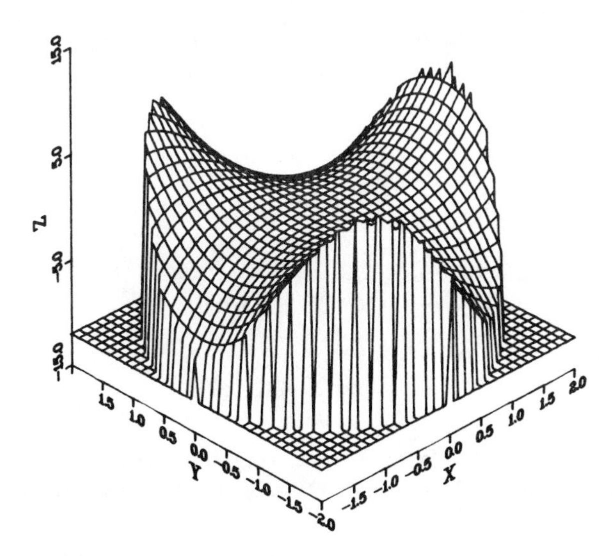

Figure 5.15 The catastrophe germ D_{-4} is shown. Graphics courtesy of N. A. Pope.

The two- and threefold degenerate critical points are found by requiring one of the eigenvalues of the stability matrix to vanish. When this occurs, the determinant of (5.44;2) is zero:

$$\det F_{ij} = 4(a^2 - y^2 - x^2) = 0 \tag{5.47}$$

The "bad set" is the circle

$$y^2 + x^2 = a^2 \tag{5.48}$$

This means that whenever a critical point lies on this circle, it is doubly or triply degenerate (fourfold if $a = 0$).

We can now use the two variables x and y to give a parametric representation for the separatrix in control parameter space. The parametric representation for $a(x, y)$ is given in (5.48), and the parametric representations for b, c can be determined from (5.44;1a) and (5.44;1b):

$$\begin{aligned} -b &= 2xy + 2ax \\ -c &= x^2 - y^2 + 2ay \end{aligned} \tag{5.49}$$

As usual, there is an easier way to approach this problem, and it involves scaling. The scaling relations for (5.44) are identical to those for (5.14):

$$\begin{matrix} x \to \lambda x \\ y \to \lambda y \end{matrix} \Rightarrow \begin{matrix} a \to \lambda a \\ b \to \lambda^2 b \\ c \to \lambda^2 c \end{matrix} \tag{5.50}$$

Therefore it is sufficient to determine the b–c cross section of the separatrix in the planes $a = +1$, $a = 0$, $a = -1$, and then reconstruct the entire separatrix through these scaling relations. For $a > 0$ we determine how the circular "bad set" parameterizes the b–c separatrix by making the substitutions [cf. (5.21), (5.22)]

$$\begin{aligned} y &= -a \cos\theta \\ x &= a \sin\theta \end{aligned} \tag{5.51}$$

Then

$$\begin{aligned} \frac{b}{a^2} &= \sin 2\theta - 2\sin\theta \\ \frac{c}{a^2} &= \cos 2\theta + 2\cos\theta \end{aligned} \tag{5.52}$$

The symmetry $\theta \to -\theta \Rightarrow (x, y) \to (-x, +y)$ and $(a, b, c) \to (+a, -b, +c)$ has already been noted. The curve (5.52) is easily plotted and is shown in Fig. 5.16.

The slice $a = 0$ may be obtained easily from (5.48) and (5.49) and is shown also in Fig. 5.16. Finally, the $a = -1$ slice may be obtained from (5.51) and (5.52)

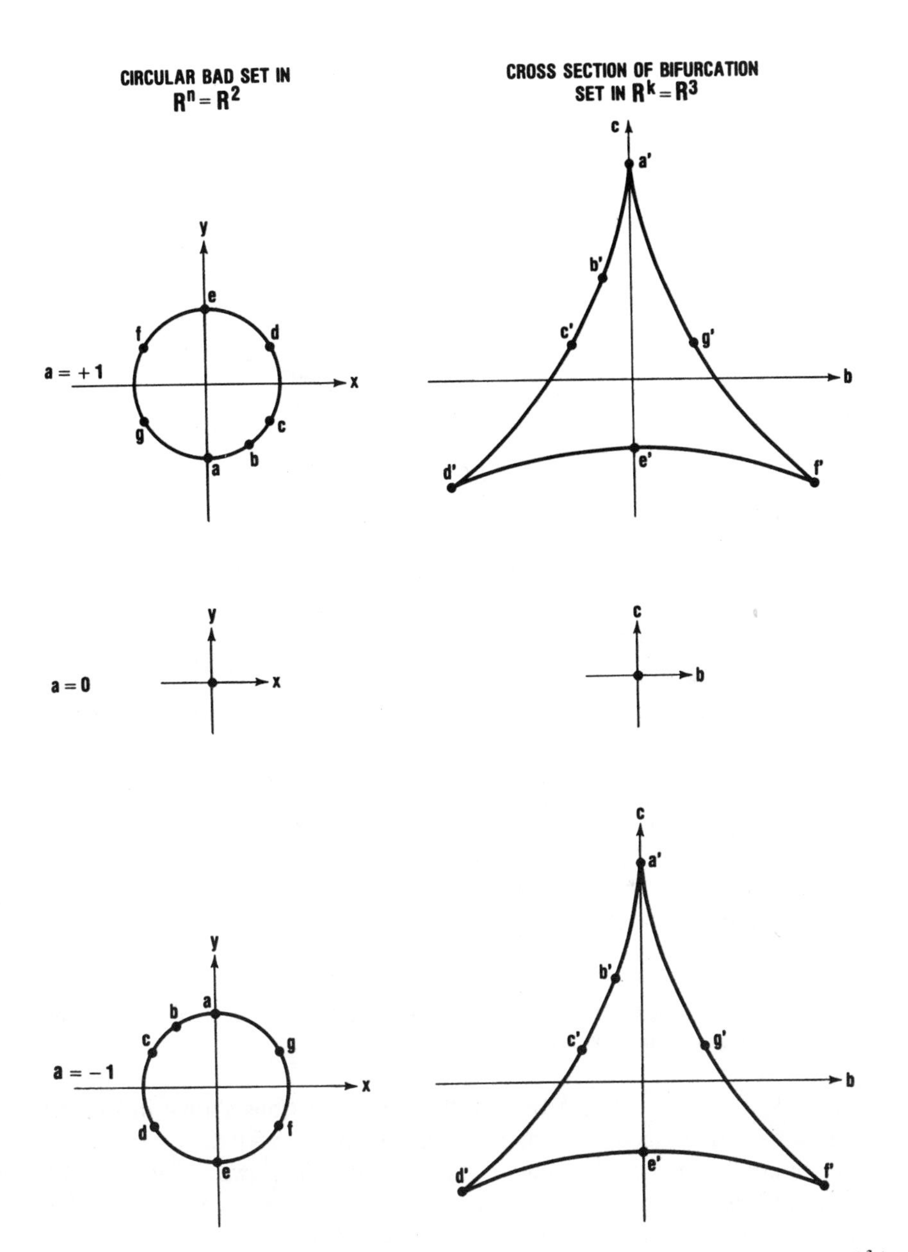

Figure 5.16 The relationship between the circular "bad set" and the separatrix of D_{-4} in $\mathbb{R}^3$ is shown for $a = +1$; $a = 0$; and $a = -1$.

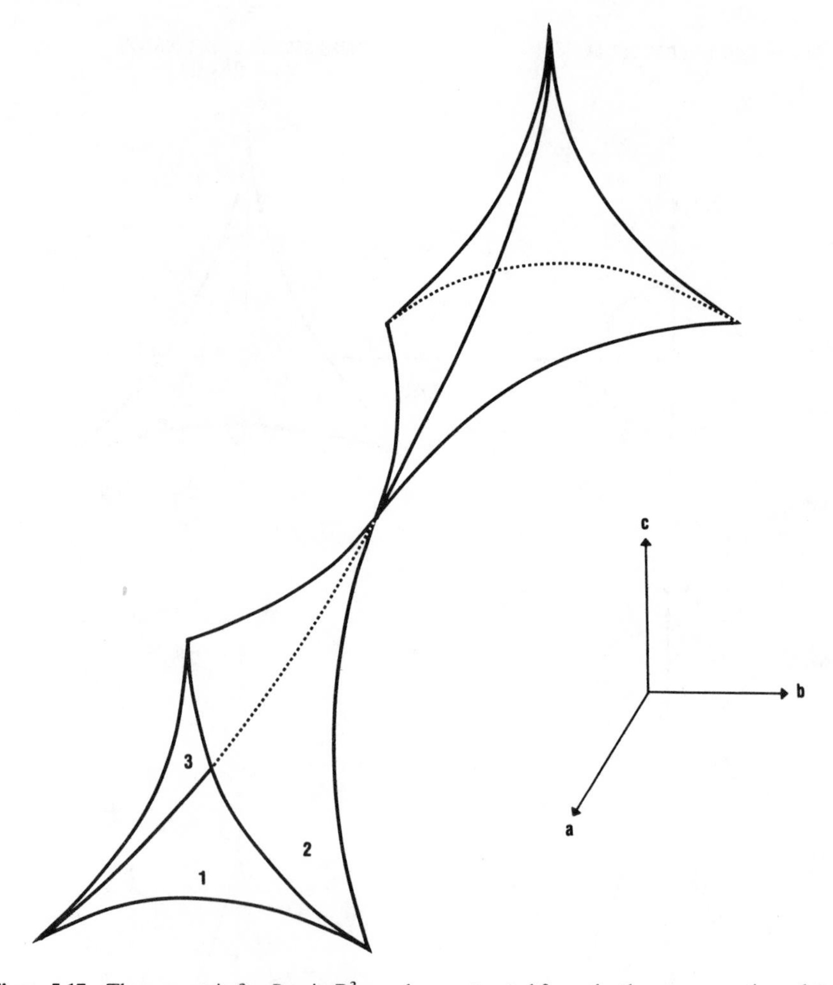

Figure 5.17 The separatrix for D_{-4} in $\mathbb{R}^3$ may be constructed from the three cross sections shown in Fig. 5.16 by using the scaling relations (5.50).

and it, too, is shown in Fig. 5.16. The three cross sections shown in Fig. 5.16 may now be strung together using the scaling relations (5.50) to give the separatrix, or bifurcation set, of the catastrophe D_{-4} which is shown in Fig. 5.17. This surface divides control parameter space $\mathbb{R}^3$ into three open regions which describe functions of three qualitatively different types.

To determine the qualitative properties of functions parameterized by points in each of these open regions, we investigate the properties of functions parameterized by points along the line $a = +1$, $b = 0$ in $\mathbb{R}^3$ (Fig. 5.18). Along this line the equations (5.44), which determine the critical points, are explicitly

$$2xy + 2x = 0 \tag{5.53a}$$

$$x^2 - y^2 + 2y + c = 0 \tag{5.53b}$$

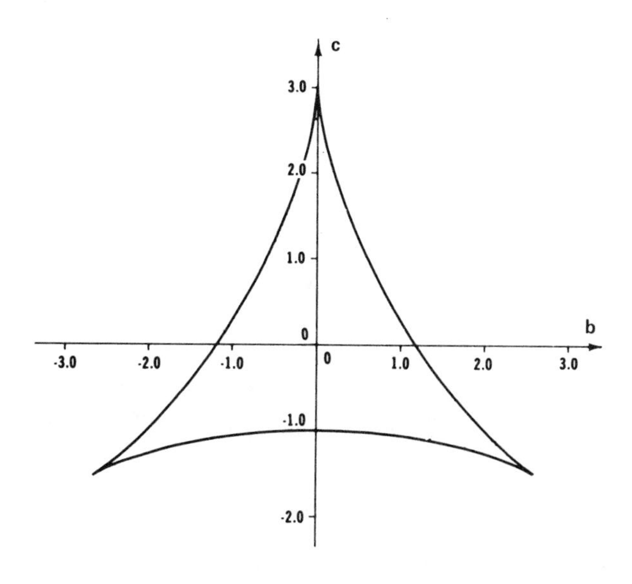

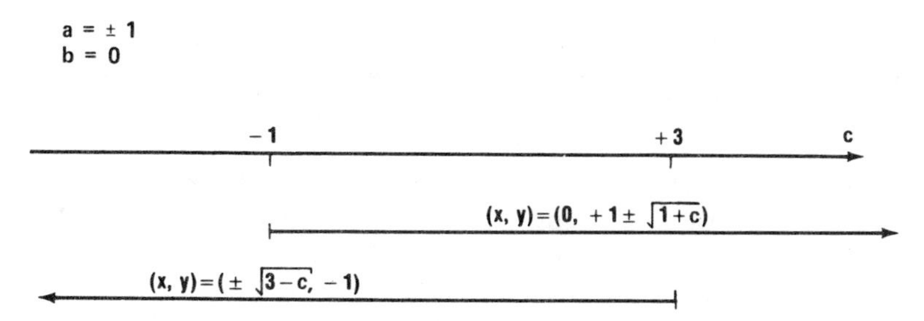

Figure 5.18 In the plane $a = +1$ the structurally stable Morse functions parameterized by the two open regions may be determined by setting $b = 0$ and varying c.

From (5.53a) the critical points must have as coordinates either $x = 0$ or $y = -1$. If $x = 0$, we have, from (5.53b),

$$x = 0 \xrightarrow{(5.53b)} y = +1 \pm \sqrt{1 + c} \tag{5.54}$$

There is a real pair of critical points for $c > -1$. If $y = -1$, we have from (5.53b),

$$y = -1 \xrightarrow{(5.53b)} x = \pm\sqrt{3 - c} \tag{5.55}$$

This real pair of critical points exists for $c < +3$. The locations and range of c values for which the critical points (5.54) and (5.55) exist is summarized in Fig. 5.18.

Once the locations of the critical points are known, their stability properties can be determined from the stability matrix (5.44;2). At the critical points (5.54), the stability matrix is

$$F_{ij} = 2\begin{bmatrix} y + 1 & 0 \\ 0 & -y + 1 \end{bmatrix}, \qquad x = 0, \quad y = +1 \pm \sqrt{1 + c} \qquad (5.56).$$

Thus the x and y axes are principal directions, with the eigenvalues in the x and y principal directions equal to $2(1 + y)$ and $2(1 - y)$, respectively. The x eigenvalue of the lower critical point changes sign when y goes through -1, while the y eigenvalue is zero when $y = +1$. These are the points at which the critical points (5.54) cross the "bad set" (5.48), where we expect interesting things to happen. The locations and stability properties of these critical points are shown in Fig. 5.19.

The stability properties and principal axes of the critical points (5.55) are determined similarly. From the stability matrix we find

$$F_{ij} \xrightarrow{(5.55)} 2\begin{bmatrix} 0 & \pm\sqrt{3 - c} \\ \pm\sqrt{3 - c} & 2 \end{bmatrix} \Rightarrow \begin{array}{l} \lambda_1 = 2(1 + \sqrt{1 + (3 - c)}) \\ \lambda_2 = 2(1 - \sqrt{1 + (3 - c)}) \end{array} \qquad (5.57)$$

The principal directions are no longer along the x and y axes, so they must be determined by a simple matrix diagonalization. The results are shown in Fig. 5.19. These two roots intersect the "bad set" (5.48) at the point $(x, y) = (0, -1)$ when $c = +3$. At this point the eigenvalue $\lambda_2 = 0$, and we expect more interesting things to happen.

From our previous experience in Sec. 6 we can see from Fig. 5.19 immediately what happens as c increases along the line $a = +1$, $b = 0$. For $c < -1$ there is a pair of saddle points. As c increases through -1, a fold catastrophe occurs at $(x, y) = (0, 1)$, with a saddle and a local minimum (inside the circle) coming into existence. The x direction is the "good" direction and the y direction is the "bad" direction at this fold catastrophe. As c increases further toward $+3$, the new saddle on the x axis moves away to the north, the local minimum moves across the circle toward the point $(x, y) = (0, -1)$, and the two original saddles move toward the same point along the line $y = -1$. As the three critical points approach the point $(0, -1)$, the principal directions of the two saddles rotate toward the x and y axes. The y eigenvalues all approach $+4$ and the x eigenvalues approach zero. As c passes through $+3$, the two saddles and the local minimum collide at $(x, y) = (0, -1)$ in a dual cusp catastrophe. For $c > +3$, the two saddles and local minimum have combined to form a single saddle (along $x = 0$, $y < -1$). A second saddle exists along $x = 0$, $y > +1$.

From this discussion it is clear what types of functions are parameterized by each of the three open regions shown in Fig. 5.17. The interior of the triangular region parameterizes functions which have three saddles and one local minimum (if $a > 0$) or local maximum (if $a < 0$). The saddles exist outside the circular "bad set" (5.48), and the minimum or maximum exists inside the "bad set."

The region outside the triangular domain in Fig. 5.17 parameterizes functions with two saddles, both outside the circular "bad set."

This study has been carried out on the symmetry plane $b = 0$ to make both the calculations simple and the figures pretty. When the point in control parameter space moves through the separatrix shown in Fig. 5.17, a fold or cusp catastrophe occurs depending on whether the point moves through one of the three surfaces or one of the three edges. Upon moving through a surface (assume $a > 0$), a minimum collides with and annihilates one of the three saddles. The saddle which is annihilated depends on which surface the control point moves through. This is easily determined by inspection of Fig. 5.16. The saddles in Fig. 5.19 and the surfaces in Fig. 5.17 have corresponding numbering.

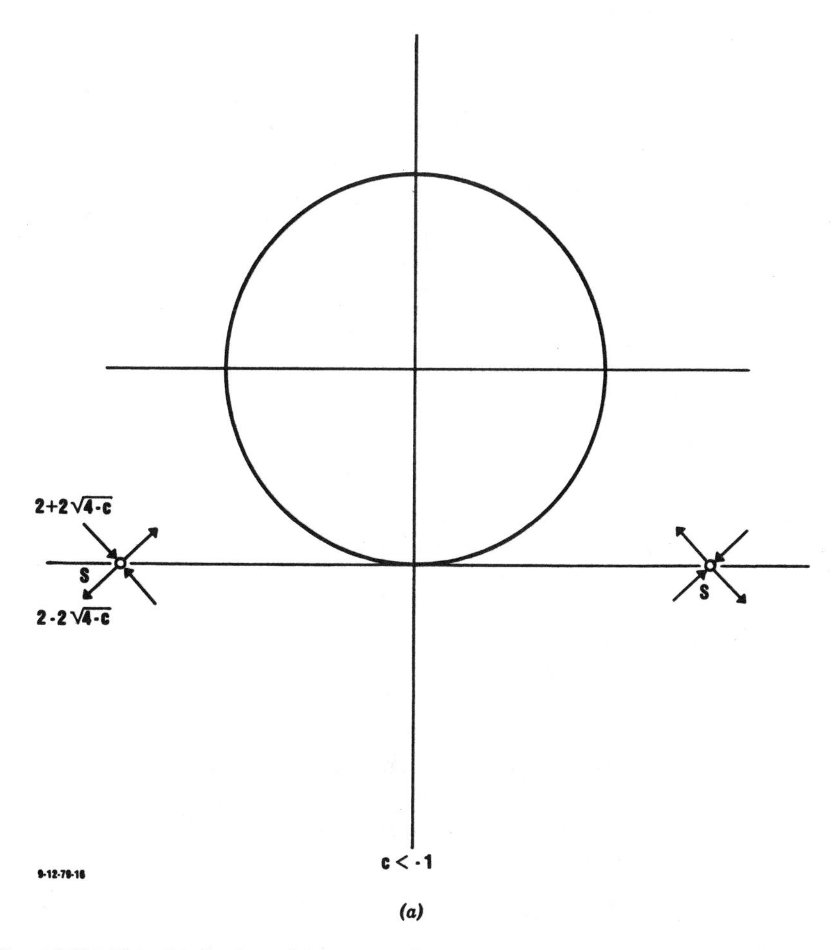

Figure 5.19 The critical points of the catastrophe D_{-4} are shown in the x–y plane as a function of the control parameter c for $a = +1, b = 0$.

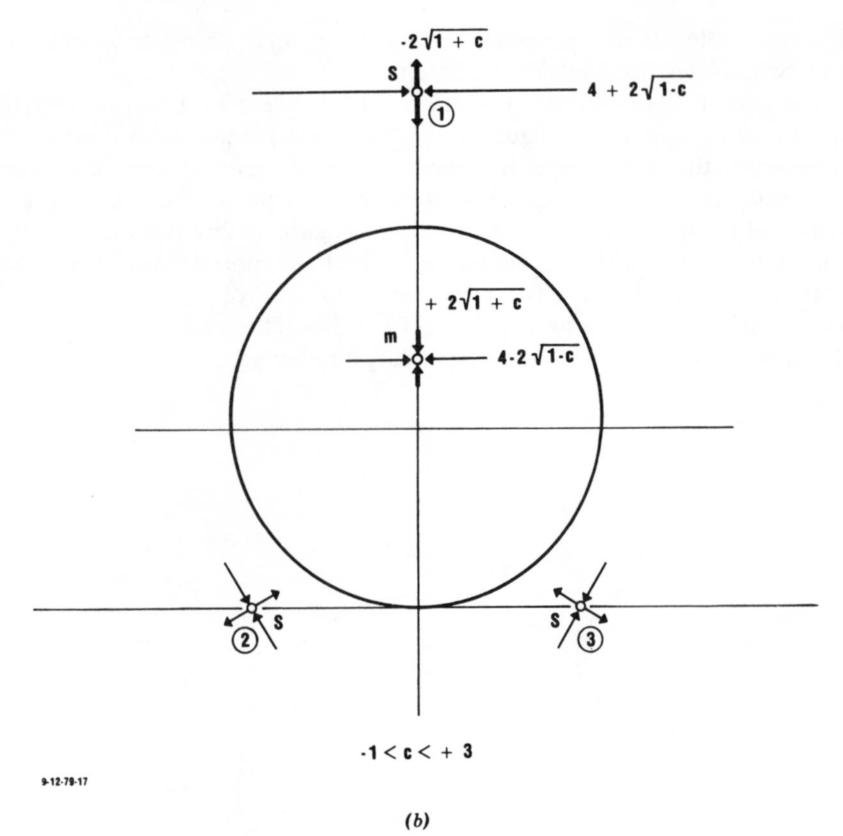

(b)

Figure 5.19 (*Continued*)

The canonical forms of the catastrophes which occur for $(a, b, c) = (1, 0, -1)$ at $(x, y) = (0, +1)$ can be determined as in Sec. 6. At this point,

$$F(x, y; 1, 0, -1) = x^2y - \tfrac{1}{3}y^3 + (x^2 + y^2) - y \tag{5.58}$$

Expanding this function about the degenerate critical point (0, 1) using $x = X$, $y = (y - 1) + 1 = Y + 1$ gives

$$F(X, Y) = X^2Y - \tfrac{1}{3}Y^3 + 2X^2 - \tfrac{1}{3} \tag{5.59}$$

It is clear that X is the "good" direction and Y the "bad" direction. We can attempt to find a canonical form by introducing a nonlinear axis-preserving transformation (5.31). The canonical form we seek is $2x'^2 + k'y'^3$. We can attempt to satisfy the equation

$$2X^2 + X^2Y - \tfrac{1}{3}Y^3 \stackrel{?}{=} +2x'^2 + k'y'^3 \tag{5.60}$$

This is for all intents and purposes identical to (5.32), so the result must be the same, modulo a few sign changes (i.e., $k' = -1/3$). The argument following

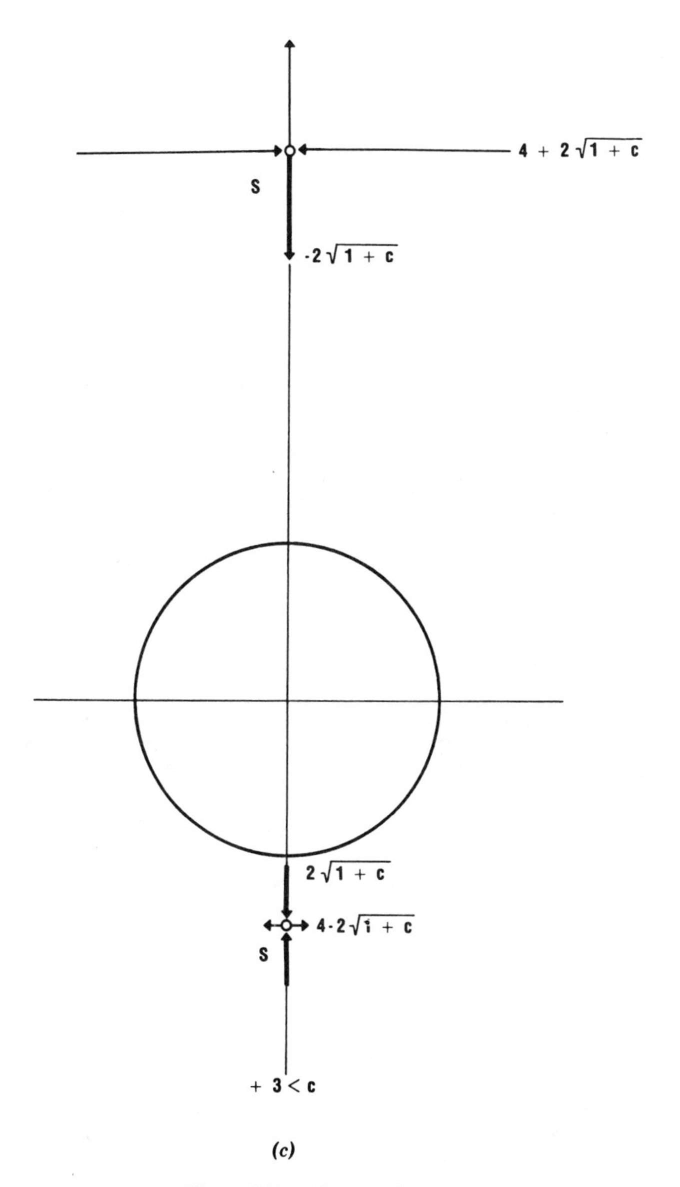

(c)

Figure 5.19 (*Continued*)

(5.60) is identical to that following (5.32); the counting argument is summarized in Table 5.1, and the appropriate canonical form, analogous to (5.35), is

$$F(X, Y) \doteq +x''^2 - y''^3 \tag{5.61}$$

At the control point $(a, b, c) = (1, 0, +3)$ we expect a cusp catastrophe at $(x, y) = (0, -1)$. At this point

$$F(x, y; 1, 0, +3) = x^2y - \tfrac{1}{3}y^3 + (y^2 + x^2) + 3y \tag{5.62}$$

If we expand this function about this critical point using $x = X$, $y = (y + 1) - 1 = Y - 1$, we find

$$F(X, Y) = X^2Y - \tfrac{1}{3}Y^3 + 2Y^2 - \tfrac{5}{3} \tag{5.63}$$

This is the same as (5.36), except for a sign change. Thus it is clear that Y is the "good" direction and X the "bad" direction. We can expect a canonical form $k'x'^4 + 2y'^2$, so we attempt to satisfy the equation

$$2Y^2 + X^2Y - \tfrac{1}{3}Y^3 \overset{?}{=} 2y'^2 + k'x'^4 \tag{5.64}$$

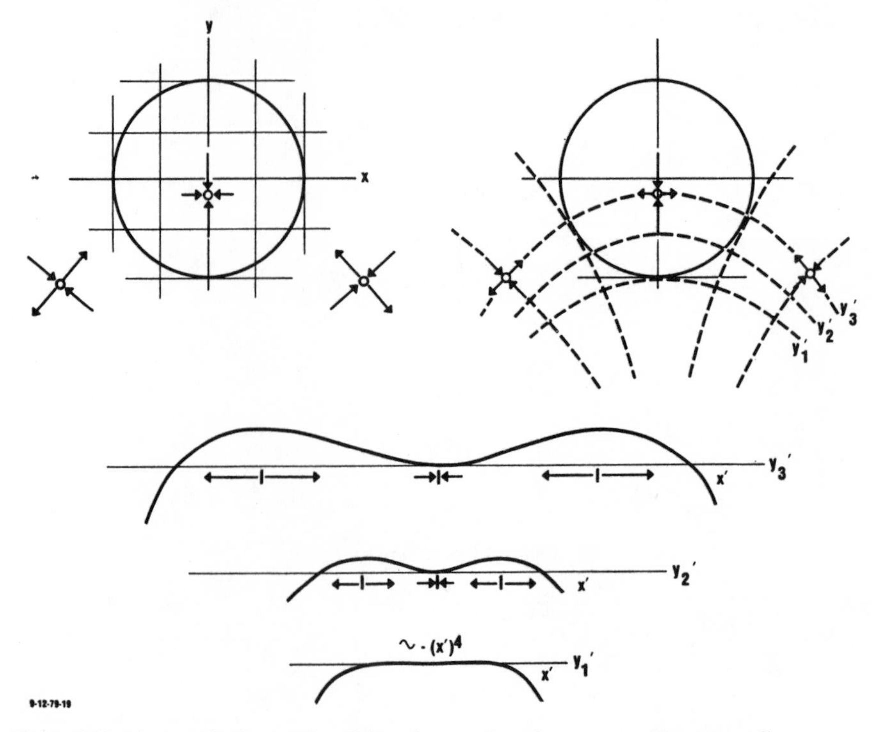

Figure 5.20 A smooth change of variables from a Cartesian to a curvilinear coordinate system shows clearly how the dual cusp catastrophe is associated with the D_{-4} catastrophe.

The arguments following (5.64) differ in no essential way from those following (5.38). We find again $k' = -\frac{1}{8}$, so that a dual cusp catastrophe A_{-3} occurs. This is clear by inspection of Fig. 5.19. The counting argument is summarized in Table 5.2, and the final canonical form, analogous to (5.43), is

$$F(X, Y) \doteq -x''^4 + y''^2 \tag{5.65}$$

The curvilinear change of variables from the Cartesian coordinates X, Y to x', y' is shown in Fig. 5.20.

8. The Bifurcation Set and the Maxwell Set

In the preceding sections we have computed the bifurcation sets $\mathscr{S}_B$ for the elementary catastrophes of control dimension $k \leq 3$. It is at the bifurcation set that the qualitative nature of functions in a family of functions changes because equilibria are either created or destroyed.

There is another set of points in $\mathbb{R}^k$ which represent structurally unstable functions and which is sometimes of importance in applications. This is called the **Maxwell set** $\mathscr{S}_M$. On the Maxwell set the critical values of a function at two or more critical points are equal. This critical value degeneracy is generally removed by an arbitrary perturbation. The Maxwell set consists of components of dimension less than k in $\mathbb{R}^k$.

The Maxwell set is of importance for the description of physical systems whose state is governed by the global minimum of a potential. In such cases there is a qualitative change in the nature of the system when the control parameters are varied so that the global minimum value jumps from one local minimum to another. The components of the Maxwell set of primary interest in applications are those which describe critical value degeneracy of the global minima.

9. The Clausius–Clapeyron Equations

The equations which determine the Maxwell set $\mathscr{S}_M$ are called the **Clausius–Clapeyron equations.** These equations can be derived from (5.3). We first assume that for $c = c^0 \in \mathbb{R}^k$ there are t isolated critical points $x^{(1)}, \ldots, x^{(t)} \in \mathbb{R}^n$ at which the potential has degenerate critical values:

$$\begin{gathered} V^{(1)} = V^{(2)} = \cdots = V^{(t)} \\ V^{(p)} = V(x^{(p)}, c^0), \qquad 1 \leq p \leq t \end{gathered} \tag{5.66}$$

Under a small change in the values of the control parameters, the pth critical value changes by

$$\delta^{(1)} V^{(p)} = V_\alpha^{(p)} \delta c^\alpha \tag{5.67}$$

The requirement that all t critical values change by the same amount imposes $t-1$ constraints on the variations δc^{α} of the k control parameters. This system of $t-1$ equations can be written

$$[V^{(p)} - V^{(p+1)}]_{\alpha}\delta c^{\alpha} = 0, \qquad p = 1, 2, \ldots, t-1 \tag{5.68}$$

In matrix form this becomes

$$\begin{bmatrix} V_1^{(1)} - V_1^{(2)} & V_2^{(1)} - V_2^{(2)} & \cdots & V_k^{(1)} - V_k^{(2)} \\ V_1^{(2)} - V_1^{(3)} & V_2^{(2)} - V_2^{(3)} & \cdots & V_k^{(2)} - V_k^{(3)} \\ \vdots & \vdots & \ddots & \vdots \\ V_1^{(t-1)} - V_1^{(t)} & V_2^{(t-1)} - V_2^{(t)} & \cdots & V_k^{(t-1)} - V_k^{(t)} \end{bmatrix} \begin{bmatrix} \delta c^1 \\ \delta c^2 \\ \vdots \\ \delta c^k \end{bmatrix} = 0 \tag{5.68'}$$

These Clausius–Clapeyron equations determine the $k-(t-1)$-dimensional tangent plane to the Maxwell set for t degenerate critical values in a k-parameter family of potentials.

A set of $k+1-t$ basis vectors in this tangent plane at c^0 can be determined by an elementary algorithm which is valid almost everywhere on the Maxwell set $\mathscr{S}_M$. This is done by setting

$$B_j = \begin{bmatrix} \delta c^1 \\ \vdots \\ \delta c^{t-1} \\ 0 \\ \vdots \\ 0 \end{bmatrix} + C_j \delta s_j, \qquad C_j = \begin{bmatrix} 0 \\ \vdots \\ 0 \\ 0 \\ 1 \\ 0 \end{bmatrix}, \qquad t \le j \le k \tag{5.69}$$

where the column vector C_j has zeros in all rows except the jth row and the δs_j are a set of $k-t+1$ independent parameters. The $k+1-t$ vectors B_j obey an equation of the form

$$MB_j = C_j \delta s_j \tag{5.70}$$

where M is a $k \times k$ matrix obtained by augmenting the $(t-1) \times k$ matrix appearing in (5.68′) with a $(k+1-t) \times k$ matrix consisting of zeros and ones such that

$$\begin{aligned} M_{ij} &= V_j^{(i)} - V_j^{(i+1)}, & 1 \le i \le t-1 \\ M_{ij} &= \delta_{ij}, & t \le i \le k \\ & & 1 \le j \le k \end{aligned} \tag{5.71}$$

If we let

$$N_{ij} = (-)^{i+j} \det(\hat{M}_{ji}) \tag{5.72}$$

where $\hat{M}_{ji}$ is the $(k-1) \times (k-1)$ matrix obtained by removing the jth row and ith column from M, then

$$MN = I_k(\det M) \tag{5.73}$$

so that (5.73) can be inverted when $\det M \neq 0$ to give

$$B_j \simeq NC_j \delta s_j \tag{5.74}$$

In short, the column vector B_j has matrix elements proportional to the cofactors of M:

$$B_j \simeq \begin{bmatrix} N_{1j} \\ N_{2j} \\ \vdots \\ N_{kj} \end{bmatrix} = \begin{bmatrix} (-)^{j+1} \det(\hat{M}_{j1}) \\ (-)^{j+2} \det(\hat{M}_{j2}) \\ \vdots \\ (-)^{j+k} \det(\hat{M}_{jk}) \end{bmatrix} \tag{5.75}$$

These vectors span the tangent plane when $\det M \neq 0$. If $\det M = 0$, a spanning set may be determined by augmenting the $(t-1) \times k$ matrix in (5.68′) in a different way. Equation (5.75) is equivalent to the set

$$\frac{\partial c_\alpha}{\partial s_j} = (-)^{\alpha+j} \det(\hat{M}_{j\alpha}), \qquad 1 \leq \alpha \leq k, \quad t \leq j \leq k \tag{5.76}$$

for the separatrix in $\mathbb{R}^k$.

Example 1. The fold catastrophe A_2 has no Maxwell set since isolated critical points, when they exist, have different critical values. If we were to regard the point $a = 0$ as the Maxwell set because the critical values become degenerate at a degenerate critical point, then for higher catastrophes the Maxwell set would include the bifurcation set as a subset.

Example 2. The point $a < 0$, $b = 0$ is on the Maxwell set of the cusp catastrophe

$$V(x; a, b) = \tfrac{1}{4}x^4 + \tfrac{1}{2}ax^2 + bx \tag{5.77}$$

At this point

$$\begin{aligned} x^{(1)} &= -\sqrt{-a} \\ x^{(2)} &= +\sqrt{-a} \end{aligned} \tag{5.78}$$

At these two points $V^{(1)} = V^{(2)} = -a^2/4$ and under the identification $(a, b) = (c^1, c^2)$ so that $V_1 = \partial V/\partial a = \frac{1}{2}x^2$ and $V_2 = \partial V/\partial b = x$,

$$\begin{aligned} V_1^{(1)} &= -\frac{a}{2} \qquad V_2^{(1)} = -\sqrt{-a} \\ V_1^{(2)} &= -\frac{a}{2} \qquad V_2^{(2)} = +\sqrt{-a} \end{aligned} \tag{5.79}$$

The Clausius–Clapeyron equations (5.68′) reduce to

$$(0, -2\sqrt{-a})\begin{pmatrix}\delta c^1 \\ \delta c^2\end{pmatrix} = 0 \tag{5.80}$$

and the solution reduces to

$$B_2 \simeq \begin{pmatrix}-2\sqrt{-a} \\ 0\end{pmatrix}\delta s_2 \tag{5.81}$$

or δa arbitrary, $\delta b = 0$. This is certainly the hard way to determine that the Maxwell set $\mathscr{S}_M$ of A_3 consists of the half-line $a \leq 0, b = 0$, but it is the easy way to determine $\mathscr{S}_M$ for higher catastrophes.

Remark. The Clausius–Clapeyron equations (5.68′) are particularly easy to treat when $k = t$, for then the system of partial differential equations (5.76) reduces to a system of ordinary differential equations.

10. Summary

In this chapter we have demonstrated and exploited the power of the "Crowbar Principle." Briefly, this says that you get the most information by looking at the points with the worst behavior.

For a function, where most points $x \in \mathbb{R}^n$ are noncritical, we found that it was the critical points which organized the global qualitative topography (Sec. 1).

For a family of functions, where most points $c \in \mathbb{R}^k$ parameterize Morse functions, we found that it was the separatrix parameterizing the non-Morse functions which organized the qualitative properties of the family (Sec. 2). Within an open region in $\mathbb{R}^k$ away from the separatrix, small changes in the control parameters produce only small changes in the location of critical points, their principal directions, and the associated forces (the eigenvalues); thus perturbations produce no changes in the qualitative nature of the functions parameterized by that region in $\mathbb{R}^k$.

In the case of a non-Morse critical point at $(x^0; c^0)$ it was clear that the focal point for the crowbar was the corresponding catastrophe germ (Sec. 3). Perturbation of a function at a non-Morse critical point yields no qualitative surprises as far as the Morse part of the canonical form is concerned. Qualitative changes arise only through perturbations of the catastrophe germ. Since both the germs and their perturbations are canonical, the separatrices need be studied only once.

In Secs. 3–7 we have constructed the separatrices in the control parameter spaces for the elementary catastrophes A_2, A_3, A_4, D_{+4}, and D_{-4}. The separatrix in $\mathbb{R}^k$ consists of manifolds of dimensions $k-1, k-2, \ldots$. These separatrices are shown explicitly in Fig. 5.3 (a point), Fig. 5.4 (two fold curves and a point), Fig. 5.6 (for A_4), Fig. 5.11 (for D_{+4}), and Fig. 5.17 (for D_{-4}). In each instance, and for the remaining elementary catastrophes also, the separatrix divides

$\mathbb{R}^k$ into a number of open regions. Each open region is connected and parameterizes functions of qualitatively the same type. The open regions are dense in the sense that their closure contains $\mathbb{R}^k$. This means simply that any function in a catastrophe family (whether or not it has degenerate critical points) can be approximated arbitrarily closely by Morse functions.

The separatrix itself parameterizes functions with a hierarchy of catastrophes. For example, in Fig. 5.17 the 2-dimensional components describe fold (A_2) catastrophes, while the 1-dimensional curves which separate (are the intersections of) the 2-dimensional components parameterize cusps ($A_{\pm 3}$) and the point at the origin which separates (is the intersection of) the six curves parameterizes the germ D_{-4} with a fourfold degenerate critical point.

The actual reduction of a function parameterized by a point on the separatrix to canonical form (2.3b) has been carried out. For the cases $D_{\pm 4}$, at first sight it would seem obvious that fold catastrophes are present, because $D_{\pm 4}$ involve terms of degree 3, but it is not obvious how cusp catastrophes come about. (How do you get a quartic from a cubic?) However, the actual reduction to canonical form, involving as it does a nonlinear change of variables, dispels these intuitive insights.

For the sake of completeness we have studied another kind of separatrix in $\mathbb{R}^k$. This is the Maxwell set, which parameterizes functions with degenerate critical values (Sec. 8). The bifurcation set $\mathscr{S}_B$ and the Maxwell set $\mathscr{S}_M$ are determined, respectively, by the bifurcation equations and the Clausius–Clapeyron equations (Sec. 9):

$$\begin{aligned} \mathscr{S}_B: \qquad \nabla V &= 0 \\ \det V_{ij} &= 0 \\ \mathscr{S}_M: \qquad \nabla V &= 0 \\ V^{(p)} - V^{(p+1)} &= 0 \\ [V_\alpha^{(p)} - V_\alpha^{(p+1)}]\delta c^\alpha &= 0 \end{aligned}$$

In applications the important separatrix depends on the convention adopted (Chapter 8, Sec. 2). Under the delay convention $\mathscr{S}_B$ is important; under the Maxwell convention $\mathscr{S}_M$ is important.

We remark in closing that scaling transformations have been an extremely useful tool for constructing and studying the separatrices and their properties.

CHAPTER 6

Geometry of the Fold and the Cusp

In Chapter 4 we studied the effects of perturbations on the canonical forms (2.1), (2.2), and (2.3). In Sec. 2 of that chapter we saw that perturbations had no qualitative effect on the Implicit Function form, and we indicated how the value of the function and its gradient at a point depend on the Taylor series coefficients of the perturbing function. In Sec. 3 of that chapter we saw that perturbations had no qualitative effect on the Morse form either, and we saw how the location of the critical point, the value of the function at the critical point ("critical value"), and the eigenvalues depend on the Taylor series coefficients of the perturbing function. In the subsequent sections of Chapter 4 we saw how perturbations do have a qualitative effect on functions with degenerate critical points. We were silent about how the qualitative properties depended on the Taylor series coefficients of the canonical perturbation (treated in Chapter 5) and about how the locations of the critical points, the critical values, and the eigenvalues depend on the control parameters. Since catastrophe germs and their perturbations are canonical, the geometric studies carried out in this chapter are canonical and need not be done more than once.

1. Geometry of the Fold (A_2)

The A_2, or fold, catastrophe is defined by

$$A_2: \quad F(x; a) = \tfrac{1}{3}x^3 + ax \tag{6.1a}$$

The qualitative properties of $F(x; a)$ have already been described (see Fig. 5.3) but are repeated here for the sake of making this description self-contained. For $a < 0$ there are two critical points, for $a = 0$ there is one doubly degenerate critical point, and for $a > 0$ there are no critical points (Fig. 6.1*a*). The bifurcation set consists of the single point $a = 0$.

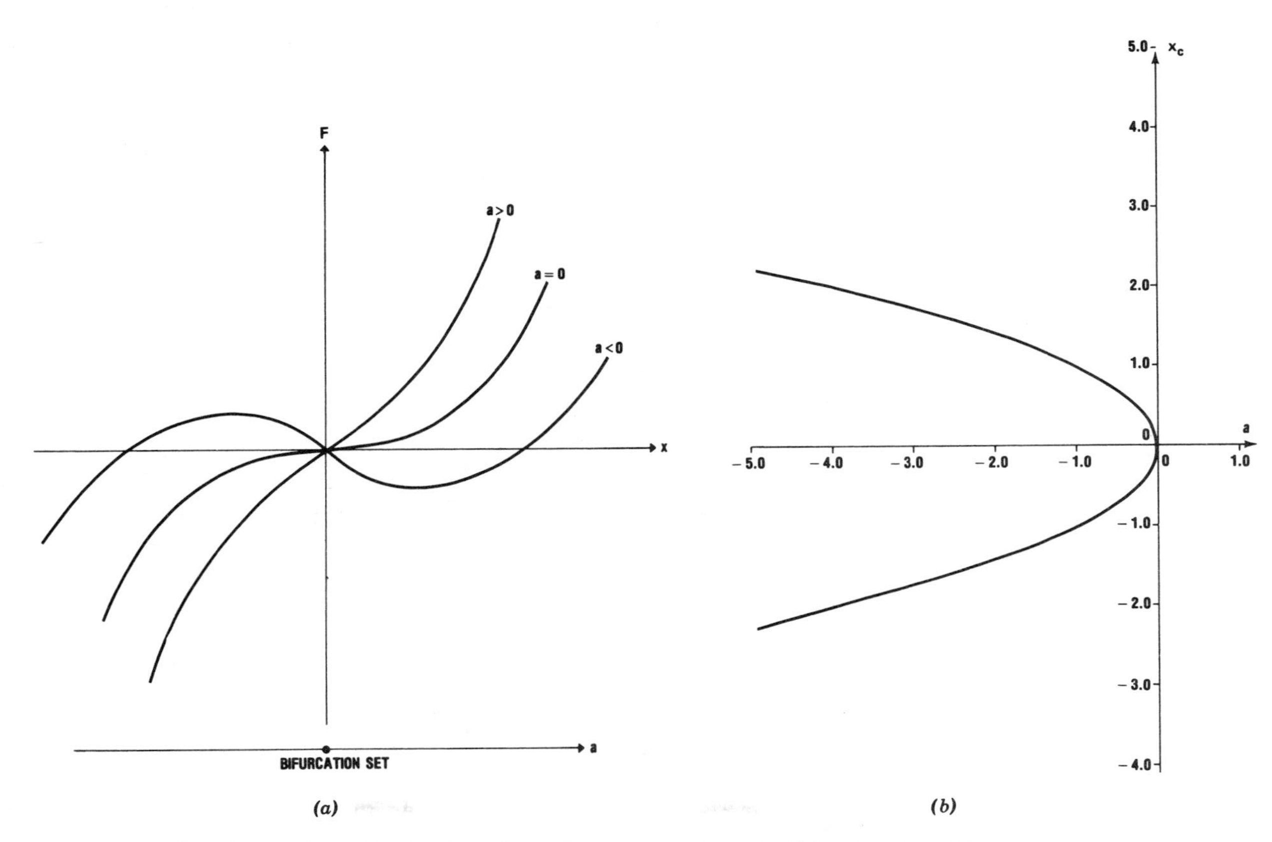

Figure 6.1 (*a*) Functions in the family $F(x; a)$ are shown for $a > 0$, $a = 0$, $a < 0$. (*b*) The location of the critical points of $F(x; a)$ are shown as a function of a.

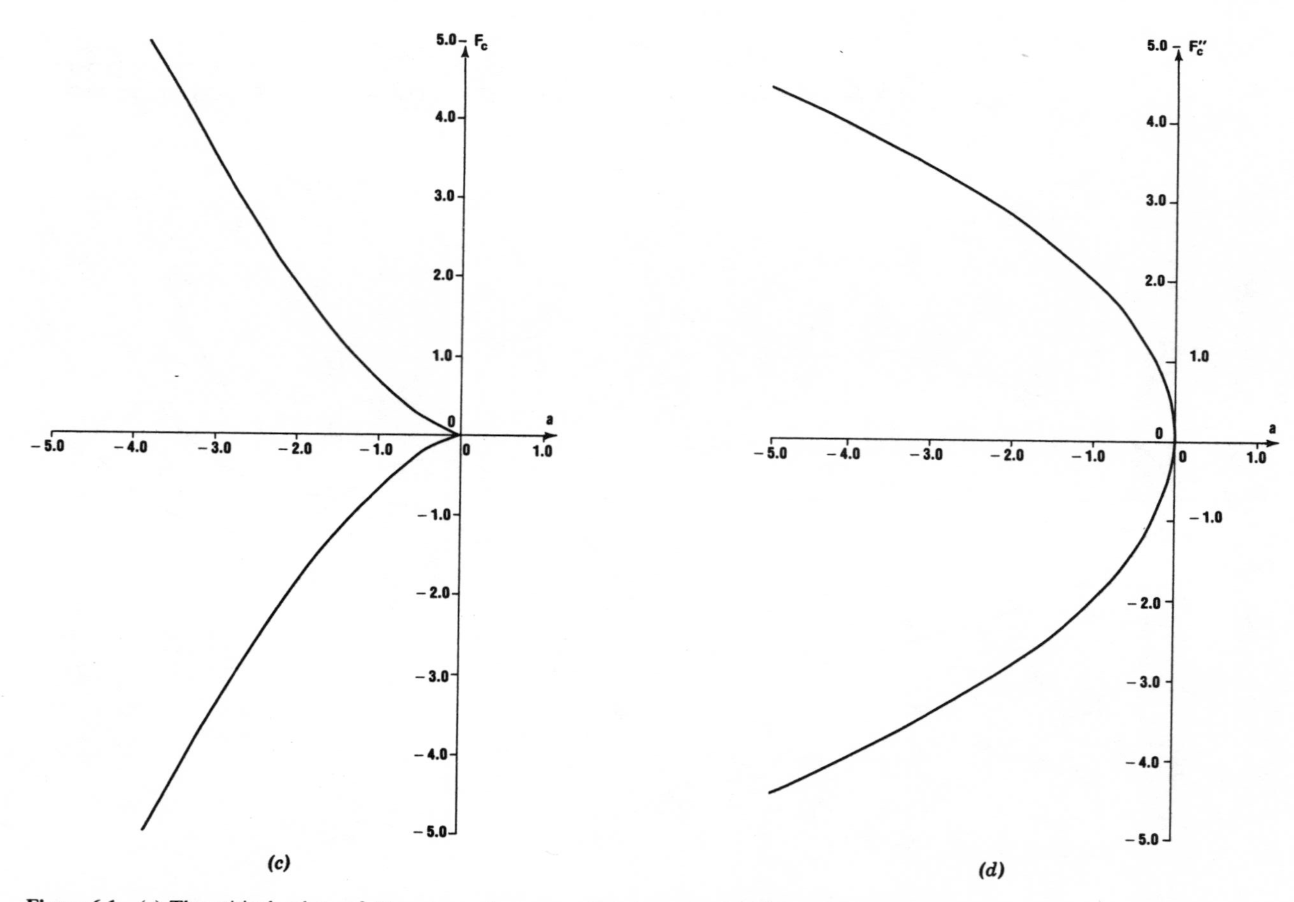

Figure 6.1 (*c*) The critical values of $F(x; a)$ are shown as a function of a. (*d*) The critical curvature of $F(x; a)$ is shown as a function of a.

The location of the critical points is determined from

$$\nabla F = x^2 + a = 0 \tag{6.1b}$$

The critical manifold determined from (6.1b) is shown in Fig. 6.1*b*.

The value of $F(x; a)$ at the critical points is determined from (6.1a) and (6.1b):

$$\begin{aligned} &\text{at} \quad x = -\sqrt{-a}, \qquad F(x; a) = \tfrac{2}{3}|a|^{3/2} \\ &\text{at} \quad x = +\sqrt{-a}, \qquad F(x; a) = -\tfrac{2}{3}|a|^{3/2} \end{aligned} \tag{6.1c}$$

The critical values are shown in Fig. 6.1*c*.

The eigenvalues of the stability matrix (the 1×1 matrix d^2F/dx^2) at the critical points are determined from (6.1b) and are

$$\begin{aligned} &\text{at} \quad x = -\sqrt{-a}, \qquad d^2F/dx^2 = -2|a|^{1/2} \\ &\text{at} \quad x = +\sqrt{-a}, \qquad d^2F/dx^2 = +2|a|^{1/2} \end{aligned} \tag{6.1d}$$

Relations (6.1d) are shown in Fig. 6.1*d*.

2. Geometry of the Cusp (A_{+3})

The A_{+3}, or cusp, catastrophe is defined by

$$A_{+3}: \quad F(x; a, b) = \tfrac{1}{4}x^4 + \tfrac{1}{2}ax^2 + bx \tag{6.2a}$$

The point $(a, b) \in \mathbb{R}^2$ parameterizes functions with one or three isolated critical points, except on the separatrix shown in Fig. 5.4. The function (6.2a) is shown in Fig. 6.2*a* for various values of the control parameters (a, b). Inside the cusp-shaped region, $F(x; a, b)$ has three isolated critical points; outside it has one.

On the boundary it has an isolated critical point and a doubly degenerate critical point. At the origin it has a triply degenerate critical point.

The location of the critical points is determined from

$$\nabla F = x^3 + ax + b = 0 \tag{6.2b}$$

Equation (6.2b) defines a 2-dimensional manifold in the 3-dimensional space whose coordinate axes are x–a–b. This critical manifold is shown in Fig. 6.2*b*. The curved fold lines $(a/3)^3 + (b/2)^2 = 0$ in the control plane are shadows of the "folded-over" part of the manifold (6.2b). The dashed line $a < 0, b = 0$ in the control parameter plane is the Maxwell set or "nonlocal separatrix" which separates the functions for which the left-hand minimum is deeper than the right-hand minimum from those with the opposite property.

The critical values of F are determined by solving the cubic equation (6.2b) for the critical points, and evaluating (6.2a) at these critical points. The resulting "critical set" (not a manifold) is shown in Fig. 6.2*c*. The projection of this critical set into the control parameter plane clearly reveals the local and nonlocal separatrices described above.

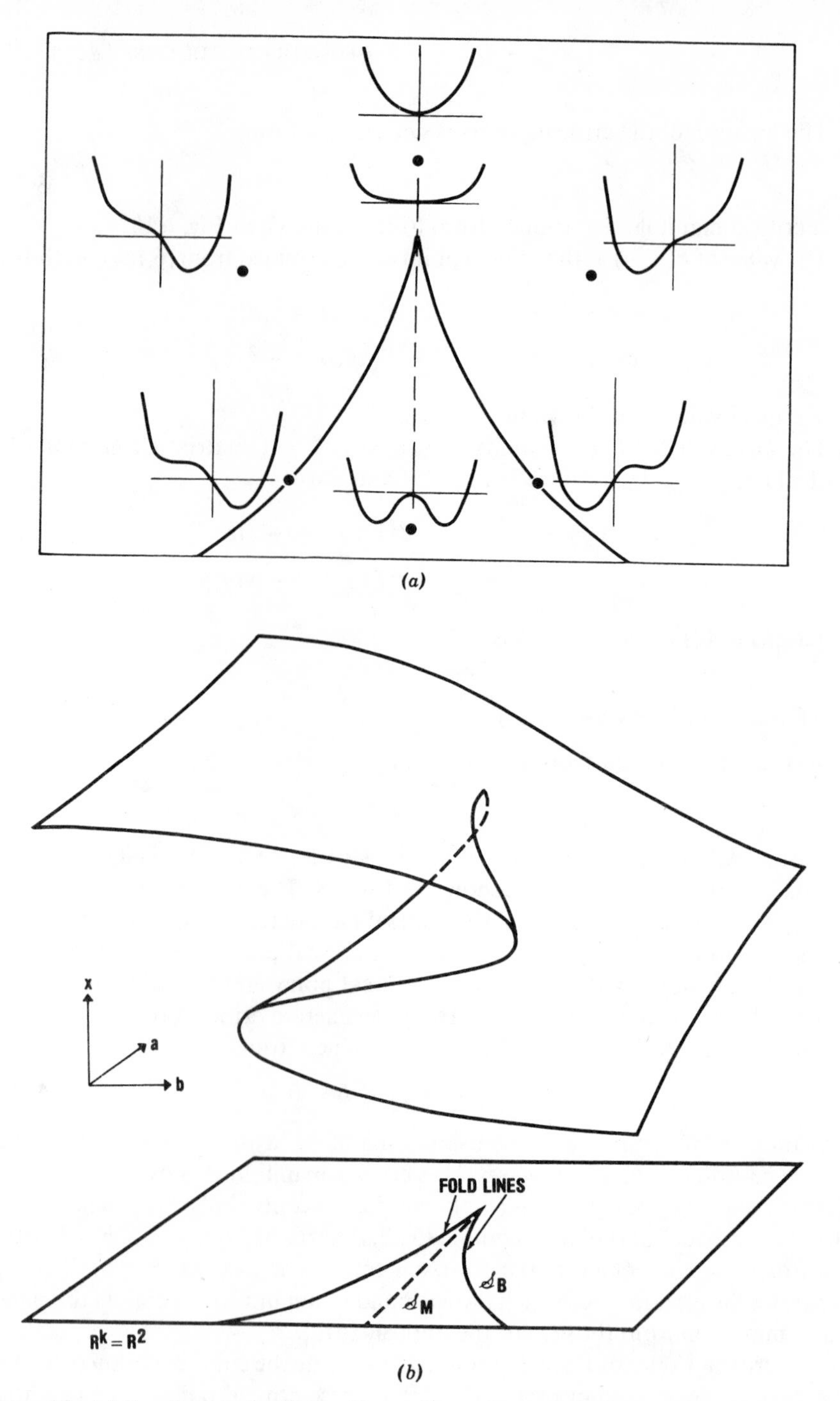

Figure 6.2 (*a*) Functions in the family $F(x;\ a,\ b)$ are shown for various values of the control parameters $(a,\ b)$. (*b*) The 2-dimensional cusp-catastrophe manifold, solution of $\nabla F = 0$, is shown embedded in $\mathbb{R}^3 = \mathbb{R}^1 \otimes \mathbb{R}^2$. The projection of this manifold down onto the control plane $\mathbb{R}^2$ shows the fold lines.

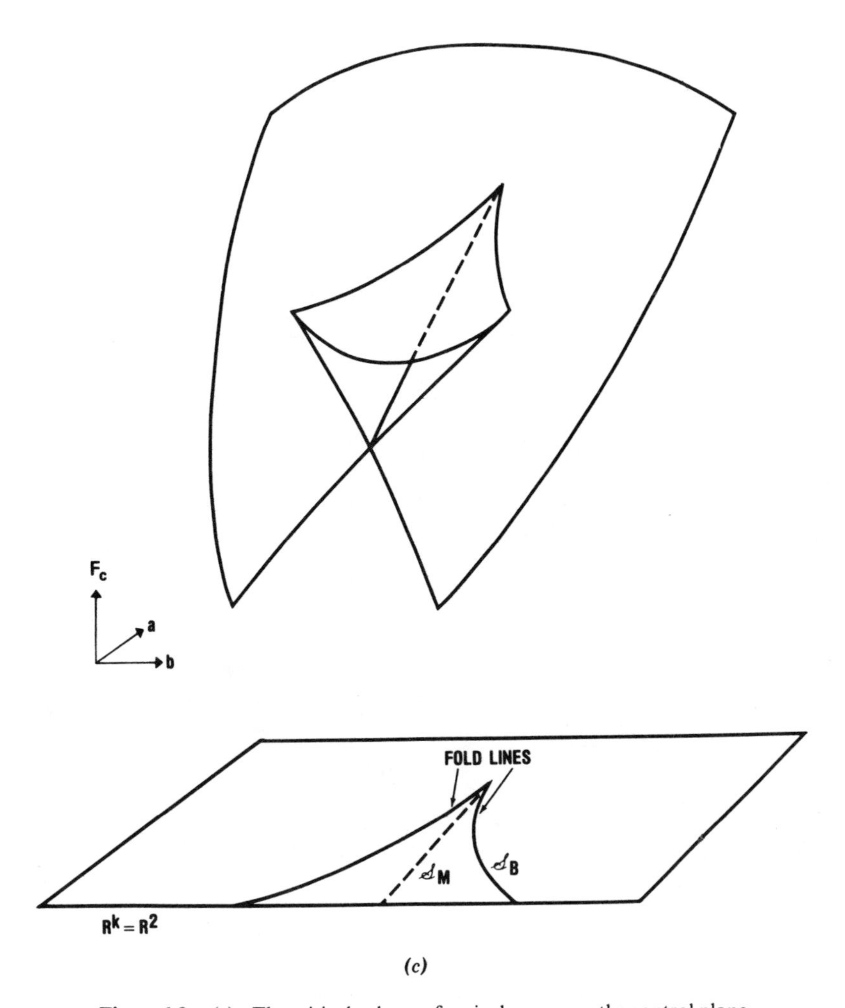

Figure 6.2 (*c*) The critical value surface is shown over the control plane.

The eigenvalues of the stability matrix of F are determined by determining the critical points from (6.2b) and computing $d^2F/dx^2 = 3x^2 + a$ at the critical points. The resulting "stability set" is shown in Fig. 6.2*d*.

Figs. 6.2*a*–*d* for the cusp are the analogs of Figs. 6.1*a*–*d* for the fold. Figs. 6.1*b*–*d* are quantitative in that explicit expressions (6.1b–d) have been derived for the curves shown in these figures. However, the corresponding Figs. 6.2*b*–*d* are qualitative in that only recipes have been given for the construction of these surfaces. Since the cusp catastrophe arises quite frequently in scientific applications, it would be useful to devise quantitative expressions for the surfaces shown in Figs. 6.2*b*–*d*.

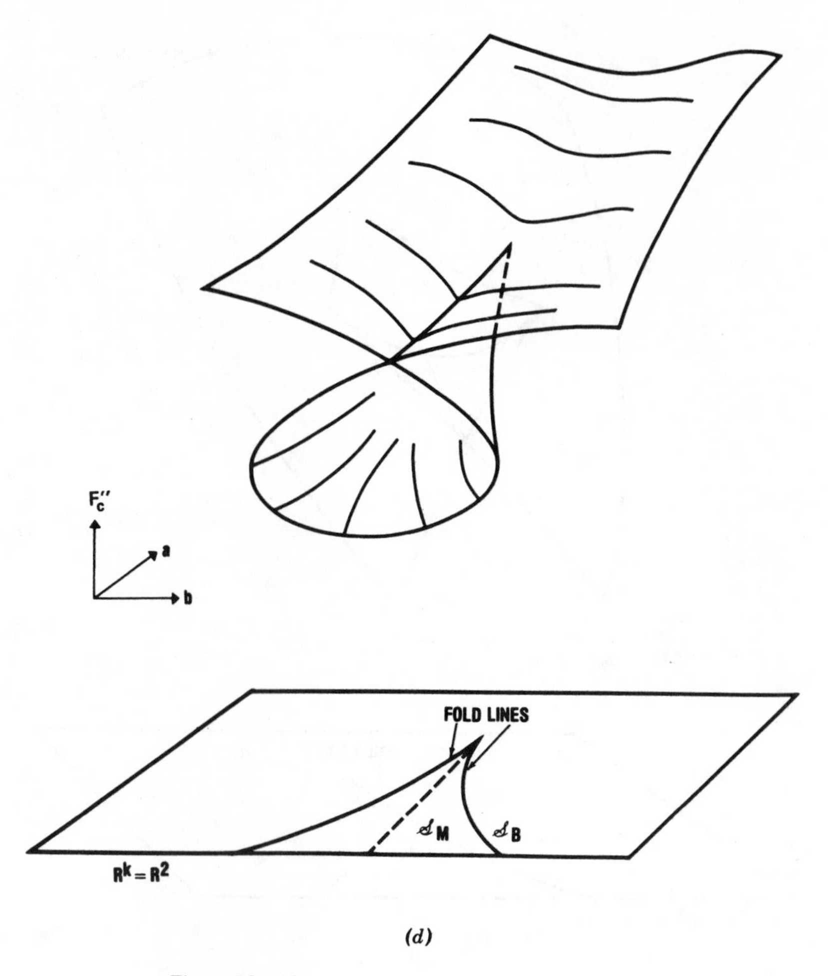

(d)

Figure 6.2 (*d*) The critical curvature surface is shown.

This can be done by using scaling relations. From the experiences encountered in Chapter 5, we note the similarity transformation

$$\begin{aligned} x &\to \lambda x \\ a &\to \lambda^2 a \\ b &\to \lambda^3 b \\ F &\to \lambda^4 F \end{aligned} \tag{6.3}$$

We can now give quantitative expressions for the intersections of the planes $a = +1$, $a = 0$, $a = -1$ with the three surfaces shown in Figs. 6.2*b*–*d*. Once quantitative expressions are available in each plane, the numerical value of any point on these surfaces is determined from the scaling relations (6.3). We now illustrate how this process works.

In the planes $a = +1$, $a = 0$, $a = -1$ the critical points of F are given as functions of b alone:

$$x^3 + x + b = 0 \qquad (6.4,\ a = +1)$$

$$x^3 \qquad + b = 0 \qquad (6.4,\ a = 0)$$

$$x^3 - x + b = 0 \qquad (6.4,\ a = -1)$$

Equations (6.4) need not be solved for x in terms of b. Rather, they can be plotted for b as a function of x. These quantitative graphs are shown in Fig. 6.3. When these graphs are strung together, Fig. 6.2b is the result. The x, a, and b axes in Fig. 6.2b scale as indicated in (6.3).

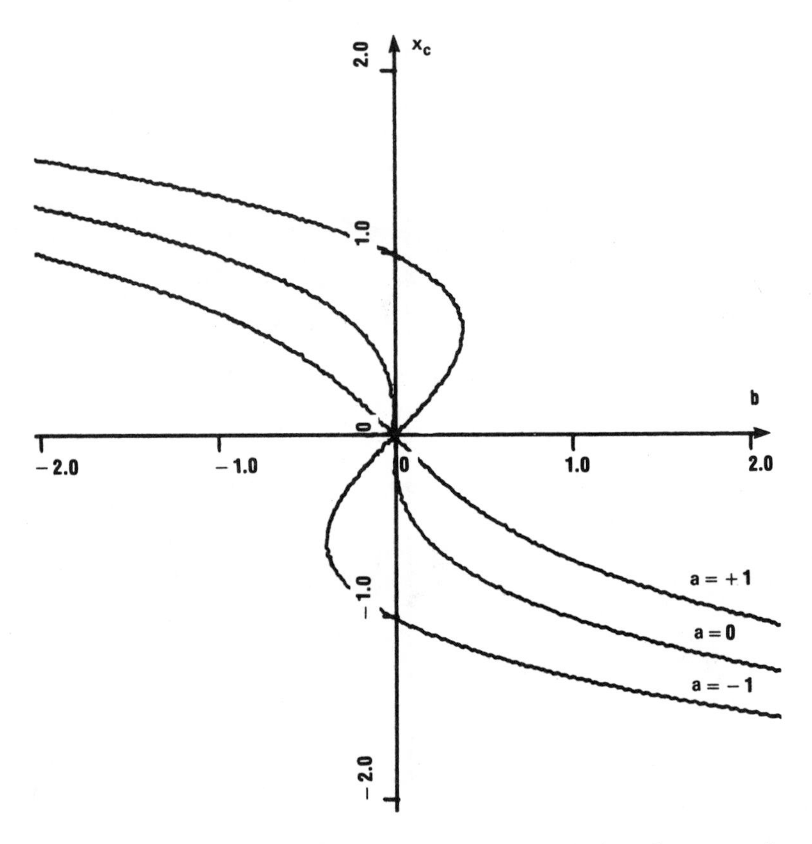

Figure 6.3 The critical points of F are shown as a function of b in the three planes $a = +1$, $a = 0$, and $a = -1$.

Example 1. Find the critical points when $(a, b) = (-3, 1)$.

Solution. Let x_0, b_0 be the points in the $a = -1$ plane which scale to the appropriate points in the $a = -3$ plane. Then

$$\begin{aligned} -3 &= \lambda^2(-1) \Rightarrow |\lambda| = \sqrt{3} \\ 1 &= \lambda^3 b_0 \Rightarrow b_0 = 3^{-3/2} = 0.19 \end{aligned} \tag{6.5}$$

For $b_0 = 0.19$ the three critical values of x, as read from Fig. 6.3, are $x_{01} = 0.88$, $x_{20} = 0.20$, $x_{30} = -1.08$.

The scaling relations give

$$\begin{aligned} x_1 &= \lambda x_{10} = 1.53 \\ x_2 &= \lambda x_{20} = 0.35 \\ x_3 &= \lambda x_{30} = -1.88 \end{aligned} \tag{6.6}$$

That these are the critical values may be verified by substituting into (6.2b).

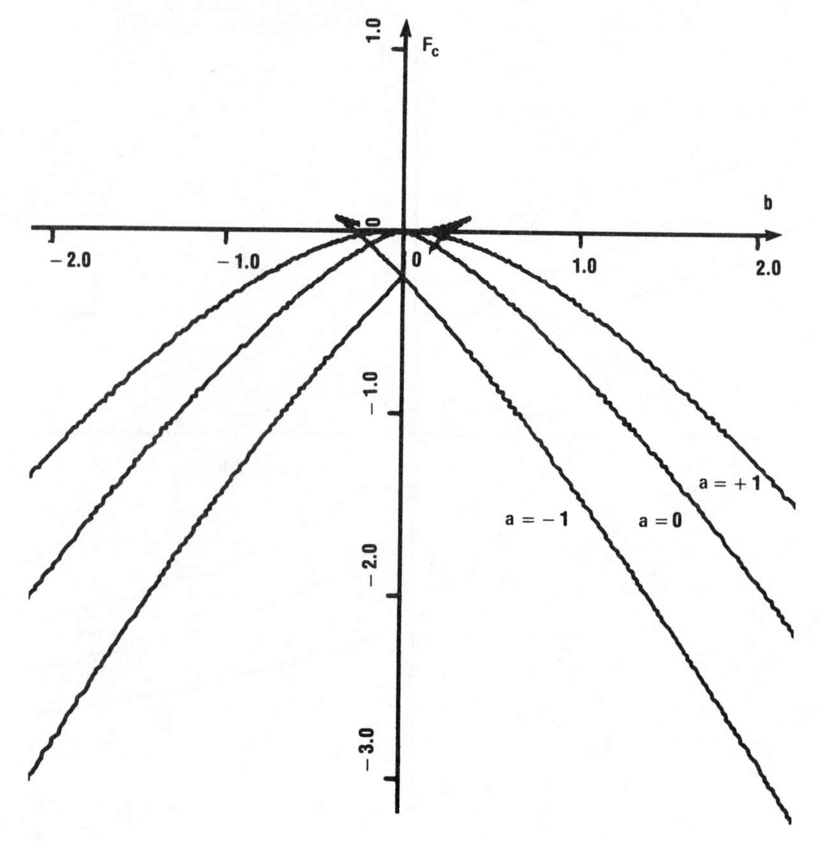

Figure 6.4 The critical values of F are shown as a function of b in the three planes $a = +1$, $a = 0$, and $a = -1$.

In the planes $a = +1, a = 0, a = -1$, the critical values of F can be determined from (6.4) and Fig. 6.3. Specifically for each value of b, the critical points are determined from Fig. 6.3 and the values of F at these points are determined from (6.2a) with $a = +1, 0, -1$. The critical values in the three planes a = constant are shown in Fig. 6.4.

Example 2. Find the critical values when $(a, b) = (-3, 1)$.

Solution. The critical values for $(a, b) = (-1, 0.19)$ are 0.02, -0.07, and -0.45. Since the critical values scale like $\lambda^4 = (\sqrt{3})^4$ the critical values for $(a, b) = (-3, 1)$ are 0.18, -0.63, and -4.05. They can also be computed directly from (6.6).

The curvature at the critical points can be determined in the same way. The critical points are determined again from Fig. 6.3, and the curvature at each such point is determined from $d^2F/dx^2 = 3x^2 + a$. The results are shown in Fig. 6.5 for the three planes $a = +1, a = 0, a = -1$.

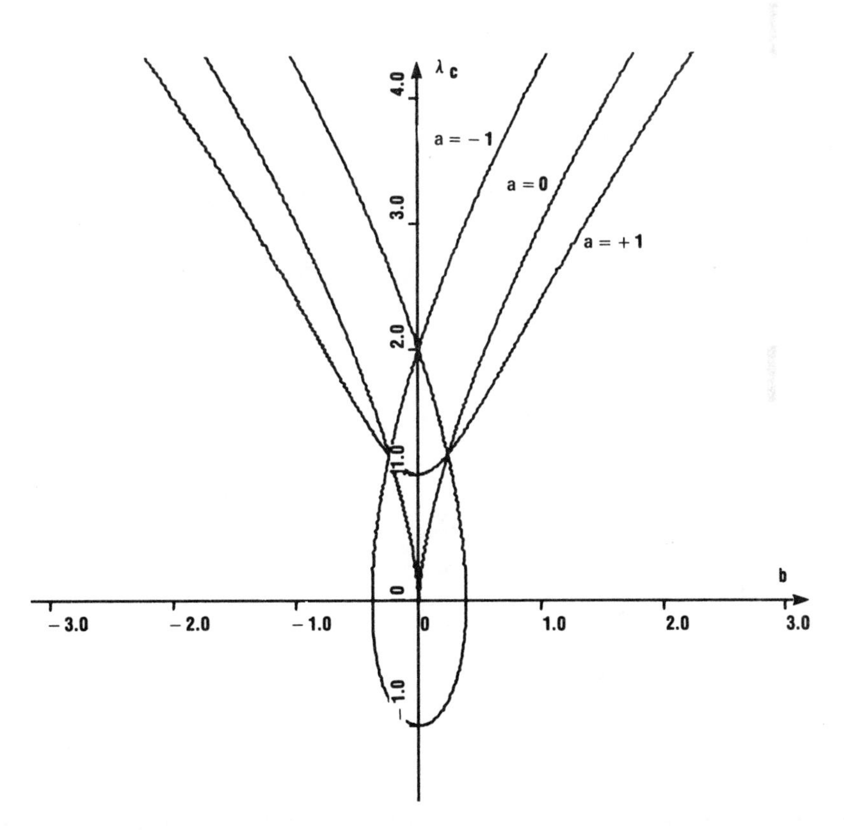

Figure 6.5 The critical curvature of F is shown as a function of b in the three planes $a = +1$, $a = 0$, and $a = -1$.

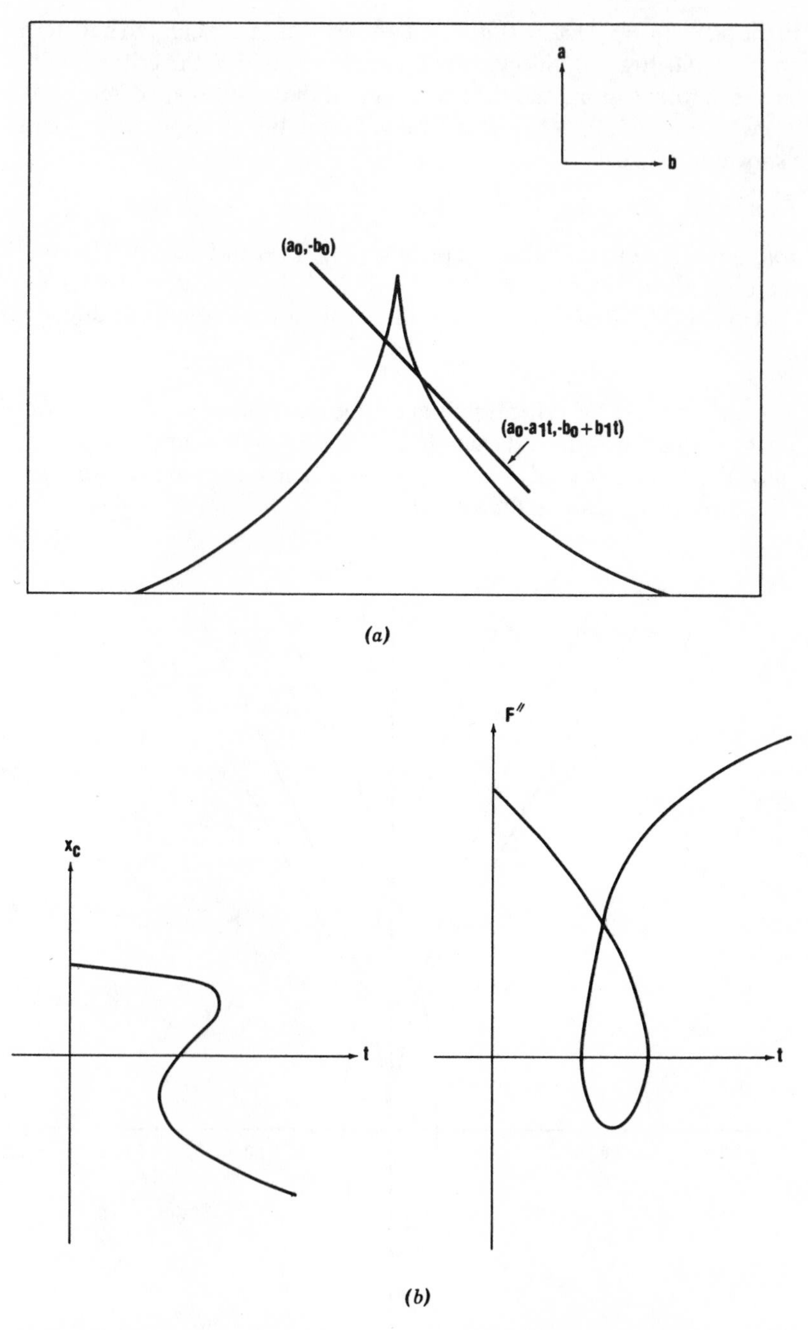

Figure 6.6 The critical curvatures along any path in control parameter space can easily be determined using the scaling relations (6.3) and the curvatures in the appropriate plane $a = +1$, $a = 0$, or $a = -1$. This is done here for a straight-line trajectory in control parameter space.

Remark. The surfaces shown in Figs. 6.2*b*–*d* may be constructed from the curves given in Figs. 6.3–6.5, respectively, and the scaling relations given in (6.3).

Example 3. Find the curvature at the critical points of the function F parameterized by points on the straight line trajectory

$$\begin{aligned} b(t) &= -b_0 + b_1 t \\ a(t) &= -a_1 t \end{aligned} \tag{6.7}$$

in control parameter space which is shown in Fig. 6.6*a*.

Solution. For each $t > 0$ the scale factor $\lambda(t) = |a_1 t|^{1/2}$ is determined. The eigenvalues are determined from the appropriate point $b_0 = b(t)/\lambda^3(t)$ in the plane $a = -1$. The eigenvalues then scale like $\lambda^2(t)$. The process can be repeated for $t = 0, t < 0$. The eigenvalues computed in this way are shown in Fig. 6.6*b* for $b_0 = 1, a_1 = 2, b_1 = 3$.

3. Geometry of the Dual Cusp (A_{-3})

The A_{-3}, or dual cusp, catastrophe is defined by

$$A_{-3}: \quad F(x; a', b') = -\tfrac{1}{4}x^4 - \tfrac{1}{2}a'x^2 - b'x \tag{6.8a}$$

We note that for $(a', b') = (a, b)$ (6.8a) is the negative of (6.2a). As a result, (6.8b) is identical to (6.2b). For the dual cusp the analogs of Figs. 6.2 for the cusp are easy to construct. The critical points of the cusp and dual cusp are determined by the same equation, so Fig. 6.2*b* is valid for both. Because (6.8*a*) is the negative of (6.2a), the analogs of Figs. 6.2*a*, *c*, and *d* are obtained by reflecting these figures in the control-parameter plane.

Although the cusp and dual cusp are mathematically simply related, their physical implications are significantly different (cf. Chapter 11). For the cusp there is always at least one global minimum. As the control parameters change, the system may move from one minimum to another.

In going from the cusp to the dual cusp, maxima and minima are interchanged. The dual cusp has a local minimum when the control parameters lie inside the cusp-shaped region in the control parameter plane. There is no global minimum. If the control parameters move outside the fold lines, a system described by such a potential has no minima at all. Catastrophe results.

4. Summary

The geometric properties of the fold and the cusp catastrophes have been discussed. In particular we have described again the qualitative nature of the functions parameterized by various regions in control parameter space. The geometric properties of particular interest to us are the locations of the critical points, the values of the function at the critical points, and the curvature of the function at its critical points. We have presented these three geometric properties in a quantitative way. Since these properties are canonical, they are invariants of all folds and cusps. In particular they are valid for folds and cusps which arise from higher catastrophes (Chapter 5, Secs. 5–7).

CHAPTER 7

Catastrophe Organization

In Chapter 5 we have seen how the control parameter space $\mathbb{R}^k$ for the lower dimensional catastrophes ($k \leq 3$) is partitioned into a small number of open regions by a separatrix. Points within each open region represent Morse functions which are qualitatively similar. The separatrices are "thin" (measure zero) and consist of components of dimensions $0, 1, 2, \ldots, k - 1$. The component of dimension 0 consists of the origin in $\mathbb{R}^k$ representing the catastrophe germ with a $(k + 1)$-fold degenerate critical point at the origin in $\mathbb{R}^l$ (state variable space). The components of dimension 1 represent catastrophe germs with $(k + 1 - 1)$-fold degenerate critical points somewhere in state variable space, and so on.

One way to determine the structure and the equations for the separatrix in $\mathbb{R}^k$ is to determine the equations of the 1-dimensional curves of catastrophe germs of control dimension $k - 1$ which emanate from the origin in $\mathbb{R}^k$ of the catastrophe with control dimension k. This can easily be done by using scaling arguments. These computational methods are illustrated in Secs. 1 and 2.

These computational methods are difficult to use for determining the components of the separatrix of dimension greater than 1. In addition, they do not easily provide answers to the questions in which we are interested. As a result, an alternative diagrammatic technique for representing catastrophes is developed in Secs. 3 and 4. Using this diagrammatic technique it is a simple matter to determine the complete "spectrum" of components of the separatrix as well as the dimension of each component. Moreover, it is just as easily possible to determine the type of structurally stable Morse functions that can be encountered

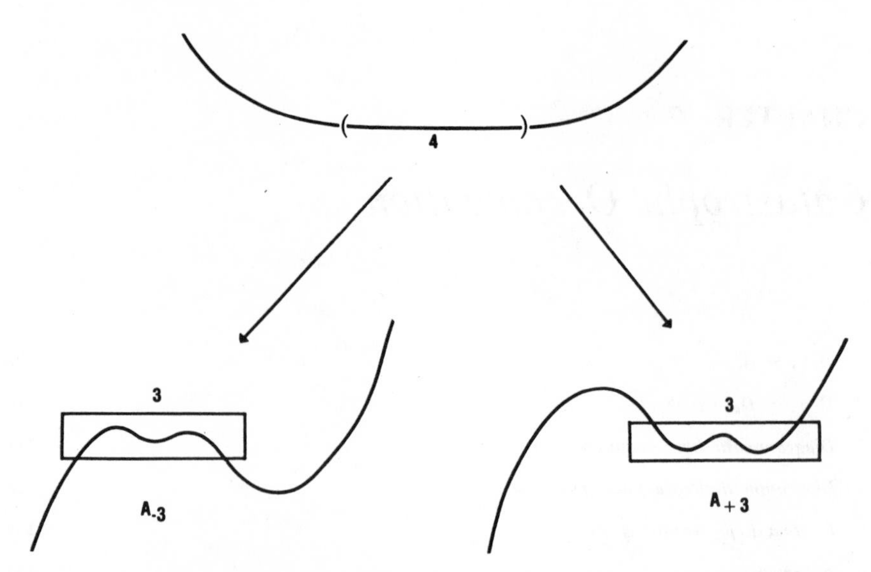

Figure 7.1 The occurrence nearby of catastrophes of type A_{+3} and A_{-3} strongly suggests that they are organized in a particular way by some higher catastrophe, which must be at least A_4.

in a perturbation of any catastrophe germ, as well as whether the open sets representing these functions have any particular component of the separatrix in common.

In Sec. 5 we introduce another scheme for representing the properties of a catastrophe. This scheme is "dual" to the diagrammatic representation of Secs. 3 and 4, and is particularly useful for providing a contour description of a catastrophe. This is a particularly convenient representation for map readers.

In Sec. 6 we discuss which and how catastrophes of control dimension k are related to ("abut") catastrophes whose control dimension is smaller by 1.

In short, higher catastrophes organize lower catastrophes, and the hierarchy of related catastrophes is organized in a very definite way. This is the view from the "top down." Conversely, there is a "bottom up" view. The presence nearby of two or more specific catastrophes, say A_{+3} and A_{-3}, strongly suggests that they are organized by a higher catastrophe, which is A_4 (cf. Fig. 7.1). We might be impelled to search for, and to find, this higher organizing catastrophe germ. This, in turn, would imply nontrivial relations between the original cusp and dual cusp catastrophes A_{+3} and A_{-3}.

1. $A_{k+1} \rightarrow A_k$

Emanating from the origin in the control parameter space $\mathbb{R}^k$ for the catastrophe germ $A_{k+1} = x^{k+2}$ are two lines which parameterize catastrophe germs $A_k = \tilde{x}^{k+1}$. The least painful way to determine the equations of these lines in control parameter space is through dimensional analysis, and in particular, through scaling arguments. We illustrate these arguments by first applying them to a familiar case: the fold lines emanating from the origin $(0, 0) \in \mathbb{R}^2$ for the control parameter space of the cusp.

A general function in the neighborhood of the cusp catastrophe germ has the form

$$F(x; a, b) = \tfrac{1}{4}x^4 + \tfrac{1}{2}ax^2 + bx$$

If we assume $dF/dx = 0$ has two roots at $x = -\lambda$ and a third root elsewhere, then

$$\begin{aligned} &A_3 \to A_2 \\ &\tfrac{1}{4}x^4 + \tfrac{1}{2}ax^2 + bx \to \tfrac{1}{4}(x + \lambda)^3(x + \alpha) + \text{constant} \\ &\qquad \tfrac{1}{4}x^4 + \tfrac{1}{4}(3\lambda + \alpha)x^3 + \tfrac{1}{4}(3\lambda^2 + 3\lambda\alpha)x^2 \\ &\qquad + \tfrac{1}{4}(\lambda^3 + 3\lambda^2\alpha)x + \text{constant} \end{aligned} \tag{7.1}$$

In the neighborhood of $x = -\lambda$, the function behaves like $(\alpha - \lambda)\tilde{x}^3$, $\tilde{x} = x + \lambda$. Equating coefficients of x^j leads to a parametric representation for the control parameters along the fold lines

$$\begin{aligned} a &= -3\lambda^2 \\ b &= -2\lambda^3 \end{aligned} \Rightarrow \left(\frac{a}{3}\right)^3 + \left(\frac{b}{2}\right)^2 = 0 \tag{7.2}$$

The general procedure for determining the 1-dimensional curves $A_{k+1} \to A_k$ is similar:

$$\begin{aligned} &A_{k+1} \to A_k \\ &\frac{1}{k+2}x^{k+2} + \sum_{j=1}^{k} \frac{1}{j} a_j x^j = \frac{1}{k+2}(x + \lambda)^{k+1}(x + \alpha) + \text{constant} \end{aligned} \tag{7.3}$$

Term by term comparison leads to

$$\frac{k+2}{j} a_j = \binom{k+1}{j-1}\lambda^{k+1-(j-1)} + \binom{k+1}{j}\lambda^{k+1-j}\alpha \tag{7.4}$$

The indicial condition $a_{k+1} = 0$ provides the relation between the k-fold degenerate root at $x = -\lambda$ and the isolated root at $x = -((k + 1)\alpha + \lambda)/(k + 2)) = k\lambda$:

$$a_{k+1} = 0 = (k + 1)\lambda + \alpha \Rightarrow \alpha = -(k + 1)\lambda \tag{7.5}$$

The lines of A_k catastrophes emanating from the A_{k+1} catastrophe germ x^{k+2} have the parametric representation

$$a_j = (-)\binom{k+1}{j-1}(k + 1 - j)\lambda^{k-j+2}, \qquad j = 1, 2, \ldots, k \tag{7.6}$$

The parameter λ may be eliminated to obtain equations involving only the control parameters

$$\lambda = \left[\frac{-a_j}{\binom{k+1}{j-1}(k+1-j)} \right]^{1/(k-j+2)} \tag{7.7}$$

The result (7.7) reduces to (7.2) for $k = 2$.

Lines of A_{k-1} catastrophes emanate from each germ A_k in exactly the same way that lines of A_k emanate from each germ A_{k+1}. Therefore 2-dimensional surfaces of A_{k-1} emanate from each germ A_{k+1}. Continuing in this way, it is possible to build up the entire separatrix of A_{k+1} in $\mathbb{R}^k$.

2. $D_{k+1} \to D_k$

Similar arguments hold for the catastrophe germs D_k, and these arguments may be treated by similar techniques. Specifically, the germ

$$D_{k+1}: \quad x^2y + y^k \tag{7.8}$$

is represented by the origin in $\mathbb{R}^k$. This germ has a $(k+1)$-fold degenerate critical point at the origin $(0, 0) \in \mathbb{R}^2$ in state variable space. 1-dimensional curves emanate from the origin in control parameter space and parameterize catastrophe germs with a k-fold degenerate critical point somewhere in $\mathbb{R}^2$. This degenerate critical point may have either one "bad" direction, in which case the corresponding germ is $A_k(\tilde{x}^{k+1})$, or it may have two "bad" directions, in which case the germ is D_k.

The lines of germs D_k emanating from the D_{k+1} germ are obtained by a scaling argument similar to that of Sec. 1. Assume that the k-fold degenerate critical point occurs at $(x, y) = (-\lambda_1, -\lambda_2)$. Then we can equate the general perturbation of D_{k+1} (left) with the germ of D_k around $(-\lambda_1, -\lambda_2)$ (right):

$$\begin{aligned} &x^2y + y^k + \sum_{j=1}^{k-2} a_j y^j + a_{k-1}x + a_k x^2 \\ &\quad = (x+\lambda_1)^2(y+\lambda_2) + (y+\lambda_2)^{k-1}(y+\alpha) + \text{constant} \end{aligned} \tag{7.9}$$

Upon expansion and comparison of coefficients, we find

$$\begin{aligned} x:&\quad a_{k-1} = 2\lambda_1\lambda_2 \\ x^2:&\quad a_k = \lambda_2 \\ xy:&\quad 0 = 2\lambda_1 \\ y^j:&\quad a_j = \binom{k-1}{j-1}\lambda_2^{k-j} + \binom{k-1}{j}\lambda_2^{k-j-1}\alpha \end{aligned} \tag{7.10}$$

Comparison (7.10) shows clearly that $\lambda_1 = 0$ and $\lambda_2 = \lambda$ may be used as the scaling factor. An indicial condition is supplied by the requirement that the coefficient of y^{k-1} be zero, since we have used x^2 instead of y^{k-1} as a universal perturbation term. This requires the relation

$$0 = (k - 1)\lambda + \alpha \tag{7.11}$$

from which we easily determine

$$a_j = -\binom{k}{j}(k - 1 - j)\lambda^{k-j} \tag{7.12}$$

The result is that the 1-parameter curve in the control parameter space $\mathbb{R}^k$ of D_{k+1} whose coordinates $(a_1, \ldots, a_k)$ represent the perturbation

$$a_1 y + a_2 y^2 + \cdots + a_{k-2} y^{k-2} + a_{k-1} x + a_k x^2 \tag{7.13}$$

with a k-fold degenerate critical point of type D_k has the parametric representation

$$\begin{aligned} a_j &= -\binom{k}{j}(k - 1 - j)\lambda^{k-j}, \qquad j = 1, 2, \ldots, k - 2 \\ a_{k-1} &= 0 \\ a_k &= \lambda \end{aligned} \tag{7.14}$$

This critical point has coordinates $(x, y) = (0, -\lambda)$ in $\mathbb{R}^2$.

Emanating from the origin of the control parameter space $\mathbb{R}^k$ for the catastrophe D_{k+1}, there are 1-dimensional curves of catastrophes of type D_k and there may in addition be 1-dimensional curves of catastrophes of type A_k. This is the case when $k + 1$ is odd, and it is also the case for catastrophes of type D_{-p}, p even, but it is not true for catastrophes of the type D_{+p}, p even. In order to see which lower dimensional catastrophes are "thrown off" by some particular higher dimensional catastrophe, it is clear that we could use a simpler computational algorithm than the one used in this and the previous section. We turn our attention now to developing such an algorithm.

3. Diagrammatic Representations: $l = 1$

In the preceding two sections we have discussed the questions of which curves (surfaces, etc.) of catastrophes emanate from a particular catastrophe germ and how to determine their equations. However, we have not really confronted the basic questions of interest. These questions are: what do the separatrices in control parameter space look like? What do they separate? What are the non-Morse functions on each component of the separatrix? And what are the qualitative natures of the Morse functions parameterized by points in each of the open regions of the control parameter space $\mathbb{R}^k$?

These questions are difficult to answer when $k > 3$. We must, therefore, resort to methods other than those of Chapter 5 for treating these questions. In this section we will introduce a diagrammatic technique for discussing these questions in the case of catastrophes involving a single bad state variable ($l = 1$). In the following section we consider catastrophes with $l = 2$ by the same diagrammatic method. A complementary method for approaching these questions is described in Sec. 5.

To illustrate the diagrammatic technique, we first consider catastrophes of type A_k. Under a universal perturbation, the k-fold degenerate critical point will split up into at most k isolated critical points. The location of these critical points can be indicated by circles placed along the x axis, since catastrophes of type A_k involve only one state variable. The local minima are indicated by placing a + sign within the circle, because the curvature is positive at a local minimum. Local maxima are indicated by ⊖ (Fig. 7.2).

Figure 7.2 provides the basis for a diagrammatic representation of the catastrophes A_k. We transform to this representation by making the following considerations:

1. It is not necessary to exhibit the x axis in the diagrammatic representation.
2. It is convenient to indicate by arrows which maxima can "flow into" which minima.
3. It is not necessary to indicate specifically the flow direction, since the maxima and minima are specifically identified.

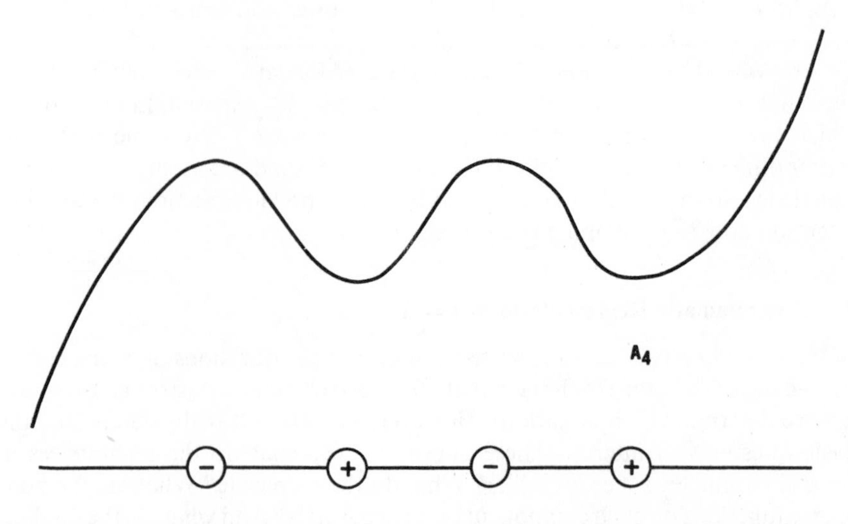

Figure 7.2 The qualitative features of a catastrophe function are adequately represented by indicating the positions of all critical points with a circle and also indicating the Morse i-saddle type of each isolated critical point. For $l = 1$ only maxima ⊖ and minima ⊕ can occur.

These considerations are summarized here:

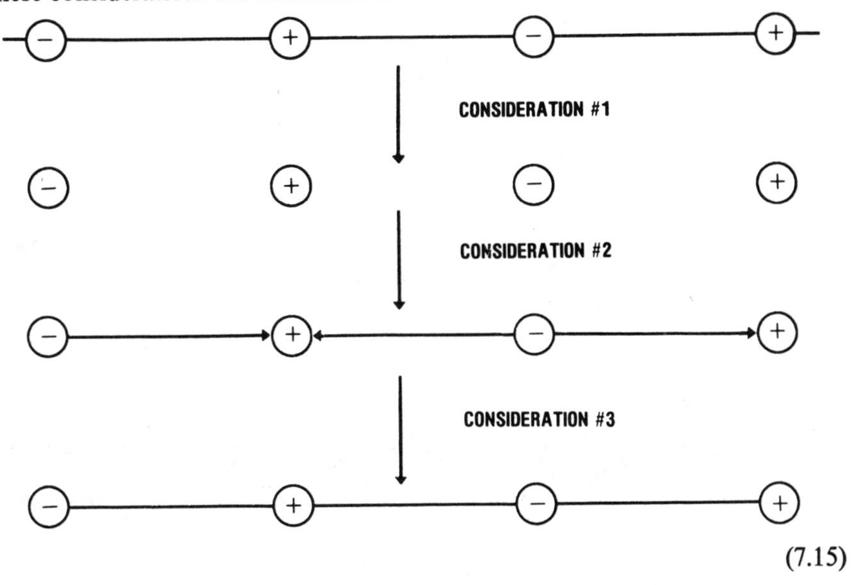

(7.15)

The lowest diagram is the diagrammatic representation of A_k ($k = 4$ in the case shown).

For k even the diagrams

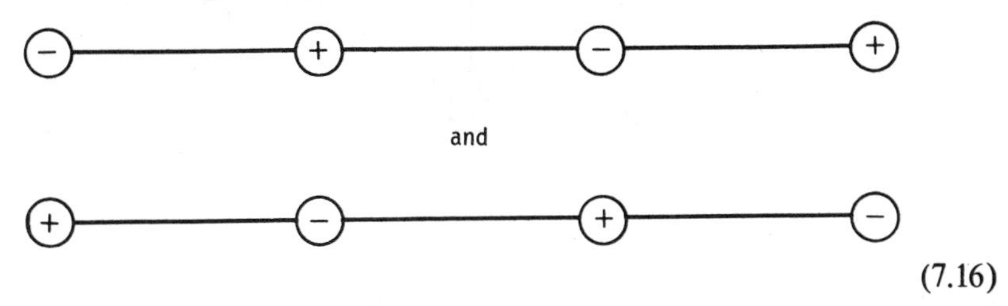

(7.16)

represent qualitatively similar functions (under $x \to -x$). Alternatively, one diagram can be transformed into the other by a rotation in the x–y plane. For k odd the diagrams

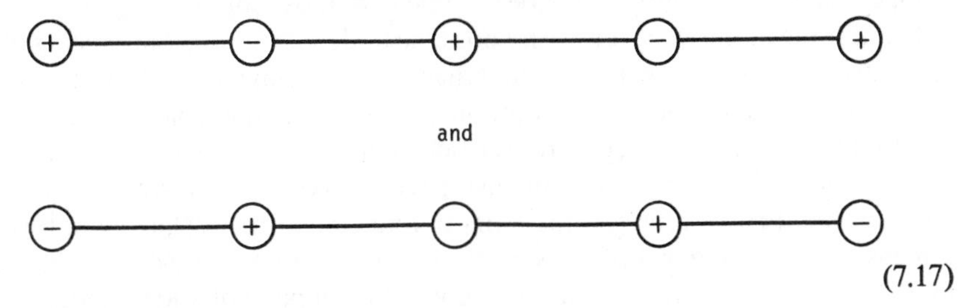

(7.17)

represent qualitatively different functions ($A_{\pm k} = \pm x^{k+1} + \cdots$).

If we now consider a 1-parameter curve in the control parameter space $\mathbb{R}^k$, then we will typically encounter only non-Morse critical points of the fold type. This will occur when the 1-dimensional curve intersects a $(k-1)$-dimensional component of the separatrix. At such an intersection a local maximum and a local minimum collide and annihilate each other. This can happen in $k-1$ distinct ways, as indicated below for A_4:

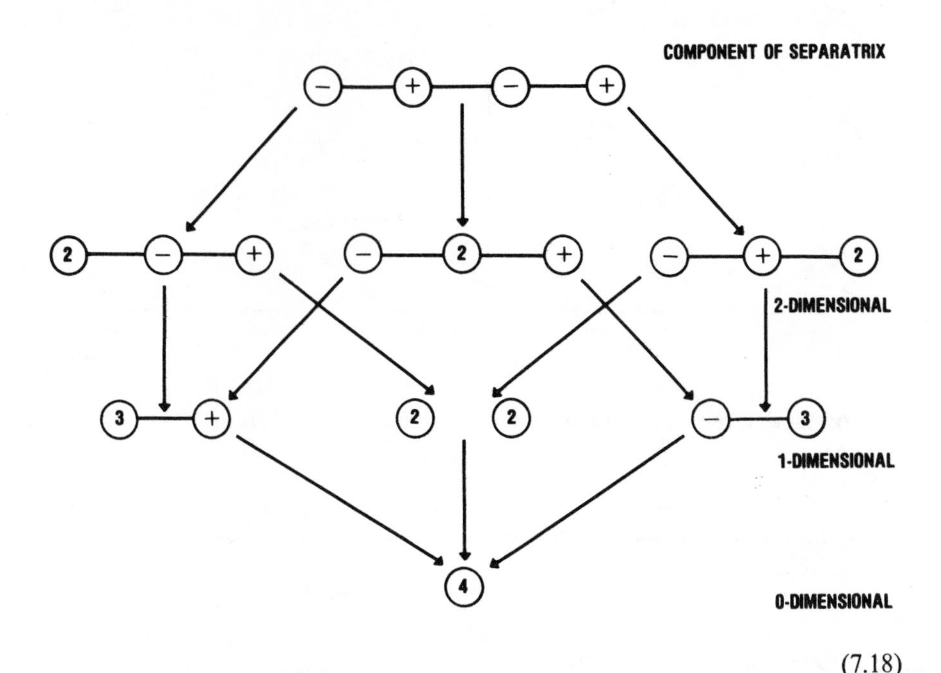

(7.18)

If we consider 2-dimensional surfaces in $\mathbb{R}^k$, these will intersect $(k-1)$-dimensional components of the separatrix in 1-dimensional curves of folds, and $(k-2)$-dimensional components of the separatrix in isolated points. These isolated intersection points represent either cusp catastrophes or pairs of fold catastrophes. These possibilities have been indicated above for A_4.

Continuing in this way, it is easily possible to determine all components of the separatrix as well as how they fit together. From the diagram above for A_4 we see that there is a 0-dimensional component, the origin in $\mathbb{R}^3$, parameterizing the function x^5 with fourfold degenerate critical point at the origin $x = 0$. Emanating from the origin in $\mathbb{R}^3$ are two 1-dimensional curves of cusp catastrophes and one 1-dimensional curve representing pairs of fold catastrophes. There are three 2-dimensional components of the separatrix representing fold catastrophes. The intersection of the "outer" surfaces to give a line of pairs of folds is indicated as clearly in (7.18) as it is in Fig. 5.6.

The open regions in control parameter space which describe distinct types of structurally stable families of Morse functions, and which are separated from each other by components of the separatrix, can also be determined by these diagrammatic methods. We illustrate how this may be done for A_4. When a 1-parameter curve in control parameter space $\mathbb{R}^3$ starts in the open region with the maximum number of isolated critical points and goes through a 2-dimensional component of the separatrix, the associated fold catastrophe involves the collision and disappearance of two critical points. This can be represented diagrammatically by

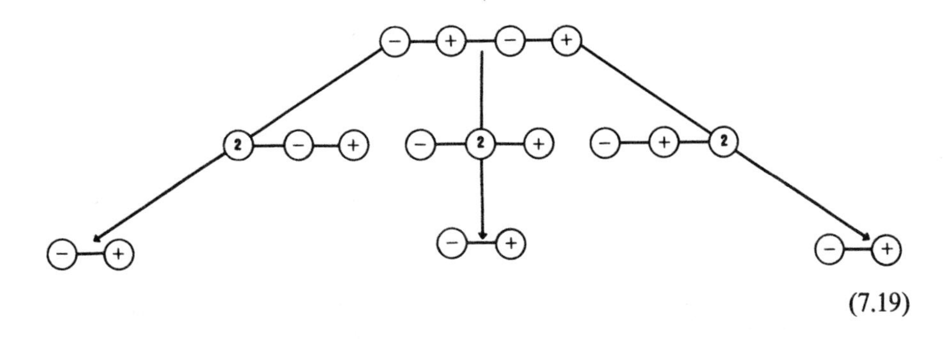

(7.19)

Thus passage through any 2-dimensional component leads to an open region in $\mathbb{R}^3$ describing functions with two isolated critical points. From such a region we can pass through another fold catastrophe to an open region representing functions with no critical points at all, as indicated by the following sequence:

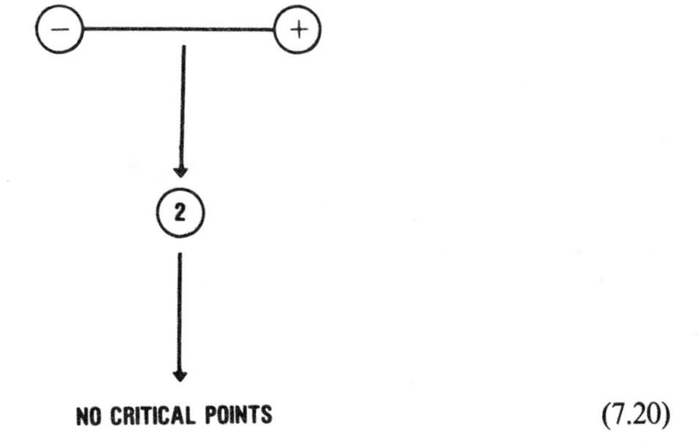

(7.20)

Such a region can also be entered from the region with four isolated critical points by passage through the curve representing a pair of fold catastrophes, as indicated by the following sequence:

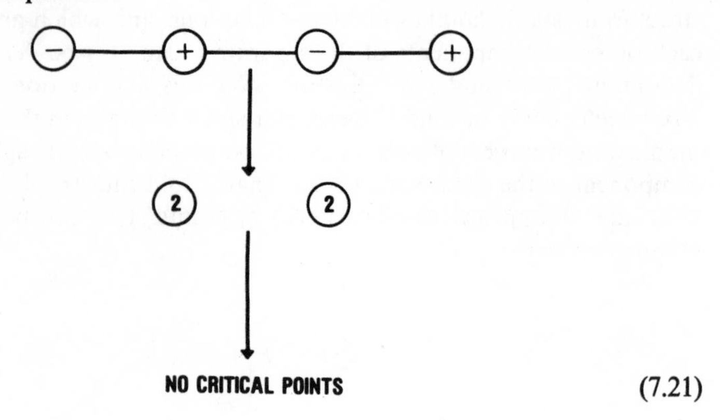

(7.21)

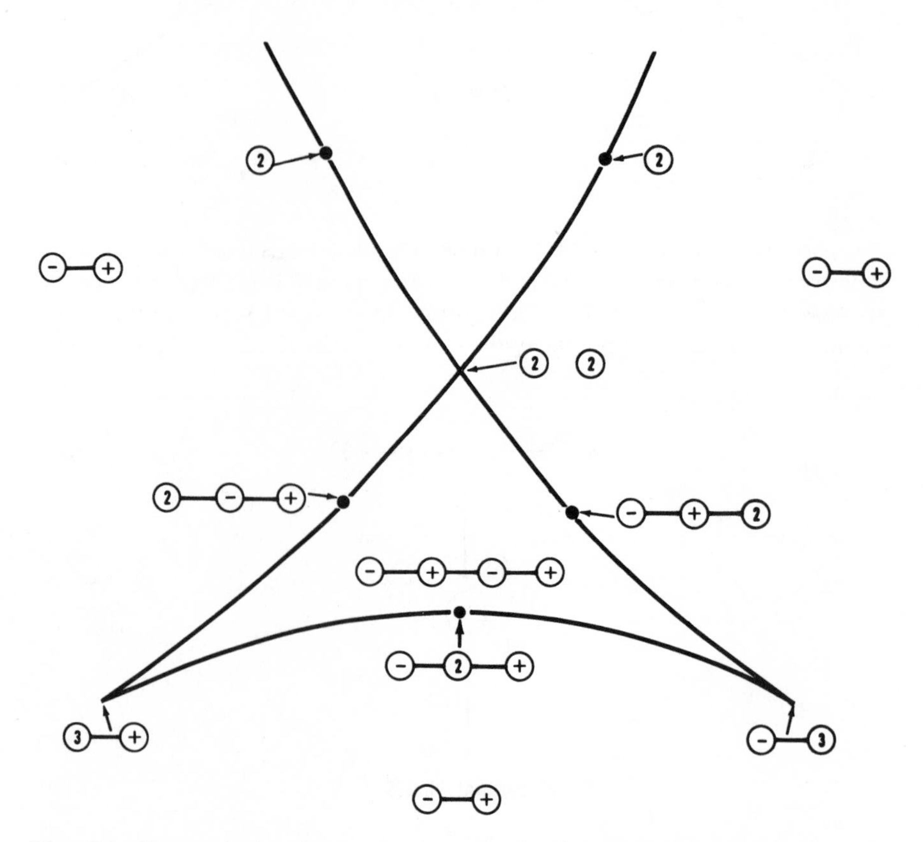

Figure 7.3 The organization of the control parameter space $\mathbb{R}^3$ for A_4 into open regions representing structurally stable functions, and into components of the separatrix of various dimensions, can be carried out using the diagrammatic technique.

In short, $\mathbb{R}^3$ is divided into three open regions by the separatrix, and these describe stable families of functions with four, two, and zero isolated critical points. The representations of the control parameter space $\mathbb{R}^3$ for A_4 given in Fig. 5.6 and in this section are synthesized in Figs. 7.3 and 7.4.

Remark. This diagrammatic construction for determining the open regions in control parameter space $\mathbb{R}^k$ can be called the **method of contraction** because it proceeds by "pulling together" "adjacent" critical points in fold catastrophes.

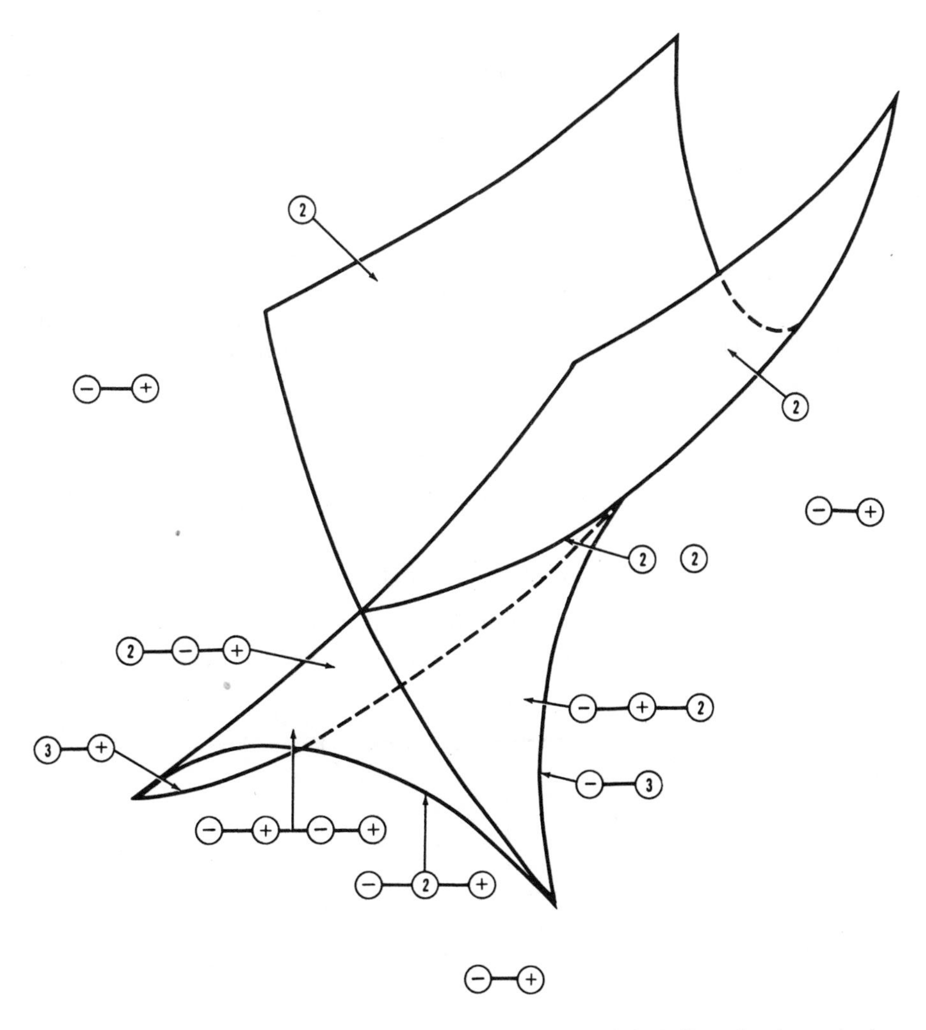

Figure 7.4 The relation between the diagrammatic technique and the 3-dimensional organization of the control space for A_4 is shown.

4. Diagrammatic Representations: $l = 2$

The diagrammatic representation can be extended with minor modifications to the description of catastrophes involving two "bad" state variables. We summarize here the rules for constructing diagrams:

1. Isolated critical points are distributed in the x–y plane.
2. The local minima, saddles, and local maxima are represented by:

   ⊕ local minimum
   ○ saddle
   ⊖ local maximum

3. Critical points are connected by a line if a flow line of the gradient system connects them. A maximum may flow to a saddle or a minimum, and a saddle may flow to a minimum. Saddles cannot be connected.*
4. The degeneracy of a non-Morse critical point is represented by a number:

   ② twofold degenerate
   ③ threefold degenerate

5. Fold catastrophes can occur only by "contracting" a saddle with either a maximum or a minimum.
6. We indicate all possible diagrammatic representations of a catastrophe obtained by choosing a perturbation leading to the maximum number of isolated critical points.
7. The components of the separatrix, and the structurally stable Morse families represented by the open sets, can be determined by the method of contraction.

Remark. The catastrophes A_k can be considered as functions of two state variables by adding a quadratic term in y:

$$F(x, y) = x^{k+1} + \text{perturbation in } x \pm y^2$$

If we choose to add $+y^2$, the 1-dimensional minima in x become minima in the x–y plane while the local maxima in the x direction become saddles in the x–y plane, so that

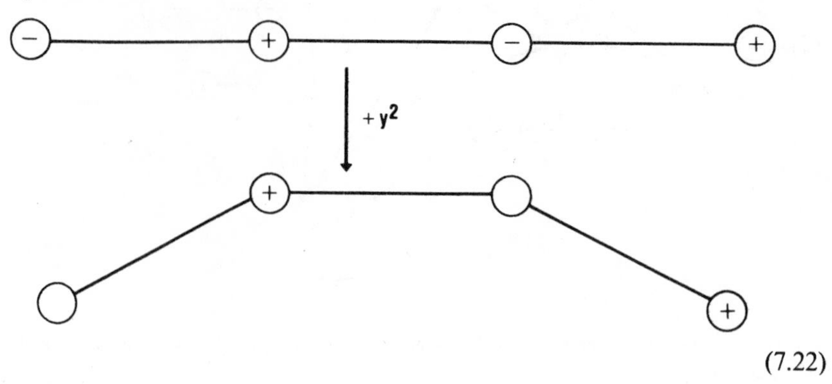

(7.22)

* Saddle connections are structurally unstable.

A 1-dimensional chain with k isolated critical points (alternating minima and saddles $(A_k + y^2)$ or alternating maxima and saddles $-(A_k + y^2)$) represents a catastrophe of type A_k.

For the series of catastrophes D_k we have seen that $D_k \neq D_{-k}$ if k is even, but $D_{-k} \doteq -D_k$ if k is odd. We therefore treat these three cases separately.

D_{2k}:

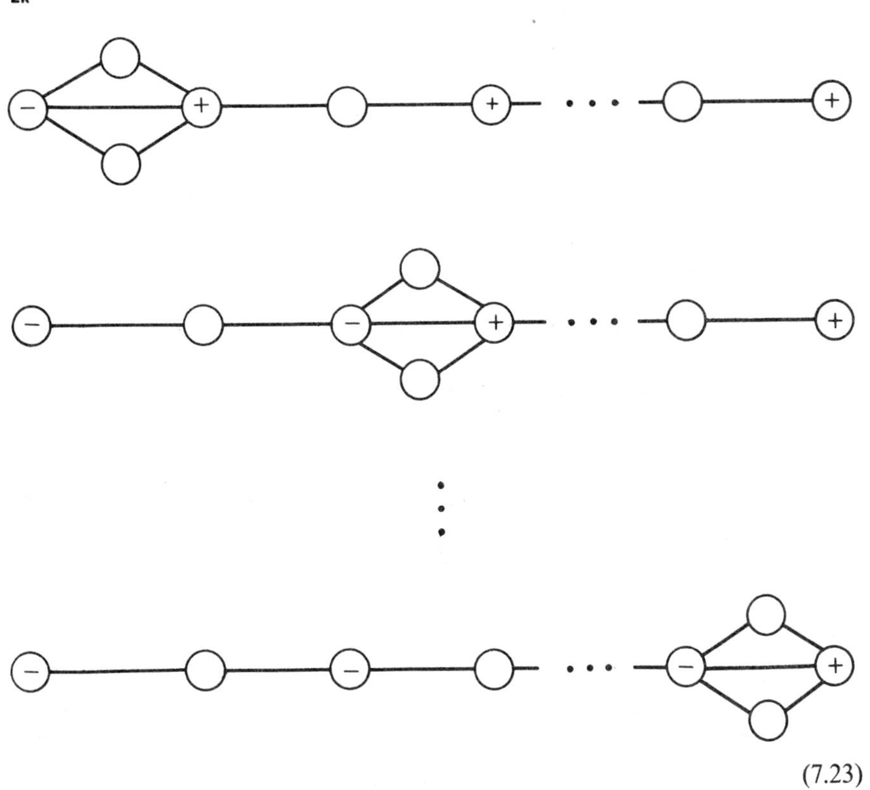

(7.23)

The general diagram has the structure

$-A_{+p}$ A_{+q}

$p + q + 2 = 2k$
p, q odd

(7.23g)

D_{-2k}:

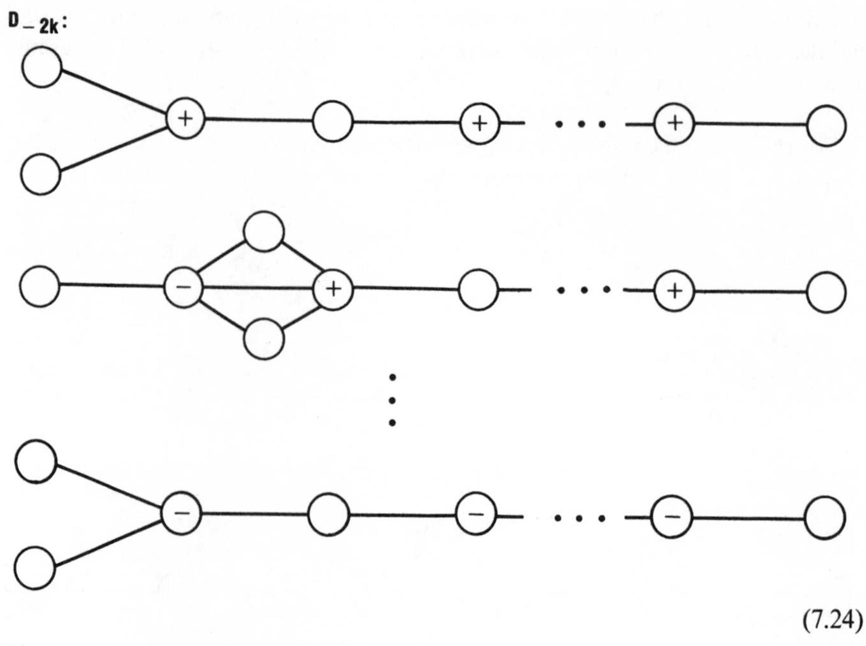

(7.24)

The general diagram has the structure

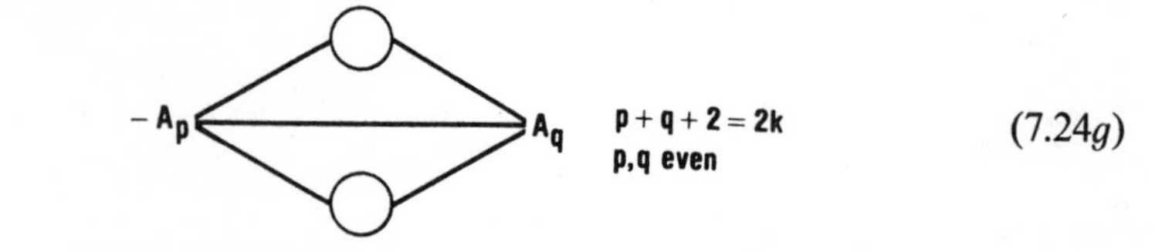

(7.24g)

D_{2k+1}:

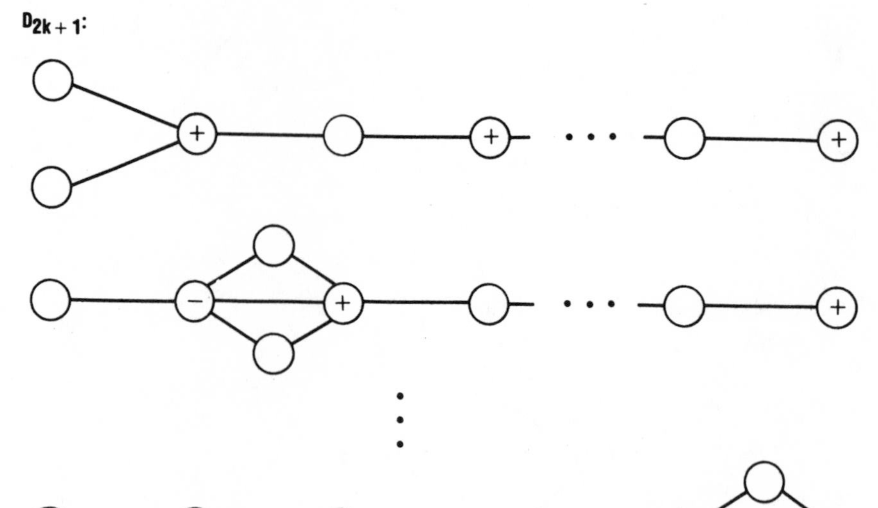

(7.25)

The general diagram has the structure

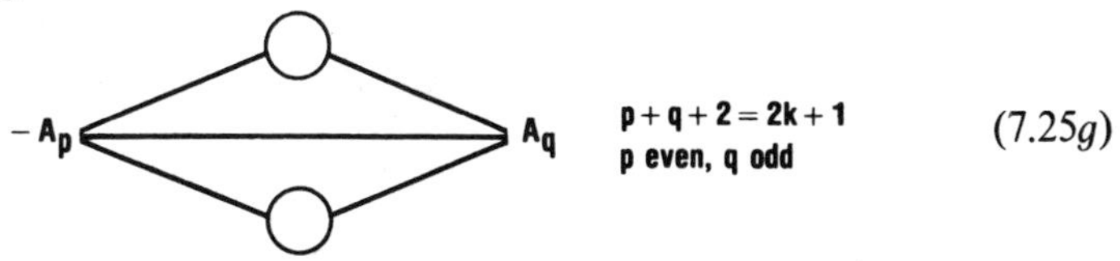

(7.25g)

The class of diagrams with p odd, q even represent $-D_{2k+1} \doteq D_{-(2k+1)}$, as can be seen by rotating the general diagram through 180°.

The diagrams for the exceptional catastrophes $E_{\pm 6}$, E_7, E_8 are known and presented in Table 7.1, where we have also collected the diagrams for the catastrophes D_k as well as the catastrophes of type A_k. The latter are presented in terms of their diagrams in the x–y plane, following the convention of (7.22).

The decomposition of the control parameter space into open sets and the components of the separatrix of various dimensions can now easily be determined by using the information summarized in the diagrammatic representations shown in Table 7.1. The dimensionality of a component of the separatrix is equal to $k - \sum (\mu_i - 1)$, where the sum extends over degenerate critical points, and μ_i is the degeneracy of the ith degenerate critical point [cf. (7.18)]. The open sets parameterize classes of structurally stable Morse functions which can be obtained by contraction. Two different open sets are contiguous through some piece of the separatrix if both nondegenerate diagrams can be contracted to the degenerate diagram describing that component of the separatrix.

Example 1: D_{+4}. From Table 7.1 the open component of $\mathbb{R}^3$ describing functions with the maximum number of isolated critical points has the diagram

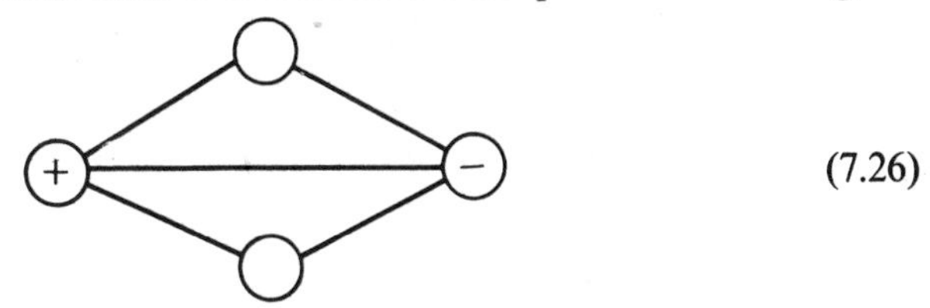

(7.26)

Other components of the open set are obtained by contracting a maximum with a saddle or a minumum with a saddle:

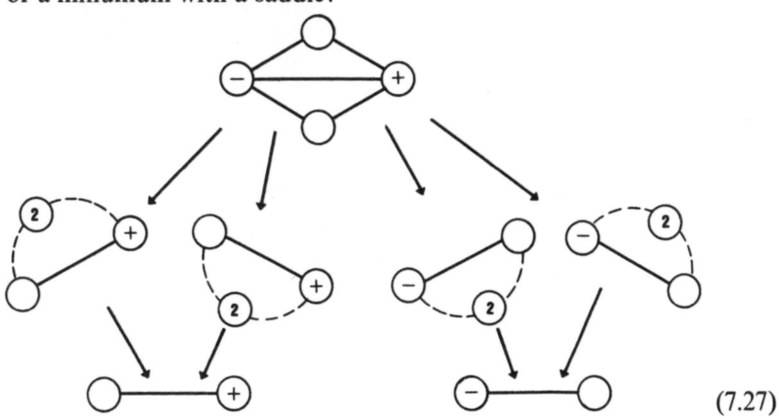

(7.27)

Table 7.1 Diagrammatic and Contour Representations of the Elementary Catastrophes

Catastrophe	Diagrammatic Representation	Contour Representation
$A_{+2k} + y^2$		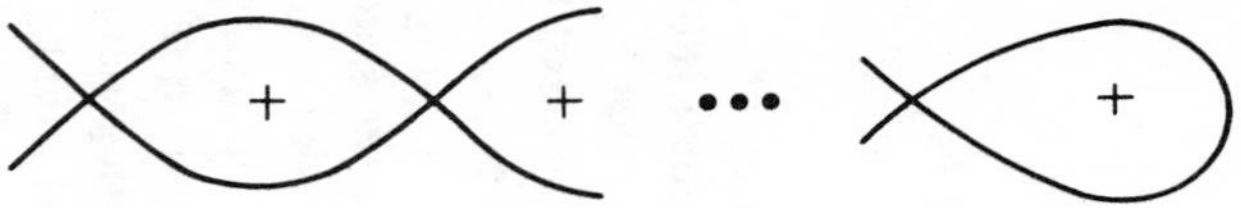
$A_{-2k} + y^2$		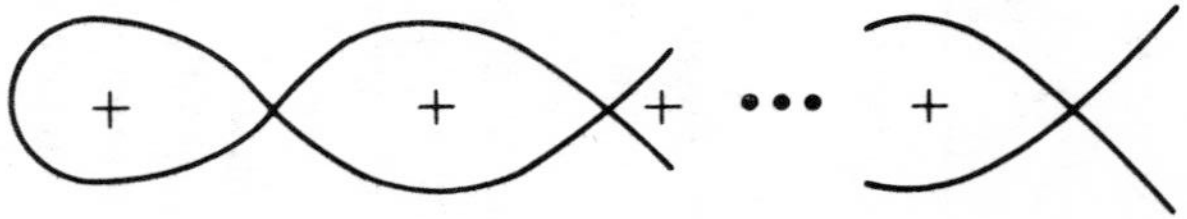
$A_{+(2k+1)} + y^2$		
$A_{-(2k+1)} + y^2$		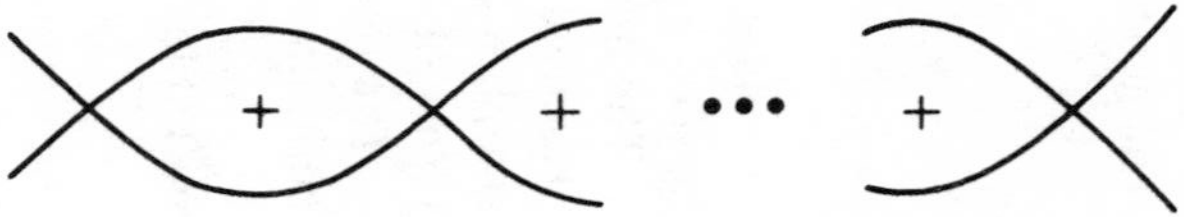

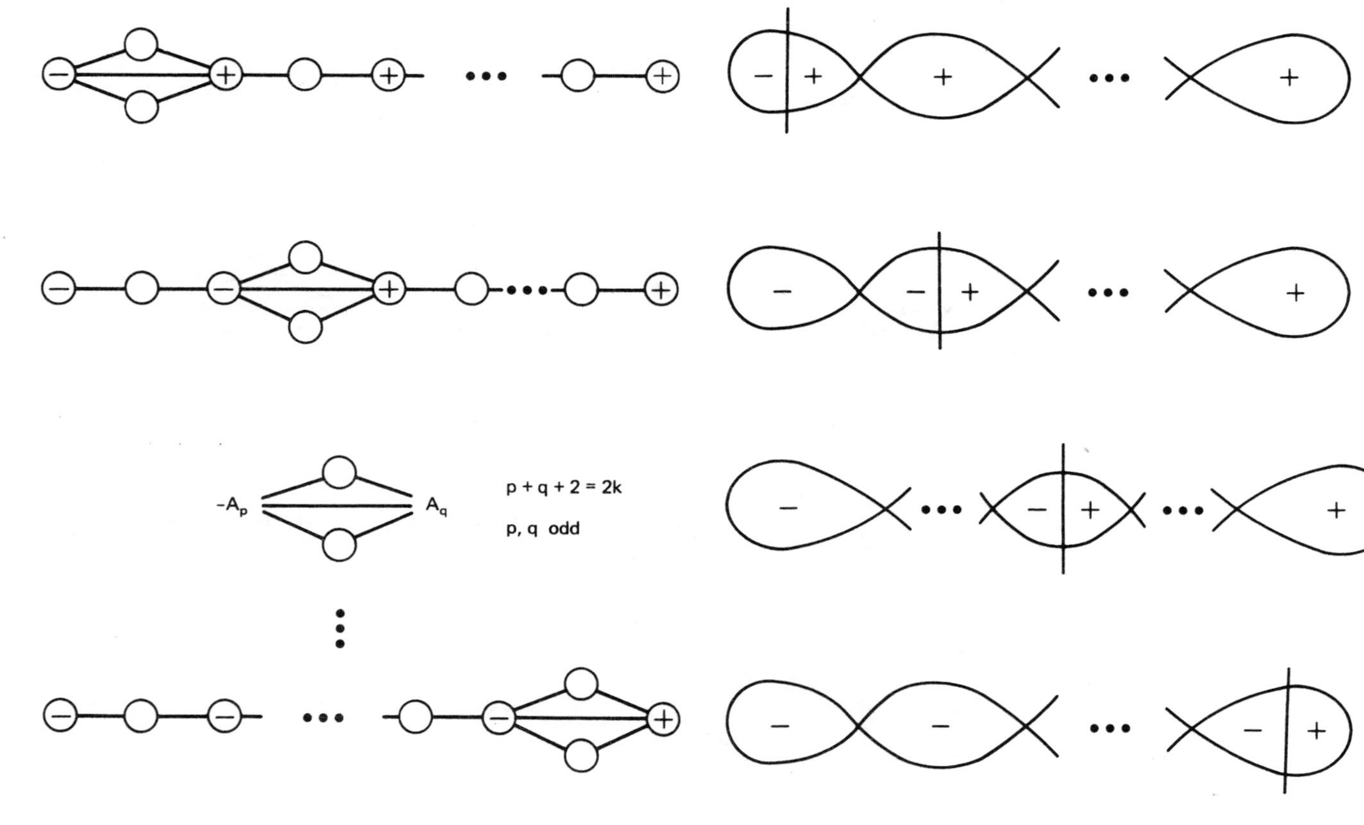

(Continued)

Table 7.1 (*Continued*)

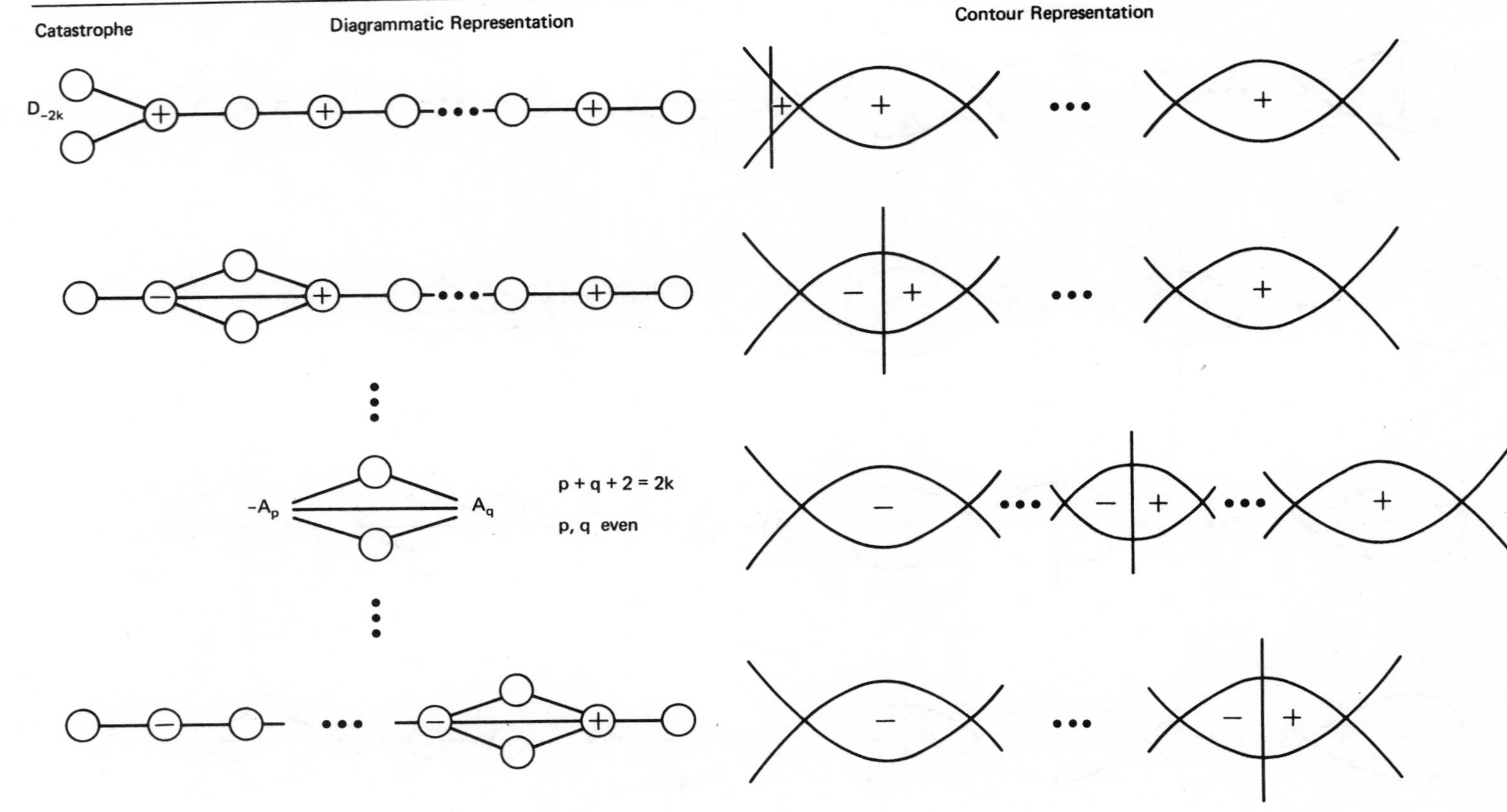

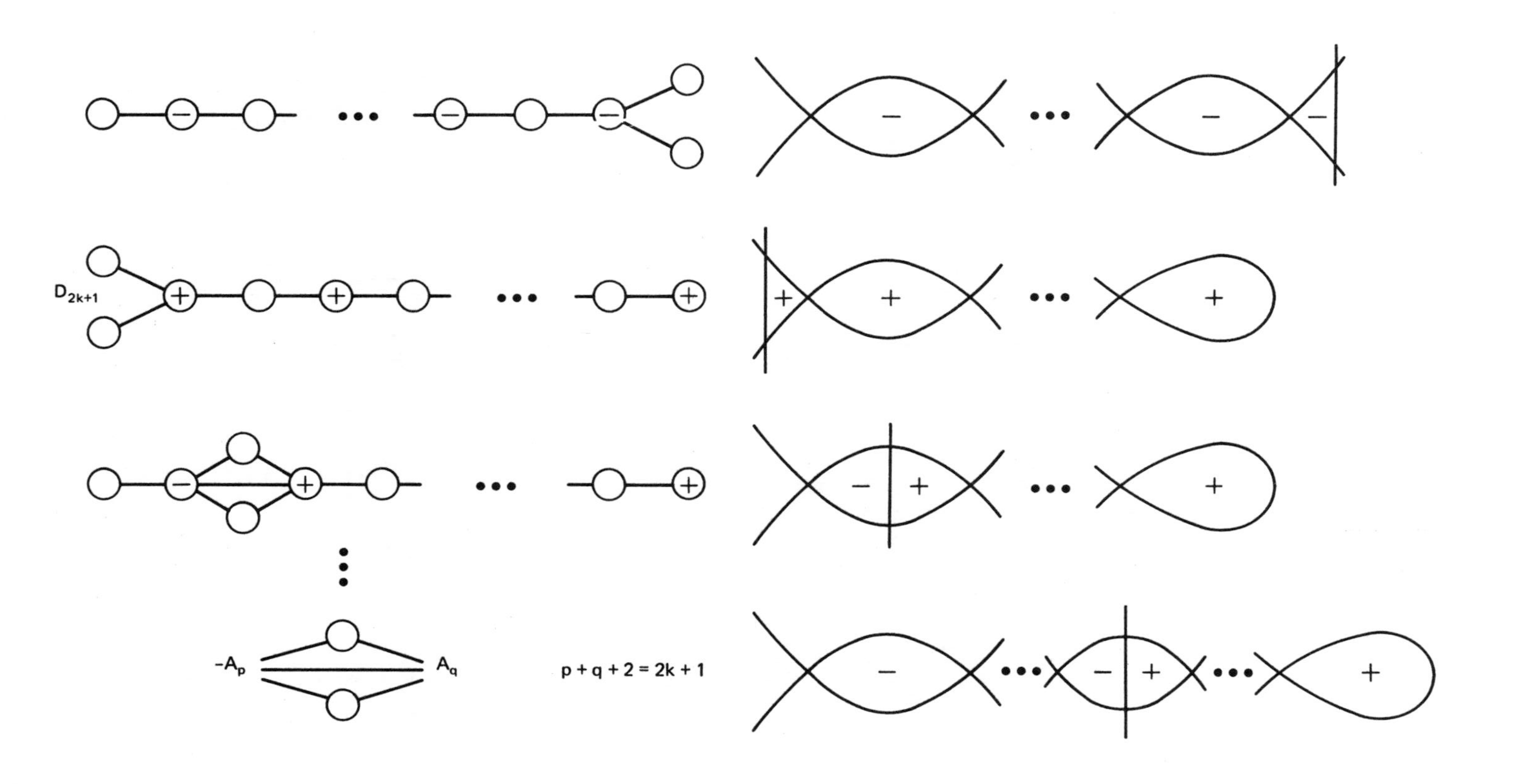

(*Continued*)

Table 7.1 (*Continued*)

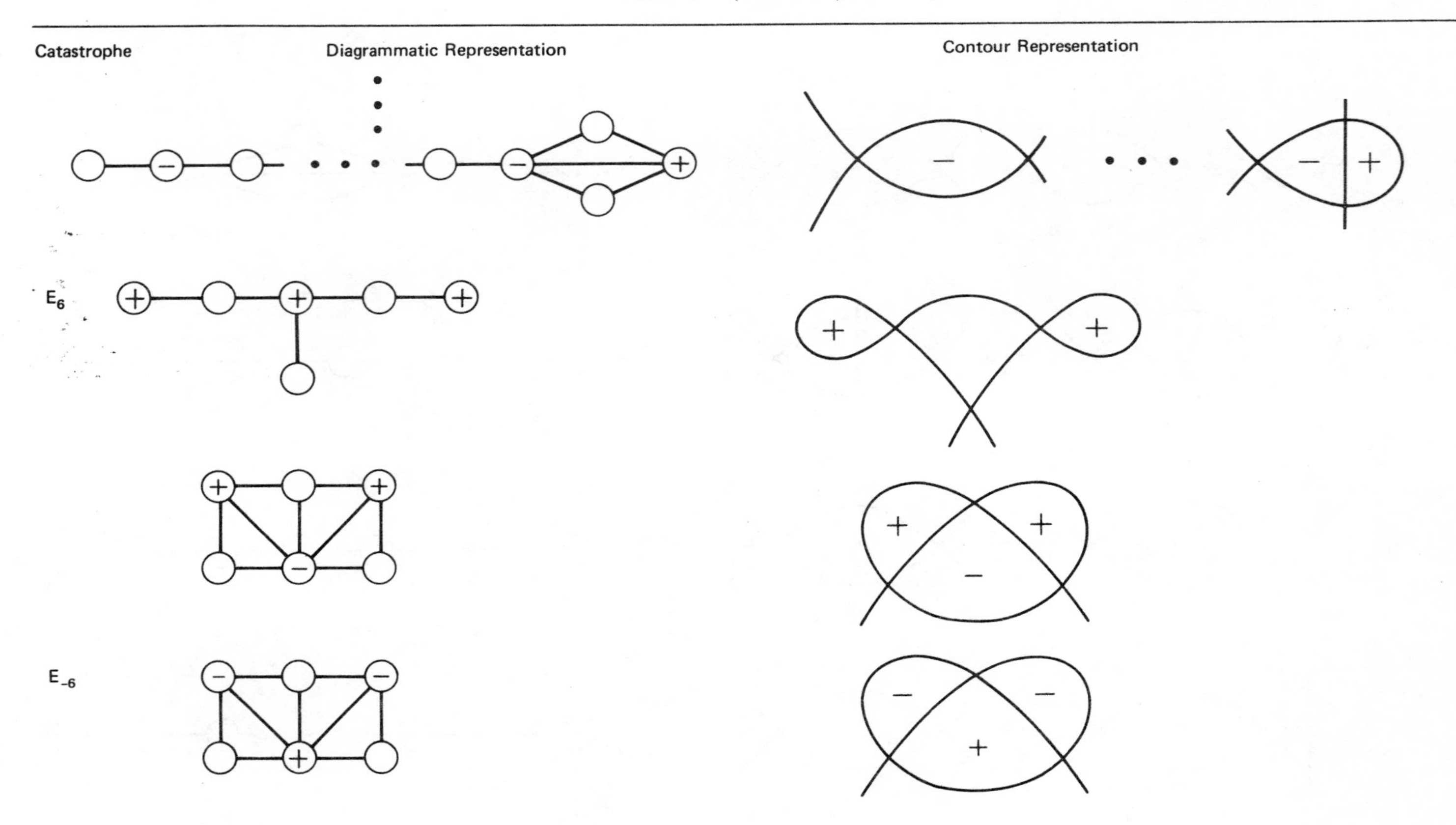

E_7

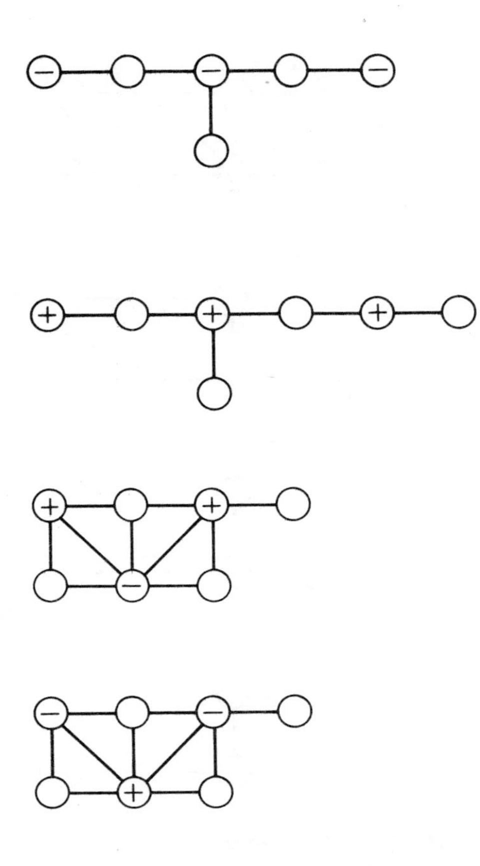

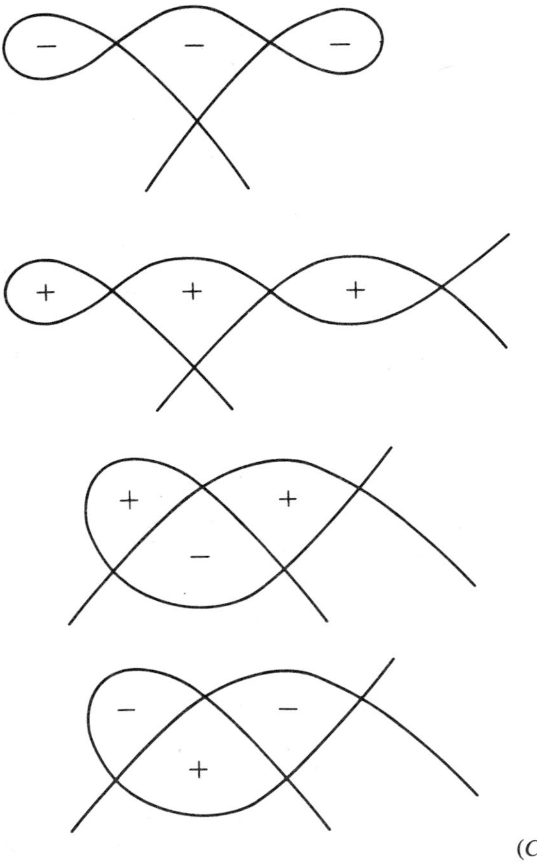

(Continued)

Table 7.1 (*Continued*)

Catastrophe	Diagrammatic Representation	Contour Representation
	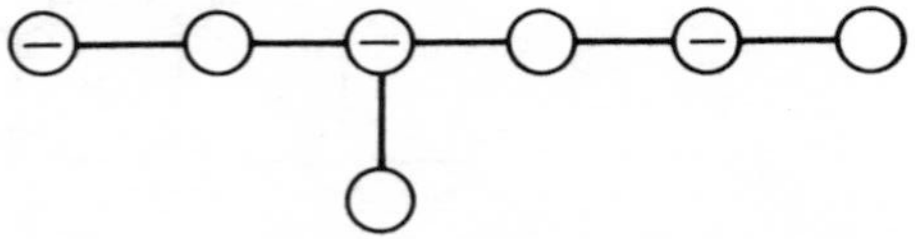	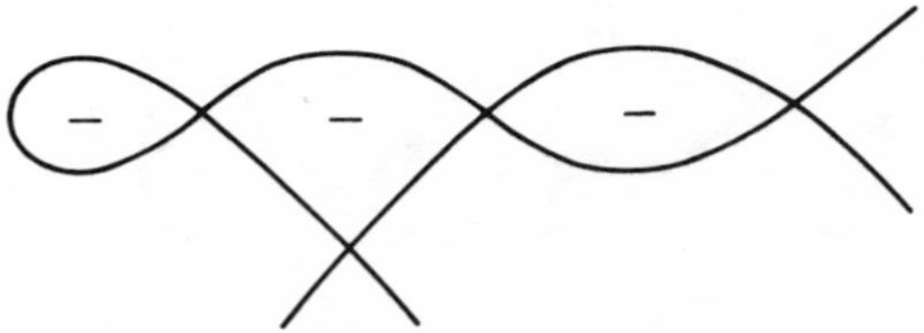
E_8	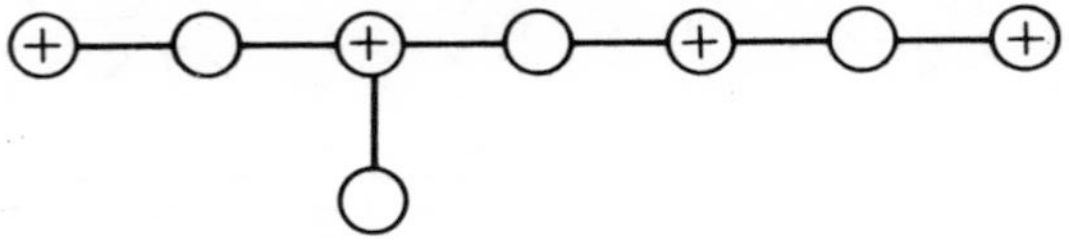	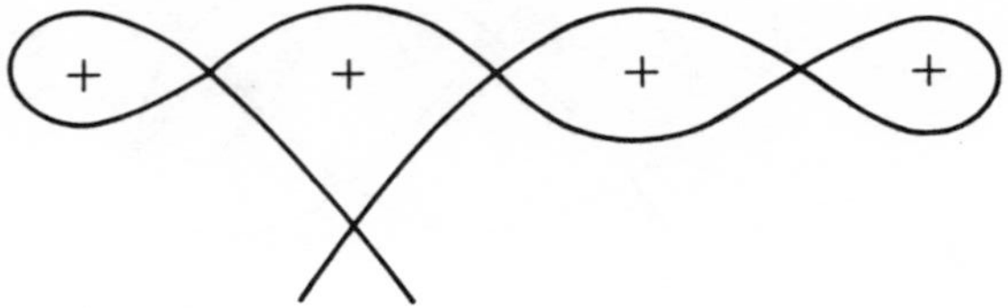
	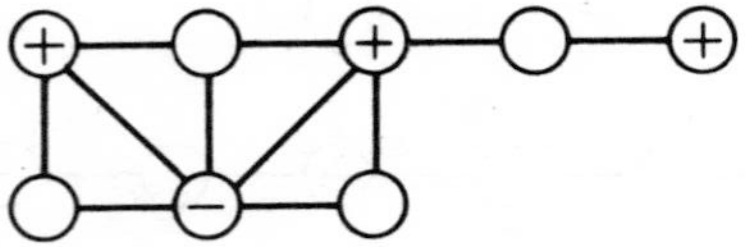	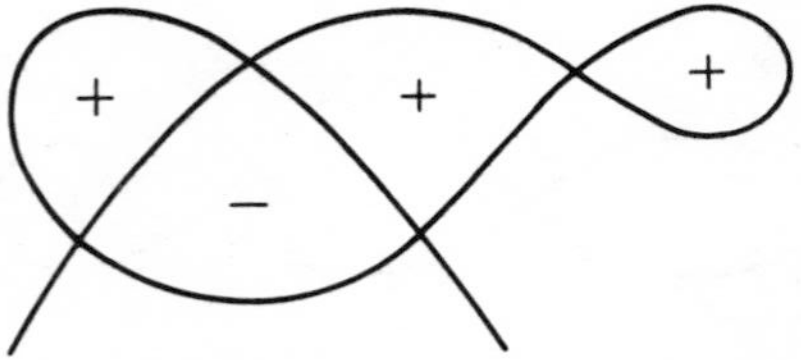
	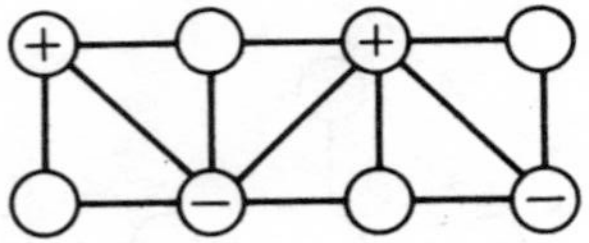	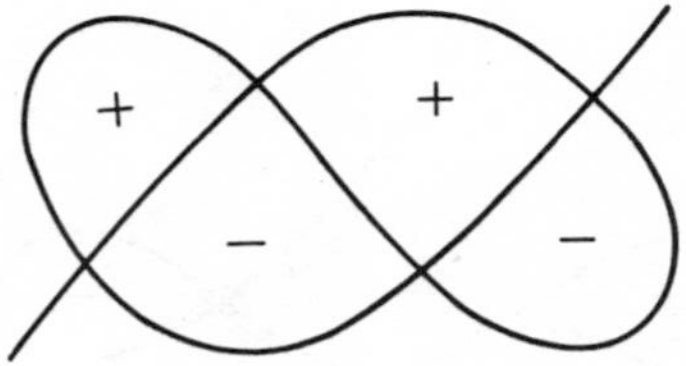

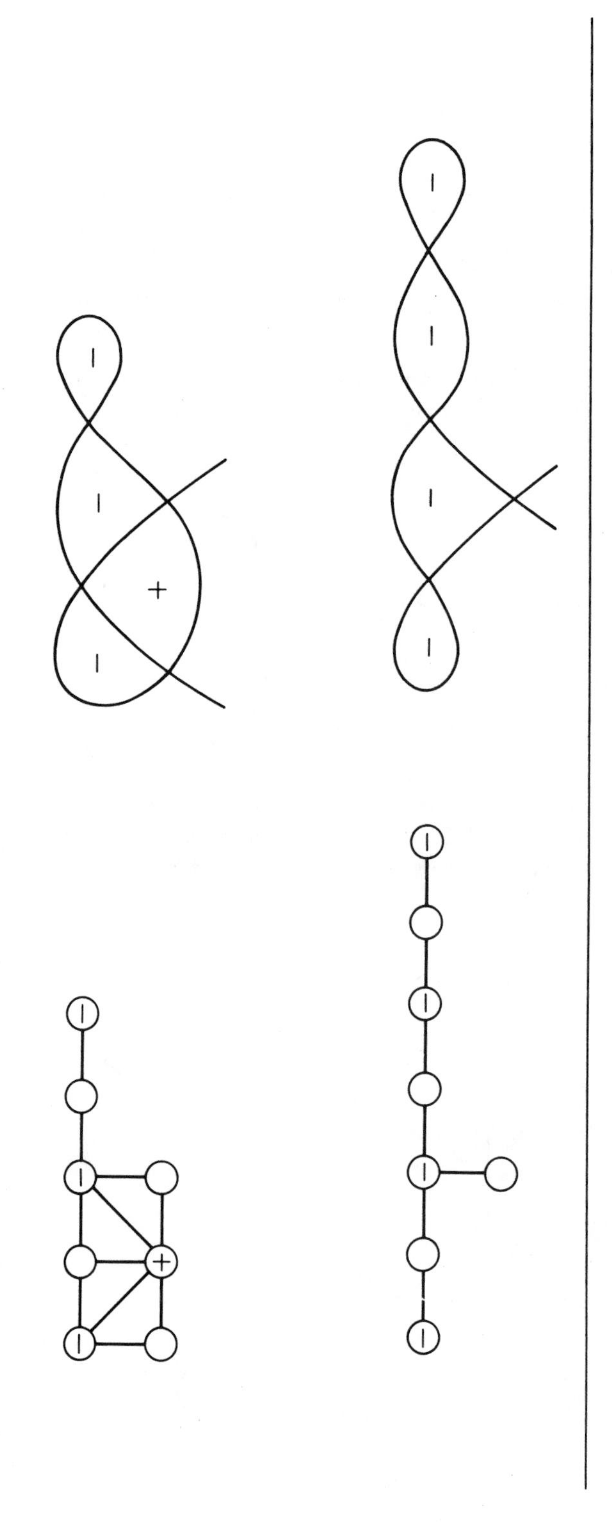

The remaining two critical points may be contracted in the same way to give an open family of functions with no critical points. A curve of cusps A_{+3} can be obtained by contracting two saddles with the maximum. Another curve of cusps, this time A_{-3}, is obtained by contracting two saddles with the minimum. Two curves in the separatrix corresponding to the component ② ② represent self-intersections. These elementary considerations lead us to a decomposition of $\mathbb{R}^3$ topologically equivalent to that given in Fig. 5.11, and determined by more laborious means.

Example 2: D_{-4}. From Table 7.1 the open components of $\mathbb{R}^3$ describing functions with the maximum number of isolated critical points have these diagrams:

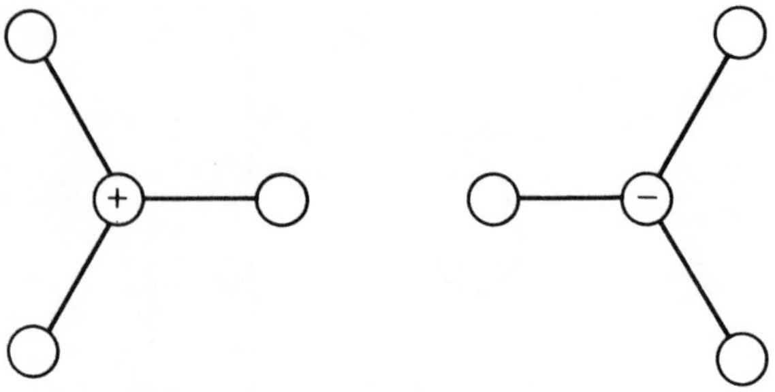

(7.28)

These diagrams immediately suggest a rotational symmetry of the type previously encountered (Chapter 5, Sec. 7). 2-dimensional fold components of the separatrix are determined by contracting an extremum with a saddle:

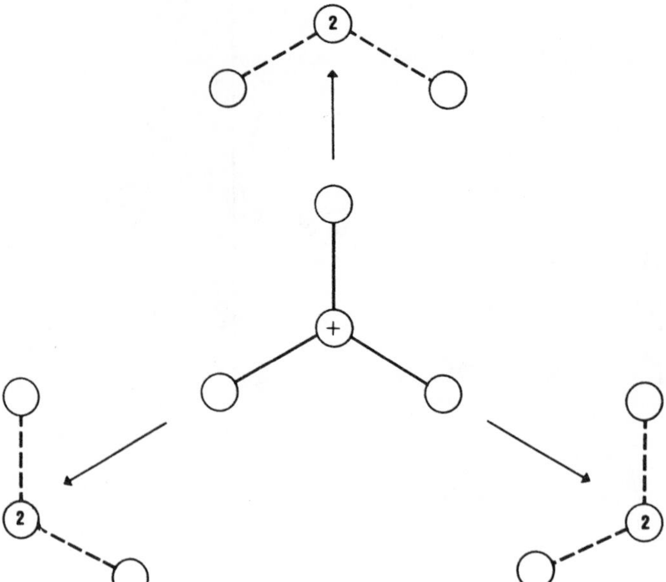

(7.29)

1-dimensional cusp components are determined by contracting three critical points:

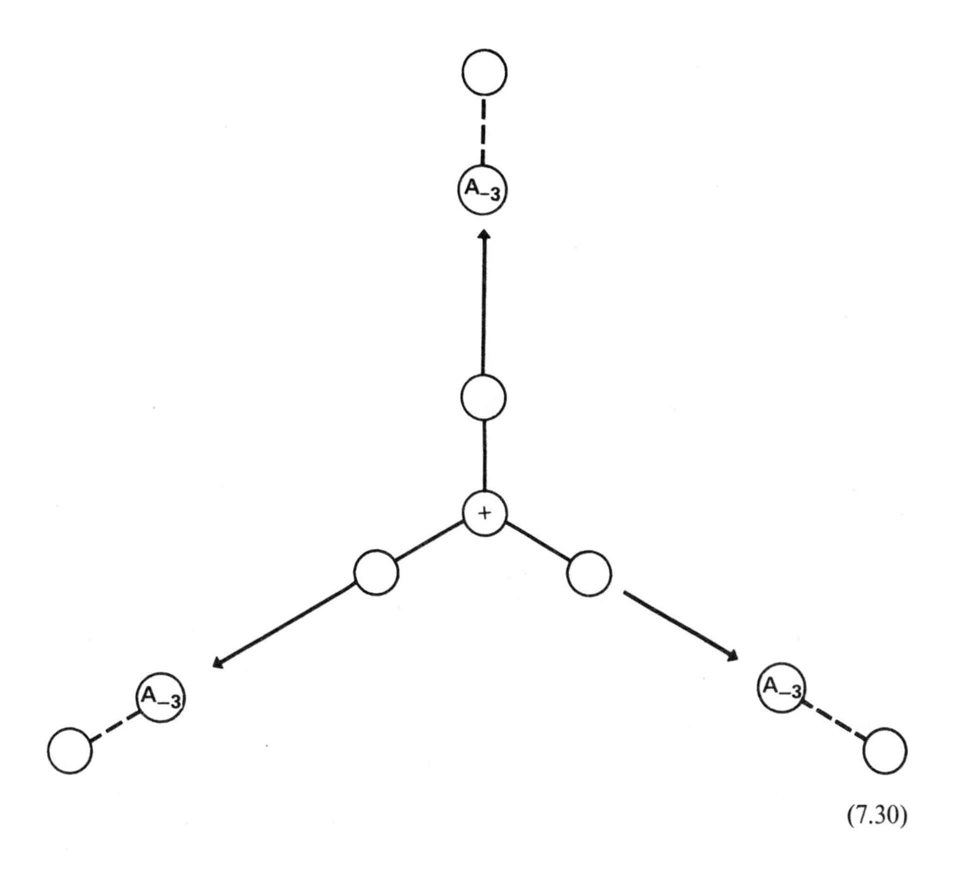

(7.30)

The open set representing structurally stable functions with fewer than four critical points is obtained by continuing the fold contractions shown in (7.29) and (7.30). In short, there are three open regions characterizing functions of the following three types:

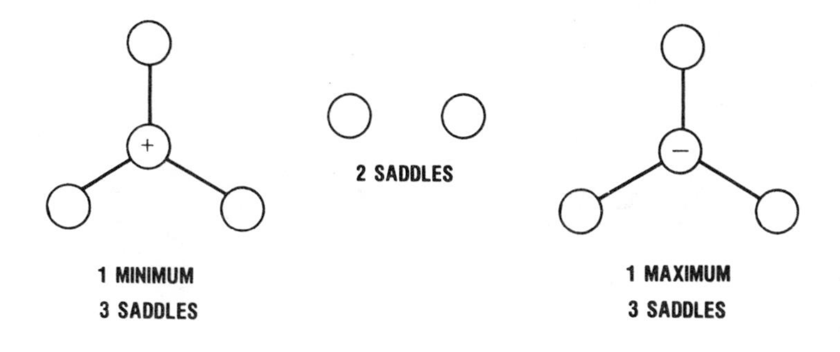

The open region describing functions with two saddles is contiguous with both of the other two open regions through 2- (and 1- and 0-) dimensional components of the separatrix. However, the two distinct open sets describing functions with four critical points are not contiguous except at the origin, since the only contraction they have in common is ④:

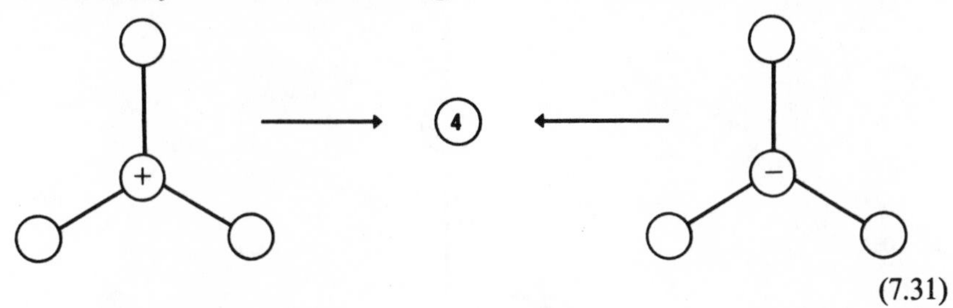

(7.31)

These considerations lead easily and immediately to a decomposition of the control parameter space $\mathbb{R}^3$ of D_{-4}, which is topologically equivalent to that shown in Fig. 5.17.

5. Contour Representations

An alternative representation for perturbed catastrophe germs can be given in terms of contour maps. To construct this representation, it is convenient to choose the origin and scale in $\mathbb{R}^1(\mathbb{R}^n \otimes \mathbb{R}^k \to \mathbb{R}^1)$ and the control parameters so that all maxima have critical value $+1$, all saddles have critical value 0, and all minima have critical value -1. This can always be done for catastrophes depending on 1 or 2 state variables.

These choices are convenient because the intersection of a saddle $f(x, y) = xy$ with the plane $z = 0$ has the form of an x. The zero-level contour then provides a convenient skeleton around which the remaining contours may be indicated, if desired (we do not). The contour representation may be constructed directly from the diagrammatic representation, as several examples should make clear.

Example 1: A_{2k}. First we write down the diagrammatic representation, which in $\mathbb{R}^2$ is

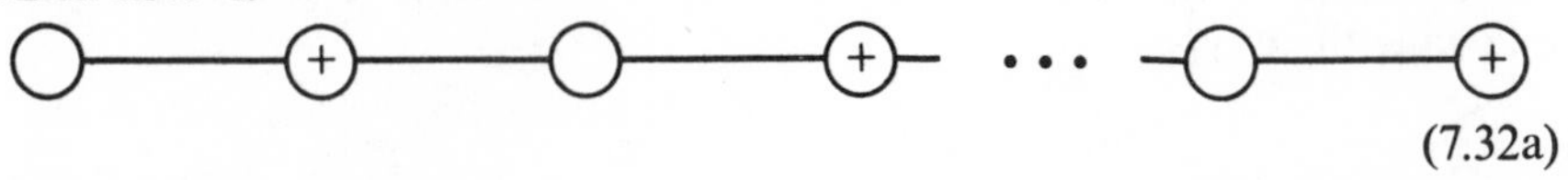

(7.32a)

The intersection of the function represented by (7.32a) with the $z = 0$ plane then has the form

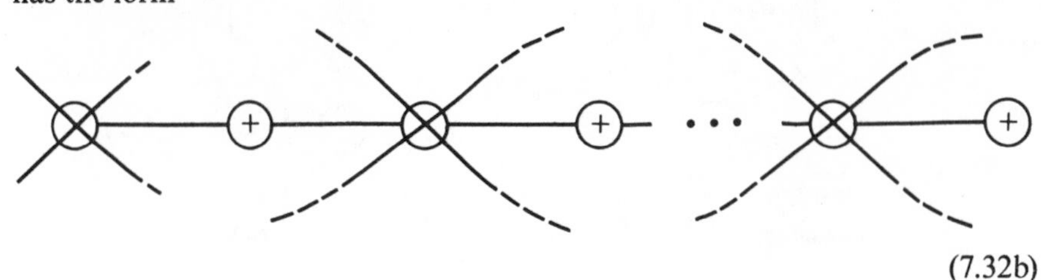

(7.32b)

The local minima must be encircled by the zero-level contours, so the contours can be completed:

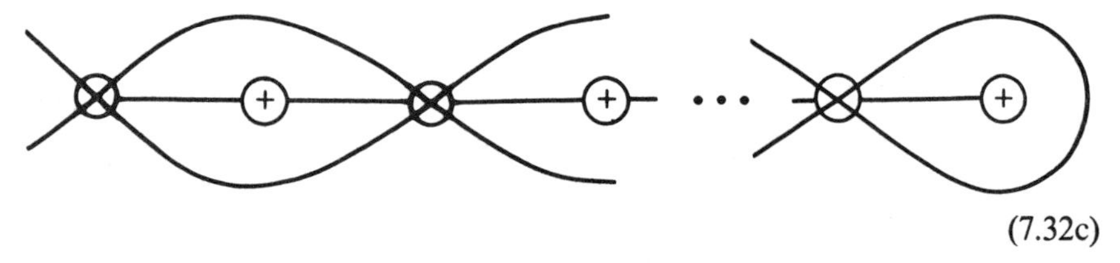

(7.32c)

Finally there is no point in retaining either the circles representing the locations of the critical points, or the flow lines connecting "adjacent" critical points. The contour representation for A_{2k} then simplifies to

(7.32d)

The saddle connections in these contours are structurally unstable.

Example 2: A_{2k+1}. The diagram for this catastrophe function is

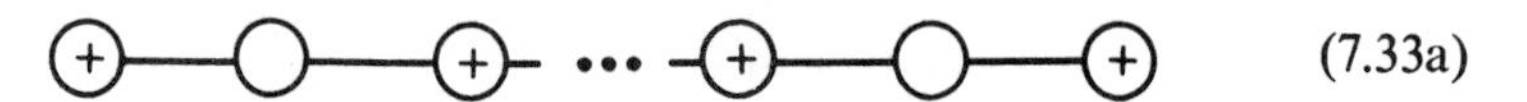

(7.33a)

Following steps *a–d* of the previous example, the associated contour representation is

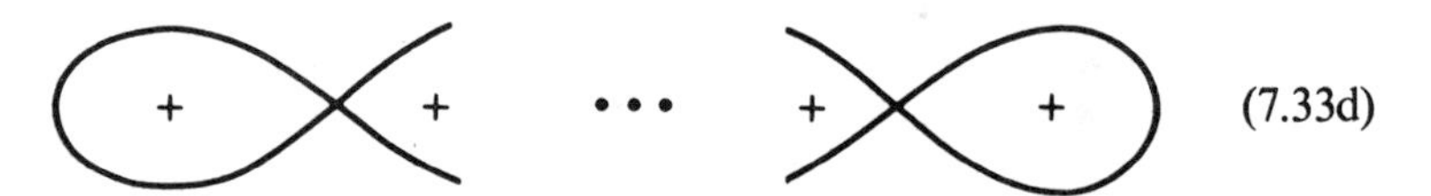

(7.33d)

Example 3. Two of the open sets for D_{2k} describing families with $2k$ isolated critical points have diagrammatic representations given by

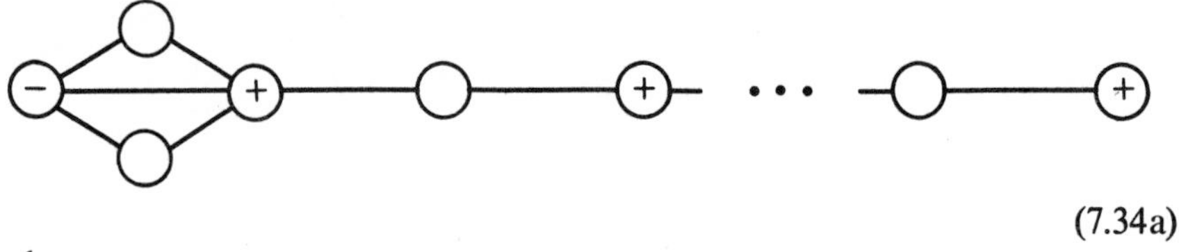

(7.34a)

and

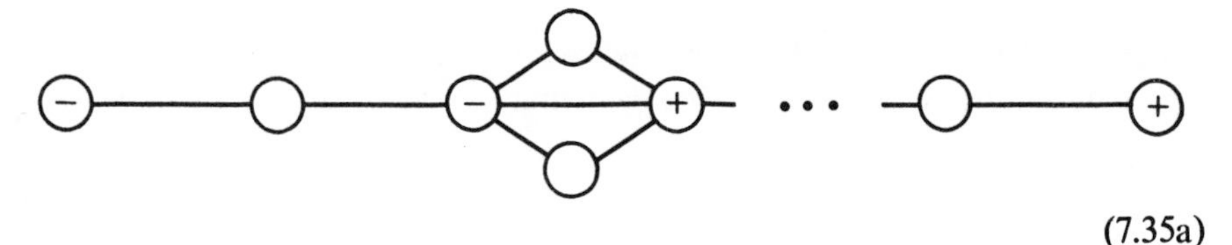

(7.35a)

The corresponding contour representatives are

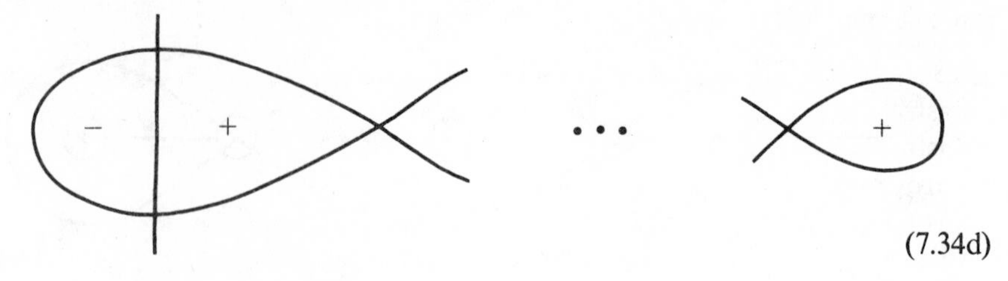

(7.34d)

and

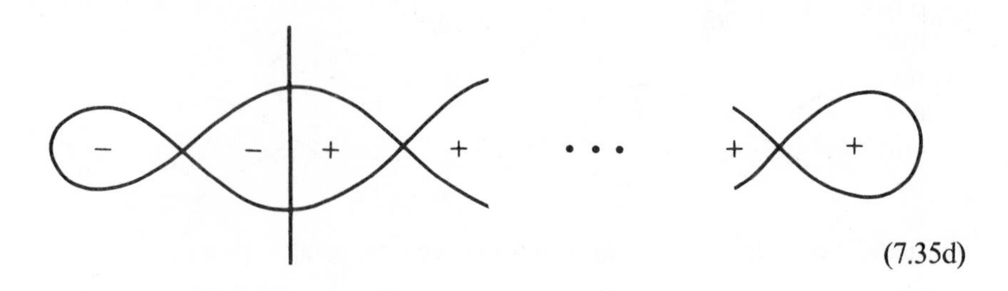

(7.35d)

Example 4. One possible diagram for D_{-2k} and the associated contour plot are

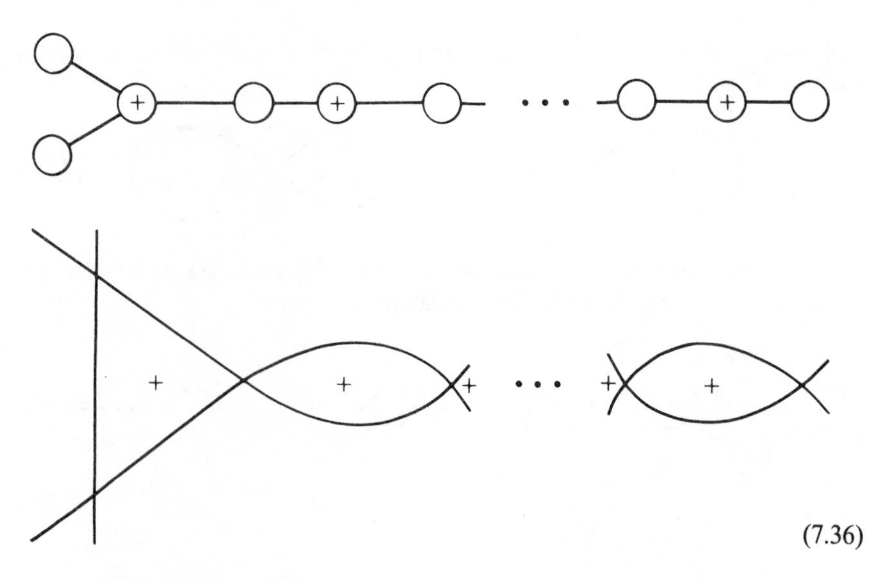

(7.36)

The zero-level contour plot associated with each type of structurally stable family of functions for each of the elementary catastrophes is also given in Table 7.1.

6. Abutment

We have seen in Sec. 1 that there exist 1-dimensional curves of catastrophes of type A_k emanating from the origin in the control parameter space $\mathbb{R}^k$ for the catastrophe A_{k+1}. In Sec. 2 we saw that there are 1-dimensional curves of catastrophes of type $D_k (k \geq 4)$ and of type A_k emanating from the origin of the control parameter space $\mathbb{R}^k$ for the catastrophe D_{k+1}. These lower dimensional catastrophes are said to **abut** the higher dimensional catastrophes.

We can now pose the question: Given an elementary catastrophe of control dimension k, what catastrophes of control dimension $k - 1$ abut this catastrophe? This question is most simply answered by using the diagrammatic representation for the original catastrophe. To illustrate how this question is answered, and how much information may be squeezed out of diagrammatic techniques, we consider which catastrophes abut E_8.

In the control parameter space $\mathbb{R}^7$ for E_8 there are numerous open sets describing structurally stable functions with an even number of critical points. However, there are only five open sets parameterizing functions with the maximum number 8 of isolated critical points (cf. Table 7.1). We first consider one particular open set whose diagram describes four minima and four saddles. By manipulating the control parameters, we can cause seven of the eight critical points to become degenerate.

Thus for the particular open set of E_8 under consideration, we have the three abutments:

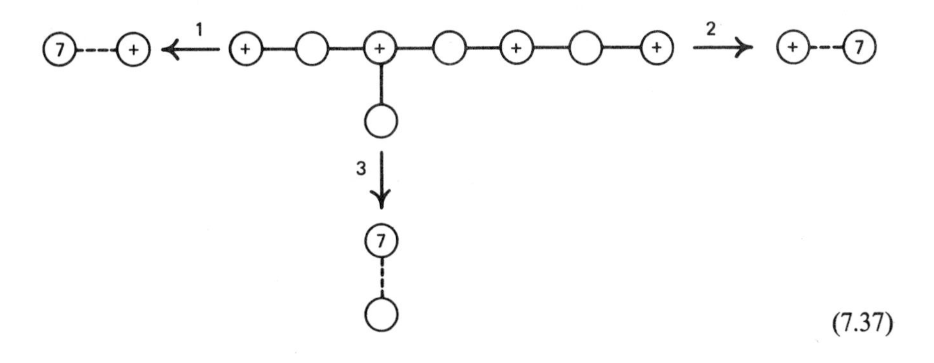

(7.37)

For abutment 1 it is clear that one morsification of the sevenfold degenerate critical point has the diagram

Abutment 1 — morsify — E_7

(7.38i)

A morsification associated with abutment 2 is

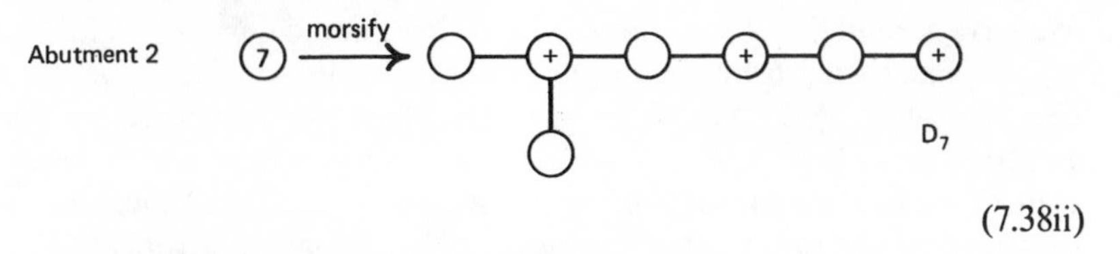

(7.38ii)

Finally a morsification associated with abutment 3 is

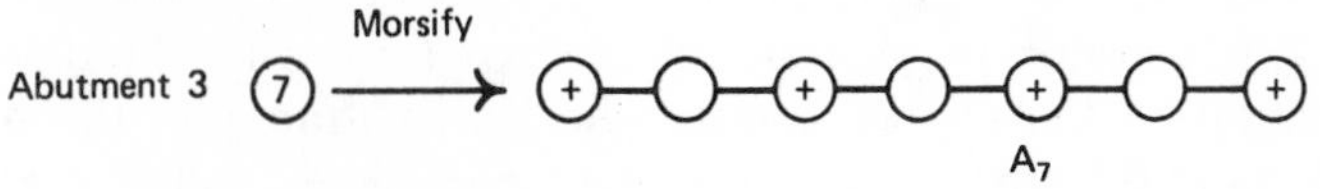

Morsify

Abutment 3 (7) ⟶ (+)—○—(+)—○—(+)—○—(+)

A_7

(7.38iii)

From this diagrammatic technique we can tell not only which lower dimensional catastrophes abut E_8, but also which open sets in these catastrophe control spaces abut a particular open set in the control space of E_8.

The remaining four open sets for E_8 parameterizing functions of type E_8 with a maximal number of isolated critical points can be dispatched similarly. The results are presented in Table 7.2.

Abutments for the other elementary catastrophes may be determined in the same way. We present in Fig. 7.5 an abutment diagram for all the elementary catastrophes. For some purposes it is useful to have a slightly clearer abutment diagram which conveys somewhat less information. Such a diagram is given in Fig. 7.6. The catastrophes $T_{3,3,3}$, $T_{2,4,4}$, and $T_{2,3,6}$ will be discussed in Chapter 17.

Table 7.2 Maximal Morsifications of E_8 and their Abutments.

Morsification	Abutments
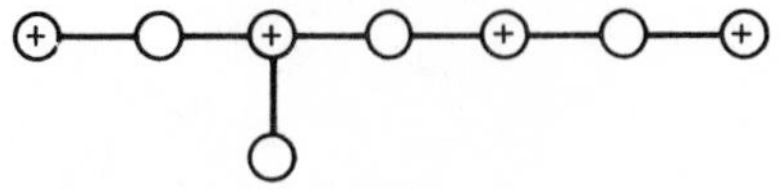 (+)—○—(+)—○—(+)—○—(+), with ○ attached below the second (+)	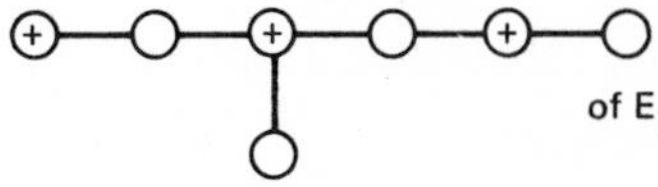 (+)—○—(+)—○—(+)—○, with ○ attached below the second (+): of E_7
	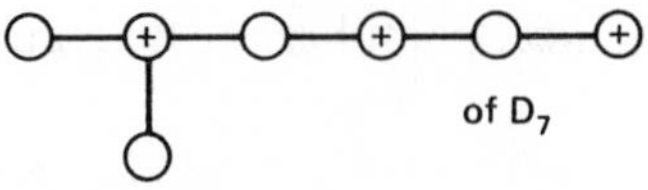 ○—(+)—○—(+)—○—(+), with ○ attached below the first (+): of D_7
	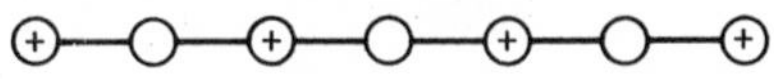 (+)—○—(+)—○—(+)—○—(+): of A_7

Table 7.2 (*Continued*)

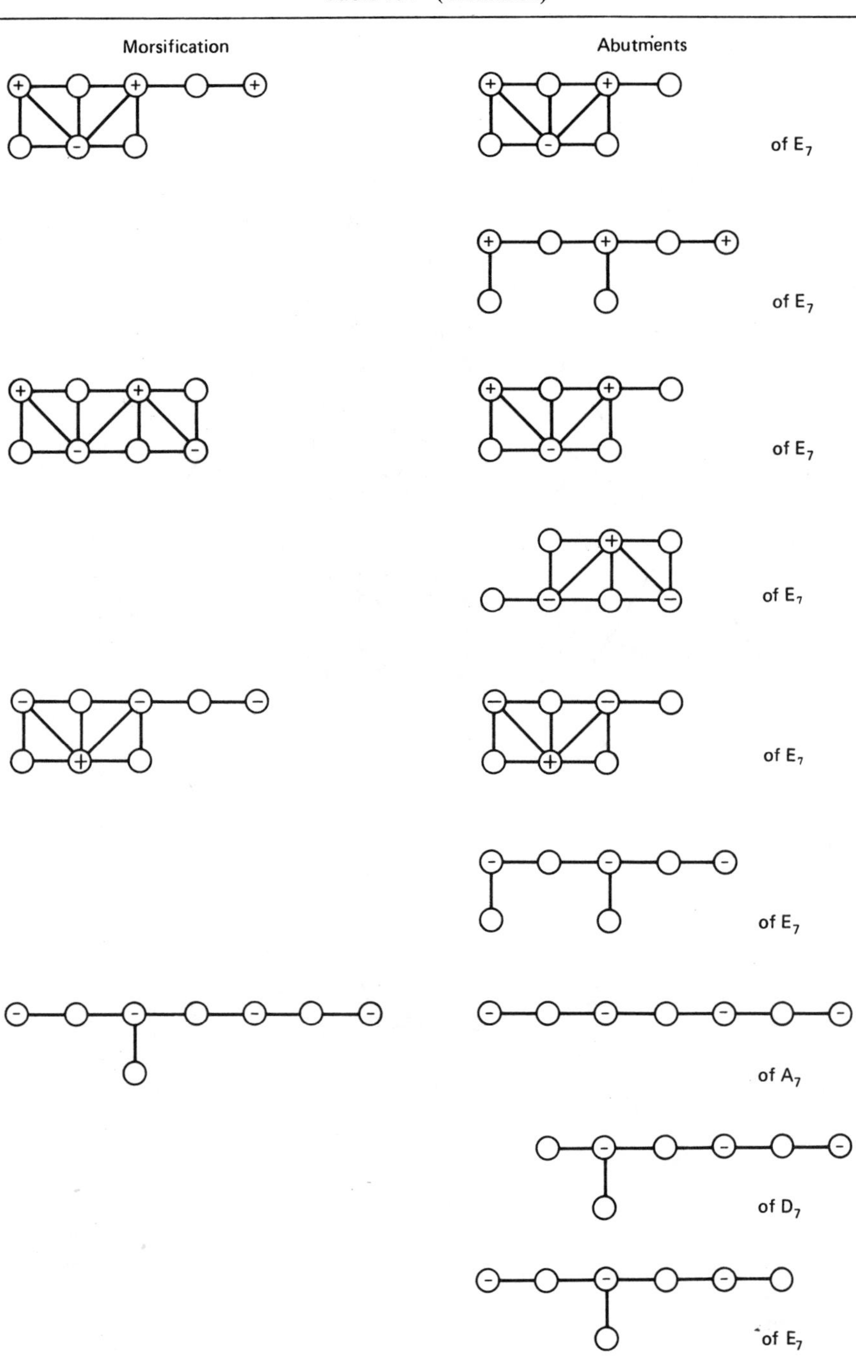

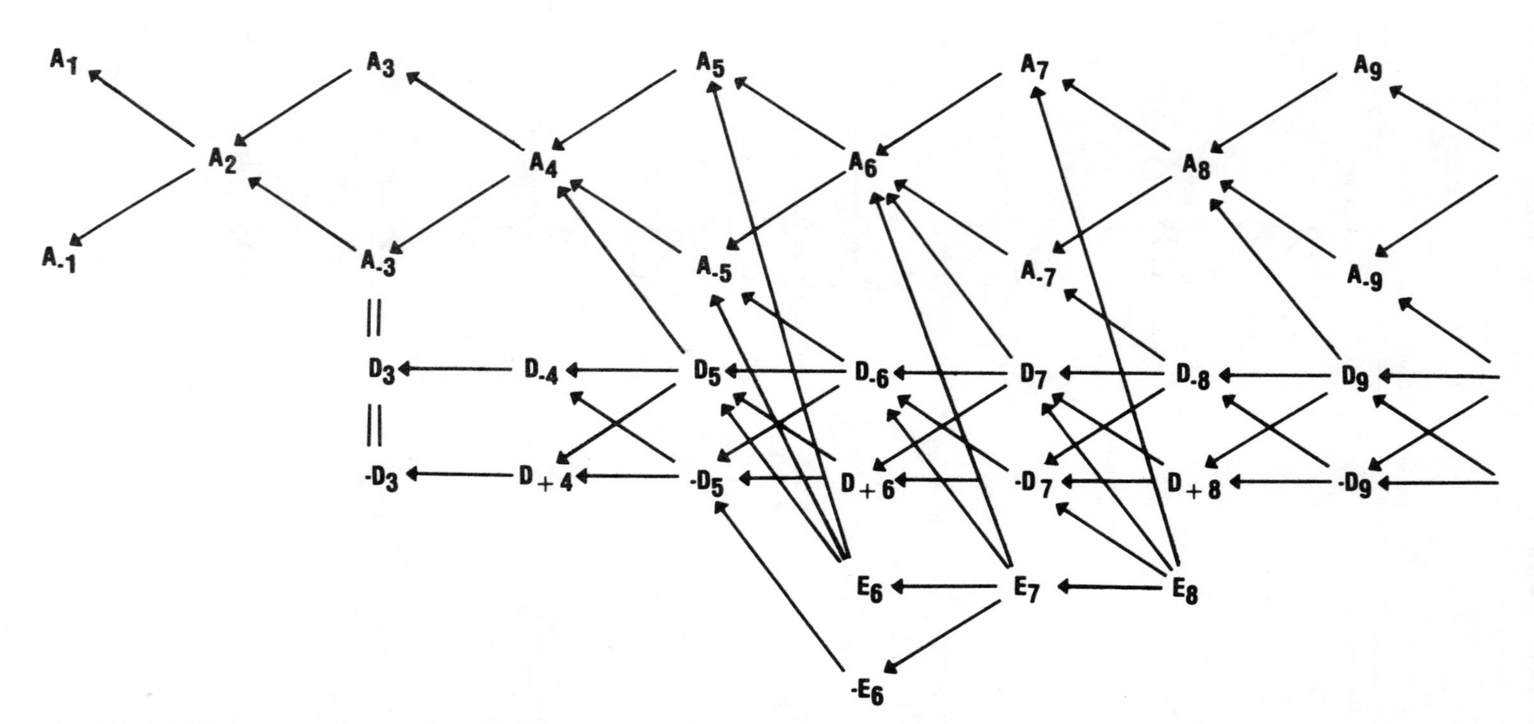

Figure 7.5 This diagram describes which catastrophes, of control dimension $k - 1$ abut each of the elementary catastrophes of control dimension k.

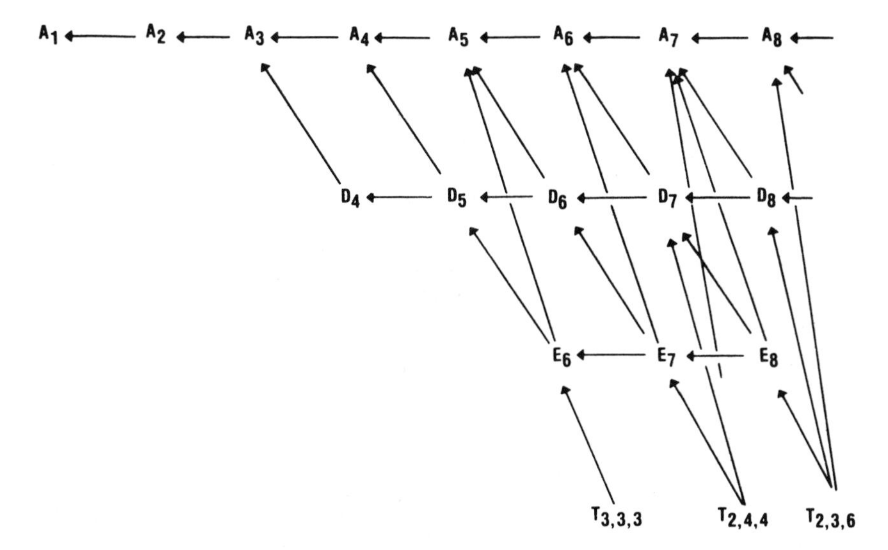

Figure 7.6 Arnol'd showed that the abutment diagram for the complex elementary catastrophes (x complex, y complex) corresponds to the Coxeter–Dynkin diagram for some of the simple Lie groups over the field of complex numbers. The catastrophes $T_{p,q,r}$ are not elementary (cf. Chapter 17, Sec. 1).

7. Summary

When F is a catastrophe function, the equations $\nabla F = 0$, det $F_{ij} = 0$ define a separatrix in the space of control parameters. Points in this separatrix represent structurally unstable functions with one or more non-Morse critical points. The separatrix consists of components of dimensions $0, 1, 2, \ldots, k - 1$ in $\mathbb{R}^k$. The separatrix divides $\mathbb{R}^k$ up into a finite number of open sets representing structurally stable functions. These open sets are dense in $\mathbb{R}^k$ when F is an elementary catastrophe.

In Secs. 1 and 2 we have seen how the equations of the 1-dimensional components of the separatrix can be determined using scaling arguments. Using such arguments, we can (in principle) determine the equations defining each of the components of the separatrix.

Since this is at best a tedious process and, in fact, does not provide answers to the important questions (how are catastrophes organized?), we have developed a much simpler and more incisive computational technique in Secs. 3 and 4. With this diagrammatic representation for catastrophe functions we are able not only to determine all the components of the separatrix and how they fit together, but also all the distinct open regions in $\mathbb{R}^k$ which parameterize the structurally stable functions. In fact, we are also able to determine which component(s) of the separatrix are common boundaries between any pair of open regions. This may be done by the "method of contraction."

A contour representation for each of the catastrophes is also possible. We have described such a representation in Sec. 5.

In Sec. 6 we returned to the original question: Given a catastrophe of control dimension k, what curves of catastrophes of control dimension $k - 1$ emanate from the origin in $\mathbb{R}^k$. This question is easily answered using the diagrammatic representation for catastrophe functions. We have shown how this question may be answered simply by inspection, using Table 7.1. The results for all the elementary catastrophes have been summarized in Figs. 7.5 and 7.6.

As catastrophes are canonical, so also are their organizations, described in this chapter. Each catastrophe germ acts as an organizing center about which a hierarchy of lower catastrophes revolve in canonical orbit and pay canonical homage.

Notes

The diagrammatic and contour methods described in this chapter were developed originally by Arnol'd [1, 2, 3] for the complex forms of the elementary catastrophes. The restriction of these techniques to real forms was carried out by Gusein-Zade [4] and A'Campo [5, 6]. An excellent survey of this work is given in Callahan [7]. I am indebted to Prof. J. Edwin Clark for bringing this last reference to my attention.

References

1. V. I. Arnol'd, Normal Forms for Functions Near Degenerate Critical Points; the Weyl Groups for A_k, D_k, E_k and Lagrangian Singularities, *Funct. Anal. Appl.* **6**, 245–272 (1973).
2. V. I. Arnol'd, Remarks on the Stationary Phase Method and Coxeter Numbers, *Russian Math. Surveys* **28**, 19–48 (1973).
3. V. I. Arnol'd, Critical Points of Smooth Functions, *Proc. Int. Cong. Math.*, Vancouver, 1974, pp. 19–39.
4. S. M. Gusein-Zade, Dynkin Diagrams for Singularities of Functions of Two Variables, *Funct. Anal. Appl.* **8**, 295–300 (1974).
5. N. A'Campo, Le Groupe de Monodromie du Deploiement des Singularités Isolées de Courbes Planes. I, *Math. Ann.* **213**, 1–32 (1975).
6. N. A'Campo, Le Groupe de Monodromie de Deploiement des Singularités Isolées de Courbes Planes. II, *Proc. Int. Cong. Math.*, Vancouver, 1974, pp. 395–404.
7. J. Callahan, Singularities and Plane Maps II: Sketching Catastrophes, *Math. Monthly* **84**, 765–803 (1977).

CHAPTER 8

Catastrophe Conventions

Two conventions are widely adopted in Catastrophe Theory. These are the Delay Convention and the Maxwell Convention. They are, as it were, dei ex machina. These conventions are not intrinsic to Catastrophe Theory. Rather, they provide the means by which the canonical mathematics of Elementary Catastrophe Theory is made available to applications in the physical and engineering sciences.

Why external conventions are required before Elementary Catastrophe Theory can be applied is the subject of Sec. 1. The two conventions are stated in Sec. 2, where we also see that these are the two extremes in a continuum of possibilities. These two conventions are merely stopgap measures to allow the results of Elementary Catastrophe Theory to be applied where the (so far mostly unavailable) results of Catastrophe Theory are called for. In Sec. 3 we discuss an equation which is of importance in many situations, which allows us to make more than qualitative statements about the regions of validity of these two conventions, and which allows us to interpolate between them. We emphasize the inadequacy of these conventions in confronting other situations of physical interest by discussing a simple example in Sec. 4.

1. The Need for Conventions

In Chapter 1 we reviewed the program of Catastrophe Theory. We saw also that this program for describing how the qualitative properties of systems of equations change as the (control) parameters in these equations change, has not yet been carried out in general. In fact, it has only been carried out in the simplest case: for the equilibria of gradient systems. In the intervening six chapters we discussed properties of the elementary catastrophes.

Many physical systems can in fact be described by potentials depending on control parameters. For such systems the results of Elementary Catastrophe Theory are immediately applicable. However, such systems must of necessity be static, because we have squeezed out all dynamical considerations (all time derivatives) by the time we have reached Step 7 in Chapter 1. Therefore a discussion of the time evolution of the equilibria of a gradient system does not fall within the scope of Elementary Catastrophe Theory. In order to discuss dynamics, we must impose some assumption of a physical or intuitive nature, such as quasistatic evolution or adiabaticity (i.e., all time derivatives are very small).

Even when a system can be described by a potential, it is not necessarily true that the system state is completely characterized by the point that minimizes the potential. A case in point is the 1-dimensional harmonic oscillator with potential $V(x) = \frac{1}{2}k(x - a)^2$. The point $x = a$ minimizes this potential, and does completely describe the (static) classical harmonic oscillator. However, the quantum mechanical harmonic oscillator is described by a probability distribution which is centered at a in the ground state, but whose width is determined by the coefficient of the diffusion term $(-\hbar^2/2m)(\partial^2/\partial x^2)$ in Schrödinger's equation. Many systems of physical interest are described by distribution functions, and these distribution functions $P(x)$ are often simply related to potential functions $V(x)$ (e.g., $P(x) \simeq e^{-V(x)/D}$). In fact, distribution functions generally occur when the differential equation describing a system involves second derivatives (diffusion terms). However, by the time we have reached Step 2 in Chapter 1, we have assumed away all space derivatives of any order. Therefore a discussion of systems described by distribution functions also falls outside the scope of Elementary Catastrophe Theory. In order to discuss such systems, we must impose some external assumptions of a physical or intuitive nature on the system we study. For example, we may suppose that the noise level is either very small or very large compared to internal barrier heights.

In short, the mathematics of Elementary Catastrophe Theory described in Chapters 2–7 is strictly applicable only at the level of Step 7 in Table 1.1. In order to apply these mathematics at some earlier level in Table 1.1, external assumptions of a more highly structured nature are required the earlier we go in Table 1.1.

2. The Conventions

The state of a physical system governed by a potential $V(x; c)$ is described (at least in part) by that point $x \in \mathbb{R}^n$ which minimizes the potential. Changing external conditions change the values of the control parameters c; changing c, in turn, changes the shape of the potential $V(x; c)$. As the shape of the potential changes, the original global minimum in which the system state sits may become a metastable local minimum (because some faraway minimum assumes a lower value), or it may even disappear. Under these conditions the system state must jump from one local minimum to another. When, and to which minimum, the jump occurs are the subjects of two commonly applied conventions.[1]

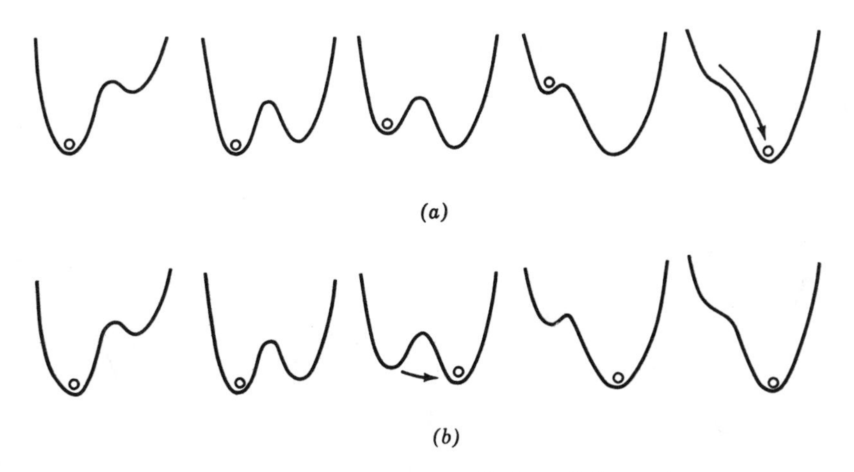

Figure 8.1 (*a*) Delay Convention. (*b*) Maxwell Convention.

Delay Convention. The system state remains in a stable or metastable equilibrium state until that state disappears (Fig. 8.1*a*).

Maxwell Convention. The system state is the one that globally minimizes the potential (Fig. 8.1*b*).

In the case of gradient dynamical systems

$$\frac{dx_i}{dt} = -\frac{\partial}{\partial x_i} V(x; c) \tag{8.1}$$

it is customary to apply the Delay Convention when $dc_\alpha/dt \ll 1$. This is because the potential changes shape very slowly, and the system state has no way of knowing when a far-distant minimum becomes globally stable. If there is a locally stable equilibrium at $(x^0(c(t)), c(t))$, then $dx^0/dt = 0$ remains the equilibrium condition until the potential loses its stability at x^0. Only then does the system seek a new local minimum.

In the case that a system is described by a distribution function, the point at which the system state leaves a metastable equilibrium and moves to a stable equilibrium or to a lower metastable equilibrium depends on the noise level in the system (Fig. 8.2). If ΔE represents the characteristic height of the potential barrier separating the metastable state from a nearby stable state, and $\mathcal{N}$ represents the characteristic noise level, then the Delay Convention is observed if $\mathcal{N}/\Delta E \ll 1$ and the Maxwell Convention is observed if $\mathcal{N}/\Delta E \simeq 1$.

These two conventions represent extremes in a continuum of possibilities. The barrier height ΔE may be determined directly from the potential $V(x; c)$. The appropriate description of noise must be in terms of partial differential equations, and $\mathcal{N}$ is determined in terms of the coefficients of the second partial

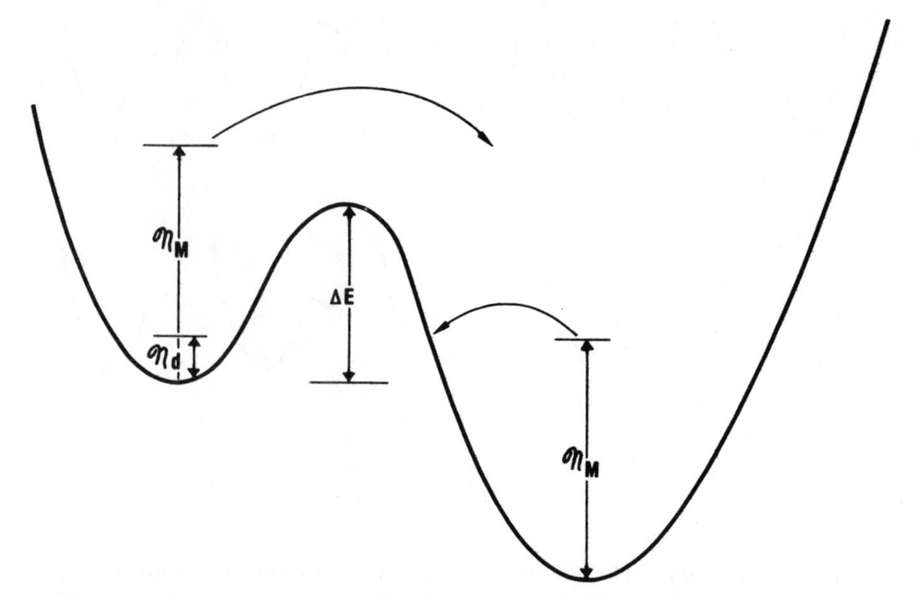

Figure 8.2 The ratio of noise level $\mathcal{N}$ to barrier height ΔE determines which of the two conventions is more appropriate. $\mathcal{N}_d/\Delta E \ll 1$: Delay Convention. $\mathcal{N}_M/\Delta E \simeq 1$: Maxwell Convention.

derivatives in the appropriate equation. A small noise level implies small fluctuations, and these, in turn, can sense the shape of the potential only locally, so distant minima are not found (Delay Convention). Large fluctuations can find distant deeper minima and correspond to the Maxwell Convention.

The applications of these two conventions in the case of the cusp catastrophe can be interpreted in terms of the critical value surface shown in Fig. 8.3. Under the Delay Convention the system state moves along a sheet until that sheet disappears. Then it jumps to the sheet below. Under the Maxwell Convention the system state moves continuously on the lowest sheet over the appropriate point in control parameter space. This involves moving from one sheet to another as the control parameter crosses the half-line $a < 0, b = 0$.

Definition. The set of points in the space of control parameters at which transition occurs from one local minimum to another is called the **bifurcation set**.

Remark 1. The bifurcation set depends on the convention adopted. For the cusp catastrophe A_3 the bifurcation set under the Delay Convention consists of the two fold lines. Under the Maxwell Convention the bifurcation set consists of the half-line $a < 0, b = 0$. The bifurcation set under the Delay Convention can in general be determined from the condition $\det|\partial^2 V/\partial x_i \partial x_j| = 0$. Since this condition depends on derivatives, it is a local condition, and the corresponding

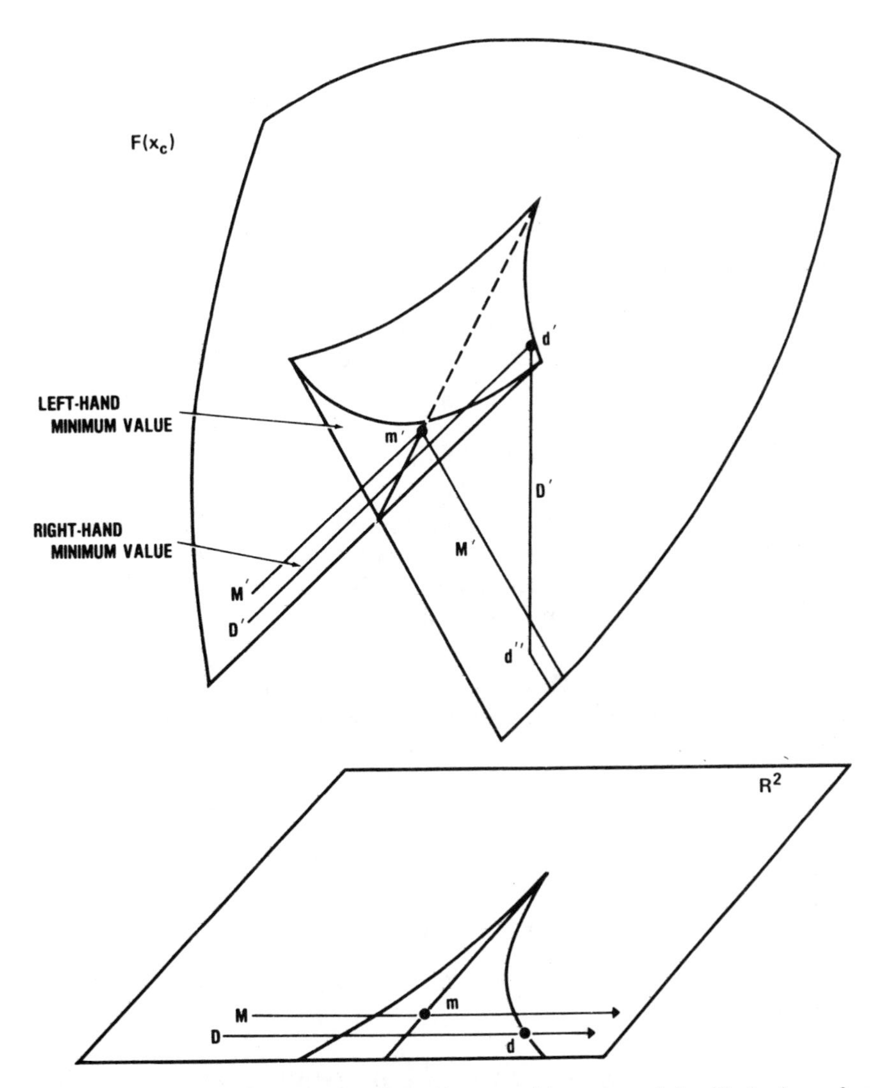

Figure 8.3 The conventions may be interpreted in terms of the catastrophe's critical value surface.

bifurcation set is called a **local bifurcation set**. The bifurcation set under the Maxwell Convention can be determined by the Clausius–Clapeyron equations. This bifurcation set is called a **nonlocal bifurcation set**.

Remark 2. The transition from one local minimum to another is called a phase transition. Phase transitions represent qualitative changes in the properties of a system. Phase transitions, their properties, and their characteristics will be discussed more fully in Chapter 10.

3. Which Convention to Use

Between the two extremes of the Delay Convention and the Maxwell Convention lies a partial differential equation which must be solved in terms of the changing control parameters if any "Ansatz Convention" is to be avoided. In this intermediate regime the bifurcation set becomes a "fuzzy set." More precisely, it is not generally possible to define a bifurcation set for some physical process if neither of these conventions is involved.

We bring these vague remarks down to the realm of concreteness by considering a situation which arises frequently in physical applications. In this type of situation the system is described by a probability distribution function P which is related to a potential function V through a Fokker–Planck equation:[2]

$$\frac{\partial}{\partial t} P = \nabla \cdot (P \nabla V) + \nabla^2 (DP) \tag{8.2}$$

The potential V is a function of a state variable $x \in \mathbb{R}^n$ and k control parameters $c \in \mathbb{R}^k$. If the values of the control parameters depend on time, then V is time-dependent. The distribution function P depends on x, on c, and on t.

We first solve (8.2) in two important cases.

A ASYMPTOTIC TIME-INDEPENDENT LIMIT. In this case the control parameters are fixed, the potential is time-independent, and we look for the stationary ($t \to \infty$) probability distribution function by setting $\partial P/\partial t = 0$. Then (8.2) reduces to

$$0 = \nabla \cdot (P \nabla V + \nabla (DP)) \tag{8.3}$$

The solution to (8.3) is

$$P(x; c) = N e^{-V(x;c)/D} \tag{8.4}$$

when D is a constant. If the diffusion coefficient $D(x)$ is not constant, then the solution is

$$P(x; c) = N D^{-1}(x) \; EXP - \left\{ \int^x D^{-1}(x') \nabla V(x'; c) \, dx' \right\} \tag{8.5}$$

The term N in (8.4) is an appropriate normalization constant.

B DIFFUSION-FREE LIMIT. In this case the diffusion coefficient D is zero, so that (8.2) assumes the form of a conservation equation with current $j(x, t)$:

$$\frac{\partial P}{\partial t} + \nabla \cdot j = 0$$

$$j(x; t) = -P \nabla V \tag{8.6}$$

It may easily be verified[3] that a solution of (8.6) is given by

$$P(x, t) = \delta(x - x_c(t)) \tag{8.7}$$

where $x_c(t)$ is any critical point of V (i.e., $\nabla V = 0$ when $x = x_c$). We discuss which critical point to choose at the end of this section.

We now return to a discussion of the qualitative properties of the Fokker–Planck equation (8.2). The right-hand side of this equation consists of two terms, a "drift" and a "diffusion" term. Roughly speaking, the drift term $\nabla(P\nabla V)$ coaxes the distribution function to move toward the nearby local minimum. The diffusion term $\nabla^2(DP)$ plays two roles: (1) it describes the width of the distribution which is centered around a local minimum, and (2) it describes the probability that a fluctuation will carry the system from a metastable minimum to a distant global minimum.

The occurrence of two terms on the right-hand side of the Fokker–Planck equation (8.2) means that this equation is a two-time-scale equation. This means that the qualitative features of (8.2) occur on two quite distinct time scales. The fast time scale T_1 is associated with the relaxation back to a local minimum after perturbation. The slow time scale T_2 is associated with the passage from a metastable minimum to a global minimum. We consider below the relaxation problem (fast time scale) and the first passage time problem (slow time scale).

Relaxation Problem. We consider for simplicity a 1-dimensional parabolic potential

$$V(x) = \tfrac{1}{2}kx^2$$

The time-independent probability distribution function (8.4) associated with this potential is $P(x) = Ne^{-kx^2/2D}$. If we perturb the system so that the initial probability is centered around $x = a$ ($a \neq 0$) rather than around $x = 0$, so that

$$P(x, t = 0) = Ne^{-k(x-a)^2/2D}$$

then the probability distribution function will evolve in time like

$$P(x, t) = Ne^{-k[x-a(t)]^2/2D} \tag{8.8}$$

where

$$a(t) = ae^{-kt} = ae^{-t/T_1} \tag{8.9}$$

The relaxation time $T_1 = 1/k = (d^2V/dx^2)^{-1}$.

More generally, if a potential V has a local minimum at $x = x^0$ and the initial probability distribution is centered around a nearby point,

$$P(x, t = 0) = Ne^{-V(x^0 + \delta x)/D} \tag{8.10}$$

then this distribution function will relax to the time-independent probability distribution function $P(x; t \to \infty) = Ne^{-V(x^0)/D}$ with time scales commensurate with the reciprocals of the eigenvalues λ_i of the stability matrix $V_{ij}(x^0)$. Therefore the fast time scale T_1 associated with relaxation is given by

$$T_1 = \operatorname*{Max}_{i=1,\cdots} 1/\lambda_i \tag{8.11}$$

The relaxation time diverges as a catastrophe germ is approached (critical slowing down).

First Passage Time Problem. Diffusion equations of the Fokker–Planck type have a very deep connection with probability theory.[4] This connection can be made through a study of random walks or Markov processes. We will exploit this connection here to determine some qualitative features of the Fokker–Planck equation without actually solving it.

In this approach we consider a (biased) random walk starting at a point x in the presence of a potential $V(x)$ and a "randomness" term D. Eventually the random walk will terminate on some boundary. For concreteness we consider only the 1-dimensional case, where the boundary consists of the points $x = a$, $x = b$, and the random walk starts from some interior point x: $a < x < b$. Each random walk starting from x will terminate after some time τ, which is a random variable. The mean termination time $T(x) = \langle\tau\rangle_{\text{ens}}$ for a random walk starting at x can be computed by the methods of probability theory. If the random walk starts from a point x with a probability $P_0(x)$, then the mean termination time for a random walk is given by

$$T_{fp} = \int_a^b P_0(x)T(x)\,dx \tag{8.12}$$

The first passage time T_{fp} is the slow time scale T_2 of the Fokker–Planck equation (8.2) for the cusp catastrophe.

Although the probability distribution function $P(x, t)$ of (8.2) is governed by a partial differential equation, the function $T(x)$ [first passage time function for a delta function distribution $P_0(y) = \delta(y - x)$] satisfies a closely related ordinary differential equation:

$$\frac{\partial P}{\partial t} = \frac{\partial}{\partial x}\left(\frac{dV}{dx}P\right) + \frac{\partial^2}{\partial x^2}(DP) \tag{8.13FPE}$$

$$-1 = -\frac{dV}{dx}\frac{dT}{dx} + D\frac{d^2T}{dx^2} \tag{8.13FPT}$$

Since a random walk beginning at the boundaries $x = a$ or $x = b$ must immediately terminate, the first passage time function $T(x)$ is subject to the boundary conditions

$$T(a) = T(b) = 0 \tag{8.14}$$

These two boundary conditions determine the two constants of integration for the second-order differential equation (8.13FPT).

Before solving the first passage time problem for the cusp catastrophe, we shall gain some insight into its meaning by studying two simpler situations.

Case 1. V = 0. In this case the equation for $T(x)$ reduces to

$$D\frac{d^2T}{dx^2} = -1 \tag{8.15}$$

The solution to (8.15), subject to the boundary conditions (8.14), is

$$T(x) = \frac{(b-x)(x-a)}{2D} \tag{8.16}$$

The random walk takes longest to terminate when $x = (a + b)/2$ ("in the middle").

Case 2. V = αx. In this case the equation for $T(x)$ looks more difficult than it is. In fact, it reduces to two simpler first-order ordinary differential equations:

$$D\frac{dT'}{dx} - \alpha T' = -1 \tag{8.17a}$$

$$T' = \frac{dT}{dx} \tag{8.17b}$$

Equation (8.17a) has the solution

$$T'(x) = Ce^{\alpha x/D} + \frac{1}{\alpha} \tag{8.18}$$

where the constant of integration remains to be determined by the boundary conditions. This is simply done as follows:

$$\int_a^b \frac{dT}{dx}\,dx$$

$$\begin{array}{ccc} T(b) - T(a) & = & \displaystyle\int_a^b \left(Ce^{\alpha x'/D} + \frac{1}{\alpha}\right) dx' \\ \| & & \| \\ 0 & = & \displaystyle C\frac{e^{\alpha b/D} - e^{\alpha a/D}}{\alpha/D} + \frac{b-a}{\alpha} \end{array} \tag{8.19}$$

This equation determines C. The function $T(x)$ can be determined similarly, replacing the upper limit b by the interior point x:

$$T(x) - T(a) = C\frac{e^{\alpha x/D} - e^{\alpha a/D}}{\alpha/D} + \frac{x-a}{\alpha}$$

$$T(x) = \frac{1}{\alpha}\left\{(x-a) - (b-a)\frac{e^{\alpha x/D} - e^{\alpha a/D}}{e^{\alpha b/D} - e^{\alpha a/D}}\right\} \tag{8.20}$$

The competition between the drift ($V' = \alpha$) and the diffusion (D) terms can be seen in (8.20). To see clearly the effects of this competition, we look at two particular limits:

1. $\alpha \to 0$, $D =$ constant, $\zeta = \alpha/D \to 0$. In this limit the drift is less important than the diffusion term, and

$$DT(x) = \zeta^{-1}\left\{(x - a) - (b - a)\frac{e^{\zeta x} - e^{\zeta a}}{e^{\zeta b} - e^{\zeta a}}\right\}$$

$$\xrightarrow{\zeta \to 0} \tfrac{1}{2}(b - x)(x - a) \tag{8.21}$$

We recover the result of Case 1.

2. $\alpha =$ constant, $D \to 0$, $\zeta = \alpha/D \to \infty$. In this limit the diffusion is less important than the drift, and

$$\alpha T(x) = \left\{(x - a) - (b - a)\frac{e^{\zeta x} - e^{\zeta a}}{e^{\zeta b} - e^{\zeta a}}\right\}$$

$$\xrightarrow{\zeta \to \infty} (x - a) - (b - a)e^{-\zeta(b - x)} \tag{8.22}$$

In this case the probability that the random walk ends at $x = b$ is very small unless $\zeta(b - x) \sim 1$. The random walk terminates at $x = a$, and the time taken $T(x)$ is related to the distance covered $(x - a)$ by $\alpha T = (x - a)$. It is therefore reasonable to interpret α as the drift velocity.

We return now to the problem of solving (8.13FPT) for the first passage time function $T(x)$ for a general 1-dimensional potential $V(x)$. For simplicity we assume that the diffusion coefficient D is a constant. Equation (8.13FPT) is a first-order ordinary differential equation in the function $T'(x) = dT/dx$. As such it may be integrated immediately by introducing an integrating factor $\Lambda(x)$. The result is

$$\Lambda T'(y) = -\int^{y} D^{-1}\Lambda(z)\, dz + C_1 \tag{8.23}$$

where $\Lambda(z) = e^{-V(z)/D}$. The constant C_1 can be determined by integration as in (8.19). The result is

$$0 = T(b) - T(a) = \int_a^b T'(y)\, dy$$

$$= C_1 \int_a^b \Lambda^{-1}(y)\, dy - \int_a^b dy\, \Lambda^{-1}(y) \int^{y} D^{-1}\Lambda(z)\, dz \tag{8.24}$$

There remains now only the problem of carrying out these integrals. To do this, we will assume that $V(x)$ is a function with two local minima, as shown in Fig. 8.4. We are particularly interested in the process of diffusion from the metastable

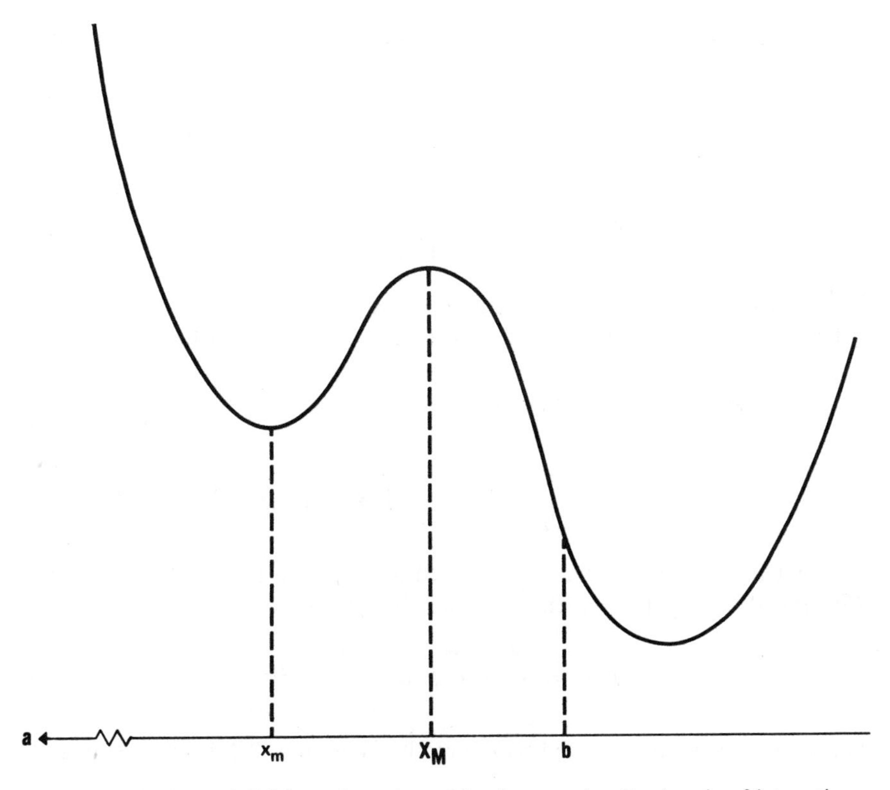

Figure 8.4 The integral (8.24) can be estimated by decomposing the domain of integration as shown, and then by applying Laplace's method for estimating the values of the integrals.

minimum to the stable minimum, so we take the random walk boundaries a, b as shown.

When D is small, the first term on the right-hand side of (8.24) is easily estimated:

$$C_1 \int_a^b e^{V(y)/D}\, dy = C_1 \int_0^{b-a} e^{[V(a)+(y+a)V'(a)+\cdots]/D}\, d(y+a)$$

$$\simeq C_1 e^{V(a)/D} \int_0^{b-a} e^{(y+a)V'(a)/D}\, d(y+a) \qquad (8.25)$$

The upper limit can be extended to ∞ without appreciably altering the value of the integral.

The second integral can be estimated similarly. The maximum contribution to the integral

$$I = \int_a^b dy \int^y dz\, D^{-1} e^{V(y)-V(z)/D} \qquad (8.26)$$

occurs for the values y, z, $y > z$, which maximize $V(y) - V(z)$. This is maximum for $y = X_M$, $z = x_m$. The result is

$$I = D^{-1}e^{V(X_M)-V(x_m)/D} \iint e^{(1/2)(y-X_M)^2V''(X_M)-(1/2)(z-x_m)^2V''(x_m)/D}\,dy\,dz$$

$$\xrightarrow{D\text{ small}} e^{V(X_M)-V(x_m)/D}\frac{2\pi}{|V''(X_M)V''(x_m)|^{1/2}} \tag{8.27}$$

The results (8.24) and (8.26) determine the value of C_1.

The first passage time $T(x)$ is determined by integrating $T'(x)$:

$$T(x) - T(a) = \int_a^x T'(y)\,dy$$

$$= C_1 \int_a^x e^{V(y)/D}\,dy - \int_a^x dy \int^y dz\, D^{-1}e^{V(y)-V(z)/D} \tag{8.28}$$

The first term in (8.28) can be estimated as in (8.25). In fact, the result is no different. By using the elementary result that $\int_a^x + \int_x^b = \int_a^b$, we find

$$T(x) = \int_x^b dy \int^y dz\, D^{-1}e^{V(y)-V(z)/D} \tag{8.29}$$

This integral can be estimated by the methods used in (8.27). The result is

$$T(x) = \frac{2\pi}{|V''(X_M)V''(x_m)|^{1/2}}\, e^{V(X_M)-V(x_m)/D}, \qquad x < x_m$$

$$= \frac{(2\pi D)^{1/2}}{V'(x)|V''(X_M)|^{1/2}}\, e^{V(X_M)-V(x)/D}, \qquad x_m < x < X_M \tag{8.30}$$

These estimates are clearly not valid in the neighborhood of a catastrophe germ. Comparable results hold in the n-dimensional case.

For the sake of comparison, the time scale T_1 for relaxation to a local minimum and the time scale T_2 for diffusion from a metastable minimum to a global minimum are given by

$$T_1 = 1/\lambda_1$$

$$T_2 = \frac{2\pi}{|\lambda_1\lambda_2|^{1/2}}\, e^{\Delta V/D} \tag{8.31}$$

where λ_1, λ_2 are the curvatures (d^2V/dx^2) at the local minimum and the local maximum, respectively, and $\Delta V = V(X_M) - V(x_m)$. In general $T_2 \gg T_1$ when $\Delta V/D \gg 1$. The evolution of a probability distribution governed by a Fokker–Planck equation is shown schematically in Fig. 8.5. It is clear that we can identify

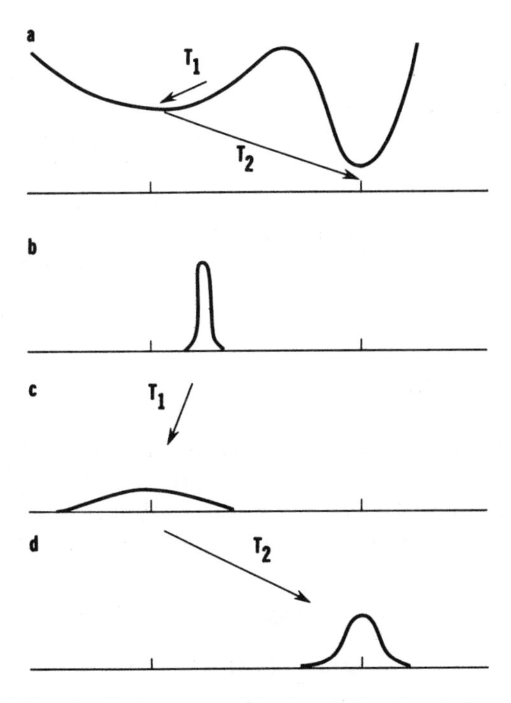

Figure 8.5 A probability distribution centered in the neighborhood of one local minimum will move "asymptotically" on a time scale T_1 to a Gaussian distribution with appropriate curvature centered on that local minimum. On the time scale $T_2 \gg T_1$ the probability distribution will move asymptotically to its time-independent form $P(x) \simeq e^{-V(x)/D}$.

the barrier height ΔV with ΔE (Fig. 8.2) and D with $\mathcal{N}$. In this sense the noise term D is a generalized temperature.

We are now in a position to discuss the applicability of conventions for systems described by diffusion equations.[5] The rate at which the critical points $x_c(t)$ of $V(x; c(t))$ move is comparable with dc/dt, so the conventions can be expressed in terms of the time derivatives of the control parameters. The result is as follows:

Delay Convention.

$$T_1^{-1} \gg \frac{dc}{dt} \gg T_2^{-1}$$

Maxwell Convention.

$$T_2^{-1} \gg \frac{dc}{dt} \tag{8.32}$$

No convention is applicable if $dc/dt \gg T_1^{-1}$ or if dc/dt is comparable with either T_1^{-1} or T_2^{-1}.

Remark. As $D \to 0$, $T_2 \to \infty$ and $T_2^{-1} \to 0$. If the control parameters change at all, $dc/dt > T_2^{-1}$ so that the Maxwell Convention is never applicable. In the absence of diffusion (no fluctuations) only the Delay Convention may be applicable. We can now answer the question left hanging in (8.7). The critical point to be chosen is the one which the system is in initially (at time $t = 0$). The system will remain at that critical point as the control parameters change value until that critical point disappears. The new critical point which the system evolves to must be determined by solving an equation of motion, which in this case is (1.6) or (8.1).

4. The Inadequacy of Conventions

The foregoing discussions have made clear that the Delay and Maxwell conventions are merely stopgap measures until more progress is made in the program of Catastrophe Theory. To illustrate the ways in which these two conventions may be inadequate, we consider a simple example.

Example 1. A classical system obeys the damped equation of motion

$$\frac{d^2x}{dt^2} = \frac{F(x)}{m} - \gamma\dot{x} \tag{8.33}$$

where the force is derived from $F = -\nabla V$ and

$$V(x) = \frac{x^6}{6} + \frac{a_1 x^4}{4} + \frac{a_2 x^3}{3} + \frac{a_3 x^2}{2} + a_4 x \tag{8.34}$$

The system is initially in the left-hand minimum (Fig. 8.6) and remains there as the control parameters change until this minimum disappears. To which minimum does the system state move?

The Delay Convention is silent on this point. If the equation of motion were

$$\frac{dx}{dt} = \frac{F(x)}{m} - \gamma\dot{x} \tag{8.35}$$

then the system would move to the middle minimum because at this minimum $F = 0$ and therefore $dx/dt = 0$. But the actual equation of motion (8.33) is second order, so that the system's inertia carries it through this point, and perhaps over the local maximum separating the middle- and right-hand minima. In the "viscous limit" $\gamma \to \infty$, equation (8.33) behaves like (8.35), and the system winds up in the middle minimum. In the slightly damped limit (γ small) the system may stabilize in either of the two minima. In fact, the minimum in which the system ultimately comes to rest depends on the value of the damping parameter γ, as shown in Fig. 8.6.

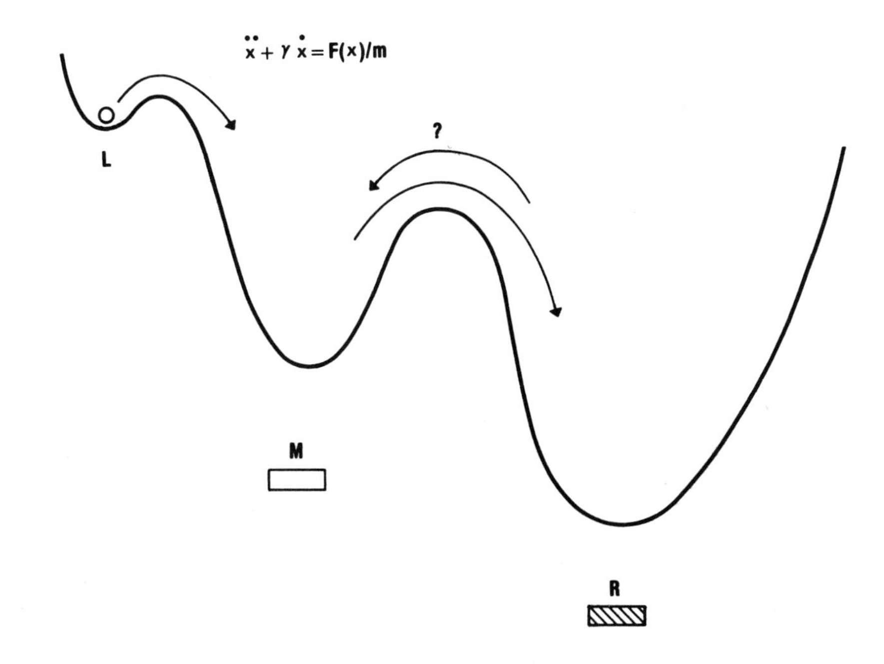

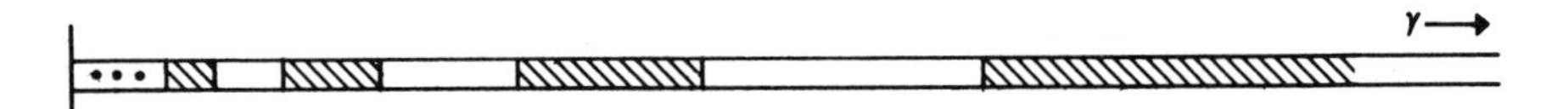

Figure 8.6 The final state of the system (8.33) whose metastable minimum disappears at time $t = 0$ may be in either the central or the right-hand minimum, depending on the value of the damping constant, as shown.

5. Summary

Very few systems of practical interest are strictly governed by the equilibria of gradient dynamical systems. In general either the control parameters depend on time or the state variables are subject to fluctuations, or both. Elementary Catastrophe Theory is not strictly applicable to the description of such systems. In many important instances the imposition of conventions allows us to lift the results of Elementary Catastrophe Theory from the level of Step 7 (Table 1.1) to some higher level. The conventions are ad hoc, the price is an incompleteness in description, and the rewards may be enormous.

Notes

The introduction of these two conventions into Catastrophe Theory was first made by Thom.[1] The use of a Fokker–Planck equation as an interpolating mechanism between these two conventions was first made by Gilmore.[5]

References

1. R. Thom, *Stabilité Structurelle et Morphogénèse*, New York: Benjamin, 1972.
2. M. C. Wang and G. E. Uhlenbeck, On the Theory of the Brownian Motion II, *Rev. Mod. Phys.* **17**, 323–342 (1945).
3. J. D. Jackson, *Classical Electrodynamics*, New York: Wiley, 1962.
4. R. L. Stratonovich, *Topics in the Theory of Random Noise*, New York: Gordon and Breach, 1963.
5. R. Gilmore, Catastrophe Time Scales and Conventions, *Phys. Rev* **A20**, 2510–2515 (1979).

CHAPTER 9

Catastrophe Flags

Elementary Catastrophe Theory will eventually become a useful tool in the bag of tricks of most working scientists and engineers. The presence of an elementary catastrophe can be recognized by the occurrence of a non-Morse critical point in a family of potentials. However, such a point may not be recognized immediately for one reason or another. For example, (1) the potential may be a very complicated function or (2) it may not even be known accurately. Worse yet, (3) the system may not even be a gradient system, or worst of all, (4) we may have not the foggiest idea of the equations that describe the system properly. Yet catastrophes do occur in realistic situations, and it is important to recognize their presence.

This is not generally too difficult, as catastrophes have characteristic fingerprints and often even wave flags[1] to gain our attention. This chapter is devoted to a description of some of the standard catastrophe flags.

The first five flags occur when the physical control parameters can move into a region of control parameter space in which the corresponding potential has more than one minimum. They all occur together, with the possible exception of hysteresis, which does not occur if the Maxwell Convention is obeyed.

The remaining three flags can occur even when the potential does not have multiple minima. This fact can be extremely important when studying the properties of systems where the sudden jump to another state is not desired, for example, the collapse of a building, the qualitative change in global weather patterns, or the explosive transformation of some chemical system.

Once one of these catastrophe flags has suggested the presence of a catastrophe, the control parameters may be varied in a search for the relevant remaining flags. The others will occur under suitable conditions. The location of the catastrophe will often provide, at low cost, an enormous amount of information about the description of the physical system. For the four cases of increasing ignorance outlined above, the presence and type of catastrophe could suggest:

1. A simplified model potential depending on only the appropriate state variables and control parameters; or
2. The appropriate germ of a potential which, in turn, could suggest the physical process that is actually occurring; or
3. Indicate the appropriate type of equations (dynamical, diffusive, etc.) for the system, and how a potential might occur in such equations; or
4. Dispense with the need for equations altogether, if the inferences to be drawn from such equations depend primarily on the canonical geometry of the associated catastrophe.

Although catastrophes arise from the qualitative study of equations, there is a feedback effect which sometimes allows us to determine qualitative consequences even if we do not know the equations, provided we recognize the presence and type of catastrophe.

1. Modality

This means that the physical system has two or more distinct physical states in which it may occur. In other words, the potential describing the system has more than one local minimum for some range of the external control parameters.

Example. The cusp catastrophe becomes bimodal when the control parameters lie within the cusp-shaped region shown in Fig. 9.1.

2. Inaccessibility

This means that the system has an equilibrium state which is unstable—a Morse i-saddle ($i > 0$). Such equilibria are unstable because infinitesimal perturbations exist which decrease the value of the potential. Whenever the potential V has more than one local minimum, it must have at least one i-saddle ($i > 0$) which is an unstable equilibrium.

Example. The two sheets over the cusp-shaped region in Fig. 9.1 representing the locally stable minima are separated by the middle inaccessible sheet representing an unstable local maximum.

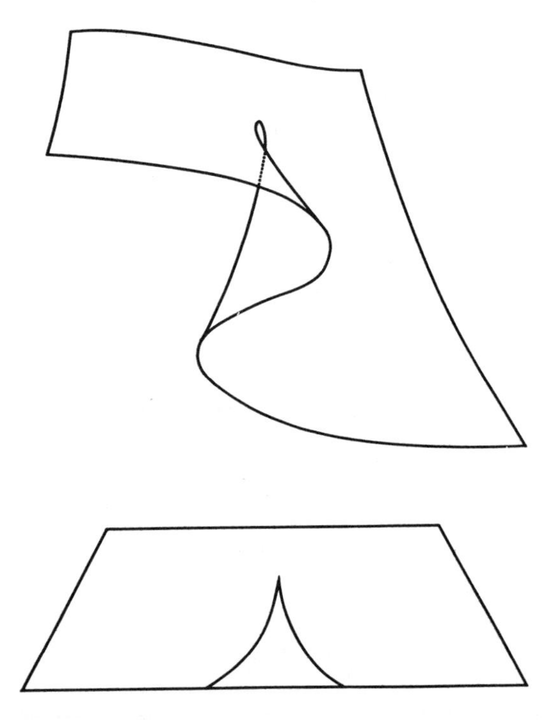

Figure 9.1 The stability properties of the critical points of the cusp catastrophe functions are easily determined from the cusp catastrophe manifold.

3. Sudden Jumps

If either the Maxwell Convention or the Delay Convention is adopted, then a small change in the value of the control parameter may result in a large change ("sudden jump") in the value of the state variable (Fig. 8.1) as the system jumps from one local minimum to another. If the Maxwell Convention is adopted, this sudden jump is accompanied by a smooth but nondifferentiable change in the value of the potential. If the Delay Convention is adopted, the jump from one disappearing local minimum to a global or some other local minimum is accompanied by a discontinuous change in the value of the potential.

The occurrence of sudden jumps is independent of the convention adopted, if any. The transition from the neighborhood of one local minimum to another represents a large change in the value of the state variable which often occurs on a fairly rapid time scale.

Example. A sudden jump in the value of the state variable occurs as the system state jumps from one sheet of the cusp catastrophe manifold to the other (Fig. 9.2), even when neither of the two extreme conventions is applicable.

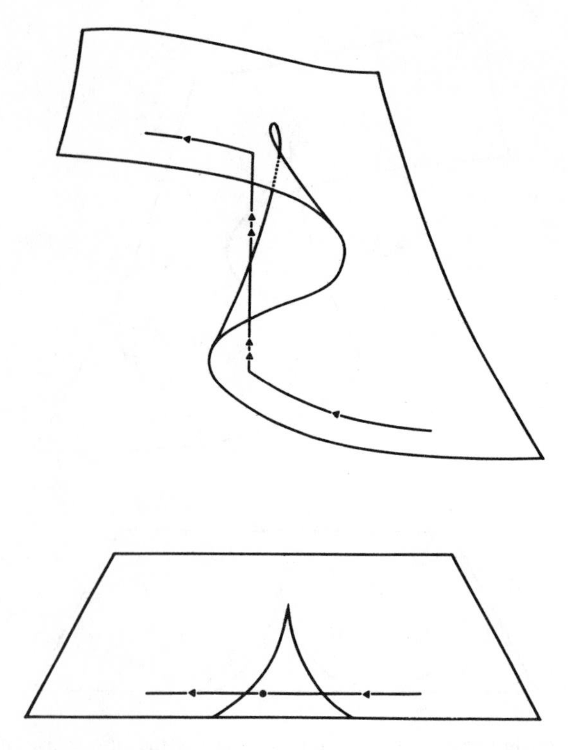

Figure 9.2 A slow change in the control parameters may lead to a sudden change in the state variable even if no catastrophe convention is observed.

4. Divergence

A finite change in the value of the control parameters leads to a finite change in the equilibrium value of the state variables. Usually a small perturbation in the initial values of the control parameters will lead to only a small change in the initial and final values of the state variables. However, in the neighborhood of a non-Morse critical point, small changes in the control parameter initial values may lead to large changes in the state variable final values.

We have already discussed the mathematical instability of non-Morse germs against perturbations (Chapter 5). The instability of physical processes against perturbation of the control parameter trajectory is called divergence.

Example. The cusp catastrophe exhibits divergence. Assume that some physical process changes the control parameters (a, b) from (positive, ε) to (negative, ε) (ε small and constant). Then a perturbation of the initial values of the control parameters which changes the sign of b has little effect

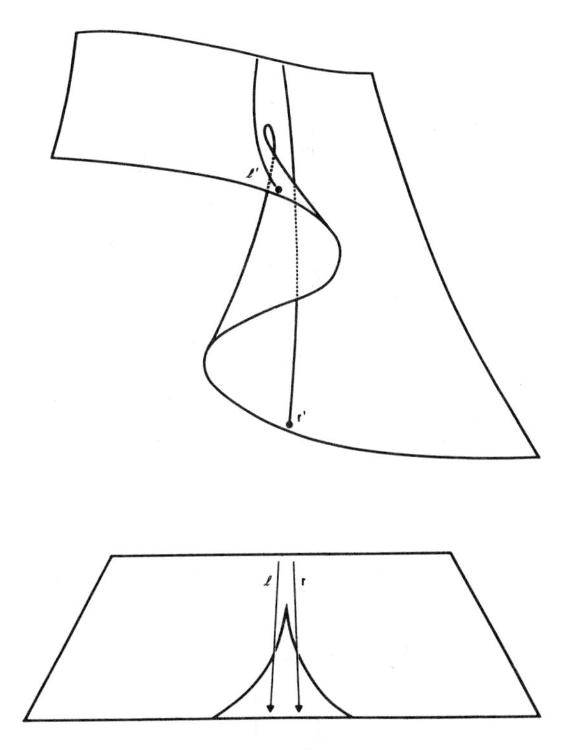

Figure 9.3 Two nearby paths in control parameter space may lead to widely divergent final values of the state variables.

on the initial value of the state variable but a profound effect on its final value (Fig. 9.3).

5. Hysteresis

Hysteresis occurs whenever a physical process is not strictly reversible. That is, the jump from local minimum 1 to local minimum 2 does not occur over the same point in control parameter space as the reciprocal jump from local minimum 2 to local minimum 1. Hysteresis fails to occur only when the Maxwell Convention is obeyed.

Example. Hysteresis occurs for the cusp catastrophe as illustrated in Fig. 9.4. We have adopted the Delay Convention and assumed that the control parameters sweep out the loop $a < 0$ and b first increasing, then decreasing and crossing both fold lines. The time-dependent response of the state variable follows the characteristic hysteresis loop.

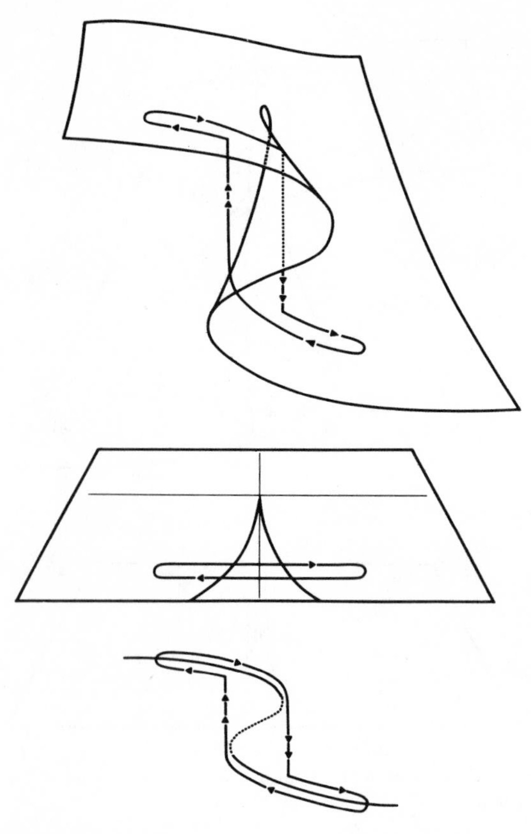

Figure 9.4 Hysteresis occurs when the jump from one sheet to another does not occur for the same values of the control parameters as the reciprocal jump.

6. Divergence of Linear Response

Under a slight change in the values of the control parameters ($c \to c^0 + \delta c^0$) the location of an equilibrium will move slightly ($x \to x^0 + \delta x^0$). The relation between the response of the equilibrium to a change in the controls can be obtained by making a Taylor series expansion of $V(x; c)$ in powers of $(x - x^0)$, $(c - c^0)$ and neglecting all but the linear terms of $\nabla V = 0$ [cf. (5.2)]. The linear response is given by

$$\begin{aligned}\delta x_j^0 &= -(V^{-1})^{jk} V_{k\alpha} \delta c_\alpha^0 \\ &= \chi_{j\alpha}(x^0; c^0) \delta c_\alpha^0\end{aligned} \tag{9.1}$$

The linear response of δx_j^0 to δc_α^0 is given through the susceptibility tensor $\chi_{j\alpha}$, which is expressed in terms of the second derivatives of the potential at a stable equilibrium. As the equilibrium approaches a non-Morse critical point, $\det V_{ij} \to 0$ so that its matrix inverse V^{-1} has some matrix elements which

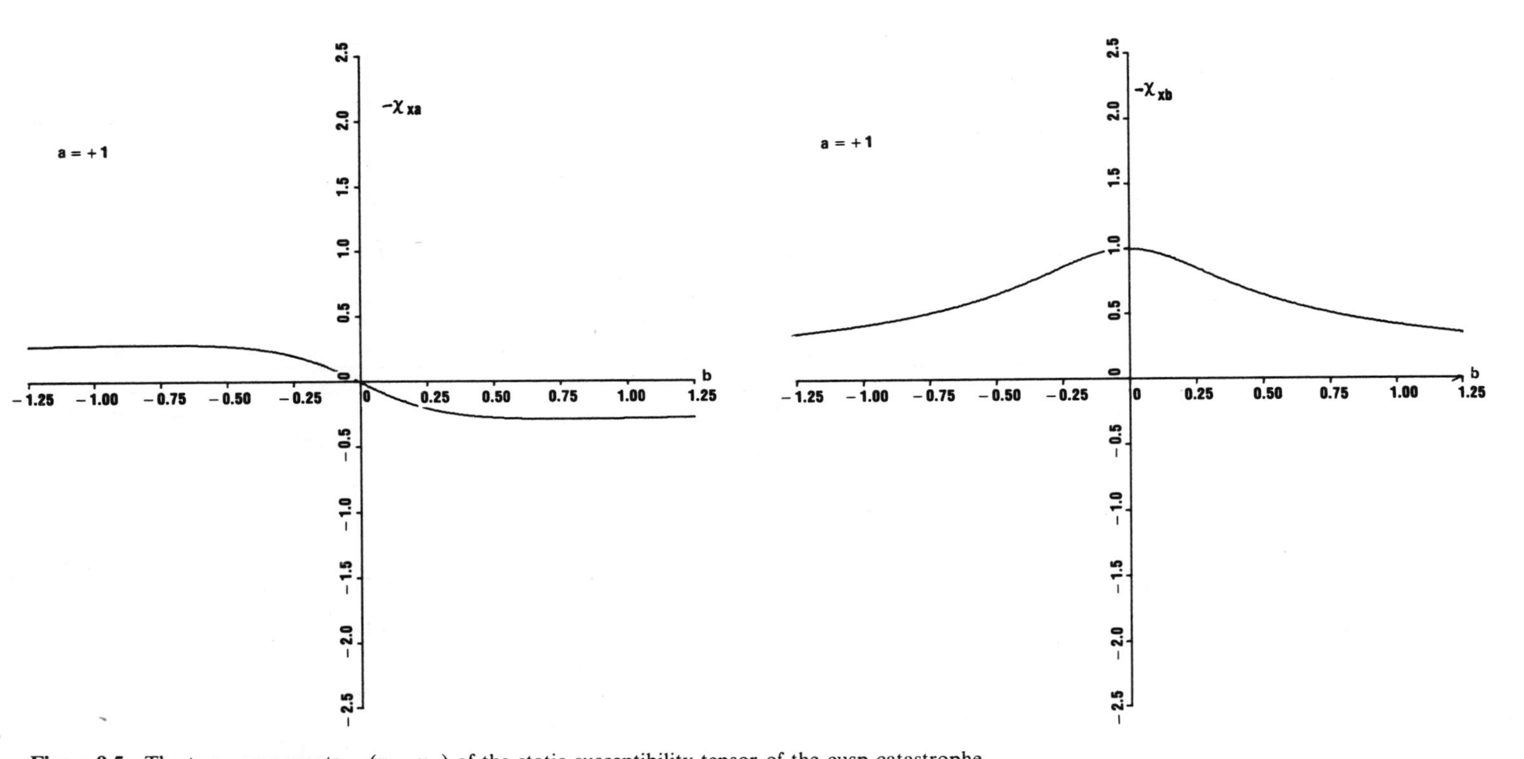

Figure 9.5 The two components $-(\chi_{xa}, \chi_{xb})$ of the static susceptibility tensor of the cusp catastrophe are shown as a function of the control parameter b in the three planes $a = +1$, $a = 0$, and $a = -1$.

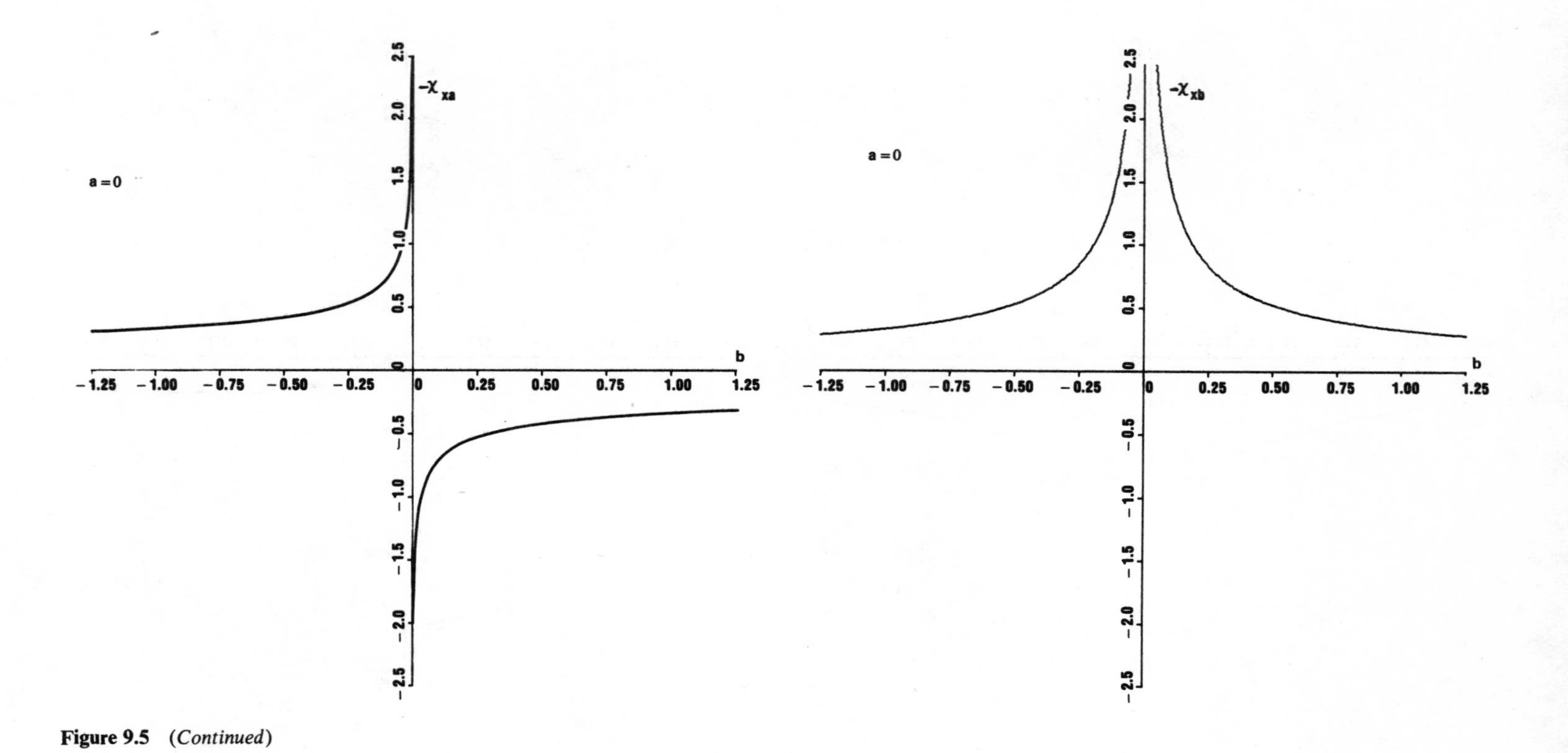

Figure 9.5 *(Continued)*

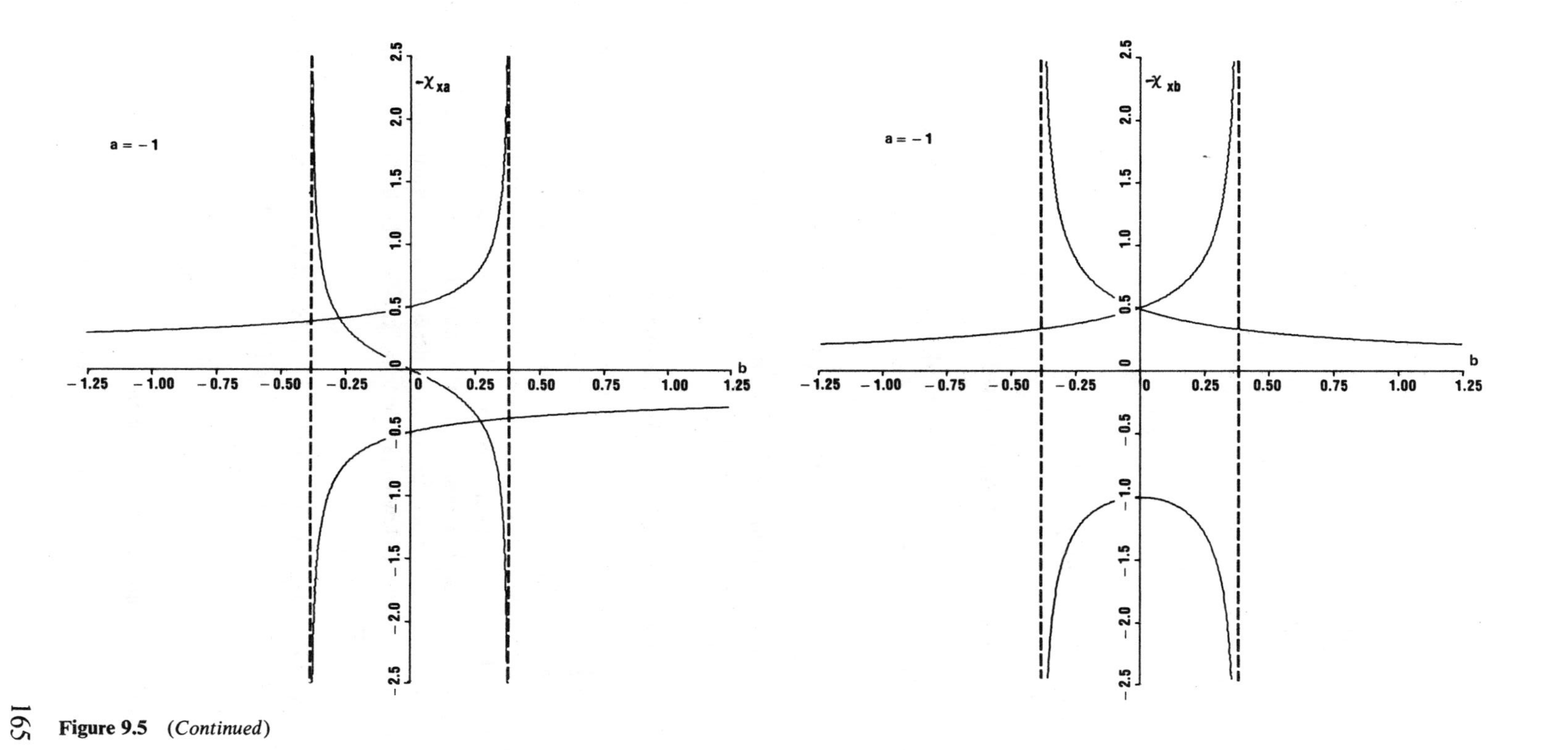

Figure 9.5 *(Continued)*

become exceedingly large. Therefore the linear response function $\chi_{j\alpha}$ diverges as a degenerate critical point is approached.

The divergence of $\chi_{j\alpha}$ tells us not only that we are in the neighborhood of a degenerate critical point, but also what the "bad" state variables are. The "bad" state variables correspond to those directions in which the state variable response to a small control perturbation diverges. This is seen most easily by assuming that V has canonical Morse form at x^0. Then $V_{jk} = \lambda_j \delta_{jk}$ so that (9.1) simplifies to

$$\delta x_j^0 = -\left(\frac{V_{j\alpha}}{\lambda_j}\right)\delta c_\alpha^0 \qquad \text{(no sum on } j) \tag{9.2}$$

As $\lambda_j \to 0$, the response of δx_j^0 to a nonzero perturbation of control parameters diverges.

Example. For the cusp catastrophe $V(x; a, b) = x^4/4 + ax^2/2 + bx$, the curvatures $V'' = 3x^2 + a$ at the equilibria are shown in Figs 6.2 and 6.5. The function $\partial^2 V/\partial x\, \partial a = x_c$ can be determined from Figs. 6.2 and 6.3, while $\partial^2 V/\partial x\, \partial b = 1$. For the cusp the static suspectibility tensor is

$$(\chi_{xa}, \chi_{xb}) = \left(\frac{-x}{3x^2 + a}, \frac{-1}{3x^2 + a}\right) \tag{9.3}$$

The two components of this tensor are shown for the three paths with b increasing and $a = +1$, $a = 0$, and $a = -1$ in Fig. 9.5.

7. Critical Slowing Down/Mode Softening

In the previous section we have danced around the question of how (in time) a physical system responds to a perturbation. Such a discussion must depend on the details of the time dynamics, that is, on some assumed equation of motion. Dynamical assumptions have been suppressed by the time we have reached the level of approximation of Step 7, Table 1.1. However, two distinctly different dynamical situations are frequently encountered in physical systems, and are worth being discussed in detail. These are the "gradient dynamical system" and the "gradient Newtonian system," each of which carries its own catastrophe fingerprint.

A GRADIENT DYNAMICAL SYSTEM. The equations governing such a system have been discussed in Chapter 1 [cf. (1.6)]:

$$\frac{dx_i}{dt} = -\frac{\partial V}{\partial x_i} \tag{9.4D}$$

If $(x^0; c^0)$ is an equilibrium, then

$$V(x; c) = \text{constant} + \tfrac{1}{2}\delta x_i \delta x_j V_{ij} + \mathcal{O}(3) \tag{9.5}$$

The equations of motion in the neighborhood of the equilibrium are

$$\frac{dx_i}{dt} = -V_{ij}\delta x_j + \mathcal{O}(2) \tag{9.6D}$$

By neglecting the terms of order 2 and higher, the dynamical equations reduce to a simple system of linear equations. The equilibrium at x^0 is stable if all the eigenvalues of the stability matrix $V_{ij}(x^0; c^0)$ are positive. The normal modes have time dependences of the form $e^{-\lambda_i t}$, where $\lambda_1, \ldots, \lambda_n$ are the eigenvalues of V_{ij}, so that $1/\lambda_i$ is the characteristic relaxation time of the ith eigenmode.

As the bifurcation set (Delay Convention) is approached, det $V_{ij} \to 0$ so that one or more of the eigenvalues approaches zero. The relaxation time of the corresponding mode becomes larger and larger. That is, as we approach a non-Morse critical point, it becomes more and more difficult for (at least) one of the modes to relax back to zero. This lengthening of the relaxation time scale is called **critical slowing down**.

B GRADIENT NEWTONIAN SYSTEMS. Gradient Newtonian systems differ from gradient dynamical systems (9.4D) in having the first time derivatives replaced by second derivatives (9.4N):

$$\frac{d^2x_i}{dt^2} = -\frac{\partial V}{\partial x^i} \tag{9.4N}$$

$$\frac{d^2x_i}{dt^2} = -V_{ij}\delta x_j + \mathcal{O}(2) \tag{9.6N}$$

The dynamical response in the neighborhood of an equilibrium (9.5) is given by (9.6N). A normal-mode transformation may be performed on (9.6N) as on (9.6D). When x^0 is a local minimum, all eigenvalues of V_{ij} are positive and all modes have a periodic time dependence of the form $e^{i\omega_j t}$, $\omega_j^2 = \lambda_j$.

As the bifurcation set (Delay Convention) is approached, det $V_{ij} \to 0$ so that one or more of the oscillation frequencies ω_j approaches zero. This diminishing of the oscillation frequency for certain modes is called **mode softening**.

Example. For the cusp catastrophe the stability matrix is simply $V'' = 3x^2 + a|_{x_c}$, where x_c is the value of a critical point. The canonical curvature for the cusp has already been plotted in Figs. 6.2d and 6.5. In the case of gradient dynamical systems the relaxation time $1/V''$ is shown in Fig. 9.6 for $a = +1$, $a = 0$, and $a = -1$ as a function of b for the equilibria. In the case of gradient Newtonian systems, the oscillation frequency $(V'')^{1/2}$ is shown in Fig. 9.7 for $a = +1$, $a = 0$, and $a = -1$ as a function of b for the stable equilibria. Scaling relations hold for both Figs. 9.6 and 9.7.

Remark 1. The eigenvector of V_{ij} associated with the eigenvalue λ_j which approaches zero gives the "bad" direction associated with the non-Morse critical point of V.

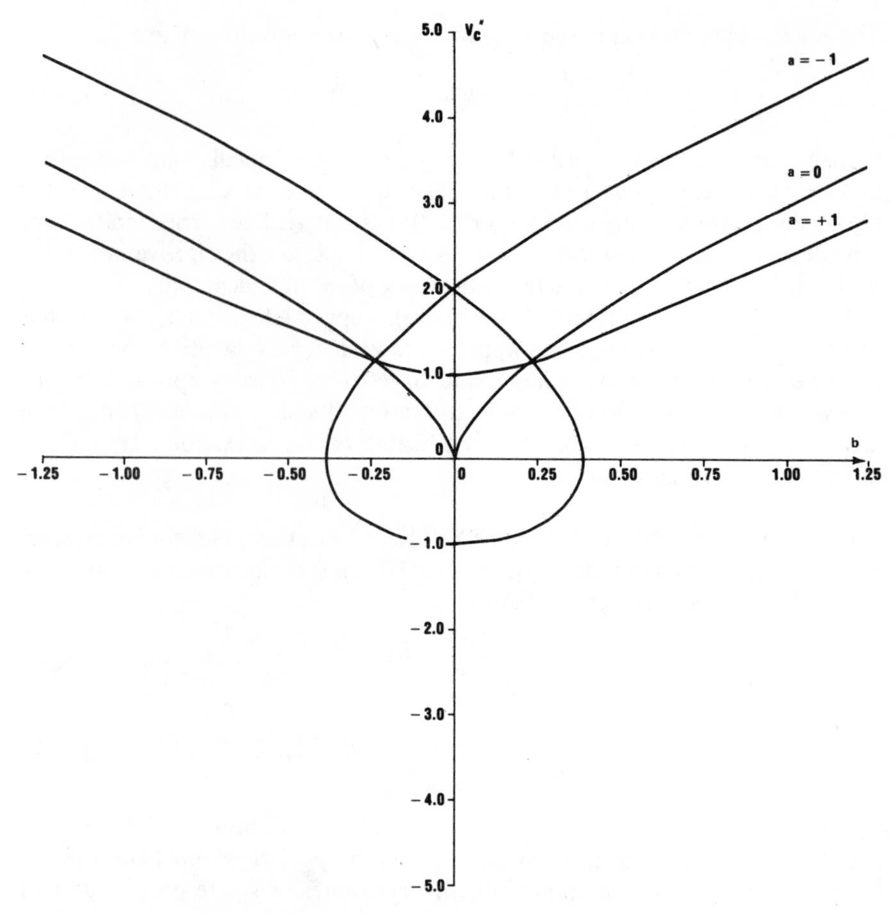

Figure 9.6 The critical curvature and local relaxation times are shown as a function of b in the three planes $a = +1$, $a = 0$, and $a = -1$.

Remark 2. Both the gradient dynamical system and the gradient Newtonian system may be viewed as limits of a slightly more complex classical equation of motion:

$$m \frac{d^2 x_i}{dt^2} = -\frac{\partial V}{\partial x_i} - \gamma \frac{dx_i}{dt} \tag{9.7}$$

Here the term $-\partial V/\partial x_i$ may be interpreted as the conservative force derivable from a potential $V(x; c)$, and $-\gamma\, dx_i/dt$ as a dissipative frictional force. Both forces act on a particle of mass m in a space of dimension $\mathbb{R}^n$ (state variable space). Then (9.7) is just Newton's equation of motion. The gradient dynamical system is obtained in the limit $m/\gamma \to 0$ and the gradient Newtonian system is obtained in the reverse limit $\gamma/m \to 0$ [cf. (8.2) and Fig. 8.6]. These two limits may be referred to vividly as the "molasses" limit and the "conservative" limit.

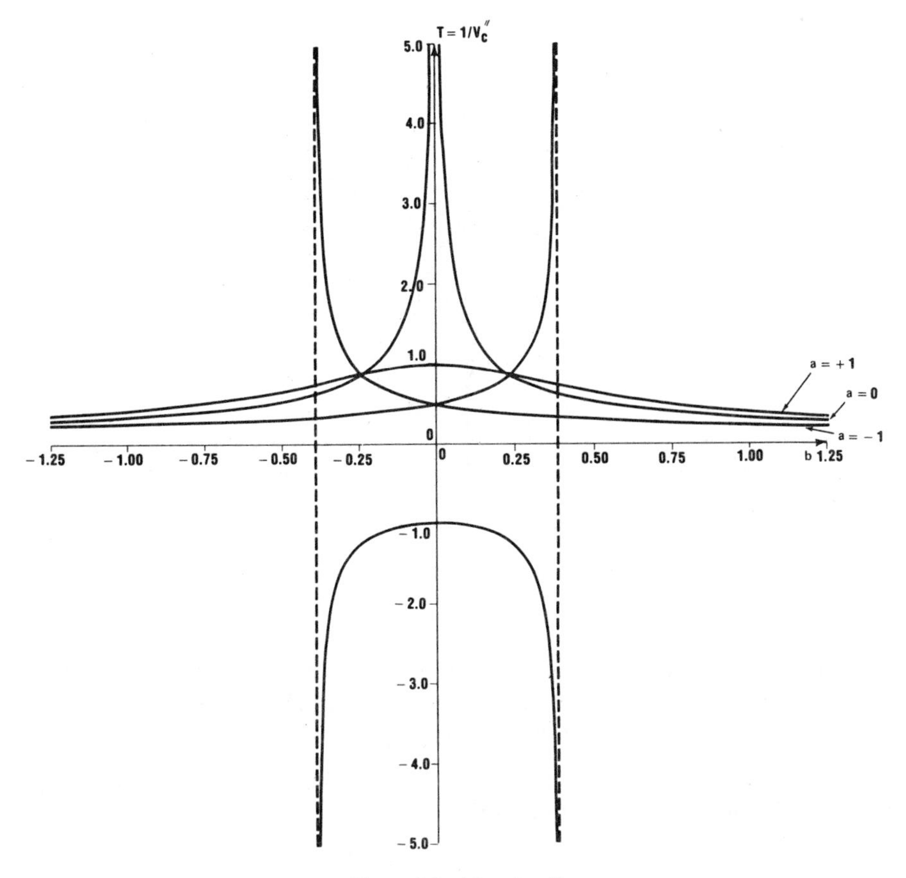

Figure 9.6 (*Continued*)

Remark 3. Although gradient dynamical systems and gradient Newtonian systems represent limit points of the more general possibility (9.7) in much the same way that Maxwell's Convention and the Delay Convention are two extremes in a range of possibilities, the critical dynamical properties of the system (9.7) are well defined within this range ($0 < \gamma/m < \infty$). To see this, we linearize the potential according to (9.5) and assume a time dependence of the form $\delta x_j(t) = \delta x_j e^{i\omega t}$. The result is

$$[(-m\omega^2 + i\omega\gamma)\delta_{ij} + V_{ij}]\delta x_j = 0 \tag{9.8}$$

This equation has a nontrivial solution only when the determinant of the matrix within square brackets is zero. As we approach a degenerate critical point, $\det V_{ij} \to 0$ so that at least one complex root ω approaches zero. This is the general case intermediate between critical slowing down and mode softening. As above, the nontrivial eigenvector of (9.8) associated with the vanishing eigenvalue describes the "bad" direction in state variable space.

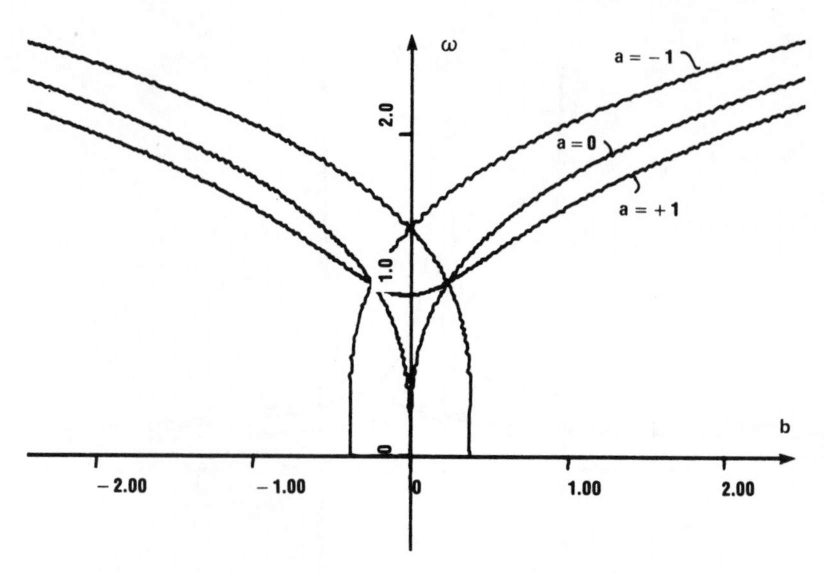

Figure 9.7 The normal mode frequencies are shown as a function of b in the three planes $a = +1$, $a = 0$, and $a = -1$.

Remark 4. If the time dynamics is represented by the operator $f(D)$, $D = d/dt$, then the equation of motion around an equilibrium can be written

$$[f(D) + V_{ij}]\delta x_j(t) = 0 \tag{9.9}$$

If a time dependence $\delta x_j(t) = \delta x_j e^{\lambda t}$ is assumed, and if $f(D)$ is a linear operator, then (9.9) reduces to the matrix equation

$$[f(\lambda)\delta_{ij} + V_{ij}]\delta x_j = 0 \tag{9.10}$$

If f is a polynomial without constant term, then as a non-Morse critical point is approached, det $V_{ij} \to 0$ and at least one of the roots λ of (9.10) also approaches zero. As usual, the associated eigenvector(s) describes the "bad" direction(s).

8. Anomalous Variance

A physical system may be defined by a probability $P(x)$ defined over the space of state variables rather than by an isolated point ("distribution") in the space of state variables. Such a system is characterized by the moments of the probability distribution function

$$\begin{aligned} 1 &= \int P(x)\,dx \\ \langle x_i \rangle &= \int x_i P(x)\,dx \\ \langle x_i x_j \rangle &= \int x_i x_j P(x)\,dx \\ &\;\vdots \end{aligned} \tag{9.11}$$

Of the moments, the most important are the mean value (first moment) $\langle x_i \rangle$ and the covariance

$$\begin{aligned}\langle \Delta x_i \Delta x_j \rangle &= \langle (x_i - \langle x_i \rangle)(x_j - \langle x_j \rangle) \rangle \\ &= \langle x_i x_j \rangle - \langle x_i \rangle \langle x_j \rangle \end{aligned} \tag{9.12}$$

If such a physical system is also associated with a potential $V(x; c)$ defined over the space $\mathbb{R}^n$ of state variables x and depending on control parameters $c \in \mathbb{R}^k$, then the time-independent probability distribution function is often related to the potential function in a simple exponential way:

$$P(x; c) = Ne^{-V(x;c)/D} \tag{9.13}$$

where N is a normalization constant and D is a positive diffusion constant [cf. (8.4)].

It is often the case that D is small. In this case the deepest minimum of $V(x; c)$ is emphasized. If the deepest minimum is a Morse critical point at $x = x^0$, then

$$P(x; c) \simeq Ne^{-V_{ij}(x-x^0)_i(x-x^0)_j/D} \tag{9.14}$$

The variance at this critical point can be computed using (9.12). The computations are somewhat simpler when V is taken in canonical Morse form in the neighborhood of x^0. This can be done because the operation of diagonalizing $V_{ij}\Delta x_i \Delta x_j$ commutes with the operation of taking the exponential. Then for

$$P(x; c) \simeq Ne^{-\Sigma \lambda_i (x_i - x_i^0)^2/D} \tag{9.15}$$

the variance is

$$\langle \Delta x_i \Delta x_j \rangle = \tfrac{1}{2}\delta_{ij} \frac{D}{\lambda_j} \tag{9.16}$$

Since D is small, the variance is small unless one of the eigenvalues of V_{ij} is small.

This suggests that the variance may become large ("anomalous") in the neighborhood of a non-Morse critical point. In fact, two quite distinct types of anomalies may occur as the control parameters c_α are changed.

1 The minimum at x^0 may become degenerate by colliding with one or more other critical points. In this case the "bad" state variable(s) and the type of degenerate critical point can be determined as follows. Since the covariance matrix (9.12) is real and symmetric, it may be diagonalized by a real orthogonal transformation. This transformation also diagonalizes V_{ij}. If the variance of Δx_1 becomes anomalously large after the principal axis transformation, x_1 is the "bad" state variable. More generally, we have, according to (2.3),

$$P(x; c^0) \simeq Ne^{-f_{NM}/D} \otimes e^{-f_M/D} \tag{9.17}$$

The Morse function gives terms with normal small variance. The non-Morse germs which appear in (9.17) must have the property that the functions $e^{-f_{NM}/D}$

are normalizable. The only simple catastrophe germs which have this property are $A_{+(2n-1)} = +x^{2n}$. For these germs $\langle x \rangle = 0$, and the variance can be expressed in terms of gamma functions:

$$\langle x^2 \rangle = \frac{\int_{-\infty}^{+\infty} x^2\, e^{-x^{2n}/D}\, dx}{\int_{-\infty}^{+\infty} e^{-x^{2n}/D}\, dx} = D^{1/n} \frac{\Gamma(3/2n)}{\Gamma(1/2n)} \tag{9.18}$$

2 A distant metastable minimum at x' may assume a lower value than the minimum at x^0. For $|V(x^0; c) - V(x'; c)| \gtrsim 10D$ the probability distribution is sharply peaked near x^0 or x', depending on whether $V(x^0; c) - V(x'; c)$ is negative or positive. Then $\langle x_i \rangle \simeq x_i^0$ or x_i' and $\langle \Delta x_i^2 \rangle \sim \frac{1}{2}D/\lambda_i^0$ or $\frac{1}{2}D/\lambda_i'$, where λ_i^0 or λ_i' is the appropriate eigenvalue. However, as the deepest minimum passes from x^0 to x' as the control parameters change, the mean value $\langle x_i \rangle$ passes rapidly from x_i^0 to x_i'. When $V(x^0; c) = V(x'; c)$,

$$\begin{aligned} \langle x_i \rangle &\simeq \tfrac{1}{2}(x^0 + x')_i \\ \langle \Delta x_i^2 \rangle &\simeq [\tfrac{1}{2}(x^0 - x')_i]^2 \end{aligned} \tag{9.19}$$

In this case the variance may be much greater than in Case 1.

Example. We assume that a probability distribution

$$\begin{aligned} P(x; a, b) &= Ne^{-V(x; a, b)/D} \\ V(x; a, b) &= \frac{x^4}{4} + \frac{ax^2}{2} + bx \end{aligned} \tag{9.20}$$

is associated with each point (a, b) in the control parameter plane $\mathbb{R}^2$ of the cusp catastrophe. In Fig. 9.8 we show four paths in the control plane, as well as the average $\langle x \rangle$ and variance $\langle \Delta x^2 \rangle$ as a function of position along each path. These are described now.

Path 1 ($a = +1$, b increasing). The eigenvalue λ of the stability matrix d^2V/dx^2 is shown as a function of b for $a = +1$, $a = 0$, and $a = -1$ in Fig. 6.5. The curvature is minimized in the neighborhood of $b = 0$, so the variance is largest at $b = 0$.

Path 2 ($a = 0$, b increasing). As $b \to 0$, the curvature also goes to zero, and the quartic term plays an increasingly important role in keeping the probability distribution more or less (less than more) sharply peaked around $x = 0$. The variance at $(a, b) = (0, 0)$ is given by (9.18)

$$\begin{aligned} \langle \Delta x^2 \rangle &= \sqrt{4D}\, \frac{\Gamma(\frac{3}{4})}{\Gamma(\frac{1}{4})} \\ &= 0.33799\sqrt{4D} \end{aligned} \tag{9.21}$$

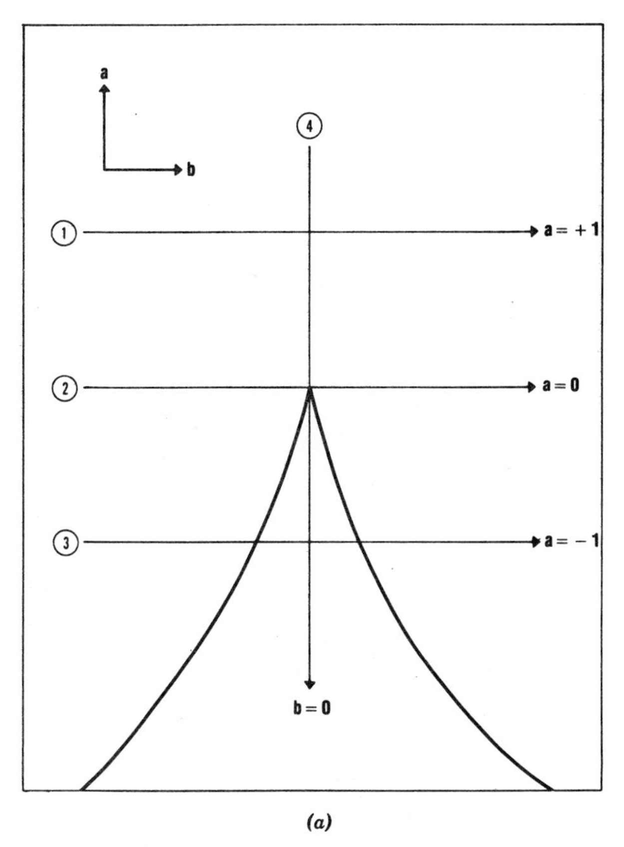

(a)

Figure 9.8 $\langle x \rangle$ and $\langle \Delta x^2 \rangle$ are shown as a function of distance along each of the four paths in the control parameter plane $\mathbb{R}^2$ of the cusp catastrophe. The expectation values are taken with respect to the time-independent probability distribution (9.20).

Since D is assumed small ($D \ll 1$), $\sqrt{4D} \gg D$ and there is an anomalous variance in the neighborhood of $(a, b) = (0, 0)$.

Path 3 ($a = -1$, *b increasing*). For b outside the cusp-shaped region there is only one minimum and, for D small, $\langle x \rangle = x_c$ to a good approximation. For b inside the critical region V has three critical values, but only the deeper of the two minima is of any importance when $|b| \gtrsim 10D$. As b passes through the value zero, the deeper minimum passes from the positive root of $dV/dx = 0$ to the negative root. There is a fast transition of $\langle x \rangle$ from positive to negative values, in accordance with the discussion of Case 2 above. This fast transition is associated with a very large anomaly in the variance.

Path 4 (*a decreasing*, $b = 0$). In this case yet another situation occurs. For $a > 0$, V has a minimum at $x = 0$, and we find $\langle x \rangle = 0$, $\langle x^2 \rangle = D/2a$. As

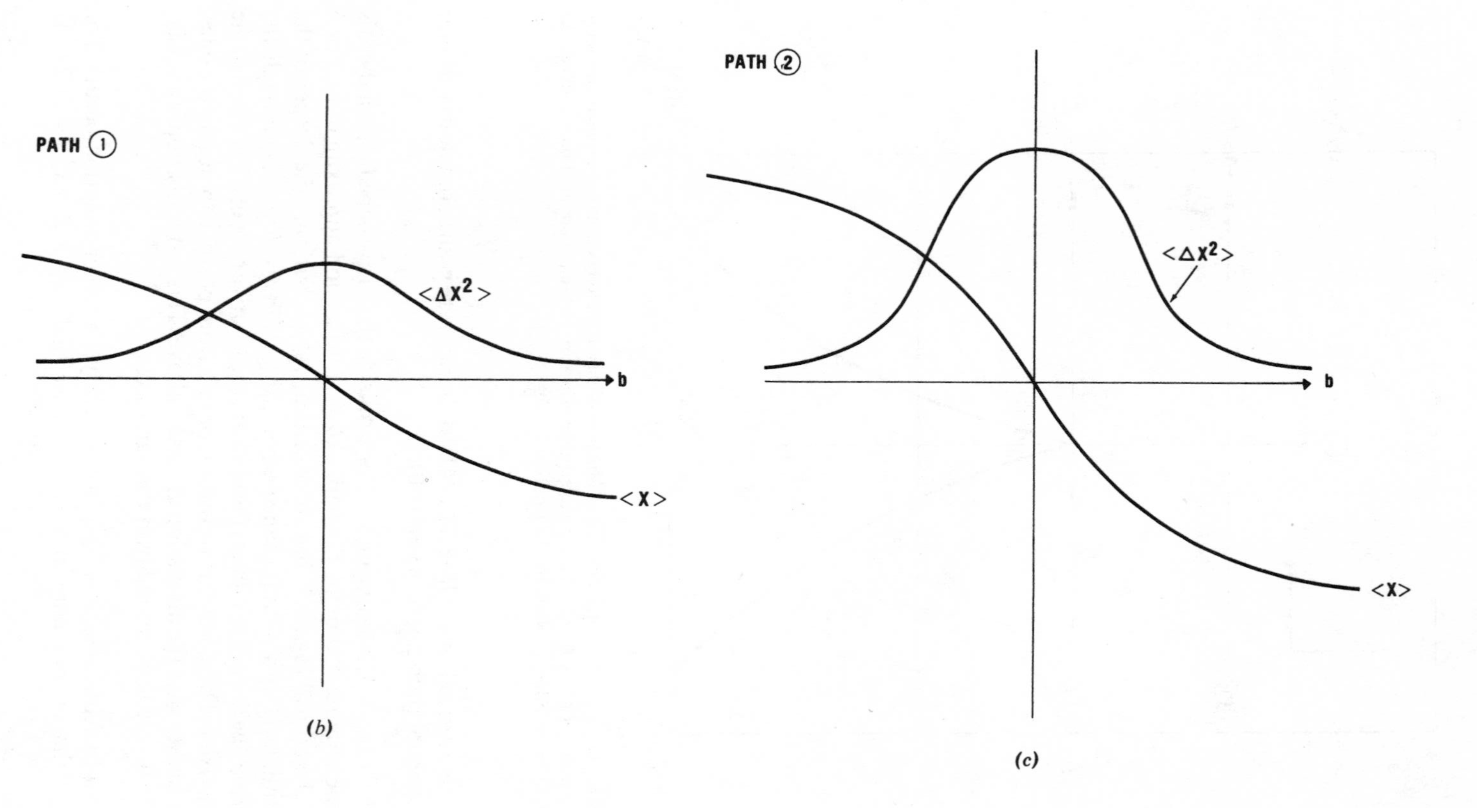

Figure 9.8 (*Continued*)

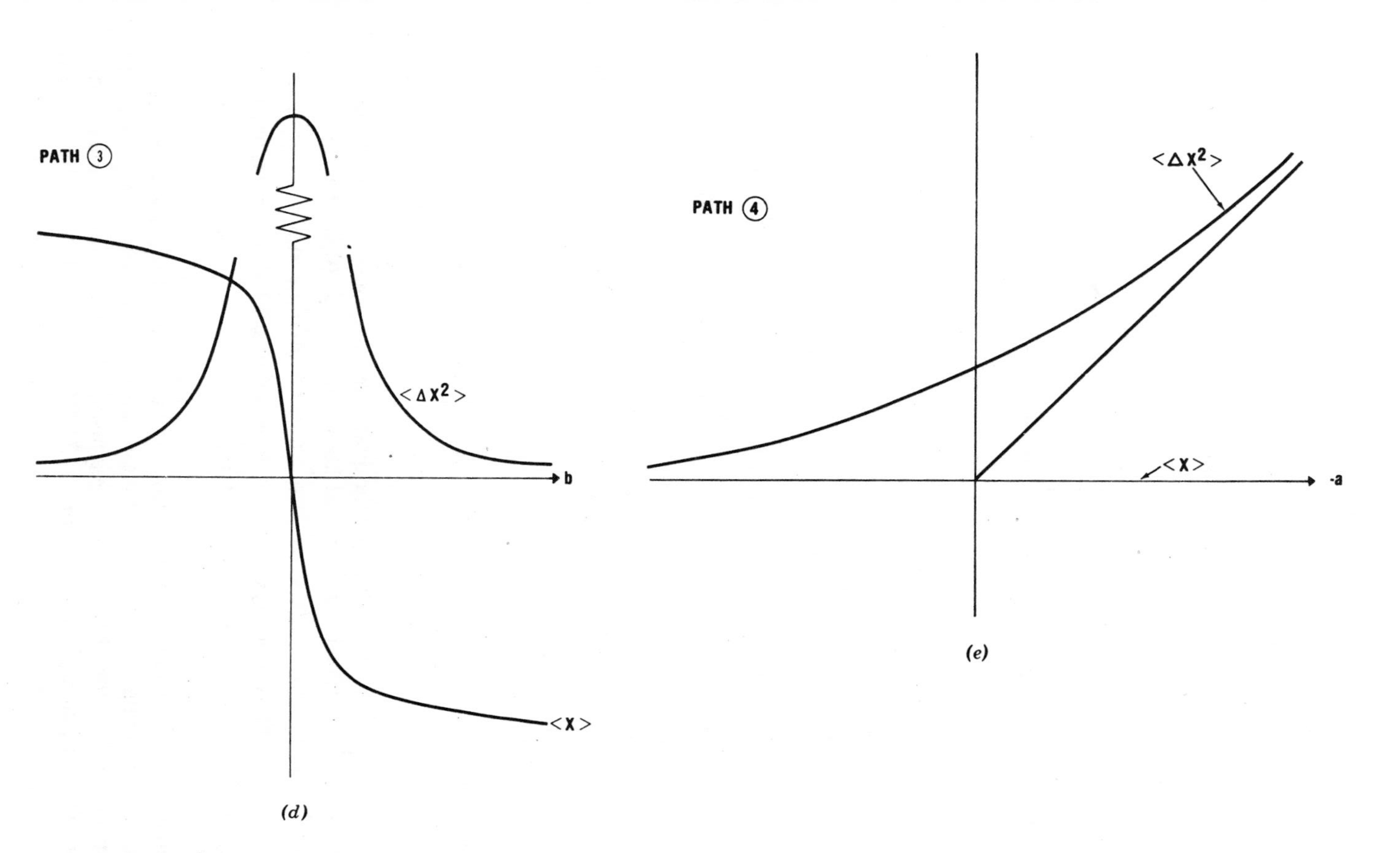

Figure 9.8 (*Continued*)

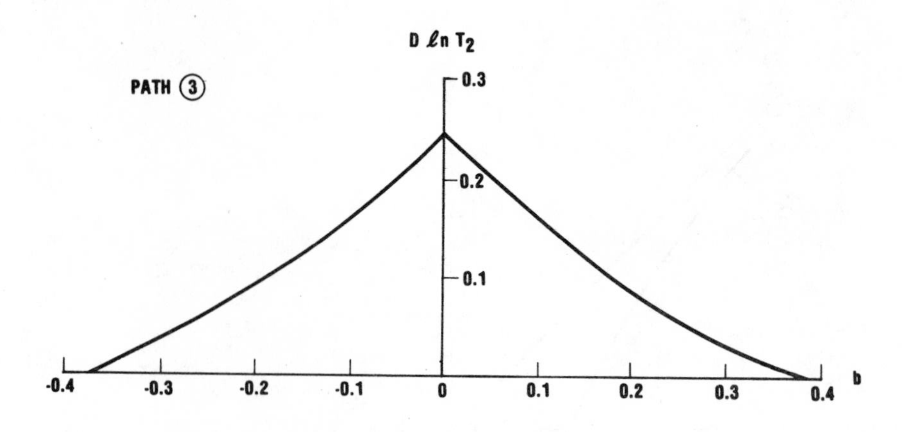

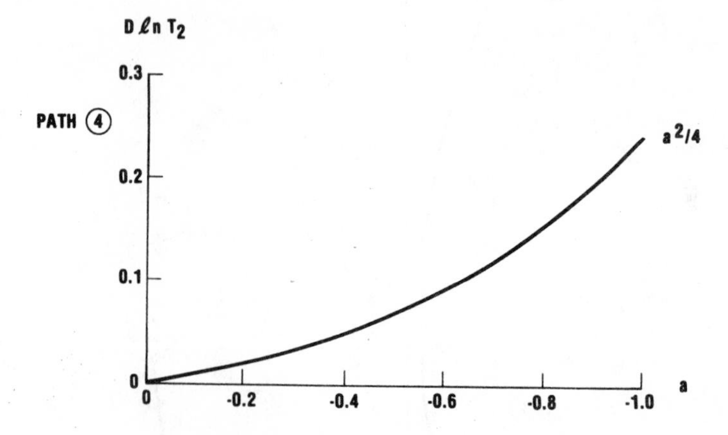

Figure 9.9 The tunneling time scale T_2 is shown for the bimodal paths 3 and 4 of Fig. 9.8.

$a \to 0^+$, the variance increases, and the quartic term becomes more important in truncating the variance, as described above. At $(a, b) = (0, 0)$ the variance is given by (9.21). For $a < 0$, V has two symmetric minima at $x = -\sqrt{-a}$ and $x = +\sqrt{-a}$. By symmetry, $\langle x \rangle = 0$ and the variance is given approximately by $(2\sqrt{-a}/2)^2 = |a|$, in accordance with (9.19).

Remark 1. Throughout this section we have ignored the question of time scale by assuming the asymptotic time-independent form for the probability distribution function. This may not be a justifiable assumption, so the description of the variances given in Cases 1 and 2 by (9.16) and (9.19) must be modified accordingly. The variance (9.16) will be observed on a time scale T_1 and the variance (9.19) will be observed on a time scale T_2 (8.31) if the system probability distribution function obeys an equation of diffusion type. The time scale T_1 is shown in Fig. 9.6 for Paths 1, 2, and 3. The time scale T_2 for equilibration between two isolated minima exists only in the case of Paths 3 and 4, and is shown for these paths in Fig. 9.9. The smaller anomalous variance (9.16) occurs on the shorter time scale T_1, the larger (9.19) on the longer time scale T_2. In many instances it happens that the time scale on which an experiment is done lies between T_1 and T_2. In this case the variance associated with any single experiment is of the type (9.16). To obtain large anomalous variances of type (9.19) it is necessary to average over many experiments. We illustrate by considering the cyclic path ($a = -1, b(t) = -b_0 \cos \omega t$) in control parameter space (Fig. 9.10*a*). If $T_1^{-1} \gg \omega \gg T_2^{-1}$, then the system follows a hysteresis cycle, as shown in Figs. 9.4 and 9.10*b*. The mean value $\bar{x} = \langle x \rangle$ and variance

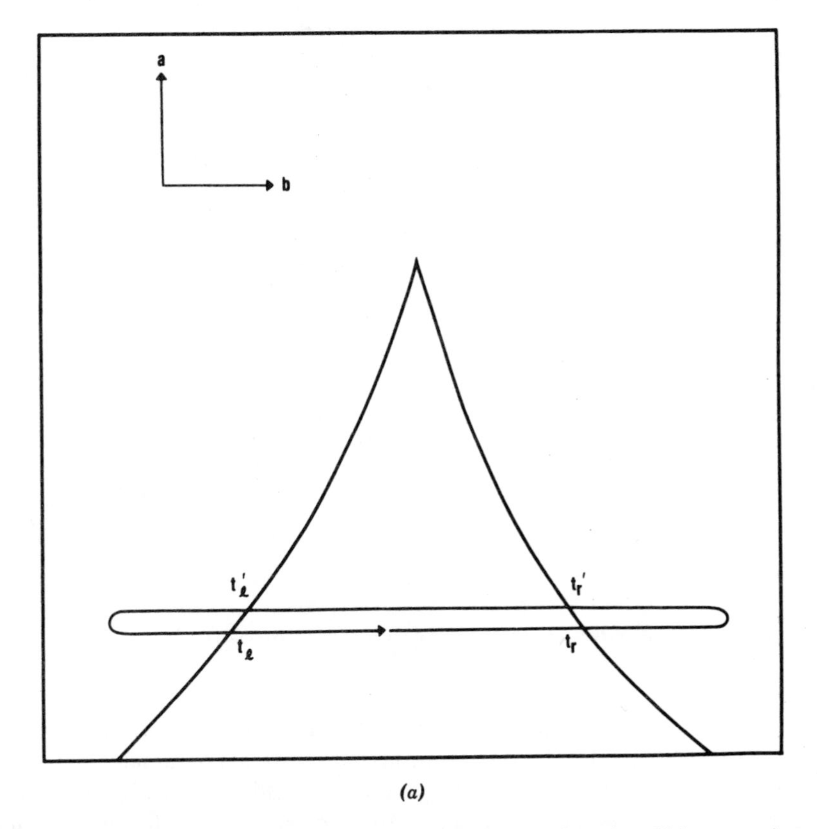

(*a*)

Figure 9.10 Behavior of mean value and variance on the time scale under which an experiment is carried out, and the methods used for data processing are illustrated in this sequence. (*a*) A cyclic path is followed in the control parameter space of the cusp catastrophe. The experimental time scale ω^{-1} obeys $T_2 \gg \omega^{-1} \gg T_1$.

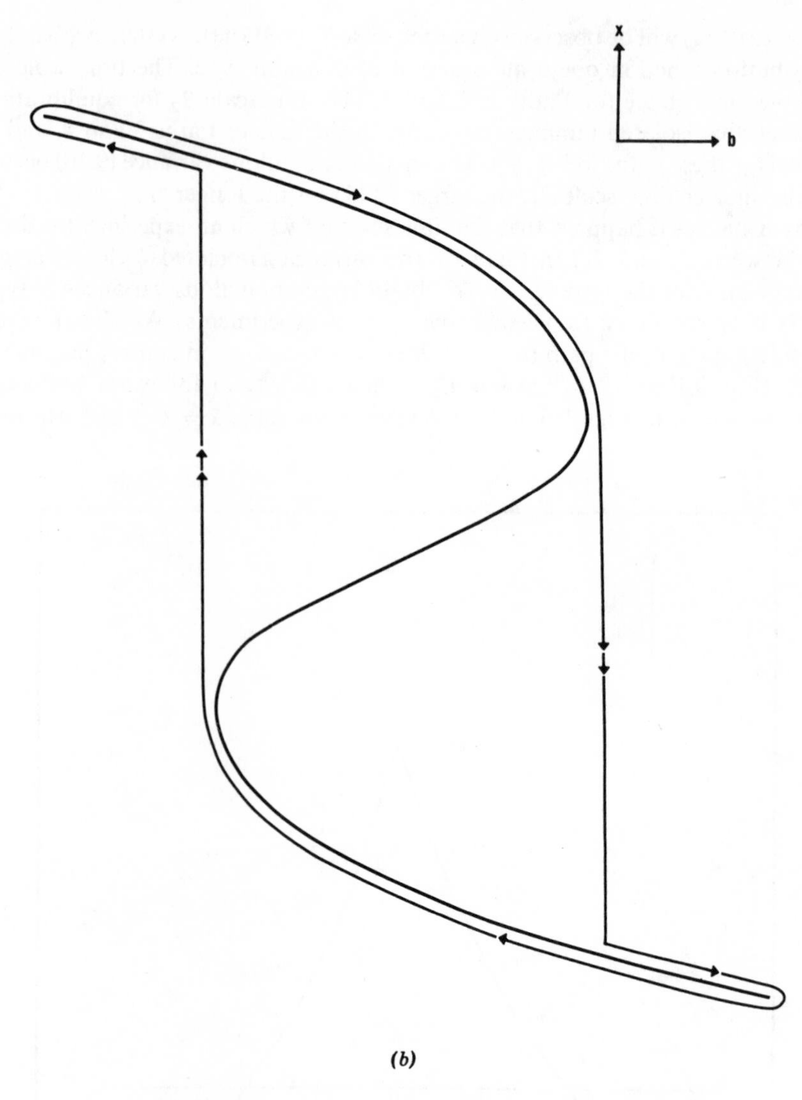

(b)

Figure 9.10 (*b*) The system exhibits a hysteresis cycle.

$\langle (x - \bar{x})^2 \rangle$ are shown as functions of time over the course of one cycle $\tau = 2\pi/\omega$ in Fig. 9.10*c*. We can plot these as functions of the control parameter *b*, provided we distinguish between paths with *b* increasing and those with *b* decreasing (Fig. 9.10*d*). Anomalous variance of type (9.16) occurs only at the fold line and is associated with time scale T_1. However, if we compute the mean and variance by averaging over all experiments, where each single experiment is a time interval of length π/ω with $\dot{b} \geq 0$ or $\dot{b} \leq 0$, then we compute the anomalous variance

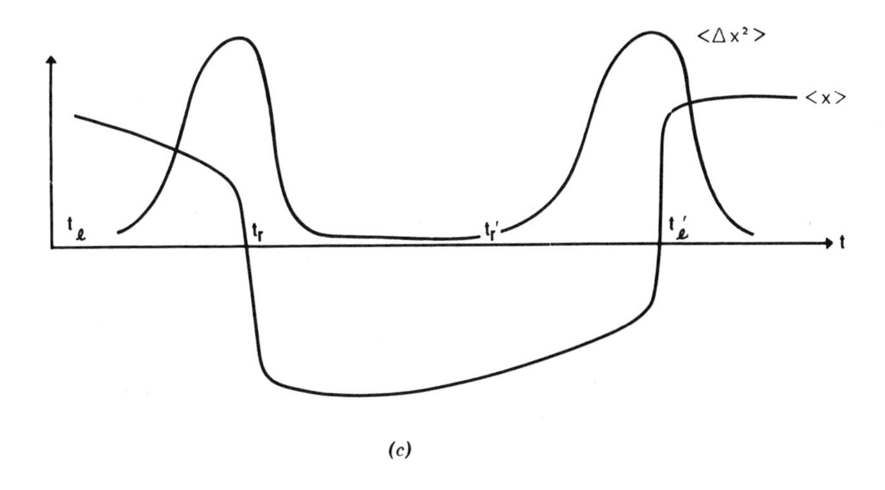

(c)

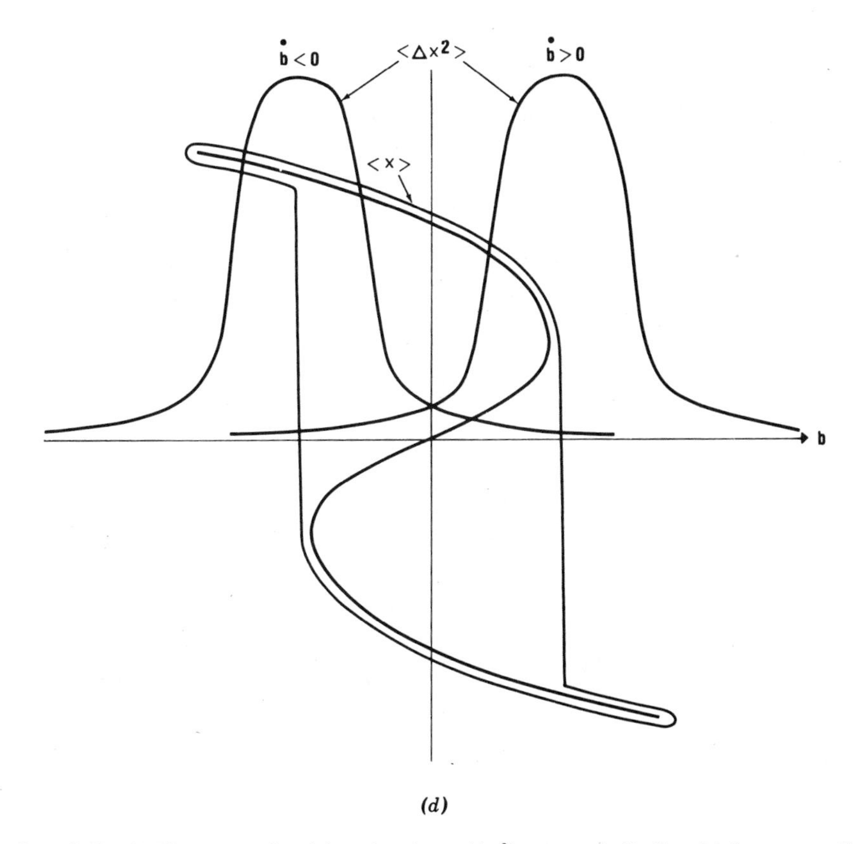

(d)

Figure 9.10 (*c*) The mean value $\langle x \rangle$ and variance $\langle \Delta x^2 \rangle$ vary periodically, with frequency $\omega/2\pi$. (*d*) If the data are segregated according to whether b is increasing or decreasing, the anomalous variance has the form shown. For b increasing, the variance is anomalously large near the right-hand fold line. When b is decreasing, the variance becomes anomalously large near the left-hand component of the bifurcation set.

associated with time scale T_2, which occurs at the bifurcation set associated with the Maxwell Convention. This is shown in Fig. 9.10*e*.

Remark 2. Distribution functions exhibit the phenomenon of critical slowing down as the bifurcation set under the Delay Convention is approached. This can be seen most easily from (8.9).

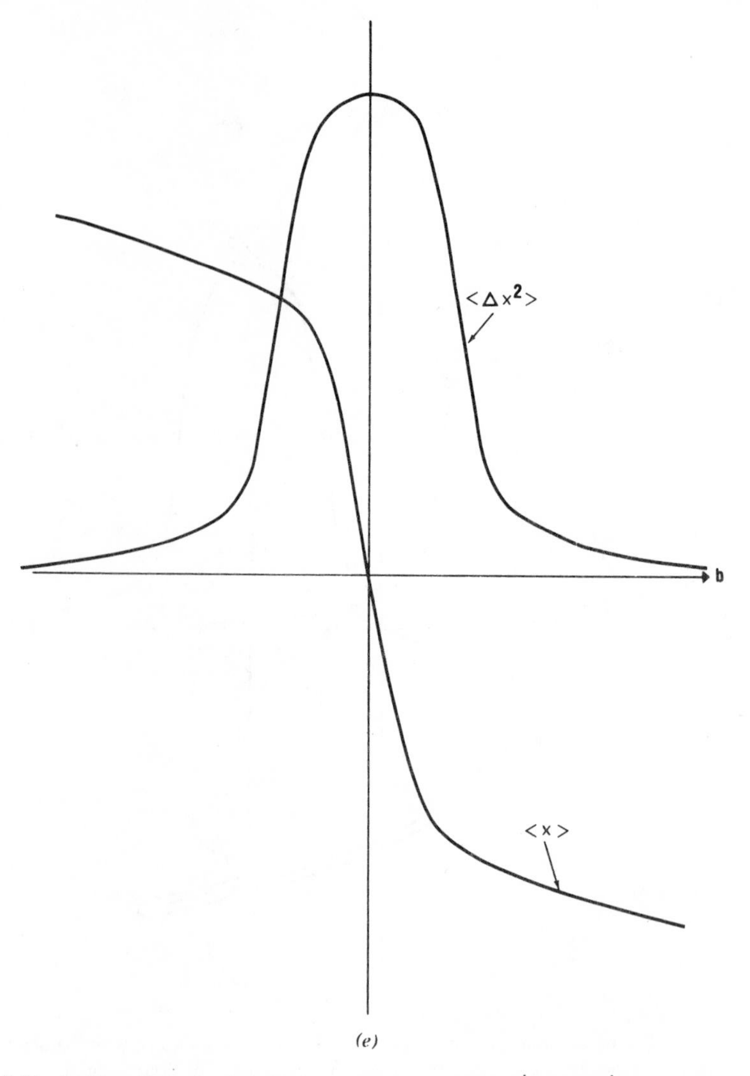

Figure 9.10 (*e*) If no distinction is made between data taken with $\dot{b} > 0$ and $\dot{b} < 0$, $\langle x \rangle$ is no longer bimodal, and the variance becomes huge around the Maxwell set. This excessive variance swamps out even the anomalously large variance associated with the bifurcation set.

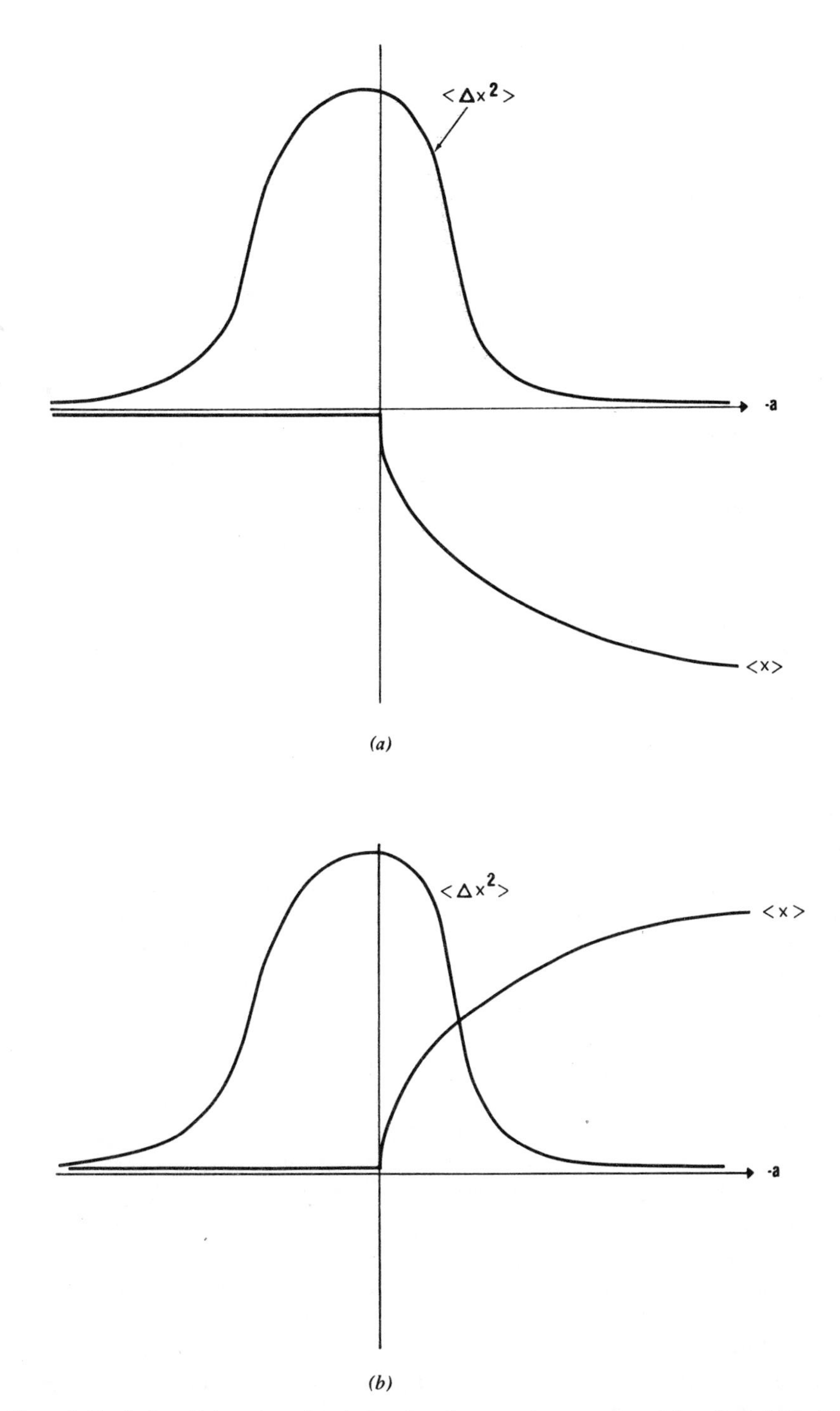

Figure 9.11 In usual laboratory situations, when the controls are made to follow Path 4 (Fig. 9.8), the physical system settles into either the left-hand minimum or the right-hand minimum for $a \lesssim -1$. The mean and variance in these two different minima are shown.

Remark 3. When Path 4 (Fig. 9.8) is followed in a laboratory experiment, the following usually occurs. As a becomes negative, the system chooses either the left-hand minimum or the right-hand minimum. The associated probability distribution function is then $P_l(x) \simeq Ne^{-V_l(x)/D}$ or $P_r(x) \simeq Ne^{-V_r(x)/D}$, where $V_l(x) = |a|(x + \sqrt{-a})^2$ and $V_r(x) = |a|(x - \sqrt{-a})^2$. The corresponding mean and variance are shown in Fig. 9.11. Once one of the two minima has been chosen, the tunneling time T_2 to the other minimum is generally too long to occur on a laboratory time scale. If a number of different experiments are averaged over, then the mean and variance are as given in Fig. 9.8*e*. This phenomenon occurs, for example, when a sample of iron is cooled below its Curie temperature. Among repeated experiments, the magnetization is "up" about as often as it is "down." The variance around the average "up" value is small, as is the case for the "down" direction. If all experiments are averaged over, the average magnetization is zero and the variance huge. The lack of precision in the magnetization direction is itself an example of the phenomenon of divergence (Sec. 4).

Remark 4. In the field of classical thermodynamics, anomalous variance associated with a cusp germ is called **critical opalescence**.

9. Summary

Recognizing the presence and type of catastrophe is often an important input for the proper description of a physical system. This input can be used either to save work (since catastrophes are canonical) at one end of the spectrum, or to provide qualitative and even quantitative information when not even a mathematical description (i.e., equations) exists at the other end of the spectrum. We have discussed a number of features which may occur in physical systems to suggest the presence of a catastrophe.

The first five (modality, inaccessibility, sudden jumps, divergence, and hysteresis) generally occur in conjunction with each other. These depend on the accessibility to the physical system of a region in control parameter space in which the potential has more than one local minimum. Hysteresis may not be observed if the Maxwell Convention is obeyed, but even in this case it is sometimes possible, by careful experimental technique, to observe it (supercooling, superheating).

The remaining three catastrophe flags (divergence of linear response, critical slowing down/mode softening, and anomalous variance) may be observed even when the potential has only one local minimum. This is an important property of these three catastrophe flags, and can be used to establish both the critical values and safety margins on the control parameters in the numerous cases in which a sudden "catastrophe jump" would be a disaster.

Notes

Catastrophe fingerprints were first systematically discussed by Zeeman.[1]

References

1. E. C. Zeeman, Catastrophe Theory, *Sci. American* **234** (4), 65–83 (1976). Published in original form in: E. C. Zeeman, *Catastrophe Theory, Selected Papers 1972–1977*, Reading: Addison-Wesley, 1977, p. 18.

PART 2

Applications of Elementary Catastrophe Theory

In which we illustrate how Elementary Catastrophe Theory has been and can be applied in several scientific and engineering disciplines.

CHAPTER 10

Thermodynamics

Our first series of applications of Catastrophe Theory to problems of scientific interest is in the area of thermodynamics. We discuss three distinct problems in this chapter, which is organized like a sandwich. The three meaty problems (Secs. 2–6, Secs. 7–10, and Secs. 11–15) are surrounded by two lighter weight sections. In Sec. 1 we review the Ehrenfest classification of phase transitions as applied to some elementary catastrophe functions. In Sec. 16 we summarize a number of questions which are raised by looking at thermodynamics from a Catastrophe Theory point of view.

In Secs. 2–6 we relate the critical point of a fluid to the germ of a cusp catastrophe, at first qualitatively (Secs. 2, 4) and then quantitatively (Sec. 5). Starting from the cusp catastrophe manifold and making simplest possible identifications between the mathematical and physical state variables and control parameters, it is possible to derive the van der Waals equation of state by making four more or less reasonable assumptions. Unfortunately the van der Waals equation becomes inaccurate in the neighborhood of the critical point (Sec. 6).

If it is possible to go from the A_{+3} catastrophe to the equation of state for a substance in the neighborhood of its critical point, it should also be possible to go from the A_{+5} catastrophe to the equation of state for a substance in the neighborhood of its triple point. The details of this identification are given in Secs. 7–10.

The third of the three major topics in this chapter is not so much an application of Catastrophe Theory as a reformulation of classical thermodynamics in the spirit of Catastrophe Theory. In this reformulation we consider the equation of state of a substance as an n-dimensional manifold in the $2n$-dimensional space $\mathbb{R}^n \otimes \mathbb{R}^n$ of n intensive and n extensive thermodynamic variables. The equation of state manifold is identified with the critical manifold of a family of potentials $\mathscr{U}$ depending on n control parameters (extensive thermodynamic variables) and n state variables (intensive thermodynamic variables). This viewpoint leads to a very geometric interpretation of equilibrium thermodynamics: the critical manifold is a Riemannian manifold. The computation of thermodynamic partial derivatives is greatly simplified by studying susceptibility tensors on the n-dimensional tangent space to the critical manifold in $\mathbb{R}^n \otimes \mathbb{R}^n$. This formulation, in addition, invites extension from equilibrium thermodynamics to non-equilibrium thermodynamics, at least in the neighborhood of the critical manifold (Sec. 16).

1. General Description of Phase Transitions

The classical theory of phase transitions fits naturally within the framework of Elementary Catastrophe Theory. We present such a formulation in this section.

To describe a physical system, we introduce a general family of potentials $V(x; c)$ depending on n state variables or order parameters $x \in \mathbb{R}^n$ and k control parameters $c \in \mathbb{R}^k$. We further assume that the state of the physical system is described by the value of x which minimizes the potential, at least locally. The study of such a physical system then reduces to a study of the equilibrium and stability properties of the potentials $V(x; c)$:

$$\begin{aligned} \frac{\partial V}{\partial x_i} &= 0 \qquad \text{equilibrium} \\ \frac{\partial^2 V}{\partial x_i \, \partial x_j} &> 0 \qquad \text{local stability} \end{aligned} \tag{10.1}$$

as well as the critical values on the stable equilibrium branches.

In general, for almost all $c \in \mathbb{R}^k$ the potential $V(x; c)$ will have only isolated critical points $x^{(1)}, \ldots, x^{(p)}, \ldots$. The first of equations (10.1), which consists of n equations of state, can be used to determine both the locations of the isolated critical points as well as their dependence on the control parameters c [cf. (5.2)]:

$$x^{(p)} = x^{(p)}(c), \qquad p = 1, \ldots \tag{10.2}$$

The local stability of the pth critical point is determined by evaluating the stability matrix at $(x^{(p)}(c), c)$.

A phase transition occurs when the point $x \in \mathbb{R}^n$ describing the state of a physical system jumps from one critical branch to another.

Phase transitions can occur when the control parameters are varied. Usually the control parameters are assumed or constrained to depend on only a single parameter (e.g., time), and this parameter is used to describe a curve in control parameter space:

$$\begin{aligned} c_\alpha &\to c_\alpha(s), \qquad s \in \mathbb{R}^1 \\ x^{(p)}(c_\alpha) &\to x^{(p)}(c_\alpha(s)) \to x^{(p)}(s) \\ V^{(p)}(x^p(c); c) &\to V^{(p)}(x_i^{(p)}(s); c_\alpha(s)) \to V^{(p)}(s) \end{aligned} \tag{10.3}$$

A phase transition will occur when the curve $c_\alpha(s) \in \mathbb{R}^k$ crosses an appropriate component of the appropriate separatrix in $\mathbb{R}^k$. If the Delay Convention is adopted, phase transitions will occur when the curve passes through the component of the bifurcation set $\mathscr{S}_B$ on which local minima are created and destroyed. If the Maxwell Convention is adopted, phase transitions will occur when the curve passes through the component of the Maxwell set $\mathscr{S}_M$ on which two or more global minima are degenerate.

It is often convenient to classify phase transitions by their order[1] (Ehrenfest classification). Assume that a phase transition occurs because the system state jumps from the pth critical branch to the qth critical branch at $c_\alpha(s^0)$ when s increases through s^0. The transition is of **order m** if

$$\lim_{\varepsilon \to 0} \frac{d^i}{ds^i} V^{(p)}(s)\bigg|_{s^0 - \varepsilon} = \lim_{\varepsilon \to 0} \frac{d^i}{ds^i} V^{(q)}(s)\bigg|_{s^0 + \varepsilon}, \qquad i = 0, 1, \ldots, m - 1 \tag{10.4}$$

but the equality is not satisfied for $i = m$. The phase transition is also termed **local** or **soft** if

$$\lim_{\varepsilon \to 0} [x^{(p)}(s^0 - \varepsilon) - x^{(q)}(s^0 + \varepsilon)] = 0 \tag{10.5}$$

Otherwise it is called **nonlocal** or **hard**. Phase transitions encountered in Nature are usually zeroth, first, or second order, or else do not conform to the Ehrenfest classification scheme.

2 Ginzburg–Landau Second-Order Phase Transitions

An extremely useful mathematical model for describing second-order phase transitions is the Ginzburg–Landau model.[1] Reduced to its essence, the model is as follows. An order parameter x is introduced to describe the state of a physical system. The total energy of the system is expressed as the sum of a kinetic energy term ($\sim(dx/dt)^2$) and a potential energy term $V(x)$. This potential depends on external control parameters as well as the state variable x. The system ground state is determined by minimizing the sum of these two terms. In the static case ($dx/dt = 0$) the system state is determined by the value of x which minimizes $V(x)$. Often some symmetry principle is present or is invoked which restricts the form of the potential to be an even function of x. The Ginzburg–Landau potential has the form

$$V(x; a) = \tfrac{1}{4}x^4 + \tfrac{1}{2}ax^2 \tag{10.6a}$$

This potential has a unique minimum at $x = 0$ when $a \geq 0$, and two equally deep minima at $x = \pm\sqrt{-a}$ when $a < 0$ (Fig. 10.1). A second-order phase transition occurs as a goes through zero.

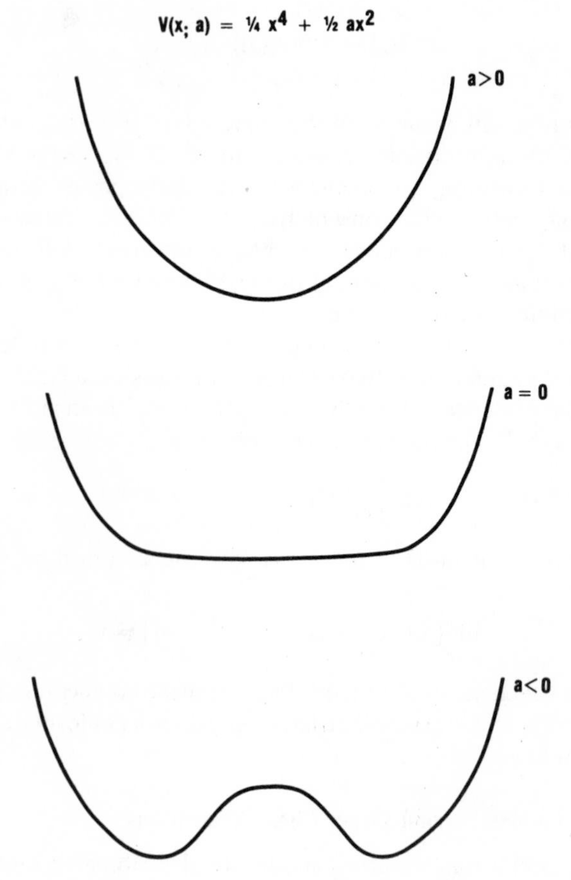

Figure 10.1 The potential $V(x; a)$ has a unique minimum at $x = 0$ when $a \geq 0$ and two equally deep minima at $x = \pm(-a)^{1/2}$ when $a \leq 0$.

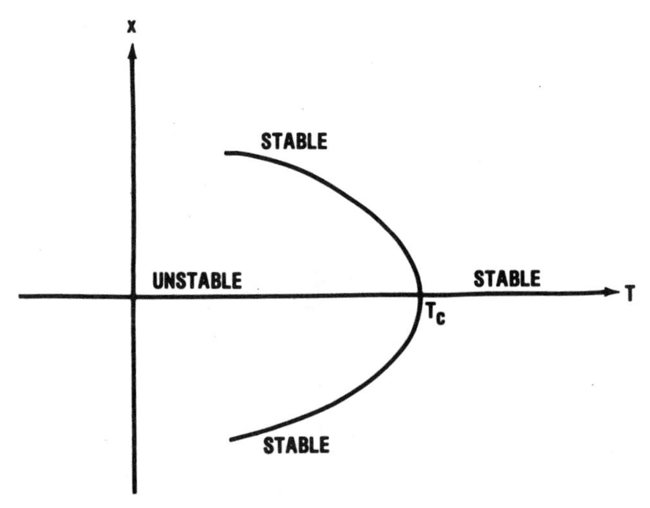

Figure 10.2 The cross section of the cusp catastrophe manifold by the symmetry plane $b = 0$ has the form of a trident. Below the critical temperature T_c two stable states exist. Above there is only one.

The Ginzburg–Landau phase transition is often used to describe second-order thermodynamic phase transitions. In this case the control parameter a is identified with $T - T_c$ in the neighborhood of the critical temperature T_c. The bifurcation diagram for the Ginzburg–Landau description of a second-order thermodynamic phase transition is the standard trident, shown in Fig. 10.2.

Experimentally, second-order phase transitions are subject to some "peculiarities" of behavior which are not anticipated from the simple model (10.6a). For example, a second-order phase transition will simply disappear under an arbitrarily small symmetry-breaking perturbation. The phase transition may reappear at a far-distant point as either a zeroth- or a first-order phase transition. This behavior is related to the structural instability of second-order phase transitions (cf. Remark 5 below).

To give a complete discussion of the Ginzburg–Landau model (10.6a), including the effects of symmetry-breaking terms, it is necessary to study the most general perturbation of this potential. This is the cusp catastrophe

$$V(x; a, b) = \tfrac{1}{4}x^4 + \tfrac{1}{2}ax^2 + bx \tag{10.6b}$$

We now illustrate the general concepts described in Sec. 1 in a series of three examples involving the cusp catastrophe.

Example 0. We adopt the Delay Convention and follow the curve 0 in Fig. 10.3. The separatrix of importance is the bifurcation set $\mathscr{S}_B$, consisting of the two fold lines. A phase transition occurs at the point marked with an $\times$. At this point the system state variable jumps from the disappearing right-hand minimum

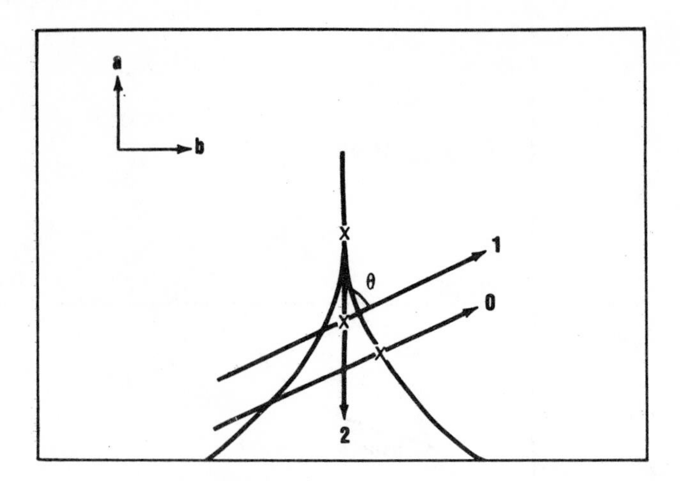

Figure 10.3 The order of a phase transition depends on the convention adopted and the path followed. A zeroth-order phase transition occurs on path 0 when the Delay Convention is adopted. A first-order phase transition occurs on path 1 when the Maxwell Convention is adopted. A second-order transition occurs on path 2. In each case, an × marks the location of the appropriate discontinuity.

$x^{(r)} = \sqrt{-a/3}$ to the global left-hand minimum at $x^{(l)} = -2\sqrt{-a/3}$. The critical values on the left- and right-hand branches are

$$V^{(r)} = \frac{a^2}{12} \qquad V^{(l)} = -\frac{2a^2}{3} \tag{10.7}$$

Since the change in the value of the potential is nonzero at ×,

$$\Delta V|_{\mathscr{S}_B} = V^{(l)} - V^{(r)} = -\tfrac{3}{4}a^2 \tag{10.8}$$

this phase transition is zeroth order. The magnitude of this discontinuity depends only on the point at which the curve 0 crosses the fold line and is independent of path direction at this point.

Example 1. We now adopt the Maxwell Convention and follow curve 1 in Fig. 10.3. The separatrix of importance is now the Maxwell set $\mathscr{S}_M$, consisting of the half-line ($a < 0, b = 0$). A phase transition occurs at the point marked with an ×. At this point the system state variable jumps from the right-hand minimum $x^{(r)} = +\sqrt{-a}$ to the left-hand minimum $x^{(l)} = -\sqrt{-a}$. The potential assumes equal critical values $V^{(r)} = V^{(l)} = -a^2/4$, but there is a discontinuity in the first derivative along curve 1 at the Maxwell set. This discontinuity is computed by evaluating

$$\frac{d}{ds}V(x; a, b) = \frac{\partial V}{\partial x}\frac{dx}{ds} + \frac{\partial V}{\partial a}\frac{da}{ds} + \frac{\partial V}{\partial b}\frac{db}{ds} \tag{10.9}$$

on the Maxwell set. The derivatives da/ds, db/ds may be interpreted as direction cosines, where $db/da = (db/ds)/(da/ds) = \tan\theta$, as shown in Fig. 10.3. Since $\partial V/\partial x = 0$,

$$\begin{aligned} \Delta \frac{dV}{ds}\bigg|_{\mathscr{S}_M} &= (\tfrac{1}{2}x^2 \cos\theta + x \sin\theta)^{(l)} - (\tfrac{1}{2}x^2 \cos\theta + x\sin\theta)^{(r)} \\ &= -2\sqrt{-a}\sin\theta \end{aligned} \tag{10.10}$$

This first-order phase transition has a discontinuity in the first derivative which depends not only on the location of the phase transition point, but also on the direction of the curve as it crosses the Maxwell set.

Example 2. We now follow curve 2 in Fig. 10.3. Along this curve $s = a$ and

$$\begin{aligned} V(s) &= 0, & s &\geq 0 \\ V(s) &= -\frac{s^2}{4}, & s &\leq 0 \end{aligned} \tag{10.11}$$

The phase transition point is marked by $\times$ in Fig. 10.3. There is no discontinuity in either the zeroth or the first derivative at the origin, but there is a discontinuity in the second derivative, for

$$\Delta \frac{d^2V}{ds^2} = \frac{d^2V^{(1)}}{ds^2} - \frac{d^2V^{(0)}}{ds^2} = -\frac{1}{2} \tag{10.12}$$

where the superscript (1) indicates that the potential is evaluated on either of the nonzero solution branches ($s < 0$) and the superscript (0) indicates the zeroth branch ($s > 0$) as $s \to 0^{\pm}$. The phase transition along curve 2 is second order.

Remark 1. The second-order phase transition is local but the zeroth- and first-order phase transitions are nonlocal.

Remark 2. Phase transitions of order greater than 2 cannot be described by typical paths in the control parameter space of the cusp. They can be described by paths in the control parameter space of higher dimensional catastrophes.

Remark 3. Not every phase transition fits neatly into the Ehrenfest classification scheme described above.

Remark 4. Under the Maxwell Convention the bifurcation set of the cusp catastrophe consists of a line ($a < 0, b = 0$) of first-order phase transitions ending in a second-order phase transition at the boundary point ($a = 0, b = 0$). This is a useful catastrophe flag.

Remark 5. The nonintuitive behavior of second-order phase transitions under perturbations, as described at the beginning of this section, has a very natural interpretation within the context of Elementary Catastrophe Theory.

The Ginzburg–Landau potential (10.6a) is the most general 1-parameter family of even functions containing a non-Morse germ. This family can only exhibit second-order phase transitions. Breaking the symmetry allows more general perturbations to occur. The most general perturbation of the Ginzburg–Landau potential (10.6a) is the cusp catastrophe (10.6b). Under an arbitrarily small symmetry-breaking perturbation the path in control parameter space will miss the cusp point $(a, b) = (0, 0)$, so the second-order phase transition will be destroyed. However, if the Maxwell Convention is obeyed and the perturbed path crosses the Maxwell set ($a < 0, b = 0$) at a point generally not near $a = 0$, a first-order phase transition will occur. If the Delay Convention is observed and the path crosses the appropriate fold curve [again, generally far from ($a = 0$, $b = 0$)] a zeroth-order phase transition will occur.

3. Topological Remarks

The critical manifold $\nabla V = 0$ may have several critical points $x^{(p)}$, $p = 1, 2, \ldots$, over a fixed point c^0 in control parameter space (Fig. 10.4). Since the critical

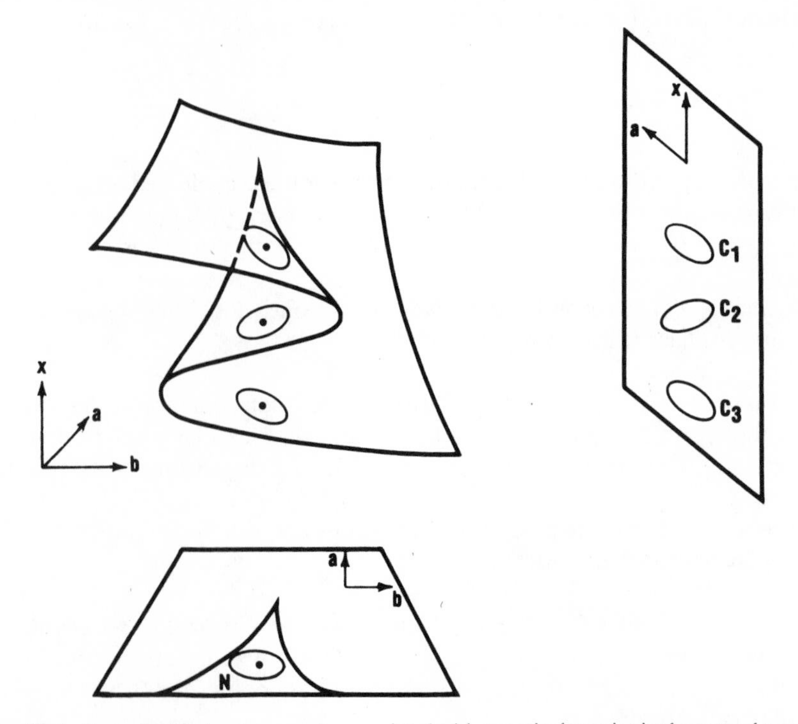

Figure 10.4 Several critical points may be associated with a particular point in the control parameter space $\mathbb{R}^k$. The neighborhoods of these isolated critical points in the k-dimensional critical manifold may be parameterized by a neighborhood $N \subset \mathbb{R}^k$ of the control point. They may also be parameterized by neighborhoods $C_1, C_2, C_3, \ldots$ in some k-dimensional subspace of $\mathbb{R}^n \otimes \mathbb{R}^k$.

manifold is a k-dimensional manifold, it is convenient but by no means necessary to use the k control parameters in order to parameterize the k-dimensional neighborhood of each critical point. This parameterization fails as a critical point approaches the bifurcation set.

Alternative parameterizations are possible, including those using one or more of the state variables. This idea is illustrated in Fig. 10.4. There are three distinct critical points over the point (a^0, b^0) in the control plane. Neighborhoods of these three isolated critical points may be coordinatized by a neighborhood of (a^0, b^0) in the control plane. An alternative, and for some purposes more convenient, parameterization is given in terms of the state variable x and one of the two control parameters. We have shown a parameterization using the pair $(x; a)$ in Fig. 10.4. This parameterization is convenient because there is a global 1–1 correspondence between points in the critical manifold and points in the x–a plane.

In applications where the minimum value of the potential determines the state of a system, the stable critical manifold depends on the convention adopted. In Fig. 10.5*a* we show the manifold of stable and metastable minima, used when the Delay Convention is adopted. This is a "surgered manifold" because the portion of the critical manifold describing critical points which are not local minima has been removed. The boundary of this manifold is the curve $(x; a, b) = (\lambda; -3\lambda^2, 2\lambda^3)$ which projects down onto the fold lines in the control plane, and to the parabola $(x; a) = (\lambda; -3\lambda^2)$ in the x–a plane.

When the Maxwell Convention is adopted, the critical manifold is a "soldered manifold" because the portion of the manifold $\nabla V = 0$ describing unstable critical points and metastable minima has been removed and replaced by a flat piece which interpolates between the two minima. The physical interpretation for a point on the flat part of the critical manifold is as follows. If $x^{(l)}$ and $x^{(r)}$ are the locations of the two equally deep minima for $a < 0$, $b = 0$, then x $(x^{(l)} \leq x \leq x^{(r)})$ describes the fraction of the system in the state corresponding to the minimum at $x^{(l)}$ (say, liquid state) and the fraction in the state corresponding to the minimum at $x^{(r)}$:

$$\frac{x - x^{(l)}}{x^{(r)} - x^{(l)}} = \text{fraction of system in state corresponding to minimum at } x^{(r)}$$

$$\frac{x^{(r)} - x}{x^{(r)} - x^{(l)}} = \text{fraction of system in state corresponding to minimum at } x^{(l)} \tag{10.13}$$

Thus the state variable x interpolates between the two pure states $x = x^{(l)}$ and $x = x^{(r)}$ which can exist at $a < 0$, $b = 0$. For $x^{(l)} < x < x^{(r)}$ the two states can coexist. It is clear that the control parameters a, b fail to provide a coordinate system for neighborhoods of points in the flat part of the critical manifold. In this case it becomes necessary to use a mixture of state variables and control parameters to uniquely parameterize the critical manifold (cf. Fig. 10.5*b*). The parabola

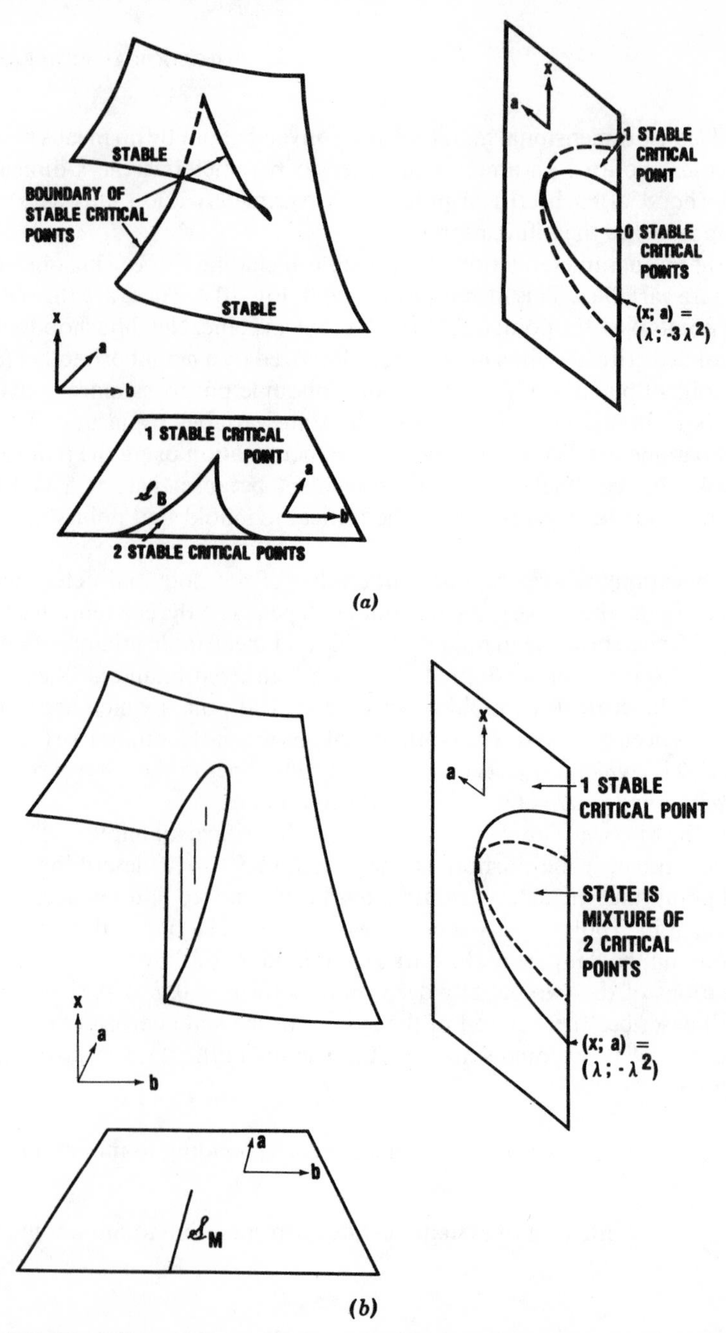

Figure 10.5 The subset of the critical manifold representing stable states is not technically a manifold under either the Delay Convention or the Maxwell Convention. (*a*) Under the Delay Convention the unstable surface of the cusp catastrophe manifold is removed. The boundary projects down onto the cusp-shaped fold curve in the (a, b) plane and onto the parabola $a = -3x^2$ in the $(x; a)$ plane. (*b*) Under the Maxwell Convention the equation of state surface has the form shown. The flat portion of this surface projects down onto the Maxwell set $\mathscr{S}_M$ in the (a, b) control plane, while the boundary projects onto the parabola $a = -x^2$ in the $(x; a)$ plane.

$a = -x^2$ in the $(x; a)$ parameterizing space is a boundary for pure phases. Outside this parabola the system is in a pure phase. Inside this parabola the system coexists in a mixture of two phases, as just described.

The Maxwell set in the control plane is the half-line ($a < 0, b = 0$). In general the Maxwell set $\mathscr{S}_M$ is determined by the Clausius–Clapeyron equations (5.68), and consists of manifolds of various dimensions in $\mathbb{R}^k$. When the critical manifold is uniquely parameterized by a mixture of state variables and control parameters, as in Fig. 10.5*b*, the Maxwell set in this space is empty, and submanifolds such as the parabola in Fig. 10.5*b* bound the regions in the parameterizing space in which various phases can coexist simultaneously.

The linear response of a system to changes in the value of the control parameters has been analyzed in (5.2). However, the infinitesimal linear response theory summarized in that equation can be visualized topologically by constructing the tangent space to the point $(x^0; c^0)$ in the k-dimensional critical manifold in $\mathbb{R}^n \otimes \mathbb{R}^k$. The tangent space is a k-dimensional linear vector space. If we define $\mathbf{e}_1, \ldots, \mathbf{e}_n$ to be basis vectors for $\mathbb{R}^n$ at the origin in $\mathbb{R}^n$, and $\mathbf{f}_1, \ldots, \mathbf{f}_k$ to be the basis vectors in $\mathbb{R}^k$ at the origin, then basis vectors in the tangent space can be determined from the linear response function computed in (5.2). The linear response function is a susceptibility tensor, so (5.2) may be written

$$\delta x^i = \chi^i_\alpha \delta c^\alpha, \qquad \chi^i_\alpha = -(V^{-1})^{ij} V_{j\alpha} \tag{10.14}$$

The susceptibility tensor χ^i_α is defined on the critical manifold.

To determine a set of k basis vectors which span the tangent space, we move the basis vectors $(\mathbf{e}_1, \ldots, \mathbf{e}_n)$ and $(\mathbf{f}_1, \ldots, \mathbf{f}_k)$ to the point $(x^0; c^0)$ in $\mathbb{R}^n \otimes \mathbb{R}^k$ and form a general linear combination of these basis vectors which lies in the tangent space. Since the infinitesimal vector δv defined by

$$\delta v = \delta x^i \mathbf{e}_i + \delta c^\alpha \mathbf{f}_\alpha \tag{10.15}$$

lies in this space when δx^i and δc^α are related by (10.14), we can choose as k linearly independent basis vectors the set

$$| f_\alpha \rangle = \mathbf{e}_i \chi^i_\alpha + \mathbf{f}_\alpha, \qquad \alpha = 1, 2, \ldots, k \tag{10.16}$$

This choice is particularly convenient if we choose the control parameters $c \in \mathbb{R}^k$ to parameterize neighborhoods of points in the critical manifold. If a mixture of state variables and control parameters is used, other choices of basis vectors may be more convenient.

In order to find a complete set of basis vectors for the space $\mathbb{R}^n \otimes \mathbb{R}^k$ at the point $(x^0; c^0)$, it is necessary to find n additional vectors which are linearly independent of the vectors $| f_\alpha \rangle$ defined in (10.16). This is particularly easy to do if an inner product, or some notion of orthogonality, is available. The only available real symmetric second-order tensor is the matrix of mixed second partial derivatives of V. If we define an inner product at $(x^0; c^0)$ by

$$\begin{aligned} (\mathbf{e}_i, \mathbf{e}_j) &= V_{ij} \\ (\mathbf{e}_i, \mathbf{f}_\beta) &= V_{i\beta} \\ (\mathbf{f}_\alpha, \mathbf{f}_\beta) &= V_{\alpha\beta} \end{aligned} \tag{10.17}$$

then the n vectors $|e_i\rangle = \mathbf{e}_i$, $i = 1, 2, \ldots, n$, are orthogonal to the k vectors $|f_\alpha\rangle$ for

$$\begin{aligned}\langle e_i | f_\alpha \rangle &= (\mathbf{e}_i, \mathbf{f}_\alpha + \mathbf{e}_j(-V^{-1})^{jm} V_{m\alpha}) \\ &= V_{i\alpha} - V_{ij}(V^{-1})^{jm} V_{m\alpha} = 0\end{aligned} \tag{10.18}$$

In short, the k vectors $|f_\alpha\rangle$ form a system of "natural" basis vectors in the tangent space and the n vectors $|e_i\rangle$ span the orthogonal complementary subspace at $(x^0; c^0)$. With respect to this choice of basis vectors, the metric matrix at $(x^0; c^0)$ assumes the block-diagonal form

$$G = \left[\begin{array}{c|c} V_{ij} & \bigcirc \\ \hline \bigcirc & V_{\alpha\beta} - V_{\alpha r}(V^{-1})^{rs} V_{s\beta} \end{array}\right] \tag{10.19}$$

The $n + k$ vectors $|e_i\rangle$, $|f_\alpha\rangle$ fail to span $\mathbb{R}^n \otimes \mathbb{R}^k$ on the bifurcation set $\mathscr{S}_B$.

Example. For the cusp catastrophe

$$\begin{aligned} V(x; a, b) &= \tfrac{1}{4}x^4 + \tfrac{1}{2}ax^2 + bx \\ V_{xx} &= 3x^2 + a \\ V_{xa} &= x \\ V_{xb} &= 1 \end{aligned} \tag{10.20}$$

The susceptibility tensor at a point $(x; a, b)$ in the critical manifold is determined from

$$\begin{aligned} \delta x &= -\frac{1}{3x^2 + a}(x, 1)\begin{pmatrix}\delta a \\ \delta b\end{pmatrix} \\ \chi_{1\alpha} &= -\frac{1}{3x^2 + a}(x, 1), \qquad \alpha = 1, 2 \end{aligned} \tag{10.21}$$

If $\mathbf{f}_1$, $\mathbf{f}_2$ span the control parameter space and $\mathbf{e}_1$ spans the state variable space, then

$$\begin{aligned} |f_1\rangle &= \mathbf{f}_1 - \frac{x}{3x^2 + a}\mathbf{e}_1 \\ |f_2\rangle &= \mathbf{f}_2 - \frac{1}{3x^2 + a}\mathbf{e}_1 \end{aligned} \tag{10.22}$$

span the tangent plane at $(x; a, b)$, as shown in Fig. 10.6. In terms of the displacements δa, δb the metric tensor in the tangent plane is

$$g_{\alpha\beta} = -\begin{pmatrix} x \\ 1 \end{pmatrix}\frac{1}{3x^2 + a}(x, 1) \tag{10.23}$$

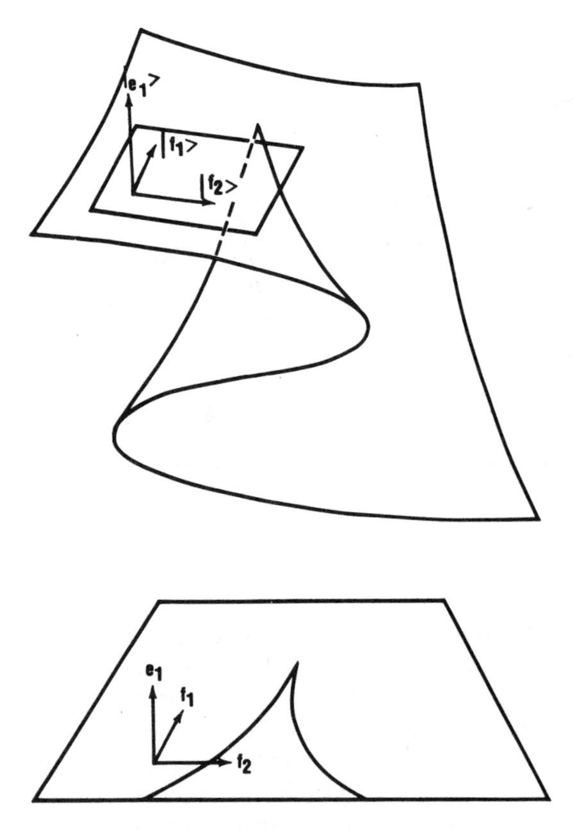

Figure 10.6 The tangent plane to the cusp catastrophe manifold is spanned by the basis vectors $|f_1\rangle$ and $|f_2\rangle$, constructed from the basis vectors $\mathbf{e}_1$, $\mathbf{f}_1$, $\mathbf{f}_2$ spanning the state variable and control parameter space $\mathbb{R}^1 \otimes \mathbb{R}^2$, as shown in (10.22).

If the mixed system $(x; a)$ is used to parameterize the critical manifold instead of the control parameters (a, b), the susceptibility tensor is determined by rearranging (5.2) $(V_{ij}\delta x^j + V_{i\alpha}\delta c^\alpha = 0)$:

$$\begin{aligned} (3x^2 + a)\delta x + x\delta a + \delta b &= 0 \\ \delta b = -(x, 3x^2 + a)&\begin{pmatrix} \delta a \\ \delta x \end{pmatrix} \end{aligned} \tag{10.24}$$

Basis vectors in the tangent plane and the metric tensor in this new coordinate system can be determined from this susceptibility tensor.

4. Critical Point of a Fluid

Many simple substances have sharply defined sublimation, melting, and boiling temperatures which change as a function of the pressure. The solid-gas (sublimation), solid-liquid (melting), and liquid-gas (boiling) coexistence curves for

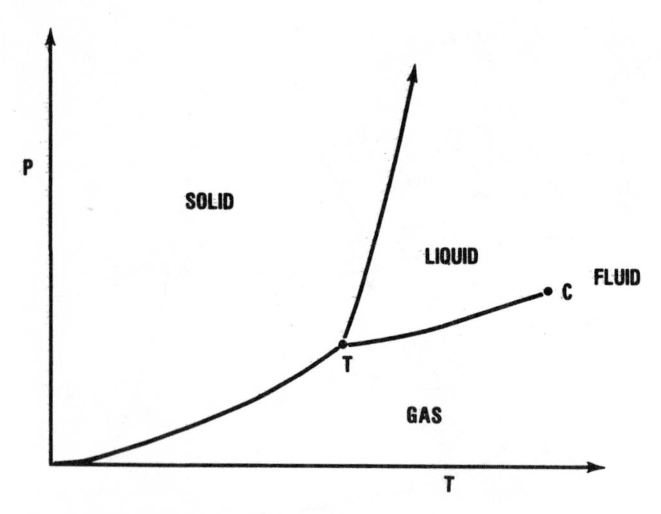

Figure 10.7 The phase portrait for a simple substance possesses a critical point C and a triple point T.

such substances are summarized in a phase diagram. A phase diagram for a simple substance is shown in Fig. 10.7. The three coexistence curves are determined by Clausius–Clapeyron relations. The two points of particular interest in this diagram are the **triple point** T and the **critical point** C. At the triple point a substance can exist at equilibrium simultaneously in the solid, liquid, and gas phases. At the critical point the difference between the liquid and gas phases ceases to exist. The three coexistence lines are lines of first-order phase transitions, and the boundary point C of the liquid-gas coexistence curve describes a second-order phase transition. Beyond C the substance is said to be in a "fluid" state.

Warning. The physical critical point C is associated with a mathematical degenerate critical point.

We focus our attention first on the critical point C. This is the boundary point for a line of first-order phase transitions. Comparison of Fig. 10.7 with Fig. 10.3 suggests strongly that a cusp catastrophe is centered around the point C, that the physical variables P, T are related to the mathematical control parameters $a = a(P, T)$, $b = b(P, T)$, and that the liquid-vapor coexistence curve can be identified with the Maxwell set $\mathscr{S}_M$ of the cusp catastrophe ($a < 0$, $b = 0$). If such an identification is reasonable, a number of catastrophe flags should be waving.

1. *Modality.* The fluid is bimodal in the neighborhood of the liquid-gas coexistence curve, having well defined liquid and gas states.
2. *Inaccessibility.* There is no observable physical counterpart for this catastrophe fingerprint.

3. *Sudden Jumps.* A slight change in the temperature and pressure which moves the control point from one side of the coexistence curve to the other will produce a large change in the volume. The change in volume is sudden if considered only as a function of the control parameters P and T. For a fixed value of control parameters lying on the coexistence curve the change in volume is gradual (continuous) when considered as a function of the total internal energy of the system.

4. *Divergence.* On decreasing P, T from above to below P_c, T_c the final state of the substance may be liquid or gas, depending on a slight perturbation of the initial conditions. From another point of view, the substance can be transformed from its liquid to its gas state either "discontinuously" by crossing the coexistence curve, or continuously by performing an end run around the cusp point C, as indicated in Fig. 10.8.

5. *Hysteresis.* A careful experimentalist can obtain a nonreproducible hysteresis cycle by first raising the temperature and superheating the liquid phase, and after evaporation (if nonexplosive), cooling the gas below the condensation point. If C is a cusp point, the coexistence curve will be surrounded by two fold lines, beyond which the substance no longer is bimodal. These fold lines determine the limits to superheating and its inverse, supersaturation. These fold lines are called **spinodal lines** (Fig. 10.9).

6. *Divergence of Linear Response.* A single component fluid has a number of linear response functions. One of these is the isothermal compressibility $\kappa_T = -V^{-1}(\partial V/\partial P)_T$. This response function diverges as the coexistence curve is approached (Fig. 10.10).

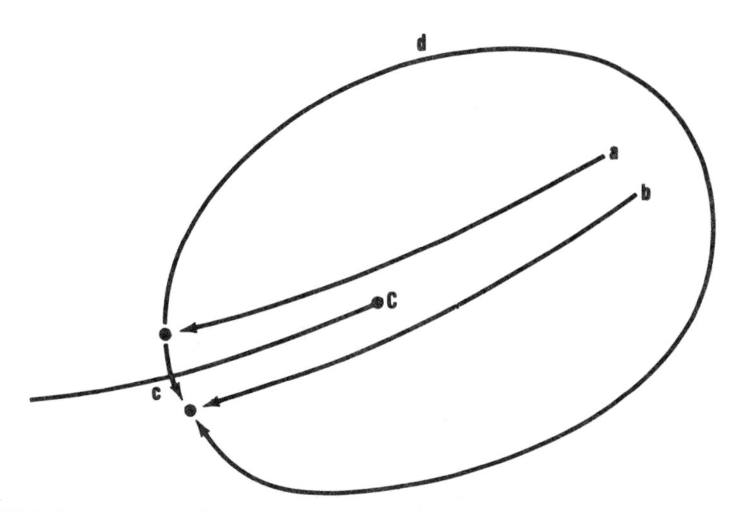

Figure 10.8 Nearby paths a, b can lead from nearby initial states to final states which are drastically different (cf. Fig. 9.3). The system may be taken from the liquid to the gas state by a path c on which a first-order phase transition occurs or along a path d on which no phase transition occurs.

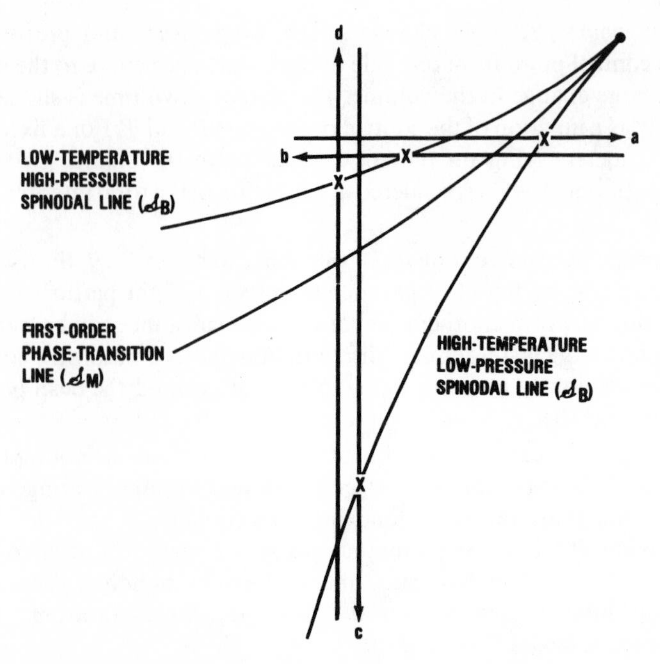

Figure 10.9 If the critical point C is associated with a cusp catastrophe, then it is surrounded by fold curves, called spinodal lines. At fixed pressure the system may be superheated above the boiling temperature (defined by $\mathcal{S}_M$), but not above the high-temperature spinodal line ($\times$ on path a). Reversing direction, supersaturation cannot proceed below the low-temperature spinodal line ($\times$ on path b). The spinodal lines $\mathcal{S}_B$ may be traced out by holding T fixed and varying P, or by choosing other paths across the first-order phase transition line.

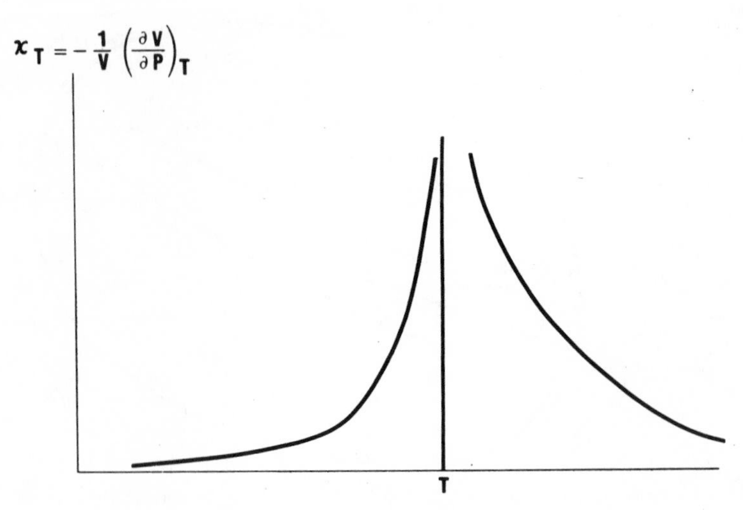

Figure 10.10 Linear response functions such as the isothermal compressibility diverge as the coexistence curve is approached.

7. *Critical Slowing Down/Mode Softening.* From dimensional considerations, the speed of sound in a fluid depends on $\kappa^{-1/2}$, where κ is a compressibility, but not necessarily isothermal. As κ diverges, the speed of sound falls to zero.

8. *Anomalous Variance.* When a fluid condenses from its gas to its liquid state, small droplets are formed in the gaseous region, which then fall to the bottom of the containing vessel because the liquid is denser than the gas. In the reverse (boiling) transition, small bubbles of the gas are formed in the liquid phase and are then forced into the gas phase by these same density considerations. As the critical point is approached from below, the difference in the density of the two fluid states approaches zero, the time scale on which droplets fall and bubbles rise increases, and the droplet/bubble sizes increase to the point at which light is very strongly scattered. This phenomenon, due to the anomalously large expectation values for the radii of bubbles and droplets, is called **critical opalescence**.

5. Fluid Equation of State: van der Waals Equation

It is remarkable that we can infer the existence of so many physical effects in real substances simply by comparing the phase diagram in the neighborhood of a critical point with the Maxwell set of the cusp catastrophe, and then looking for catastrophe fingerprints. It would be more remarkable still if we could find a quantitative description (equation of state) of a fluid in the neighborhood of its critical point rather than being satisfied with the qualitative identifications made in the previous section.

The first step in the transformation of our qualitative understanding into a quantitative understanding is to identify the mathematical state variable x with the appropriate physical order parameter. The first and most obvious guess would be to identify x with V (volume). However, this would be a bad guess for the following reason. On crossing the coexistence curve from the liquid to the gas phase, the value of V increases, but on crossing the cusp catastrophe Maxwell set $\mathscr{S}_M$ in the same direction the critical value giving the global minimum decreases; jumping from positive ($b < 0$) to negative ($b > 0$). The next most obvious thing to do would be to identify x with V^{-1}, which is proportional to the density ρ of the substance. We adopt this identification.

There now remain two technical difficulties. The first is that the mathematical cusp occurs at $(x; a, b) = (0; 0, 0)$, while the physical cusp occurs at $(\rho_c; P_c, T_c)$. The second is that the mathematical parameters $(x; a, b)$ are dimension-free, while the physical parameters $(\rho; P, T)$ are not. Both difficulties may be overcome by introducing the scaled dimension-free physical quantities $(\rho_r; P_r, T_r)$ defined by

$$\frac{\rho - \rho_c}{\rho_c} = \rho_r - 1, \qquad \frac{P - P_c}{P_c} = P_r - 1 = p - 1, \qquad \frac{T - T_c}{T_c} = T_r - 1 = t - 1 \tag{10.25}$$

The subscript r means "reduced" (measured with respect to its value at the critical point).

The simplest possible relation between the mathematical control parameters (a, b) and the physical control parameters $(p - 1, t - 1)$ is a linear relation:

$$\begin{bmatrix} a \\ b \end{bmatrix} = \begin{bmatrix} A & B \\ C & D \end{bmatrix} \begin{bmatrix} p - 1 \\ t - 1 \end{bmatrix} \tag{10.26}$$

This assumption requires the coexistence curve to be a straight line, which is not a bad assumption in the neighborhood of the critical point. This line is defined by

$$0 = b = C(p - 1) + D(t - 1) \tag{10.27}$$

The reduced slope m of this line is

$$m = \frac{dp}{dt} = \frac{T_c}{P_c} \lim \frac{dP}{dT} = -\frac{D}{C} \tag{10.28}$$

where the limit is taken approaching the critical point along the coexistence curve.

The linear identification (10.26) between the control parameters and the linear identification $x = \rho_r - 1$ between the state variables leads to the cusp equation for critical points:

$$(\rho_r - 1)^3 + (Ap + Bt - A - B)(\rho_r - 1) + (Cp + Dt - C - D) = 0 \tag{10.29}$$

After some algebra we find

$$(\rho_r - 1)^3 + p(A\rho_r + C - A) - [\rho_r(A + B) + (D + C - B - A)] = -t[B\rho_r + D - B] \tag{10.30}$$

and after yet more algebra we have

$$(A - C)P_r V_r - AP_r + (D + C - B - A + 1)V_r + (B + A - 3) - \frac{1}{V_r^2} + \frac{3}{V_r} = BT_r + (D - B)V_r T_r \tag{10.31}$$

To this messy equation we can apply the following sequence of unjustified but simplifying assumptions:

1. The coefficient of the term $V_r T_r$ is zero.

This assumption is motivated by our knowledge of the perfect gas law ($PV = nRT$). Since identification with canonical catastrophes is only a local property and reduction to perfect gas form (as $T \to \infty$) is not local, there is no compelling reason to make this assumption.

2. The coefficient of V_r is zero.
3. The constant term is zero.

The only motivation for assumptions 2 and 3 is that (10.31) simplifies considerably [cf. (10.34) below].

These assumptions lead to the following constraints on the linear transformation (10.26):

$$\begin{aligned} 1. &\Rightarrow & D - B &= 0 \\ 2. &\Rightarrow & D + C - B - A + 1 &= 0 \\ 3. &\Rightarrow & A + B - 3 &= 0 \\ 1. + 2. &\Rightarrow & C - A + 1 &= 0 \end{aligned} \tag{10.32}$$

With these assumptions the linear transformation (10.26) and the equation of state (10.31) simplify to

$$\begin{bmatrix} a \\ b \end{bmatrix} = \begin{bmatrix} 3 - D & D \\ 2 - D & D \end{bmatrix} \begin{bmatrix} p - 1 \\ t - 1 \end{bmatrix} \tag{10.33}$$

$$P_r V_r - (3 - D)P_r + \frac{3}{V_r} - \frac{1}{V_r^2} = DT_r \tag{10.34}$$

We can now make one of two additional simplifying assumptions.

4a. The left-hand side of (10.34) factorizes into the form

$$\left(P_r + \frac{\alpha_r}{V_r^2}\right)(V_r - \beta_r) = DT_r \tag{10.35}$$

For this to occur, $D = \frac{8}{3}, \alpha_r = 3, \beta_r = +\frac{1}{3}$. Equation (10.35) leads immediately to the van der Waals equation under the substitutions $P_r = P/P_c$, and so on:

$$\left(P + \frac{\alpha}{V^2}\right)(V - \beta) = nRT \tag{10.36}$$

where

$$\begin{aligned} \alpha &= 3P_c V_c^2 \\ \beta &= \tfrac{1}{3} V_c \\ nR &= \frac{8}{3}\frac{P_c V_c}{T_c} \end{aligned} \tag{10.37}$$

Here n is the number of moles, R is the gas constant, and α, β, R scale like n^2, n^1, n^0, respectively.

4b. If we force the coefficients in (10.34) to reproduce the reduced slope m of the coexistence curve at the critical point [cf. (10.28)], then $D = 2m/(m - 1)$ and (10.34) no longer factors into the van der Waals form.

The two assumptions 4a and 4b are equivalent, provided the reduced slope $m = 4$. This leads us to pose the following question.

Question 1. What is the reduced slope $m = (T_c/P_c)\lim(dP/dT)$ of a simple fluid near its critical point?

6. Predictions of the van der Waals Equation

The van der Waals equation (10.36) is remarkable for its staying power. Proposed originally in 1873,[2] it provides a good qualitative description for a simple fluid in the neighborhood of its critical point. As an equation of state, it can be pressed to give quantitative relations between the three parameters V, P, T of a fluid near its critical point. Computations of this type have been carried out by countless generations of students.

We will view some of these computations in the context of Catastrophe Theory. The van der Waals equation is the critical manifold of the cusp catastrophe. Once a point on this manifold has been fixed, we can investigate the linear response of the system to small perturbations using the methods described in Sec. 3. Instead of using the susceptibility tensor as given in (10.21), we will find it more useful here (and especially in Secs. 11–16) to treat the state variables and control parameters on an equal footing by writing (5.2) ($V_{ij}\delta x^j + V_{i\alpha}\delta c^\alpha = 0$) for the cusp catastrophe:

$$(3x^2 + a)\delta x + x\delta a + \delta b = 0 \tag{10.38}$$

where $x = \rho_r - 1 = V_c/V - 1$, $a = \frac{1}{3}(P/P_c - 1) + \frac{8}{3}(T/T_c - 1)$,

$$\delta x = -\frac{V_c}{V^2}\delta V \quad \text{and} \quad \begin{pmatrix} \delta a \\ \delta b \end{pmatrix} = \begin{pmatrix} \frac{1}{3} & \frac{8}{3} \\ -\frac{2}{3} & \frac{8}{3} \end{pmatrix} \begin{pmatrix} \delta p \\ \delta t \end{pmatrix} \tag{10.39}$$

All the linear response functions, which are just thermodynamic partial derivatives, may be determined directly from (10.38). For example, at constant volume $\delta x = 0$ and

$$\left(\frac{V_c}{V} - 1\right)(\tfrac{1}{3}\delta p + \tfrac{8}{3}\delta t) + (-\tfrac{2}{3}\delta p + \tfrac{8}{3}\delta t) = 0 \tag{10.40}$$

so that

$$\left(\frac{dp}{dt}\right)_V = \frac{T_c}{P_c}\left(\frac{dP}{dT}\right)_V = \frac{8V_c}{3V - V_c} \tag{10.41}$$

The remaining thermodynamic derivatives may be computed similarly.

Next we use the arc length s to parameterize a path between the cusp point $(a, b) = (0, 0)$ and the point (a_0, b_0) in the control plane (Fig. 10.11). Then for s sufficiently small,

$$s = \frac{-x^3}{a_0 x + b_0} \tag{10.42}$$

If $b_0 \neq 0$, then in the neighborhood of the cusp point

$$s = -x^3/b_0 \tag{10.43}$$

This can be translated into physical terms using the transformation (10.26) and the relation $x = \rho_r - 1 = -v/(1 + v)$, where $v = V_r - 1$. Then

$$s \simeq v^3/b_0 \tag{10.44}$$

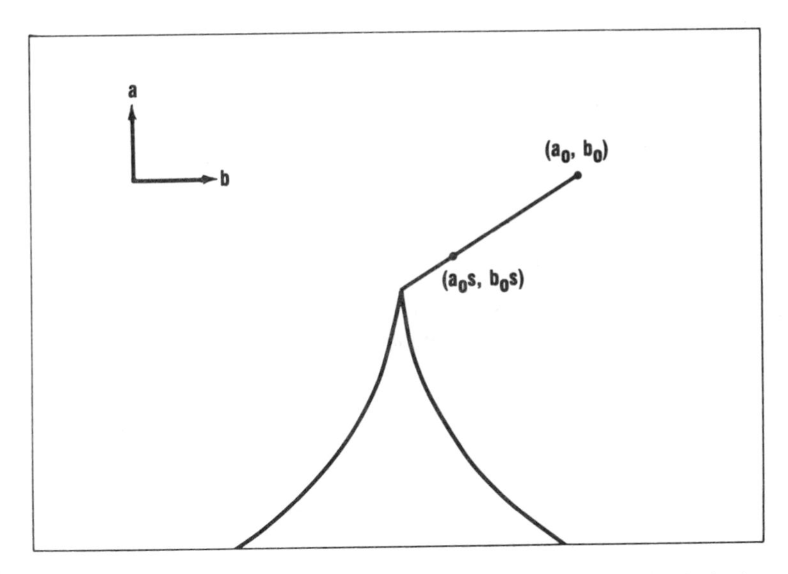

Figure 10.11 Directional derivatives on the cusp catastophe manifold have canonical values which may be identified with thermodynamic properties of the van der Waals equation of state.

Along any straight-line trajectory passing through the critical point the quantities $P_r - 1$ and $T_r - 1$ are proportional. For any path that is not approximately parallel to the coexistence curve, $P_r - 1$ and $T_r - 1$ are approximately proportional to $(V_r - 1)^3$.

Third, we consider a path which crosses the coexistence curve at an angle ψ at a reduced temperature $T_r < 1$ (Fig. 10.12). At this point $P_r - 1 = -(D/C)(T_r - 1)$

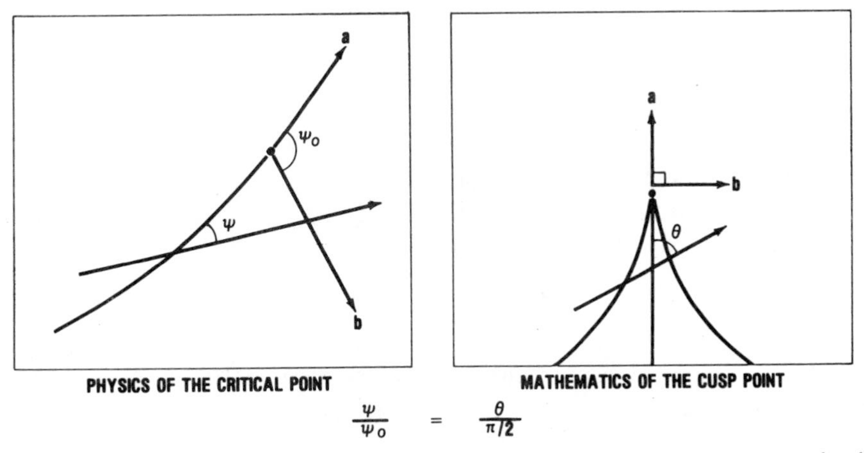

Figure 10.12 The neighborhoods of the critical point and the cusp catastrophe germ are related by an affine transformation.

and $a = C^{-1}(AD - BC)(1 - T_r) \rightarrow -4(1 - T_r)$ for the van der Waals fluid. The angle ψ in the phase plane must be related to the angle θ in the cusp control plane (Fig. 10.3) by observing that the affine transformation (10.26) does not preserve angles. The angle between the coexistence line (a axis) and the complementary axis in the phase plane is determined by elementary geometry and defined by

$$\cos \psi_0 = \frac{(1, -B/A) \cdot (1, -D/C)}{\sqrt{1 + (B/A)^2}\sqrt{1 + (D/C)^2}} \tag{10.45}$$

For the van der Waals fluid this is 158.84°, so that the internal energy change on crossing the coexistence curve at angle ψ is [cf. (10.10)]

$$\Delta \frac{d}{ds}(U/U_c) = -2\sqrt{4(1 - T_r)}\sin(0.5666\psi) \tag{10.46}$$

Fourth, we can ask how the liquid-gas density difference varies along the coexistence curve. From the canonical geometry of the cusp, $\Delta x = \Delta\rho = 2\sqrt{-a} \sim (T_c - T)^{1/2}$.

The quantitative inadequacies of the van der Waals equation of state show up most clearly when its predictions, particularly the last, are confronted with experimental data. In Fig. 10.13 we reproduce Guggenheim's plot[3] of the reduced temperature T/T_c against ρ/ρ_c for eight substances in two phases along the coexistence curve. The data fall essentially on a single universal curve, suggesting that the law of corresponding states is valid for these gases. The **law of corresponding states** says that the thermodynamic state functions for each of these gases are isomorphic to a single universal function of dimension-free reduced quantities.

These experimental data reveal two important features.

1. The mean density of the liquid and gas phases along the coexistence curve depends linearly on the distance from the critical point:

$$\frac{1}{2}\left(\frac{\rho_l}{\rho_c} + \frac{\rho_g}{\rho_c}\right) = 1 + \frac{3}{4}(1 - T_r) \tag{10.47}$$

This is known as the **law of rectilinear diameters**.

2. The density difference between the liquid and the gas phases along the coexistence curve depends on $(1 - T_r)^\beta$ where $\beta \simeq \frac{1}{3}$:

$$\frac{\rho_l}{\rho_c} - \frac{\rho_g}{\rho_c} = \frac{7}{2}(1 - T_r)^{1/3} \tag{10.48}$$

The first of these two features suggests another question that may be answered by experiment.

Question 2. Does the law of rectilinear diameters hold above the critical point? That is, if we make a linear extrapolation of the coexistence curve through

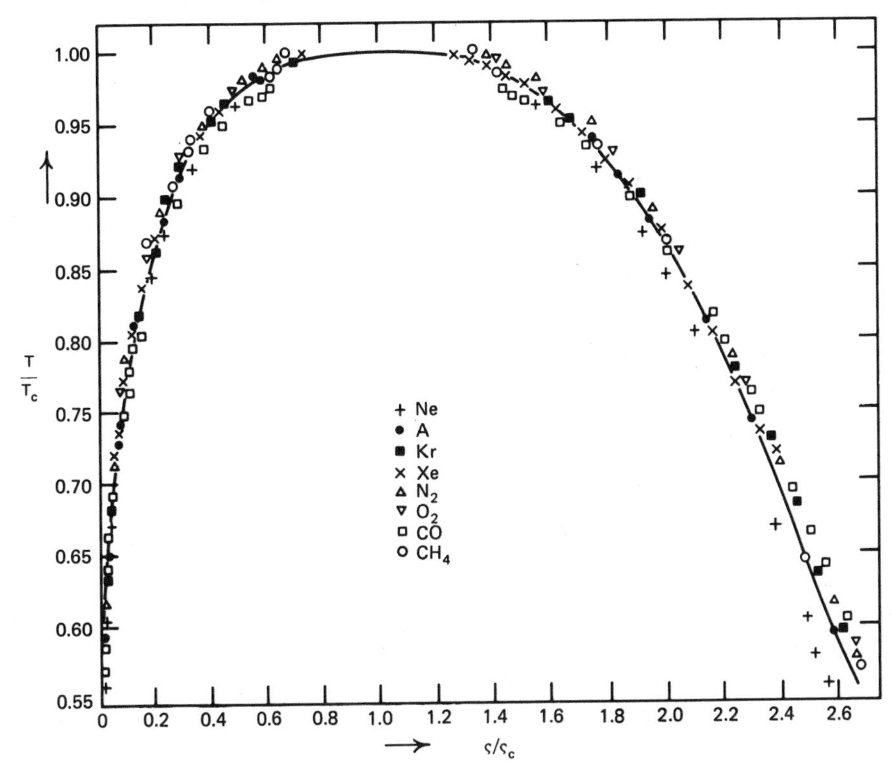

Figure 10.13 Guggenheim plotted the reduced density against the reduced temperature for the liquid and gas phases of eight simple substances. The plot indicated a "law of corresponding states" for these substances. Reprinted with permission from E. A. Guggenheim, *J. Chem. Phys.* **13** (1945), p. 257. Copyright 1945 by American Institute of Physics.

the critical point, does the fluid density above T_c obey (10.47) with the left-hand side replaced by ρ_f/ρ_c (and $T_r > 1$)?

The second feature (10.48) points clearly to a limitation of the van der Waals equation. This equation predicts a $\frac{1}{2}$ power law dependence of $\Delta\rho$ on $T_c - T$, whereas experiments reveal a $\frac{1}{3}$ power law dependence. This problem raises a third question.

Question 3. Is it possible to reproduce the experimental data by identifications of the mathematical parameters $(x; a, b)$ with the physical parameters $(\rho_r; P_r, T_r)$ different from those given in Sec. 5?

The answer is no for the cusp catastrophe. The identification $x \sim |\rho_r - 1|^{3/2}$ is not smooth. Alternative identifications of the control parameters, such as $a \sim [A(p-1) + B(t-1)]^{2/3}$, are unsatisfactory because the mathematical controls fail to provide a coordinate system at the critical point.

In short, the cusp catastrophe can be used to construct an equation of state for a fluid in the neighborhood of its critical point. But the van der Waals equation and all equations like it derived from mean field theory arguments or from cusp catastrophe arguments fail quantitatively because they predict a $\frac{1}{2}$ power law dependence of the density difference along the coexistence curve instead of the measured $\frac{1}{3}$ power law dependence.

7. Ginzburg–Landau First-Order Phase Transitions

Ginzburg and Landau have also proposed an extremely useful model for describing first-order phase transitions.[1] This model is as described in Sec. 2, except that now a potential of sixth degree is involved:

$$V(x; a, b, c, d) = \tfrac{1}{6}x^6 + \tfrac{1}{4}ax^4 + \tfrac{1}{3}bx^3 + \tfrac{1}{2}cx^2 + dx \tag{10.49}$$

This function has the following scaling properties:

$$\begin{aligned} x \to x' &= \lambda x \\ a \to a' &= \lambda^2 a \\ b \to b' &= \lambda^3 b \Rightarrow V \to V' = \lambda^6 V \\ c \to c' &= \lambda^4 c \\ d \to d' &= \lambda^5 d \end{aligned} \tag{10.50}$$

In the Ginzburg–Landau formulation a symmetry principle restricts the form of the A_{+5} potential above to be an even function of x. The symmetry-restricted potential will be denoted simply by $V(x; a, c)$.

The Ginzburg–Landau potential has a 2-dimensional control space $\mathbb{R}^2$. As usual, it is useful to determine how this space is partitioned into open subsets by the bifurcation set $\mathscr{S}_B$ and the Maxwell set $\mathscr{S}_M$. The critical set is determined by

$$\frac{dV}{dx} = x(x^4 + ax^2 + c) = 0 \tag{10.51}$$

The point $x = 0$ is a critical point, as is always the case for symmetry-restricted potentials. The remaining quartic factor may be solved for x^2:

$$x^2 = -\frac{a}{2} \pm \sqrt{\left(\frac{a}{2}\right)^2 - c} \tag{10.52}$$

If c is negative, one of the two solutions for x^2 is positive and the other is negative. There are then two real solutions for x. If c is positive but less than $(a/2)^2$, then both solutions in (10.52) are positive if $-a/2$ is positive, both negative if $-a/2 < 0$. If $c > (a/2)^2$, x has no real roots. As a result, the a–c control plane $\mathbb{R}^2$ is partitional into open regions parameterizing functions with 1, 3, or 5 critical points, as shown in Fig. 10.14. The critical set is shown in Fig. 10.15, where we have not

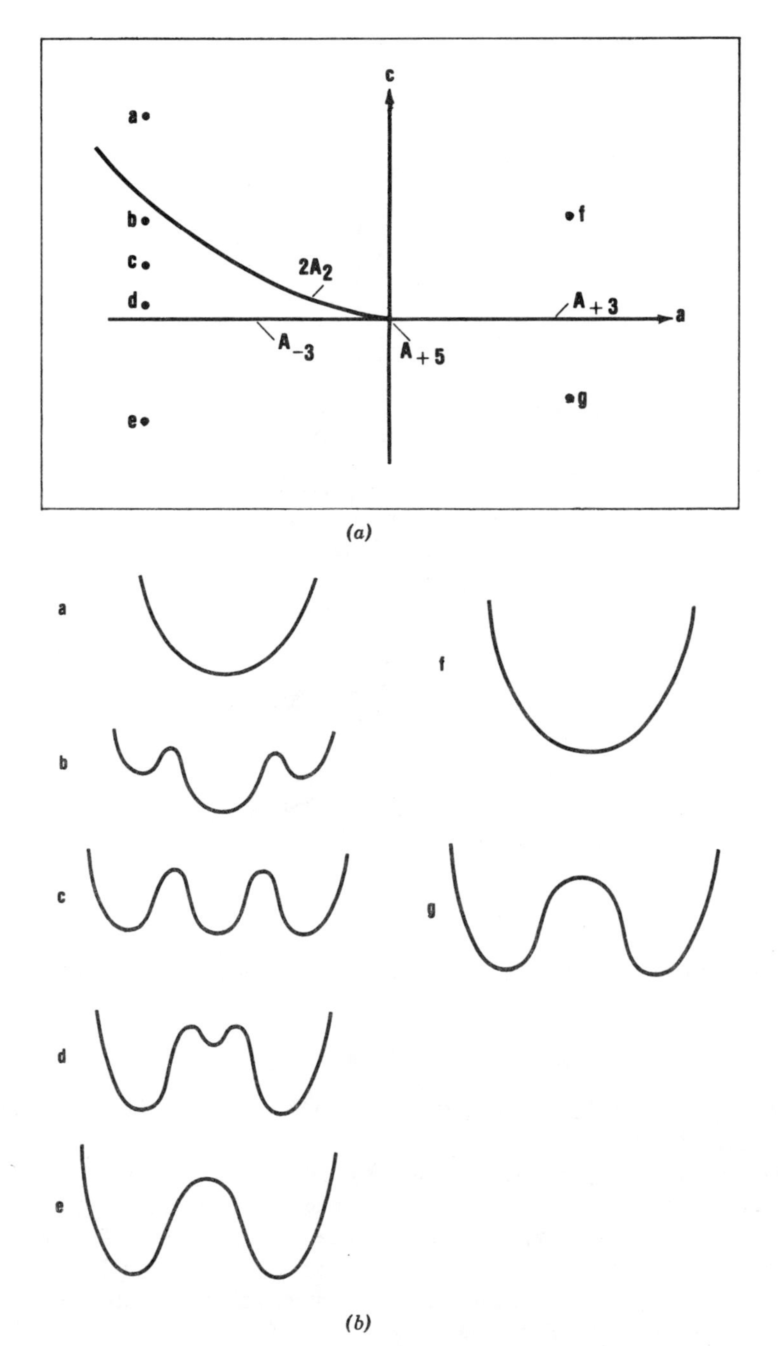

Figure 10.14 (*a*) The a–c control plane of the symmetry-restricted A_{+5} catastrophe is partitioned into three open regions representing functions with 1, 2, or 3 minima. (*b*) Potentials associated with several points in the control plane are shown.

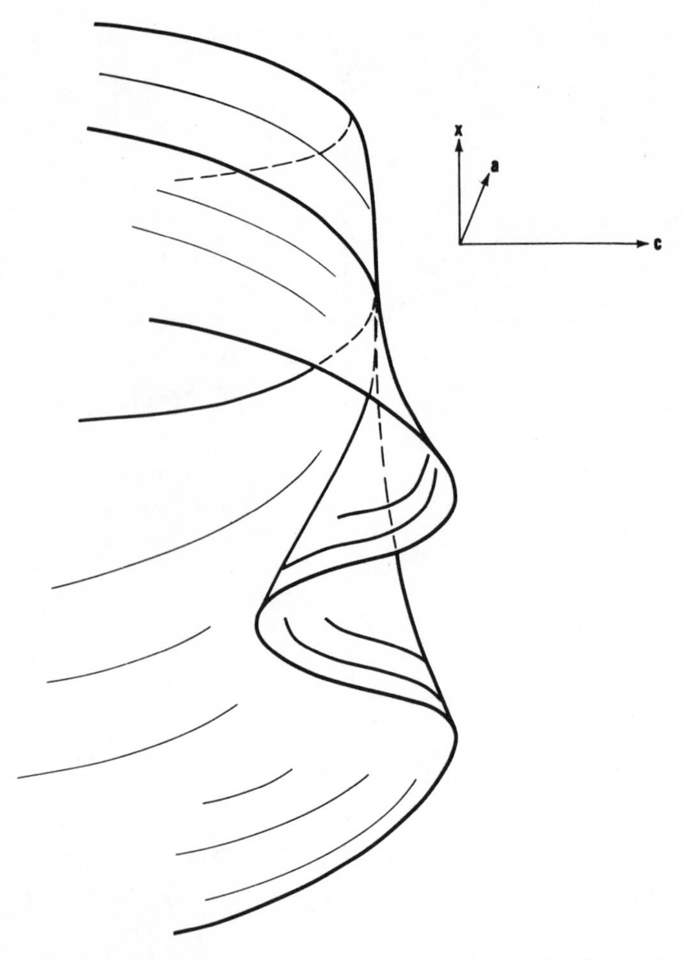

Figure 10.15 The surface of critical points $\nabla V(x; a, c) = 0$ is shown in $\mathbb{R}^3$. The plane $x = 0$ is not explicitly shown.

specifically indicated the plane $x = 0$ in $\mathbb{R}^1 \otimes \mathbb{R}^2$ as part of the critical set for purposes of clarity.

The bifurcation set can be computed by the usual means of setting second derivatives equal to zero. However, it can just as easily be seen as the "shadow" of the critical set (Fig. 10.15) in the control plane $\mathbb{R}^2$. This consists of the half-lines $a > 0, c = 0$ (A_{+3}), $a < 0, c = 0$ (A_{-3}), the curve $a < 0$, $c = (a/2)^2$ $(2A_2)$, and the point $a = 0, c = 0$ (A_{+5}). The qualitative shapes of the potentials associated with several points are also shown in Fig. 10.14. On passing from a to b, two symmetric fold catastrophes are encountered. At point c on the Maxwell set the three local minima are equally deep, as the critical values of the side minima approach zero from above. On passing from d to e a dual cusp catastrophe A_{-3} is encountered. On passing from point f to g a standard cusp catastrophe is encountered.

The locations of the three minima with equal critical value 0 are $x = 0$, $x^2 = -3a/4$. The Maxwell set is defined by:

$$\mathscr{S}_M: \quad a < 0, \qquad c = \tfrac{3}{16}a^2 \tag{10.53}$$

Phase transitions of the zeroth, first, and second order may be encountered in this symmetry-restricted Ginzburg–Landau potential, as discussed below.

0. If we assume the Delay Convention, then zeroth-order phase transitions are encountered on crossing the components of $\mathscr{S}_B$ with $a < 0$ in a suitable direction. For example, on crossing the curve $c = (a/2)^2$ following path 0a in Fig. 10.16, the change in the value of the potential is

$$\Delta V = V(0; a, c) - V\left(\pm\sqrt{\frac{-a}{2}}; a, c\right) = -\frac{1}{6}\left(-\frac{a}{2}\right)^3 \tag{10.54}$$

On crossing the half-line $a < 0$, $c = 0$ along the path 0b, the change in potential is

$$\Delta V = V(\pm\sqrt{-a}; a, 0) - V(0; a, 0) = -\tfrac{1}{12}(-a)^3 \tag{10.55}$$

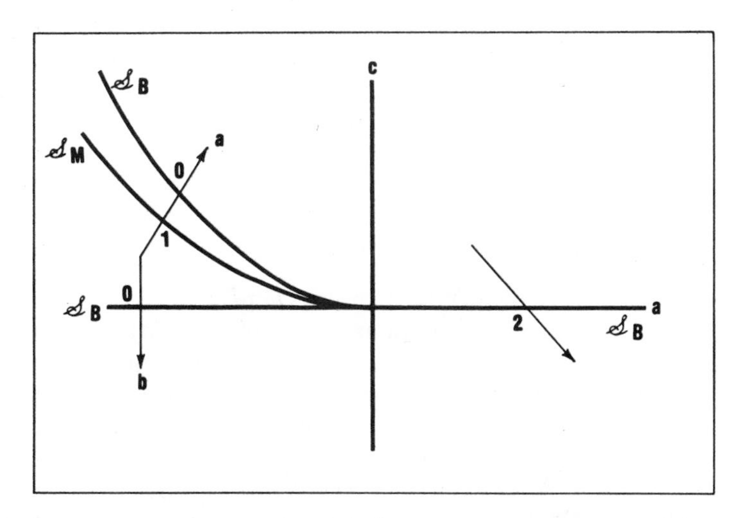

Figure 10.16 The Ginzburg–Landau symmetry-restricted potential exhibits zeroth-, first-, and second-order phase transitions in much the same way that the catastrophe A_{+3} exhibits such phase transitions.

1. If we adopt the Maxwell Convention, then a first-order phase transition occurs as we cross $\mathscr{S}_M$. For Path 1 shown in Fig. 10.16, the change in the first derivative of the potential is

$$\Delta \frac{dV}{ds} = \frac{d}{ds} V \bigg|_{0;\, a,\, c} - \frac{d}{ds} V \bigg|_{\pm \sqrt{-3a/4};\, a,\, c}$$
$$= -\left\{\frac{1}{4}\left(-\frac{3a}{4}\right)^2 \frac{da}{ds} + \frac{1}{2}\left(-\frac{3a}{4}\right)\frac{dc}{ds}\right\} \tag{10.56}$$

where da/ds, dc/ds are the direction cosines of the path in $\mathbb{R}^2$ as it crosses $\mathscr{S}_M$. The spinodal lines for this first-order phase transition are the two curves of zeroth-order phase transitions discussed above.

2. A second-order phase transition will occur as we cross the bifurcation set along Path 2. The magnitude of the discontinuity of the second derivative is direction dependent and is given by

$$\Delta \frac{d^2V}{ds^2} = -\frac{1}{2a}\left(\frac{dc}{ds}\right)^2 \tag{10.57}$$

where the direction cosine is evaluated on the bifurcation set.

8. Tricritical Points

Symmetry breaking in the Ginzburg–Landau model for first-order phase transitions (Sec. 7) can be studied by the same methods used for the study of symmetry breaking in the canonical model for second-order phase transitions (Sec. 2). The variety of phenomena encountered is much richer since the catastrophe A_{+5} has a larger control parameter space than does A_{+3}. The first studies of symmetry breaking for this potential were carried out for just the single odd term dx. Accordingly the family of potentials we study is

$$V(x; a, c, d) = V(x; a, c) + dx$$
$$= \tfrac{1}{6}x^6 + \tfrac{1}{4}ax^4 + \tfrac{1}{2}cx^2 + dx \tag{10.58}$$

We shall call the potential (10.58) the "tricritical potential."

The tricritical potential is of interest in the area of thermodynamics, where the Maxwell Convention is usually adopted. In particular it is of interest to determine the Maxwell set $\mathscr{S}_M$ of this potential. This has already been done in the previous section in the a–c plane $d = 0$ in the control space $\mathbb{R}^3$. In this plane there are two equally deep minima below the positive half-line ($a \geq 0$, $c = 0$) and also below the curve ($a \leq 0$, $c = +\frac{3}{16}a^2$). These two curves are themselves boundaries for the Maxwell set in $\mathbb{R}^3$, the curve ($a > 0$, $c = 0$) parameterizing second-order phase transitions and the curve ($a < 0$, $c = +\frac{3}{16}a^2$) parameterizing first-order phase transitions. Along this latter curve the potential has three equally deep minima. This is therefore called the triple-point curve. The boundary of these boundaries [i.e., $(a, c, d) = (0, 0, 0)$] is called a "tricritical point"; it is an A_{+5} catastrophe germ. The Maxwell set of $V(x; a, c, d)$ in $\mathbb{R}^3$ is most easily determined by determining how the shape of the potential changes along straight lines in $\mathbb{R}^3$ with (a, c) constant and d increasing. For the four points shown in the a–c plane $d = 0$ in Fig. 10.17*a*, we indicate the shape of the potential as a function of increasing d. From these figures we conclude that a first order phase transition occurs along lines 1 and 2, when the global minimum jumps from the right hand local minimum to the left-hand local minimum as d increases through zero. Two first order phase transitions can occur along line 3 as the global minimum jumps first from the right-hand to the central local minimum at $d = -d_0$, and then from the central to the left-hand local minimum at $d = +d_0$. Along line 4, sufficiently far above the curve of triple points, no phase transitions of any kind occur as a function of increasing d.

These simple considerations reveal that the part of the a–c plane $d = 0$ below the lines of $A_{\pm 3}$ catastrophes parameterize first-order phase transitions involving

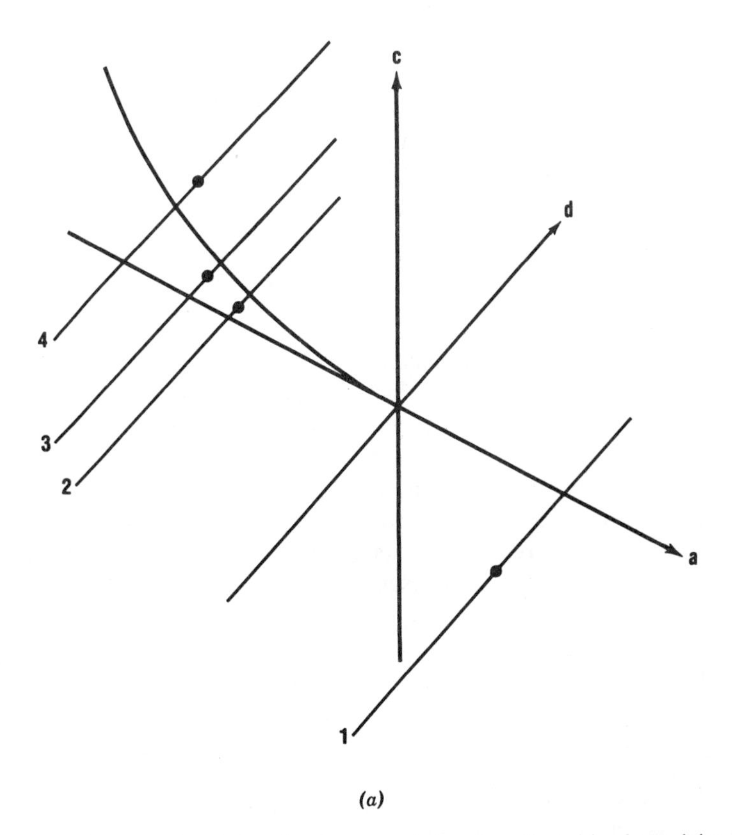

(a)

Figure 10.17 The properties of the tricritical potential are investigated by determining how the shape changes for fixed control parameters (a, c) as a function of increasing control parameter d.

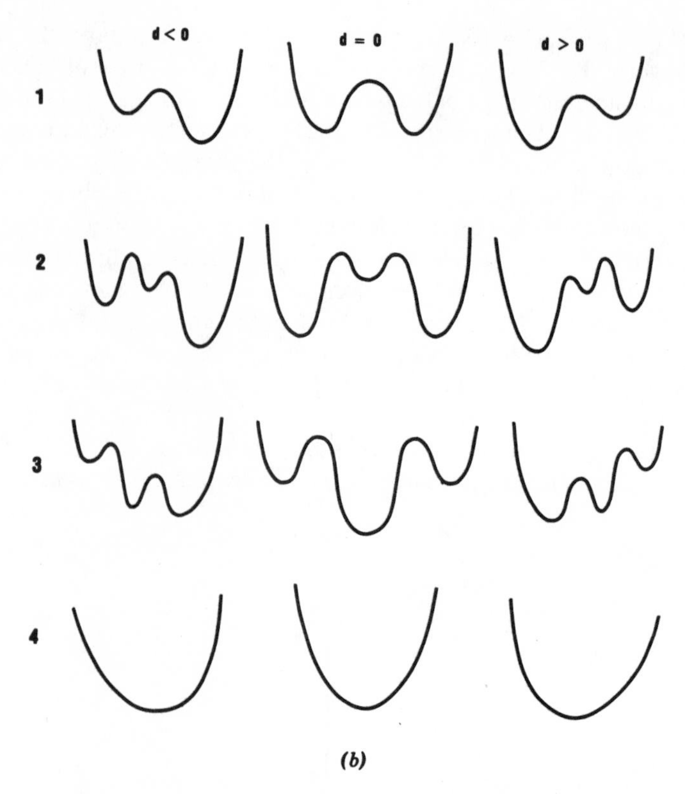

(b)

Figure 10.17 (*Continued*)

jumps between the left- and right-hand minima at $x^{(l)}$, $x^{(r)}$. Above the curve of triple points and located symmetrically about the a–c plane $d = 0$ are two "wings" of first order phase transitions involving a side minimum and the "middle minimum" at $x^{(m)}$. These wings cannot extend upward indefinitely and must therefore have a boundary. The curves bounding these wings must parameterize A_{+3} catastrophes for reasons indicated in Fig. 10.18. Since the boundary curve is part of the bifurcation set, its equation can easily be determined in parametric form by the methods described in (5.9) → (5.12). The result is

$$\begin{aligned} a &= -\tfrac{10}{3}\lambda^2 \\ c &= 5\lambda^4, \qquad \lambda \in \mathbb{R}^1, \lambda \neq 0 \\ d &= -\tfrac{8}{3}\lambda^5 \end{aligned} \tag{10.59}$$

where the cusp germ is $(x - \lambda)^4$.

The 2-dimensional wings of first-order phase transitions may be determined by integrating the Clausius–Clapeyron equations. However, it is simpler first to reduce the dimensionality of the problem by 1 by restricting to the plane $a = -1$.

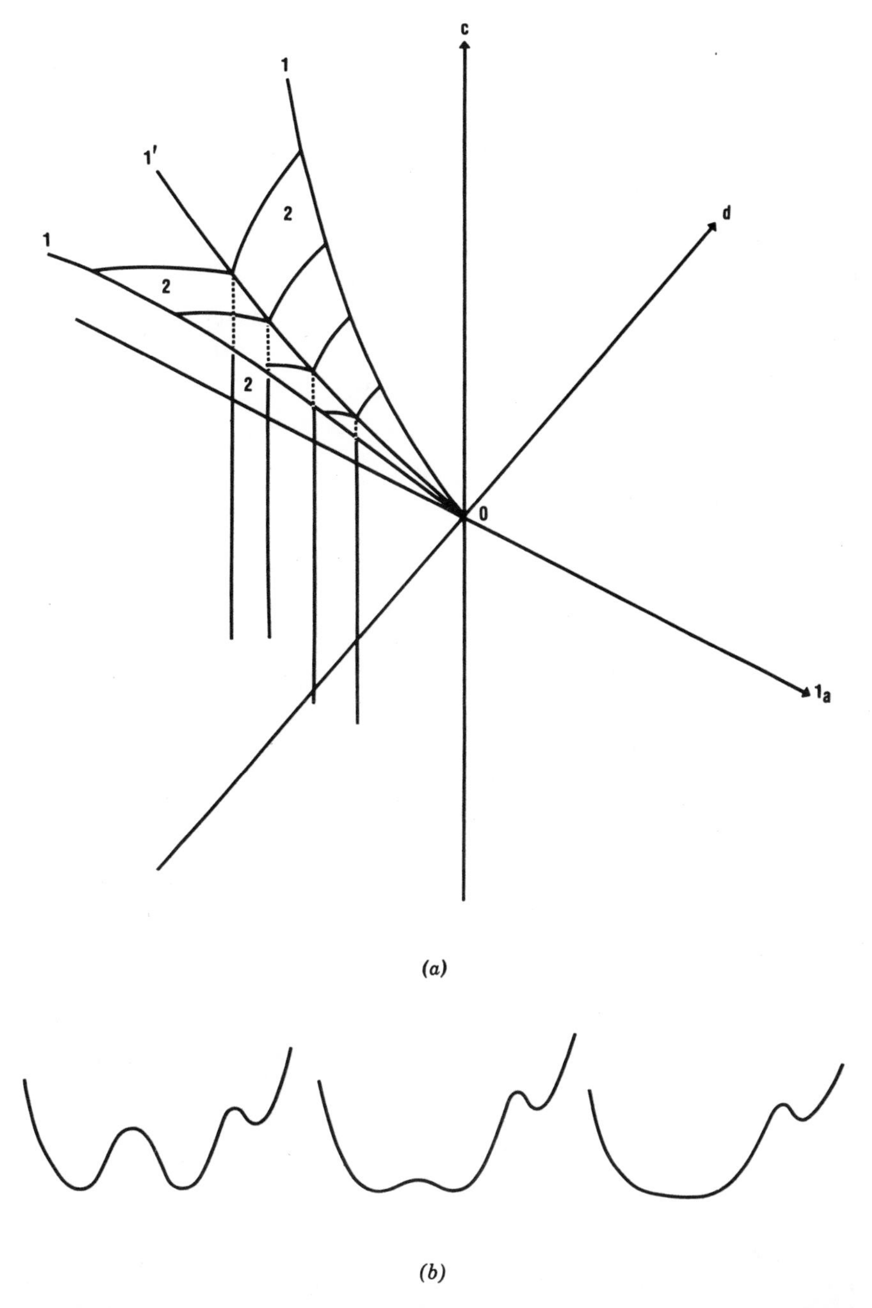

Figure 10.18 (*a*) The Maxwell set of the tricritical potential consists of (part of) the symmetry plane $d = 0$ together with two wings. The boundaries of these wings parameterize cusp catastrophes A_{+3}. (*b*) Potentials associated with points in one of the wings change shape as shown as the bounding curve 1 is approached.

The intersection of the 2-dimensional wings with this plane gives a 1-dimensional curve whose determination is relatively simple. The full wings are easily recovered by the scaling relations (10.50). The intersection curve is shown in Fig. 10.19*a*.

The topology of the Maxwell set $\mathscr{S}_M$ in $\mathbb{R}^3$ for the tricritical potential $V(x; a, c, d)$ is as shown in Fig. 10.18.

0. There is a 0-dimensional component $(a, c, d) = (0, 0, 0)$, the tricritical point, parameterizing an A_{+5} catastrophe germ.

1. There are three 1-dimensional components parameterizing second-order phase transitions. These are the curves $(a > 0, c = 0, d = 0)$ and the two curves whose parametric representation is given in (10.59).

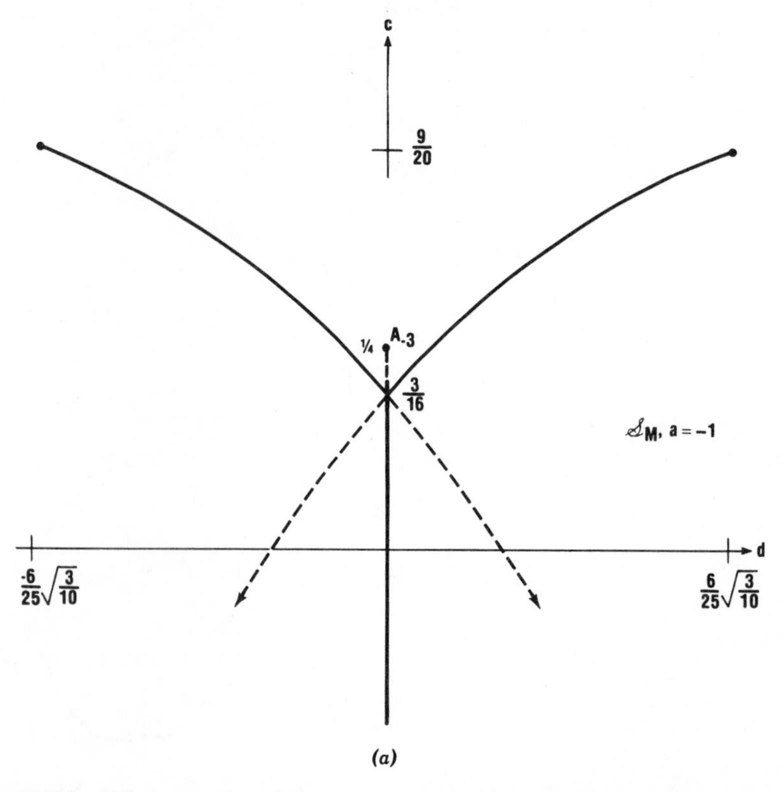

Figure 10.19 (*a*) The intersection of the Maxwell set $\mathscr{S}_M$ of the tricritical potential with the plane $a = -1$ has the form shown. Heavy lines indicate control parameter values for which the two deepest minima are equally deep (stable Maxwell set). Their dashed continuations indicate control parameter values for which these two equally deep minima are metastable with respect to the third minimum (metastable Maxwell set). The metastable Maxwell set is of no concern for physical applications.

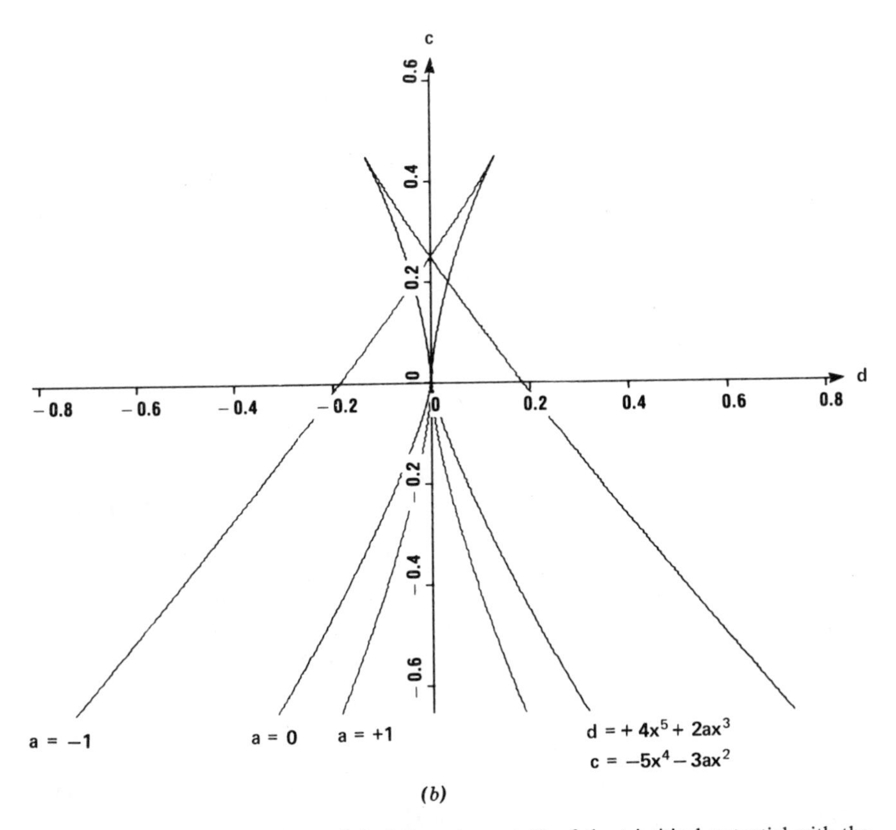

Figure 10.19 (*b*) The intersections of the bifurcation set $\mathscr{S}_B$ of the tricritical potential with the planes $a = -1, 0, +1$ have the form shown. The entire bifurcation set may be reconstructed from these cross sections and the scaling relations (10.50).

1′. There is one 1-dimensional curve parameterizing first-order phase transitions. Its equation is $(a < 0, c = +\frac{3}{16}a^2, d = 0)$.

2. There are three 2-dimensional separatrices parameterizing functions for which the two deepest of the three local minima have degenerate critical values. These surfaces intersect in the triple point curve 1′. The boundaries of these three components are the curves 1 and 1′ described above. These three separatrices can be extended through curve 1′; if so, they parameterize functions with degenerate metastable critical values.

The separatrix $\mathscr{S}_B$ in $\mathbb{R}^3$ can be constructed from the three cross sections $a = -1$, $a = 0$, $a = +1$ shown in Fig. 10.19*b* and a scaling relation. These are obtained from the parametric representation

$$\begin{aligned} c(x, a) &= -5x^4 - 3ax^2 \\ d(x, a) &= +4x^5 + 2ax^3 \end{aligned} \tag{10.60}$$

9. Maxwell Set for A_{+5}

The most general perturbation for the symmetry-restricted Ginzburg–Landau model (Sec. 7) for first-order phase transitions is the catastrophe A_{+5} (10.49), which has a 4-dimensional control parameter space. The bifurcation set $\mathscr{S}_B$ may be determined as was the bifurcation set for A_4 (Chapter 5, Sec. 5). Of more interest is the Maxwell set $\mathscr{S}_M$. This may be determined by integrating the Clausius–Clapeyron equations.

The components of the Maxwell set of particular interest are those that parameterize functions which have two or three equally deep global minima. Roughly but accurately speaking, it takes one control parameter (i.e., function of the controls) to produce each critical value degeneracy. Thus there are three 3-dimensional components ($3 = 4 - (2 - 1)$) of $\mathscr{S}_M$ on which two of the three minima are degenerate, and one 2-dimensional component ($2 = 4 - (3 - 1)$) parameterizing functions with three equally deep minima. It is this "triple point" surface that we will be particularly concerned with here. Unfortunately the Clausius–Clapeyron equations are relatively easy to integrate only in the 1-dimensional case, and the triple point surface is two dimensional. This suggests that clever methods are called for to reduce the dimensionality of the problem by 1. The clever method involves the scale transformation (10.50).

Our first impulse is to build on what we have already derived, that is, to scale Fig. 10.18. However, we cannot do this because this Maxwell set has been determined for $b = 0$ [cf. (10.58)]. Instead, we shall fix $a =$ constant in (10.49), and then use the scaling relations to go back from $\mathscr{S}_M$ in $\mathbb{R}^3$ ($a =$ constant) to $\mathscr{S}_M$ in $\mathbb{R}^4$. The potential A_{+5} has no more than two minima for $a > 0$, so we fix $a = -1$ and determine the 1-dimensional triple point curve in $\mathbb{R}^3$.

This is done as follows. We let $(c^1, c^2, c^3, c^4) = (-1, b, c, d)$ and $x^{(1)}, x^{(2)}, x^{(3)}$ be the locations of the three equally deep minima. Then the Clausius–Clapeyron equations reduce to

$$\begin{aligned}
\delta c^2 &= (-)^{1+2} \det\begin{bmatrix} M_3^1 & M_4^1 \\ M_3^2 & M_4^2 \end{bmatrix} \delta s \\
\delta c^3 &= (-)^{1+3} \det\begin{bmatrix} M_2^1 & M_4^1 \\ M_2^2 & M_4^2 \end{bmatrix} \delta s \\
\delta c^4 &= (-)^{1+4} \det\begin{bmatrix} M_2^1 & M_3^1 \\ M_2^2 & M_3^2 \end{bmatrix} \delta s
\end{aligned} \tag{10.61}$$

where

$$M_\alpha^i = V_\alpha^{(i)} - V_\alpha^{(i+1)}$$

and

$$V_\alpha^{(i)} = \left.\frac{\partial V}{\partial c^\alpha}\right|_{x = x^{(i)}}, \qquad i = 1, 2, 3 \qquad \alpha = 2, 3, 4$$

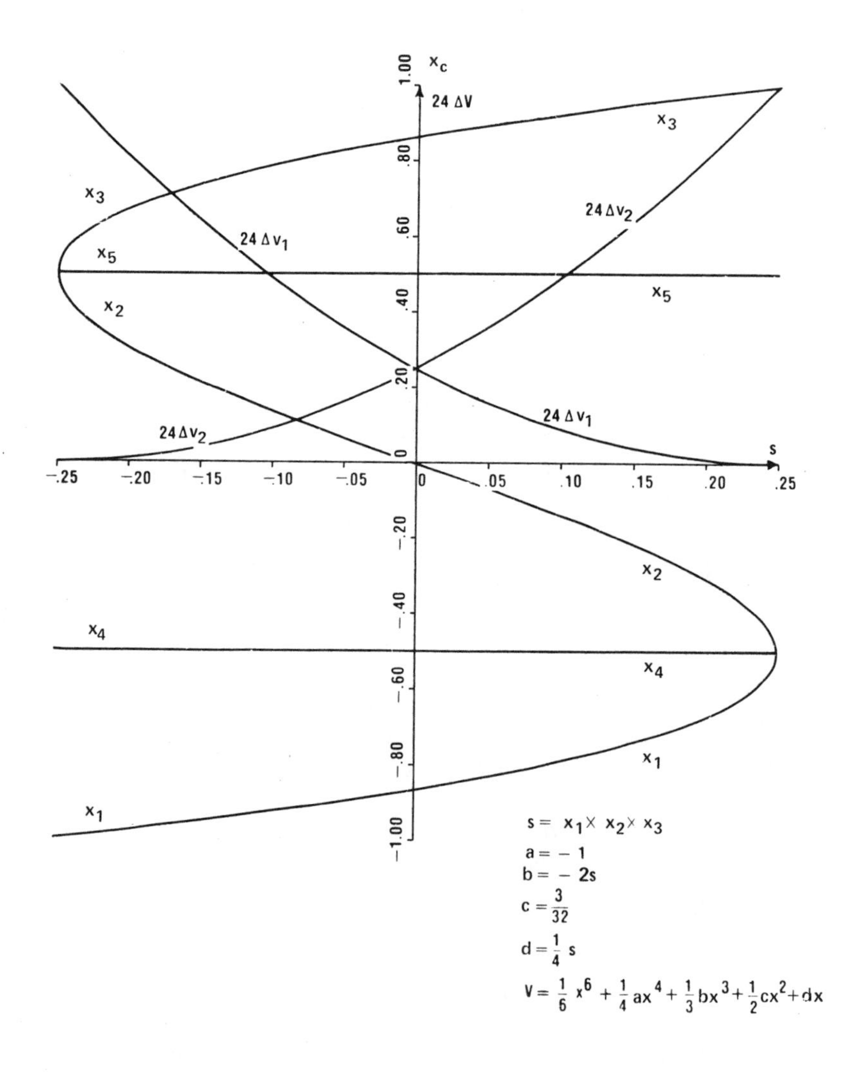

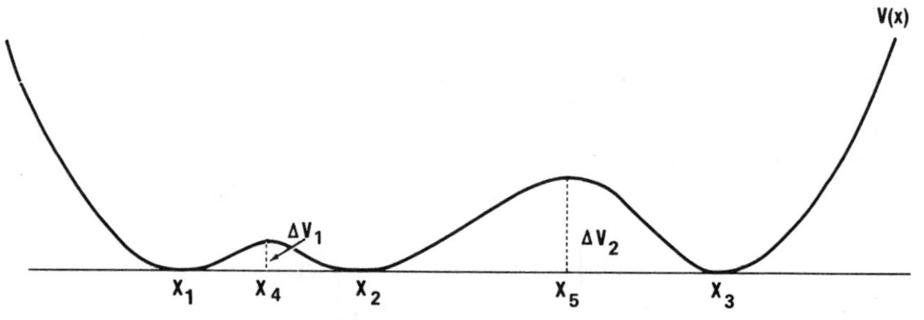

Figure 10.20 The Clausius–Clapeyron equations (10.61) have been integrated numerically to give a parametric representation for the locations of the three minima $x^{(1)}$, $x^{(2)}$, $x^{(3)}$ and the corresponding control parameter values $a = -1$, b, c, d.

The displacements of the three minima are determined using (5.2):

$$\delta x^{(i)} = -\sum_{\alpha=2}^{4} \frac{1}{V_{xx}^{(i)}} V_{x\alpha}^{(i)} \delta c^{\alpha}$$

where

$$\begin{aligned} V_{xx}^{(i)} &= \left.\frac{\partial^2 V}{\partial x^2}\right|_{x=x^{(i)}} \\ V_{x\alpha}^{(i)} &= \left.\frac{\partial^2 V}{\partial x \partial c^{\alpha}}\right|_{x=x^{(i)}} \end{aligned} \tag{10.62}$$

These equations have been integrated numerically using a small step length δs, and the results are presented in Fig. 10.20. Above any value of s we read the values of the three control parameters b, c, d for which the potential (10.49) has three equally deep minima as well as the locations $x^{(1)}, x^{(2)}, x^{(3)}$ of these critical points. The initial conditions for integrating this first-order ordinary differential equation were obtained from Fig. 10.16.

10. Triple Point

The reader who has grasped the "Crowbar Principle" cannot fail to notice that we did not exploit this principle when we discussed Fig. 10.7. We studied the critical point ($\sim x^4$), but it is the triple point ($\sim x^6$) which cries out to be treated with this principle.

However, there are difficulties. Whereas the critical point lies in the P–T plane and is characterized qualitatively by a catastrophe germ, the tricritical point, characterized qualitatively by the germ x^6, does not lie in the P–T plane. Rather, it is the triple point which lies in this plane, and this is not a catastrophe germ. The catastrophe germ itself is not accessible by varying P and T; it is only the triple point and the coexistence curves which are accessible.

To determine an equation of state for a substance in the neighborhood of the triple point, analogous to the van der Waals equation for the neighborhood of the critical point, we may proceed as in Sec. 5:

1. Identify the mathematical state variable x with the physical quantity ρ through $x \simeq (\rho - \rho_0)$.
2. Assume the simplest possible relationship between the four mathematical control parameters c^{α} and the two physical control parameters P, T:

$$\begin{bmatrix} a \\ b \\ c \\ d \end{bmatrix} = \begin{bmatrix} A_P^1 & A_T^1 & A^1 \\ A_P^2 & A_T^2 & A^2 \\ A_P^3 & A_T^3 & A^3 \\ A_P^4 & A_T^4 & A^4 \end{bmatrix} \begin{bmatrix} P \\ T \\ 1 \end{bmatrix} \tag{10.63}$$

Remark. Geometrically, we are considering the 2-dimensional physical P–T control plane $\mathbb{R}^2$ as embedded in the 4-dimensional mathematical (a, b, c, d) control parameter space $\mathbb{R}^4$. The plane $\mathbb{R}^2$ will then intersect the 2-dimensional component of the Maxwell set $\mathscr{S}_M$ in a point (the triple point) and the three 3-dimensional components (parameterizing potentials with two equally deep global minima) in three 1-dimensional curves (the coexistence curves). The embedding of physical $\mathbb{R}^2$ in mathematical $\mathbb{R}^4$ requires 12 parameters, given in the 4×3 matrix in (10.63).

With 14 free parameters (12 from the embedding, ρ_0, and an overall energy scale) one can "fit an elephant." However, there are several constraints on these parameters that can be immediately determined from the phase portrait of any substance (Fig. 10.21).

1. The mean density at the triple point may be determined.
2. Two of the three density differences at the triple point are independent.
3. Two of the three angles $\theta_1, \theta_2, \theta_3$ are independent, as is the slope of any one of the coexistence curves at T.

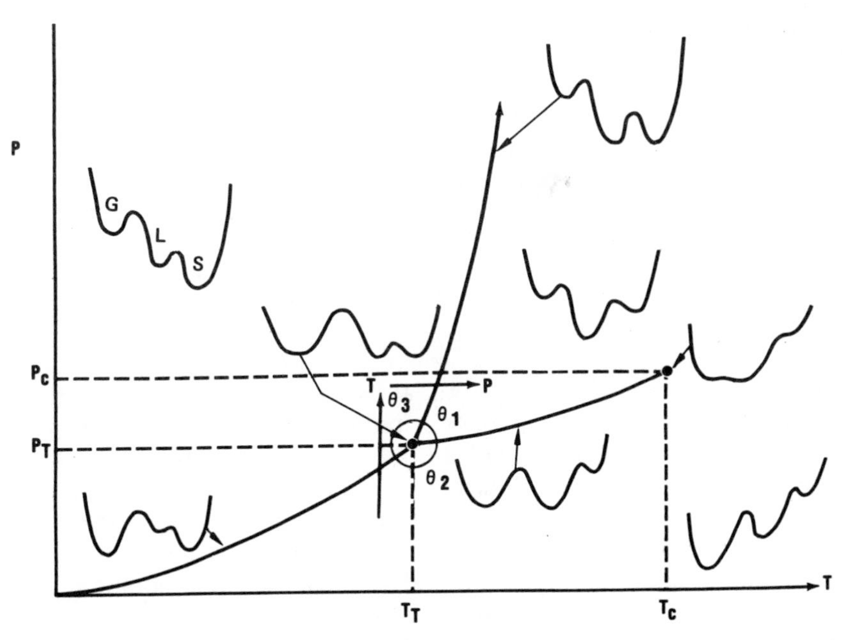

Figure 10.21 If the state of a substance near its triple point is governed by a potential—analog of the van der Waals potential—then the qualitative properties of the potential in various parts of the phase plane are as shown.

4. Latent heats along two independent paths near the triple point are independent.
5. The ratios P_C/P_T, T_C/T_T are independent.

The remaining 4 ($= 14 - 10$) degrees of freedom may be determined by fitting latent heat or density difference data along the coexistence curves.

It would be of great interest to determine how well a best fit embedding of physical $\mathbb{R}^2$ into mathematical $\mathbb{R}^4$ can reproduce available experimental data for well-studied simple physical substances. These considerations suggest the following two questions.

Question 4. What is the equation of state for a substance near its triple point, analog of the van der Waals equation of state for a substance near its critical point?

Question 5. Can the principle of corresponding states be extended to include the solid as well as the liquid and gas phases of a substance?

Remark. The van der Waals equation of state is an inadequate representation for a fluid because it predicts incorrect critical exponents as the cusp germ x^4 is approached. Since the germ x^6 does not lie in the physical P–T plane $\mathbb{R}^2$, it is anticipated that the classical exponents predicted by this A_{+5} analog of the A_{+3} van der Waals equation will provide a generally adequate representation for a substance in the neighborhood of its triple point.

11. Catastrophe Topology and Thermodynamics

Our discussions dealing with thermodynamics so far have been phenomenological. That is, we have been interested in determining equations that could reproduce experimental data to a greater or a lesser extent. We now turn in another direction to see whether Catastrophe Theory might be of any help in a conceptual formulation of thermodynamics.

A system in thermodynamic equilibrium is described by specifying the values of the appropriate thermodynamic variables. These variables fall broadly into two classes: those that are extensive, or proportional to the total system mass ($\sim M^1$) such as entropy and volume, and those that are intensive, or independent of the total system mass ($\sim M^0$) such as temperature and pressure. Extensive and intensive thermodynamic variables occur in conjugate pairs. In Table 10.1 we present a list of commonly occurring conjugate thermodynamic variables. It is convenient to adopt covariant (i_α) and contravariant (E^α) notation for the description of intensive and Extensive thermodynamic variables.

The n extensive variables E^α and the n intensive conjugate variables i_α describing a system *in thermodynamic equilibrium* are not all independent. In fact, only n of these variables can assume independent values. Once values are assigned to n of

Table 10.1 Conjugate Pairs of Extensive and Intensive Thermodynamic Variables

Extensive			Intensive
Entropy	S	T	Temperature
Volume	V	$-P$	Pressure
Mole number	N	μ	Chemical potential
Strain	$S_{\mu\nu}$	$T_{\mu\nu}$	Stress
Magnetization	$\mathbf{M}$	$\mathbf{H}$	Magnetic field
Polarization	$\mathbf{P}$	$\mathbf{E}$	Electric field
Electric charge	Q	Φ	Electric potential
Angular momentum	$\mathbf{J}$	$\mathbf{\Omega}$	Angular velocity
Black hole event horizon surface area	A	κ	Black hole surface gravity

these variables, the remaining n variables are determined. For example, if we regard the n extensive variables as independent variables, the n intensive variables are uniquely determined. Thus

$$i_\alpha = i_\alpha(E^\beta) \tag{10.64}$$

Relations of this type are called equations of state. Of course, we may choose any* n of the $2n$ thermodynamic variables to be independent. Then the remaining n dependent thermodynamic variables may be expressed as functions of the independent variables by manipulating the equations of state (10.64).

More generally, it is useful to regard the equations of state as a description of how the n-dimensional equilibrium manifold is embedded in a $2n$-dimensional space $\mathbb{R}^n \otimes \mathbb{R}^n$. The coordinates in this Euclidean space are the values of the n extensive and the n intensive thermodynamic variables. Specification of values for n of the $2n$ thermodynamic quantities determines the equilibrium state of the system uniquely, and therefore the corresponding point on the equation of state manifold.

This discussion suggests a very natural connection with Catastrophe Theory. The critical manifold is determined by a zero gradient condition. In the space $\mathbb{R}^n \otimes \mathbb{R}^k$ of n state variables and k control parameters this manifold has dimension k. It is useful to identify the equation of state manifold (thermodynamics) with the critical manifold (Catastrophe Theory). Since the equation of state manifold is an n-dimensional manifold embedded in a $2n$-dimensional space, we must choose n of the thermodynamic variables as control parameters (i.e., independent variables) and the remaining n dependent variables as state variables. The equation of state manifold is then determined by a zero gradient condition,

* This must be qualified slightly when several phases coexist in thermodynamic equilibrium.

where the gradient of some potential is taken with respect to the state variables.

There remain the details of choosing appropriate state variables and control parameters, an appropriate function of these whose critical manifold may be identified with the equation of state manifold of physical substances, and an appropriate convention to adopt. We will adopt the Maxwell Convention in conformity with our previous studies of phase transitions (Secs. 5 and 10). This done, we realize that the equation of state manifold is a soldered manifold. As a consequence the intensive and extensive variables do not occur on a completely equal footing. For example, systems in which more than one phase exists in equilibrium are described by flat portions of the equation of state manifold (cf. Fig. 10.5*b*). Points on this flat piece of the soldered manifold can be uniquely parameterized by the control parameters only if the control parameters are judiciously chosen. Otherwise they must be described by a mixture of state variables and control parameters.

To illustrate these remarks we will often discuss the thermodynamics of a simple single-component substance in this and the succeeding sections. The state of such a substance is characterized by the two pairs of conjugate thermodynamic variables $S, T; V, P$. The equation of state is a 2-dimensional manifold embedded in the space $\mathbb{R}^4$ whose coordinates are $S, T; V, P$. Any two of these four variables may be chosen as independent, and the equation of state (soldered) manifold may be projected down into the control space $\mathbb{R}^2$. In Fig. 10.22 we show four such projections for four different choices of the two control parameters: (a) two extensive variables (S, V); (b), (b′) an extensive and a nonconjugate intensive variable; and (c) two intensive variables (T, P). In this case the flat pieces (describing coexistence of phases) of the equation of state manifold can be uniquely parameterized by the extensive variables. This is true for a general thermodynamic system. It is therefore natural to identify the extensive thermodynamic variables with the system control parameters, in which case the state variables must be identified with the intensive thermodynamic variables.

We will therefore study the critical manifold $\nabla_i \mathscr{U}(i_\alpha; E^\beta)$ of some family of potential functions $\mathscr{U}$ depending on n intensive state variables i_α and n extensive control parameters E^β.

Remark. Thermodynamics was transformed from a study of processes to a study of states by America's greatest natural philosopher, J. W. Gibbs. In a Gibbsian formulation of thermodynamics the state of a system is determined by some kind of a thermodynamic "potential." The potential used depends on the choice of independent variables. The potentials commonly used to describe a single component substance are $U(S, V)$ (internal energy), $G(T, P)$ (Gibbs free energy), $A(T, V)$ (Helmholtz free energy), and $H(S, P)$ (enthalpy).* Under conditions of thermodynamic equilibrium these functions are uniquely defined once their two arguments have been specified. Therefore they are not potentials

* Functions depending on conjugate pairs (S, T), (V, P) are not used because they cannot be obtained from the four functions above by Legendre transformations.

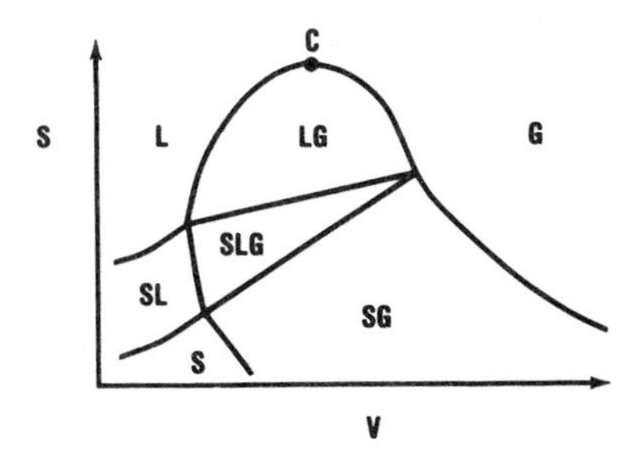

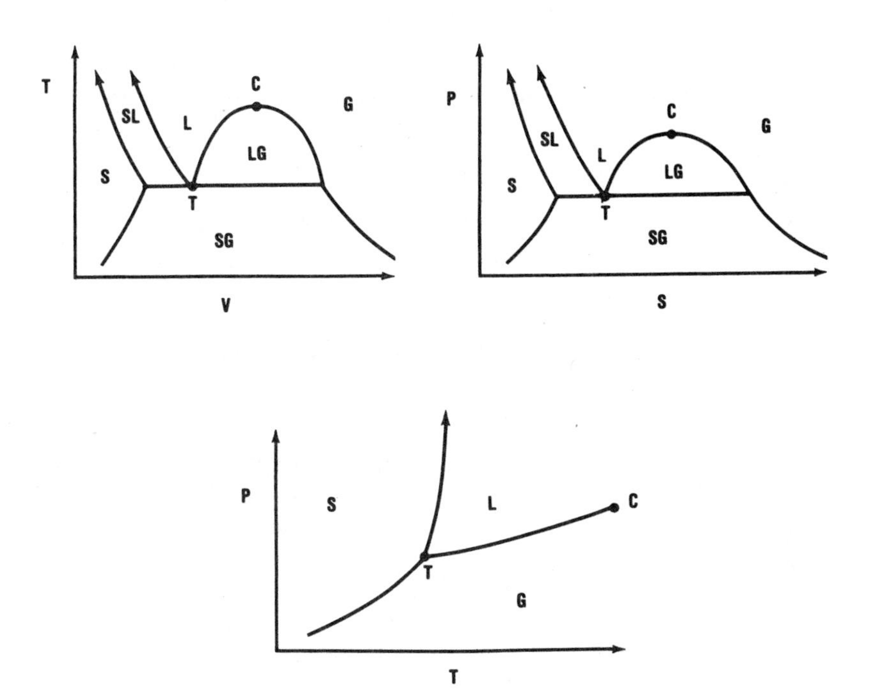

Figure 10.22 The 2-dimensional equation of state manifold in $\mathbb{R}^4$ for a simple substance can be projected down onto any 2-dimensional subspace. Projections onto those subspaces spanned by nonconjugate pairs of thermodynamic variables are shown here. Solid lines indicate phase boundaries.

because the system state is not determined by a zero gradient condition. Rather, they are generating functions because their gradients determine the conjugate thermodynamic variables:

$$\left(\frac{\partial U}{\partial S}\right)_V = T, \qquad \left(\frac{\partial U}{\partial V}\right)_S = -P \tag{10.65}$$

In fact, it is this nonzero gradient condition which is responsible for the Legendre transformations which exist among these thermodynamic generating functions. More generally, all thermodynamic properties of a system can be obtained by taking various partial derivatives of its thermodynamic "potentials," so the term "thermodynamic generating function" is more appropriate.

The identification of the Catastrophe Theory formulation of equilibrium thermodynamics with the classical Gibbsian formulation is now made as follows. The critical points of the potential $\mathscr{U}(i_\alpha; E^\beta)$ are determined through the zero gradient condition

$$\nabla_i \mathscr{U}(i_\alpha; E^\beta) = 0, \qquad 1 \leq \alpha, \beta \leq n \tag{10.66}$$

These n equations of state can be solved to determine the intensive variables i_α as functions of the extensive variables at each local minimum (other critical points are not interesting):

$$i_\alpha = i_\alpha^{(r)}(E^\beta), \qquad r = 1, \ldots \tag{10.67}$$

At each local minimum the value of the potential $\mathscr{U}$ is a well-defined function of the extensive control parameters:

$$U^{(r)}(E^\beta) = \mathscr{U}(i_\alpha^{(r)}(E^\beta); E^\beta) \tag{10.68i}$$

The system state is determined by the global minimum condition:

$$U(E^\beta) = \min_{r=1,\ldots} U^{(r)}(E^\beta) \tag{10.68ii}$$

This minimum condition defines the intensive variables through (10.67). The Gibbsian thermodynamic generating function $U(E^\beta)$ is equal to the potential function $\mathscr{U}(i_\alpha; E^\beta)$ evaluated on the critical manifold [defined by (10.66)] and chosen in accordance with the Maxwell Convention.

In the classical Gibbsian formulation of thermodynamics the extensive and conjugate intensive thermodynamic variables are related through the gradient of the generating function

$$i_\alpha = \frac{\partial U}{\partial E^\alpha} \tag{10.69}$$

Therefore the identification is completed by requiring (cf. (5.3; 1))

$$i_\alpha = \left. \frac{\partial \mathscr{U}(i; E)}{\partial E^\alpha} \right|_{CM} \tag{10.70}$$

where the subscript indicates that the derivative is evaluated on the critical manifold. The two definitions (10.66) and (10.70) of the intensive variables impose some constraints on the structure of the potential $\mathscr{U}(i; E)$.

Phase transitions occur on the "flat" part of the critical manifold. These flat parts project down into the Maxwell set in the control space. Therefore the

Maxwell set depends on our choice of control parameters. The Clausius–Clapeyron equations (5.68) may be used to determine the Maxwell set in the control space. If the system has t distinct phases in thermodynamic equilibrium, the Clausius–Clapeyron equations are determined from

$$\mathscr{U}_{\alpha}^{(1)}\delta E^{\alpha} = \mathscr{U}_{\alpha}^{(2)}\delta E^{\alpha} = \cdots = \mathscr{U}_{\alpha}^{(t)}\delta E^{\alpha} \tag{10.71}$$

All derivatives are evaluated on the critical manifold. Since the extensive variables uniquely describe the state of the system, their variations are arbitrary, and the Clausius–Clapeyron equations reduce to

$$i_{\alpha}^{(1)} = i_{\alpha}^{(2)} = \cdots = i_{\alpha}^{(t)} \tag{10.72}$$

In other words, in a system with t phases in thermodynamic equilibrium, each intensive thermodynamic variable ($T, P, \ldots$) assumes the same value in each phase. This is a transcription of a Newtonian equilibrium condition to thermodynamic systems. For this reason it has been customary to consider intensive thermodynamic variables as generalized forces and extensive thermodynamic variables as generalized displacements.

Of course, when a different collection of thermodynamic variables (p intensive and q extensive variables, where $p + q = n$) is used as control variables, the Maxwell set is different, and the Clausius–Clapeyron equations (5.68) provide a different set of conditions for the coexistence of phases.

12. Metric Geometry and Thermodynamics[4,5]

Changes in the state variables ($i_{\alpha} \to i_{\alpha} + \delta i_{\alpha}$) and in the control parameters ($E^{\beta} \to E^{\beta} + \delta E^{\beta}$) produce changes in the value of the potential $\mathscr{U}$:

$$\mathscr{U}(i_{\alpha} + \delta i_{\alpha}; E^{\beta} + \delta E^{\beta}) = \mathscr{U}^{(0)} + \delta^{(1)}\mathscr{U} + \delta^{(2)}\mathscr{U} + \cdots \tag{10.73}$$

Expressions for the first- and second-order terms $\delta^{(1)}\mathscr{U}$ and $\delta^{(2)}\mathscr{U}$ are given in (5.1). When the control parameters are varied sufficiently slowly ("quasistatically") so that the system remains in thermodynamic equilibrium, the displacements δi_{α} and δE^{β} are related as in (5.2). Under these conditions the variation in the value of the potential $\mathscr{U}$ on the critical manifold (= generating function U) is given by

$$U(E^{\beta} + \delta E^{\beta}) = U^{(0)} + \delta^{(1)}U + \delta^{(2)}U + \cdots \tag{10.74}$$

Expressions for the first and second-order terms $\delta^{(1)}U$ and $\delta^{(2)}U$ are given in (5.3). Specifically, the relation between the variations in the potential $\mathscr{U}$ and the generating function U are

$$\begin{array}{rl} \delta^{(1)}\mathscr{U} = & \mathscr{U}^{r}\delta i_{r} + \mathscr{U}_{\alpha}\delta E^{\alpha} \\ & \big\downarrow \text{CM} \\ \delta^{(1)}U = & \qquad\quad U_{\alpha}\delta E^{\alpha} \end{array} \tag{10.75i}$$

and

$$\begin{aligned}\delta^{(2)}\mathscr{U} &= \tfrac{1}{2}\mathscr{U}^{rs}\delta i_r \delta i_s + \mathscr{U}^r_\beta \delta i_r \delta E^\beta + \tfrac{1}{2}\mathscr{U}_{\alpha\beta}\delta E^\alpha \delta E^\beta \\ \delta^{(2)}U &= \qquad\qquad\qquad\qquad\qquad\qquad \tfrac{1}{2}U_{\alpha\beta}\delta E^\alpha \delta E^\beta\end{aligned} \tag{10.75ii}$$

where

$$U_{\alpha\beta} = \mathscr{U}_{\alpha\beta} - \mathscr{U}^r_\alpha (\mathscr{U}^{-1})_{rs} \mathscr{U}^s_\beta \tag{10.76}$$

Here $\mathscr{U}^r = \partial\mathscr{U}/\partial i_r$, $\mathscr{U}^r_\beta = \partial^2\mathscr{U}/\partial i_r \partial E^\beta$, and so on.

The expressions for the first- and second-order terms $\delta^{(1)}U$ and $\delta^{(2)}U$ are closely related to the first and second laws of thermodynamics. Under the identifications (10.75i) the first-order term may be written

$$\begin{aligned}\delta^{(1)}U &= U_\alpha \delta E^\alpha \\ &= T\,dS - P\,dV + \mu_j\,dN_j \cdots\end{aligned} \tag{10.77i}$$

This is the first law of thermodynamics. The second law of thermodynamics may be reformulated as a stability condition, specifically,

$$\frac{\partial^2 U}{\partial E^\alpha\,\partial E^\beta}\delta E^\alpha \delta E^\beta = U_{\alpha\beta}\delta E^\alpha \delta E^\beta \geq 0 \tag{10.77ii}$$

where the equality holds only if all displacements δE^α are zero. In short, the second law of thermodynamics is equivalent to the requirement that the metric tensor $U_{\alpha\beta}$ be positive definite.

Remark. The first derivatives U_α of the generating function and $\mathscr{U}_\alpha$ of the potential on the critical manifold are equal: $U_\alpha = \mathscr{U}_\alpha|_{CM}$. This is not true of the second derivatives. Since the critical manifold is defined by a minimum condition $\mathscr{U}^{rs}$ is positive definite on this manifold. So also is its inverse $(\mathscr{U}^{-1})_{rs}$. Therefore $\mathscr{U}^r_\alpha(\mathscr{U}^{-1})_{rs}\mathscr{U}^s_\beta$ is positive definite, so that $\mathscr{U}_{\alpha\beta}$ is "more positive definite" than $U_{\alpha\beta}$ in the sense that

$$(\mathscr{U}_{\alpha\beta} - U_{\alpha\beta})\delta E^\alpha \delta E^\beta \geq 0 \tag{10.78}$$

and equality implies that all displacements δE^α are zero.

The change in $U(E)$ produced by a change in the extensive parameters $E \to E + \delta E$ is

$$U = U^{(0)} + U_\alpha \delta E^\alpha + \tfrac{1}{2}U_{\alpha\beta}\delta E^\alpha \delta E^\beta + \cdots \tag{10.79}$$

where $U_\alpha = i_\alpha$. This expression leads us to a number of important conclusions. First, classical thermodynamics may be formulated as a constrained variational principle by looking for minima of the constrained function

$$\delta(U - \lambda_\alpha E^\alpha) = \delta(U^0 + \tfrac{1}{2}U_{\alpha\beta}\delta E^\alpha \delta E^\beta + \cdots) = 0 \tag{10.80}$$

Comparison with (10.69) leads to an immediate identification of the Lagrange multipliers λ_α with the intensive variables i_α:

$$\lambda_\alpha = i_\alpha \tag{10.81}$$

at the constrained minimum.

Next, by computing $\partial U/\partial E^{\alpha}$, we find

$$\begin{aligned}\frac{\partial U}{\partial E^{\alpha}} &= U_{\alpha} + U_{\alpha\beta}\delta E^{\beta} \\ &= i_{\alpha} + \delta i_{\alpha}\end{aligned} \tag{10.82}$$

Therefore the positive-definite metric tensor $U_{\alpha\beta}$ can be interpreted as a susceptibility tensor, since it gives the equilibrium response δi_{α} of an intensive variable i_{α} to a change in the values δE^{β} of the extensive variables E^{β}:

$$\delta i_{\alpha} = U_{\alpha\beta}\delta E^{\beta} \tag{10.83}$$

Since $U_{\alpha\beta}$ is positive definite, it is nonsingular, so it can be inverted. Thus the response of the extensive variables to a change in intensive variables is

$$\delta E^{\beta} = U^{\beta\alpha}\delta i_{\alpha}, \qquad (U_{\alpha\beta}U^{\beta\gamma} = \delta_{\alpha}{}^{\gamma}) \tag{10.84}$$

The change $\delta^{(2)}U$ can now be written in three convenient ways:

$$\begin{aligned}2\delta^{(2)}U &= U_{\alpha\beta}\delta E^{\alpha}\delta E^{\beta} \\ &= \delta i_{\alpha}\delta E^{\alpha} \\ &= \delta i_{\alpha}\delta i_{\beta}U^{\alpha\beta}\end{aligned} \tag{10.85}$$

and (10.79) can be written in the suggestive form

$$U = U^{(0)} + i_{\alpha}\delta E^{\alpha} + \tfrac{1}{2}\delta i_{\alpha}\delta E^{\alpha} + \cdots \tag{10.86}$$

Remark. The susceptibility tensor describing the infinitesimal linear response of the state variables to infinitesimal changes in the control parameters is given for the general case by (5.2) and for the particular case under study by (10.76). Since these susceptibilities must be equal, we find

$$U_{\alpha\beta} = (\mathscr{U}_{\alpha\beta} - \mathscr{U}^{r}_{\alpha}(\mathscr{U}^{-1})_{rs}\mathscr{U}^{s}_{\beta}) = -(\mathscr{U}^{-1})_{\alpha s}\mathscr{U}^{s}_{\beta} \tag{10.87}$$

on the critical manifold. This expression can be written in the simpler form

$$\mathscr{U}_{\alpha\beta} = (\mathscr{U}^{r}_{\alpha} - \delta^{r}_{\alpha})(\mathscr{U}^{-1})_{rs}\mathscr{U}^{s}_{\beta} \tag{10.87$'$}$$

Equations (10.70) and (10.87) are the only two constraints imposed on the thermodynamic potential $\mathscr{U}$ in order to reproduce the thermodynamics of the classical generating function U.

Example. The thermodynamic response functions for the simple single-component substance which are most easily measured are

$$\begin{aligned}C_V &= T\left(\frac{\partial S}{\partial T}\right)_V, & C_P &= T\left(\frac{\partial S}{\partial T}\right)_P \\ \beta_S &= -\frac{1}{V}\left(\frac{\partial V}{\partial P}\right)_S, & \beta_T &= -\frac{1}{V}\left(\frac{\partial V}{\partial P}\right)_T \\ \Gamma_V &= T\left(\frac{\partial S}{\partial P}\right)_V, & \alpha_P &= \frac{1}{V}\left(\frac{\partial V}{\partial T}\right)_P\end{aligned} \tag{10.88}$$

Here C_V, C_P are the heat capacities at constant volume and pressure, β_S and β_T are the adiabatic (constant S) and isothermal (constant T) compressibilities, Γ_V is the constant-volume heat of pressure variation, and α_P is the thermal expansion coefficient.

How are these thermodynamic response functions related to the metric tensors $U_{\alpha\beta}$, $U^{\alpha\beta}$? We determine the covariant components first by expressing the response of the intensive variables to a change in the extensive variables:

$$\begin{bmatrix} di_1 \\ di_2 \end{bmatrix} = \begin{bmatrix} dT \\ -dP \end{bmatrix} = \begin{bmatrix} A & B \\ B & C \end{bmatrix} \begin{bmatrix} dS \\ dV \end{bmatrix} = U_{\alpha\beta} \begin{bmatrix} dE^1 \\ dE^2 \end{bmatrix} \tag{10.89}$$

At constant volume ($dV = 0$) we find

$$\begin{aligned} dT = A\,dS \Rightarrow \frac{1}{A} &= \left(\frac{\partial S}{\partial T}\right)_V \\ -dP = B\,dS \Rightarrow \frac{1}{B} &= -\left(\frac{\partial S}{\partial P}\right)_V \end{aligned} \tag{10.90}$$

At constant entropy ($dS = 0$) we find

$$\begin{aligned} dT = B\,dV \Rightarrow \frac{1}{B} &= \left(\frac{\partial V}{\partial T}\right)_S \\ -dP = C\,dV \Rightarrow \frac{1}{C} &= -\left(\frac{\partial V}{\partial P}\right)_S \end{aligned} \tag{10.91}$$

Although the adiabatic expansion coefficient $\alpha_S = (1/V)(\partial V/\partial T)_S$ is not one of the response functions listed in (10.88), it is closely related to Γ_V by the symmetry $U_{\alpha\beta} = U_{\beta\alpha}$. The covariant susceptibility tensor is

$$U_{\alpha\beta} = \begin{bmatrix} \dfrac{T}{C_V} & -\dfrac{T}{\Gamma_V} \\ -\dfrac{T}{\Gamma_V} & \dfrac{1}{V\beta_S} \end{bmatrix} \tag{10.92}$$

The contravariant susceptibility tensor $U^{\alpha\beta}$ is somewhat easier to compute:

$$\begin{bmatrix} dE^1 \\ dE^2 \end{bmatrix} = \begin{bmatrix} dS \\ dV \end{bmatrix} = \begin{bmatrix} \left(\dfrac{\partial S}{\partial T}\right)_P & -\left(\dfrac{\partial S}{\partial P}\right)_T \\ \left(\dfrac{\partial V}{\partial T}\right)_P & -\left(\dfrac{\partial V}{\partial P}\right)_T \end{bmatrix} \begin{bmatrix} dT \\ -dP \end{bmatrix} = U^{\alpha\beta} \begin{bmatrix} di_1 \\ di_2 \end{bmatrix}$$

$$U^{\alpha\beta} = \begin{bmatrix} \dfrac{C_P}{T} & V\alpha_P \\ V\alpha_P & V\beta_T \end{bmatrix} \tag{10.93}$$

Since the covariant and contravariant metric tensors $U_{\alpha\beta}$ and $U^{\alpha\beta}$ are matrix inverses, $(U^{\alpha\beta})^{-1} = U_{\alpha\beta}$ and

$$\frac{1}{\frac{C_P}{T} V\beta_T - (V\alpha_P)^2} \begin{bmatrix} V\beta_T & -V\alpha_P \\ -V\alpha_P & \frac{C_P}{T} \end{bmatrix} = \begin{bmatrix} \frac{T}{C_V} & -\frac{T}{\Gamma_V} \\ -\frac{T}{\Gamma_V} & \frac{1}{V\beta_S} \end{bmatrix} \tag{10.94}$$

Only three of the six response functions listed in (10.88) are independent.[5]

The existence of a positive-definite metric tensor $U_{\alpha\beta}$ on the critical manifold leads to a number of useful statements about the general properties of thermodynamic systems. These statements take the form of inequalities and equalities. Although it is possible to discuss these properties directly on the equation of state manifold, it may be somewhat easier to discuss them in the n-dimensional linear vector space which is tangent to the critical manifold at an equilibrium point.

The construction of this tangent space has been discussed in Sec. 3. Suppose that $\mathbf{e}^\alpha$ span the state variable space $\mathbb{R}^n$ and $\mathbf{f}_\beta$ span the control parameter space $\mathbb{R}^n$. Then a general point in $\mathbb{R}^n \otimes \mathbb{R}^n$ is described by a vector $i_\alpha \mathbf{e}^\alpha + E^\beta \mathbf{f}_\beta$. The n vectors tangent to the critical manifold are then

$$|v_\beta\rangle = \mathbf{f}_\beta + \mathbf{e}^\alpha U_{\alpha\beta} \tag{10.95}$$

by (10.16). The vectors $|v_\beta\rangle$ may be considered as linear combinations of the vectors $\mathbf{e}^\alpha, \mathbf{f}_\beta$, defined at the origin in $\mathbb{R}^n \otimes \mathbb{R}^n$, and parallel transported to a point on the critical manifold. The vectors $|v_\beta\rangle$ span the tangent space.

An infinitesimal displacement δv with coordinates $(\delta i_\alpha; \delta E^\beta)$ in the equilibrium manifold may be identified with an infinitesimal vector $\delta v = \delta E^\beta |v_\beta\rangle$ in the tangent space, for

$$\delta E^\beta |v_\beta\rangle = \delta E^\beta(\mathbf{f}_\beta + U_{\beta\alpha}\mathbf{e}^\alpha) = \delta i_\alpha \mathbf{e}^\alpha + \delta E^\beta \mathbf{f}_\beta \tag{10.96}$$

In some instances it may be desirable to write the infinitesimal vector (10.96) in terms of the displacements δi_α. This can be done by introducing a second set of basis vectors $|v^\alpha\rangle$ in the tangent space, so that

$$\begin{array}{c} \delta v \\ /\!/ \qquad \backslash\!\backslash \\ \delta E^\beta |v_\beta\rangle = \delta i_\alpha |v^\alpha\rangle \end{array} \tag{10.97}$$

The relation between the basis vectors $|v_\beta\rangle$ and the basis vectors $|v^\alpha\rangle$ is easily determined from the relation between components:

$$\delta i_\alpha = U_{\alpha\beta}\delta E^\beta \Leftrightarrow |v^\alpha\rangle U_{\alpha\beta} = |v_\beta\rangle \tag{10.98}$$

An inner product in the tangent space can be obtained by identifying $(\delta v, \delta v)$ with $2\delta^{(2)}U$:

$$\begin{array}{c} (\delta v, \delta v) \\ /\!/ \qquad\qquad \backslash\!\backslash \\ \delta i_\alpha \delta i_\beta \langle v^\alpha | v^\beta \rangle \;\; \| \;\; \langle v_\alpha | v_\beta \rangle \delta E^\alpha \delta E^\beta \\ \delta i_\alpha \delta E^\beta \langle v^\alpha | v_\beta \rangle \end{array} \tag{10.99}$$

Comparison with (10.85) leads to the identifications

$$\begin{aligned}\langle v_\alpha | v_\beta \rangle &= U_{\alpha\beta} \\ \langle v_\alpha | v^\beta \rangle &= \delta_\alpha^\beta \\ \langle v^\alpha | v^\beta \rangle &= U^{\alpha\beta}\end{aligned} \tag{10.100}$$

In short, the basis vectors $|v^\alpha\rangle$ are dual to the basis vectors $|v_\beta\rangle$.

Remark. The properties of the metric in the tangent space have previously been studied by Weinhold.[5] The connection between Weinhold's dual set of basis vectors $|i_\alpha\rangle$, $|E^\beta\rangle$ and those used here is

$$\begin{aligned}|v_\alpha\rangle &\leftrightarrow |i_\alpha\rangle \\ |v^\beta\rangle &\leftrightarrow |E^\beta\rangle\end{aligned} \tag{10.101}$$

The existence of a positive-definite inner product in a vector space leads to inequalities based on the Schwartz inequality and to both inequalities and equalities based on the Bessel inequality.

Schwartz Inequality. If **u**, **v** are arbitrary vectors in a linear vector space with positive-definite inner product (,), then

$$(\mathbf{u}, \mathbf{v})^2 \leq (\mathbf{u}, \mathbf{u})(\mathbf{v}, \mathbf{v}) \tag{10.102}$$

Particular choices for **u** and **v** lead to the inequalities below:

$$\begin{array}{ccr} \mathbf{u} & \mathbf{v} & \\ |v^\alpha\rangle & |v^\beta\rangle & (U^{\alpha\beta})^2 \leq U^{\alpha\alpha}U^{\beta\beta} \\ |v_\alpha\rangle & |v^\beta\rangle & \delta_\alpha^\beta \leq U_{\alpha\alpha}U^{\beta\beta} \\ |v_\alpha\rangle & |v_\beta\rangle & (U_{\alpha\beta})^2 \leq U_{\alpha\alpha}U_{\beta\beta} \end{array} \tag{10.103}$$

Application. For the single component fluid, the metric tensors $U_{\alpha\beta}$, $U^{\alpha\beta}$ are given in (10.92) and (10.93). The Schwartz inequalities (10.102) lead to the following inequalities among the response functions (10.88):

$$\begin{aligned} \alpha_P^2 &\leq \frac{C_P \beta_T}{TV} & 1 &\leq \frac{C_P}{C_V} \\ \frac{1}{\Gamma_V^2} &\leq \frac{1}{TVC_V\beta_S}, & 1 &\leq \frac{\beta_T}{\beta_S} \end{aligned} \tag{10.104}$$

Bessel Inequality. If $\mathbf{u}_1, \mathbf{u}_2, \ldots$ is a system of orthonormal vectors in a space with positive-definite inner product, and **w** is an arbitrary vector, then

$$(\mathbf{w}, \mathbf{w}) \geq \sum_i (\mathbf{u}_i, \mathbf{w})^2 \tag{10.105}$$

The equality holds if the vector **w** lies in the space spanned by the vectors $\mathbf{u}_i$, or in particular if the vectors $\mathbf{u}_i$ span the linear vector space. Since $|v_\alpha\rangle, |v^\beta\rangle$ form a dual system, we can choose $\mathbf{u}_1 = |v_1\rangle/(U_{11})^{1/2}$ and $\mathbf{u}_2 = |v^2\rangle/(U^{22})^{1/2}$. Then $\mathbf{u}_1, \mathbf{u}_2$ are orthonormal. Particular choices for $\mathbf{w}$ lead to the inequalities below:

$$\begin{array}{ll} \mathbf{w} & \\ |v^1\rangle & U^{11} \geq \dfrac{1}{U_{11}} + \dfrac{(U^{12})^2}{U^{22}} \\ |v_2\rangle & U_{22} \geq \dfrac{(U_{12})^2}{U_{11}} + \dfrac{1}{U^{22}} \\ |v^3\rangle & U^{33} \geq 0 + \dfrac{(U^{32})^2}{U^{22}} \end{array} \tag{10.106}$$

Application. For the single component substance $n = 2$, so $\mathbf{u}_1, \mathbf{u}_2$ span the tangent plane and the inequalities become equalities. These are

$$\begin{array}{ll} \dfrac{1}{\beta_S} = \dfrac{TVC_V}{\Gamma_V^2} + \dfrac{1}{\beta_T}, & \beta_T = \beta_S + \dfrac{TV\alpha_P^2}{C_P} \\ \dfrac{1}{C_V} = \dfrac{TV\beta_S}{\Gamma_V^2} + \dfrac{1}{C_P}, & C_P = C_V + \dfrac{TV\alpha_P^2}{\beta_T} \end{array} \tag{10.107}$$

13. Thermodynamic Partial Derivatives

The bête noire of generations of thermodynamics students has been the computation of thermodynamic partial derivatives. Just how does one go about computing these creatures?

The classical method is to introduce a number of additional thermodynamic generating functions related to the internal energy U by Legendre transformations. Many thermodynamic partial derivatives can then be computed by noting the equality of mixed second partial derivatives of these generating functions. Choosing the correct generating function amounts to making inspired guesses and borders on black magic. Not all thermodynamic partial derivatives can be computed by this process.

The process of computing thermodynamic partial derivatives can be simplified when we realize that they are nothing more than susceptibilities, determining the response of one thermodynamic variable to a change in another when certain others are held fixed. The tangent space to the critical manifold contains all the information about the response of a system at equilibrium to a small change in the thermodynamic variables. The points in this n-dimensional tangent space can be parameterized by any n of the $2n$ thermodynamic variables $(i_\alpha; E^\beta)$, provided those chosen are independent. Computing thermodynamic partial derivatives amounts to changing independent coordinates describing points in the tangent space. This in turn is simply a problem in the algebra of linear vector spaces.

We illustrate a simple general procedure first by a simple example for $n = 2$.

Example. $(\partial E^1/\partial E^2)_{i_2} = ?$

Solution. The susceptibility tensor may be written

$$\delta E^\alpha = U^{\alpha\beta}\delta i_\beta \tag{10.108i}$$

In matrix form this is

$$\begin{bmatrix} \delta E^1 \\ \delta E^2 \end{bmatrix} = \begin{bmatrix} A & B \\ B & C \end{bmatrix} \begin{bmatrix} \delta i_1 \\ \delta i_2 \end{bmatrix} \tag{10.108ii}$$

This equation is "unsymmetric" in the sense that the extensive and intensive variables are treated differently. This equation

$$\begin{bmatrix} 1 & 0 \\ 0 & 1 \end{bmatrix} \begin{bmatrix} \delta E^1 \\ \delta E^2 \end{bmatrix} = \begin{bmatrix} A & B \\ B & C \end{bmatrix} \begin{bmatrix} \delta i_1 \\ \delta i_2 \end{bmatrix} \tag{10.108iii}$$

is more symmetric. We now decide which two of the four thermodynamic variables we would like to be independent. For the purpose of computing $(\partial E^1/\partial E^2)_{i_2}$, δE^2 and δi_2 are useful independent displacements. We bring the independent displacements to one side of the equation and place the dependent responses on the other side, simply by exchanging the appropriate columns of the 2×2 matrices above:

$$\begin{bmatrix} 1 & -A \\ 0 & -B \end{bmatrix} \begin{bmatrix} \delta E^1 \\ \delta i_1 \end{bmatrix} = \begin{bmatrix} 0 & B \\ -1 & C \end{bmatrix} \begin{bmatrix} \delta E^2 \\ \delta i_2 \end{bmatrix} \tag{10.108iv}$$

This equation is then solved for the dependent displacements as a function of the independent displacements by matrix inversion and multiplication:

$$\begin{bmatrix} \delta E^1 \\ \delta i_1 \end{bmatrix} = \begin{bmatrix} \dfrac{A}{B} & \dfrac{B^2 - AC}{B} \\ \dfrac{1}{B} & -\dfrac{C}{B} \end{bmatrix} \begin{bmatrix} \delta E^2 \\ \delta i_2 \end{bmatrix} \tag{10.108v}$$

The result is $(\partial E^1/\partial E^2)_{i_2} = A/B$. For a 1-component substance we see from (10.93) that $(\partial S/\partial V)_P = (C_P/T)/(V\alpha_P)$ and $(\partial V/\partial S)_T = \beta_T/\alpha_P$. These derivatives cannot be obtained from Maxwell relations.

In Table 10.2 we summarize the distinct susceptibility tensors that can occur for a system described by two intensive thermodynamic variables and two extensive thermodynamic variables. The independent variables may be: two intensive variables, two extensive variables, one extensive and a nonconjugate intensive variable, or a conjugate pair of variables. The susceptibility tensors are

Table 10.2 Susceptibility Tensors for an Equilibrium System Described by Two Intensive and Two Extensive Variables

$$\begin{bmatrix} \delta E^1 \\ \delta E^2 \end{bmatrix} = \begin{bmatrix} A & B \\ B & C \end{bmatrix} \begin{bmatrix} \delta i_1 \\ \delta i_2 \end{bmatrix}$$

$$\begin{bmatrix} \delta E^1 \\ \delta i_1 \end{bmatrix} = \begin{bmatrix} \dfrac{A}{B} & \dfrac{B^2 - AC}{B} \\ \dfrac{1}{B} & -\dfrac{C}{B} \end{bmatrix} \begin{bmatrix} \delta E^2 \\ \delta i_2 \end{bmatrix}$$

$$\begin{bmatrix} \delta E^1 \\ \delta i_2 \end{bmatrix} = \begin{bmatrix} \dfrac{AC - B^2}{C} & \dfrac{B}{C} \\ -\dfrac{B}{C} & \dfrac{1}{C} \end{bmatrix} \begin{bmatrix} \delta i_1 \\ \delta E^2 \end{bmatrix}$$

$$\begin{bmatrix} \delta i_1 \\ \delta i_2 \end{bmatrix} = \frac{1}{AC - B^2} \begin{bmatrix} C & -B \\ -B & A \end{bmatrix} \begin{bmatrix} \delta E^1 \\ \delta E^2 \end{bmatrix}$$

symmetric only when the independent displacements are either all extensive or all intensive.

The general prescription for computing thermodynamic partial derivatives is as follows:

1. Write down the susceptibility tensor in matrix form:

$$\delta^{\alpha}_{\beta}\,\delta E^{\beta} = U^{\alpha\beta}\,\delta i_{\beta}$$

2. Decide which displacements are independent. Bring them to the right-hand side of this equation, and put the dependent displacements on the left-hand side. This is done by exchanging the appropriate matrix columns.

3. Solve for the dependent displacements as a function of the independent displacements by matrix inversion and multiplication.

4. A change of basis to linear combinations of the independent displacements may be carried out.

5. The individual matrix elements of the resulting susceptibility tensor are the appropriate thermodynamic partial derivatives.

Example. For a simple magnetic 1-component substance, compute $(\partial S/\partial T)_{M,\sigma}$. Here we consider only magnetic fields in the z direction, and the derivative is computed along the coexistence curve.

Solution. The three conjugate pairs of variables describing this system are (S, T), $(V, -P)$, (M, H). If the slope of a coexistence curve is m, then along this curve

$$\frac{dP}{dT} = m \Leftrightarrow 0 = d\sigma = -dP + m\,dT \tag{10.109}$$

We therefore must choose as independent displacements $dT, d\sigma, dM$. Since $d\sigma$ is a linear combination of $-dP$ and dT, we begin by choosing $dT, -dP, dM$ as independent displacements. Here follow the steps of the algorithm described above:

STEP 1.

$$\begin{bmatrix} 1 & 0 & 0 \\ 0 & 1 & 0 \\ 0 & 0 & 1 \end{bmatrix} \begin{bmatrix} dE^1 \\ dE^2 \\ dE^3 \end{bmatrix} = \begin{bmatrix} A & B & C \\ B & D & E \\ C & E & F \end{bmatrix} \begin{bmatrix} di_1 \\ di_2 \\ di_3 \end{bmatrix} \tag{10.110i}$$

STEP 2.

$$\begin{bmatrix} 1 & 0 & -C \\ 0 & 1 & -E \\ 0 & 0 & -F \end{bmatrix} \begin{bmatrix} dE^1 \\ dE^2 \\ di_3 \end{bmatrix} = \begin{bmatrix} A & B & 0 \\ B & D & 0 \\ C & E & -1 \end{bmatrix} \begin{bmatrix} di_1 \\ di_2 \\ dE^3 \end{bmatrix} \tag{10.110ii}$$

STEP 3.

$$\begin{bmatrix} dS \\ dV \\ dH \end{bmatrix} = \begin{bmatrix} A - \dfrac{C^2}{F} & B - \dfrac{CE}{F} & \dfrac{C}{F} \\ B - \dfrac{CE}{F} & D - \dfrac{E^2}{F} & \dfrac{E}{F} \\ -\dfrac{C}{F} & -\dfrac{E}{F} & \dfrac{1}{F} \end{bmatrix} \begin{bmatrix} dT \\ -dP \\ dM \end{bmatrix} \tag{10.110iii}$$

STEP 4.

$$\begin{bmatrix} dT \\ d\sigma \\ dM \end{bmatrix} = \begin{bmatrix} 1 & 0 & 0 \\ m & 1 & 0 \\ 0 & 0 & 1 \end{bmatrix} \begin{bmatrix} dT \\ -dP \\ dM \end{bmatrix} \tag{10.110iv}$$

STEP 5.

$$\begin{bmatrix} dS \\ dV \\ dH \end{bmatrix} = \begin{bmatrix} A - \frac{C^2}{F} & B - \frac{CE}{F} & \frac{C}{F} \\ B - \frac{CE}{F} & D - \frac{E^2}{F} & \frac{E}{F} \\ -\frac{C}{F} & -\frac{E}{F} & \frac{1}{F} \end{bmatrix} \begin{bmatrix} 1 & 0 & 0 \\ -m & 1 & 0 \\ 0 & 0 & 1 \end{bmatrix} \begin{bmatrix} dT \\ d\sigma \\ dM \end{bmatrix} \tag{10.110v}$$

$$\left(\frac{\partial S}{\partial T}\right)_{\sigma, M} = \left(A - \frac{C^2}{F}\right) - m\left(B - \frac{CE}{F}\right) \tag{10.110vi}$$

The matrix elements A, B, C, etc. are easily computed. For example, $B = -(\partial S/\partial P)_{T,H} = (\partial V/\partial T)_{P,H}$ is one of the response functions which is easily measured in the laboratory.

Remark. This algorithm is simpler, more transparent, and of more widespread applicability for the computation of thermodynamic partial derivatives (thermodynamic response functions) than methods using Maxwell relations.

Table 10.3 Metric Tensors for an Equilibrium System Described by Two Intensive and Two Extensive Variables

$$(\delta i_1, \delta i_2)\begin{bmatrix} A & B \\ B & C \end{bmatrix}\begin{bmatrix} \delta i_1 \\ \delta i_2 \end{bmatrix}$$

$$(\delta E^2, \delta i_2)\begin{bmatrix} \frac{A}{B^2} & \frac{B^2 - AC}{B^2} \\ \frac{B^2 - AC}{B^2} & C\frac{AC - B^2}{B^2} \end{bmatrix}\begin{bmatrix} \delta E^2 \\ \delta i_2 \end{bmatrix}$$

$$(\delta i_1, \delta E^2)\begin{bmatrix} \frac{AC - B^2}{C} & 0 \\ 0 & \frac{1}{C} \end{bmatrix}\begin{bmatrix} \delta i_1 \\ \delta E^2 \end{bmatrix}$$

$$(\delta E^1, \delta E^2)\begin{bmatrix} \frac{C}{AC - B^2} & \frac{-B}{AC - B^2} \\ \frac{-B}{AC - B^2} & \frac{A}{AC - B^2} \end{bmatrix}\begin{bmatrix} \delta E^1 \\ \delta E^1 \end{bmatrix}$$

Further, Maxwell relation techniques compute the components of the susceptibility tensor one matrix element at a time, while the algorithm described above yields all the components of the susceptibility tensor simultaneously.

The metric tensor is equal to the susceptibility tensor only when the independent displacements are all of the same type, either extensive or intensive. When the independent displacements are mixed, the susceptibility tensor is generally not symmetric (Table 10.2), but the metric tensor must remain symmetric. It is a simple matter to compute the metric tensor in a mixed coordinate system by setting

$$2\delta^{(2)}U = \delta i_\alpha \delta E^\alpha \tag{10.111}$$

and then using the susceptibility tensor, computed as described by the algorithm above, to write this expression in terms of the independent displacements.

Example. Compute the metric tensor in a coordinate system in which δE^1 and δi_2 are the independent displacements.

Solution. The dependent displacements $\delta i_1, \delta E^2$ may be expressed in terms of δE^1, δi_2 using the inverse of the appropriate susceptibility tensor given in Table 10.2:

$$\begin{aligned}
\delta i_1 \delta E^1 + \delta i_2 \delta E^2 &= \left[\frac{1}{A}\delta E^1 - \frac{B}{A}\delta i_2\right]\delta E^1 + \delta i_2\left[\frac{B}{A}\delta E^1 + \left(\frac{AC - B^2}{A}\right)\delta i_2\right] \\
&= (\delta E^1, \delta i_2)\begin{bmatrix} \frac{1}{A} & 0 \\ 0 & \frac{AC - B^2}{A} \end{bmatrix}\begin{bmatrix} \delta E^1 \\ \delta i_2 \end{bmatrix}
\end{aligned} \tag{10.112}$$

Table 10.3 for metric tensors is the analog of Table 10.2 for susceptibility tensors.

14. Alternative Variational Representations

The thermodynamic variables most easily subject to experimental control are not always extensive. It is useful to consider catastrophe-theoretic formulations of equilibrium thermodynamics in which the thermodynamic variables which are most easily constrained in a given experimental situation are chosen as the control parameters. For example, for a single component substance we might want to choose (T, V) or (T, P) rather than (S, V) as the control parameters.

A variational formulation of thermodynamics based on a mixed set of extensive and intensive thermodynamic variables as control parameters and their conjugate variables as state variables may readily be obtained from the formulation given in Sec. 11. To the potential $\mathscr{U}(i_\alpha; E^\beta)$ we add a function f of the intensive and extensive variables. It is our objective to choose the function f to reproduce a variational formulation of thermodynamics with a predetermined mixture of extensive and intensive thermodynamic variables as control parameters. Further, we look for the *simplest* function f that will do this job. This task can be accomplished by expanding $\mathscr{U} + f$ up to second order about a point on the critical manifold:

$$(\mathscr{U} + f) = (\mathscr{U} + f)^0 + \delta^{(1)}(\mathscr{U} + f) + \delta^{(2)}(\mathscr{U} + f) + \cdots \quad (10.113)$$

$$(\mathscr{U} + f)^0 = \mathscr{U}(i_\alpha^{(0)}; E_{(0)}^\beta) + f(i_\alpha^{(0)}, E_{(0)}^\beta) \quad (10.114.0)$$

$$\delta^{(1)}(\mathscr{U} + f) = (\mathscr{U}^r + f^r)\delta i_r + (\mathscr{U}_\beta + f_\beta)\delta E^\beta \quad (10.114.1)$$

$$\delta^{(2)}(\mathscr{U} + f) = \tfrac{1}{2}(\mathscr{U} + f)^{rs}\delta i_r \delta i_s + (\mathscr{U} + f)_\beta^r \delta i_r \delta E^\beta + \tfrac{1}{2}(\mathscr{U} + f)_{\alpha\beta}\delta E^\alpha \delta E^\beta \quad (10.114.2)$$

The critical manifold is determined by the zero gradient condition, where the gradient is taken with respect to the state variables. Thus (10.114.1) can be used to determine the function f by requiring that the appropriate n of the $2n$ first derivatives vanish. The remaining n nonzero derivatives are then uniquely determined by the simplicity requirement.

Example 1. We illustrate these remarks by considering a potential depending on two intensive variables and two extensive variables. Then $\mathscr{U} = \mathscr{U}(i_1, i_2; E^1, E^2)$. Now suppose we wish to use i_1, E^2 as controls. The first derivatives appearing in (10.114.1) are now

$$(\mathscr{U} + f)^1, \qquad (\mathscr{U} + f)^2, \qquad (\mathscr{U} + f)_1, \qquad (\mathscr{U} + f)_2 \quad (10.115)$$

Further, $\mathscr{U}^r = \partial\mathscr{U}/\partial i_r = 0$ on the critical manifold. Since we want to use i_1, E^2 as control parameters, i_2 and E^1 must be state variables. Therefore we require

$$\begin{aligned} \frac{\partial(\mathscr{U} + f)}{\partial i_2} &= \qquad \frac{\partial f}{\partial i_2} = 0 \\ \frac{\partial(\mathscr{U} + f)}{\partial E^1} &= \mathscr{U}_1 + \frac{\partial f}{\partial E^1} = i_1 + \frac{\partial f}{\partial E^1} = 0 \end{aligned} \quad (10.116)$$

As a result, f is independent of i_2 and proportional to E^1. Specifically

$$f = -i_1 E^1 \quad (10.117)$$

is the simplest function that performs this job. The remaining two first derivatives (i.e., with respect to the new control parameters i_1 and E^2) are

$$\begin{aligned} \frac{\partial}{\partial i_1}(\mathscr{U} - i_1 E^1) &= \frac{\partial \mathscr{U}}{\partial i_1} - E^1 = -E^1 \\ \frac{\partial}{\partial E^2}(\mathscr{U} - i_1 E^1) &= \frac{\partial \mathscr{U}}{\partial E^2} - 0 = i_2 \end{aligned} \quad (10.118)$$

The general situation is now clear. In order to transform to a variational description of equilibrium thermodynamics depending on a mixed bag of control variables with the conjugate thermodynamic variables as state variables:

1. Add to $\mathscr{U}$ one term $-i_{\alpha'}E^{\alpha'}$ for each intensive variable $i_{\alpha'}$ which is to be considered as a control parameter ($f = -\sum_{``r"} i_{\alpha'}E^{\alpha'}$). Here "$r$" means the sum is restricted.
2. The function $\mathscr{U}' = \mathscr{U} + f$ so obtained has zero gradient with respect to the new state variables.
3. The first derivatives with respect to the control parameters are

$$\frac{\partial}{\partial E^{\beta'}}(\mathscr{U} + f) = i_{\beta'}$$

$$\frac{\partial}{\partial i_{\alpha'}}(\mathscr{U} + f) = -E^{\alpha'} \tag{10.119}$$

$(i_{\alpha'}, E^{\beta'})$ are the controls.

The critical manifold defined by the zero gradient condition above is the same as the critical manifold defined by $\nabla_i \mathscr{U}(i; E) = 0$.

What kind of variational principle does the prescription above lead to? Are we looking for a minimum, some saddle, or a maximum of $\mathscr{U}' = \mathscr{U} + f$? This question is resolved by investigating the expression $\delta^{(2)}\mathscr{U}'$:

$$\delta^{(2)}\mathscr{U}' = \delta^{(2)}\mathscr{U} - \sum_{``r"} \delta i_{\alpha'}\delta E^{\alpha'} \tag{10.120}$$

The matrix of mixed second partial derivatives of f with respect to the state variables is identically zero since each term in the sum comprising f is a product of a state variable and a control parameter. The matrix of mixed second partial derivatives of $\delta^{(2)}\mathscr{U}$ is a symmetric $n \times n$ submatrix of the $2n \times 2n$ matrix

$$\left[\begin{array}{c|c} \mathscr{U}^{rs} & \mathscr{U}^{r}_{\ \beta} \\ \hline \mathscr{U}^{s}_{\alpha} & \mathscr{U}_{\alpha\beta} \end{array}\right] \tag{10.121}$$

which is positive definite. Therefore $\delta^{(2)}\mathscr{U}'$ is positive definite with respect to the n new state variables. Thus the system thermodynamic state is determined by minimizing the function $\mathscr{U}'$ with respect to the new state variables for fixed value of the new control parameters.

Next, what are the properties of the metric tensor determined from $\delta^{(2)}U'$, where U' is the value of $\mathscr{U}'$ on the critical manifold? This is most easily determined from

$$\begin{aligned} \delta^{(2)}U' &= \delta^{(2)}U + \delta^{(2)}f \\ &= \tfrac{1}{2}\delta i_{\alpha}\delta E^{\alpha} - \sum_{``r"} \delta i_{\alpha'}\delta E^{\alpha'} \end{aligned} \tag{10.122}$$

We have seen already that $\delta^{(2)}U$ is positive definite. If we were to use as control parameters only intensive variables, the "restricted" sum above would reduce to

$$\delta^{(2)}U' = \tfrac{1}{2}\delta i_\alpha \delta E^\alpha - \sum_{\alpha'=1}^{n} \delta i_{\alpha'} \delta E^{\alpha'} = -\tfrac{1}{2}\delta i_\alpha \delta E^\alpha \qquad (10.123)$$

Thus $\delta^{(2)}U'$ would be negative definite. It follows that there is one sign change in the signature of the metric tensor obtained from $\delta^{(2)}U'$ for each term included in the restricted sum for f. Therefore the only choices of control parameters which lead to (positive or negative)-definite metrics on the critical manifold are all extensive or all intensive thermodynamic variables.

Example 2. Continuing Example 1, we consider

$$\mathscr{U}'(E^1, i_2; i_1, E^2) = \mathscr{U}(i_1, i_2; E^1, E^2) - i_1 E^1 \qquad (10.124)$$

The critical manifold is determined by minimizing $\mathscr{U}'$ with respect to E^1 and i_2. On the critical manifold $E^1 = E^1(i_1, E^2)$, $i_2 = i_2(i_1, E^2)$, and

$$U'(i_1, E^2) = \mathscr{U}'[E^1(i_1, E^2), i_2(i_1, E^2); i_1, E^2] \qquad (10.125)$$

where the global minimum is chosen. Then $\delta^{(2)}U' \geq 0$ for $\delta i_1 = 0$ and $\delta^{(2)}U' \leq 0$ for $\delta E^2 = 0$. The susceptibilities may be obtained from the derivatives of U' or from Table 10.2. The indefinite metric may be determined from $\delta^{(2)}U'$ in the coordinate system (i_1, E^2) or in any other coordinate system following the method outlined in (10.112).

In Fig. 10.23 we apply these arguments to a potential $\mathscr{U}(T, P; S, V)$ for a simple 1-component substance. The properties of the resulting functions are summarized.

Remark. Transformations of the type carried out above ($\mathscr{U} \to \mathscr{U}' = \mathscr{U} + f$, $f = -\sum_{``r"} i_\alpha E^\alpha$) are called **Legendre transformations**. It is not possible to find a Legendre transformation leading to a potential for which a conjugate pair of thermodynamic variables occurs among the control parameters (hence also among the state variables). Many thermodynamic partial derivatives can be constructed using the Maxwell relations (equality of mixed second partial derivatives) after choice of the appropriate thermodynamic generating function. Thermodynamic partial derivatives employing a conjugate pair of thermodynamic variables as independent variables [such as $(\partial S/\partial P)_V$] cannot be so computed because of the lack of a Legendre transformation, and must instead be computed from the susceptibility tensor, as summarized in Table 10.2.

Of fundamental importance for the description of a system in thermodynamic equilibrium is the equation of state. We have chosen to regard this as an n-dimensional (soldered) manifold in the space $\mathbb{R}^n \otimes \mathbb{R}^n$ of dimension $2n$ by considering the appropriate intensive and extensive thermodynamic variables on an

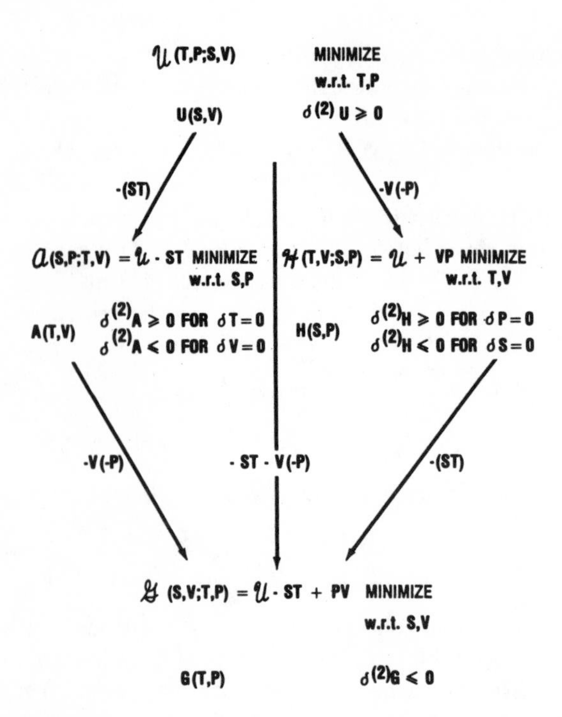

Figure 10.23 The relationships among, and properties of, thermodynamic potentials and their generating functions for a system with two extensive variables are as shown. A single manifold in $\mathbb{R}^{2n} = \mathbb{R}^4$ is represented by the various equations $\nabla(\text{potential}) = 0$, where the gradient is taken with respect to the state variables of the potential.

equal footing. To construct a variational formulation for determining the equation of state manifold, we have broken the symmetry which exists among the intensive and extensive thermodynamic variables by choosing n of them as the independent controls, and the remaining n as state variables with respect to which some potential must be minimized. The critical manifold (Maxwell Convention) was then identified with the equation of state manifold. The projection of the critical manifold in $\mathbb{R}^n \otimes \mathbb{R}^n$ down into the control space $\mathbb{R}^n$ then determines the Maxwell set. Different projections, required by different choices of control parameters, lead to different Maxwell sets. For any projection, the Maxwell set is determined by the Clausius–Clapeyron equations.

We have originally chosen the extensive thermodynamic variables as control parameters because they uniquely describe the state of a system in thermodynamic equilibrium. For this choice of control parameters the Maxwell set is empty and the Clausius–Clapeyron equations lead to the condition (10.71) requiring each intensive variable to have the same value in all phases that coexist.

Other choices of controls may lead to nonempty Maxwell sets in the control parameter space. The components of the Maxwell set are determined as usual by the Clausius–Clapeyron equations. The components of the Maxwell set have maximal dimension when projected into the control space of intensive thermodynamic variables.

The state of a system containing c components is specified by the $c + 1$ pairs of thermodynamic variables (S, T), (N_1, μ_1), (N_2, μ_2), $\ldots$, (N_c, μ_c). The equation of state is a $(c + 1)$-dimensional manifold in the space $\mathbb{R}^{2(c+1)}$. The state is uniquely determined by the extensive variables $(S, V, N_2, \ldots, N_c)$. If ν phases coexist in thermodynamic equilibrium, the potential describing the system has ν equally deep minima. The Maxwell set is empty in the control space of extensive variables but maximal in the control space of intensive variables. In the latter case the Clausius–Clapeyron equations for ν equally deep minima provide $(\nu - 1)$ constraints on the intensive control parameters. Therefore there are only

$$1 + c - (\nu - 1) = c + 2 - \nu \tag{10.126}$$

independent intensive control parameters (Gibbs phase rule). The system state can be uniquely prescribed by specifying in addition the values of $(\nu - 1)$ appropriate extensive thermodynamic variables.

To illustrate these remarks, we again consider a single component substance and choose the two intensive variables (T, P) as controls. If we assume that two phases coexist in thermodynamic equilibrium, the Clausius–Clapeyron equations reduce to

$$\left[\frac{\partial \mathscr{G}^{(1)}}{\partial T} - \frac{\partial \mathscr{G}^{(2)}}{\partial T}\right]\delta T + \left[\frac{\partial \mathscr{G}^{(1)}}{\partial P} - \frac{\partial \mathscr{G}^{(2)}}{\partial P}\right]\delta P = 0 \tag{10.127}$$

Since $\partial \mathscr{G}/\partial T = \partial G/\partial T = -S$ and $\partial \mathscr{G}/\partial(-P) = -V$ (cf. (10.119)], this reduces to

$$[-S^{(1)} + S^{(2)}]\delta T + [V^{(1)} - V^{(2)}]\delta P = 0 \tag{10.128}$$

or

$$\frac{dP}{dT} = \frac{S^{(1)} - S^{(2)}}{V^{(1)} - V^{(2)}} = \frac{\Delta S}{\Delta V} \tag{10.129}$$

where ΔS is the entropy change in converting the system from phase 1 to phase 2 and ΔV is the associated volume change. The system state is uniquely determined by specifying the value of the one independent control parameter, such as position along the coexistence curve, and one extensive thermodynamic variable, either S or V.

The system state at the triple point is uniquely specified by $0(=1 + 2 - 3)$ intensive parameters and 2 extensive variables.

15. Fluctuations

If g_{ij} is a positive-definite symmetric $n \times n$ matrix and g^{ij} is its inverse ($g^{ij}g_{jk} = \delta^i_k$), then

$$\langle x^i x^j \rangle = \frac{\int \cdots \int x^i x^j e^{-(1/2)g_{rs}x^r x^s}\, dx^1 \cdots dx^n}{\int \cdots \int e^{-(1/2)g_{rs}x^r x^s}\, dx^1 \cdots dx^n} = g^{ij} \tag{10.130}$$

This result can be proved as follows:

1. $\langle x^i x^j \rangle$ are the components of a second-rank contravariant tensor which is symmetric. Ditto for g^{ij}.
2. This equality is true in one particular coordinate system: that in which $g_{..} = g^{..} = I_n$.
3. Therefore it is true in all coordinate systems.

Fluctuations in the intensive and extensive variables δi_α and δE^β will occur about any point $(i_\alpha;\, E^\beta)$ in the critical manifold. The fluctuations may be equilibrium fluctuations, in which case the displacements δi_α and δE^β are related by a susceptibility tensor, or they may be nonequilibrium fluctuations, in which case the displacements δi_α and δE^β are not so related. Nonequilibrium fluctuations occur about the point $(i_\alpha;\, E^\beta)$ in $\mathbb{R}^n \otimes \mathbb{R}^n$, but equilibrium fluctuations are constrained to occur in the n-dimensional tangent space to the critical manifold at $(i_\alpha;\, E^\beta)$. Nonequilibrium and equilibrium fluctuations are described, respectively, by the probability distributions $\mathscr{P}$ and P defined by

$$\begin{aligned} \mathscr{P}(\delta i_\alpha;\, \delta E^\beta) &\simeq e^{\delta S/k} \simeq e^{-\delta^{(2)}\mathscr{U}/kT} \\ P(\delta E^\beta) &\simeq e^{\delta S/k} \simeq e^{-\delta^{(2)}U/kT} \end{aligned} \tag{10.131}$$

The positive-definite expressions $\delta^{(2)}\mathscr{U}$ and $\delta^{(2)}U$ are given in (10.75ii). These expressions, in conjunction with (10.130), lead to simple expressions for the expectation values of nonequilibrium and equilibrium fluctuations. In the nonequilibrium case

$$\begin{bmatrix} \langle \delta i_r \delta i_s \rangle & \langle \delta i_r \delta E^\beta \rangle \\ \langle \delta E^\alpha \delta i_s \rangle & \langle \delta E^\alpha \delta E^\beta \rangle \end{bmatrix}_{NE} = kT \begin{bmatrix} \mathscr{U}^{rs} & \mathscr{U}^r_\beta \\ \mathscr{U}^s_\alpha & \mathscr{U}_{\alpha\beta} \end{bmatrix}^{-1} \tag{10.132}$$

Therefore the mixed second partial derivatives of the potential $\mathscr{U}$ can be expressed in terms of the nonequilibrium fluctuations.

In the equilibrium case $\delta^{(2)}U = \frac{1}{2}U_{\alpha\beta}\delta E^\alpha \delta E^\beta$ and

$$\begin{aligned} \frac{\langle \delta E^\alpha \delta E^\beta \rangle}{kT} &= U^{\alpha\beta} \\ \frac{\langle \delta i_\alpha \delta E^\beta \rangle}{kT} &= \delta^\beta_\alpha \\ \frac{\langle \delta i_\alpha \delta i_\beta \rangle}{kT} &= U_{\alpha\beta} \end{aligned} \tag{10.133}$$

Example. For a single component system we have immediately, from (10.133),

$$\begin{bmatrix} \langle \delta S \delta S \rangle & \langle \delta S \delta V \rangle \\ \langle \delta V \delta S \rangle & \langle \delta V \delta V \rangle \end{bmatrix} = kT \begin{bmatrix} \dfrac{C_P}{T} & V\alpha_P \\ V\alpha_P & V\beta_T \end{bmatrix} \tag{10.134i}$$

$$\begin{bmatrix} \langle \delta T \delta S \rangle & \langle \delta T \delta V \rangle \\ \langle -\delta P \delta S \rangle & \langle -\delta P \delta V \rangle \end{bmatrix} = kT \begin{bmatrix} 1 & 0 \\ 0 & 1 \end{bmatrix} \tag{10.134ii}$$

$$\begin{bmatrix} \langle \delta T \delta T \rangle & \langle -\delta T \delta P \rangle \\ \langle -\delta P \delta T \rangle & \langle \delta P \delta P \rangle \end{bmatrix} = kT \begin{bmatrix} \dfrac{T}{C_V} & -\dfrac{T}{\Gamma_V} \\ -\dfrac{T}{\Gamma_V} & \dfrac{1}{V\beta_S} \end{bmatrix} \tag{10.134iii}$$

16. Additional Questions

It is well known that geometrical optics is the short wavelength ("$\lambda \to 0$") limit of the wave optics which is described by Maxwell's equations, and that classical mechanics is the "$h \to 0$" limit of quantum mechanics. So also it is known that statistical mechanics is the basis of thermostatics. Thus thermodynamics may be regarded as the "$k \to 0$" limit of statistical mechanics. We have used quotes to indicate that our limits have not been precisely defined because the three quantities λ, h, k are not dimensionless. These limits can be precisely defined by forming dimensionless ratios of these quantities, and then taking the limit as these dimensionless ratios go to zero. We will not do that here.

It is also known that much of wave optics can be recovered from geometrical optics in terms of a path integral (Huyghens') construction, and that much of quantum mechanics can be recovered from classical mechanics using a Feynman path integral approach. It is thus natural to ask if statistical mechanics can be recovered from thermodynamics by a natural inversion process somewhat akin to a path integral formulation.

Question 6. How much of statistical mechanics can be reconstructed from thermodynamics from a path integral approach? Is $S(\mathscr{U})/k$ the analog of the action $i\Delta t\mathscr{L}(q, \dot{q})/\hbar$? Does the Bloch equation play the same role vis-à-vis the transition (thermodynamics $\to$ statistical mechanics) as the Schrödinger equation does in the transition (classical mechanics $\to$ quantum mechanics)? How closely analogous is the density operator ρ to the scattering matrix S?

In a Hamiltonian formulation,

$$\frac{dq_i}{dt} = \frac{\partial \mathscr{H}}{\partial p_i}, \qquad \frac{dp_i}{dt} = -\frac{\partial \mathscr{H}}{\partial q_i} \tag{10.135}$$

classical mechanics receives an elegant interpretation in terms of symplectic geometry in $\mathbb{R}^n \otimes \mathbb{R}^n$. In the present variational formulation of thermodynamics the equations analogous to (10.135) are

$$i_{\beta'} = \frac{\partial \mathcal{U}'}{\partial E^{\beta'}}, \qquad -E^{\alpha'} = \frac{\partial \mathcal{U}'}{\partial i_{\alpha'}} \tag{10.136a}$$

$$0 = \frac{\partial \mathcal{U}'}{\partial E^{\beta''}}, \qquad 0 = \frac{\partial \mathcal{U}'}{\partial i_{\alpha''}} \tag{10.136b}$$

where $i_{\alpha'}$, $E^{\beta'}$ are the control parameters, $i_{\alpha''}$, $E^{\beta''}$ are the state variables, and $\mathcal{U}'$ is obtained from $\mathcal{U}$ as in (10.119). The relation between (10.135) and (10.136a) suggests the following question.

Question 7. Is there an intrinsic geometric structure in $\mathbb{R}^n \otimes \mathbb{R}^n$ associated with this variational formulation of thermodynamics? Is this geometry associated with the symplectic or orthogonal group $Sp(2n)$ or $SO(2n)$ or some related real form?

LeChatelier's principle is, like the second law of thermodynamics, enormously powerful but often vaguely stated. LeChatelier's principle may be stated as follows: An external force which disturbs the equilibrium of a system induces processes in the system which tend to weaken the effects of this force. This principle has been regarded by some authors as a statement within the context of equilibrium thermodynamics. When interpreted in this way, LeChatelier's principle is equivalent to the stability requirement on $\delta^{(2)}U$ which has been used to derive the positive definiteness of the metric tensor $U_{\alpha\beta}$. In point of fact, LeChatelier's principle is used to derive thermodynamic inequalities such as we have derived from the Schwartz and Bessel inequalities (10.102–10.107). The ensemble of such inequalities demands the positive definiteness of $U_{\alpha\beta}$.

I prefer to regard LeChatelier's principle as a statement within the context of nonequilibrium thermodynamics, and particularly about the approach to equilibrium of perturbed systems. M. Lax has pointed out[6] that LeChatelier's principle may be regarded even more generally as a stability condition describing the return of a system to its equation of state manifold, independent of whether this manifold is an equilibrium or some nonequilibrium steady-state manifold.

When a system in thermodynamic equilibrium is perturbed slowly, the susceptibility tensor is $U_{\alpha\beta}$. However, if the perturbation is carried out on a time scale faster than the system response time, the system response is "larger." This is the context of LeChatelier's principle. In short, the nonequilibrium response ($\mathcal{U}_{\alpha\beta}$) of a system to perturbation is greater than the equilibrium response ($U_{\alpha\beta}$) because the tensor $\mathcal{U}_{\alpha\beta} - U_{\alpha\beta}$ is positive definite [cf. (10.78)]. These considerations suggest the following question.

Question 8. Is LeChatelier's principle a statement about the positive definiteness of $\mathcal{U}_{\alpha\beta} - U_{\alpha\beta}$ rather than the positive definiteness of $U_{\alpha\beta}$? Is it a principle of nonequilibrium thermo*dynamics* ($\mathcal{U}_{\alpha\beta} - U_{\alpha\beta}$) rather than equilibrium thermo*statics*? How are the time constants describing relaxation of a perturbed system from its initial nonequilibrium response to its final steady-state equilibrium response related to the positive-definite tensors $\mathcal{U}_{\alpha\beta}$, $U_{\alpha\beta}$?

If the equation of state manifold of some system is derived from some variational principle (1.7)

$$\nabla V = 0 \tag{10.137.0}$$

then it is not unreasonable to assume that its time evolution off the critical manifold is given by a gradient system equation of the form (1.6)

$$\frac{dx_i}{dt} = -\frac{\partial V(x; c)}{\partial x_i} \tag{10.137.1}$$

In particular, the approach to the critical manifold may then be described by linearized equations of the type

$$\frac{dx_i}{dt} = -V_{ij}\delta x_j + \mathcal{O}(2) \tag{10.137.2}$$

Near the critical manifold the higher order terms may be neglected, and the time scale characteristic of the system is the reciprocal of the smallest eigenvalue of the positive-definite matrix V_{ij}.

When a system is displaced from thermodynamic equilibrium, then its regression back to the critical manifold is determined by the appropriate $n \times n$ submatrix of the $2n \times 2n$ matrix of mixed second partial derivatives of $\mathcal{U}$. For example, if the intensive parameters are held fixed and the extensive variables perturbed, then the appropriate gradient system equations are

$$\begin{aligned}\frac{dE^\alpha}{dt} &= -\frac{\partial}{\partial E^\alpha}(\mathcal{U} - i_\beta E^\beta)\\ &= -\frac{\partial}{\partial E^\alpha}(\mathcal{U}_\beta \delta E^\beta + \tfrac{1}{2}\mathcal{U}_{\beta\alpha}\delta E^\beta \delta E^\gamma + \cdots - i_\beta E^\beta)\\ &= -\mathcal{U}_{\alpha\beta}\delta E^\beta + \mathcal{O}(2)\end{aligned} \tag{10.138}$$

It is possible to prevent a system from reaching thermodynamic equilibrium by maintaining flows and gradients in the system. For example, if the two ends of an iron bar are maintained at different temperatures, the temperature gradient is responsible for a heat flow ($dS/dt \neq 0$) through the bar. Similarly, if an electrical potential difference is maintained between the two ends of this same bar, a current (dQ/dt) will flow through the bar. There are also cross effects. When two or more gradients are maintained simultaneously, the flows are proportional to both gradients. That is, near the critical manifold the generalized currents J^α are related to the generalized forces X_β by a susceptibility tensor L:

$$J^\alpha = L^{\alpha\beta} X_\beta \tag{10.139}$$

where

$$J^{\alpha} \simeq \frac{dE^{\alpha}}{dt}$$
$$X_{\beta} \simeq \nabla i_{\beta} \tag{10.140}$$

In the absence of magnetic fields the matrix $L^{\alpha\beta}$ is a real symmetric positive-definite matrix

$$L^{\alpha\beta} = L^{\beta\alpha} \quad \text{(Onsager reciprocity relations)}^{7}$$

This symmetry relation for steady-state behavior near equilibrium, compared to the symmetry relation $\mathscr{U}_{\alpha\beta} = \mathscr{U}_{\beta\alpha}$ describing nonsteady-state approach to thermodynamic equilibrium, suggests our closing question.

Question 9. How is the $n \times n$ matrix L of phenomenological coefficients related to the $2n \times 2n$ matrix (10.121) of mixed second partial derivatives of $\mathscr{U}$, evaluated on the critical manifold?

17. Summary

Our first substantive applications of Catastrophe Theory have been in the area of thermodynamics. We have considered explicitly three applications. These have been: (1) the derivation of the classical equation of state of a fluid in the neighborhood of its critical point (Secs. 2–6); (2) the discussion of the equation of state of a substance in the neighborhood of its triple point (Secs. 7–10); and (3) an alternative formulation for thermodynamics (Secs. 11–15). These applications have been surrounded by a general description of the Ehrenfest classification of phase transitions according to their order (Sec. 1), and a summary of questions raised by the attempted synthesis of Catastrophe Theory with thermodynamics through Riemannian geometry (Sec. 16).

Comparison of the Ginzburg–Landau canonical model for second-order phase transitions (Sec. 2) with the phase portrait of a fluid in the neighborhood of its critical point (Sec. 4) strongly suggests the presence of a cusp catastrophe underlying the liquid-gas first-order phase transition. Most of the flags indicative of an underlying catastrophe (Chapter 8) can be observed waving along the liquid-vapor coexistence curve, and in particular in the neighborhood of the critical point (Sec. 4). The identification of the critical point and coexistence curve with the cusp catastrophe can be made quantitative through affine linear identifications of the state variables ρ, x and control parameters (P_r, T_r), (a, b). A set of four reasonable assumptions leads from the cusp catastrophe manifold to the van der Waals equation of state (Sec. 5). This classical equation is quantitatively incorrect because it predicts incorrect values for certain critical exponents (Sec. 6). It is not possible, by making modifications of the identifications between the mathematical and the physical state variables or control parameters, to construct an equation with the appropriate critical exponents.

The triple point is of somewhat more interest than the critical point. In an attempt to construct an equation of state for a substance in the neighborhood of its triple point, analog of the van der Waals equation for a fluid near its critical point, we first discuss the canonical Ginzburg–Landau model for first-order phase transitions (Sec. 7). This symmetry-restricted model can be extended "by a half-step" to a discussion of tricritical points (Sec. 8), or "by a full step" to a discussion of the catastrophe A_5, and in particular to its Maxwell set. Then the phase portrait for a simple substance is the intersection of the physical P–T plane $\mathbb{R}^2$ with the Maxwell set $\mathcal{S}_M$ of A_5 in the mathematical control space $\mathbb{R}^4$. The general embedding of physical $\mathbb{R}^2$ in mathematical $\mathbb{R}^4$ is given explicitly (Sec. 10), but no embeddings for specific substances have yet been carried out.

In Sec. 11 we discuss the difference between the "potentials" characterizing systems in thermodynamic equilibrium and the potentials and families of potentials which are the subject of Catastrophe Theory. This distinction, and the natural identification of the equation of state of a substance with an n-dimensional manifold in the space $\mathbb{R}^n$ (intensive variables) $\otimes$ $\mathbb{R}^n$ (extensive variables), leads to a rather natural variational formulation of equilibrium thermodynamics in the spirit of Catastrophe Theory. In this formulation a real potential $\mathcal{U}$ is introduced. This is a function of n intensive state variables and n extensive control parameters. The equation of state is identified with the critical manifold.

In this formulation the equilibrium and stability conditions on families of functions, manifested in terms of the first and second derivatives, are equivalent to the first and second laws of thermodynamics. More specifically, the first law of thermodynamics is equivalent to the absence of infinitesimal terms in the intensive state variables from the expression for the first-order change in the potential at equilibrium

$$\delta^{(1)}\mathcal{U} = \mathcal{U}^r \delta i_r + \mathcal{U}_\alpha \delta E^\alpha \to 0 + U_\alpha \delta E^\alpha$$

The second law of thermodynamics is equivalent to a stability requirement, a positive-definiteness requirement on the second-order terms in the expansion of U on the critical manifold

$$\delta^{(2)} U = \tfrac{1}{2} U_{\alpha\beta} \delta E^\alpha \delta E^\beta \geq 0$$

The positive definiteness of the metric tensor $U_{\alpha\beta}$ means that the equation of state surface is a Riemann surface. Le Chatelier's principle appears to be a statement about the relative sizes of $\mathcal{U}_{\alpha\beta}$ and $U_{\alpha\beta}$:

$$(\mathcal{U}_{\alpha\beta} - U_{\alpha\beta}) \delta E^\alpha \delta E^\beta \geq 0$$

The positive definiteness of the metric tensor $U_{\alpha\beta}$ leads, by way of the Schwartz and Bessel inequalities, to a number of thermodynamic inequalities and equalities.

Linear response theory is most easily carried out in the tangent space to the critical manifold at a point. Any point in the tangent space ($\simeq \mathbb{R}^n$) in $\mathbb{R}^n \otimes \mathbb{R}^n$ may be determined by specifying n of the $2n$ coordinates. In this way susceptibility tensors and their matrix elements, the thermodynamic partial derivatives, may easily be computed. In particular, the metric tensor $U_{\alpha\beta}$ is a susceptibility tensor (Sec. 13).

If intensive and extensive variables are to be treated on an equal footing, then a Catastrophe Theory formulation of thermodynamics ought to be available using a mixed set of intensive and extensive thermodynamic variables as control parameters and a conjugate set as the state variables. Such formulations are presented in Sec. 14.

In a first attempt to go beyond purely classical thermodynamics, we consider both nonequilibrium fluctuations ($k \neq 0$) and equilibrium fluctuations, which are restricted to lie in the tangent plane to the critical manifold. The fluctuation correlation quantities are given by the matrices of mixed partial derivatives of the potential $\mathscr{U}$ and generating function U (Sec. 15).

Two considerations suggest that the current catastrophe theoretic formulation of thermodynamics is a valid and useful extension of the classical Gibbsian formulation. The first is the multiplicity of interpretations of the matrix $U_{\alpha\beta}$:

1. It determines $\delta^{(2)}U$ (Sec. 11).
2. It is a positive-definite metric tensor (Sec. 12).
3. It is a susceptibility tensor (Sec. 13).
4. It describes how the tangent space sits in $\mathbb{R}^n \otimes \mathbb{R}^n$.
5. It is a fluctuation correlation function (Sec. 15).
6. It puts a lower bound on the nonequilibrium response $\mathscr{U}_{\alpha\beta}$.

The second is that it raises more questions than it answers. In particular, the current formulation might be a suitable vehicle for extending thermodynamics from equilibrium to nonequilibrium configurations. There is "enough room" in $\mathbb{R}^n \otimes \mathbb{R}^n$ to step off the critical manifold in some direction not lying in the tangent space. There is not enough room to do so in the standard Gibbsian formulation of equilibrium thermodynamics, because it is formulated in $\mathbb{R}^n$. How to bridge the gap between equilibrium and nonequilibrium thermodynamics is the subject of the questions raised in Sec. 16.

References

1. L. D. Landau and E. M. Lefschitz, *Statistical Physics*, London: Pergamon, 1958.
2. J. D. van der Waals, Ph.D. Thesis, Leiden, 1873.
3. E. A. Guggenheim, The Principle of Corresponding States, *J. Chem. Phys.* **13**, 253–261 (1945).
4. Ref. 1, p. 61, eq. (21.2).

5. (a) F. Weinhold, Metric Geometry of Equilibrium Thermodynamics, *J. Chem. Phys.* **63**, 2479–2483 (1975).
(b) II. Scaling, Homogeneity, and Generalized Gibbs-Duhem Relations, *J. Chem. Phys.* **63**, 2484–2487 (1975).
(c) III. Elementary Formal Structure of a Vector-Algebraic Representation of Equilibrium Thermodynamics, *J. Chem. Phys.* **63**, 2488–2495 (1975).
(d) IV. Vector-Algebraic Evaluation of Thermodynamic Derivatives, *J. Chem. Phys.* **63**, 2496–2501 (1975).
(e) V. Aspects of Heterogeneous Equilibrium, *J. Chem. Phys.* **65**, 559–564 (1976).
6. M. Lax, private communication.
7. L. Onsager, Reciprocal Relations in Irreversible Processes. I, *Phys. Rev.* **37**, 405–426 (1931); Reciprocal Relations in Irreversible Processes. II, *Phys. Rev.* **38**, 2265–2279 (1931).

CHAPTER 11

Structural Mechanics

Many large-scale engineering structures can be described by a potential. The minimum value of the potential governs the state of the structure. The state is described by a point in some space—the structure state space. As the load which the structure (bridge, building) carries becomes heavier, the potential changes. Under severe loading, collapse may ensue. Collapse occurs when the locally stable state for which the system was designed loses its stability. Equilibrium, stability, and loss thereof are the hallmarks of Catastrophe Theory. We should therefore expect Catastrophe Theory to have many positive contributions to make to the study of engineering systems. This chapter is designed to illustrate the types of contributions Catastrophe Theory can bring to such engineering studies.

In this chapter we study a number of simple examples: Euler buckling (A_{+3}), collapse of a shallow arch (A_{-3}), and stability exchange (A_2). The methods of Catastrophe Theory enable us to determine the sensitivity of the critical or failure load to both imperfections and dynamical loading in the form of oscillations. Several severe imperfection sensitivities are encountered.

The methods of Catastrophe Theory can be used to study compound systems exhibiting several failure modes. It is a disquieting result that an "optimized" compound system composed of several "irreducible" structural elements may exhibit unexpected failure modes with severe imperfection sensitivity when strong coupling is present. This occurs in the collapse of a propped cantilever, for example.

Real-world structures are constructed from numerous individual elements. Failure studies have traditionally been carried out by means of bifurcation analyses. The standard algorithms (Chapter 4) for computing universal perturbations now allow an organized approach to the study of optimized systems, the sensitivity of such systems to imperfections, and the spectrum of their failure modes.

1. Systems Governed by a Potential

In order to describe the properties of a design, model, or actual physical structure, it is necessary first to introduce a coordinate system $x_1, x_2, \ldots, x_n$. These coordinates may be called order parameters. It is also useful to introduce an additional set of parameters $c_1, c_2, \ldots, c_k$. These parameters may represent the loads (external forces) on the system, the imperfections which occur in the manufacture of the component elements of the structure, and the imperfections which occur during construction of the system.

The total energy ξ of such a system may often be represented as the sum of a kinetic energy and a potential energy:

$$\xi = KE + PE \tag{11.1}$$

It is also often possible to represent the kinetic energy as a positive-definite quadratic form in the generalized velocities $\dot{x}_i$ and the potential energy in closed form as a function of the state variables $x \in \mathbb{R}^n$ and control parameters $c \in \mathbb{R}^k$:

$$\begin{aligned} KE &= \tfrac{1}{2} M_{ij} \dot{x}_i \dot{x}_j \geq 0 \\ PE &= V(x; c) \end{aligned} \tag{11.2}$$

We will first study conservative systems which are static. For such structures $KE = 0$ and $\xi = V$. In particular we will be concerned with:

1. Equilibrium configurations
2. Stability of equilibrium configurations
3. Loss of stability

If we assume that the configurations of the structure (in order parameter space) are determined by the minima of ξ, then the above three considerations translate into:

1. Equilibrium: $\nabla V = 0$
2. Stability: $V_{ij} \simeq M_k^n, \quad k = 0$
3. Loss of stability: $\det V_{ij} = 0$

These considerations make it clear that Catastrophe Theory must be a useful tool for describing the equilibrium, stability, and failure properties of physical structures.

A structure is usually designed to have certain characteristics. With perfect materials and worksmanship, these characteristics can be realized. Although we live in the best of all possible worlds, perfect materials and worksmanship are a chimera. This means that the constructed system will fail to realize its design characteristics. Of course we cannot tell beforehand by how much a system will fail to meet its design objectives, but we can tell beforehand how sensitive a design may be to imperfections that will occur.

The approach we take is as follows. The potential $V(x; c)$ for the ideal system is expanded about the designed equilibrium state which is stable for small external loads. It is usual to choose the state variables x_i to be zero in the ideal configuration. The Taylor series expansion for the ideal system then has the form

$$V(x; F) = \tfrac{1}{2}V_{ij}(F)x_i x_j + \text{higher order terms} \tag{11.3}$$

The constant term is unimportant, and has been dropped. The linear term is absent by the assumption $\partial V/\partial x_i = 0$ at $x = 0$. The leading Taylor series terms are therefore the second-degree terms. The Taylor series coefficients are functions of the control parameters $c \in \mathbb{R}^k$. In the absence of imperfections the only controls are the externally applied forces F. The designed stable equilibrium state remains locally stable as the load increases from zero as long as V_{ij} remains positive definite. The condition

$$\det V_{ij}(F) = 0 \tag{11.4}$$

determines the critical load $F = F_p$ at which the perfect system will fail.

When the germ of the potential $V(x; F)$ at $F = F_p$ is known, the methods of Chapter 4 can be used to determine the form of the most general perturbation of the ideal potential $V(x; F)$. This perturbation can be used to model all imperfections which arise through faulty worksmanship and the use of substandard building materials.

The potential $V(x; F, \varepsilon)$ representing the imperfect system can be studied in the same way as the potential $V(x; F)$ representing the perfect system. The critical load F_c at which the imperfect system fails is determined from

$$\det V_{ij}(F, \varepsilon) = 0 \tag{11.5}$$

It is natural to expect the imperfect system to have a lower carrying capacity ($F_c \leq F_p$) than the perfect system. (Otherwise why hire an architect?) The use of Catastrophe Theory allows the reduction in load carrying capability to be put into quantitative form. For the models which are considered below, we find

$$F_c = F_p - k|\varepsilon|^p \tag{11.6}$$

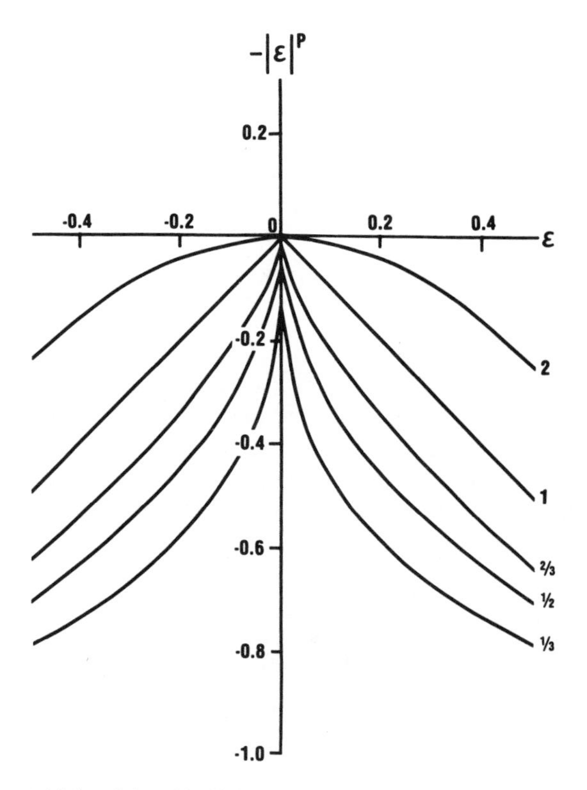

Figure 11.1 The sensitivity of the critical load F_c depends dramatically on the imperfection ε when the power p is small, where $F_c = F_p - k|\varepsilon|^p$.

where k is some positive constant, p is a positive rational fraction, and ε is some imperfection parameter. The imperfection sensitivity is shown in Fig. 11.1 for various values of the power p. The smaller p, the more drastic is the imperfection sensitivity.

Civil structures are made to house people and carry traffic. People and traffic pump kinetic energy into such structures. This dynamic loading of a system is also responsible for lowering the load carrying capacity of a structure. In short, the average kinetic energy ΔE added to a system by traffic noise may be considered as a (dynamic) imperfection parameter. We can determine the imperfection sensitivity of a structure to dynamic loading as follows. At any load F the critical points $x^{(0)}, x^{(1)}, \ldots$ of V are determined from $\nabla V(x; F, \varepsilon) = 0$. The critical values $V^{(i)} = V(x^{(i)}; F, \varepsilon)$ are determined at each critical point. If $x^{(0)}$ represents the locally stable equilibrium and $x^{(1)}$ represents the lowest nearby Morse 1-saddle, then the dynamical imperfection sensitivity is determined from

$$\Delta E = V^{(1)} - V^{(0)} \tag{11.7}$$

Physically, this means that the system remains in the locally stable state at $x^{(0)}$ for zero or small oscillations ($V^{(0)} + \Delta E < V^{(1)}$) until the kinetic energy pumped into the system is sufficiently large that the system can be "bounced" over the barrier of height $V^{(1)} - V^{(0)}$ into some distant configuration. Dynamical imperfection sensitivities obtained from (11.7) have the form (11.6) with $\varepsilon \to \Delta E$. For the systems studied in this chapter, the dynamical imperfection sensitivity is more severe than the static imperfection sensitivity.

Remark. Two structures with identical potential functions $V(x; F, \varepsilon)$ may have different kinetic energy functions. Their behavior under static loading will then be identical, but they may react differently under dynamic loading.

Remark. Even if a system is conservative, important perturbations of it may not be conservative. These include wind and rain loading, and some kinds of dynamic loading, such as a train going over a bridge. These caveats should be kept in mind to prevent the erroneous impression from occurring that stability and imperfection sensitivity analyses are completely reducible to Elementary Catastrophe Theory algorithms.

2. Euler Buckling of a Beam

As a first concrete illustration of the methods discussed in the previous section, we consider the Euler buckling of a compressed beam. Assume that a force F is applied to one end of an ideal incompressible beam, as shown in Fig. 11.2. For F small, the beam will remain unbent. For F very large, the beam will be very bent. We are interested in the details of what happens for intermediate values of F.

It is convenient to express the shape of the beam in a Fourier series expansion

$$y(x) = \sum_{j=1}^{\infty} a_j \sin \frac{j\pi x}{l} \tag{11.8}$$

Here $y(x)$ is the horizontal deflection of the beam as a function of distance, x, between the two ends, which are a distance l apart. The Fourier coefficients a_j determine the beam shape. They therefore play the role of the system state variables. The externally applied force F plays the role of a control parameter.

The fixed length L of the incompressible beam is given in terms of the parameter l and the deflection $y(x)$ by the expression

$$L = \int_0^l \sqrt{1 + \left(\frac{dy}{dx}\right)^2}\, dx \tag{11.9}$$

Equation (11.9) serves as a constraint on the parameter l and the state variables $a_1, a_2, \ldots$. This constraint can be made more explicit by writing

$$\begin{aligned} L &= \int_0^l \left\{1 + \frac{1}{2}\left(\frac{dy}{dx}\right)^2 - \frac{1}{8}\left(\frac{dy}{dx}\right)^4 + \cdots\right\} dx \\ &= l + \frac{1}{2}\sum_{j=1}^{\infty} a_j^2 \left(\frac{j\pi}{l}\right)^2 \frac{l}{2} + \text{higher order terms} \end{aligned} \tag{11.10}$$

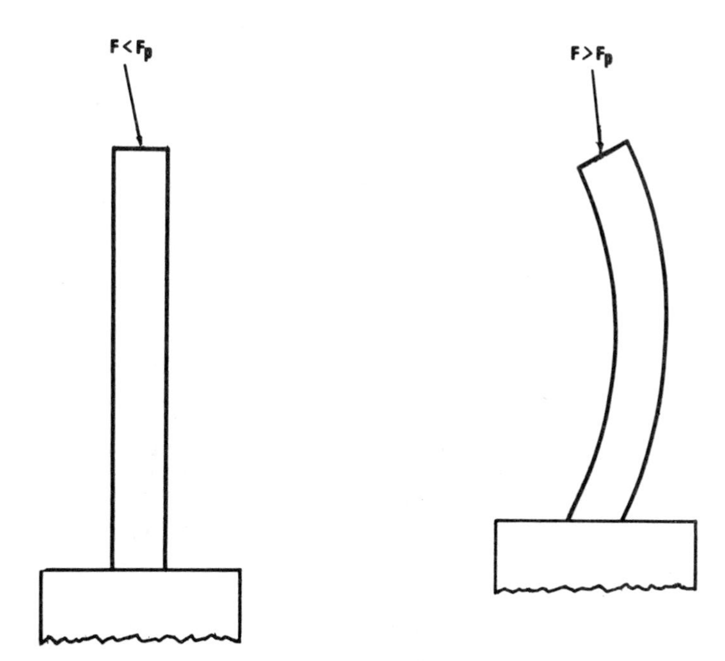

Figure 11.2 Under small loads the beam will remain unbent. Under large loads the beam will bend. Buckling was first systematically studied by Euler.[2]

Of the higher order terms, the only one of importance for our future purpose is the quartic term in a_1: $-l(3/2^{+6})(\pi/l)^4a_1^4$.

The potential energy stored in the curved beam is proportional to the square of the curvature of the beam, integrated over the beam length:

$$\frac{B}{2}\int_0^l\left(\frac{d^2y}{dx^2}\right)^2 dx = \frac{B}{2}\frac{l}{2}\sum_{j=1}^{\infty}\left(\frac{j\pi}{l}\right)^4 a_j^2 \tag{11.11}$$

Here B is a constant called the bending stiffness.[1] The work done by the external force is

$$W = \int_L^l F\cdot dx = -F(L-l) \tag{11.12}$$

The potential representing the energy of the perfect static beam is the sum of (11.11) and (11.12):

$$\begin{aligned} V_p(a_j; F) &= \frac{Bl}{4}\sum_{j=1}^{\infty}\left(\frac{j\pi}{l}\right)^4 a_j^2 - \frac{Fl}{4}\sum_{j=1}^{\infty}\left(\frac{j\pi}{l}\right)^2 a_j^2 + \text{higher order terms} \\ &= \frac{l}{4}\sum_{j=1}^{\infty}\left(\frac{j\pi}{l}\right)^2\left[B\left(\frac{j\pi}{l}\right)^2 - F\right]a_j^2 + \text{higher order terms} \end{aligned} \tag{11.13}$$

In deriving (11.13), we have neglected the weight of the beam.

The state of the beam is determined by minimizing the potential $V_p(a_j; F)$. This function is positive definite for $F < F_1 = B(\pi/l)^2$. As a function of increasing load F, the first non-Morse critical point is encountered at $F = F_1$. If we remain on the unstable undeflected branch, additional non-Morse critical points are encountered at $F_2, F_3, \ldots$, where $F_j = B(j\pi/l)^2$. The function $V_p(a_j; F)$ is a Morse function for all values of F except $F = F_j (j = 1, 2, \ldots)$. At these values $V_p(a_j; F)$ is a non-Morse function in the state variable a_j, but is Morse in all remaining state variables.

In order to describe the state of the beam for $F > F_1$ (the so-called post-buckling state), it is necessary to look at the terms of higher than second degree in the single state variable a_1. In compensation, all other state variables may be neglected. The potential describing the state of the beam is now effectively

$$V_p(a_j; F) \rightarrow V(a_1; F) = \frac{l}{4}\left(\frac{\pi}{l}\right)^2 (F_1 - F)a_1^2 + \frac{3Fl}{2^6}\left(\frac{\pi}{l}\right)^4 a_1^4 + \cdots \tag{11.14}$$

For $F > F_1$ the amplitude of the first Fourier coefficient is determined by

$$a_1^2 = \frac{8}{3F}\frac{F - F_1}{(\pi/l)^2}, \qquad F \geq F_1 \tag{11.15}$$

If the beam is constrained so that it cannot buckle into the lowest energy configuration ($j = 1$), then it will buckle at a higher value of $F(=F_2)$ into the next higher ($j = 2$) bending configuration. More generally, if the first $j - 1$ buckling modes are forbidden by constraints, buckling will occur into the jth mode whose shape is $\sin j\pi x/l$ at $F_j = B(j\pi/l)^2$. The values of a_j for $F > F_j$ are shown in Fig. 11.3.

We have restricted our considerations so far only to the static properties of perfect beams. We next consider the effects of perturbations on the beam shape. The most general perturbation of the potential (11.13) includes also a linear term, so that

$$V_i(a_1; F, \varepsilon) = \varepsilon a_1 + \frac{l}{4}\left(\frac{\pi}{l}\right)^2 (F_1 - F)a_1^2 + \frac{3Fl}{2^6}\left(\frac{\pi}{l}\right)^4 a_1^4 \tag{11.16}$$

That is, the most general imperfection in the beam can be modelled by a warp of nonzero magnitude. The equilibrium values for a_1 are determined by the zero gradient condition $\nabla_1 V_i(a_1; F, \varepsilon) = 0$. These values of $a_1(F)$ are shown for various values of ε in Fig. 11.4. These curves are easily seen to be intersections of the cusp catastrophe manifold $x^3 + ax + b = 0$ with planes $b =$ constant. The stability of the beam along these equilibrium paths is easily determined from the stability properties of the cusp catastrophe.

If the beam is not static, but possesses a kinetic energy ΔE entirely in the lowest mode, then a_1 will oscillate about its static value $a_1(F)$. The range of oscillation of the order parameter a_1 is shown in Fig. 11.5 for several values of the imperfection parameter ε.

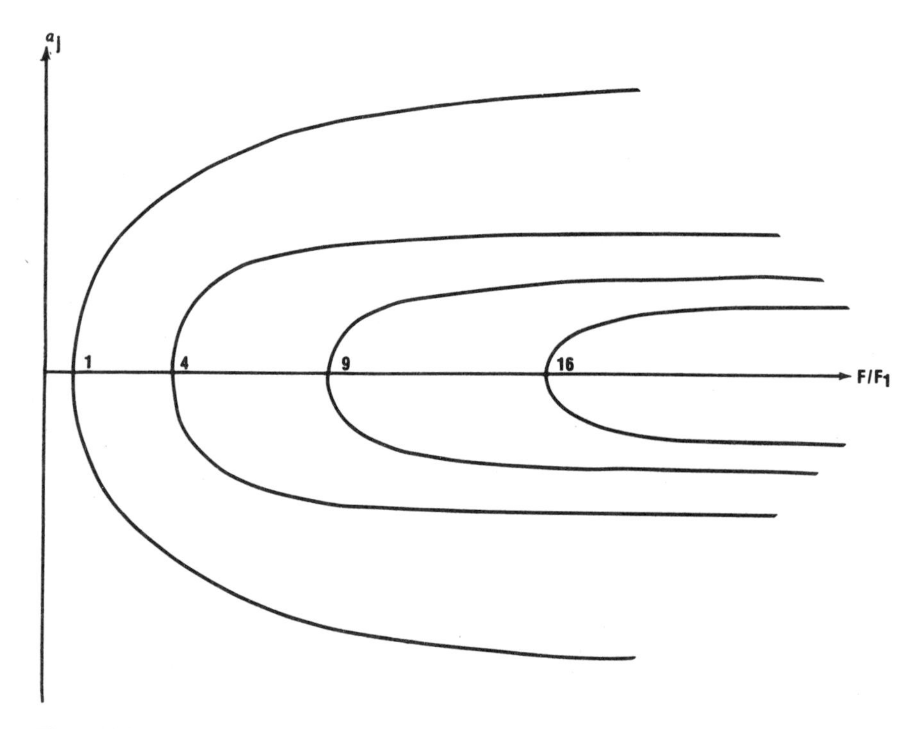

Figure 11.3 Under a large force F, one of the early Fourier coefficients a_j may be nonzero, as shown.

Euler buckling of a loaded beam is a second-order phase transition. The buckling failure is "soft" because the postbuckling path is connected continuously to the prebuckling path. Structures exhibiting soft postbuckling behavior do not collapse when the critical load is exceeded—they merely sag. Therefore it is useful to employ some sort of safety criterion to determine the limits of safe loading. For many practical purposes it is useful to declare a beam unsafe when the beam deflection amplitude a_1 exceeds a preassigned safe value s: $|a_1| > s > 0$. Since the deflection amplitude is determined by the zero gradient condition, the maximum safe deflection determines the maximum safe load F_s through

$$\nabla_1 V(a_1 = s; F_s, \varepsilon) = 0 = \varepsilon + \frac{l}{2}\left(\frac{\pi}{l}\right)^2 (F_1 - F_s)s + \frac{3F_s l}{2^4}\left(\frac{\pi}{l}\right)^4 s^3$$
$$F_s = F_c(s) - k(s)\varepsilon \tag{11.17}$$

where

$$F_c(s) = \frac{F_1}{1 - \frac{3}{8}(\pi s/l)^2}$$
$$k(s) = \frac{(2/|s|l)(l/\pi)^2}{1 - \frac{3}{8}(\pi s/l)^2} \tag{11.18}$$

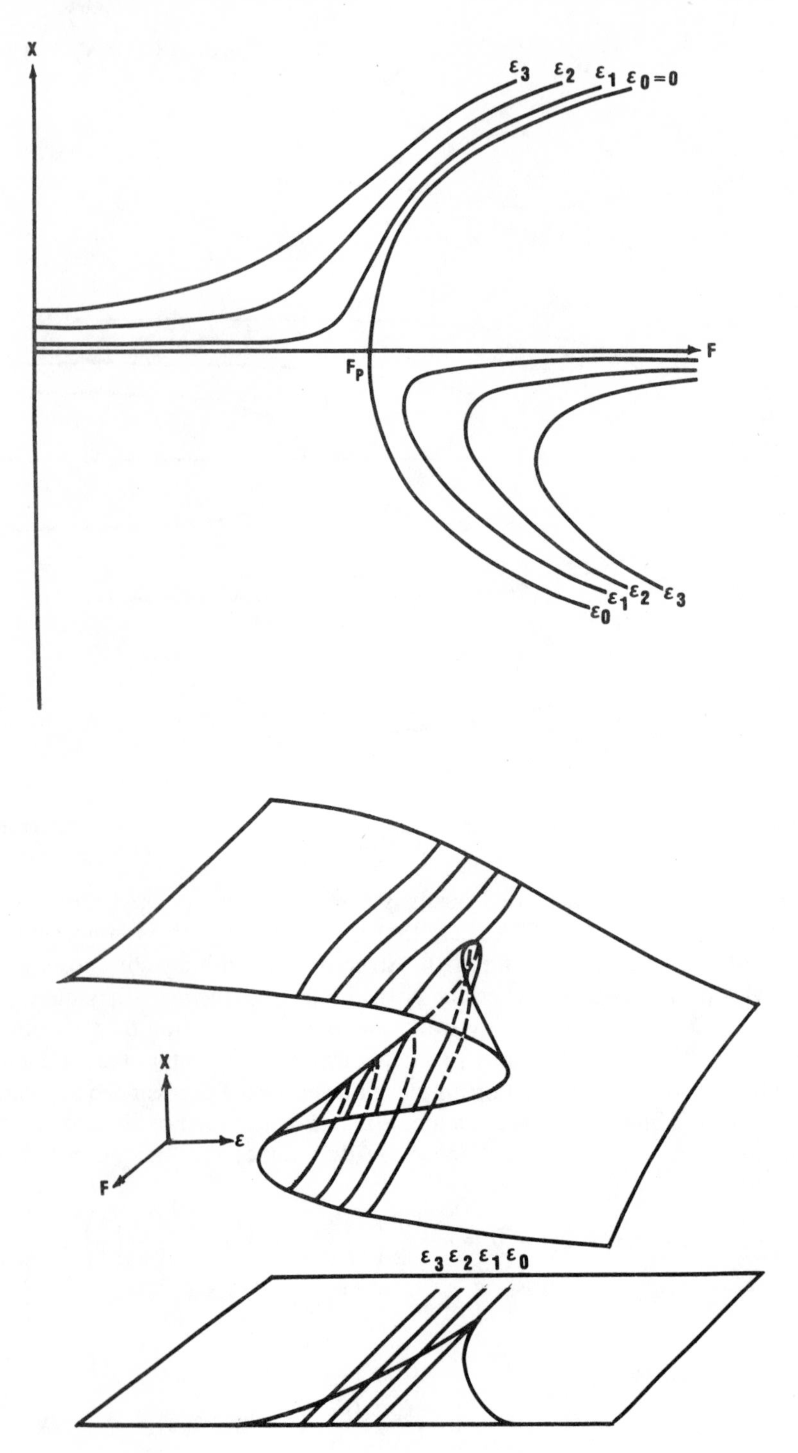

Figure 11.4 The equilibrium configuration of the imperfect beam depends on the magnitude of the loading force F and of the imperfection parameter ε.

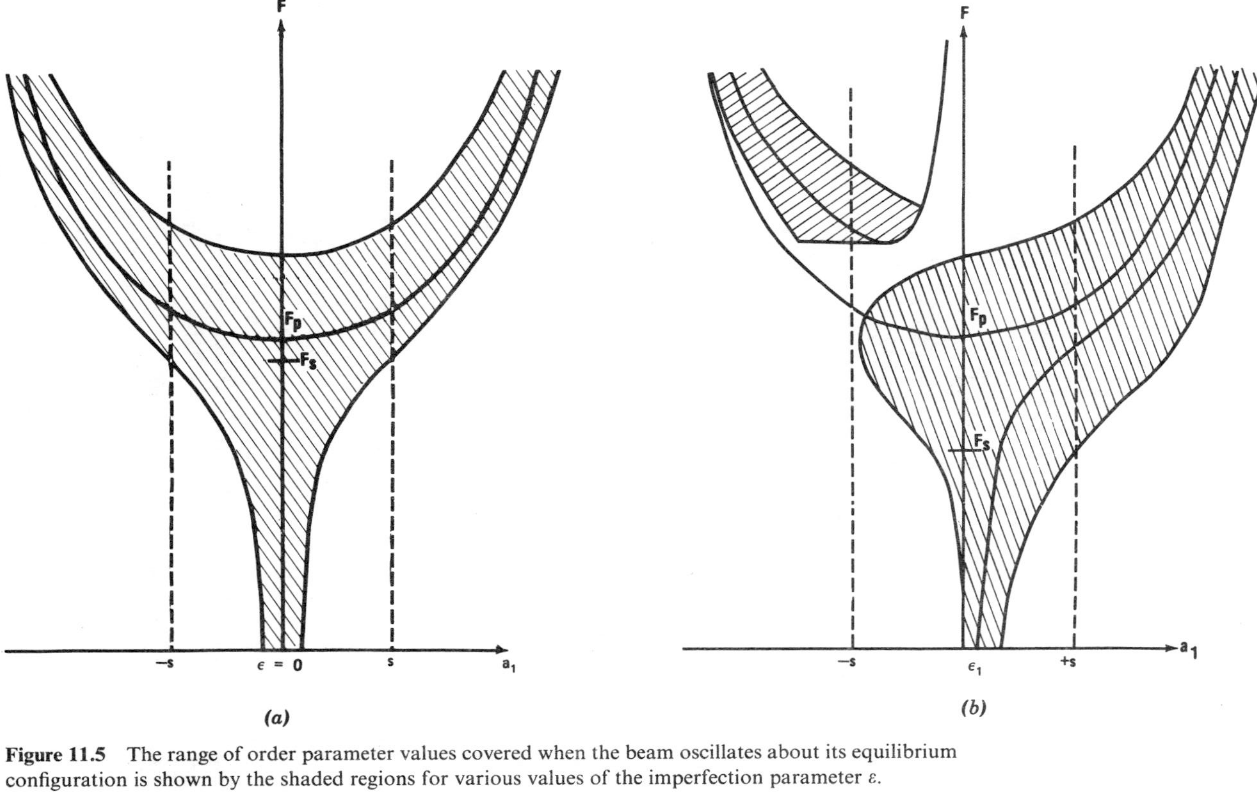

Figure 11.5 The range of order parameter values covered when the beam oscillates about its equilibrium configuration is shown by the shaded regions for various values of the imperfection parameter ε.

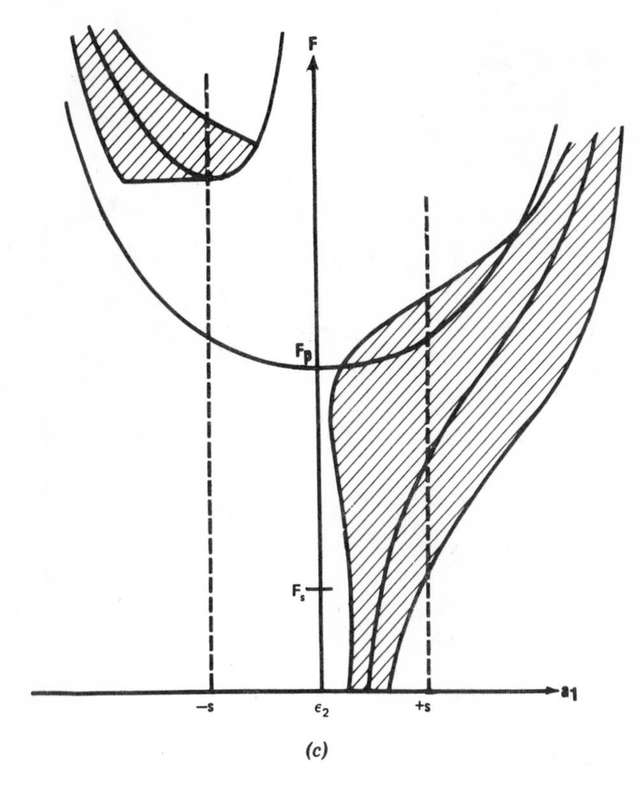

Figure 11.5 (*Continued*)

The imperfection sensitivity of the safe load is fairly soft, depending on the first power of the imperfection parameter ε. Here $F_c(s)$ is the safe load in the absence of imperfections. For sufficiently large imperfections no load is safe.

We can define the maximum safe load carrying capacity for an oscillating beam in a similar way. The oscillations occur about the static equilibrium positions, as shown in Fig. 11.5. The maximum safe load carrying capacity may be defined as the value of F at which the deflection amplitude reaches s. For a perfect system this load is determined from

$$\Delta E = \frac{l}{4}\left(\frac{\pi}{l}\right)^2 (F_1 - F_s)s^2 + \frac{3F_s l}{2^6}\left(\frac{\pi}{l}\right)^4 s^4 \tag{11.19}$$

This provides the following linear relation between the kinetic energy ΔE and the maximum safe load:

$$F_s = \frac{F_1 - (\Delta E/l)(2l/\pi s)^2}{1 - 3(\pi s/4l)^2} \tag{11.20}$$

Buckling beams, and other structural systems exhibiting an A_{+3} catastrophe, have very mild safe loading sensitivity toward both imperfections and kinetic energy.

3. Collapse of a Shallow Arch

Buckled beams are not necessarily poor building components. The shallow arch (Fig. 11.6) is extremely useful for bridging gaps. For small vertical loads the arch is undeformed, while for large loads the arch will collapse. Once again we are interested in the details of what happens for intermediate values of the load. In particular, we will attempt to determine the critical load as well as its sensitivity to both imperfections and dynamical loading.

It is convenient to assume that the shallow arch has been obtained by applying an axial force to an incompressible beam, as shown in Fig. 11.2. Then its equilibrium shape, in the absence of loading forces and imperfections, is

$$y(x) = a_1^0 \sin \frac{\pi x}{l} \tag{11.21}$$

Once again the arch shape under loading may be determined by a Fourier series analysis. The calculations can be carried through as before in an infinite-dimensional state space in which the state variables are the Fourier series coefficients a_j. However, this was unnecessary for the study of Euler buckling. In that case we were able to use a finite-dimensional state space $\mathbb{R}^n$, where in fact $n = 1$. This is because of the high energy cost for buckling in each successively higher mode. In the case of the collapse of a shallow arch, the infinite-dimensional state space may also be replaced by a finite-dimensional space. It is sufficient to retain only the two leading Fourier coefficients.

$$y(x) = a_1 \sin \frac{\pi x}{l} + a_2 \sin \frac{2\pi x}{l} \tag{11.21$'$}$$

These two coefficients are not independent, but related by the fixed-length condition

$$\begin{aligned} L = l &+ \frac{l}{4}\left[\left(\frac{\pi a_1}{l}\right)^2 + \left(\frac{2\pi a_2}{l}\right)^2\right] \\ &- \frac{l}{8}\left[\frac{3}{8}\left(\frac{\pi a_1}{l}\right)^4 + \frac{3}{2}\left(\frac{\pi a_1}{l}\right)^2\left(\frac{2\pi a_2}{l}\right)^2 + \frac{3}{8}\left(\frac{2\pi a_2}{l}\right)^4\right] \end{aligned} \tag{11.22}$$

The relation between L, l, and a_1^0 is obtained by setting $a_1 \to a_1^0, a_2 \to 0$ in (11.22). Terms of degree greater than 4 have not been retained. This is because the single control parameter F can be used to annihilate only one Taylor series coefficient. (The Fourier series in x becomes, after integration, a Taylor series in the Fourier coefficients.)

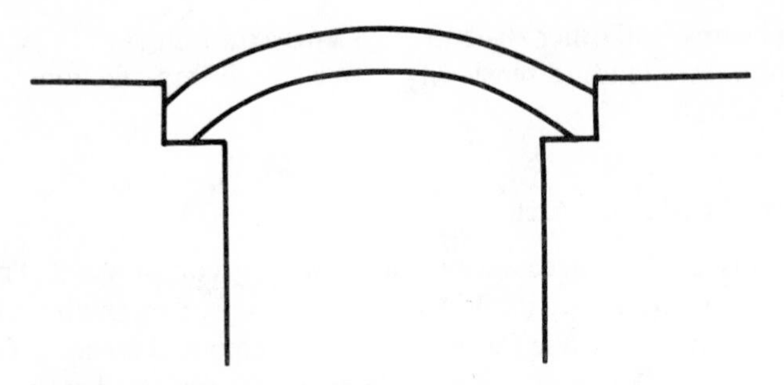

Figure 11.6 The shallow arch is a useful structural element.

The energy stored in the shallow arch is the same as that stored in the bent beam, and given by (11.11). The work done by the external load F is $F(a_1^0 - a_1)$. The potential describing the static perfect shallow arch is therefore

$$V_p(a_1, a_2; F) = \frac{Bl}{4}\left[\left(\frac{\pi}{l}\right)^4 a_1^2 + \left(\frac{2\pi}{l}\right)^4 a_2^2\right] - F(a_1^0 - a_1) \qquad (11.23)$$

This constrained problem involving two variables a_1, a_2 can be transformed to an unconstrained problem involving a single variable by using the constraint equation to solve for a_1 in terms of a_2 or vice versa. We note that this constraint equation is invariant under $a_1 \to \pm a_1, a_2 \to \pm a_2$. A solution for a_1 can be given as a function of a_2 as follows:

$$a_1 = \pm f_1(a_2) \qquad (11.24)$$

where

$$a_1^0 = f_1(0)$$

Here the two choices of sign represent quite distinct physical situations. The positive sign represents the normal arch configuration. The negative sign represents an inverted arch. The normal arch is a useful structural element, the inverted arch is practically useless. If we eliminate a_2 in favor of a_1, a sign choice must be made because of the linear term present in (11.23).

Positive and negative values of the second harmonic amplitude a_2 represent a different physical situation (Fig. 11.7). For fixed a_1, the potential $V_p(a_1, a_2; F)$ is clearly invariant under $a_2 \to -a_2$. It is therefore useful to eliminate $a_1(a_1 > 0)$ in favor of a_2. The corresponding unconstrained potential then assumes the form

$$V_p(a_2; F) = A_0 + \tfrac{1}{2}A_2 a_2^2 + \tfrac{1}{4}A_4 a_2^4 \qquad (11.25)$$

$$A_0 = \frac{Bl}{4}\left(\frac{\pi}{l}\right)^4 (a_1^0)^2$$

$$\tfrac{1}{2}A_2 = 3Bl\left(\frac{\pi}{l}\right)^4 - \frac{2F}{a_1^0} \qquad (11.26)$$

$$\tfrac{1}{4}A_4 < 0$$

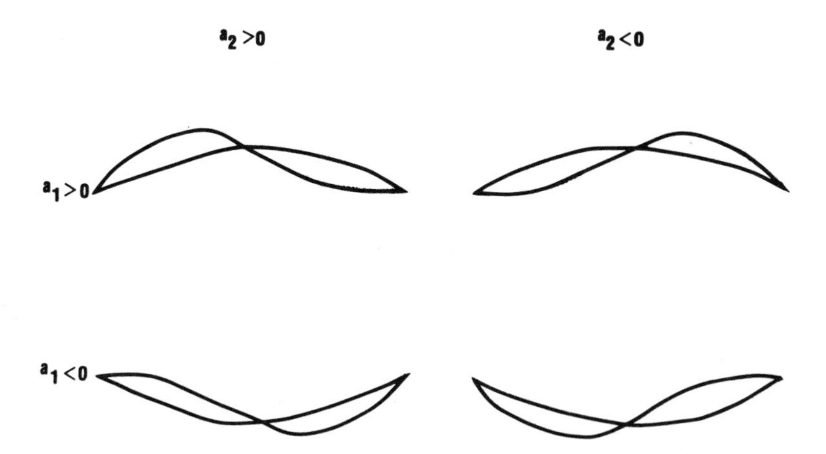

Figure 11.7 The two sign choices for a_1 represent drastically different physical situations. The two sign choices for a_2 represent distinct but equivalent collapse modes.

The constant term is unimportant. The linear and other odd terms are absent because of the invariance present in the problem. The quadratic term shows that the arch retains its shape for $F < F_p = \frac{3}{2}Bla_1^0(\pi/l)^4$. The quartic term shows that the perfect shallow arch collapses ($a_1 > 0 \rightarrow a_1 < 0$) for $F > F_p$, since there are no nearby stable equilibria. The potential $V_p(a_2; F)$ exhibits a catastrophe of type A_{-3} for $F = F_p$.

Mathematically, the collapsing arch is described by a dual cusp catastrophe. Initially there are two unstable equilibria with $a_2 \neq 0$ near the stable equilibrium at $a_2 = 0$. The two unstable equilibria represent the "snap-through" positions. As F is increased, smaller and smaller values of $|a_2|$ bring the arch into the snap-through position. As the critical load is approached, the snap-through positions occur at $|a_2| \rightarrow 0$. The slightest perturbation will cause collapse of the arch.

Physically, the collapse comes about as follows. The vertical force applied at the center of the shallow arch tries to push the center down. In order to lower the center, a higher harmonic must be added into the arch shape. This can only be done at the cost of bending energy. This cannot take place until the energy gain in work done $[F(a_1^0 - a_1)]$ exceeds the energy cost in deforming the arch. Once the arch begins to move downward, there is no mechanism preventing it from snapping past the unstable position into the inverted position. This latter position is useless as a structural support position. The perfect shallow arch may therefore be considered as having collapsed once the critical load F_p has been exceeded.

Imperfections of a shallow arch may occur because of inhomogeneities in the building materials, because of displacement of the loading point, or for a thousand other reasons. The most general perturbation of the cusp catastrophe germ $\pm x^4$ has the form $\frac{1}{2}\varepsilon_2 x^2 + \varepsilon_1 x$. Therefore the potential associated with an imperfect shallow arch has the form

$$V_i(x; F, \varepsilon_1, \varepsilon_2) = \varepsilon_1 x + \tfrac{1}{2}(F_p - F + \varepsilon_2)x^2 - \tfrac{1}{4}x^4 \tag{11.27}$$

For simplicity of description we have written $a_2 \simeq x$ and have chosen the proportionality constant and the scale of F to give simple coefficients in (11.27).

As the external load is increased, the control parameters follow a linear trajectory in the cusp control parameter space. The system will remain in the locally stable state corresponding to the middle sheet of the dual cusp catastrophe manifold until the bifurcation set is reached. The imperfection sensitivity of the shallow arch therefore has the form

$$F_c - F_p + \varepsilon_2 = -k|\varepsilon_1|^{2/3}$$

or

$$F_c = F_p - \varepsilon_2 - k|\varepsilon_1|^{2/3} \tag{11.28}$$

The shallow arch is much more sensitive to symmetry-breaking imperfections ($\varepsilon_1 \neq 0$) than it is to symmetry-preserving imperfections. It is therefore adequate, here and below, to concentrate on the symmetry-breaking imperfections. The most general symmetry-breaking imperfection is equivalent to a displacement of the applied force by a distance ε_1 from the symmetry plane.

The $\frac{2}{3}$ power law dependence of the failure load on the imperfection parameter was shown in a series of experiments by Roorda.[3] In these experiments the imperfection was simply the offset f/L of the applied force from the symmetry plane of the shallow arch. The experimental data are shown in Fig. 11.8. The experimental arrangement is illustrated in Fig. 11.8*a*. The state variable $\theta(P)$ is shown as a function of the applied force P in Fig. 11.8*b*. Both stable and unstable equilibria are shown. Stable equilibria occur only on the central branch. These equilibria disappear at the point P^M/P^C of horizontal tangency. This point defines the failure load at which snap-through takes place. The scaled failure load P^M/P^C is shown as a function of the scaled imperfection parameter f/L in Fig. 11.8*c*. The cusp is offset due to the imperfections intrinsic to the model. It is important to observe how sensitive the carrying capacity is to imperfections. In this experiment a $\frac{1}{4}\%$ imperfection is responsible for a 6% reduction in the arch load carrying capacity.

We now consider the reduction in the carrying capacity caused by dynamical loading. For the perfect system subjected to oscillations carrying kinetic energy ΔE in the failure mode, the critical load is determined through

$$\Delta E = \tfrac{1}{2}(F_p - F)x^2 - \tfrac{1}{4}x^4 \tag{11.29}$$

As a result

$$F_c = F_p - 2|\Delta E|^{1/2} \tag{11.30}$$

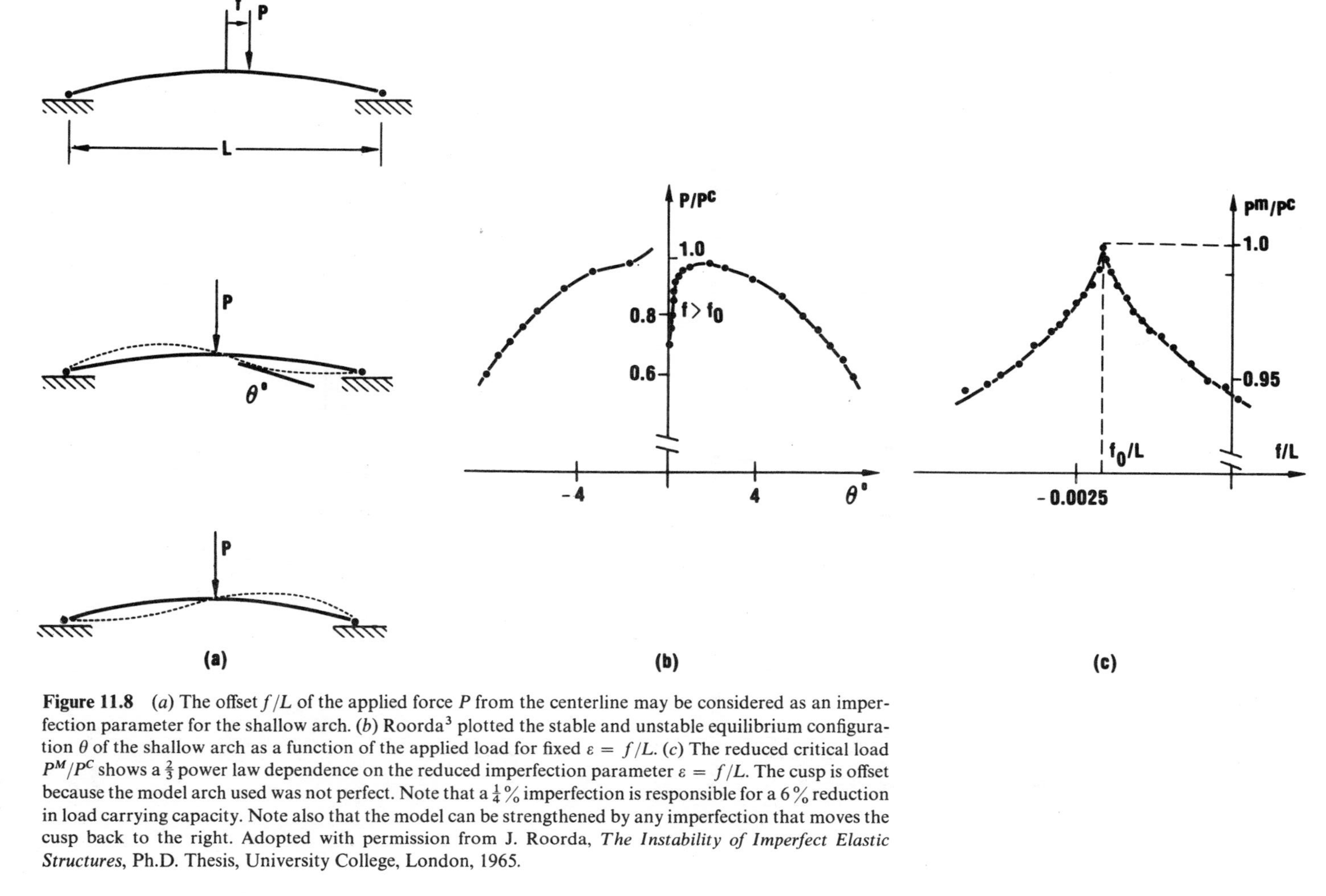

Figure 11.8 (*a*) The offset f/L of the applied force P from the centerline may be considered as an imperfection parameter for the shallow arch. (*b*) Roorda[3] plotted the stable and unstable equilibrium configuration θ of the shallow arch as a function of the applied load for fixed $\varepsilon = f/L$. (*c*) The reduced critical load P^M/P^C shows a $\frac{2}{3}$ power law dependence on the reduced imperfection parameter $\varepsilon = f/L$. The cusp is offset because the model arch used was not perfect. Note that a $\frac{1}{4}$% imperfection is responsible for a 6% reduction in load carrying capacity. Note also that the model can be strengthened by any imperfection that moves the cusp back to the right. Adopted with permission from J. Roorda, *The Instability of Imperfect Elastic Structures*, Ph.D. Thesis, University College, London, 1965.

The reduction in load carrying capacity of the imperfect system can be determined from scaling arguments. For example,

$$\begin{gathered} x \to \lambda x \\ |F_c - F_p| \to \lambda^2 |F_c - F_p| \Rightarrow \Delta E \to \lambda^4 \Delta E \\ |\varepsilon_1| \to \lambda^3 |\varepsilon_1| \end{gathered} \tag{11.31}$$

This surface may be constructed from the canonical properties of the cusp catastrophe. The collapse surface in F–ΔE–ε_1 space is shown in Fig. 11.9. The entire surface can be reconstructed from a knowledge of the scaling relation (11.31) and any single cross section. It is clear that the shallow arch is even more sensitive to dynamical "imperfections" than it is to symmetry-breaking imperfections.

Remark. The difference between the cusp and the dual cusp catastrophe germs is nothing more than a sign—mathematically. Physically, there is a world of difference. The cusp catastrophe is globally stable, while the dual cusp catastrophe is globally unstable. For the former, there is always some stable state. For the latter, stable states exist only within the cusp-shaped region in the control plane. For systems described by the cusp catastrophe we must make a subjective estimate for the safe loading criterion. In the case of Euler buckling

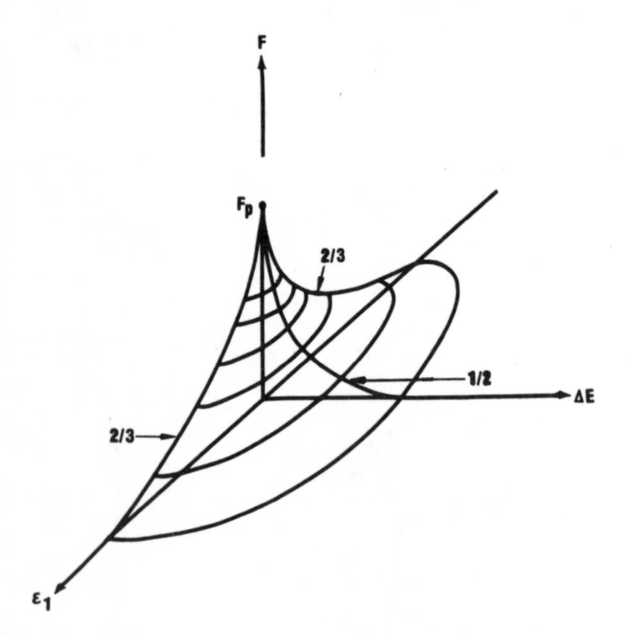

Figure 11.9 The critical surface for the shallow arch in F–ΔE–ε_1 space has the form shown. The extreme sensitivity to static imperfections is overshadowed by the even more extreme sensitivity to dynamic imperfections.

we required that the order parameter a_1 remain less than some maximum safe deflection s. For systems described by dual cusp catastrophes no such subjective criterion is required—we have an objective criterion. When the critical load limit is exceeded, the system collapses.

4. Exchange of Stability

In the two previous examples the potential describing the perfect system has been invariant under the symmetry $x \to -x$, where x is the system order parameter (that is, the first or second Fourier coefficient). This is not strictly true unless we restrict ourselves to expansions (11.10) involving only the first two Fourier coefficients. However, we made this approximation based on the rapidly increasing bending energy in the successively higher bending modes.

Since the perfect systems were described by an even potential $V_p(x; F)$ depending on a single control parameter F, only one Taylor series coefficient could be annihilated $[\sim(F_p - F)x^2]$, so we were inexorably led to a consideration of either the cusp ($\sim x^4$) or the dual cusp ($\sim -x^4$) catastrophe.

In the absence of symmetry the Taylor series expansion of the potential $V_p(x; F)$ describing some perfect system will have the form

$$V_p(x; F) = V_0 + V_1 x + \tfrac{1}{2}V_2 x^2 + \frac{1}{3!} V_3 x^3 + \cdots \tag{11.32}$$

In general we choose the order parameter x so that the perfect system has an equilibrium at $x = 0$. Then $V_1 = 0$. The constant term is unimportant, and may be removed by a shift of origin. The quadratic, cubic, and higher order terms are generally nonzero. If the system fails as a function of increasing applied force F at $F = F_p$, then we may write the potential in the form

$$V_p(x; F) = \tfrac{1}{2}(F_p - F)x^2 + \tfrac{1}{3}x^3 + \cdots \tag{11.33}$$

by a change in scale in the x and F axes. The quartic and higher order terms may be neglected. This form for the potential of a perfect system is valid in the neighborhood of the critical load F_p. Two perfect systems described by a potential equivalent to (11.33) are shown in Fig. 11.10*a*.

The critical points associated with V_p are determined as usual from

$$\frac{d}{dx} V_p = 0 = x\{(F_p - F) + x\} \tag{11.34}$$

The location of the critical points, together with their stability type, is shown in Fig. 11.11. There is clearly an exchange of stability as the two critical points $x_1(F) = 0$, $x_2(F) = F - F_p$ pass through each other.

The potential describing a corresponding imperfect system is

$$V_i(x; F, \varepsilon_1, \varepsilon_2, \dots) = V_p(x; F) + p(x)$$
$$p(x) = \varepsilon_1 x + \tfrac{1}{2}\varepsilon_2 x^2 + \tfrac{1}{3}\varepsilon_3 x^3 + \cdots \tag{11.35}$$

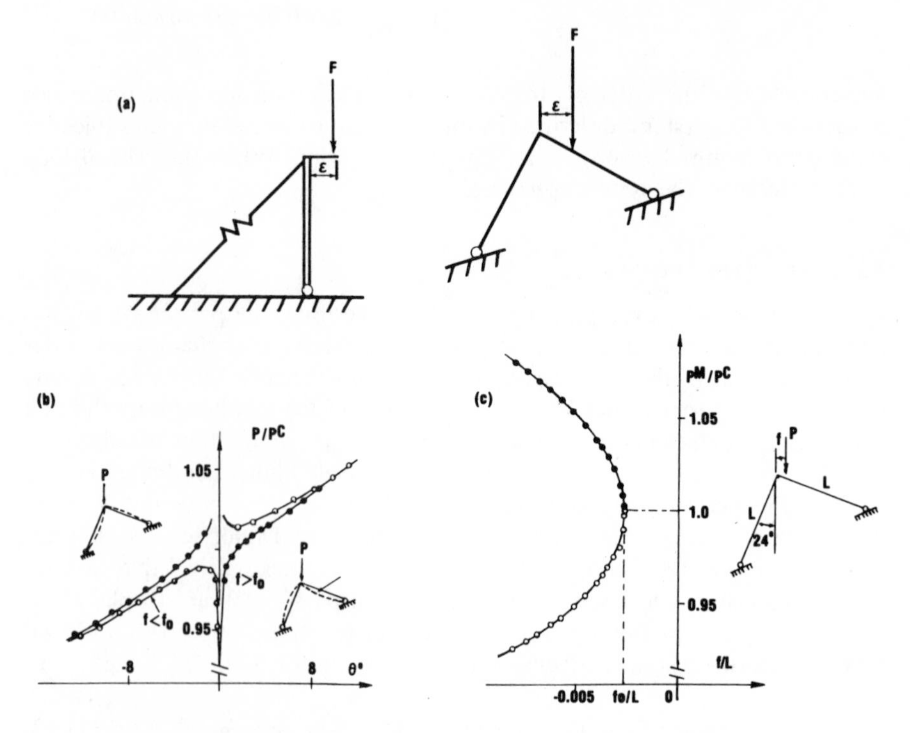

Figure 11.10 (*a*) Two perfect systems which exhibit an exchange of stability are the propped cantilever and the rigid jointed frame. Imperfections are modeled by offsetting the load. (*b*) The stable and unstable equilibria for an imperfect rigid jointed frame were determined experimentally by Roorda.[3] (*c*) The experimentally determined imperfection sensitivity exhibits a $\frac{1}{2}$ power law behavior. Adopted with permission from J. Roorda, *The Instability of Imperfect Elastic Structures*, Ph.D. Thesis, University College, London, 1965.

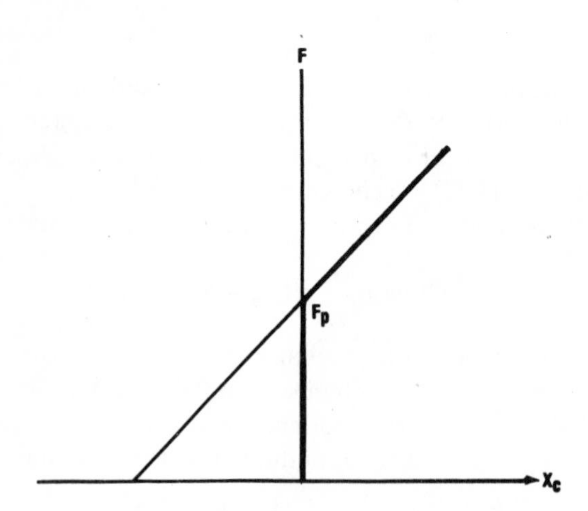

Figure 11.11 The two critical points for a system which exhibits an exchange of stability are shown as a function of the applied load F. The stable critical point is shown as a heavy line.

Of course the results of Chapter 4 tell us all about the most general perturbation of the germ x^3: it is a linear term. In short, (11.35) may be transformed, by a suitable nonlinear transformation, to the canonical form $x^3 + ax$. What is true from a mathematical point of view may not be suitable from a physical point of view. Such a nonlinear transformation will produce a complicated nonlinear relationship between the physical loading parameter F and the imperfection parameters $\varepsilon_1, \varepsilon_2, \ldots$. Instead of carrying out the nonlinear transformation which brings (11.35) into canonical form, we proceed as follows. We neglect terms of degree exceeding 2 in the perturbation expansion. We keep the quadratic term only to show that it is less important than the linear term. Eventually we will drop it also. The potential for the imperfect system then takes the form

$$V_i(x; F, \varepsilon_1) = \varepsilon_1 x + \tfrac{1}{2}(F'_p - F)x^2 + \tfrac{1}{3}x^3 \tag{11.36}$$

where $F'_p = F_p + \varepsilon_2$. The critical points are determined by

$$\frac{dV_i}{dx} = \varepsilon_1 + (F'_p - F)x + x^2 = 0 \tag{11.37}$$

Equation (11.37) represents a 2-dimensional manifold embedded in the space $\mathbb{R}^3$ whose coordinate axes are x–F–ε_1. This surface is shown in Fig. 11.12*a*.

The equilibrium states of the imperfect system are then the intersections of this manifold with the planes $\varepsilon_1 =$ constant. Three such intersections are shown in Fig. 11.12*b*. The stability properties of the critical points are easily determined. All points on the upper sheet of the critical manifold represent locally stable critical points of V_i, all those on the lower sheet represent unstable critical points.

The behavior of the imperfect system as a function of increasing load F depends critically on the sign of the imperfection ε_1. For $\varepsilon_1 < 0$ there are two critical points for all values of F. For $\varepsilon_1 > 0$ there is a region in which V_i has no critical points at all. This occurs for

$$-\varepsilon_1 + \left(\frac{F'_p - F}{2}\right)^2 < 0 \tag{11.38}$$

The sensitivity of this system to imperfections is therefore

$$F_c = F_p + \varepsilon_2 - 2(\varepsilon_1)^{1/2}, \qquad \varepsilon_1 > 0 \tag{11.39a}$$

$$F_c = \text{“}\infty\text{”} \qquad \varepsilon_1 < 0 \tag{11.39b}$$

In the case $\varepsilon_1 \leq 0$ the locally stable equilibrium exists for all values of the external force F. This situation is similar to the situation for Euler buckling. A safety criterion may be introduced. In this case the system is declared unsafe when a critical load forces the state value x to exceed some preassigned safe value $s: |x| > s$. This is a subjective criterion.

In the case $\varepsilon_1 > 0$ there is no need for a subjective criterion. The locally stable equilibrium ceases to exist at the critical load F_c defined by (11.39a).

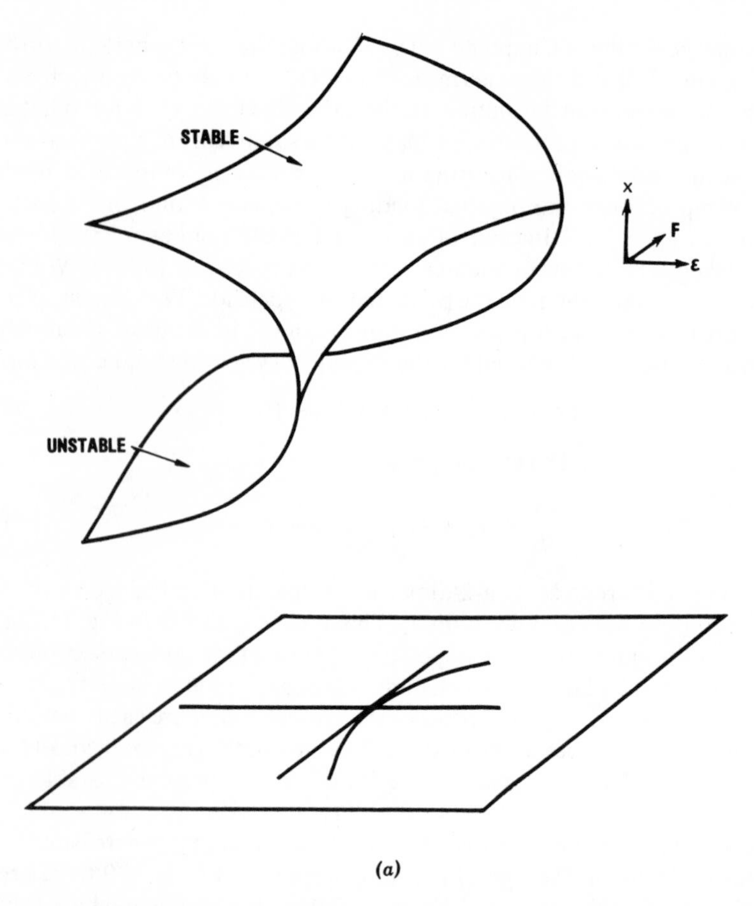

Figure 11.12 (*a*) The equilibrium manifold for the imperfect propped cantilever has the form shown in the state variable control parameter space x–F–ε. All points on the upper part of this surface represent locally stable states. Those on the lower sheet represent locally unstable states. The upper and lower branches have a separatrix which projects down onto the parabola $\varepsilon = (F - F'_p/2)^2$ in the control parameter space.

We see here a very weak dependence of imperfection sensitivity on ε_2 but a strong dependence of imperfection sensitivity on ε_1. Thus ε_1 is more important than ε_2; ε_2 more important than ε_3; etc. It is for this reason that all but the linear perturbations can be neglected on physical grounds.

The $\frac{1}{2}$ power law dependence of the failure load on the imperfection parameter was shown by Roorda for a system which undergoes an exchange of stability in the absence of imperfections. The stable and unstable critical points for a system depicted in Fig. 11.10*a* are shown in Fig. 11.10*b*. The imperfection sensitivity is shown in Fig. 11.10*c*. Parabolic contact with the vertical tangent clearly indicates the $\frac{1}{2}$ power law dependence. The failure locus is not tangent to

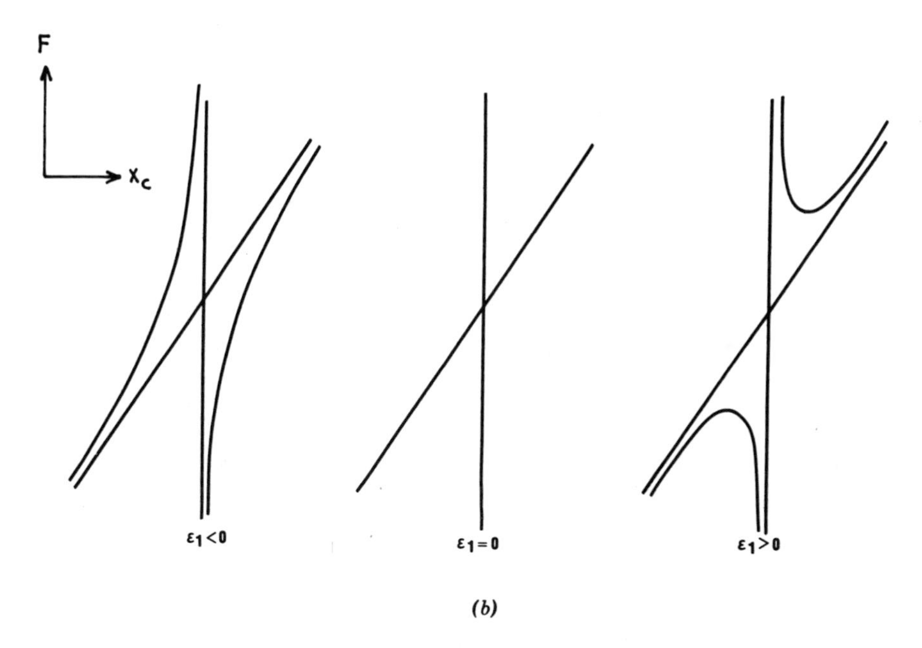

Figure 11.12 (*b*) The pre- and postbuckling paths are obtained as intersections of the equilibrium manifold with planes ε_1 = constant. The stability properties along these paths are determined directly from the equilibrium manifold.

the axis $f/L = 0$ because of imperfections intrinsic to the model. Once again there is a dramatic imperfection sensitivity, with a $\frac{1}{4}\%$ imperfection leading to a 5% reduction in the failure load.

When the potential (11.36) has two equilibria, the unstable equilibrium acts as a separatrix between the domain of attraction of the locally stable state and the abyss. The energy difference ΔV between the local maximum and the local minimum is

$$\Delta V = \tfrac{1}{6}[(F - F_p)^2 - 4\varepsilon_1]^{3/2} \tag{11.40}$$

If the system is subject to dynamic loading ΔE as a dynamic "imperfection," then the relation between the failure load F_c, the imperfection ε_1, and the dynamic loading factor ΔE is given by

$$F_c = F_p - [4\varepsilon_1 + (6\Delta E)^{2/3}]^{1/2} \tag{11.41}$$

In the absence of dynamic loading, the sensitivity of the failure load reduces to the $\frac{1}{2}$ power law dependence given in (11.39a). However, in the absence of imperfections, the dynamic loading sensitivity has the very drastic $\frac{1}{3}$ power law dependence on the energy ΔE. These imperfection sensitivities are shown in Fig. 11.13.

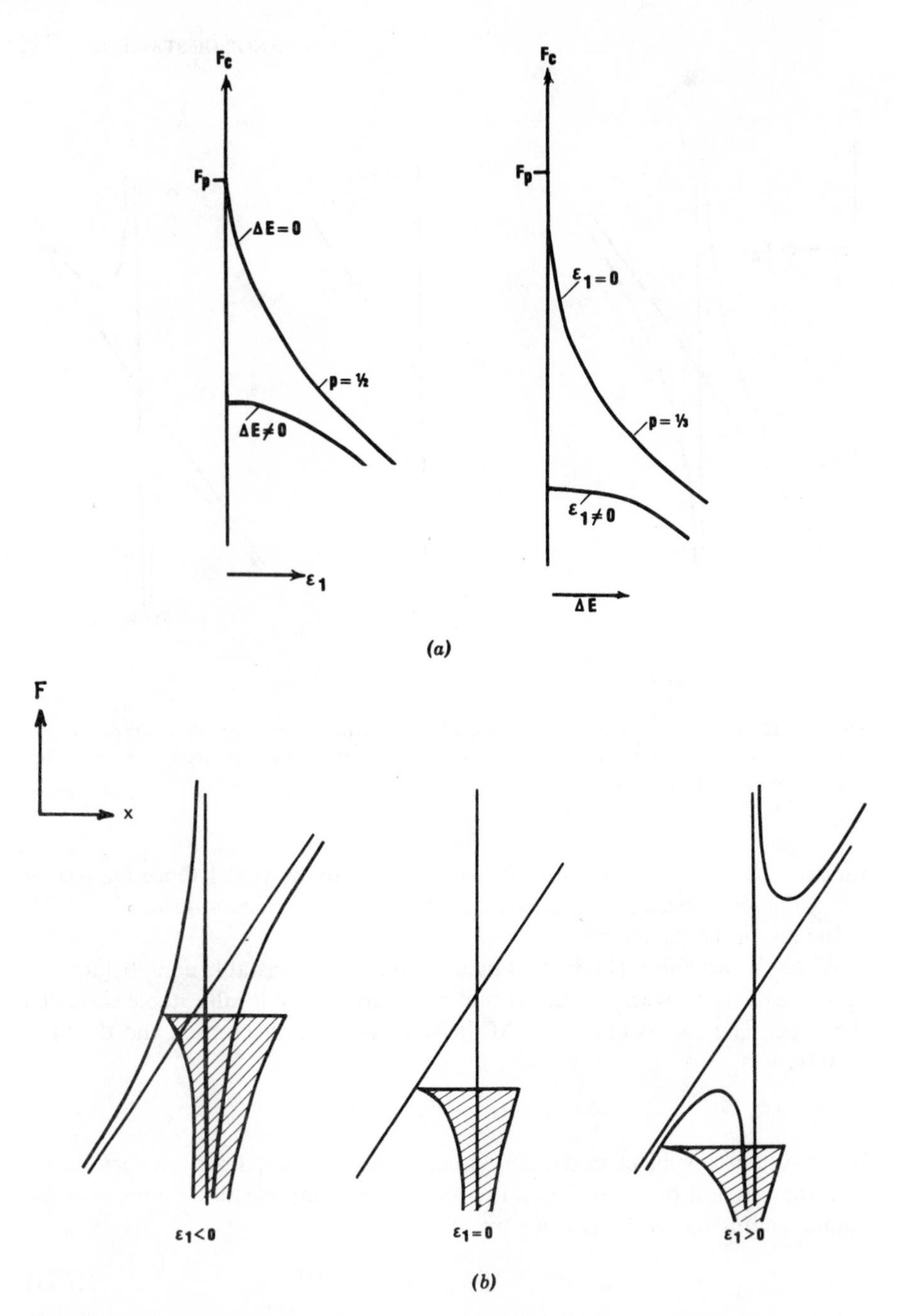

Figure 11.13 (*a*) The imperfection sensitivity of systems exhibiting an exchange of stability is extremely severe. Pure static and pure dynamic imperfections exhibit a $\frac{1}{2}$ and a $\frac{1}{3}$ power law dependence, respectively. (*b*) Not only can dynamic imperfections decrease the carrying capacity of imperfect systems which exhibit an exchange of stability (as for $\varepsilon_1 \geq 0$), but they can also shake a stable system over the edge into oblivion ($\varepsilon_1 < 0$). The shaded areas indicate the extent of oscillations about the stable equilibrium. Stability is lost when the oscillation crosses the unstable local maximum.

For imperfect systems with a finite critical load F_c given by (11.39a) it is reasonable to expect dynamic fluctuations to reduce the carrying capacity. It is also reasonable to expect a very drastic sensitivity of the perfect system on dynamical factors. The potential describing the perfect system is both extremely flat as well as globally unstable near $F \simeq F_p$, so that very small fluctuations can send it over the edge. It may be surprising that dynamical oscillations can transform the "stable" case $\varepsilon_1 < 0$ into an unstable case. Fluctuations larger than the critical value difference can bounce the system out of its smugly stable state, over the potential barrier, to the bottomless pits below. This transformation from stable to unstable by dynamical processes is disquieting, and should be a source of contemplation.

5. Compound Systems

In the previous sections we have considered the very elementary ("irreducible") components out of which more complicated systems are constructed. For the moment the most complicated compound system which we choose to consider will be constructed from two components. We shall assume that the two components with order parameters x and y are perfect and also invariant under $x \to \pm x$, $y \to \pm y$. We also assume that the compound system is constructed perfectly, so that it also has symmetries

$$V_p(x, y; F) = V_p(\pm x, \pm y; F) \tag{11.42}$$

Then the leading terms in a Taylor series expansion of this ideal potential have the form

$$\begin{aligned} V_p(x, y; F) = {} & \tfrac{1}{2}(F_1 - F)x^2 + \tfrac{1}{2}(F_2 - F)y^2 + \tfrac{1}{4}\sigma_1 x^4 + \tfrac{1}{4}\sigma_2 y^4 \\ & + \tfrac{1}{2}cx^2y^2 + \text{higher order terms} \end{aligned} \tag{11.43}$$

We will neglect terms of degree 6 and higher. The symbols $\sigma_1 = \pm 1$, $\sigma_2 = \pm 1$ are introduced to take into account the possibilities that the two individual components fail by means of $A_{\pm 3}$ catastrophes.

Before jumping into a detailed analysis of the critical points of this perfect potential, it is useful first to determine qualitatively its global stability properties. This can be done by neglecting the quadratic terms as unimportant compared to the quartic terms for large values of $r^2 = x^2 + y^2$ (assuming that the terms of degree larger than 4 are zero). For example, if the quartic contribution to V_p is

$$\begin{aligned} V_p & \xrightarrow{r \text{ large}} \tfrac{1}{4}x^4 + \tfrac{1}{4}y^4 - \tfrac{1}{2} \cdot 3x^2y^2 \\ & = \tfrac{1}{4}(x^2 - y^2)^2 - x^2y^2 \end{aligned} \tag{11.44}$$

then V_p is stable in the "direction" $x^2 - y^2$ and unstable in the "direction" $x^2 = y^2$ (i.e., $x^2 - y^2 = 0$). These stability results are summarized in Table 11.1.

Table 11.1 Global Stability Properties for the Potential $V(x, y) = \frac{1}{4}\sigma_1 x^4 + \frac{1}{4}\sigma_2 y^4 + \frac{1}{2}cx^2y^2$

σ_1	σ_2	Condition	$x^2 - y^2$	x^2y^2	"Morse Type"
+1	+1	$-1 < c$	+	+	M_0^2
		$c < -1$	+	−	M_1^2
+1	−1	Arbitrary			M_1^2
−1	+1				
−1	−1	$+1 < c$	−	+	M_1^2
		$c < +1$	−	−	M_2^2

The local stability properties of the perfect potential (11.43) can now be determined in the usual way. The zero gradient condition leads to the pair of coupled nonlinear equations

$$\frac{\partial V}{\partial x} = x\{(F_1 - F) + \sigma_1 x^2 + cy^2\} = 0 \tag{11.45a}$$

$$\frac{\partial V}{\partial y} = y\{(F_2 - F) + \sigma_2 y^2 + cx^2\} = 0 \tag{11.45b}$$

The stability matrix is

$$V_{ij} = \begin{bmatrix} F_1 - F + 3\sigma_1 x^2 + cy^2 & 2cxy \\ 2cxy & F_2 - F + 3\sigma_2 y^2 + cx^2 \end{bmatrix} \tag{11.46}$$

The set of equations (11.45) has up to nine solutions, as follows:

1. $x = 0,\quad y = 0,\qquad V_{ij} = \begin{bmatrix} F_1 - F & 0 \\ 0 & F_2 - F \end{bmatrix}$

2, 3. $F_1 - F + \sigma_1 x^2 = 0,\quad y = 0,\qquad V_{ij} = \begin{bmatrix} -2(F_1 - F) & 0 \\ 0 & F_2 - F + cx^2 \end{bmatrix}$

4, 5. $x = 0,\quad F_2 - F + \sigma_2 y^2 = 0,\qquad V_{ij} = \begin{bmatrix} F_1 - F + cy^2 & 0 \\ 0 & -2(F_2 - F) \end{bmatrix}$

6–9. $F_1 - F + \sigma_1 x^2 + cy^2 = 0,\quad F_2 - F + \sigma_2 y^2 + cx^2 = 0,\qquad V_{ij} = \begin{bmatrix} 2\sigma_1 x^2 & 2cxy \\ 2cxy & 2\sigma_2 y^2 \end{bmatrix}$

The last set of solutions exists if the quadratic equations represent real conics (ellipses and/or hyperbolas) and these two conics intersect.

We shall call the solution set $(x, y) = (0, 0)$ the **zeroth branch**, or **trunk**. This solution exists for all values of the load parameter F, by symmetry. The solution sets which exist in the x–F plane $y = 0$ (#2, 3) and the y–F plane $x = 0$ (#4, 5) will be called **primary branches**, when they exist. The remaining four symmetric solutions ($x \to \pm x, y \to \pm y$) will be called **secondary branches.** Primary branches bifurcate from the zeroth branch. Secondary branches bifurcate from the primary branch.

We shall determine the local stability properties of V_p by performing an elementary bifurcation analysis. We assume that $F_2 > F_1 > 0$. For $F = 0$, the solution at $(x, y) = (0, 0)$ is stable. This solution remains stable as F is increased until $F = F_1$. At this point a primary branch $(x, y) = (\pm\sqrt{\sigma_1(F - F_1)}, 0)$ bifurcates from the zeroth branch. This primary branch exists only for $F \geq F_1$ or $F \leq F_1$, depending on the catastrophe type $A_{\pm 3}$. At $F = F_2$ a second primary branch bifurcates from the zeroth branch. On this primary branch $(x, y) = (0, \pm\sqrt{\sigma_2(F - F_2)})$.

Along the zeroth branch the x and y axes are principal axes and one eigenvalue of V_{ij} changes sign at each bifurcation. Along the primary branches the x and y axes are also principal directions. The first primary branch ($x \neq 0, y = 0$) represents stable Morse critical points M_0^2 if $\sigma_1 = +1$, unstable Morse saddles M_1^2 if $\sigma_1 = -1$. The second primary branch ($x = 0, y \neq 0$) represents unstable Morse saddles M_1^2 if $\sigma_2 = +1$, M_2^2 if $\sigma_2 = -1$.

If the secondary branches exist, they must bifurcate from the primary branches, by arguments involving conic sections. The simplest way to locate these bifurcations is to search for zeros in the eigenvalues of the stability matrix along the primary branches. Bifurcation can occur from the first primary branch at

$$F = \frac{F_2 - c\sigma_1 F_1}{1 - c\sigma_1} \tag{11.47a}$$

if this value of F lies along the branch. Similarly, bifurcations can occur from the second primary branch at

$$F = \frac{F_1 - c\sigma_2 F_2}{1 - c\sigma_2} \tag{11.47b}$$

provided this value of F lies along the branch ($x = 0, y \neq 0$).

Two interesting bifurcation diagrams are shown for the perfect potential (11.43) in Fig. 11.14. In the first of these a secondary branch is "emitted" from one of the primary branches and "absorbed" by the other. In the second of these the two primary branches both turn upward but the secondary branches turn down and are unstable. The descending unstable (M_1^2) secondary branches are potential barriers separating the stable (M_0^2) critical point at ($x \doteq 0, y = 0$) from the global instability ($c < -1$). For $F_2 \gg F_1$ or $F_1 \gg F$ these saddles are far removed from the zeroth branch, and the barrier height is substantial. For $F_2 - F_1 \simeq 0$ and $F_1 - F$ small, this is no longer the case.

An imperfection analysis can be carried out for the perfect systems represented by the potential (11.43). For $F_2 - F_1 \gg 0$ these analyses reduce to the analyses of the cusp catastrophes, carried out in Secs. 2 and 3. The situation is more

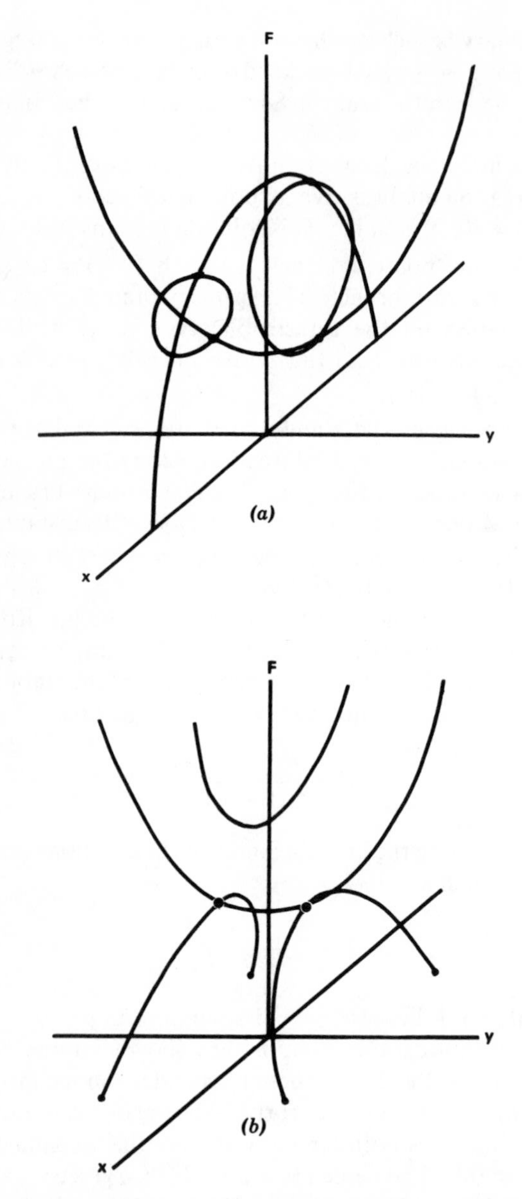

Figure 11.14 The potential (11.43) has several interesting bifurcation diagrams, depending on the coefficients of the quartic terms. (*a*) Secondary branches "emitted" from one primary branch are "absorbed" by the other primary branch. (*b*) The secondary branches are Morse saddles M_1^2. These saddles separate the locally stable equilibrium at $(x, y) = (0, 0)$ from the global instability which exists for large r.

complicated when $F_2 \simeq F_1$. In this case, although both bifurcations are "soft" (A_{+3}), the secondary branches turn down if $c < -1$, presenting the possibility not only of extreme, but worse, of unexpected, imperfection sensitivity. This unexpected sensitivity arises through the strong coupling of the load-matched failure modes of the two component subsystems.

Remark. In terms of imperfection sensitivity, the fact that an "unsafe" system can result from the combination of two "safe" systems should be a cause for pause.

6. Engineering Optimization

Many component parts are involved in the construction of a bridge or a building. For example, an incompressible beam and a shallow arch might be incorporated into a bridge design as shown in Fig. 11.15. If the bridge must be capable of carrying a maximum load F_M, then the beam must not buckle, and the arch must not collapse, for $F < F_M$. In fact, it is prudent design practice to use beams and arches that are safe for loads in excess of F_M, say by 50%. In addition, it is the design philosophy of optimization procedures to use structural elements

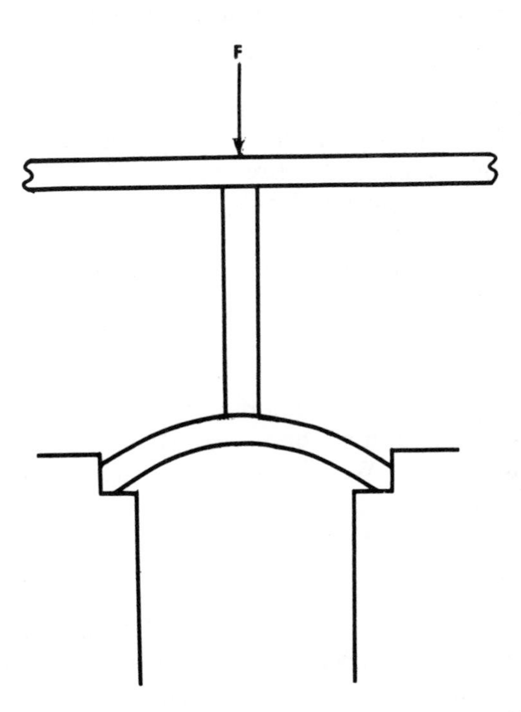

Figure 11.15 Useful structures are composed of many irreducible structural building blocks. The failure modes of a compound system may be more severe than the failure modes of its constituent elements.

which all fail at the same load. After all, what is the point of using a heavy beam that can carry a load of $3F_M$ before buckling, if the arch has already collapsed at $1.5F_M$? Heavy beams cost money, so a saving is involved in using a beam that buckles also at $1.5F_M$.

7. Dangers of Engineering Optimization

Such optimization procedures can sometimes be dangerous.[4] For example, suppose a compound system is constructed from n irreducible components, each of which fails by means of a cusp catastrophe $A_{\pm 3}$. Then the potential representing the perfect system has the form

$$V(x_1, x_2, \ldots, x_n; F) = \sum_{i=1}^{n} \tfrac{1}{2}(F - F_i)x_i^2 + \sum_{i=1}^{n} \pm\tfrac{1}{4}x_i^4 \tag{11.48}$$

If the lowest critical load occurs at $F = F_1$, and only one mode fails at this load, the catastrophe is of type $A_{\pm 3}$, and the imperfection sensitivity is as described above.

In an optimized system $F_1 = F_2 = \cdots = F_n(=F_p)$. At this ideal critical load everything fails simultaneously. The germ of the potential has the form $\pm x_1^4 \pm x_2^4 \pm \cdots \pm x_n^4$. The universal perturbation of this germ is

$$P(x) = \sum P_{i_1, i_2, \ldots, i_n} x_1^{i_1} x_2^{i_2} \cdots x_n^{i_n} \tag{11.49}$$

where no exponent i_j exceeds 2. The imperfect system $V_p(x; p) + P(x)$ may now have as many as $(3)^n$ critical points for various values of the load F and the perturbation parameters $P_{i_1, i_2, \ldots, i_n}$. Even if each individual component has only one stable critical point for $F < F_p$, the compound system may have numerous critical points for $F < F_p$ if the individual components are strongly coupled together. The drastic imperfection sensitivity of compound systems is due to symmetry-breaking imperfections which connect the locally stable branch at $(x_1, \ldots, x_n) = (0, \ldots, 0)$ to an unstable branch which can exist near the locally stable branch for $F < F_p$ because of strong coupling between modes.

Remark. Catastrophe germs of the form $\pm x_1^4 \pm \cdots \pm x_n^4$ are called **multiple cusp catastrophes** or ***n*-fold cusp catastrophes**. Multiple cusp catastrophes are not elementary catastrophes. We will encounter the double cusp catastrophe in the following section.

8. Propped Cantilever. C_{4v}

The propped cantilever shown in Fig. 11.16 exhibits the type of imperfection sensitivity for a compound system described in the previous section. If the springs along the x and y axes have spring constants k_1 and k_2, then the potential describing the perfect system has the symmetry

$$V_p(\pm x, \pm y; F) = V_p(x, y; F) \tag{11.50}$$

A Taylor series expansion of V_p can contain only even powers of x and y.

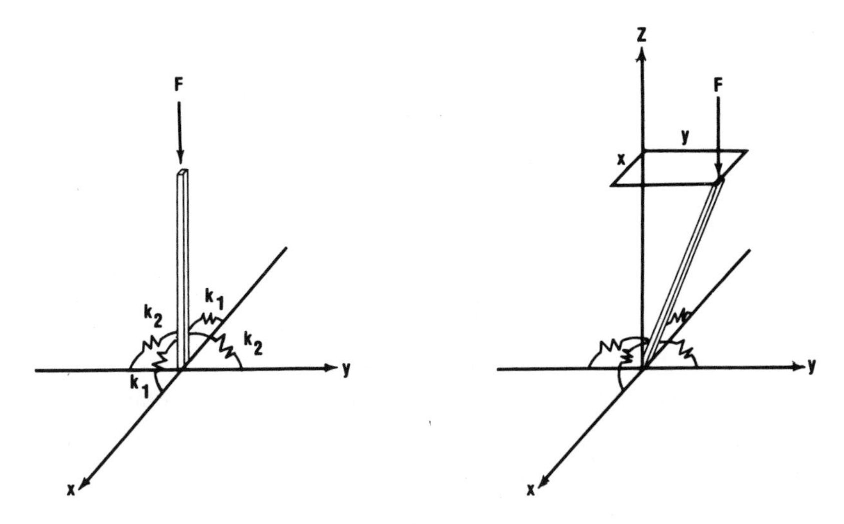

Figure 11.16 The potential describing this perfect propped cantilever has symmetry group C_{4v} when all springs have the same spring constant k.

If $k_1 = k_2$, then V_p is invariant under a larger symmetry group. This group is C_{4v}. This is the group of rotations through $\pi/2$ about the z axis and reflections in the four planes containing the z axis and the lines $x = 0$, $x - y = 0$, $y = 0$, $x + y = 0$. The effects of the eight group operations on the coordinates (x, y) of any point in the x–y plane are

	E	C_4	C_4^2	C_4^3	σ_{v1}	σ_{v2}	σ_{d1}	σ_{d2}
x	x	y	$-x$	$-y$	x	$-x$	y	$-y$
y	y	$-x$	$-y$	x	$-y$	y	x	$-x$

(11.51)

The invariance of V_p under the operations of the group C_{4v} means in particular that the coefficients of the terms $x^p y^q$ in its Taylor series expansion about $(x, y) = (0, 0)$ must be zero if p or q is odd, and the coefficient of $x^p y^q$ must equal the coefficient of $x^q y^p$.

This statement can be tightened considerably. The Taylor series expansion of V_p can contain only terms of the form $(x^2 + y^2)$ and $x^2 y^2$, and products of these functions:

$$V_p(x, y; F) = f(x^2 + y^2, x^2 y^2; F) \tag{11.52}$$

The proof of this uses group theoretical techniques, but as the result is useful and the proof not difficult, we present it here. The Taylor series expansion of V_p contains linear terms in x and y, quadratic terms, terms of degree 3, 4, ... The linear vector space of terms of degree d has dimension $d + 1$. We may choose as basis vectors in this space the monomials $x^d, x^{d-1}y, \ldots, y^d$. Each operation in

C_{4v} maps a term of degree d into a term of degree d, by homogeneity arguments. For example, the operation C_4 has the effect

$$\begin{array}{cccc} & x^2 & xy & y^2 \\ C_4 & \downarrow & \downarrow & \downarrow \\ & y^2 & (y)(-x) & (-x)^2 \end{array} \tag{11.53}$$

This may be written in matrix form as follows:

$$\begin{bmatrix} x^2 \\ xy \\ y^2 \end{bmatrix}' = \begin{bmatrix} 0 & 0 & 1 \\ 0 & -1 & 0 \\ 1 & 0 & 0 \end{bmatrix} \begin{bmatrix} x^2 \\ xy \\ y^2 \end{bmatrix} \tag{11.54}$$

The 3×3 matrix describing the action of the group operation C_4 on the 3-dimensional linear vector space spanned by the degree 2 monomials is called a **matrix representation** of C_4. The matrix representatives of the other group operations in this space and all other spaces spanned by monomials of various degree can be similarly computed.

For the purpose at hand the matrices themselves are of less importance than their traces (characters). These characters can be used to compute the number of linearly independent polynomial invariants which exist in any $(d + 1)$-dimensional subspace of terms of degree d. This number is

$$\frac{1}{8} \sum_{g \in C_{4v}} \chi^d(g) \tag{11.55}$$

Here χ^d is the trace of the $(d + 1) \times (d + 1)$ matrix representative of the group element g. The characters for various values of d are shown in Table 11.2. This table repeats in the obvious way beyond $d = 7$. The number of independent

Table 11.2 Character Table for Representations of the Group C_{4v} in $d + 1$-Dimensional Spaces Spanned by Degree d Monomials of the Form $x^p y^q$, $p + q = d$

d	E	C_2	$2C_4$	$2\sigma_v$	$2\sigma_d$	Number[a]
0	1	1	1	1	1	1
1	2	−2	0	0	0	0
2	3	3	−1	1	1	1
3	4	−4	0	0	0	0
4	5	5	1	1	1	2
5	6	−6	0	0	0	0
6	7	7	−1	1	1	2
7	8	−8	0	0	0	0

[a] This column gives the number of linearly independent functions of degree d which are invariant under C_{4v}.

invariants of any degree is shown in the right-hand column of this table. The invariant of degree 0 is the constant term. The invariant of degree 2 is $x^2 + y^2 = I_2$. The two invariants of degree 4 are $(x^2 + y^2)^2 = I_2^2$ and $x^2y^2 = I_4$.

All invariants have the form

$$I_2^p I_4^q, \qquad p \geq 0, q \geq 0 \tag{11.56}$$

where p and q are nonnegative integers. To show this, it is sufficient to show that the number of invariants of degree d is exactly equal to the number of invariants of degree d which can be written in the form $I_2^p I_4^q$. From Table 11.2 the number of invariants of degree d is

$$\begin{gathered} 0 \text{ if } d \text{ is odd} \\ \left[\frac{n}{4}\right] + 1 \text{ if } d \text{ is even} \end{gathered} \tag{11.57}$$

where $[x]$ means "greatest integer in." On the other hand, the number of non-negative integer solutions (p, q) for

$$2p + 4q = d, \qquad p \geq 0, \quad q \geq 0, \quad p, q \text{ integer} \tag{11.58}$$

is also given by (11.57). Therefore I_2 and I_4 are the only two functionally independent invariants under C_{4v}, and we have

$$V_p(x, y; F) = \sum_{p,q \geq 0} A_{p,q}(F) I_2^p I_4^q = f(I_2, I_4; F) \tag{11.59}$$

as claimed. We have assumed that the functions V_p, f are real and analytic.

This digression has not allowed us to write down explicitly the potential representing the perfect propped cantilever shown in Fig. 11.16. This can be done by direct computation. Assume that (x, y) represents the position of the end of the cantilever. For simplicity we choose the length of the cantilever to be 1. Then the potential is

$$V_p(x, y; F) = \tfrac{2}{2}k\theta_1^2 + \tfrac{2}{2}k\theta_2^2 + Fz \tag{11.60}$$

We have neglected the weight of the cantilever in this expression. Here θ_1, θ_2 are the deflection angles in the x and y directions, and $z = \sqrt{1 - x^2 - y^2}$. The angles θ_i can be expressed in terms of x, y:

$$\theta_1 = \sin^{-1} x = x + \frac{x^3}{6} + \cdots \tag{11.61}$$

Up to fourth order the expansion of V_p is

$$V_p(x, y; F) = F + (k - \tfrac{1}{2}F)(x^2 + y^2) + \left(\frac{k}{3} - \frac{F}{8}\right)(x^4 + y^4) - \tfrac{1}{4}Fx^2y^2 \tag{11.62}$$

The critical load occurs for $F = 2k$. At the critical load the coefficients of x^4 and y^4 are $k/12$ and that of x^2y^2 is $-k/2$. The coefficient of the coupling term is sufficiently negative so that the system is globally unstable.

Unstable collapse modes occur in the direction $x = \pm y$. In these directions the potential reduces to ($\pm x = t$, $\pm y = t$)

$$V_p(t; F) = F + (2k - F)t^2 + \left(\frac{2k}{3} - \frac{F}{2}\right)t^4 \tag{11.63}$$

At the cusp point the coefficient of t^4 is negative. Therefore the strong coupling between two modes, each of which exhibits an A_{+3} buckling mode and is therefore relatively safe, can lead to an A_{-3} failure mode with its drastic imperfection sensitivity. In fact, the imperfection sensitivity to perturbations of the form $\varepsilon_1(x \pm y)$ has the form

$$F_c = F_p - k' |\varepsilon_1|^{2/3} \tag{11.64}$$

while the sensitivity to dynamical loading has the canonical $\frac{1}{2}$ power dependence

$$F_c = F_p - k''(\Delta E)^{1/2} \tag{11.64$'$}$$

where $F_p = 2k$. The combined sensitivity to static ($\varepsilon_1 \neq 0$) and dynamic ($\Delta E \neq 0$) imperfections is described by the surface shown in Fig. 11.9.

Remark. If $k_1 \ll k_2$, the only possible failure mode involves the A_{+3} catastrophe in the x direction. If $k_1 \gg k_2$, only the A_{+3} catastrophe in the y direction can occur. In either case the failure and imperfection studies reduce to 1-dimensional problems. When k_1 and k_2 are made equal in accordance with the design philosophy of engineering optimization, a second degree of freedom is added to the failure study. Instead of one cusp there are now two. Instead of 3 postbuckling paths there are now 3^2. Worse, more than one exists for $F < F_p$. These represent saddles separating the locally stable state $(x, y) = (0, 0)$ from the global instability which exists for large r^2.

9. Propped Cantilever. C_{3v}

The propped cantilever shown in Fig. 11.17 differs from that shown in Fig. 11.16 in the symmetrical disposition of the springs. The perfect system is invariant

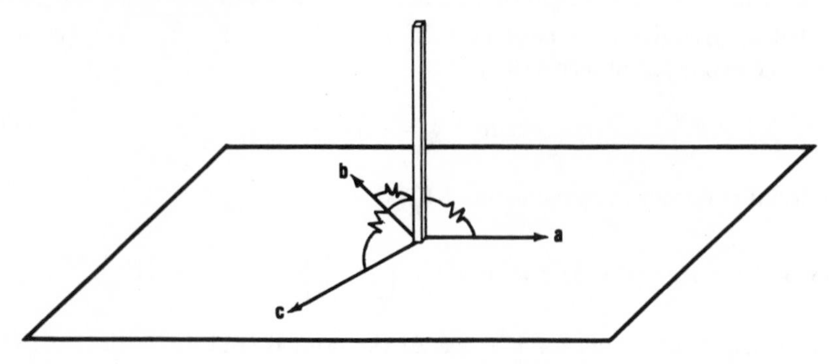

Figure 11.17 The potential describing this perfect propped cantilever has symmetry group C_{3v}.

under rotations through $2\pi/3$ radians as well as reflection in each of the three planes containing one spring. These six operations $(E, C_3, C_3^2; \sigma_1, \sigma_2, \sigma_3)$ form a group called C_{3v}. The potential $V_p(x, y; F)$ describing the perfect system must be invariant under the group C_{3v}.

The polynomial functions which are invariant under C_{3v} may be computed by the methods used in the preceding section. We first introduce the three vectors

$$\begin{aligned} a &= x \\ b &= -\tfrac{1}{2}x + \frac{\sqrt{3}}{2}y \\ c &= -\tfrac{1}{2}x - \frac{\sqrt{3}}{2}y \end{aligned} \tag{11.65}$$

which are shown in Fig. 11.18. It is clear that

$$a + b + c = 0 \tag{11.65$'$}$$

These vectors are linearly dependent and span the state variable plane $\mathbb{R}^2$. Any two of these three vectors may be chosen as basis vectors. The action of the elements of C_{3v} on these vectors is as follows:

(11.66)

	E	C_3	C_3^2	σ_1	σ_2	σ_3
a	a	b	c	a	c	b
b	b	c	a	c	b	a
c	c	a	b	b	a	c

The action of any group element on the $d + 1$ linearly independent terms of degree d may be determined by choosing a and b as basis vectors in $\mathbb{R}^2$. Then

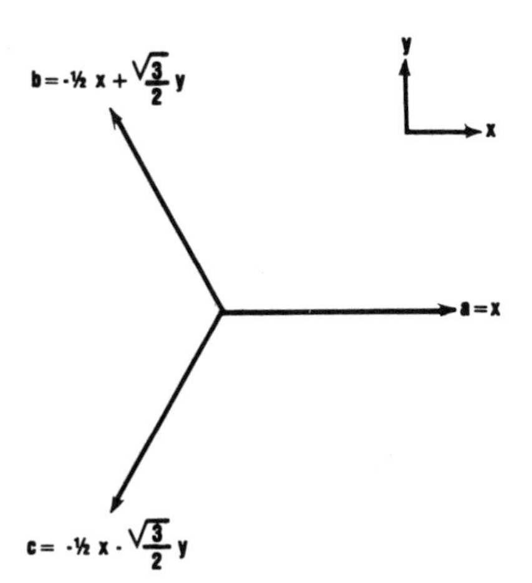

Figure 11.18 Matrix representatives for the operators of C_{3v} are particularly easy to construct when the group elements are allowed to act on the overcomplete system of vectors a, b, c.

basis vectors in the space of terms of degree d have the form $a^p b^q$, $p + q = d$. For example, under C_3 we have

$$\begin{aligned} & \quad C_3 \\ a^4 &\to b^4 \\ a^3 b &\to b^3(-a - b) \\ a^2 b^2 &\to b^2(-a - b)^2 \\ ab^3 &\to b(-a - b)^3 \\ b^4 &\to (-a - b)^4 \end{aligned} \tag{11.67}$$

The 5×5 matrix representative of C_3 is then determined from

$$\begin{bmatrix} a^4 \\ a^3b \\ a^2b^2 \\ ab^3 \\ b^4 \end{bmatrix}' = \begin{bmatrix} 0 & 0 & 0 & 0 & 1 \\ 0 & 0 & 0 & -1 & -1 \\ 0 & 0 & 1 & 2 & 1 \\ 0 & -1 & -3 & -3 & -1 \\ 1 & 4 & 6 & 4 & 1 \end{bmatrix} \begin{bmatrix} a^4 \\ a^3b \\ a^2b^2 \\ ab^3 \\ b^4 \end{bmatrix} \tag{11.68}$$

The trace of this matrix representative is -1.

The matrix representatives of all the group operations in each of the $(d + 1)$-dimensional spaces of homogeneous polynomials of degree d may be computed. The characters of these matrix representatives are given in Table 11.3. The

Table 11.3 Character Table for Representations of the Group C_{3v} in $(d + 1)$-dimensional Spaces Spanned by Degree d Monomials of the Form $x^p y^q$, $p + q = d$

d	E	$2C_3$	3σ	Number[a]
0	1	1	1	1
1	2	−1	0	0
2	3	0	1	1
3	4	1	0	1
4	5	−1	1	1
5	6	0	0	1
6	7	1	1	2
7	8	−1	0	1
8	9	0	1	2
9	10	1	0	2
10	11	−1	1	2
11	12	0	0	2

[a] This column gives the number of linearly independent functions of degree d which are invariant under C_{3v}.

character table repeats itself in the obvious way as d increases. The right-hand column contains the number of linearly independent invariant polynomials of degree d. This number is computed by a standard character analysis and is

$$\frac{1}{6}\sum_{g\in C_{3v}} \chi^d(g) \tag{11.69}$$

One invariant of degree d is clearly

$$I_d = a^d + b^d + c^d \tag{11.70}$$

Then $I_0 \simeq 1$, $I_1 = 0$ (by 11.65′), $I_2 \simeq x^2 + y^2$, $I_3 \simeq abc \simeq x^3 - 3xy^2$. We will now show that I_2 and I_3 are the only functionally independent invariants. To do this, we observe the following:

1. The number of invariants of degree d is, from the character analysis,

$$\begin{aligned} &\left[\frac{d}{6}\right] + 1, \qquad d \bmod 6 \neq 1 \\ &\left[\frac{d}{6}\right] \qquad\quad d \bmod 6 = 1 \end{aligned} \tag{11.71}$$

2. The number of nonnegative integer solutions of

$$2p + 3q = d, \qquad p \geq 0, \quad q \geq 0, \; p, q \text{ integers}$$

is $[d/6] + 1$ unless $d \bmod 6 = 1$, in which case it is $[d/6]$.

Since every invariant of degree d can be written in the form $I_2^p I_3^q$, I_2 and I_3 are the only functionally independent invariants. As a result

$$V_p(x, y; F) = f(I_2, I_3; F) \tag{11.72}$$

The single control parameter F can be used to annihilate a single Taylor series coefficient, so that the potential describing the perfect system is, to lowest nontrivial order,

$$V_p(x, y; F) = \tfrac{1}{2}(F_p - F)(x^2 + y^2) + (x^3 - 3xy^2) \tag{11.73}$$

The most general perturbation of the germ $x^3 - 3xy^2$ is 3-dimensional, with basis vectors x, y, $x^2 + y^2$. A perturbation involving $x^2 + y^2$ will produce a soft first-power imperfection sensitivity. It is therefore sufficient to consider imperfections of the form $\varepsilon_1 x + \varepsilon_2 y$ to determine the imperfection sensitivity of this propped cantilever. The perturbed potential has the form

$$V(x, y; F, \varepsilon_1, \varepsilon_2) = x^3 - 3xy^2 + \tfrac{1}{2}(F_p - F)(x^2 + y^2) + \varepsilon_1 x + \varepsilon_2 y \tag{11.74}$$

The critical points are determined from

$$\frac{\partial V}{\partial x} = 3x^2 - 3y^2 + (F_p - F)x + \varepsilon_1 = 0 \tag{11.75a}$$

$$\frac{\partial V}{\partial y} = -6xy + (F_p - F)y + \varepsilon_2 = 0 \tag{11.75b}$$

The stability matrix is given by

$$V_{ij} = \begin{bmatrix} 6x + (F_p - F) & -6y \\ -6y & -6x + (F_p - F) \end{bmatrix} \tag{11.76}$$

On the bifurcation set det $V_{ij} = 0$. The bifurcation set is therefore determined by the three equations (11.75a), (11.75b), and

$$(6x)^2 + (6y)^2 = (F_p - F)^2 \tag{11.75c}$$

It is convenient to give a parametric representation of this bifurcation set in terms of x and y:

$$\begin{aligned} \varepsilon_1 &= -3(x^2 - y^2) \pm 6x\sqrt{x^2 + y^2} \\ \varepsilon_2 &= \quad 6xy \pm 6y\sqrt{x^2 + y^2} \\ F_p - F &= \quad \pm 6\sqrt{x^2 + y^2} \end{aligned} \tag{11.77}$$

This bifurcation set is shown in Fig. 11.19*a*. In fact, only the smallest F component of this bifurcation set is of any physical interest, because the system collapses on this set. This component is shown in Fig. 11.19*b*.[5]

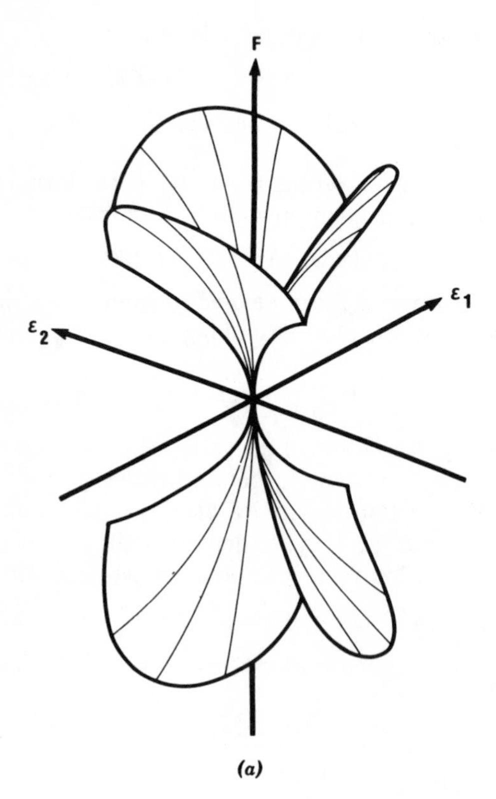

(a)

Figure 11.19 (*a*) The bifurcation set of the D_{-4} catastrophe is shown in the F-ε_1-ε_2 control parameter space.

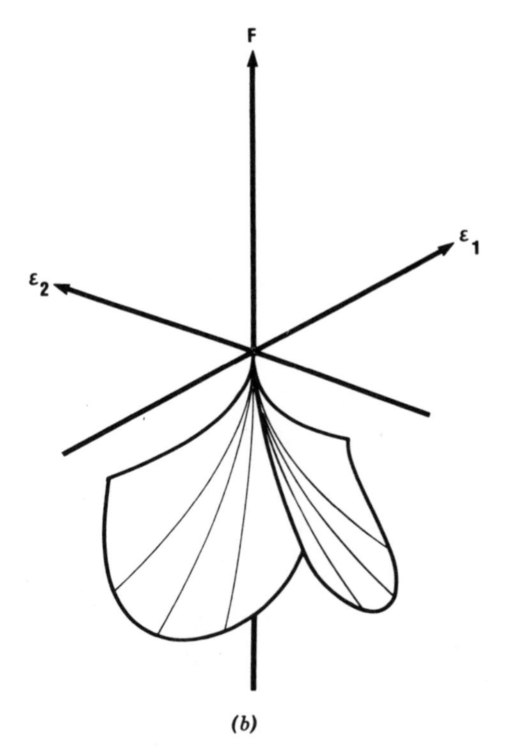

(b)

Figure 11.19 (*b*) Failure occurs the first time the bifurcation set is encountered as F increases. The failure surface is the bottom leaf of the bifurcation surface. Reprinted with permission from J. M. T. Thompson and G. W. Hunt, *J. Appl. Math. Phys.* **26** (1975), p. 596. Copyright 1975 by Birkhäuser Verlag.

Along any straight line through the origin we have $y = tx$. The parametric representation of the bifurcation set then assumes the form

$$\begin{aligned} \varepsilon_1 &= \{-3(1 - t^2) \pm 6\sqrt{1 + t^2}\}x^2 \\ \varepsilon_2 &= \{6t \pm 6t\sqrt{1 + t^2}\}x^2 \\ F_p - F &= \pm 6\sqrt{1 + t^2}\,x \end{aligned} \tag{11.78}$$

Except for certain values of t, there is a $\frac{1}{2}$ power law dependence of the failure load on the imperfection parameters $\varepsilon_1, \varepsilon_2$.

Remark. This $\frac{1}{2}$ power law imperfection sensitivity should not be too surprising. The germ $x^3 - 3xy^2$ is like the germ $x^3 + y^3$. This can be called the "double fold" catastrophe, in analogy with the "double cusp" catastrophe, $\pm x^4 \pm y^4$. Imperfection sensitivities for the double cusp have the same power law dependence as those for the single cusp: $\frac{2}{3}$ for static imperfections and $\frac{1}{2}$ for dynamic imperfections. Similarly, imperfection sensitivities for the double fold have the same power law dependence as those for the single fold: $\frac{1}{2}$ for static imperfections and $\frac{1}{3}$ for dynamic imperfections.

10. Mode Softening

When a structure nears the "breaking point" of the type described in this chapter, one of the collapse modes has a frequency that becomes very small. To see this, we consider the energy of the system exhibiting small oscillations around a Morse critical point:

$$\xi = \tfrac{1}{2}M_{ij}\dot{x}_i\dot{x}_j + \tfrac{1}{2}V_{ij}(F)x_i x_j \tag{11.79}$$

Then the linearized equations of motion are

$$M_{ij}\ddot{x}_j + V_{ij}x_j = 0 \tag{11.80}$$

The associated eigenvalue equation is

$$\det|V_{ij} - \omega^2 M_{ij}| = 0 \tag{11.81}$$

On the bifurcation set $\det V_{ij} = 0$, so that at least one eigenvalue $\omega = 0$ on the bifurcation set and, by continuity arguments, is small near the bifurcation set. This decrease in the oscillation frequency of a normal mode as the critical point is approached is called mode softening.

Mode softening can be used as a quantitative tool to determine both the type of catastrophe to be encountered, and the critical value of the load. For example, if $V \sim x^3$ at $F = F_c$, then

$$\omega^2 \sim (F_c - F)^{1/2} \tag{11.82}$$

as $F \uparrow F_c$. On the other hand, if $V \sim x^4$ at $F = F_c$, then

$$\omega^2 \sim (F_c - F)$$

as $F \uparrow F_c$. In Fig. 11.20 we indicate how plots of $\omega(F)$ against F can be used to determine the critical point of a system.

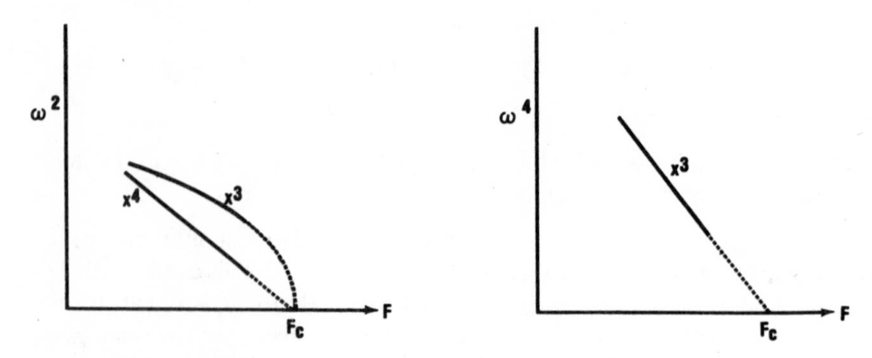

Figure 11.20 The oscillation frequency of the collapse mode decreases as the critical load is approached. A plot of ω^p versus F can determine not only the approximate failure load F_c (by extrapolation) but also the catastrophe type (through p).

Remark. As the bifurcation set is approached, the potential flattens out and in addition departs significantly from a harmonic oscillator shape. The flattening is responsible for mode softening. The departure from the harmonic oscillator shape is responsible for the occurrence of nonlinear oscillations. These contain admixtures of higher harmonics. The amplitudes of these higher harmonics increase as $F \uparrow F_c$. These nonlinear oscillations may be described by means of elliptic functions, and may be used as another diagnostic for the location of catastrophes.

11. Summary

Many structures can be described by a potential. This potential is a function of the load F imposed on the structure. The state of the structure is determined by the critical points of the potential. The stability of a state is determined by the Morse character of the potential at a critical point. The critical load that the structure can carry is determined by critical point degeneracies. In short, the study of failure loadings and imperfection sensitivities of static structures is intimately related to Catastrophe Theory. Much of this chapter is an exposition of these observations.

Our approach has been the following. The potential describing the ideal or perfect system was determined after suitable state variables had been introduced. The critical load for the perfect system was then determined by looking for values of F at which the stability matrix became singular. The germ of the potential at this non-Morse critical point then was used to determine the mathematical form of the most general perturbations of the potential describing the ideal system. Physical considerations were than used to restrict this form somewhat further. This universal perturbation was then used to determine the sensitivity of the critical load F_c to the various imperfections that were possible. These methods were also extended to study the reduction in load carrying capacity due to oscillations of the structure. This critical load reduction was determined by setting the critical value difference ΔV between the metastable local minimum and the lowest contiguous Morse 1-saddle equal to ΔE. The sensitivity of the critical load to imperfections and oscillations was found to have a power law dependence $F_c = F_p - k\,|\varepsilon|^p$ or $F_c = F_p - k\,(\Delta E)^p$, where p is as follows:

	$\lvert\varepsilon\rvert^p$	ΔE^p	Section
A_{+3}	1	1	2
A_{-3}	$\frac{2}{3}$	$\frac{1}{2}$	3, 8
A_2	$\frac{1}{2}$	$\frac{1}{3}$	4, 9

For the systems studied a severe imperfection sensitivity was accompanied by an even more severe dynamical sensitivity.

We saw also how a current design philosophy (engineering optimization, Secs. 5–7) can lead, in the presence of strong coupling, to unexpected failure modes with severe imperfection and dynamical sensitivity.

In Secs. 8 and 9 we saw how group theoretical methods could be used to do part of the Catastrophe Theory analysis of compound systems exhibiting some kind of symmetry. In Sec. 10 we saw how mode softening could be used to locate the critical load before collapse occurs.

Notes

For physical reasons it may not be desirable to allow changes of variables which mix state variables and control parameters. When this is the case, the imperfection sensitivity of a system exhibiting a non-Morse critical point may be determined by a detailed analysis using the methods outlined in Chapters 3–5. Such an analysis has been carried out by Golubitsky and Schaeffer. Their results have been presented in Ref. 6.

References

1. J. M. T. Thompson and G. W. Hunt, *A General Theory of Elastic Stability*, New York: Wiley, 1973.
2. L. Euler, *Methodes Inveniendi Lineas Curvas Maximi Minimive Proprietate Gaudentes* (Appendix, *De Curvis Elasticis*), Lausanne and Geneva: Marcum Michaelum Bousquet, 1744.
3. J. Roorda, Stability of Structures with Small Imperfections, *J. Eng. Mech. Div. Am. Soc. Civil Eng*. **91**, 87 (1965).
4. J. M. T. Thompson and G. W. Hunt, Dangers of Structural Optimization, *Eng. Optimization* **1**, 99–110 (1974).
5. J. M. T. Thompson and G. W. Hunt, Towards a Unified Bifurcation Theory, *J. Appl. Math. Phys*. **26**, 581–603 (1975).
6. M. Golubitsky and D. Schaeffer, A Theory for Imperfect Bifurcation via Singularity Theory, *Commun. Pure Appl. Math*. **32**, 21–98 (179).

CHAPTER 12

Aerodynamics

A high-performance aircraft is like a thoroughbred racehorse. Both are sleek, fast, high-spirited, and, at times, unpredictable. About the unpredictability of horses and humans (equus et egos) we have nothing to say. About the unpredictability of aircraft there is something to say. This "unpredictability" may sometimes be related to Elementary Catastrophes. When this is the case, the unpredictable becomes predictable.

In a certain class of aircraft the nonlinear equations of motion are sufficiently simple that they can be treated qualitatively. This treatment leads to an identification of the steady-state manifold with an Elementary Catastrophe.[1] Once the catastrophe has been identified and its bifurcation set located, the aircraft's qualitative dynamics are essentially determined. We shall illustrate how this procedure is carried out in this chapter.

The qualitative predictions based on the identification of the equation of state manifold with a particular catastrophe A_{+5} can be compared with detailed numerical solutions of the dynamical equations of motion. The statics of

Catastrophe Theory and the dynamics of numerical integration are compatible, even complementary. Numerical integration can be used to locate the stable components of the steady-state manifold. Catastrophe Theory can provide a fundamental understanding in terms of which to interpret the detailed results of a numerical integration.

1. Dynamical System Equations

Aerodynamics is an engineering endeavor in which the assumptions of static behavior that pervade some other engineering fields are clearly not realistic. An airplane in flight is obviously not a static object. Banks, turns, dives, rolls, and ascents involve intimate dynamical details—more often than not they are interrelated through complicated nonlinear mechanisms.

The equations of motion describing an aircraft are complicated and nonlinear. They are also first-order equations. Before jumping into a detailed study of the equations of motion for a particular aircraft, it is more useful first to sit back and consider from arm's length the general properties of systems of coupled non-linear first-order ordinary differential equations. A family of such equations has the form:

$$\frac{dx_i}{dt} = F_i(x; c), \qquad x \in \mathbb{R}^n, c \in \mathbb{R}^k \tag{12.1}$$

As usual, we will regard the n-tuple x as the system state variables and the k-tuple c as the control parameters.

One of the objects of the study of aerodynamics is the determination of how the solution(s) $x(t) \in \mathbb{R}^n$ depend on the control parameters $c(t) \in \mathbb{R}^k$. More generally, the objects of Catastrophe Theory are the determination of the numbers, types, and stability properties of the solutions of (12.1), the locations of the "bifurcation set" $\mathscr{S}_B \subset \mathbb{R}^k$ at which these numbers and types change, and an enumeration of the properties of the phase transitions that occur when the system state jumps from one stable solution to another.

There are several ways to approach this problem. If one has access to a large computer with inexpensive computer time and an efficient Runge–Kutta algorithm for integrating equations of the form (12.1), then it is only necessary to program these equations and let the machine do the rest. While such an approach is extremely economical of time (after debugging) and can lead to enormous insights, it is not entirely satisfactory for several reasons. For one thing, this approach may produce a "feeling" for how the solutions behave as a function of the control parameters, but it will not produce a fundamental insight into their properties. For another, the asymptotic steady state (if one exists) may depend in a discontinuous way on the initial conditions and/or the control parameters. Figure 12.1 illustrates how this could come about. The numerical approach is unable to determine the locations of the separatrices responsible for discontinuities except in terms of detailed numerical mappings of the initial conditions

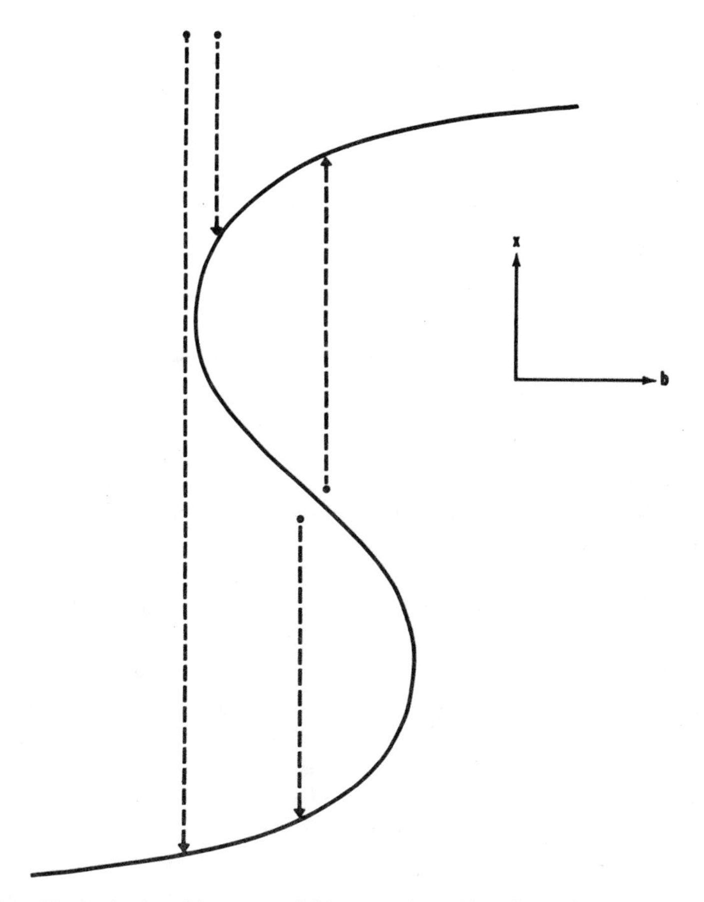

Figure 12.1 The final value of the state variable x may depend in a discontinuous way on the initial value of the state variable x or the control parameter b.

and control parameter settings in $\mathbb{R}^n \otimes \mathbb{R}^k$. Such searches are costly, even on a "free" computer.

An alternative and complementary approach for studying the behavior catastrophes of the dynamical system (12.1) is to search initially for its steady-state solutions $dx_i/dt = \dot{x}_i = 0$. (We distinguish carefully between steady-state conditions and equilibrium conditions. An aircraft may be in steady-state when flying at 13,000 m. An aircraft in equilibrium is an aircraft on the ground.) This method is complementary to the previous approach, and suffers from its own drawbacks. For one thing, (12.1) may not have any (stable) steady states at all. For another thing, it may not easily be possible to solve the system of n equations $F_i(x; c) = 0$.

In some important cases both of these objections are absent. A particularly useful case arises when all but one of the state variables $x = x_1, x_2, \ldots, x_n$ can be eliminated from the n equations $F_i = 0$, leaving a single equation of the form

$$\Phi'(x; c) = 0, \qquad x \in \mathbb{R}^1, c \in \mathbb{R}^k \tag{12.2}$$

Then we may set

$$\begin{aligned} \Phi(x; c) &= -\int^x \Phi'(y; c)dy \\ \Phi'(x; c) &= -\frac{d}{dx}\Phi(x; c) \end{aligned} \tag{12.3}$$

When this is the case, the steady states of (12.1) are identified with the critical points of a function $\Phi(x; c)$. The function Φ is called a **Lyapunov function.** A Lyapunov function plays the same role for steady-state systems as a potential function does for equilibrium systems.

When a Lyapanov function can be associated with a dynamical system, the bifurcation properties and catastrophes of the dynamical system can be partially related to the catastrophes of $\Phi(x; c)$. In the case just described the catastrophes which may occur (for Φ) are those of type A_k. This identification of dynamical system steady states with elementary catastrophes provides information about the adequacy of a polynomial approximation of $\Phi(x; c)$. Conversely, if Φ is a polynomial of degree n, then exactly $n - 2$ control parameters are required to obtain any desired perturbation of a steady-state system. If $k > n - 2$, there are redundant controls. If $k < n - 2$ there are insufficient controls.

A great deal of understanding about the properties of a dynamical system can be obtained when both the numerical and the analytical approaches outlined above are applicable. The analytical approach can be used to determine the equation of state surface in $\mathbb{R}^n \otimes \mathbb{R}^k$ for the dynamical system steady states. The numerical integration procedure can then be used to study how the dynamical system actually approaches the steady-state manifold as a function of (1) control parameter values; (2) initial conditions; and (3) time. We adopt this approach for a particular model aircraft in the present chapter.

2. Truncated Dynamical System Equations

Sometimes dynamical system equations can profitably be studied by making a Taylor series expansion about a steady-state solution of the force functions $F_i(x; c)$. If we assume the state variable coordinates are chosen so that $x = 0$ is a solution for $c = 0$, then a Taylor series expansion of (12.1) about this state assumes the form

$$\dot{x}_i = F_i + F_i^j x_j + F_i^{jk} x_j x_k + \cdots \tag{12.4}$$

Here we have employed covariant notation, so that $F_i^{jk} = \partial^2 F_i / \partial x_j \, \partial x_k$, for example. All derivatives are evaluated at the point $(x; c) = (0; c)$. For fixed value of the control parameters c, the matrix F_i^j may be regarded as a linear response function in the state variables, for it describes the change in $\dot{x}_i$ produced by a change in x_j. More generally, $F_i^{jk\cdots}$ are the components of a susceptibility tensor.

The coefficients $F_i(c), F_i^j(c), \cdots$ may be expanded in a Taylor series expansion in the control parameters c:

$$\begin{aligned} F_i(c) &= F_i(0) + F_i^{\alpha}c_{\alpha} + F_i^{\alpha\beta}c_{\alpha}c_{\beta} + \cdots \\ F_i^j(c) &= F_i^j(0) + F_i^{j\alpha}c_{\alpha} + \cdots \end{aligned} \tag{12.5}$$

These coefficients are the partial derivatives of F_i evaluated at the point $(x; c) = (0; 0) \in \mathbb{R}^n \otimes \mathbb{R}^k$. The coefficients F_i^{α} may be regarded as linear response functions in the control parameters. In the Taylor series expansion about $(0; 0)$, $F_i^j(0; 0)$ and $F_i^{\alpha}(0; 0)$ may be regarded as linear response functions in the state variables and the control parameters, and all other terms are considered nonlinear in the state variables and/or control parameters.

For some dynamical systems the intrinsic nonlinear behavior can be studied by retaining the leading nonlinear terms in the state variables. Such nonlinear systems have the form (12.4), with terms of degree 3 and higher neglected. A number of systems of this type possess one additional property. This is the existence of a constant nonsingular positive-definite symmetric metric tensor G^{li} with the property

$$x_l G^{li} F_i^{jk} x_j x_k = \tilde{F}^{ijk} x_i x_j x_k = 0 \tag{12.6}$$

For such systems one can define an "energy function" by means of the calculation

$$\begin{aligned} x_l G^{li} \dot{x}_i &= x_l G^{li} F_i + x_l G^{li} F_i^j x_j + x_l G^{li} F_i^{jk} x_j x_k \\ \frac{1}{2}\frac{d}{dt}(G^{li} x_l x_i) &= \tilde{F}^i x_i + \tilde{F}^{ij} x_i x_j + \tilde{F}^{ijk} x_i x_j x_k \end{aligned} \tag{12.7}$$

The last term is zero by assumption (12.6). The product $G^{il} F_l^j = \tilde{F}^{ij}$ is generally not symmetric. It may be decomposed into a symmetric and an antisymmetric part. The antisymmetric tensor contracts to zero against the symmetric tensor $x_i x_j$. It is therefore sufficient to take $\tilde{F}^{ij}$ as the symmetric part of $G^{il} F_l^j$. When the symmetric matrix $\tilde{F}^{ij}$ is nonsingular with inverse $\tilde{F}_{ij}$, (12.7) can be written in the form

$$\frac{d}{dt}(\tfrac{1}{2} G^{ij} x_i x_j) = \tilde{F}^{ij}(x_i + b_i)(x_j + b_j) - \tilde{F}^{ij} b_i b_j \tag{12.8}$$

where $b_j = \frac{1}{2}\tilde{F}^i \tilde{F}_{ij}$ is a function of the control parameters alone. When G^{ij} is positive definite, the quadratic form can be interpreted as an energy

$$E = \tfrac{1}{2} G^{ij} x_i x_j \tag{12.9}$$

Equation (12.8) then describes the time rate of change of the dynamical system's generalized energy E. The negative of the generalized energy function (12.9) is a useful Lyapunov function for dynamical systems of the type described above.

Remark. When the forces F_i are given by gradients, then

$$F_i \dot{x}_i \to -\frac{\partial V}{\partial x_i}\frac{dx_i}{dt} = -\frac{dV}{dt} \tag{12.10}$$

For a gradient dynamical system the Lyapunov function is simply the potential. For dynamical systems the Lyapunov function, $-E$ in the case studied here, may be called a generalized potential.

3. Linear Stability Analysis

Once the steady-state solution or solutions $\dot{x}_i = 0$ of (12.1) have been found, it is necessary to determine its (their) stability properties. This is done most easily by a standard linear stability analysis. If $x = x^0$ is a steady-state solution of (12.1) at $c = c^0$, then the equation of motion for a point $x = x^0 + \delta x$ in the neighborhood of x^0 is

$$\delta \dot{x}_i = M_i^j \delta x_j + \mathcal{O}(2) \tag{12.11}$$

$$M_i^j = \frac{\partial F_i}{\partial x_j}(x^0; c^0) \tag{12.12}$$

The terms of order 2 in δx can be dropped when the $n \times n$ stability matrix M has no eigenvalues with zero real part (Chapter 19). When the terms of order 2 can be neglected, (12.11) has a solution:

$$\delta x(t) = e^{M(t-t_0)}\delta x(t_0) \tag{12.13}$$

where δx is an $n \times 1$ column vector, and the exponential is defined by its power series expansion. The stability properties near x^0 can be determined by transforming M to Jordan canonical form (Chapter 14). In particular, if any one of the eigenvalues of M has positive real part, x^0 is an unstable steady state. If all eigenvalues of M at x^0 are in the left half-plane, a perturbation of the system will return to the stable steady state at x^0 on a time scale determined from the eigenvalue with the smallest real part:

$$T_1 \simeq \max_{1 \le i \le n} (-\text{Re}\,\lambda_i)^{-1} \tag{12.14}$$

If x^0 is a steady-state solution for $c = c^0$, then changing the value of the controls slightly $c \to c^0 + \delta c$ will change the location of the steady-state solution slightly. If all eigenvalues of $F_i^j(x^0; c^0)$ lie in the left half-plane, so also do all eigenvalues of $F_i^j(x^0 + \delta x; c^0 + \delta c)$. The open set in $\mathbb{R}^k$ containing c^0 which parameterizes locally stable steady-state solutions of (12.1) is bounded by a bifurcation set $\mathcal{S}_B \subset \mathbb{R}^k$ on which one or more eigenvalues of $F_i^j[x^0(c); c]$ have zero real part. If M_i^j (12.11) were symmetric, as is the case for gradient systems, then all eigenvalues would be real and the bifurcation set of F_i^j would be locally equivalent to the bifurcation set of some catastrophe function. However, F_i^j is not necessarily symmetric, may and often does have complex eigenvalues, so that its bifurcation set and catastrophes hold new surprises in store for us (Chapter 19).

In this chapter we will concern ourselves only with an aircraft whose loss of stability is related to one of the elementary catastrophes. For this aircraft the stability matrix on some of the stable steady-state solutions has complex conjugate roots, but it is one of the real roots which goes through zero and which is responsible for the loss of stability. The equations of motion governing the aircraft studied below have the form (12.4) (without $\cdots$). If $x = x^0$ is a steady-state solution for this set of equations, the linearized stability matrix around this point is

$$M_i^j = F_i^j + 2F_i^{jk}x_k^0 \tag{12.15}$$

4. Description of Aircraft

Before we can describe the possible steady states of an aircraft we must first determine an appropriate set of state variables. This set must be sufficiently large that our description is not a bad caricature of the aircraft, yet small enough to be useful.

We begin by choosing a body-fixed set of axes in the aircraft. The origin is taken at the aircraft center of gravity, and the axes in the directions of the aircraft principal moments of inertia. These axes are shown in Fig. 12.2.

The orientation of the body axes with respect to an inertial frame of reference is determined by the three angles (ϕ, θ, ψ). The nonzero components of the inertia tensor I_{ij} along the three principal axes are (I_{xx}, I_{yy}, I_{zz}). The components of the angular velocity vector $\boldsymbol{\omega}$, with respect to the principal axes, are (p, q, r), while the components of torque $\mathbf{T}$ are (L, M, N). The components of the velocity vector $\mathbf{v}$ of the center of mass with respect to the inertial frame are (u, v, w), while the components of force $\mathbf{F}$ are (X, Y, Z). These definitions are summarized in Fig. 12.2.

If the speed is approximately constant, it is often useful to use instead of $\mathbf{v}$ the "polar" variables $|\mathbf{v}|, \alpha, \beta$, where α is called the angle of attack and β the sideslip angle. For small angles measured in radians, we have approximately $\alpha \simeq w/|\mathbf{v}|$, $\beta \simeq v/|\mathbf{v}|$.

In order to describe the stability of an aircraft, it is in many instances sufficient to use as state variables the three components of angular velocity (p, q, r), the two "polar" angles (α, β), and the three orientation angles (ϕ, θ, ψ). Then the space of state variables $\mathbb{R}^n$ is 8-dimensional, with $(p, q, r, \alpha, \beta, \phi, \theta, \psi) = (x_1, \ldots, x_8) \in \mathbb{R}^8$.

The orientation of the aircraft is determined by the left and right aileron settings a_l, a_r, the left and right elevator settings e_l, e_r, and the rudder setting r. These angles are measured from the trim settings for straight and level flight, and are usually small. These five deflections are the controls:

$$(a_l, a_r, e_l, e_r, r) = (c_1, c_2, c_3, c_4, c_5) \in \mathbb{R}^5$$

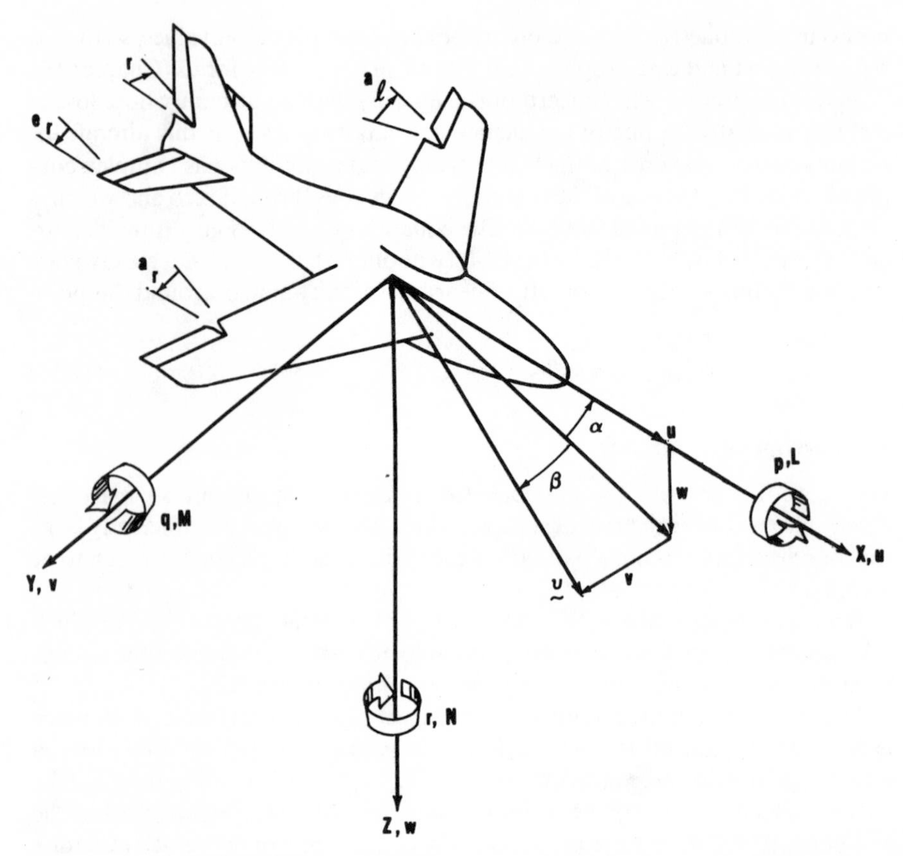

Figure 12.2 The principal body-fixed aircraft axes are shown. These axes have Euler angles (ϕ, θ, ψ) with the axes of an inertial frame. (Adapted from Refs. 1, 2.)

X, Y, Z: components of force vector
L, M, N: components of torque tensor
u, v, w: components of velocity vector with respect to inertial frame
p, q, r: components of angular velocity tensor with respect to principal axes,
α, β: angle of attack, sideslip angle
a_l, a_r: aileron trim sittings
e_l, e_r: elevator trim sittings
r: rudder trim sitting

For subsequent convenience, it is useful to set

$$\begin{aligned} a &= \tfrac{1}{2}(a_l + a_r), \qquad & \delta a &= a_l - a_r \\ e &= \tfrac{1}{2}(e_l + e_r), \qquad & \delta e &= e_l - e_r \end{aligned} \tag{12.16}$$

5. Aircraft Equations of Motion

Now that the appropriate state variables and control parameters have been identified, it is necessary to determine the structure of the equations of motion.

Newton's equations of motion are second order, so the aerodynamic equations of motion must also be second order. Since the three angular velocity components (p, q, r) are already first-order time derivations, these three state variables obey equations of the form

$$\dot{x}_i = F_i(x; c) \tag{12.17}$$

with $1 \leq i \leq 3$. The two "polar" angles have also been expressed in terms of the velocity components w, v, which are first-order time derivatives. Therefore these two angles must also obey first-order equations of motion of the type (12.17) with $4 \leq i \leq 5$. Finally the time derivatives (ϕ, θ, ψ) of the three orientation angles are related to the three components of angular velocity (p, q, r) by a transformation linear in the latter:

$$\begin{aligned} \dot{\phi} &= p + q \tan\theta \sin\phi + r \tan\theta \cos\phi \\ \dot{\theta} &= \quad\quad q \cos\phi \quad\quad - r \sin\phi \end{aligned} \tag{12.18}$$

Thus relation (12.17) for $6 \leq i \leq 8$ assumes the role of a constitutive equation. In short, all eight state variables obey a set of dynamical system equations of the form (12.1). This 8-variable family of equations depends on five control parameters.

The dynamical equations of motion for (p, q, r) $(1 \leq i \leq 3)$ are independent of the orientation angles (ϕ, θ, ψ). The dynamical equations for (α, β) $(4 \leq i \leq 5)$ depend on the angles ϕ, θ through the gravitational coupling constant scaled by the airspeed $(g/|\mathbf{v}|)$. For a jet aircraft $|\mathbf{v}|$ is large and $g/|\mathbf{v}| \to 0$, so that these coupling terms can often be neglected. For jet aircraft, then, the first five equations decouple from the last three. We have as equations of motion:

$$\dot{x}_i = F_i(x_j; c), \qquad 1 \leq i, j \leq 5 \tag{12.19a}$$

$$\dot{x}_k = F_k(x_l; c), \qquad 6 \leq k \leq 8, \qquad 1 \leq l \leq 8 \tag{12.19b}$$

We shall be concerned with solving the first set of equations (12.19a).

The system of equations (12.19a) for a symmetrical jet aircraft has one steady-state solution $x = 0 \in \mathbb{R}^5$ when $c = 0 \in \mathbb{R}^5$. To search for additional solutions we make a Taylor series expansion of the dynamical system equations as in (12.4):

$$\dot{x} = LA + LC + NL \tag{12.20}$$

Here LA are the linear aerodynamic coefficients and LC are the linear control coefficients.

$$\begin{aligned} LA &= F_i^j x_j \\ LC &= F_i^\alpha c_\alpha \end{aligned} \tag{12.21}$$

Of the nonlinear terms NL, the inertial terms IN are the most important. These terms are

$$IN = F_i^{jk} x_j x_k, \qquad 1 \leq i \neq j \neq k \leq 3$$

where

$$F_1^{23} = F_1^{32} = \frac{(I_{yy} - I_{zz})}{I_{xx}}$$
$$F_2^{31} = F_2^{13} = \frac{(I_{zz} - I_{xx})}{I_{yy}} \tag{12.22}$$
$$F_3^{12} = F_3^{21} = \frac{(I_{xx} - I_{yy})}{I_{zz}}$$

If we neglect all nonlinear terms except the inertial terms, then the nonlinear aerodynamic equations of motion assume the form

$$\dot{x}_i = \sum_\alpha F_i^\alpha c_\alpha + \sum_{j=1}^{5} F_i^j x_j + \sum_{1 \le j < k}^{3} F_i^{jk} x_j x_k, \qquad 1 \le i \le 3$$
$$\dot{x}_i = \sum_\alpha F_i^\alpha c_\alpha + \sum_{j=1}^{5} F_i^j x_j, \qquad 4 \le i \le 5 \tag{12.23}$$

This system of equations has the form and properties of the dynamical system studied in Sec. 2. Its Lyapunov function therefore has the form given in (12.8). More specifically,

$$E = \tfrac{1}{2}M\mathbf{v} \cdot \mathbf{v} + \tfrac{1}{2}I_{ij}\omega_i\omega_j \tag{12.24}$$

is the negative of the Lyapunov function. The linear stability analysis is carried out directly on (12.23), once its steady-state solutions have been obtained.

6. Steady-State Manifolds

The steady-state solutions of (12.23) are obtained by setting $\dot{x}_i = 0$. The resulting equations are

$$0 = F_i^\alpha c_\alpha + F_i^j x_j + F_i^{jk} x_j x_k, \qquad 1 \le i \le 3 \tag{12.25a}$$
$$0 = F_i^\alpha c_\alpha + F_i^j x_j, \qquad 4 \le i \le 5 \tag{12.25b}$$

The two equations (12.25b) are linear in the state variables $x_1, \ldots, x_5$. They can be solved by standard techniques to express x_4 and x_5 as linear combinations of the control parameters and the remaining state variables x_1, x_2, x_3. Under these substitutions (12.25a) reduce to

$$0 = k_1^\gamma c_\gamma + F_1'^j x_j + F_1^{23} x_2 x_3$$
$$0 = k_2^\gamma c_\gamma + F_2'^j x_j + F_2^{31} x_3 x_1 \tag{12.26}$$
$$0 = k_3^\gamma c_\gamma + F_3'^j x_j + F_3^{12} x_1 x_2$$

Here the $k_i^\gamma c_\gamma$ are linear combinations of the control parameters, the $F_i'^j$ are complicated functions of the original linear response coefficients F_i^j, and the inertial terms remain unchanged.

The system of three coupled nonlinear equations (12.26) may be solved as follows. The second and third equations of (12.26) may be solved simultaneously for x_2 and x_3 in terms of the remaining state variable $p = x_1 = x$. The solutions are ratios of quadratic functions:

$$x_2 = \frac{Q_2}{Q_1}, \qquad x_3 = \frac{Q_3}{Q_1}. \tag{12.27}$$

$$\begin{aligned} Q_1 &= (F_2'^3 + F_2^{31}x_1)(F_3'^2 + F_3^{21}x_1) - F_2'^2 F_3'^3 \\ Q_2 &= -(k_3^\gamma c_\gamma + F_3'^1 x_1)(F_2'^3 + F_2^{31}x_1) + F_3'^3(k_2^\gamma c_\gamma + F_2'^1 x_1) \\ Q_3 &= -(k_2^\gamma c_\gamma + F_2'^1 x_1)(F_3'^2 + F_3^{21}x_1) + F_2'^2(k_3^\gamma c_\gamma + F_3'^1 x_1) \end{aligned} \tag{12.28}$$

These ratios are then substituted into the first of equations (12.26) to give the following algebraic expression:

$$0 = k_1^\gamma c_\gamma Q_1^2 + F_1'^1 x_1 Q_1^2 + F_1'^2 Q_1 Q_2 + F_1'^3 Q_1 Q_3 + F_1^{23} Q_2 Q_3 \tag{12.29}$$

This is a quintic in the state variable $x = x_1$. Therefore the equation of state $\Phi' = 0$ is derived from a Lyapunov function $\Phi(x; c)$ which is a 6th-degree polynomial in the single state variable x. The associated catastrophe can be no worse than A_5, which requires four control parameters. Therefore one of the five control parameters $c \in \mathbb{R}^5$ is mathematically redundant.

Remark. The quadratic term Q_1 does not depend on any control parameters, while the quadratic terms Q_2 and Q_3 depend linearly on the controls. As a result, all terms in the expression (12.29) for Φ' depend linearly on the control parameters c_γ except those arising from the last term $F_1^{23} Q_2 Q_3$.

7. Application to a Particular Aircraft

The various assumptions leading to the equation of state (12.29) are applicable to a class of aircraft. This class includes a short stubby wing jet aircraft which has been studied extensively.[2] The aerodynamic linear response coefficients F_i^j, the control linear response coefficients F_i^γ, and the components of the moment of inertia tensor have been given by Etkin and will not be reproduced here. For this aircraft the moment of inertia component I_{xx} about the fuselage axis is much smaller than the remaining two components I_{yy}, I_{zz}. Because the wings are stubby, $I_{yy} \simeq I_{zz}$ and in particular $|I_{yy} - I_{zz}|/I_{xx} \simeq 0$, so that the term F_1^{23} in (12.29) can be neglected. As a consequence, the equation of state $\Phi'(x; c) = 0$ is a quintic in the state variable x, but linear in all the control parameters.

Although this model has been studied extensively, it is not entirely reasonable because the initial equations of motion, which have the form (12.4), are assumed to depend on the two control parameters a, e only, and not on the remaining three δa, δe, r. A better model would take these three additional control degrees of freedom into account.

For this standard model, the equation of state is

$$\Phi'(x; c) = \sum_{k=0}^{5} a_k x^k = 0 \tag{12.30}$$

$$\begin{aligned} a_5 &= -21.6 \\ a_4 &= -326.4a \\ a_3 &= 50.3e + 358.9 \\ a_2 &= 5412.6a \\ a_1 &= 11752.8e - 1525.9 \\ a_0 &= -23015a \end{aligned} \tag{12.31}$$

The angles a, e are measured in radians unless otherwise explicitly indicated. The deficiency in the model at least allows us to visualize the equation of state surface $\Phi' = 0$ as a 2-dimensional manifold embedded in the 3-dimensional space $\mathbb{R}^3$ whose coordinates are $(x; a, e)$.

The quintic equation of state can be transformed to canonical form by performing a shift of the origin:

$$x = \tilde{x} - \frac{a_4}{5a_5}$$

This transformation leads to the canonical form of the catastrophe A_5, in which the canonical coefficients are nonlinear functions of the physical control parameters. The bifurcation set for Φ can then be determined since the bifurcation set for A_5 is canonical. However, this identification is nonlinear and complicated.

8. The Bifurcation Set

A simpler way to determine the bifurcation set is in terms of a parametric representation. We set $\Phi'(x; a, e) = 0$ and $\Phi''(x; a, e) = 0$ simultaneously. The nonlinearities in these expressions occur only in the state variable x. The state variable x can thus be used to give a parametric representation for the values of the control parameters $(a(x), e(x))$ on the bifurcation set. This is carried out explicitly by writing $\Phi' = 0$, $\Phi'' = 0$ in matrix form:

$$\begin{bmatrix} x^5 & x^4 & x^3 & x^2 & x & 1 \\ 5x^4 & 4x^3 & 3x^2 & 2x & 1 & 0 \end{bmatrix} \begin{bmatrix} 0 & 0 & -21.6 \\ -326.4 & 0 & 0 \\ 0 & 50.3 & 358.9 \\ 5412.6 & 0 & 0 \\ 0 & 11752.8 & -1525.9 \\ -23015 & 0 & 0 \end{bmatrix} \begin{bmatrix} a \\ e \\ 1 \end{bmatrix} = 0 \tag{12.32}$$

Each value of x leads to a pair of simultaneous linear equations in the control parameters (a, e). The solution is unique provided the determinant is nonzero.

In Fig. 12.3 we show the parametric representation of the fold curves $(a(x), e(x))$ and the isolated cusp points. There are three disconnected components in this bifurcation set because the determinant of the appropriate 2×2 matrix determined from (12.32) has two transversal zero crossings at $x_c = \pm 2.88$. The fold curves parameterized by $x < -|x_c|$ and $x > |x_c|$ have one cusp kink each, while the fold curve parameterized by the interior segment $-|x_c| < x < +|x_c|$ has three cusp kinks.

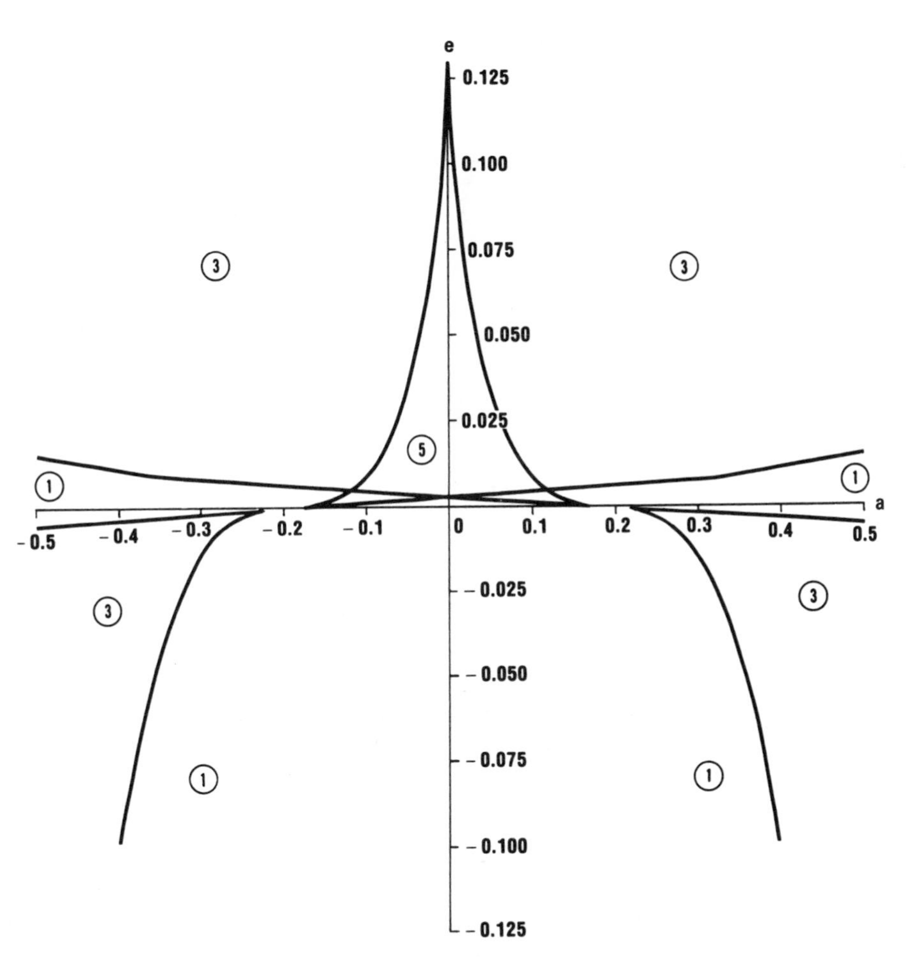

Figure 12.3 The bifurcation set for the potential $\Phi(x; a, e)$ describing a stubby-winged jet aircraft is shown in the (a, e) control plane $\mathbb{R}^2$. The fold curves bound regions with different numbers of equilibria (stable and unstable), as shown in circles.

The bifurcation set divides the control parameter plane $\mathbb{R}^2$ into disjoint open regions which parameterize functions Φ with 1, 3, or 5 critical points, as shown. Of these steady states 1, 2, or 3 are stable, respectively.

The locus of vertical tangents $(x; a(x), e(x))$, its projection, the bifurcation set $\mathscr{S}_B$ in $\mathbb{R}^2$, and the information about numbers of critical points, is sufficient to determine the shape of the equation of state manifold.

9. Catastrophe Phenomena

Slices of the equation of state surface $\Phi'(x; a, e) = 0$ can easily be determined in parametric form from the first of the matrix equations (12.32). Several intersections of planes $e =$ constant with the manifold $\Phi' = 0$ are shown in Fig. 12.4.

These cross sections provide no surprises. For some values of the control parameter a three stable steady states are possible, while for others there are two

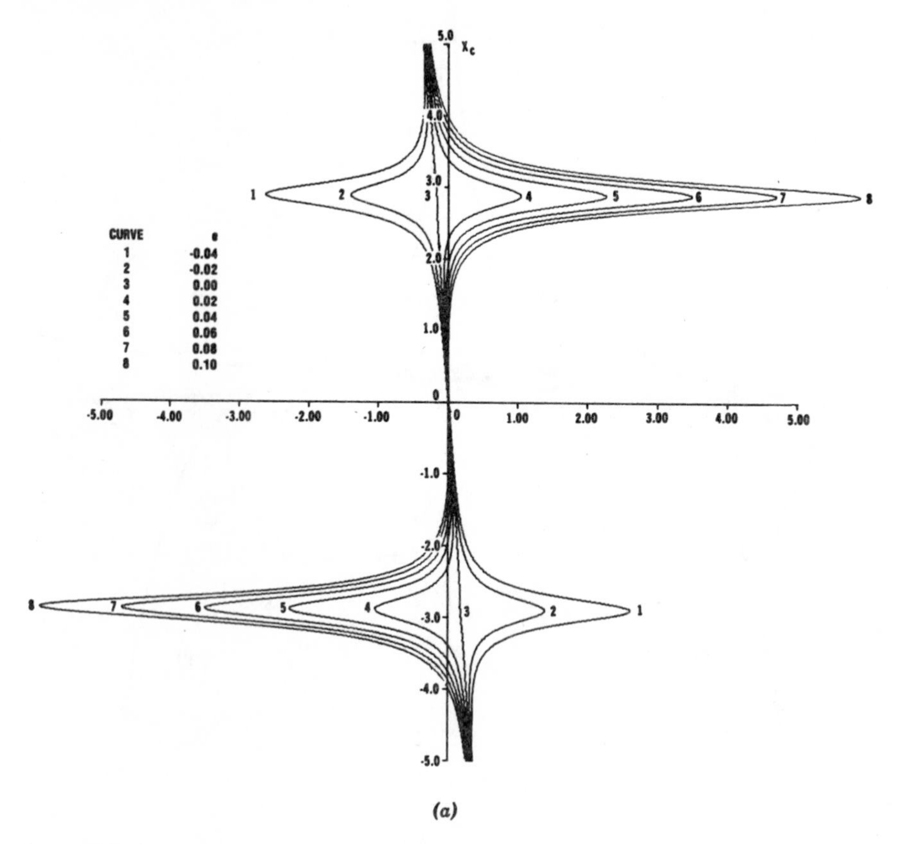

Figure 12.4 (*a*) Intersections of the equilibrium manifold $\Phi'(x; a, e) = 0$ with planes $e =$ constant are shown for several values of e.

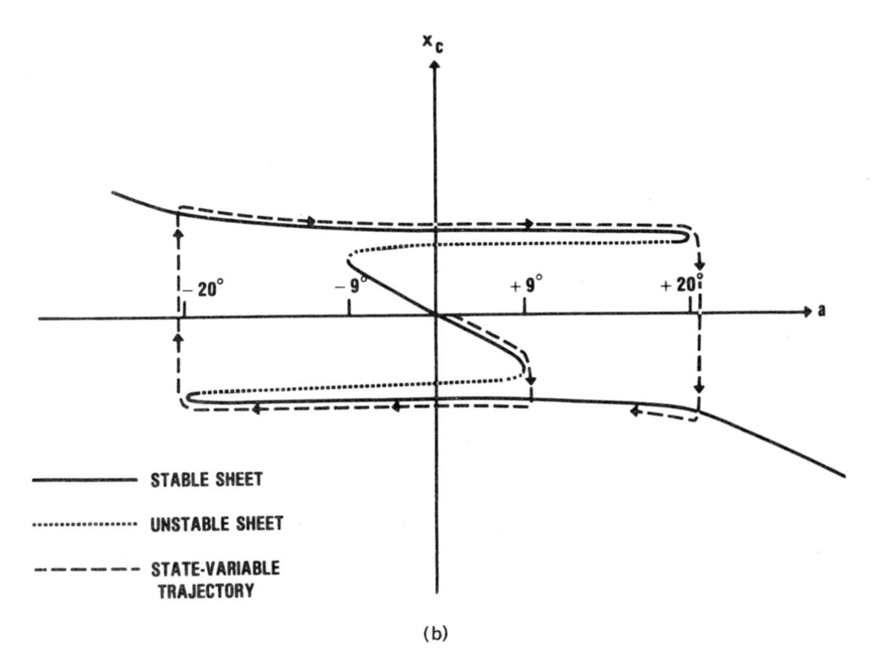

Figure 12.4 (*b*) For an initial state $(x; a, e) = (0; 0, 2°)$, slow variations of a $(0° \to 9°)$ move the state variable off the central sheet. Decreasing a leaves x on the lower sheet until $a \to -20°$, at which point the state variable jumps to the upper sheet. The central sheet remains inaccessible as long as e remains fixed.

or one. Suppose that the control parameters are set initially at $(a, e) = (0, 2°)$. If e is held fixed while a is increased slowly (on a time scale consistent with the Delay Convention), then initially the system state x will respond linearly to changes in a. This linear response breaks down for $a \simeq 9°$,[1] when a jump to the lower sheet occurs. On the lower sheet the aircraft is extremely unresponsive to changes in the aileron settings $(-20 < a < +20)$. It is as if the aircraft "goes overboard" for $a \sim 9°$, and tries to make up for it by not responding at all while on the lower sheet. If the pilot tries to return to the state $x = 0$ by decreasing a while holding e fixed, he is out of luck. The aircraft will not respond until $a \simeq -20°$, at which point it will jump to the upper sheet. Increasing a will produce essentially no response until $a \simeq +20°$, at which point the system state jumps back to the lower sheet. The middle sheet remains inaccessible for $e = 2°$. The system can be steered back to the middle sheet by changing both control parameters a and e. For example, if a is returned to zero and then e decreased below zero, the system will jump to a unique stable state. Returning e to zero will return the system to the central sheet. The system may also be returned to the central sheet without jumps (how?).

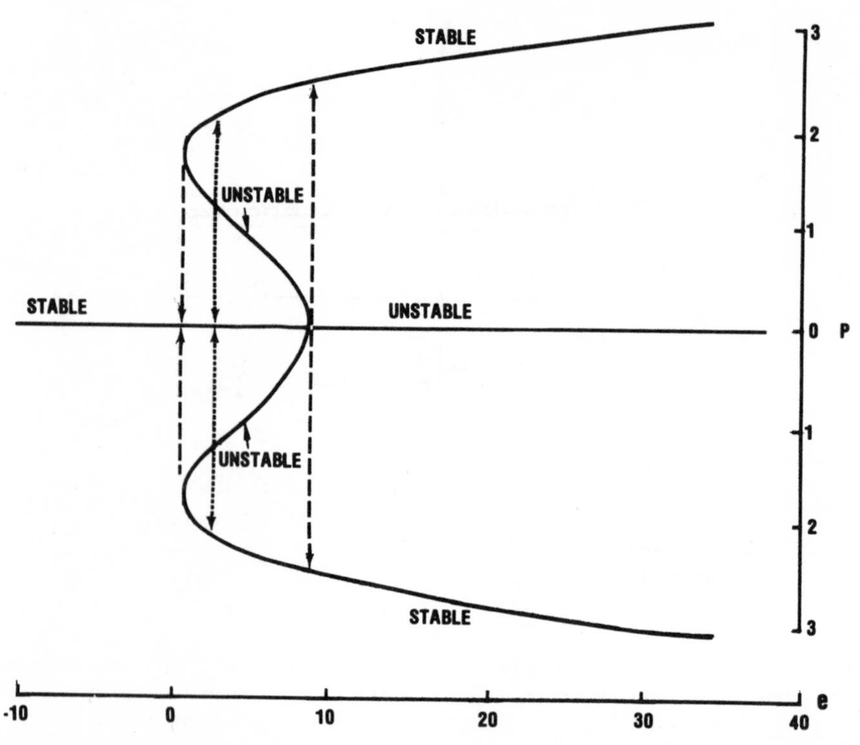

Figure 12.5 The intersection of the plane $a = 0$ with the steady-state manifold $\Phi' = 0$ has the shape shown. The stability properties of the various branches are easily determined by performing a stability analysis at any single nondegenerate point on the steady-state manifold. The Maxwell set is indicated by dotted lines, the bifurcation set by dashed lines. The latter are spinodal lines for the former. Here e is measured in degrees and $p = x_1 = x$.

The intersection of the plane $a = 0$ with the manifold $\Phi' = 0$ is shown in Fig. 12.5. For $e < 0°$ there is a unique stable stationary state at $x = 0$. The branch $x = 0$ loses its stability at $e \simeq 9°$, at which point two symmetric unstable solutions bifurcate (A_{-3}) from this branch. These two new branches turn around at $e \simeq 0°$ and undergo a stability change from unstable to stable. In the range $e > 9°$ there are two symmetric stable solutions as well as the unstable solution $x = 0$. In the range $0 < e < 9°$ there are three stable solutions and two unstable solutions.

If the Delay Convention is obeyed, a first-order phase transition will occur as a function of increasing $e(a = 0)$ at $e \simeq 9°$. In the return case, a first-order phase transition will occur at $e \simeq 0°$. Hysteresis occurs. It is impossible to say whether the jump from the branch $x = 0$ at $e \simeq 9°$ occurs to the upper or the lower branch. This depends on small random perturbations.

If the Maxwell Convention were obeyed, the value of e at which the transition from the branch $x = 0$ to one of the branches $x \neq 0$ occurs must be determined

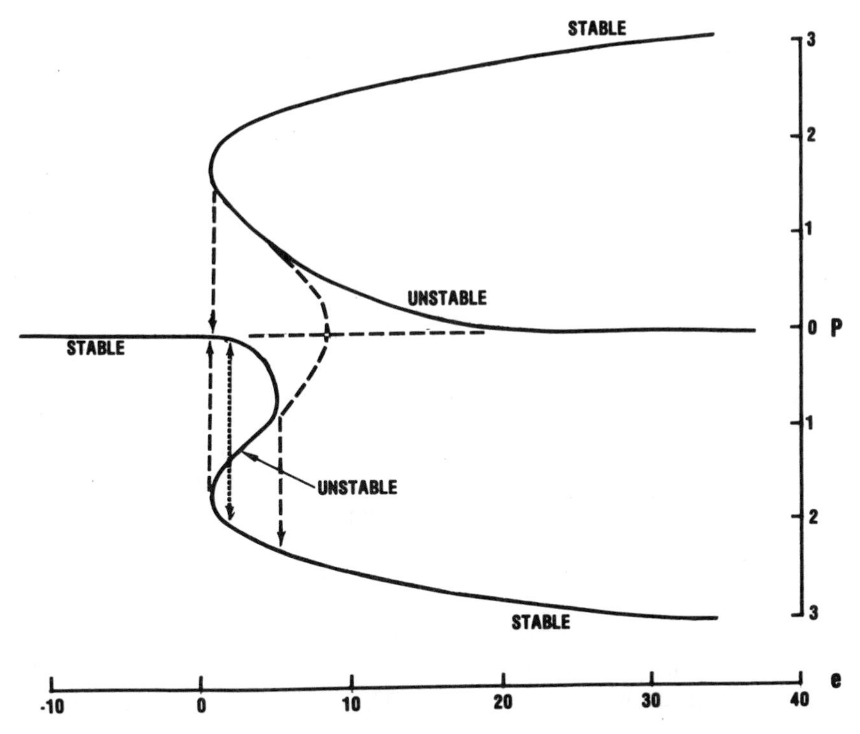

Figure 12.6 The bifurcation is structurally unstable against perturbations of the form $a \neq 0$. Heavy lines indicate intersection of a plane $a \neq 0$ with the steady-state manifold.

from the Lyapanov function $\Phi(x; a = 0, e_c)$. In this case no hysteresis is involved, but the first-order phase transition is surrounded by spinodal lines.

This 1-parameter bifurcation is unstable against perturbations of the form $a \neq 0$. The intersection of the plane $a =$ *small constant* with the manifold $\Phi' = 0$ is shown in Fig. 12.6. The imperfection sensitivity of the aircraft spinodal lines can be determined following the procedures outlined in Chapter 11.

When the control parameters a, e are specified, the state variable x may assume 1, 2, or 3 possible stable steady-state values. Which value is assumed is determined by the aircraft history. For each $(x; a, e)$ satisfying $\Phi' = 0$, there is a unique point $(x_1, x_2, x_3, x_4, x_4) \in \mathbb{R}^5$, the state space for (12.23). The values of x_2, x_3, are uniquely determined from $x_1 = x$ by (12.27). The values of x_4, x_5 are uniquely determined from x_1, x_2, x_3 by linear equations. Therefore if $\Phi' = 0$ is t-valued over (a, e), then the 2-dimensional manifold $\mathscr{M}$ in $\mathbb{R}^5 \otimes \mathbb{R}^2$ which is determined by (12.25) is also t-valued over (a, e). A linear stability analysis can be performed at each point in this manifold $\mathscr{M}$. Such a linear stability analysis yields five eigenvalues. A point $x \in \mathscr{M}$ represents a dynamically stable steady state for the system (12.19) provided all five eigenvalues have negative real parts. Loss of dynamical stability is associated with the real part of one or more eigenvalues increasing through zero. The bifurcation set $\mathscr{S}_B \subset \mathbb{R}^2$ corresponds to a real eigenvalue increasing through zero. For the most part the stable sheets of Φ' correspond to dynamically stable sheets of $\mathscr{M}$. However, there are portions of stable sheets of Φ' whose corresponding points in $\mathscr{M}$ have all real eigenvalues negative, but for which a pair of complex conjugate eigenvalues have crossed the imaginary axis and moved into the positive half-plane. The corresponding system state is unstable, but near it lies a stable but nonsteady state, called a *limit cycle*. The description of this phenomenon is beyond the scope of this chapter. It is treated more thoroughly in Chapters 19 and 20.

10. Integration of the Equations of Motion

We are now ready to return to the original problem—the integration of the equations of motion. The actual dynamical equations of motion are defined on $\mathbb{R}^8 \otimes \mathbb{R}^5$ (12.19). However, the decoupling obtained at high speeds ($g/|\mathbf{v}| \to 0$) allows us to consider the reduced dynamical system defined on $\mathbb{R}^5 \otimes \mathbb{R}^5$ (12.23), and the particular model studied allows us to consider the system defined on $\mathbb{R}^5 \otimes \mathbb{R}^2$.

The steady states for the dynamical system defined on $\mathbb{R}^5 \otimes \mathbb{R}^2$ were determined by first mapping down to $\mathbb{R}^1 \otimes \mathbb{R}^2$ [(12.25)–(12.29)], and then pulling the results back from $\mathbb{R}^1 \otimes \mathbb{R}^2$ to $\mathbb{R}^5 \otimes \mathbb{R}^2$. Now that the location of the 2-dimensional steady-state manifold $\mathscr{M}$ is known in $\mathbb{R}^5 \otimes \mathbb{R}^2$, a reasonable interpretation of the integrated equations of motion is possible.

These equations were integrated as follows.[1] The system is assumed to be in a stable steady state for $t < 0$. At $t = 0$ the control parameter values are changed. The system will then asymptotically approach a new steady state. The asymptotic steady state is one of the stable steady states over the new control parameter values. It is always possible to map out the stable steady-state manifold by starting from some initial steady state and then changing the control parameters sequentially and incrementally. This procedure amounts to little more than time-dependent linear response theory. The bifurcation set shows up clearly as a divergence in the susceptibility tensor.

A more interesting and realistic procedure involves changing the control parameters by finite values, then following the (global) response of the system. The final system state depends on the initial state $x^0 \in \mathbb{R}^5$ as well as the final values of the control parameters.

Mehra, Kessel, and Carrol have performed these integrations numerically. Their results are normalized by using the same initial conditions $(x; c) = (0; 0) \in \mathbb{R}^8 \otimes \mathbb{R}^2$ for all integrations. Thus only the control parameter final values are changed from integration to integration. Parts of the equation of state surface are accessible from this initial condition, while other parts remain in the "shadows" of unstable components of the equation of state manifold.

The time evolution of the state variables $(p, q, r, \alpha, \beta, \phi, \theta)$ is shown in Fig. 12.7.

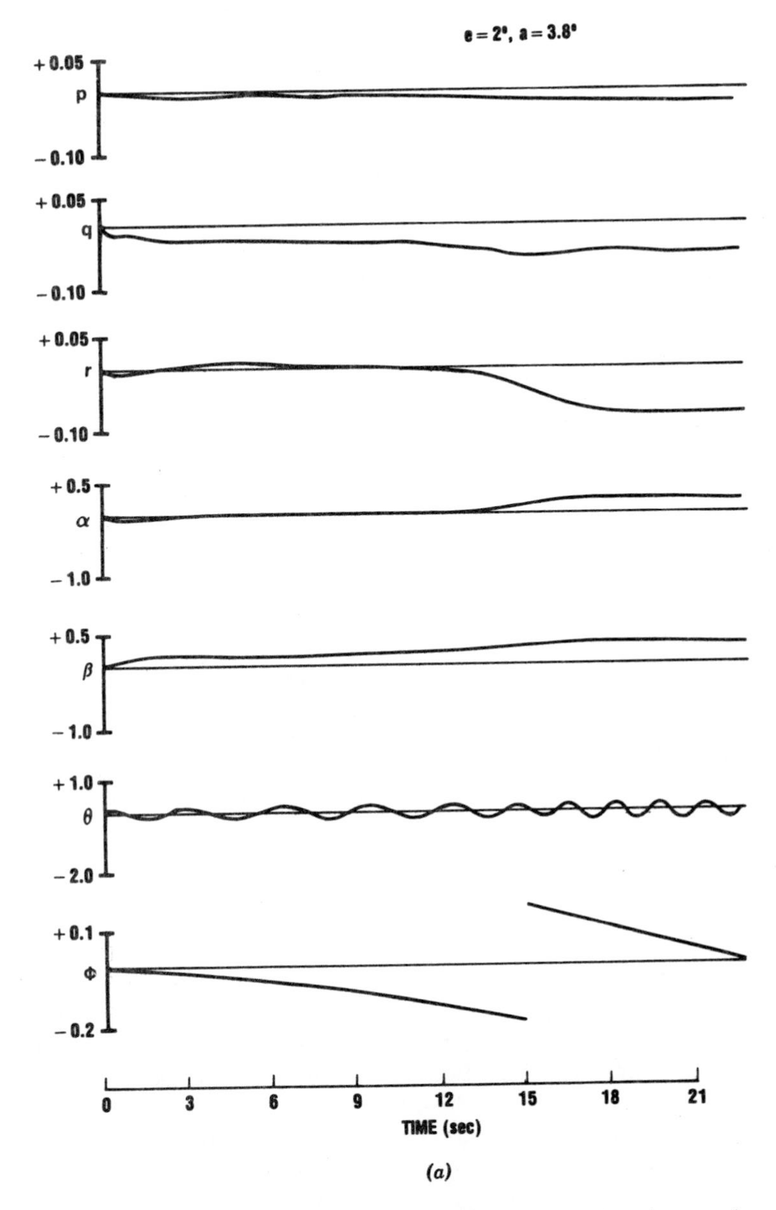

Figure 12.7 (*a*) The time evolution of seven state variables is shown when the control parameter final values carry the system from one stable sheet to another. Data from Ref. 1.

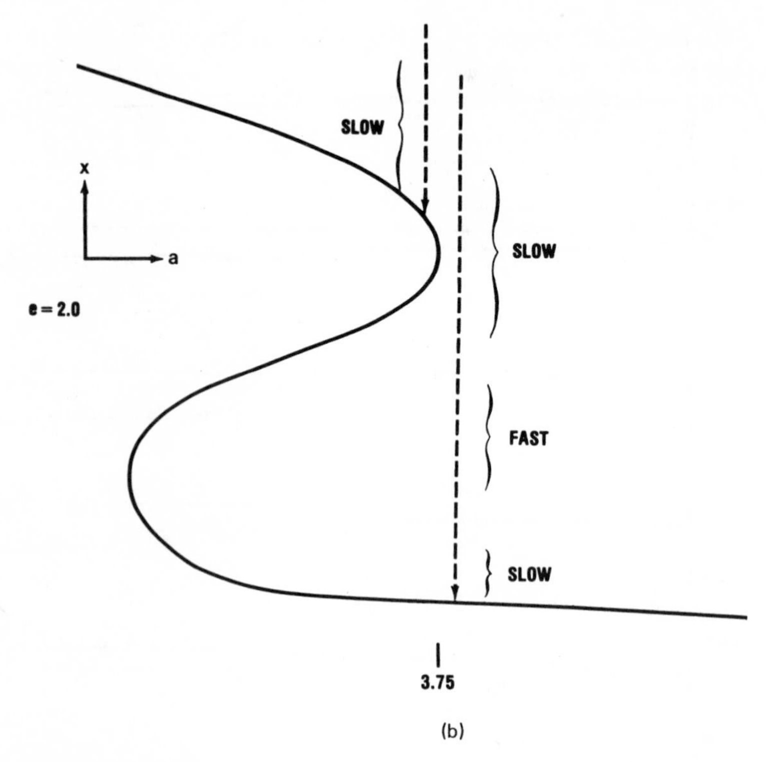

Figure 12.7 (*b*) Critical slowing down occurs as $a \to 3.75°$ because the driving force $F_i(x; c)$ approaches zero at the fold. Critical slowing down occurs whether the system steady state lies on the middle or the lower sheet.

The curves shown in this figure have $(a, e) = (3.8°, 2°)$. There is a jump from the central to the lower sheet at $(a, e) \simeq (3.75°, 2°)$. For these control parameter values, there is a stable steady state on the lower sheet with $x \simeq -2$ and another stable steady state on the upper sheet with $x \simeq 1.6$. The stable states on the upper sheet are screened behind a potential barrier which is represented by the unstable sheet, so the time-integrated solution must asymptotically approach the stable steady state corresponding to the lower sheet.

The evolution toward steady state is shown clearly in the curves of Fig. 12.7. These curves indicate that the approach toward steady state proceeds slowly at first. This is because the state point lies near the middle sheet after the control parameters have been changed. As a consequence, all forces F_i are small and the motion is slow. The point approaches but just misses the equation of state manifold at the fold. Beyond the fold, the system state moves away from the critical manifold as it continues moving toward the lower sheet. As it moves away from the critical manifold it speeds up because the forces F_i become larger. As the lower sheet is approached, the motion slows down. Eventually the stable steady state represented by $(x; a, e) \simeq (-2.0; 3.8°, 2°)$ is reached. If the control parameters were set at $(a, e) = (3.7°, 2°)$, the final system state would lie on the central sheet, and this sheet would be approached slowly.

For control parameters $e = 2°$, $a > 3.8°$ the motion of the state variables would be qualitatively similar to that shown in Fig. 12.7, except that the time scale decreases. If $e = 2°$, $a < 3.7°$, the final system state would lie on the central sheet, but the approach to the stable state would also be faster. This decrease in time scale occurs because the forces F_i are larger as the controls move away from the bifurcation set. The stretching out of the time scale for $e = 2°$ as $a \to 3.75°^{\pm}$ is an example of critical slowing down. This occurs whether the asymptotic stable state lies on the central or the lower sheet.

The five first-order coupled nonlinear dynamical equations of motion can be integrated numerically using an enormous combination of initial conditions $x^{(i)} \in \mathbb{R}^5$ and control parameter final values $c^{(f)} \in \mathbb{R}^2$. The large number of time-dependent functions so obtained would be very difficult to interpret, had we no idea of the structure of the equation of state manifold. With such a knowledge it is possible to determine the qualitative behavior of the trajectories $x(t) \in \mathbb{R}^5$ for arbitrary $(x^{(i)}; c^{(f)}) \in \mathbb{R}^5 \otimes \mathbb{R}^2$, even without integrating the equations of motion.

11. Energy as Lyapunov Function

The aircraft kinetic energy may be expressed in terms of the velocity $\mathbf{v}$ and angular velocity $\boldsymbol{\omega}$ or in terms of the state variables $(p, q, r)(=\boldsymbol{\omega})$ and (α, β) (polar velocity angles):

$$E = \tfrac{1}{2}M\mathbf{v} \cdot \mathbf{v} + \tfrac{1}{2}I_{ij}\omega_i\omega_j \tag{12.33}$$

$$\dot{E} = M\mathbf{v} \cdot \dot{\mathbf{v}} + I_x p\dot{p} + I_y q\dot{q} + I_z r\dot{r} \tag{12.34}$$

The energy function (12.33) is the particular case of (12.9) for the dynamical system currently under consideration. This function is defined on the entire equation of state manifold. On the manifold $\dot{E} = 0$. Upon approach to steady state $\dot{E}$ approaches zero asymptotically. Steady state is not obtained when $\dot{E}$ goes through zero (transversally). The value of E changes significantly when the system state jumps from one sheet to another.

The evolution of the two functions E and $\dot{E}$ is shown in Fig. 12.8. In both cases the initial conditions are $x = 0 \in \mathbb{R}^5$. In Fig. 12.8*a*, $(a, e) = (3.7°, 2°)$, so that the system state evolves toward the edge of the middle sheet. The Lyapunov function $-E$ decreases initially until $t \simeq 3$ sec. At this time $\dot{E}$ changes sign and approaches zero asymptotically while $-E$ increases to a steady-state value. In the second case, shown in Fig. 12.8*b*, $(a, e) = (3.8°, 2°)$, so that the system evolves to a steady-state value on the lower sheet. Again the Lyapunov function $-E$ decreases initially, and at $t \sim 5$ sec, the function $\dot{E}$ almost changes sign. However, it does not quite undergo a sign change, so that $-E$ continues to decrease as the state

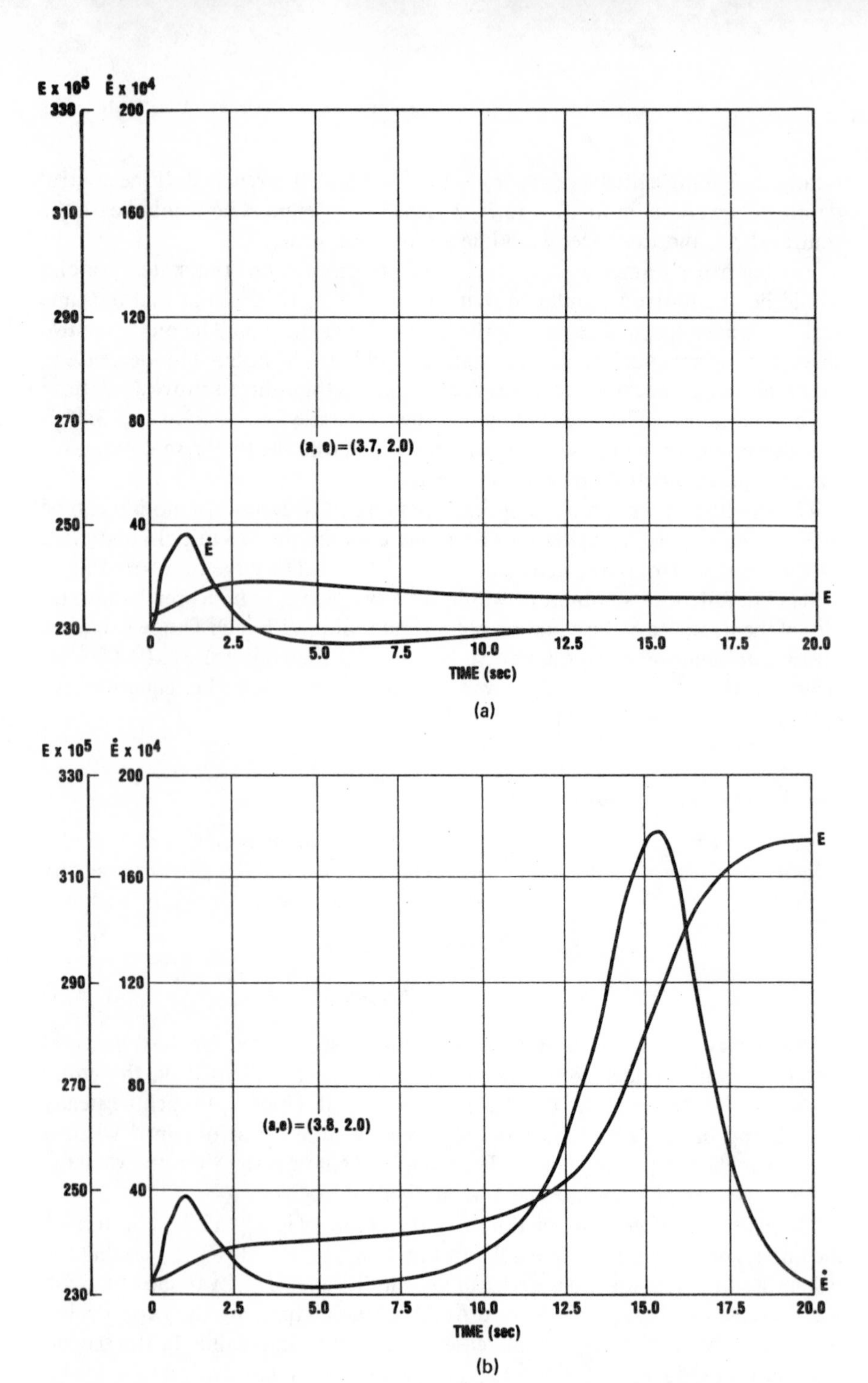

Figure 12.8 The energy E and its time derivative $\dot{E}$ are shown as functions of time for two cases. (*a*) The final steady-state value lies on the middle sheet near the fold edge. (*b*) The system just passes the fold edge ($t \simeq 5$ seconds), then sails onto the lower sheet. Energy measurements are given in British units. Data from Ref. 1.

variable sails past the edge of the equation of state manifold and heads toward the lower sheet. The system state is furthest from the equation of state manifold when $|\dot{E}|$ is maximum. Upon approach to the lower sheet, $\dot{E}$ approaches zero asymptotically while E approaches the steady-state value indicated. The Lyapunov function $-E$ has value -2.32×10^7 slug ft^2/sec^2 on the middle sheet and -3.15×10^7 slug ft^2/sec^2 on the lower sheet for $(a, e) = (3.7°, 2°)$. The central sheet no longer exists above $(3.8°, 2°)$, while the value of $-E$ on the lower sheet is essentially unchanged from its value over $(3.7°, 2°)$.

12. Summary

Our objective in this chapter has been to show that Elementary Catastrophe Theory can be used to provide a useful framework in terms of which the dynamic properties of some aircraft may be understood. We have done this to show that the methods and even the results of Elementary Catastrophe Theory are applicable for the description of, and useful in the understanding of, dynamical (nongradient) systems with suitably prescribed equations of motion. We defer until Chapter 19 a more thorough discussion of dynamical systems.

Since our approach is not restricted to the study of aircraft, we proceed in the first three sections to a general discussion of the methods used. Dynamical system equations are introduced (Sec. 1) and a useful subset is studied in Sec. 2. This subset contains linear terms and some nonlinear terms of a certain type. For this class of dynamical systems it is possible to define a generalized "potential" function. In this sense such dynamical systems begin to resemble gradient systems. Linear stability analyses are described in Sec. 3.

In Sec. 4 we introduce the spaces of state variables and control parameters required to describe a simple model aircraft. The equations of motion are described in Sec. 5. These equations simplify considerably under certain assumptions which are approximately valid for stubby winged jet aircraft. These assumptions involve neglect of the gravity terms $(g/|\mathbf{v}| \to 0)$ and all nonlinear terms except the inertial terms in the equations of motion. With these assumptions the residual equations have the properties of the dynamical system equations studied in Sec. 2. As a consequence, many of their properties are generically known. The reduction of the state variable/control parameter space in this sequence is

$$\begin{array}{l} \mathbb{R}^n \otimes \mathbb{R}^k \to \mathbb{R}^8 \otimes \mathbb{R}^5 \\ \qquad\qquad \downarrow (g/|\mathbf{v}| \to 0) \\ \qquad\qquad \mathbb{R}^5 \otimes \mathbb{R}^5 \\ \qquad\qquad \downarrow (12.25b) \\ \qquad\qquad \mathbb{R}^3 \otimes \mathbb{R}^5 \\ \qquad\qquad \downarrow (12.29) \\ \qquad\qquad \mathbb{R}^1 \otimes \mathbb{R}^5 \\ \qquad\qquad \downarrow \\ \qquad\qquad \mathbb{R}^1 \otimes \mathbb{R}^4 \end{array}$$

This last reduction occurs because the equation of state $\Phi'(x; c) = 0$ is a quintic. The worst catastrophe associated with this manifold is A_5, which has only four effective control parameters.

These results are applied to a particular aircraft in Sec. 7. The equation of state based on the aircraft parameters is presented in (12.30) and (12.31). The bifurcation set is easily derived from this equation of state (Sec. 8). Once the bifurcation set is known, the related catastrophe phenomena are easy to describe (Sec. 9).

The dynamical system equations of motion can be integrated directly. This was done by Mehra, Kessel, and Carroll.[1] The results of these numerical integrations become easy to interpret once the location and structure of the steady-state manifold has been determined. More to the point, integration of the equations of motion can be used to locate the stable components of the equation of state manifold and the bifurcation set (by projection). Conversely, knowledge of the steady-state manifold can be used to infer the solutions of the equations of motion, without performing any integrations at all. Numerical solutions of the equations of motion are presented in Sec. 10, and the time dependence of the Lyapunov function $-E$ is presented in Sec. 11 for two nearby cases on opposite sides of a bifurcation set.

We have used only two of the four control parameter degrees of freedom associated with the catastrophe A_5. This is a deficiency of the model, not the method, used. A more realistic model would have nonzero linear response coefficients for rudder deflection r and aileron and elevator deflection differences δa, δe. These additional nonzero coefficients would force changes only at stage (12.31) of our discussion, as well as the subsequent detailed analysis.

References

1. R. K. Mehra, W. C. Kessel, and J. V. Carroll, *Global Stability and Control Analysis of Aircraft at High Angles of Attack*, Cambridge: Scientific Systems, 1977.
2. B. Etkin, *Dynamics of Atmospheric Flight*, New York: Wiley, 1972.

CHAPTER 13

Caustics and Diffraction Patterns

Geometrical optics may be considered as the short wavelength limit of classical optics, itself an approximation of Maxwell's equations. We show in Sec. 1 how the variational principle which governs geometrical optics comes about from wave optics in the short wavelength limit. In Sec. 2 we investigate in more detail the integral that determines the amplitude of an optical signal at any point. The only nonzero contributions come from neighborhoods of critical points. We are thus immediately forced into a study of catastrophes. Morse critical points give finite contributions to the amplitude (Sec. 2), while non-Morse critical points give very large contributions to the amplitude (Sec. 3).

The direct connection between the large amplitude contributions (caustics) and catastrophe bifurcation sets is studied in Sec. 4. In Sec. 5 we consider what happens when the wavelength is finite. Each caustic is then approximated by a diffraction pattern. Associated with a canonical caustic is a canonical diffraction pattern. These patterns can be computed by taking the Fresnel integral of the appropriate catastrophe function.

1. Geometrical Optics

The wave-particle duality has long been a vocal focal point for philosophers and scientists. In a crude way of speaking, most things have wavelike properties when looked at very carefully. This means that the measuring apparatus has dimensions effectively comparable with the wavelength involved. However, when the wavelength is much smaller than any apparatus dimension, the system behaves for all practical purposes like a particle. That is, the system may

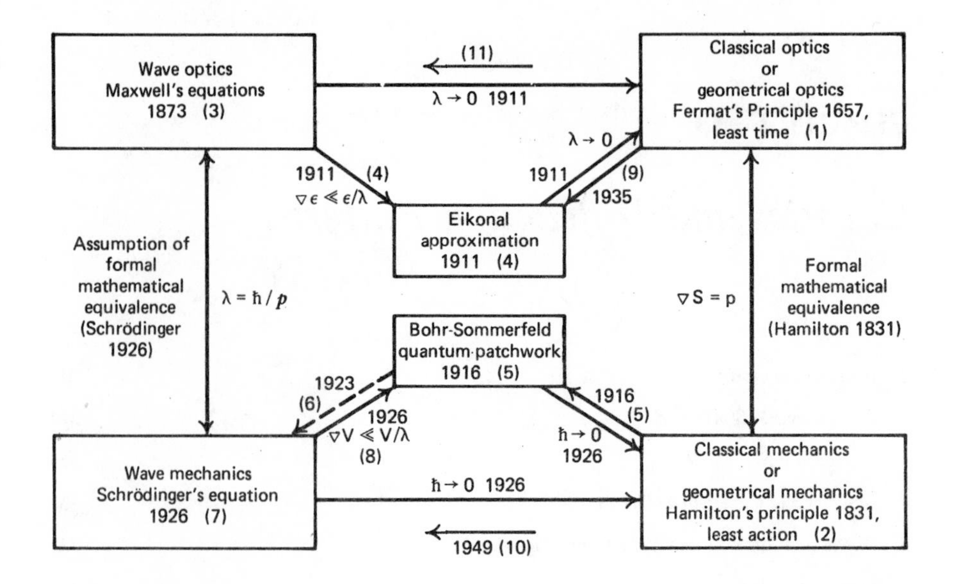

References

1. Geometrical Optics, Principle of Least Time, *Oeuvres de Fermat*, Vol. 2, Paris, 1891, p. 354.
2. Least Action Formulation of Classical Mechanics, W. R. Hamilton, *Trans. Roy. Irish Acad.* **17**, 1 (1833). *Hamilton's Mathematical Papers*, J. L. Synge, W. Conway, editors, Cambridge University Press, p. 285.
3. Wave Optics, J. C. Maxwell, *A Treatise on Electricity and Magnetism*: Oxford University Press, 1873.
4. The Eikonal Approximation, A Sommerfeld and J. Runger, *Ann. Physik* **35**, 289 (1911).
5. The Quantum Patchwork. N. Bohr, *Phil. Mag.* **26**, 1 (1913). W. Wilson, *Phil. Mag.* **29**, 795 (1915). A Sommerfeld, *Ann. Physik* **51**, 1 (1916).
6. M. Born forbids W. Heisenberg to try extending the Bohr-Sommerfeld Quantum Patchwork. Matrix Mechanics results.
7. Wave Mechanics. L. deBroglie, *Nature* **112**, 540 (1923). L. deBroglie, These, Paris (1924). L. deBroglie, *Ann. Physique* **3**, 22 (1925). P. A. M. Dirac, *Sc. Am.* **208**, 45 (1963). (Schrödinger derives the Klein-Gordon equation, solves for the hydrogen atom, never publishes). E. Schrödinger, *Ann. Physik* **79**, 361 (1926). E. Schrödinger, *Ann. Physik* **79**, 489 (1926). E. Schrödinger, *Ann. Physik* **80**, 437 (1926). E. Schrödinger, *Ann. Physik* **81**, 109 (1926).
8. The WKB Approximation. J. Liouville, *J. Math.* **2**, 16 (1837). J. Liouville, *J. Math.* **2**, 418 (1837). Lord Rayleigh, *Proc. Roy. Soc. (London)* **A86**, 207 (1912). H. Jeffreys, *Proc. London Math. Soc.* (*2*) **23**, 428 (1923). G. Wentzel, *Z. Physik* **38**, 518 (1926). H. A. Kramers, *Z. Physik* **39**, 828 (1926). L. Brillouin, *Comptes Rendus* **183**, 24 (1926).
9. Derivation of the Eikonal Approximation from Geometrical Optics, P. Frank and R. von Mises, *Die Differential- und Integralgleichungen der Mechanik und Physik*, Vol. II, Braunschweig: Friederich Viewig und Sohn (1935), Prob. 4-2.
10. Path Integral Formulation of Wave Mechanics, based on the Action Integral. R. P. Feynman, *Revs. Mod. Phys.* **20**, 267 (1948). R. P. Feynman and A. R. Hibbs, *Path Integrals and Quantum Mechanics*, McGraw Hill, New York, 1965.

propagate along well-defined particlelike trajectories even though it is described fundamentally by a wave equation. Classical mechanics and geometrical optics are formally the short wavelength limits of their parents: quantum mechanics and electromagnetic theory.

Remark 1. The short wavelength limits might just as well be called geometrical mechanics and classical optics. Their common origin as short wavelength limits of wave theories makes it not surprising that these two fields can be mathematically formulated in an essentially identical manner.

Remark 2. These essentially identical Hamiltonian formulations of classical mechanics and geometrical optics, together with the knowledge that classical optics was the short wavelength limit of wave optics, suggested to Schrödinger that, if Nature has any sense of symmetry, classical mechanics ought to be the short wavelength limit of some *new* field which he named, by analogy, wave mechanics. He then pushed this analogy hard enough to "derive" the Schrödinger equation from Maxwell's equations. The relation that exists between wave optics and wave mechanics and their short wavelength limits, classical optics and classical mechanics, is summarized in Fig. 13.1.

To illustrate why geometrical optics can be considered the short wavelength limit of wave optics, we consider below two closely related examples. In both examples light is propagating from P_1 to P_2 in 3-dimensional space (Fig. 13.2). In the first example we consider the trajectory followed by a light ray according to the laws of classical optics. In the second example we consider the propagation of light waves according to the laws of wave optics. The important point is that the path followed by light, whether ray or wave, is determined by one minimization principle.

Example 1. The path of a light ray traveling from P_1 to P_2 (Fig. 13.2*a*) is determined by Fermat's Principle of least time:

$$\delta \int_{P_1}^{P_2} dt = 0 \tag{13.1}$$

---

11. Path Integral Formulation of Maxwell's Equations, based on the Time Integral. This has not been formulated mathematically. The groundwork was laid a long time ago: C. Huyghens, (Huyghens' Principle) *Traite de la Lumiere*, 1678 T. Young, (Interference between waves) Phil. Trans. Roy, Soc. (London) xcii **12**, 387 (1802). *Young's Works*, Vol. **1**, p. 202. A. Fresnel, (Synthesis of Huyghen's, Young's Principles) *Ann. Chim. et Phys.* (2) **1**, 239 (1816). *Oevres* **1**, pp. 89 and 129.

Figure 13.1 There is a close mathematical relation between wave and geometrical optics and mechanics. Schrödinger was lead to his wave equation by "filling in the missing box."

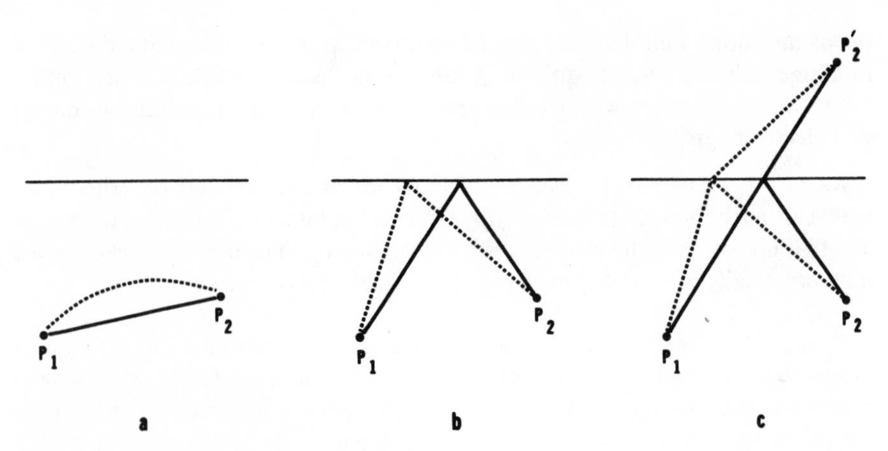

Figure 13.2 The path followed by a light ray between points P_1 and P_2 is one for which the optical path length is stationary under variation (solid line). Dotted lines indicate nearby paths which are nonstationary under variation.

If the velocity of light in some reference medium (vacuum) is c and $n(x, y, z)$ is the index of refraction of the medium through which the light propagates, then the distance traversed ds during a time interval dt is $ds/dt = c/n$, so that Fermat's Principle can be written

$$\delta \int_{P_1}^{P_2} \frac{1}{c} n \, ds = 0 \tag{13.2}$$

If n is constant, then (13.2) reduces to a statement that the path followed is the curve of minimum length connecting P_1 and P_2. If $D(P_1, P_2)$ is the length of any path connecting P_1 with P_2, then this minimization principle can be expressed as

$$\delta D(P_1, P_2) = 0 \tag{13.3}$$

where the variation is over all paths connecting these two points.

Example 2. The intensity of light of wavelength λ which is observed at P_2 is[1]

$$I(P_1, P_2) = I_0 \left| \frac{1}{\lambda} \int \psi e^{2\pi i \Phi/\lambda} \, dx \right|^2 \tag{13.4}$$

where I_0 is the intensity of the source at P_1, and the integral extends over all possible paths connecting P_1 with P_2 (path integral). If all the light reaching P_2 is reflected from the mirror M (Fig. 13.2*b*), we are interested in determining the region of the mirror which contributes most significantly to the integral (13.4) in the short wavelength limit.

In this integral Φ is the length of a path connecting P_1 to P_2 (via the mirror), and ψ is the measure on the space of path integrals. For all practical purposes we need consider only piecewise linear paths broken at the mirror. In this case the path integral is replaced by a 2-dimensional integral over the mirror surface:

$$\begin{aligned} \Phi &= R_1(x) + R_2(x) \\ \psi &= [R_1(x) + R_2(x)]^{-1} \end{aligned} \tag{13.5}$$

where $R_i(x)$ is the shortest distance between P_i and an arbitrary point $x = (x_1, x_2)$ in the mirror surface. Then (13.4) simplifies to

$$I(P_1, P_2) = I_0 \left| \frac{1}{\lambda} \iint \psi(x_1, x_2) e^{2\pi i \Phi(x_1, x_2)/\lambda} \, dx_1 \, dx_2 \right|^2 \tag{13.6}$$

We show in the following section that in the limit $\lambda \to 0$, the contribution of a neighborhood of x^0 to rapidly oscillating integrals of the type (13.6) is zero if $\nabla\Phi \neq 0$ at x^0. Therefore in the short wavelength limit, most of the light that is emitted from P_1 and detected at P_2 is reflected from the neighborhood of $x^0 \in M \simeq \mathbb{R}^2$ such that

$$\delta(R_1 + R_2) = 0 \tag{13.7}$$

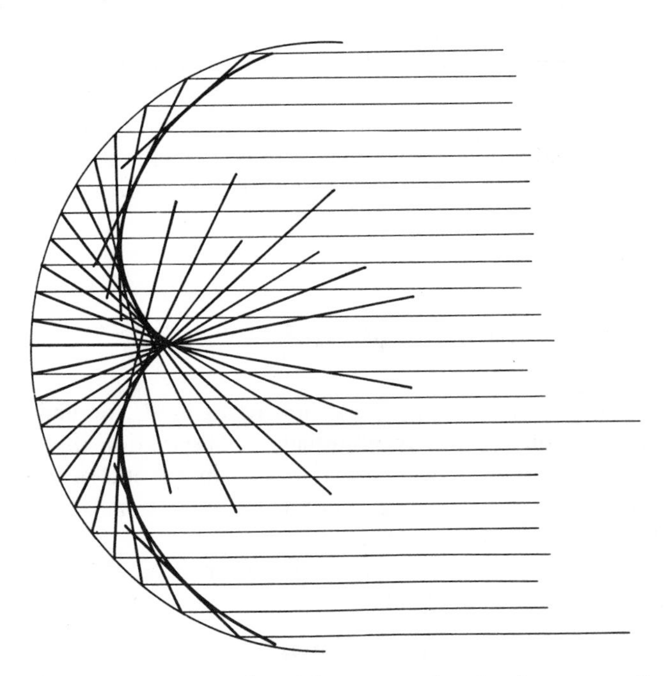

Figure 13.3 The envelope of rays reflected from a curved surface forms a caustic. The caustic formed by reflection of a parallel beam from a spherical surface looks suspiciously like a cusp catastrophe.

This result is equivalent to the Euclidean construction shown in Fig. 13.2*c*. It is useful to compare (13.3) with (13.7).

This consequence ($\nabla\Phi = 0$) of the $\lambda \to 0$ limit of wave optics is the basis for the relation between wave optics and geometrical optics shown in Fig. 13.1.

In the event that reflection occurs from a curved surface, there may be points in $\mathbb{R}^3$ through which two or more rays pass. Along envelopes, such as the one shown in Fig. 13.3, the light intensity may be significantly greater than at nearby points off the enevelope. In fact, such intensities may produce temperatures sufficiently high to burn paper or wood along the envelope. Hence these envelopes are called **caustics**.

If normal intensities are associated with typical points where $\nabla\Phi = 0$ (Morse critical points), then we may expect caustics to be associated with atypical critical points. We will make the connection between caustics and catastrophes in the following sections.

2. Stationary Phase Method

The geometrical nature of the short wavelength limit of a wave theory comes about because propagation is described by rapidly oscillating integrals which have sharply defined peaks in very narrow direction cones. To make this explicit, we imagine that a wave of wavelength λ is propagated from one 2-dimensional surface to another. We assume that the optical distance between the source point $x \in \mathbb{R}^n$ ($n = 2$) and the observation point $\Omega \in \mathbb{R}^k$ ($k = 2$) is $\Phi(x; \Omega)$, and that $\psi(x; \Omega)$ is the amplitude of the signal emitted from the source point x and received at the observation point Ω. Then the intensity of the signal received at Ω is

$$I(\Omega) = |A(\Omega)|^2$$

$$A(\Omega) = \frac{1}{\lambda} \iint \psi(x; \Omega) e^{2\pi i \Phi(x; \Omega)/\lambda}\, d^2x \tag{13.8}$$

The vector Ω may represent an observation point in $\mathbb{R}^2$, or under different geometrical arrangements, it may represent an observation direction. Since Ω is generally at the control of the observer, we shall interpret Ω as a set of control parameters without further specification of its interpretation. It is then natural to interpret x as a set of state variables.

More generally, we can consider the n-dimensional case. We set $k = 1/\lambda\!\!\!^{-} = 2\pi/\lambda$. Then the generalization of (13.8) is

$$I(\Omega) = |A(\Omega)|^2$$

$$A(\Omega) = (2\pi)^{-n/2} k^{n/2} \int \psi(x; \Omega) e^{ik\Phi(x; \Omega)}\, d^n x \tag{13.9}$$

In order to discuss the contribution to the integral (13.9) from the neighborhood of any point x^0 in the domain of integration, we make a Taylor series expansion of Φ and $\ln \psi$ about x^0:

$$\begin{aligned} \Phi(x;\Omega) &= \Phi(x^0;\Omega) + \delta x \cdot \nabla\Phi + \cdots \\ \ln \psi(x;\Omega) &= \ln \psi(x^0;\Omega) + \delta x \cdot \nabla \ln \psi + \cdots \end{aligned} \tag{13.10}$$

Then (13.9) can be written

$$A_{x^0}(\Omega) = (2\pi)^{-n/2} k^{n/2} \psi(x^0;\Omega) e^{ik\Phi(x^0;\Omega)} \times \int_{N(x^0)} e^{(ik\nabla\Phi + \nabla \ln \psi)\cdot \delta x}\, d^n x \tag{13.11}$$

where $N(x^0)$ is a small neighborhood of x^0, and we have neglected for the moment the higher order terms in the exponent. The n-fold integral in (13.11) is a product of n 1-dimensional integrals, each of which has the form

$$k^{1/2} \int_{\delta x = -\varepsilon}^{\delta x = +\varepsilon} e^{(ikA + B)\delta x}\, dx = k^{1/2} \frac{e^{(ikA+B)\varepsilon} - e^{-(ikA+B)\varepsilon}}{ikA + B} \tag{13.12}$$

By choosing ε so that $\varepsilon k A = 2\pi \times$ (integer), this reduces to

$$k^{1/2} \frac{2 \sinh \varepsilon B}{ikA + B} \xrightarrow{\varepsilon \text{ small}} \frac{2k^{1/2} B \varepsilon}{ikA + B} \xrightarrow{\lambda \to 0} 0 \tag{13.13}$$

This latter limit is valid when $A \neq 0$. As long as $\nabla\Phi(x^0;\Omega) \neq 0$, a neighborhood $N(x^0)$ can always be chosen so that each of the 1-dimensional integrals (13.12) vanishes in the limit $\lambda \to 0$.

Since contributions to the amplitude $A(\Omega)$ from neighborhoods of non-critical points vanish, the integral (13.9) can be estimated accurately in the $\lambda \to 0$ limit by considering contributions only from the neighborhoods of the critical points of $\Phi(x;\Omega)$. In these small neighborhoods the function $\psi(x;\Omega)$ may be taken as constant, so that

$$\begin{gathered} A(\Omega) \simeq \sum_{r=1} (2\pi)^{-n/2} k^{n/2} \psi(x^{(r)};\Omega) \int e^{ik\Phi(x^{(r)} + \delta x;\Omega)}\, d^n x \\ \nabla\Phi(x^{(r)};\Omega) = 0 \end{gathered} \tag{13.14}$$

This method for estimating the values of rapidly oscillating integrals by considering only neighborhoods of points where the phase $\Phi(x;\Omega)/\lambda$ varies slowly is called the **stationary phase method**.

We next consider what happens at a Morse critical point. To estimate the integral (13.9) in the neighborhood of such a point, we neglect the variation of $\psi(x;\Omega)$ throughout this neighborhood (valid unless $\psi \simeq 0$), expand $\Phi(x;\Omega)$ about x^0:

$$\Phi(x;\Omega) = \Phi(x^0;\Omega) + \tfrac{1}{2}\delta x_i \delta x_j \Phi_{ij} + \cdots \tag{13.15}$$

and neglect higher order terms. Then we perform an orthogonal transformation to diagonalize Φ_{ij} so that

$$(2\pi)^{-n/2}k^{n/2}\int\psi(x;\Omega)e^{ik\Phi(x;\Omega)}\,d^nx$$

$$\to \psi(x^0;\Omega)e^{ik\Phi(x^0;\Omega)}\prod_{j=1}^{n}\left[\left(\frac{k}{2\pi}\right)^{1/2}\int e^{ik\lambda_j(\delta x_j)^2/2}\,dx_j\right]$$

$$= \psi(x^0;\Omega)e^{ik\Phi(x^0;\Omega)}\prod_{j=1}^{n}(|\lambda_j|^{-1/2}e^{\pm i\pi/4}) \tag{13.16}$$

where the $\pm$ sign is chosen according to the sign of the eigenvalue $\lambda_j \neq 0$. We may express the contribution to a rapidly oscillating integral from the neighborhood of a Morse critical point in coordinate-free form as follows:

$$\left(\frac{k}{2\pi}\right)^{n/2}\int_{N(x^0)}\psi(x;\Omega)e^{ik\Phi(x;\Omega)} \simeq \psi(x^0;\Omega)e^{ik\Phi(x^0;\Omega)}\frac{e^{it\pi/4}}{|\det\Phi_{ij}(x^0;\Omega)|^{1/2}} \tag{13.17}$$

Here $t = (n-i) - i = n - 2i$ is the inertia of a Morse i-saddle ($t = n$ for a local minimum).

3. Degree of a Singularity

So far we have considered the contribution to a rapidly oscillating integral from the neighborhood of a noncritical point (13.13) and from a Morse critical point (13.17). In the former case the contribution was zero and in the latter case it was finite. This leads us to expect that the contribution from the neighborhood of a non-Morse critical point might diverge, and in fact to guess that such critical points are closely connected with caustics.

Suppose that x^0 is a non-Morse critical point of $\Phi(x;\Omega)$. Then by a smooth change of variables we can bring Φ to the canonical form (2.4):

$$\Phi \doteq \mathrm{CG}(l) + M_i^{n-l} \tag{13.18}$$

Then the contribution to the rapidly oscillating integral from the neighborhood of x^0 factors as follows:

$$A_{x^0}(\Omega) = \psi(x^0;\Omega)e^{ik\Phi(x^0;\Omega)}J(x^0)$$

$$\times\int\left(\frac{k}{2\pi}\right)^{n-l/2}e^{ikM_i^{n-l}}d^{n-l}x'\int\left(\frac{k}{2\pi}\right)^{l/2}e^{ik\mathrm{CG}(l)}d^lx' \tag{13.19}$$

where $J(x^0)$ is the Jacobian of the transformation which takes Φ to its canonical form (13.18) at x^0. The integral over the Morse part in (13.19) is given in (13.17), so the only contribution remaining to calculate is that due to the l-variable catastrophe germ CG(l). Such integrals can be estimated using dimensional analysis and computed using complex analysis.

We illustrate these methods in the 1-dimensional case. We assume the following forms for $\Phi(x;\Omega)$ and $\psi(x;\Omega)$:

$$\begin{aligned}\Phi(x;\Omega) &= \Phi(x^0;\Omega) + \Phi_r(x;\Omega)\\ \psi(x;\Omega) &= \sum c_j(x - x^0)^j\end{aligned} \tag{13.20}$$

Here the remainder term $\Phi_r(x;\Omega)$ is to be expanded in a Taylor series expansion around x^0. For $|x - x^0| \ll 1$ the lower degree terms $|x - x^0|^p$ vary more rapidly than the higher degree terms $|x - x^0|^q$ $(p < q)$. Since it is the rapid variation of the trigonometric terms in rapidly oscillating integrals which determines these integrals, it is sufficient to consider only the leading nonzero term in $\Phi_r(x;\Omega)$:

$$\Phi_r(x;\Omega) = \gamma(x - x^0)^p, \qquad \gamma \neq 0, \quad p > 0 \tag{13.21}$$

Then we find

$$\begin{aligned}&\left(\frac{k}{2\pi}\right)^{1/2}\int_{x^0-\varepsilon}^{x^0+\varepsilon}\psi(x;\Omega)e^{ik\Phi(x;\Omega)}\,dx\\ &\quad = e^{ik\Phi(x^0;\Omega)}\sum_{j=0} c_j\left(\frac{k}{2\pi}\right)^{1/2}\int_{x^0-\varepsilon}^{x^0+\varepsilon}(x - x^0)^j e^{ik\gamma(x-x^0)^p}\,dx\end{aligned} \tag{13.22}$$

These integrals can conveniently be estimated by making the substitution $y = k^{1/p}|\gamma|^{1/p}(x - x^0)$. The limits of integration are then

$$\pm\,\varepsilon|\gamma|^{1/p}\left(\frac{2\pi}{\lambda}\right)^{1/p} \tag{13.23}$$

If the wavelength λ and the size of the neighborhood of integration ε are such that $\lambda^{1/p} \ll \varepsilon$, then the limits of the transformed integral can conveniently be extended to $\pm\infty$ without undue violence to the values of the integrals. We then find

$$\left(\frac{k}{2\pi}\right)^{1/2}\int_{x^0-\varepsilon}^{x^0+\varepsilon}(x - x^0)^j e^{ik\gamma(x-x^0)^p}\,dx = (2\pi)^{-1/2}k^{1/2-(j+1)/p}|\gamma|^{-(j+1)/p}I_{\pm}(j,p) \tag{13.24}$$

$$I_{\pm}(j,p) = \int_{-\infty}^{+\infty} y^j e^{\pm iy^p}\,dy \tag{13.25}$$

The plus sign is chosen for $\gamma > 0$, the minus sign if $\gamma < 0$. The result (13.24) is a factorization of the integral on the left-hand side into two parts, a physical part and a geometric part. The physical part k^σ determines the amplitude dependence on the wavelength. The geometric part (13.25) can be computed by the methods of contour integration.

The complex integral

$$I = \oint z^j e^{iz^p}\,dz \tag{13.26}$$

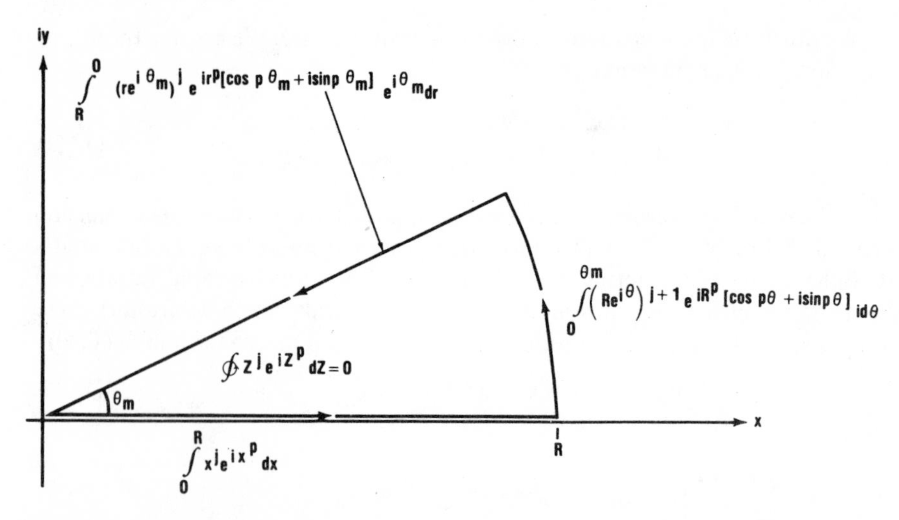

Figure 13.4 The integral (13.25) can be evaluated by computing the complex integral (13.26) along the three segments of the pie-shaped contour shown.

around the contour shown in Fig. 13.4 is zero because no poles are in the region bounded by the contour. This integral can be written as the sum of contributions from the two line segments and the arc:

$$0 = I = \int_0^R x^j e^{ix^p}\, dx + \int_0^{\theta_m} R^j e^{ij\theta} e^{iR^p[\cos p\theta + i\sin p\theta]} Re^{i\theta} i d\theta$$

$$+ \int_R^0 R^j e^{ij\theta_m} e^{iR^p[\cos p\theta_m + i\sin p\theta_m]} e^{i\theta_m}\, dR \tag{13.27}$$

By choosing θ_m in the region $0 \le p\theta_m < \pi$ we ensure that the contribution to the integral from the arc vanishes in the limit $R \to \infty$. By choosing $p\theta_m = \pi/2$ we guarantee that the third term in (13.27) is proportional to the integral over a real function:

$$\int_0^\infty x^j e^{ix^p}\, dx = e^{i(j+1/p)\pi/2} \int_0^\infty R^j e^{-R^p}\, dR \tag{13.28}$$

The substitution $y = R^p$ converts this to a standard function, so that

$$\int_0^\infty x^j e^{\pm ix^p}\, dx = e^{\pm i(\pi/2)(j+1)/p} \frac{1}{p} \Gamma\left(\frac{j+1}{p}\right) \tag{13.29}$$

where $\Gamma(n) = \int_0^\infty x^{n-1} e^{-x}\, dx$ is the standard gamma function, shown in Fig. 13.5. These expressions have order of magnitude unity for $p > 0, j \ge 0$.

Expressions (13.24) and (13.25) provide useful information about the relative

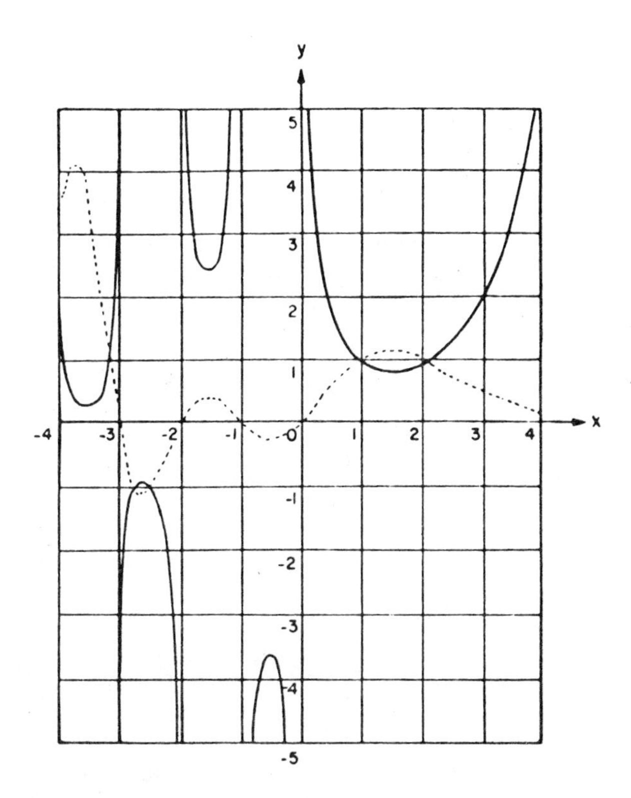

Figure 13.5 $y = \Gamma(x)$. Reprinted with permission from M. Abramowitz and I. A. Stegun, *Handbook of Mathematical Functions*, Nat. Bur. Stand. (U.S.), Applied Mathematics Series 55 (U.S. Government Printing Office, Washington, D.C., 1964). Figure 6.1, p. 255.

importance of the different pieces of $\psi(x; \Omega)$, $\Phi(x; \Omega)$ to the rapidly oscillating integral:

1. For fixed p, terms with higher j contribute less to the value of the integral. As $\lambda \to 0$, the dominant contribution comes from the leading term in the Taylor series expansion of $\psi(x; \Omega)$ around x^0.
2. The integrals $I_\pm(j = 0, p)$ have the values

$$\begin{aligned} I_\pm(0, 1) &= 0 \\ I_\pm(0, 2) &= \pi^{1/2} e^{\pm i\pi/4} \\ I_\pm(0, p) &= 2e^{\pm i\pi/2p}\Gamma\left(1 + \frac{1}{p}\right), \qquad p \text{ even} \\ &= 2\cos\frac{\pi}{2p}\,\Gamma\left(1 + \frac{1}{p}\right), \qquad p \text{ odd} \end{aligned} \tag{13.30}$$

3. The dependence of the amplitude on the wavelength at a catastrophe germ has the form

$$A \sim k^{\sigma}, \qquad \sigma = \frac{1}{2} - \frac{1}{p} \qquad \text{for } A_{p-1} \tag{13.31}$$

The remaining catastrophe germs may be treated similarly. For the catastrophe germ $D_p(x, y) = x^2y + y^{p-1}$ we can use scaling arguments $x' = k^{\alpha}x$, $y' = k^{\beta}y$ to extract the wavelength dependence from the Fresnel integral

$$k^{2/2} \iint_{-\infty}^{+\infty} e^{ik(x^2y + y^{p-1})}\, dx\, dy = k^{1-\alpha-\beta} \int_{-\infty}^{+\infty} e^{iky^{p-1}}\, d(k^{\beta}y) \int_{-\infty}^{+\infty} e^{ikx^2y}\, d(k^{\alpha}x) \tag{13.32}$$

Dimensional arguments require $\beta(p - 1) = 1$, $2\alpha + \beta = 1$, so that the integral becomes

$$k^{2/2} \iint_{-\infty}^{+\infty} e^{ikD_p(x, y)}\, dx\, dy = k^{(p-2)/2(p-1)} \iint_{-\infty}^{+\infty} e^{iD_p(x', y')}\, dx'\, dy' \tag{13.33}$$

Result. The Fresnel integral for the elementary catastrophe germs can be decomposed into the product of two factors. One (k^{σ}) depends on physics

Table 13.1 Fresnel Transform of Elementary Catastrophe Factors into the Product of Two Terms: $k^{l/2} \int_{-\infty}^{\infty} e^{ikCG(l)} d^l x = k^{\sigma} I[CG(l)]$

Germ	Canonical Form	σ	$I[CG(l)]$
$A_{\pm(p-1)}$	$\pm x^p$	$\frac{p-2}{2p}$	$[e^{\pm i\pi/2p} + e^{\pm(-)^p i\pi/2p}]\Gamma\left(1 + \frac{1}{p}\right)$
$D_{\pm(p+1)}$	$x^2y \pm y^p$	$\frac{p-1}{2p}$	$[e^{(i\pi/4)(1 \pm 1/p)} + e^{(-i\pi/4)(1 \pm (-)^{p+1}/p)}]4\Gamma(\frac{3}{2})\Gamma\left(1 + \frac{1}{2p}\right)$
$E_{\pm 6}$	$x^3 \pm y^4$	$\frac{5}{12}$	$2 \cos\frac{\pi}{6}\Gamma(\frac{4}{3}) \times 2e^{\pm i\pi/8}\Gamma(\frac{5}{4})$
E_7	$x^3 + xy^3$	$\frac{4}{9}$	$2 \cos\frac{\pi}{6}\Gamma(\frac{4}{3}) \times 3 \cos\frac{\pi}{9}\Gamma(\frac{11}{9})$
E_8	$x^3 + y^5$	$\frac{7}{15}$	$2 \cos\frac{\pi}{6}\Gamma(\frac{4}{3}) \times 2 \cos\frac{\pi}{10}\Gamma(\frac{6}{5})$

through the wavelength λ. The other (an integral) depends on geometry and is a dimensionless complex number:

$$k^{l/2} \int e^{ikCG(l)} d^l x = k^{\sigma} \int e^{iCG(l)} d^l x'$$

$$= k^{\sigma} I[CG(l)] \tag{13.34}$$

The index σ and the values of these integrals are summarized for the elementary catastrophe germs in Table 13.1.

Remark. For a rapidly oscillating integral of the form (13.9) with $\psi = 1$, the contribution from the neighborhood of a critical point of multiplicity μ has an asymptotic expansion of the form

$$A \sim \sum_{\alpha,\kappa} C_{\alpha,\kappa} \lambda^{n/2-\alpha} (\ln \lambda)^{\kappa} \tag{13.35}$$

where $0 \leq \kappa \leq n - 1$.[2]

4. Caustics and Catastrophe Germs

The results derived in the previous section and summarized in Table 13.1 show that the amplitude associated with a degenerate critical point of the phase function $\Phi(x; \Omega)$ diverges like $(1/\lambda)^{\sigma}$ as $\lambda \to 0$. Therefore the bifurcation set $\mathscr{S}_B$ of $\Phi(x; \Omega)$ determines the caustics of the associated rapidly oscillating integral. Since the family of functions $\Phi(x; \Omega)$ is parameterized by n control parameters $\Omega \in \mathbb{R}^n$, the worst catastrophe that can typically occur is n-dimensional. In general it is difficult for us to see caustics in more than two dimensions at a time because we usually look at intensity patterns on a screen ($\mathbb{R}^2$). Therefore the caustic patterns that can typically be encountered are those that can be obtained as intersections of the form $\mathbb{R}^2 \cap \mathscr{S}_B$.

If the highest dimensional catastrophe occurring in a neighborhood of Ω^0 occurs at Ω^0, then the caustics which appear on a screen $\mathbb{R}^2$ close to Ω^0 can assume only those canonical forms that can be obtained by taking 2-dimensional slices of the appropriate bifurcation set. For example, if an A_4 catastrophe germ is lurking somewhere in $\mathbb{R}^3$, then the caustics we may encounter by placing a flat screen near it in $\mathbb{R}^3$ may have the shapes shown in Fig. 13.6*a*, which is obtained directly from Fig. 5.6. Similar considerations hold for catastrophe germs D_{+4} (Fig. 13.6*b*) and D_{-4} (13.6*c*).

The canonical arrangement which exists among catastrophe germs leads directly to the same canonical arrangement among caustics, whether 3- or ($n > 3$)-dimensional. Furthermore the known canonical form for planar sections of the bifurcation sets of the elementary catastrophes, together with the scaling relations which exist among the control parameters, leads to the following possibility. From observations of caustics on two parallel planes, the plane containing the caustic germ can readily be determined in the case $n = 3$.

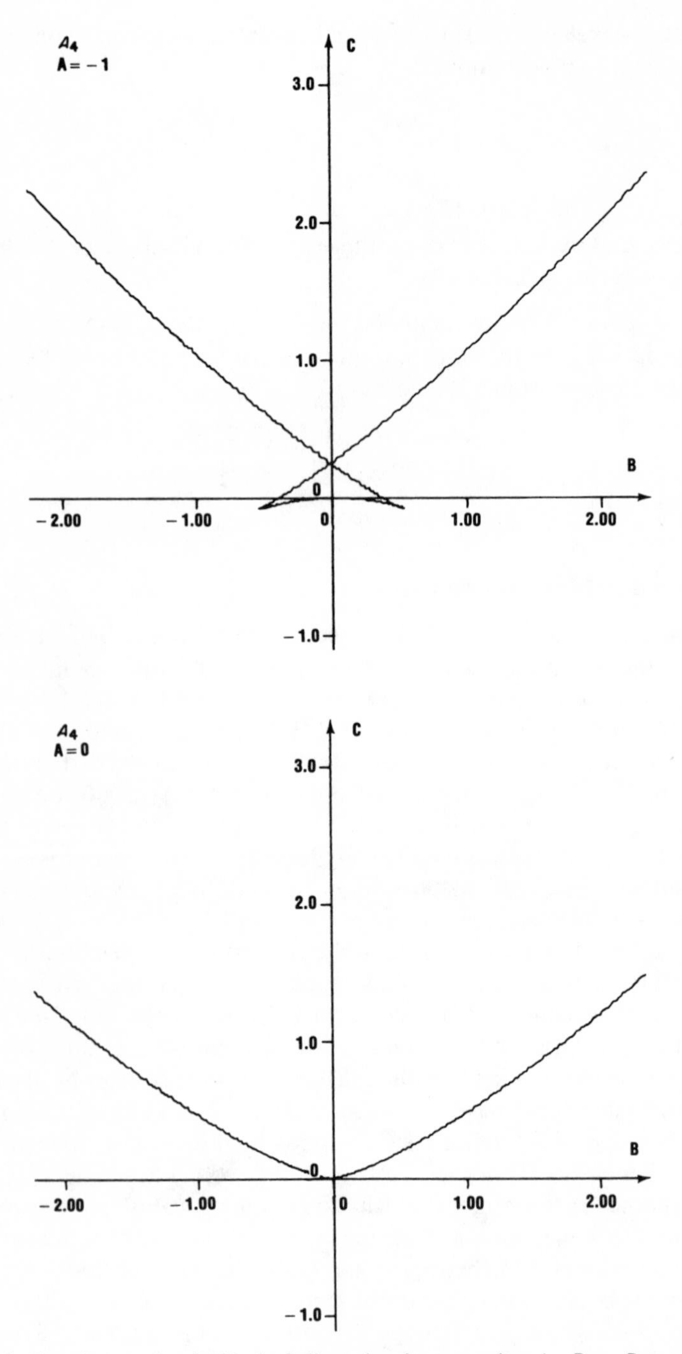

Figure 13.6 Caustics associated with the 3-dimensional catastrophes A_4, D_{+4}, D_{-4} are intersections of the appropriate bifurcation set $\mathscr{S}_B$ with a plane $\mathbb{R}^2$. The caustics shown in the figures above are obtained from Figs. 5.6, 5.11, and 5.17, respectively.

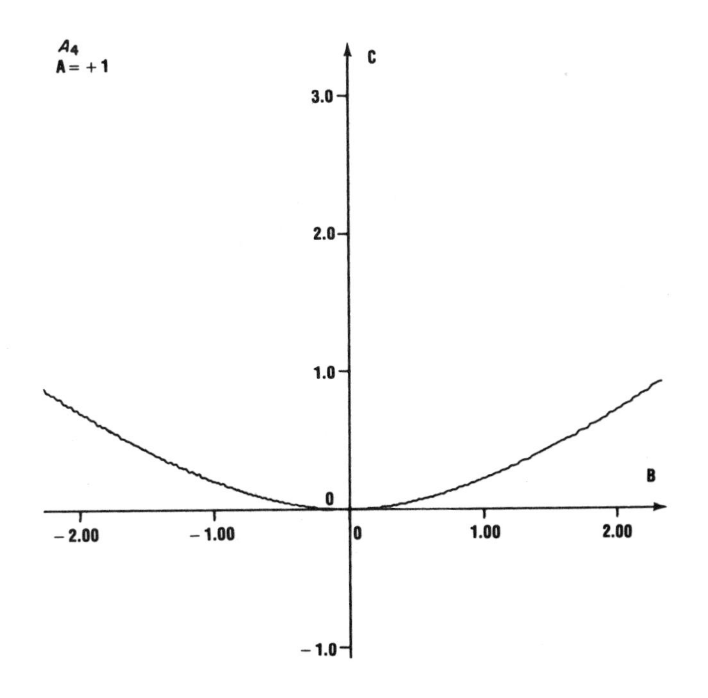

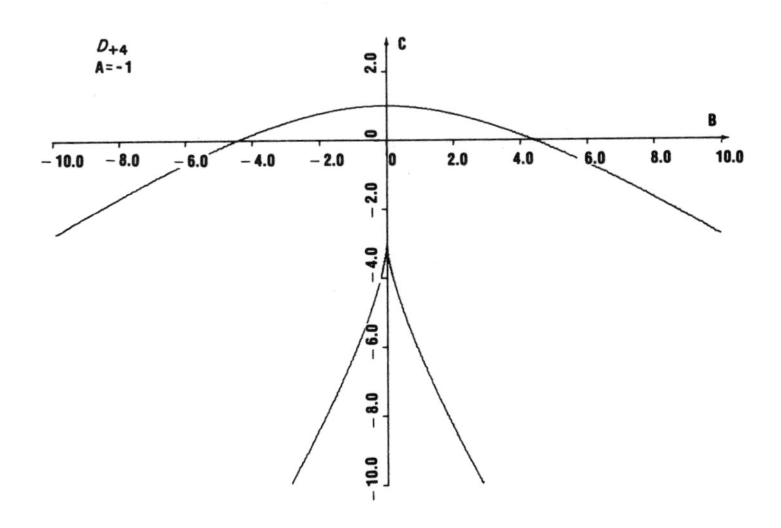

Figure 13.6 (*Continued*)

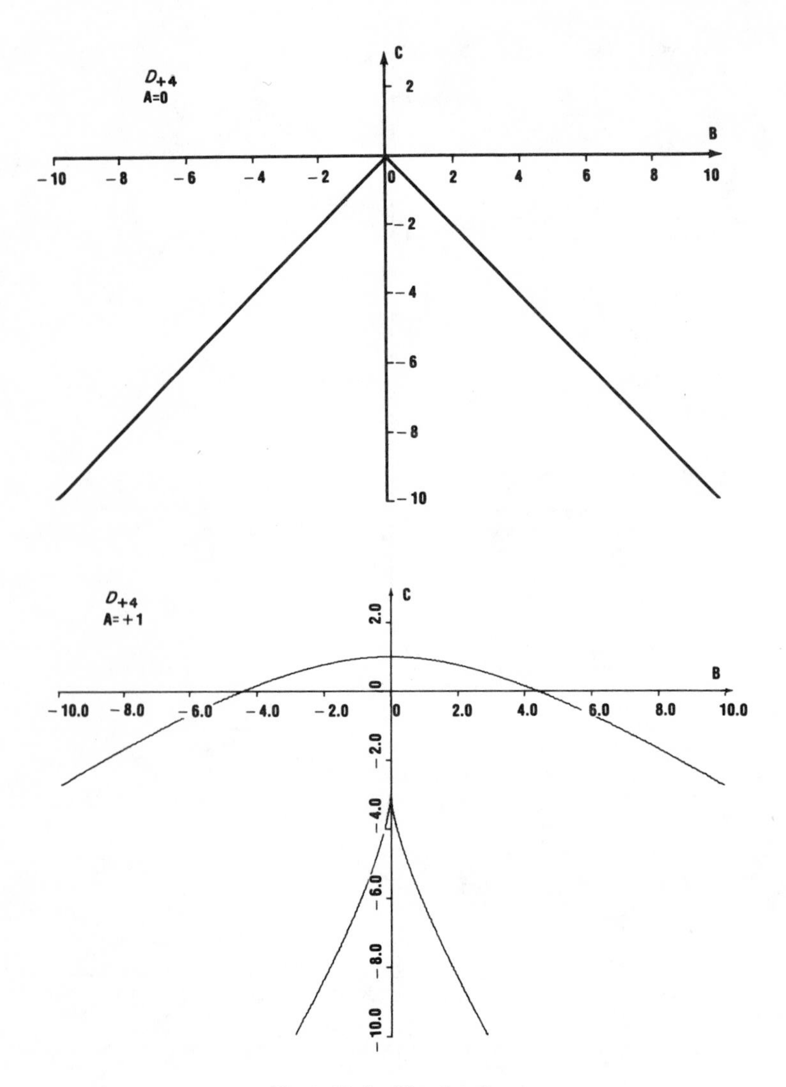

Figure 13.6 (*Continued*)

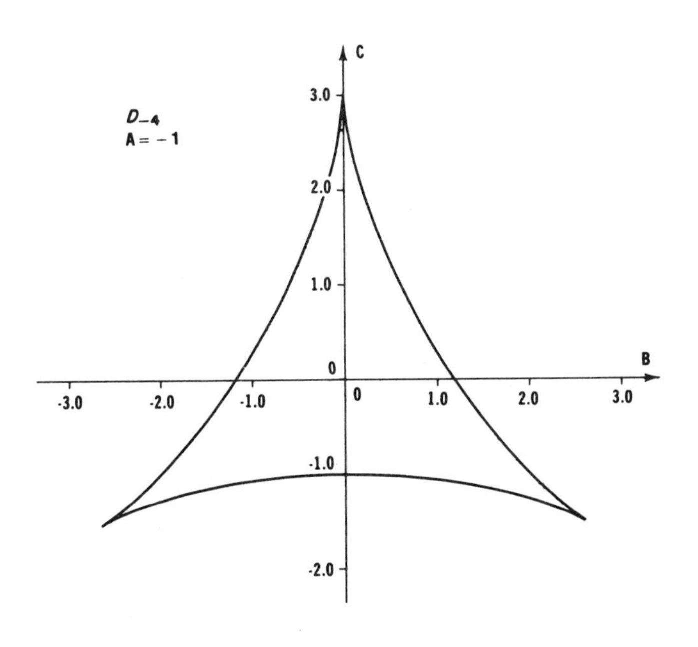

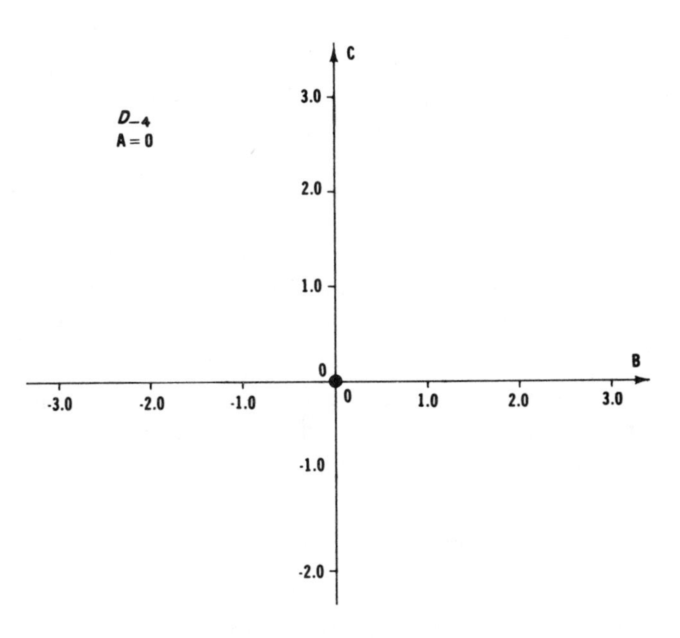

Figure 13.6 (*Continued*)

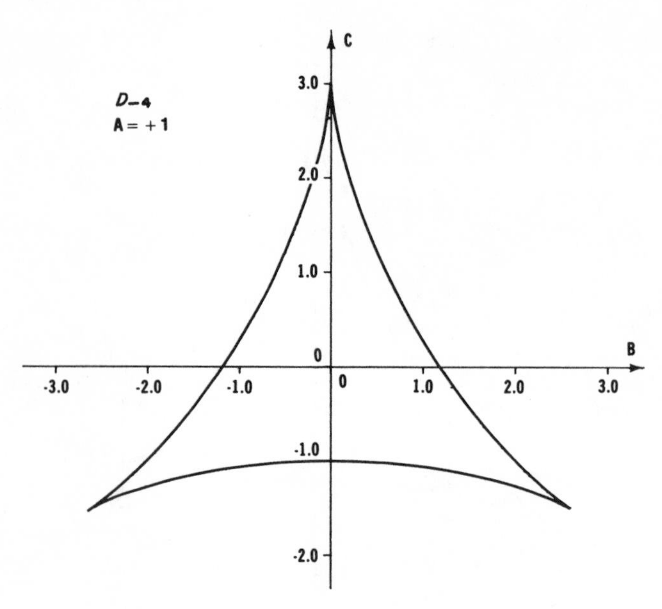

Figure 13.6 (*Continued*)

Remark 1. We can see here a close connection with our work in Chapter 10. The phase diagram of a substance in the P–T plane is the intersection $\mathbb{R}^2 \cap \mathscr{S}_M$ of the physical P–T plane with the Maxwell set of the thermodynamic potential $\mathscr{G}(E^\alpha; i_\beta)$. In this chapter the caustic diagram on a screen is the intersection $\mathbb{R}^2 \cap \mathscr{S}_B$ of the physical screen with the bifurcation set of the optical distance function $\Phi(x; \Omega)$. In these two cases we can map out the Maxwell set and the bifurcation set by changing the location and orientation of the plane $\mathbb{R}^2$ in the control parameter space of the appropriate catastrophe. In both cases there is a duality between state variables and control parameters.

Remark 2. The parallel between caustics and thermodynamics can be extended somewhat. There is a class of trigonometric integrals depending on a parameter Ω which assumes the form

$$A(\Omega) = \int e^{i[S(x) - \langle \Omega, x \rangle]/\lambda}\, d^n x \tag{13.36}$$

Here $\langle \Omega, x \rangle$ is an inner product between the two n vectors x, Ω. Such integrals are special cases of (13.9), with

$$\Phi(x; \Omega) = S(x) - \Omega \cdot x \tag{13.37}$$

The stationary phase method immediately determines the subset of points $(x; \Omega) \subset \mathbb{R}^n \otimes \mathbb{R}^n$ which contribute substantially to the integral (13.36) in the short wavelength limit. This manifold is defined by

$$\nabla_x \Phi(x; \Omega) = 0 \Leftrightarrow \frac{\partial S}{\partial x_j} = \Omega_j \tag{13.38}$$

Such n-dimensional manifolds in $\mathbb{R}^n \otimes \mathbb{R}^n$ are called **Lagrangian manifolds.** Such manifolds also occur in thermodynamics because of the relation

$$\nabla_i \mathscr{U}(i_\alpha; E^\beta) = 0 \Leftrightarrow \frac{\partial G}{\partial i_\alpha} = -E^\alpha \tag{13.39}$$

These two observations are very suggestive of a positive answer to Question 6, p. 247.

5. Diffraction Patterns and Catastrophe Functions

The picture that has now emerged of caustics is the following. The structure of a caustic is that of a 2-dimensional section of the bifurcation set of some catastrophe. The amplitude on each component of a caustic goes like $(1/\lambda)^\sigma$, where the degrees σ have been listed in Table 13.1 for the simple catastrophes. For example, the intensity on the fold part of a caustic behaves like $(1/\lambda)^{2/6}$ while that at the cusp point behaves like $(1/\lambda)^{2/4}$. These results become exact in the short wavelength limit.

These intensities also diverge in this limit. In practice, intensities may be large, but they are bounded above; wavelengths may be small, but they are bounded below. In the limit $\lambda \to$ small, $\lambda \nrightarrow 0$, the stationary phase approximation is not rigorously correct. In the neighborhood of every catastrophe germ there are functions with isolated critical points whose critical values are comparable with λ. It is therefore a bad approximation to estimate the amplitude $A(\Omega)$ by summing the contributions from these isolated critical points as in (13.14). That is, the limits on integrals (13.23) cannot be extended to $\pm\infty$ with impunity.

These considerations suggest that it is important to consider the contributions made by all stationary points in the neighborhood of a catastrophe germ in the limit where the wavelength λ is not vanishing. That is, we must consider the Fresnel integral of the universal unfolding of the appropriate catastrophe germ. We illustrate this for the germ $A_p = x^{p+1}$. The diffraction integral is

$$k^{1/2} \int_{-\infty}^{+\infty} e^{ik[(p+1)\;{}^{1}x^{p+1} + \sum_{j=1}^{p-1} a_j x^j]} = k^\sigma I[A_p; c_1, \ldots, c_{p-1}]$$

$$I[A_p; c_1, \ldots, c_{p-1}] = \int_{-\infty}^{+\infty} e^{i[(p+1)^{-1} y^{p+1} + \sum_{j=1}^{p-1} c_j y^j]}$$

$$c_j = a_j k^{1-j/(p+1)}, \quad \sigma = \frac{p-1}{2(p+1)} \tag{13.40}$$

This is the Fresnel transform of a canonical catastrophe function. The integral $I[\mathrm{CG}(l)]$ is the canonical diffraction integral evaluated at the origin in control parameter space. For the simplest catastrophe A_2 the universal unfolding is $\frac{1}{3}x^3 + tx$. For $t > 0$ there are no stationary points, and for $t < 0$ there are two. Therefore we expect the short wavelength limit $\lambda \to 0$ of the diffraction function for A_2 to be dark for $t > 0$, light for $t < 0$, and of decreasing amplitude as t decreases because the curvature of the two Morse stationary points increases with decreasing t. The regions $t > 0$ and $t < 0$ are separated by a caustic at $t = 0$. In fact,

$$\int_{-\infty}^{+\infty} e^{i[x^3/3 + tx]}\, dx = 2\,\mathrm{Ai}(t) \tag{13.41}$$

where $\mathrm{Ai}(t)$ is the Airy function (Fig. 13.7).[3] This function drops rapidly to zero for positive argument, and is a rapidly oscillating, slowly decreasing function for t negative and decreasing. From (13.40),

$$k^{1/2} \int_{-\infty}^{+\infty} e^{ik[x^3/3 + ax]}\, dx = 2k^{1/6}\,\mathrm{Ai}(ak^{2/3}) \tag{13.42}$$

Other catastrophes may be treated similarly. For example, the canonical diffraction integral $I[A_3; a, b]$ is shown in Fig. 13.8.[4] Note that cross sections orthogonal to the fold lines look like Airy functions once some background has been subtracted. The Airy function comes from the two sheets that fold over; the background from the faraway sheet. Although a dark region is associated with the fold, there is no dark region for the cusp, which has always at least one stationary point.

The classical Airy (1838) and Pearcey (1947) functions are directly related to the unfoldings of the two simplest catastrophes—they are the Fresnel transforms of these catastrophe functions. This observation provides us with a machine for constructing the higher dimensional analogs of these canonical diffraction

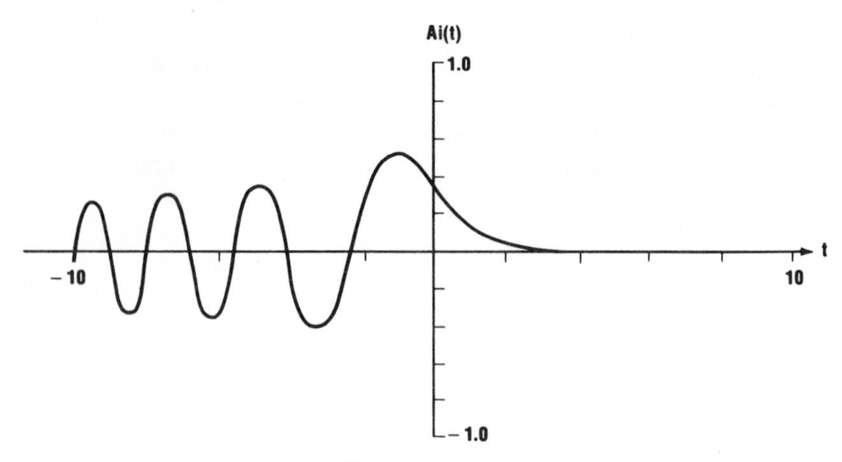

Figure 13.7 Ai(t).

patterns. The only difficulty with computing these higher dimensional canonical diffraction patterns is that they cannot be plotted and they cannot easily be visualized. It is only 2-dimensional sections of these diffraction patterns which can be observed on a screen and which can also be plotted after being computed. For example, if we observe a caustic as shown in Fig. 13.6*a*(i), then we know that this is a particular cross section of the A_4 bifurcation set with a = constant. Therefore the diffraction pattern on the $\Omega_1\Omega_2$ screen will have intensity distribution:

$$I(\Omega_1, \Omega_2) \simeq \left| \int_{-\infty}^{+\infty} e^{i[x^5/5 + ax^3 + \Omega_1 x^2 + \Omega_2 x]} \, dx \right|^2 \tag{13.43}$$

In the neighborhood of the caustics of the type shown in Fig. 13.6*a*(i), the diffraction pattern will be given by Airy or by Pearcey functions. However, if we move the $\Omega_1\Omega_2$ observation plane so that the caustic shown in Fig. 13.6*a*(ii) is observed, the associated $\lambda \to$ small $\nrightarrow 0$ diffraction pattern can no longer be expressed in terms of the classical functions, but is given by the expression (13.43) with $a = 0$.

If we looked at planar sections of the A_4 bifurcation set with b = constant or with c = constant, the diffraction patterns associated with the corresponding caustics could be expressed (locally) in terms of Airy or Pearcey functions. But as the observation plane is moved to make $b \to 0$ or $c \to 0$, the corresponding diffraction patterns would no longer be expressible in terms of these functions.

The general procedure for predicting the diffraction patterns accompanying a caustic associated with a catastrophe of control dimension k is now clear. We assume that the $\Omega_1\Omega_2$ observation plane passes a distance d from the origin in $\mathbb{R}^k$, the control parameter space. Then the k control parameters can be written as linear combinations of Ω_1 and Ω_2 (cf. (10.63)):

$$\begin{bmatrix} c_1 \\ c_2 \\ \vdots \\ c_k \end{bmatrix} = \begin{bmatrix} M_1^1 & M_1^2 & M_1^3 \\ M_2^1 & M_2^2 & M_2^3 \\ \vdots & \vdots & \vdots \\ M_k^1 & M_k^2 & M_k^3 \end{bmatrix} \begin{bmatrix} \Omega_1 \\ \Omega_2 \\ d \end{bmatrix} \tag{13.44}$$

The matrix elements $M_i^j (1 \le i \le k, 1 \le j \le 3)$ are determined from the structure of the caustics on the observation plane. Then if CG(l) is the appropriate catastrophe germ and

$$P[c_1(\Omega), \ldots, c_k(\Omega)] = P(\Omega_1, \Omega_2, d) = P(\Omega) \tag{13.45}$$

is the universal perturbation, where the coordinates c_i are expressed linearly in terms of the observation points Ω_1, Ω_2 in the plane and the distance parameter d, the canonical diffraction pattern will be

$$I(\Omega_1, \Omega_2, d) \simeq \left| \int \cdots \int_{-\infty}^{+\infty} e^{i[\mathrm{CG}(l) + P(\Omega)]} \, d^l x \right|^2 \tag{13.46}$$

This can be computed numerically for any embedding of $\mathbb{R}^2$ in $\mathbb{R}^k$.

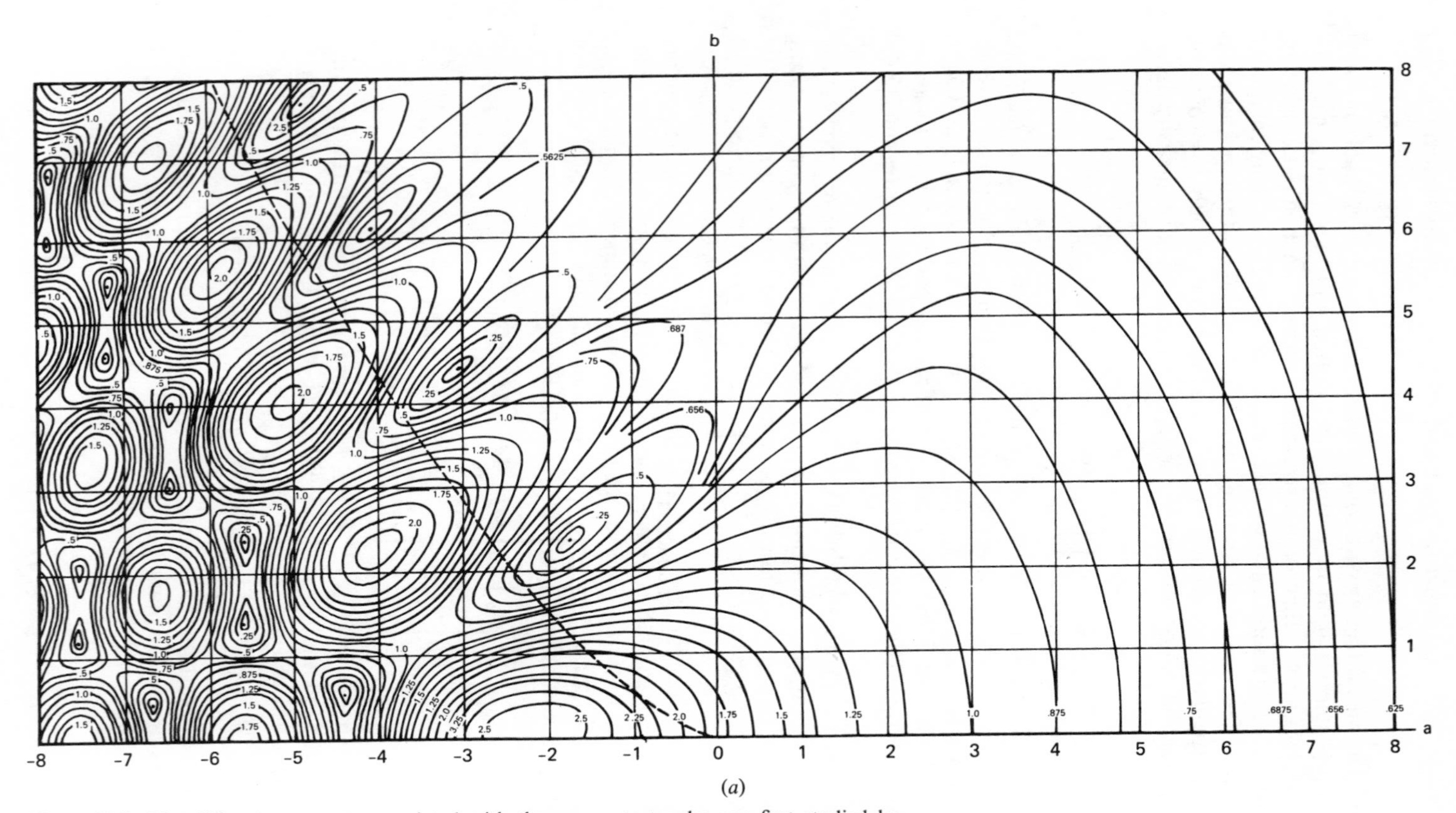

(*a*)

Figure 13.8 The diffraction pattern associated with the cusp catastrophe was first studied by Pearcey.[4] (*a*) The contour plot is from Ref. 4. Reprinted with permission from T. Pearcey, *Phil. Mag.* (7) p. 314. Copyright 1946 by Taylor and Francise Ltd.

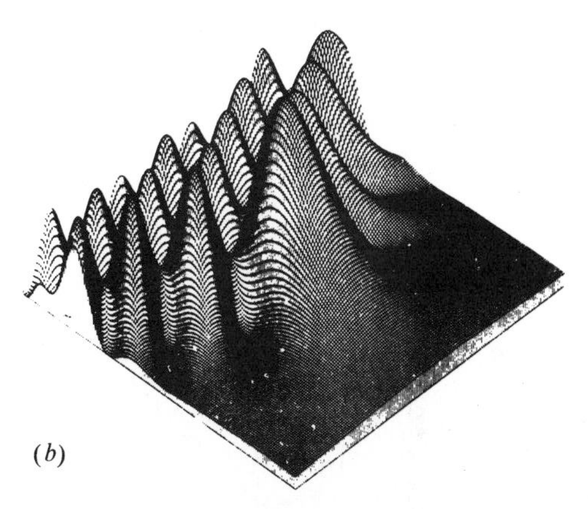

Figure 13.8 (b) The 3-dimensional plot is from Ref. 5. Reprinted with permission from H. Trinkhaus and F. Drepper, *J. Phys.* **A10** (1977), p. L14. Copyright 1977 by The Institute of Physics.

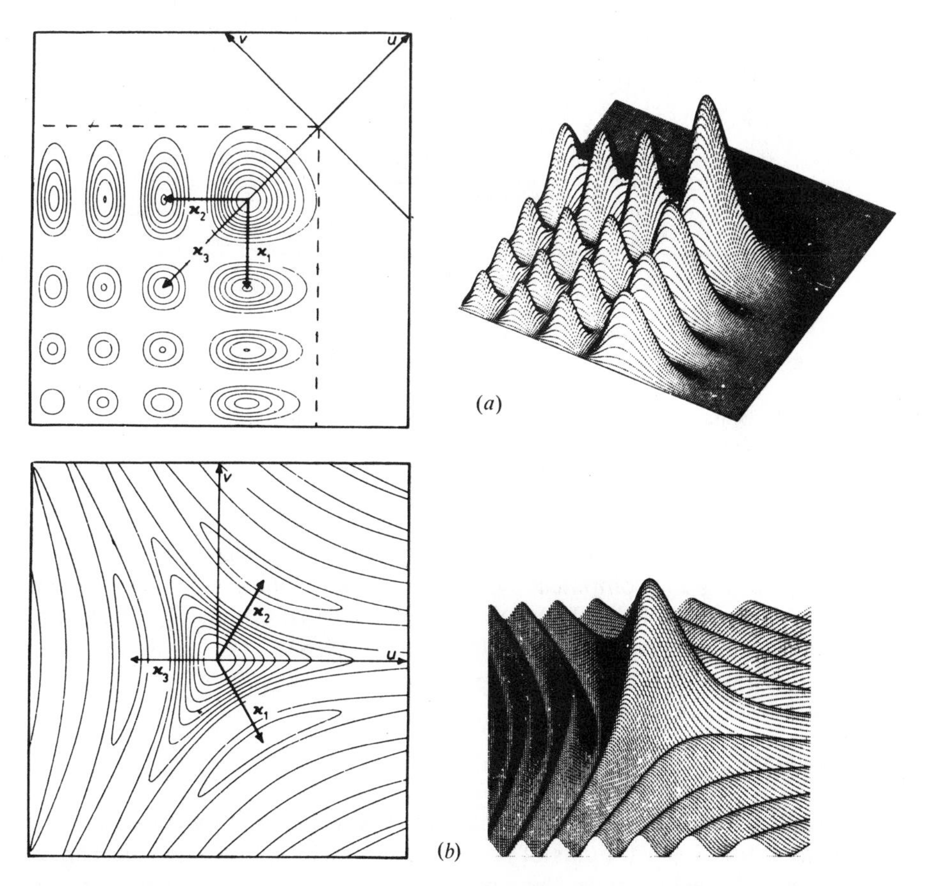

Figure 13.9 Contour plots and 3-dimensional plots of the diffraction pattern in the $a = 0$ plane of two catastrophes were made by Trinkhaus and Drepper.[5] (*a*) D_{+4}; (*b*) D_{-4}. Reprinted with permission from H. Trinkhaus and F. Drepper, *J. Phys.* **A10** (1977), p. L14. Copyright 1977 by The Institute of Physics.

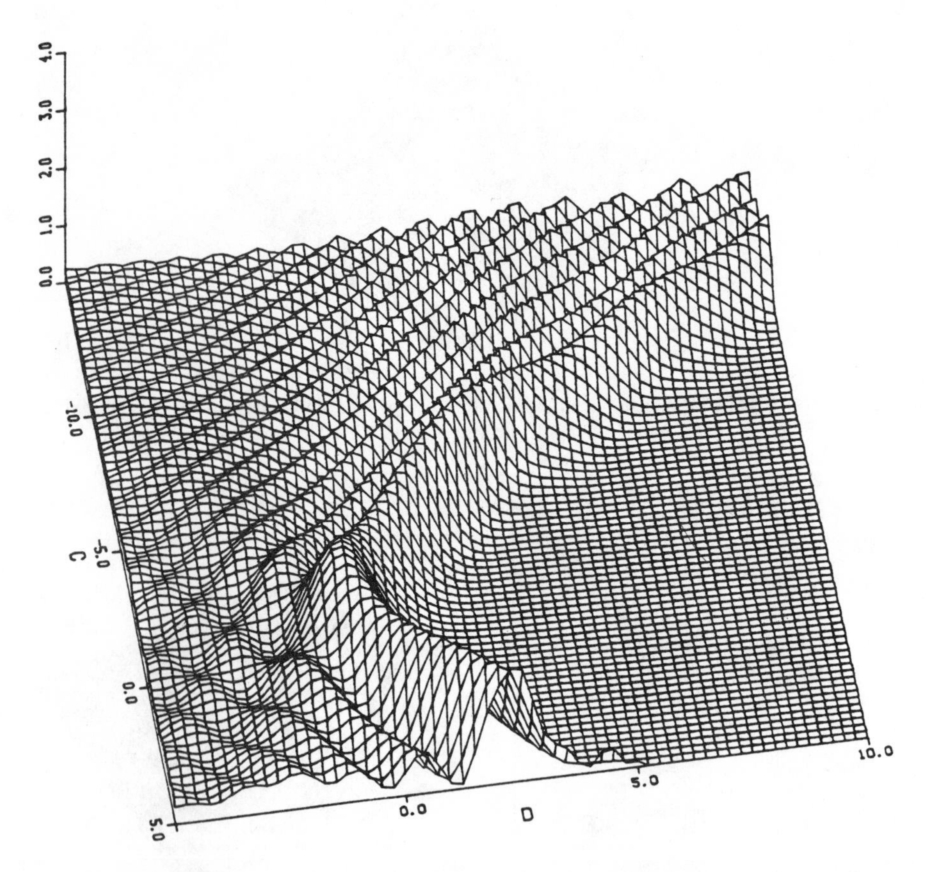

Figure 13.10 Diffraction pattern in the $b = +1$ plane of the catastrophe A_4: $x^5/5 + bx^3/3 + cx^2/2 + dx$. Graphics courtesy of N. A. Pope.

Diffraction patterns associated with the plane $a = 0$ of the catastrophes D_{+4} and D_{-4} have been computed by Trinkhaus and Drepper.[5] These are reproduced in Fig. 13.9. The diffraction patterns associated with the A_4 catastrophe have been computed by Pope. The diffraction pattern of $F(x; b, c, d) = x^5/5 + bx^3/3 + cx^2/2 + dx$ is shown in the plane $b = +1$ in Fig. 13.10.

6. Summary

Classical* optics, classical mechanics, and classical thermodynamics are the $\lambda \to 0$, $h \to 0$, and $k \to 0$ limits of their parents: wave optics, wave mechanics, and statistical mechanics. In this chapter we have seen how to construct the

* Classical = geometrical.

$\lambda \to 0$ limit of wave optics, and we have seen how to go in the reverse direction to reconstruct wave optics from its older offspring, geometrical optics.

In the opening two sections we have seen how a classical variational principle (Fermat's) is related to a wave optics variational principle, the stationary phase method. This method is a direct application of the decompositions (2.1)–(2.3):

1. The neighborhood of a noncritical point contributes $\simeq 0$ to the amplitude integral (13.9).
2. The neighborhood of a Morse critical point contributes $\simeq$finite to the amplitude integral.
3. The neighborhood of a degenerate critical point contributes $\simeq (1/\lambda)^{\sigma}$ to the amplitude integral.

These results are valid in the limit $\lambda \to 0$.

In Sec. 4 we saw that caustics on a screen $\mathbb{R}^2$ are simply 2-dimensional sections of the bifurcation set of an appropriate catastrophe. In Sec. 5 we "reconstructed" wave optics by investigating the possible diffraction patterns that can result in the neighborhood of caustics if the wavelength does not approach zero. Over a century separated the understanding of the Airy (A_2) function from that of the Pearcey (A_3) function. Canonical perturbations of catastrophes now allow us to construct a dictionary of canonical diffraction patterns, each associated with a canonical caustic. In short, the caustic is the skeleton of a diffraction pattern. This skeleton is fleshed out by the corresponding Fresnel integral.

This mathematics extends without difficulty from optics to mechanics, upon the substitution $\Phi(x; \Omega)/\lambda \to \mathscr{S}(x_i, x_f)/\hbar$, where $\mathscr{S}$ is the action of the mechanical system. In both instances complete information about the parent cannot be reconstructed from a path integral study of its classical limit. Classical optics cannot tell us that the electromagnetic field has spin 1; classical mechanics cannot tell us that the electron has spin 1/2. It remains to be seen if statistical mechanics can be reconstructed in a comparable way from classical thermodynamics and, if so, what is left out.

Notes

Thorough treatments of the relation between caustics and catastrophe theory have been made from a mathematical viewpoint by Arnol'd,[6] Duistermaat,[7] and Jänich,[8] and from the physical viewpoint by Berry.[9]

References

1. M. Born and E. Wolf, *Principles of Optics* (1st ed.), London: Pergamon, 1959.
2. V. I. Arnol'd, Critical Points of Smooth Functions and Their Normal Forms, *Russian Math. Surveys* **30**(5), 1–75 (1975).
3. G. B. Airy, On the Intensity of Light in a Neighborhood of a Caustic, *Trans. Camb. Phil. Soc.* **6**, 379–403 (1838); see also: M. Abramowitz and I. A. Stegun, *Handbook of Mathematical Functions*, Washington, D.C.: NBS, 1964.

4. T. Pearcey, The Structure of an Electromagnetic Field in the Neighborhood of a Cusp of a Caustic, *Phil. Mag.* **37**, 311–317 (1946).
5. H. Trinkhaus and F. Drepper, On the Analysis of Diffraction Catastrophes, *J. Phys.* **A10**, L11–L16 (1977).
6. V. I. Arnol'd, Wave Front Evolution and Equivariant Morse Lemma, *Commun. Pure Appl. Math.* **29**, 557–582 (1976).
7. J. J. Duistermaat, Oscillatory Integrals, Lagrange Immersions and Unfolding of Singularities, *Commun. Pure Appl. Math.* **27**, 207–281 (1974).
8. K. Jänich, Caustics and Catastrophes, *Math. Ann.* **209**, 161–180 (1974).
9. M. V. Berry, Waves and Thom's Theorem, *Adv. Phys.* **25**, 1–26 (1976).

CHAPTER 14

Jordan–Arnol'd Canonical Form

Matrices, or more generally linear systems, may be studied in exactly the same way as functions. The equation of state $\nabla V(x; c) = 0$ characterizing the roots of a family of functions $V(x; c)$ is comparable with the characteristic equation $\det M(\lambda; c) = 0$ for the eigenvalues of a family of matrices $M(\lambda; c)$. Functions with isolated critical points are stable under perturbation in exactly the same way that matrices with nondegenerate (isolated) eigenvalues are stable under perturbation.

Functions with degenerate critical points are often locally equivalent to canonical catastrophe germs. Matrices with degenerate eigenvalues are always globally equivalent to the canonical matrix germs discussed originally by Jordan. Just as catastrophe germs have canonical perturbations, so also do Jordan canonical forms have universal perturbations. These are called Jordan–Arnol'd canonical forms, for Arnol'd first studied the universal perturbations of Jordan canonical forms using the methods of Elementary Catastrophe Theory. In short, the program of Catastrophe Theory for linear systems closely parallels the program of Elementary Catastrophe Theory for functions.

In Sec. 1 we review the connection between linear systems and matrices, and introduce the Jordan canonical form appropriate when degeneracies occur. In Sec. 2 we consider arbitrary and minimal perturbations of Jordan canonical forms. Meaning is given to these considerations in the following three sections where we consider specifically the matrix analogs of the elementary catastrophes A_2, A_3, A_k. In Sec. 6 we determine the spectrum of worst possible degeneracies that can stably be encountered in k-parameter families of linear operators. In the following section we study the bifurcation set associated with each Jordan germ.

There is throughout this chapter a close parallel in spirit and even in detail with the material developed in Chapters 2–4.

1. Program of Catastrophe Theory for Linear Systems

Systems of linear equations are encountered throughout the scientific and engineering disciplines. These are systems of equations depending on n state variables $x \in \mathbb{C}^n$ which obey the following two conditions:

$$\begin{aligned} \mathscr{L}(\alpha x^{(1)} + \beta x^{(2)}) &= \alpha\mathscr{L}(x^{(1)}) + \beta\mathscr{L}(x^{(2)}) \qquad &\text{(Linear)} \\ \mathscr{L}(x) &= 0 \qquad &\text{(Equation)} \end{aligned} \tag{14.1}$$

Every linear operator on $\mathbb{C}^n$ can be written as an $n \times n$ matrix. Thus the study of canonical forms for systems of linear equations reduces to a study of canonical forms for $n \times n$ matrices. Under a similarity transformation S (change of basis transformation) every complex matrix can be brought into Jordan canonical form:

$$M \to SMS^{-1} = \begin{bmatrix} J_1(\lambda_1) & & & 0 \\ & J_2(\lambda_2) & & \\ & & \ddots & \\ 0 & & & J_k(\lambda_k) \end{bmatrix} \tag{14.2}$$

where $\lambda_1, \lambda_2, \ldots$ are the eigenvalues of M with degeneracies (multiplicities) $d_1, d_2, \ldots,$ and the Jordan matrices J_i are $d_i \times d_i$ matrices with λ_i on the major diagonal, 1 or 0 on the diagonal above the major diagonal, and 0 elsewhere, for example,

$$J(\lambda) = \begin{bmatrix} \lambda & 1 & & & \\ & \lambda & 1 & & \bigcirc \\ & & \lambda & 0 & \\ \bigcirc & & & \lambda & 1 \\ & & & & \lambda \end{bmatrix} (=\lambda^3\lambda^2) \tag{14.3}$$

The nomenclature within parentheses will be encountered in Sec. 6. We will deal here primarily with complex matrices. The reduction to the real case may be made subsequently.

A family of linear systems $\mathscr{L}(x; c)$ may depend on control parameters $c \in \mathbb{C}^k$. When this is the case, certain members of the family may be degenerate while others are not. Then the Jordan canonical form associated with a typical member in the family may be quite different from the Jordan canonical form associated with a nearby more or less degenerate member in the family. The similarity transformation taking members of the family into Jordan canonical form will then depend discontinuously on the control parameters.

It is clear that the program of Catastrophe Theory can be applied to families of linear systems. It is remarkable that the methods used for the studies of functions, their canonical forms, and their universal perturbations can be used once again to answer a similar spectrum of questions for linear systems.

Before embarking on a study in the spirit of Chapters 2–4, we first consider three examples.

Example 1. We consider the linear operator $\mathscr{L} = I_n\, d/dt - M$ on $\mathbb{C}^n$. Here M is an $n \times n$ complex matrix, $x \in \mathbb{C}^n$, and the linear system equation becomes

$$\frac{d}{dt} x = M(c)x \tag{14.4}$$

where we have assumed M depends on control parameters $c \in \mathbb{C}^k$. If the n eigenvalues $\lambda_1(c), \lambda_2(c), \ldots, \lambda_n(c)$ are distinct when $c = 0$, they will also be distinct when c is small, by continuity arguments. Therefore we conclude that if $M(0)$ has distinct eigenvalues, an arbitrary perturbation

$$M(\delta c) = M(0) + \delta M$$

$$\delta M = \delta c_\alpha \frac{\partial}{\partial c_\alpha} M(c)|_{c=0}$$

of $M(0)$ also has distinct eigenvalues.

Example 2. We now consider the linear differential equation

$$\dddot{x} + p\ddot{x} + q\dot{x} + rx = 0 \tag{14.5}$$

where the dots denote time derivatives. This equation can be put into canonical matrix form by defining

$$\begin{aligned} y_1 &= x \\ y_2 &= \dot{x} \\ y_3 &= \ddot{x} \end{aligned} \tag{14.6}$$

Then

$$\frac{d}{dt}\begin{bmatrix} y_1 \\ y_2 \\ y_3 \end{bmatrix} = \begin{bmatrix} 0 & 1 & 0 \\ 0 & 0 & 1 \\ -r & -q & -p \end{bmatrix}\begin{bmatrix} y_1 \\ y_2 \\ y_3 \end{bmatrix} \tag{14.7}$$

To determine what happens when we perturb the control parameters $(p, q, r) \in \mathbb{C}^3$, we compute the eigenvalues of the matrix representative M of the linear equation (14.5). These are determined by solving the secular equation, which is determined directly from (14.5) by the substitution $d/dt \rightarrow \lambda$:

$$\lambda^3 + p\lambda^2 + q\lambda + r = 0 \tag{14.8}$$

If the secular equation has three distinct roots, perturbation of the parameters ($p \rightarrow p + \delta p$, etc.) has no effect on the canonical structure associated with the linear equation (14.5) and, by extension, its roots. Interesting things happen only when there is a dengeneracy among the roots of (14.5). Without going into great details, we can move the center of gravity of the eigenvalues to zero by the substitution $\lambda = \lambda' - \frac{1}{3}p$, in which case the cubic equation lacks the quadratic term: $\lambda'^3 + A\lambda' + B = 0$. It is then obvious that the interesting points in the new control parameter space $(A, B) \in \mathbb{C}^2$ lie on the cusp bifurcation set.

Example 3. We consider now the system of coupled linear equations:

$$\begin{aligned} \dddot{y} + A_1\ddot{y} + B_1\dot{y} + C_1 y &= -D_1 z \\ \dot{z} + A_2 z &= -C_2\dot{y} \end{aligned} \tag{14.9}$$

What is the smallest family of coupled linear equations in $y, \dot{y}, \ddot{y}, \dddot{y}$ and $z, \dot{z}$ which contains all perturbations of the system above? We can attempt to answer this question by constructing the matrix representative of the linear system (14.9):

$$\frac{d}{dt}\begin{bmatrix} y \\ \dot{y} \\ \ddot{y} \\ z \end{bmatrix} = \begin{bmatrix} 0 & 1 & 0 & 0 \\ 0 & 0 & 1 & 0 \\ -C_1 & -B_1 & -A_1 & -D_1 \\ 0 & -C_2 & 0 & -A_2 \end{bmatrix}\begin{bmatrix} y \\ \dot{y} \\ \ddot{y} \\ z \end{bmatrix} \tag{14.10}$$

This result strongly suggests that the most general perturbation of (14.9) is obtained by the replacement

$$C_2\dot{y} \rightarrow B_2\ddot{y} + C_2\dot{y} + D_2 y \tag{14.11}$$

thus obtaining a linear system depending on two state variables $(y, z) \in \mathbb{C}^2$ and eight control parameters $(A_1, \ldots, D_2) \in \mathbb{C}^8$. In fact, a smaller family of linear systems contains (14.9) as well as every perturbation of it. This is the 6-parameter family

$$\begin{aligned} \dddot{y} + A_1\ddot{y} + B_1\dot{y} + C_1 y &= -D_1 z \\ \dot{z} + A_2 z &= -D_2 y \end{aligned} \tag{14.12}$$

in a suitable coordinate system.

The existence of a canonical form for a linear transformation was proved originally by Jordan. Jordan canonical forms have greatly simplified the analysis of linear systems. The existence of a canonical form for *families* of transformations was proved by Arnol'd.[1] Jordan–Arnol'd canonical forms lead to a canonical list of eigenvalues and their dependence on the control parameters which parameterize the members in a family of linear systems. These canonical forms also allow us to determine which distinct Jordan canonical forms occur in a family of matrices and how they are geometrically related to each other in the control parameter space.

2. Perturbation

The program of Catastrophe Theory can be implemented for linear systems by following the steps indicated in Chapter 2 and carried out in detail in Chapters 3 and 4.

In Chapter 3 we began with an arbitrary function of n state variables and looked for canonical forms for that function at degenerate critical points. These canonical forms were constructed using general nonlinear coordinate transformations. In the study of linear systems our working materials are $n \times n$ matrices instead of functions of n state variables. The analogs of degenerate critical points of functions are degenerate eigenvalues of matrices. The analogs of nonlinear transformations for functions are linear transformations for linear systems. Canonical germs at degenerate critical points have as analog canonical Jordan forms (14.2) and (14.3) for matrices with degenerate eigenvalues. In short, the reduction to canonical form in the presence of degeneracy was carried out for linear systems by Jordan long before the more difficult problem was carried out for functions by Whitney and Thom.

The calculation for matrices comparable to the reduction to canonical form for functions (done in Chapter 3) is carried out in a typical tome[2] dealing with the theory of linear vector spaces. It will not be repeated here.

On the other hand, the analog for matrices of the Chapter 4 calculations for functions (finding the smallest universal perturbation) does not appear in these same tomes. We do this below.

We recall that once a canonical germ had been determined, we wished to determine the most general perturbation of it. This was done by adding an arbitrary perturbation to it, and then "forgetting" about all those terms in the perturbation which could be obtained from the germ by a "similarity" transformation (i.e., a nonlinear coordinate transformation). We can proceed in the same way for linear systems. Given a complex $n \times n$ matrix M_0, we can add an arbitrary perturbation δM to it. The most general perturbation is a complex $n \times n$ matrix, each of whose matrix elements is small. Some small matrices δM can be obtained by applying a similarity transformation to M. These are of no interest; they can be "forgotten." The matrices that remain are analogous to the universal perturbations of the catastrophe germs, given in Table 2.2.

Before making these vague remarks more precise, it will be expedient to consider a simple example.

Example 1. In this case we set

$$M_0(\lambda) = \begin{bmatrix} \lambda & 1 & 0 \\ 0 & \lambda & 1 \\ 0 & 0 & \lambda \end{bmatrix} \tag{14.13}$$

The most general perturbation of $M_0(\lambda)$ has the form

$$\delta M = \begin{bmatrix} \delta m_{11} & \delta m_{12} & \delta m_{13} \\ \delta m_{21} & \delta m_{22} & \delta m_{23} \\ \delta m_{31} & \delta m_{32} & \delta m_{33} \end{bmatrix} \tag{14.14}$$

where each matrix element is "small." We can obtain a large collection of matrices from M_0 by performing a similarity transformation on it: $M_0 \to SM_0S^{-1}$. In general the only transformed matrices which are near M_0 are those obtained by using a similarity transformation near the identity. Then

$$\begin{aligned} SM_0S^{-1} &\to (I + \delta S)M_0(I + \delta S)^{-1} \\ &= (I + \delta S)M_0(I - \delta S + \delta S^2 - \cdots) \\ &= M_0 + [\delta S, M_0] + \cdots \end{aligned} \tag{14.15}$$

If we choose the small matrix δS to be

$$\delta S = \begin{bmatrix} A & B & C \\ a & b & c \\ \alpha & \beta & \gamma \end{bmatrix} \tag{14.16}$$

where each matrix element is "small," then

$$[\delta S, M_0] = \begin{bmatrix} -a & A - b & B - c \\ -\alpha & a - \beta & b - \gamma \\ 0 & \alpha & \beta \end{bmatrix} \tag{14.17}$$

It is clear that some matrices δS_c commute with M_0

$$\delta S_c = \begin{bmatrix} A & B & C \\ 0 & A & B \\ 0 & 0 & A \end{bmatrix} = \begin{bmatrix} \diagdown & \diagdown & \diagdown \end{bmatrix} \tag{14.18a}$$

while other matrices δS_i do not commute with M_0, for example,

$$\delta S_i = \begin{bmatrix} 0 & 0 & 0 \\ a & x' & z' \\ \alpha & \beta & y' \end{bmatrix} \tag{14.18b}$$

It is also clear that some matrices δM_i can be written as the commutator of some δS with M_0:

$$\delta M_i = \begin{bmatrix} \delta m_{11} & \delta m_{12} & \delta m_{13} \\ \delta m_{21} & \delta m_{22} & \delta m_{23} \\ 0 & -\delta m_{21} & -\delta m_{11} - \delta m_{22} \end{bmatrix} \tag{14.18c}$$

while other matrices δM_c cannot, for example,

$$\delta M_c = \begin{bmatrix} 0 & 0 & 0 \\ 0 & 0 & 0 \\ x & y & z \end{bmatrix} \tag{14.18d}$$

We can neglect all perturbations δM of M_0 which have the form (14.18c), since these are "internally" generated by a coordinate transformation (i.e., change of basis). Therefore the most general perturbation of M_0 of smallest dimension is

$$M_0 + \delta M = \begin{bmatrix} \lambda & 1 & 0 \\ 0 & \lambda & 1 \\ x & y & z + \lambda \end{bmatrix} \tag{14.19}$$

We return now to a discussion of the general problem. The infinitesimal similarity transformations $S \rightarrow I + \delta S$ produce perturbations of M_0 according to

$$\begin{aligned} (I + \delta S)M_0(I + \delta S)^{-1} &= M_0 + \delta M \\ \delta M &= [\delta S, M_0] + \mathcal{O}(2) \end{aligned} \tag{14.20}$$

The set of infinitesimal matrices δS forms a linear vector space $\mathcal{S} \simeq \mathbb{C}^{n^2}$. The subset of matrices in $\mathcal{S}$ which commutes with M_0 [cf. (14.18a)] forms a linear vector subspace $\mathcal{S}_c$ (commute) of $\mathcal{S}$, for if $\delta S_1, \delta S_2 \in \mathcal{S}_c$, then

$$[\alpha\delta S_1 + \beta\delta S_2, M_0] = \alpha[\delta S_1, M_0] + \beta[\delta S_2, M_0] = 0 \tag{14.21}$$

The subspace $\mathcal{S}_c$ is unique. The space $\mathcal{S}$ has a direct sum decomposition:

$$\mathcal{S} = \mathcal{S}_c \oplus \mathcal{S}_i, \qquad \mathcal{S}_c \cap \mathcal{S}_i = 0 \tag{14.22}$$

The subspace $\mathcal{S}_i$ [cf. (14.18b)] is not unique. The matrix structure of the space $\mathcal{S}_i$ may be chosen for convenience.

The set of infinitesimal perturbations δM of M_0 forms a linear vector space $\mathcal{P} \simeq \mathbb{C}^{n^2}$. The subset of matrices in $\mathcal{P}$ which can be written in the form $[\delta S, M_0]$ forms a linear vector subspace $\mathcal{P}_i$ [internal, cf. (14.18c)] of $\mathcal{P}$, for if $\delta M_1 = [\delta S_1, M_0]$ and $\delta M_2 = [\delta S_2, M_0]$, then

$$\begin{aligned} \alpha\delta M_1 + \beta\delta M_2 &= \alpha[\delta S_1, M_0] + \beta[\delta S_2, M_0] \\ &= [\alpha\delta S_1 + \beta\delta S_2, M_0] \end{aligned} \tag{14.23}$$

The subspace $\mathscr{P}_i$ is unique. The space $\mathscr{P}$ has a direct sum decomposition

$$\mathscr{P} = \mathscr{P}_i \oplus \mathscr{P}_c, \qquad \mathscr{P}_i \cap \mathscr{P}_c = 0 \tag{14.24}$$

The subspace $\mathscr{P}_c$ [control, cf. (14.18d)] is not unique. The matrix structure of the space $\mathscr{P}_c$ may be chosen for convenience.

The previous two paragraphs make clear that there is some sort of duality between the two spaces $\mathscr{S}$ and $\mathscr{P}$. Specifically, there is a 1–1 correspondence between the (nonunique) space $\mathscr{S}_i$ and the unique space $\mathscr{P}_i$. Suppose that, for some $\delta M \in \mathscr{P}_i$, there are two matrices $\delta S_1, \delta S_2 \in \mathscr{S}_i$ whose commutators with M_0 give δM. Then $\delta S_1 - \delta S_2 \in \mathscr{S}_i$, and

$$[\delta S_1 - \delta S_2, M_0] = \delta M - \delta M = 0 \tag{14.25}$$

implies that $\delta S_1 - \delta S_2 \in \mathscr{S}_c$. Since $\mathscr{S}_c \cap \mathscr{S}_i = 0$, $\delta S_1 - \delta S_2 = 0$. The relation between these spaces is therefore

$$\begin{array}{ccccc} & & \text{unique} & & \\ \mathscr{S} & = & \mathscr{S}_c & \oplus & \mathscr{S}_i \\ & & \updownarrow{\scriptstyle 1\text{-}1} & & \updownarrow{\scriptstyle 1\text{-}1} \\ \mathscr{P} & = & \mathscr{P}_c & \oplus & \mathscr{P}_i \\ & & & & \text{unique} \end{array} \tag{14.26}$$

Since $\mathscr{S}$ and $\mathscr{P}$ are isomorphic with $\mathbb{C}^{n^2}$ and $\mathscr{S}_i$ and $\mathscr{P}_i$ are isomorphic, there must be a 1–1 correspondence between $\mathscr{S}_c$ and $\mathscr{P}_c$. The smallest universal perturbations of M_0 live in $\mathscr{P}_c$. Thus $\mathscr{P}_c$ is constructed by the roundabout process outlined below:

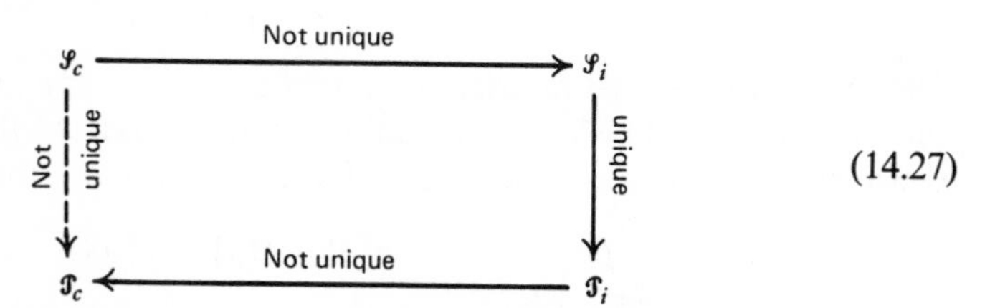

(14.27)

We illustrate this construction by several additional examples.

Example 2. We assume that M_0 has distinct eigenvalues. Without loss of generality we can take M_0 as diagonal, with eigenvalues $\lambda_1 \neq \lambda_2 \neq \cdots \neq \lambda_n$. Then we may write

$$\delta M = [\delta S, M_0] \tag{14.28}$$

$$\delta m_{ij} = \delta S_{ij}(\lambda_j - \lambda_i) \tag{14.29}$$

Since the eigenvalues are distinct, all off-diagonal matrix elements in the perturbation δM may be internally generated. The universal perturbation of a diagonal matrix with distinct eigenvalues is a diagonal matrix. The perturbation serves only to displace slightly the isolated eigenvalues, without changing the signs of their real parts* or the stability properties of M_0. In this sense a matrix with n nondegenerate (isolated) eigenvalues is analogous to a Morse function with a nondegenerate (isolated) critical point.

Example 3. We assume now that M_0 has eigenvalues $\lambda_1, \lambda_2, \ldots$ with degeneracy $d_1, d_2, \ldots$. Then M_0 may be taken in Jordan canonical form:

$$M_0 = \begin{bmatrix} J_1(\lambda_1) & & \\ & J_2(\lambda_2) & \\ & & \ddots \end{bmatrix} \tag{14.30}$$

The infinitesimal similarity transformation δS can also be written in block form:

$$\delta S = \begin{bmatrix} S_{11} & S_{12} & S_{13} & \cdots \\ S_{21} & S_{22} & & \\ \vdots & & & \ddots \end{bmatrix} \tag{14.31}$$

where S_{ij} is a $d_i \times d_j$ matrix. The commutator gives

$$\delta M_{ij} = S_{ij} J_j(\lambda_j) - J_i(\lambda_i) S_{ij} \tag{14.32}$$

This equation can be uniquely solved for the $d_i \times d_j$ matrix elements of S_{ij} in terms of the $d_i \times d_j$ matrix elements of δM_{ij} when $i \neq j$ since then $\lambda_i \neq \lambda_j$. Thus all off block diagonal perturbations δM are internal, and $\mathscr{P}_c$ can be chosen to have the structure

$$\mathscr{P}_c = \begin{bmatrix} (\delta M)_{11} & 0 & \cdots \\ 0 & (\delta M)_{22} & \\ \vdots & & \ddots \end{bmatrix} \tag{14.33}$$

Example 4. We now assume that M_0 has one k-fold degenerate eigenvalue, and that its Jordan canonical form has the structure

$$M_0 \rightarrow J(\lambda) = \begin{bmatrix} \begin{matrix} \lambda & 1 & & \\ & \lambda & & \\ & & \ddots & 1 \\ & & & \lambda \end{matrix} & & \bigcirc \\ & \begin{matrix} \lambda & 1 & \\ & \ddots & 1 \\ & & \lambda \end{matrix} & \\ \bigcirc & & \ddots \end{bmatrix} \tag{14.34}$$

* Assuming that all eigenvalues have nonzero real part.

The subspace $\mathscr{S}_c$ of matrices has the schematic form

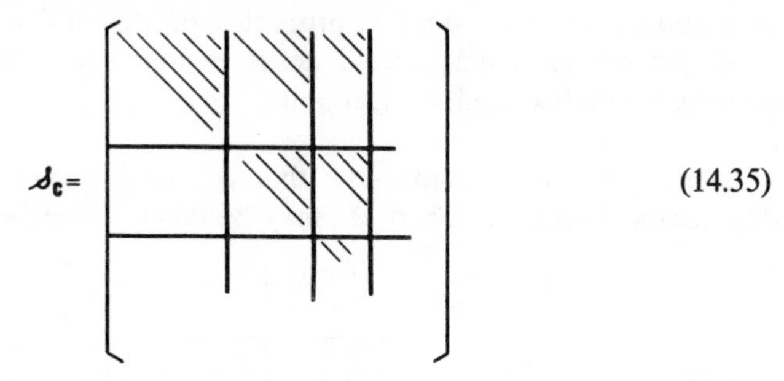

(14.35)

where the matrix elements on each piece of diagonal are equal and those not shown are zero [cf. (14.18a)]. The subspace of matrices $\mathscr{P}_c$ may be chosen to have the schematic form[1]

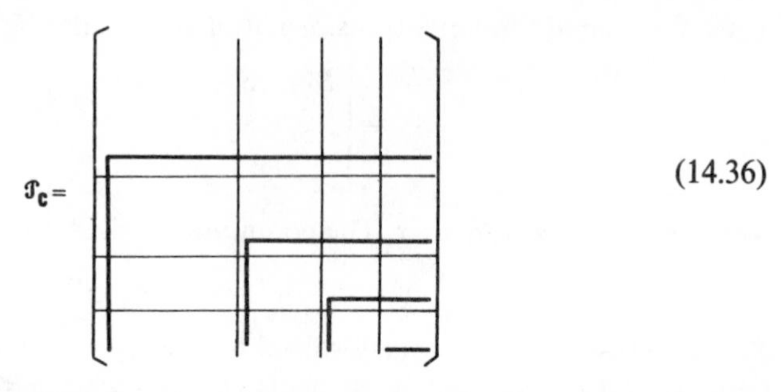

(14.36)

where the matrix elements indicated are independent and those not shown are zero [cf. (14.18d)]. The most general perturbation of the Jordan matrix (14.34) $\lambda^{n_1}\lambda^{n_2}\lambda^{n_3}\cdots(n_1 \geq n_2 \geq n_3 > \cdots > 1)$ [cf. (14.3)] has therefore dimension

$$D = n_1 + 3n_2 + 5n_3 + \cdots \qquad (14.37)$$

Example 5. The most general perturbation of the Jordan matrix

$$M_0 = \begin{bmatrix} \lambda & 1 & 0 & 0 \\ 0 & \lambda & 1 & 0 \\ 0 & 0 & \lambda & 0 \\ 0 & 0 & 0 & \lambda \end{bmatrix}$$

is

$$\delta M = \left[\begin{array}{ccc:c} 0 & 0 & 0 & 0 \\ 0 & 0 & 0 & 0 \\ \hdashline x_4 & x_3 & x_2 & x_1 \\ x_5 & 0 & 0 & x_6 \end{array}\right] \qquad (14.38)$$

Remark 1. See (14.12).

Remark 2. The reader who has clearly seen the connection between the perturbation studies of Chapter 4 and those of the present chapter may also have noticed that we have failed to carry through the analogy consistently. The nonlinear transformations of Chapter 4 have been global while the linear transformations of the present chapter have been local (infinitesimal). We have done this in the interests of simplicity. We have introduced infinitesimal similarity transformations to work with the linear vector space $\mathscr{S}$ (a Lie algebra) rather than a Lie group G of similarity transformations. Dropping this infinitesimal restriction involves making the replacements

$$\mathscr{S} \to G \qquad \text{a Lie group}$$

$$\mathscr{S}_c \to C \qquad \text{a centralizer}$$

$$\mathscr{S}_i \to G/C \qquad \text{a coset}$$

The local and the global studies produce equivalent results, although the infinitesimal version is simpler. This suggests that there should be an "infinitesimal analog" of Chapter 4. There is: it is Chapter 23. Not only does it yield the results of Chapter 4 more simply and economically, but also so transparently that powerful new statements suggest themselves.

3. Application 1: A_2

In this and the following sections we shall assume that the matrix M_0 has already been put into Jordan canonical form. We further assume that all eigenvalues of M_0 are real. It is therefore possible to restrict consideration to only real perturbations δM of M_0. In making these assumptions, we are making a restriction from the complex field to the real field: $\mathbb{C} \to \mathbb{R}$.

We choose M_0 to be the 2×2 matrix

$$M_0(\gamma) = \gamma I_2 + \begin{bmatrix} 0 & 1 \\ 0 & 0 \end{bmatrix} \tag{14.39}$$

The most general family of matrices containing the degenerate 2×2 matrix $M_0(\gamma)$ is

$$M(\gamma; x, y) = M_0(\gamma) + \begin{bmatrix} 0 & 0 \\ y & x \end{bmatrix} = \begin{bmatrix} \gamma & 1 \\ y & \gamma + x \end{bmatrix} \tag{14.40}$$

In general a nonzero perturbation $(x, y) \neq (0, 0)$ will lift the eigenvalue degeneracy, so that the Jordan canonical form of the perturbed matrix is diagonal. The eigenvalues $\lambda_\pm$ of $M(\gamma; x, y)$ are

$$\lambda_\pm - \gamma = \frac{x}{2} \pm \sqrt{\left(\frac{x}{2}\right)^2 + y} \tag{14.41}$$

The parabola $(x/2)^2 + y = 0$ forms a separatrix in the x–y control plane. In the open region $(x/2)^2 + y < 0$ the eigenvalues from a complex conjugate pair, while in the open region $(x/2)^2 + y > 0$ they are both real. On the separatrix both eigenvalues are equal to $x/2 + \gamma$.

A smooth change of variables

$$x' = \frac{x}{2}$$
$$t = \left(\frac{x}{2}\right)^2 + y \tag{14.42}$$

more easily reveals the effect of perturbations on the eigenvalues and the connection with Elementary Catastrophe Theory. The real part of the eigenvalues is given by

$$\mathrm{Re}(\lambda - \gamma - x') = \begin{cases} 0, & t \leq 0 \\ \pm\sqrt{t}, & t \geq 0 \end{cases} \tag{14.43}$$

The dependence on x' is trivial, and the dependence on t is canonical.

Remark. The smooth change of variables (14.42) may be obtained by a similarity transformation:

$$\begin{bmatrix} 0 & 1 \\ y & x \end{bmatrix} \to \begin{bmatrix} \frac{x}{2} & 1 \\ \left(\frac{x}{2}\right)^2 + y & \frac{x}{2} \end{bmatrix} = \begin{bmatrix} x' & 1 \\ t & x' \end{bmatrix} \tag{14.44}$$

4. Application 2: A_3

We now consider the 3×3 Jordan matrix

$$M_0(\gamma) = \begin{bmatrix} \gamma & 1 & 0 \\ 0 & \gamma & 1 \\ 0 & 0 & \gamma \end{bmatrix} \tag{14.45}$$

Since the matrix (14.39) is associated with A_2, we expect (14.45) to be associated with the catastrophe A_3. This canonical form-catastrophe relation is simple enough to work out easily in detail for (14.45), yet complicated enough to be representative of higher catastrophes. We shall consider this relation in some detail here.

The Jordan–Arnol'd canonical form for $M_0(\gamma)$ is

$$M(\gamma; p, q, r) = \begin{bmatrix} \gamma & 1 & 0 \\ 0 & \gamma & 1 \\ -r & -q & \gamma - p \end{bmatrix} \tag{14.46}$$

As noted previously, the secular equation for this matrix is a cubic, so a cusp catastrophe is present. A similarity transformation can be made which transforms (14.46) to a more convenient form:

$$\gamma I_3 + \begin{bmatrix} -\lambda & 1 & 0 \\ 0 & -\lambda & 1 \\ -r & -q & -p-\lambda \end{bmatrix} \to \gamma I_3 + \begin{bmatrix} c-\lambda & 1 & 0 \\ 0 & c-\lambda & 1 \\ -b & -a & c-\lambda \end{bmatrix} \tag{14.47}$$

secular equation ↓   secular equation ↓

$$(\lambda - \gamma)^3 + p(\lambda - \gamma)^2 + q(\lambda - \gamma) + r = 0 \to x^3 + ax + b = 0 \tag{14.48}$$

$$x = \lambda - \gamma - c$$

The relation between the coefficients p, q, r and a, b, c is

$$\begin{aligned} a &= \tfrac{1}{3}(3q - p^2) \\ b &= \frac{1}{(3)^3}(2p^3 - 9pq + 27r) \\ c &= -\tfrac{1}{3}p \end{aligned} \tag{14.49}$$

In short, we have the following relation between canonical forms:

$$M(\gamma; p, q, r) = M(\gamma + c; a, b) \tag{14.50}$$

One of the three parameters ($c = -p/3$) can be absorbed into the eigenvalue. It simply serves to adjust the center of gravity of the eigenvalues. The way in which the three eigenvalues split up under a perturbation is determined by the remaining two control parameters. The secular equation

$$\det M(\gamma + c - \lambda; a, b) = 0 \tag{14.51}$$

is simply the equation of state of the cusp catastrophe manifold. Inside the cusp-shaped region $M(\gamma + c; a, b)$ has three real distinct eigenvalues with average at $\gamma + c$. Outside this region there is one real eigenvalue and a complex conjugate pair. Two eigenvalues become degenerate on the fold lines. All three are degenerate at $(a, b) = (0, 0)$.

5. Application 3: A_k

We now consider the $k \times k$ Jordan matrices $M_0(\gamma)$ and their universal perturbations δM:

$$M_0(\gamma) + \delta M = M(\gamma; p_1, \ldots, p_k) = \begin{bmatrix} \gamma & 1 & & \bigcirc \\ & \gamma & 1 & \\ & & \ddots & \ddots \\ \bigcirc & & & 1 \\ & & & \gamma \end{bmatrix} - \begin{bmatrix} & \bigcirc & \\ p_k & \cdots & p_1 \end{bmatrix} \tag{14.52}$$

The secular equation

$$\det M(\gamma - \lambda; p_1, \ldots, p_k) = \sum_{j=0}^{k} p_j(\lambda - \gamma)^{k-j} = 0, \qquad p_0 = 1 \tag{14.53}$$

is a polynomial equation of degree k in the variable $x = \lambda - \gamma$. It is always possible to choose a new origin for the x axis ($x = x' - (1/k)p_1$) so that the term $(x')^{k-1}$ is absent. Such a change of origin (among functions) is effected by a similarity transformation among matrices. The resulting secular equation is simply the equation of state $\nabla V = 0$ of the catastrophe A_k. As a result, the set of points in the control parameter space $\mathbb{R}^{k-1}$ for the perturbation

$$M(x'; c_1, \ldots, c_{k-1}) = \begin{bmatrix} x' & 1 & & & \bigcirc \\ & x' & 1 & & \\ & \bigcirc & & & \\ & & & x' & 1 \\ -c_1 & -c_2 & \cdots & -c_{k-1} & x' \end{bmatrix} \tag{14.54}$$

for which the eigenvalue degeneracy is not completely lifted corresponds exactly to the bifurcation set $\mathscr{S}_B$ for A_k.

6. Degeneracies in Families of Linear Operators

Families of linear operators depending on control parameters may contain particular matrices with degenerate eigenvalues. What kinds of degeneracies can occur in a k-parameter family of linear operators? This question can be answered using the results developed in the previous sections.

In the case $k = 1$ only one doubly degenerate root can generally occur in the characteristic equation. The most degenerate member in the 1-parameter family has Jordan canonical form:

$$\alpha^2 \beta \gamma \cdots \tag{14.55}$$

In the case $k = 2$ the characteristic equation may have one triply degenerate root or two doubly degenerate roots (the remaining roots are nondegenerate).

The corresponding Jordan canonical forms are

$$\alpha^3\beta\gamma\cdots \qquad \text{and} \qquad \alpha^2\beta^2\gamma\cdots \tag{14.56}$$

These forms correspond to functions with one threefold cusp degeneracy or two twofold fold degeneracies. Below we will stop indicating explicity the non-degenerate roots.

In the case $k = 3$ we may encounter the cases

$$\alpha^4, \qquad \alpha\alpha, \qquad \alpha^3\beta^2, \qquad \alpha^2\beta^2\gamma^2 \tag{14.57}$$

That the first, third, and fourth cases occur is reasonably obvious by the following intuitive but rigorous counting argument. It takes one control to make two eigenvalues equal (analogous to the result that it takes one control for each critical point degeneracy or each critical value degeneracy). The matrix $\alpha\alpha$ is

$$\alpha\alpha = \begin{bmatrix} \alpha & 0 \\ 0 & \alpha \end{bmatrix} \tag{14.58}$$

That such matrices occur when three control parameters are present can be seen as follows. An $n \times n$ diagonal matrix λI_n has an n^2-dimensional universal perturbation. One dimension can be removed by making the perturbed matrix traceless. This corresponds to setting the center of gravity of the multiplet equal to zero, or changing the origin of the x axis to remove the term $(x_1 + \cdots + x_k)x^k$ from the potential A_k(x_i are roots).

More generally, $\alpha^p\alpha^q$ represents a $(p + q) \times (p + q)$ Jordan matrix. The upper $p \times p$ Jordan block has $+1$ along the diagonal above the major diagonal. So also does the lower Jordan block. Off-diagonal blocks are zero. The Jordan matrix $\alpha^3\alpha^2$ has the structure shown in (14.3). The matrix $\alpha^2\alpha$ is first encountered stably in a 4-parameter family, for $2 + 3 \times 1 - 1 = 4$ [cf. (14.37)]. The value of k for which Jordan matrices of the form $\alpha^p\alpha^q\alpha^r\cdots$ can first occur stably is shown in Table 14.1.

Table 14.1 Jordan Blocks of the Form $\alpha^{n_1}\alpha^{n_2}\alpha^{n_3} = \cdots = \{n_i\}$ Can First Stably Be Encountered in k-Parameter Families of Matrices as Indicated (from Ref. 1)

k	1	2	3	...	7	8	...	11	12	...	15
$\{n_i\}$	2	3	4	...	8	9	...	12	13	...	16
			1, 1	...	5, 1	6, 1	...	9, 1	10, 1	...	13, 1
					2, 2	3, 2	...	6, 2	7, 2	...	10, 2
						1, 1, 1	...	4, 1, 1	5, 1, 1	...	8, 1, 1
								3, 3	4, 3	...	7, 3
									2, 2, 1	...	5, 2, 1
											4, 4
											1, 1, 1, 1

Table 14.2 The Most Degenerate Matrices that Can Typically Be Encountered in k-Parameter Families of Matrices (after Ref. 1)

k	Jordan Form
1	α^2
2	$\alpha^3, \alpha^2\beta^2$
3	$\alpha^4, \alpha\alpha, \alpha^3\beta^2, \alpha^2\beta^2\gamma^2$
4	$\alpha^5, \alpha^2\alpha, \alpha^4\beta^2, \alpha\alpha\beta^2, \alpha^3\beta^2\gamma^2, \alpha^3\beta^3, \alpha^2\beta^2\gamma^2\delta^2$.

With the help of Table 14.1 we can construct a list of worst possible degeneracies that can typically be encountered in a k-parameter family of linear operators. This list depends only on k and not on n, the dimension of the state space on which the operator acts, except that the sum of the degeneracies cannot exceed n. A list of degenerate Jordan forms is given in Table 14.2. The Jordan–Arnol'd canonical form (universal perturbation) corresponding to each degenerate matrix (germ) can easily be constructed from (14.34) and (14.36).

7. Bifurcation Set for Jordan–Arnol'd Canonical Forms

If $M(c^0)$ is a degenerate matrix in a k-parameter family of linear operators $M(c)$, then a typical perturbation will completely lift the eigenvalue degeneracy. We may ask how the eigenvalue splitting is related to the perturbation $c^0 \to c^0 + \delta c$. In addition, a subset of perturbations of measure zero will fail to lift the degeneracy completely. We may seek to determine the structure of this subset in the space $\mathbb{R}^k$ of control parameters. Since specific Jordan "germs" have canonical perturbations, these questions can be answered canonically. In fact, these questions are completely analogous to the corresponding questions in Elementary Catastrophe Theory: what is the canonical critical manifold; what is the canonical bifurcation set?

We begin by assuming that a 1-parameter family of $n \times n$ matrices has one doubly degenerate root. The Jordan–Arnol'd canonical form for the 2×2 matrix of interest is

$$\begin{bmatrix} 0 & 1 \\ t & 0 \end{bmatrix} \tag{14.59}$$

The eigenvalues and bifurcation set are as shown in Fig. 14.1. The dashed lines indicate the complex conjugate eigenvalues, while the solid lines represent the real eigenvalues. The bifurcation set occurs at $t = 0$.

For a 2-parameter family the 3×3 Jordan–Arnol'd matrix of interest is

$$\begin{bmatrix} 0 & 1 & 0 \\ 0 & 0 & 1 \\ -b & -a & 0 \end{bmatrix} \tag{14.60}$$

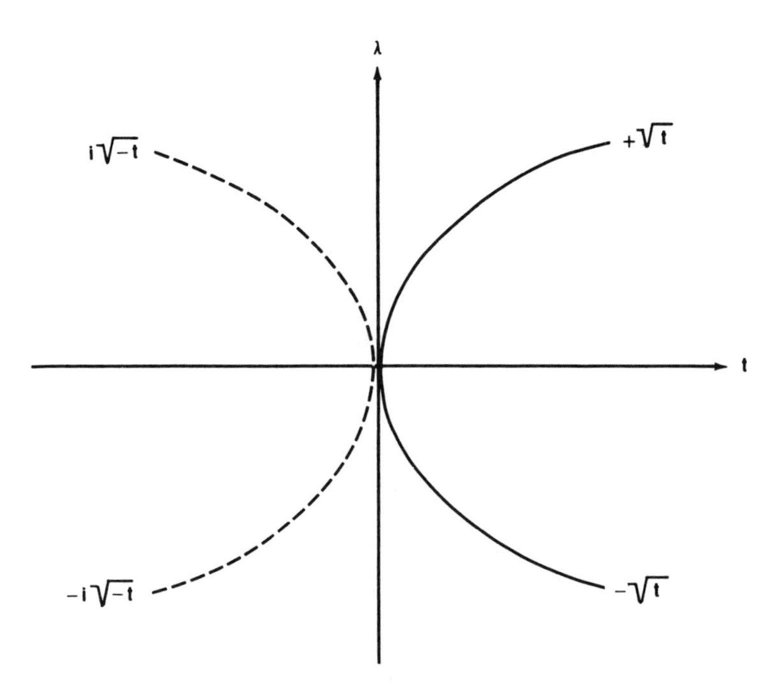

Figure 14.1 The eigenvalues of the perturbed 2×2 Jordan matrix have a canonical square root dependence on the Arnol'd universal perturbation t [cf. (14.59)]. Solid lines—real eigenvalues; dashed lines—imaginary eigenvalues.

The eigenvalues over the (a, b) plane have a close relation with the cusp catastrophe manifold, with additions as noted above. Cross sections in the planes $a = -1, 0, +1$ are shown in Fig. 14.2*b*. The bifurcation set in the a–b plane is the standard cusp. In Fig. 14.2*a* we have suppressed the imaginary parts of the complex eigenvalues.

The bifurcation sets for the Jordan germs with $k = 3$ may be determined similarly. The Jordan canonical form $\alpha^4 \simeq A_4$, so its bifurcation set is as shown in Fig. 14.3*a*. The bifurcation set of the Jordan form $\alpha^3\beta^2$ is the direct product of the bifurcation sets of the corresponding catastrophes A_3 and A_2. It is therefore as shown in Fig. 14.3*c*. The bifurcation set for $\alpha^2\beta^2\gamma^2$ is the product of three lines $\mathbb{R}^1$, and is as shown in Fig. 14.3*d*. The traceless 3-parameter perturbation of $\alpha\alpha$ is

$$\begin{bmatrix} \alpha + z & x + y \\ x - y & \alpha - z \end{bmatrix} \tag{14.61}$$

The bifurcation set is the cone $x^2 + z^2 - y^2 = 0$, shown in Fig. 14.3*b*.

Remark. In general the bifurcation set associated with the perturbation of a Jordan germ of the form $J_1(\alpha)J_2(\beta)\cdots$, where $J_1(\alpha) = \alpha^p\alpha^q\cdots$, is the direct product of the bifurcation sets associated with the Jordan form for each degenerate eigenvalue.

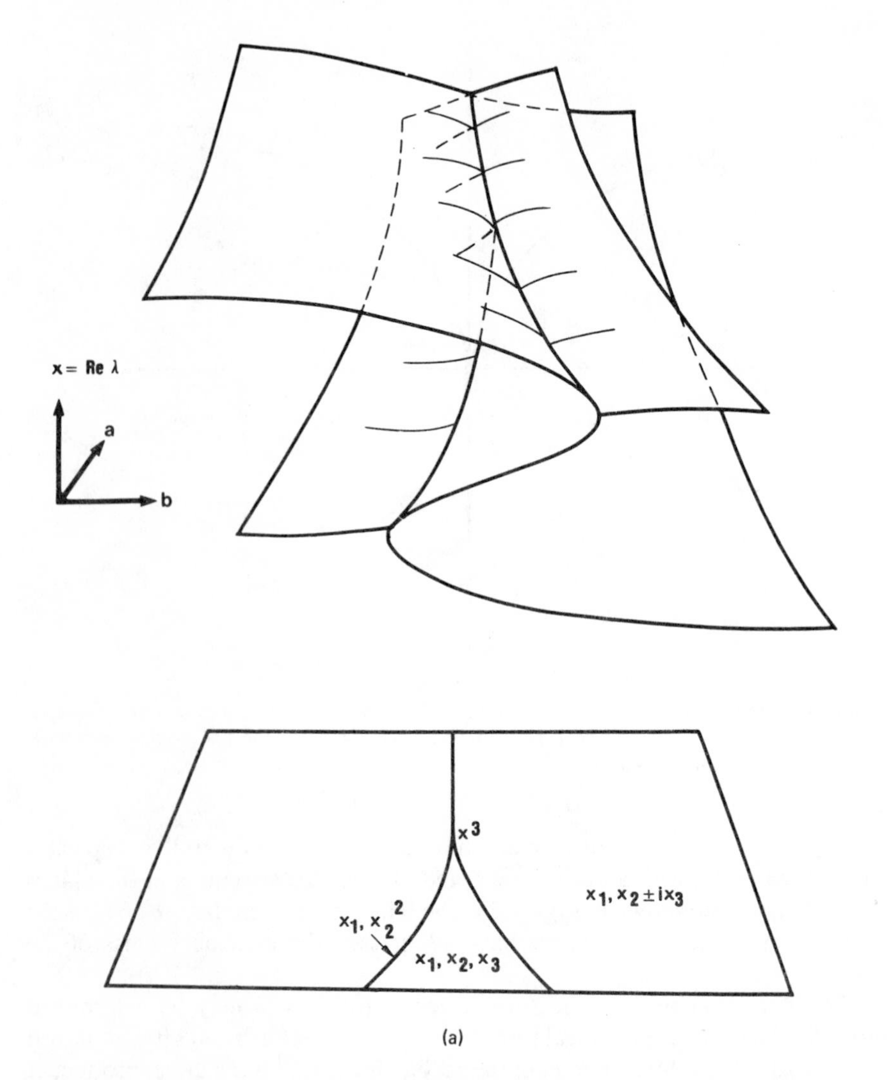

Figure 14.2 The three eigenvalues of the perturbed 3×3 Jordan matrix have a canonical dependence on the universal perturbation parameters a, b. (a) The values of the eigenvalues over the control parameter plane look like the cusp catastrophe manifold with wings attached. Only the real parts of the complex eigenvalues are shown.

The relationship between the degeneracies and perturbations of functions and of matrices suggests that there might be matrix analogs of the bifurcation set $\mathscr{S}_B$ and the Maxwell set $\mathscr{S}_M$. We have just seen that the bifurcation set describes residual degeneracies not completely lifted by perturbations. The long-term stability of a linear system of the form $dx/dt = Mx$ is governed by the eigenvalue with the largest real part. If we define the Maxwell set for the Jordan–Arnol'd canonical form as the set of points in the control parameter space on which two

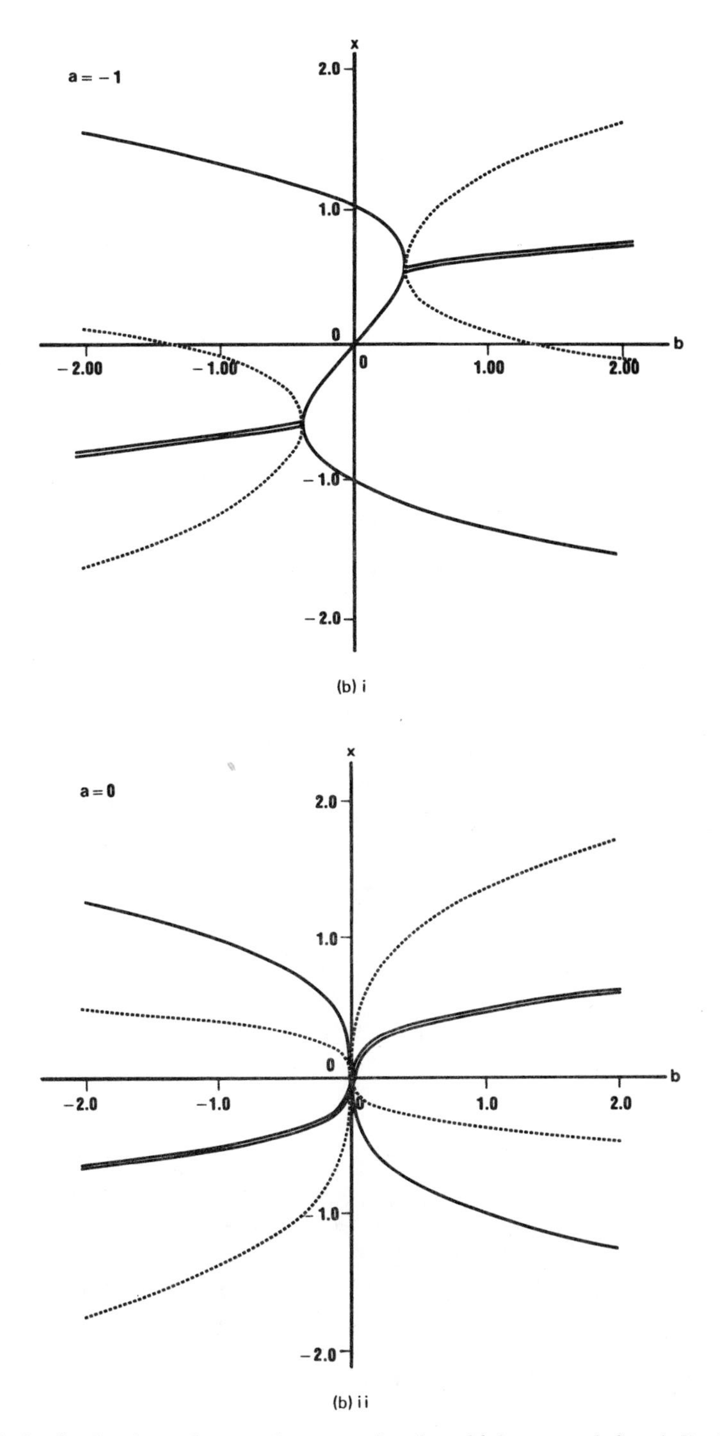

Figure 14.2 (b) The eigenvalues are shown as a function of b for $a = -1, 0, +1$. Double lines indicate the equal real parts of the complex conjugate eigenvalues. The distances of the dashed lines above and below the double lines indicate the values of the imaginary parts of the complex conjugate eigenvalues.

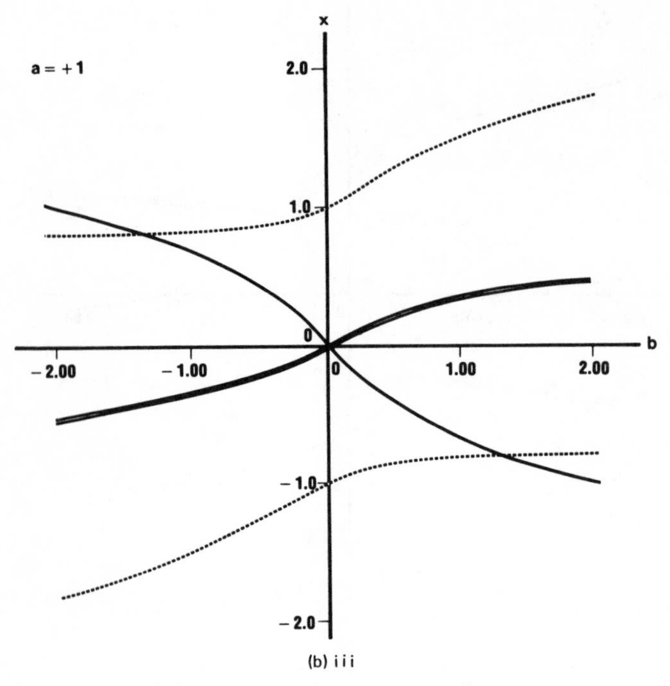

Figure 14.2 (*b*) (*Continued*)

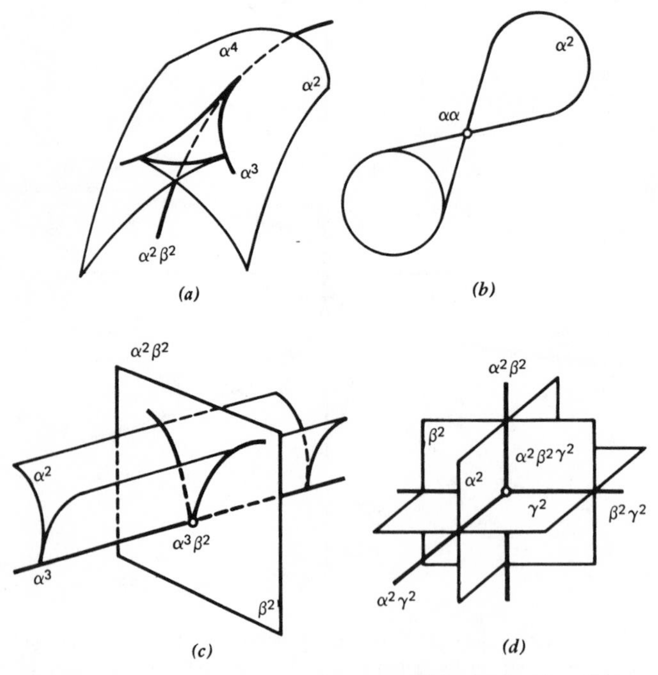

Figure 14.3 Bifurcation set in the control-parameter space $\mathbb{R}^3$ for Jordan-Arnol'd canonical forms depending on three control parameters. (*a*) α^4; (*b*) $\alpha\alpha$; (*c*) $\alpha^3\beta^2$; (*d*) $\alpha^2\beta^2\gamma^2$. (From Ref. 1.)

or more nonconjugate eigenvalues have equal real parts, then a long-time dynamical stability transformation occurs on particular components of the Maxwell set. Thus the Maxwell sets for matrices are directly analogous with the Maxwell sets for functions.

8. Summary

Linear operators may be represented by matrices acting on a linear vector space of suitable dimension. Two matrices are similar, and the corresponding operators equivalent, if both their characteristic and their minimal polynomials are equal.

Perturbation of the original linear operator produces a perturbation in the coefficients of the characteristic polynomial. If the original matrix has nondegenerate eigenvalues, the perturbed matrix also has nondegenerate eigenvalues. When there is eigenvalue degeneracy, the characteristic polynomial can be written as a product of factors $(\lambda - \alpha)^p(\lambda - \beta)^q \cdots$, some of which are degenerate. The perturbation of these separate factors can be carried out in much the same way as the perturbation of the degenerate roots of the gradient of a potential (i.e., equation of state).

The most general minimal pertubation of a matrix M_0 is what is left over after the "internal perturbations" have been removed and the center of gravity of the eigenvalues shifted. The most general perturbation of a matrix reduced to Jordan form is a block-diagonal matrix (14.33). The perturbation for each Jordan block is shown in (14.36). The Jordan–Arnol'd canonical form of a Jordan matrix is the family of matrices of minimal dimension which includes all perturbations of the original Jordan matrix.

These canonical forms provide canonical information about a given Jordan matrix under arbitrary perturbations. The effect of a perturbation on the eigenvalue spectrum is canonically determined by eigenvalue "manifolds" (really varieties) closely analogous to critical manifolds. Eigenvalue degeneracy occurs on various components of the bifurcation set, which is generally the direct product of bifurcation sets associated with the Jordan matrices for distinct eigenvalues. The most general separatrix associated with a single Jordan block is the algebraic surface

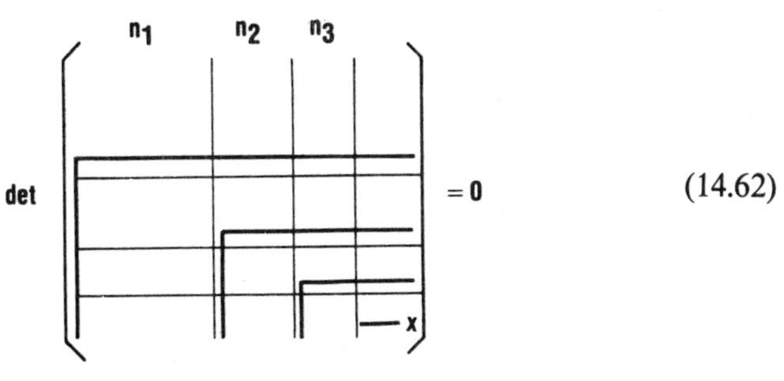

(14.62)

$$D = n_1 + 3n_2 + 5n_3 + \cdots - 1 \tag{14.63}$$

in $\mathbb{C}^D$ or $\mathbb{R}^D$. Here the x is chosen to make the matrix traceless. Each component of this bifurcation set parameterizes a matrix with degenerate eigenvalues. Jordan forms are related to each other in exactly the same way as catastrophe germs and caustics.

Notes

This chapter is based entirely on Ref. 1.

References

1. V. I. Arnol'd, On Matrices Depending on Parameters, *Russian Math. Surveys* **26**, 29–43 (1971).
2. K. Hoffman and R. Kunze, *Linear Algebra* (2nd ed.), Englewood Cliffs, N.J.: Prentice-Hall, (1971).

CHAPTER 15

Quantum Mechanics

The applications of Catastrophe Theory considered so far have been to areas of classical physics and engineering. In these areas a system may be characterized by a point in a phase or a configuration space. The coordinates and momenta which define a phase space are actually operators that commute. The actual system state may be determined by the point in this phase or configuration space which minimizes some sort of potential or Lyapunov function of these commuting operators. This minimum value condition can then lead to easy and fruitful applications of Catastrophe Theory in the classical branches of physics.

One way to exploit the results of Catastrophe Theory for the description of quantum-mechanical systems is to follow the same procedure. However, there is a difficulty. The phase space coordinates and momenta describing a quantum mechanical system are no longer commuting operators. Functions of these operators are generally operator valued (q numbers) rather than scalar valued (c numbers). As a result, this approach is not immediately applicable.

In this chapter we will consider quantum-mechanical systems that behave classically to some extent. These systems are composed of large numbers of identical interacting subsystems. The interactions among these subsystems are assumed to be of mean field type. As such, the operators which appear in the quantum-mechanical Hamiltonian can be replaced by their expectation values. The resulting c-number function then plays the role of a potential whose minimum value characterizes the system. At this point the tools of Catastrophe Theory are immediately applicable.

The systems studied in this chapter are composed of large numbers (N) of identical subsystems (atoms, molecules, nucleons), each of which has r internal degrees of freedom or energy levels. The operators describing these subsystems and the aggregate system are introduced in Sec. 1. The commutativity of the intensive collective operators suggests that all these operators are simultaneously diagonalizable in the large-N limit. This "classical limit" is computed in Sec. 2. A dynamics is assumed in Sec. 3, where we indicate explicitly the structure of the model Hamiltonians whose properties will be studied.

The first property to be studied is the ground state energy phase transition. When the coupling between the subsystems is small or nonexistent, they behave independently of each other. However, when the coupling is strong, they behave cooperatively. In this case the total system ground state is quite different from the ground state of a single isolated subsystem. The transition from single particle to cooperative behavior can be studied easily using an algorithm presented in Sec. 4. This simple algorithm involves replacing the Hamiltonian by its expectation value in the "classical limit." This produces a function of an appropriate number of state variables and control parameters. The state variables are the classical limits of the operators described in Sec. 1. The control parameters are the coupling constants which appear in the Hamiltonian. Standard Catastrophe Theory methods can then be used to determine ground state energy phase transitions. Applications of this algorithm are presented in Secs. 5-8.

Thermodynamic phase transitions can be studied by a method identical in spirit and similar in detail. Another algorithm (Sec. 9) exists for replacing the Hamiltonian operators by a potential function whose minimum is rigorously the free energy per particle. In addition to the classical limit of the Hamiltonian, this potential function contains an additive term related to system entropy. The system state variables remain unchanged while the system temperature now becomes the single control parameter. Applications of this algorithm are presented in Secs. 10 and 11. The relation between ground state phase transitions and thermodynamic phase transitions is discussed in Sec. 12.

In practice, applications of these two algorithms are implemented through bifurcation analyses. Catastrophe Theory is useful for determining which terms in the Hamiltonian are responsible for the occurrence of phase transitions and which are not. Just as functions can be "truncated" beyond terms of appropriate degree in their Taylor series expansion about a degenerate critical point, so also can Hamiltonian operators be truncated. The terms which drive a second-order phase transition are called the canonical kernel of the Hamiltonian. In Sec. 13 we see how to compute the canonical kernel of any of the Hamiltonians described in Sec. 3.

Nonequilibrium behavior as well as equilibrium behavior can benefit from a Catastrophe Theory analysis. In Sec. 14 we determine the dynamical equations of motion associated with one particular Hamiltonian. These equations are conservative, so cannot be expected to exhibit the same behavior as gradient dissipative systems. On the other hand, a potential is involved, and when degenerate critical points are present, phenomena comparable to those discussed in Chapter 9 might be anticipated. One such phenomenon, called critical elongation, is described in Sec. 14.

Dissipative phenomena are considered in Sec. 15. In particular, we consider nonequilibrium steady states, and especially the phase transitions which can occur when a system is driven far from thermodynamic equilibrium. One of the model Hamiltonians described in Sec. 3 provides a venerable laser model. The laser phase transition is of cusp type A_{+3}. Perturbations of the laser phase transition are discussed in Sec. 16. Symmetry-breaking perturbations can lead to new physical processes, one of which is called optical bistability.

1. Operators

Potential functions play a central role in classical mechanics. Operator valued functions play a central role in quantum mechanics. It is not surprising that Catastrophe Theory is a useful tool in classical physics, but it may be surprising that it is also a useful tool in quantum physics. This surprise may be mitigated to some extent by the observation that many quantum-mechanical results can be obtained directly from variational principles (Rayleigh–Ritz, Hartree, Hartree–Fock) applied to classical functions. These functions are often expectation values of operators with respect to trial wave functions. We will make the connection between quantum mechanics and classical mechanics by essentially this route.

The operators that we deal with here can be built up from boson and fermion operators. These obey the commutation and anticommutation relations

$$\begin{aligned} [b_i, b_j^\dagger] &= \delta_{ij}, & \{f_i, f_j^\dagger\} &= \delta_{ij} \\ [b_i, b_j] &= 0, & \{f_i, f_j\} &= 0 \\ [b_i^\dagger, b_j^\dagger] &= 0, & \{f_i^\dagger, f_j^\dagger\} &= 0 \end{aligned} \tag{15.1}$$

where

$$[A, B] = AB - BA \text{ and } \{A, B\} = AB + BA. \tag{15.2}$$

Here $b_i^\dagger$, b_i are the boson creation and annihilation operators for mode i, and $f_j^\dagger$, f_j are the fermion creation and annihilation operators for mode j. All particles discovered so far with integer (half odd integer) spin are bosons (fermions).

The fermion operators for a single mode act in a 2-dimensional Hilbert space with basis vectors $|0\rangle$, $|1\rangle$ as follows:

$$\begin{aligned} f^\dagger|0\rangle &= |1\rangle, \qquad & f^\dagger|1\rangle &= 0 \\ f|0\rangle &= 0, \qquad & f|1\rangle &= |0\rangle \end{aligned} \tag{15.3}$$

The operators for a single boson mode act in an infinite-dimensional Hilbert space with basis vectors $|n\rangle$, $n = 0, 1, 2, \ldots$ (Fock representation) as follows:

$$\begin{aligned} b^\dagger|n\rangle &= \sqrt{n+1}\,|n+1\rangle \\ b|n\rangle &= \sqrt{n}\,|n-1\rangle \end{aligned} \tag{15.4}$$

Bilinear combinations $b_i^\dagger b_j$, $f_i^\dagger f_j$ of boson and fermion operators close under commutation. For example,

$$[a_i^\dagger a_j, a_r^\dagger a_s] = a_i^\dagger a_s \delta_{jr} - a_r^\dagger a_j \delta_{si} \tag{15.5}$$

where $a = b$ or f. This can be verified through the following property of commutators:

$$[A, BC] = [A, B]C + B[A, C] \tag{15.6}$$

This property is called a **derivation** because it is the operator generalization of the ordinary differential operator.

Remark 1. A set of operators which closes under commutation spans a Lie algebra. The sets of operators $a_i^\dagger a_j$ $(1 \leq i, j \leq r)$ span the Lie algebra u(r) for $a = b$ or $a = f$. This Lie algebra structure will be profoundly exploited below in order to replace q numbers (operators) by c numbers (functions). This replacement is at the heart of the present series of applications of Catastrophe Theory to quantum mechanics.

Remark 2. If A, B, C are $r \times r$ matrices with $[A, B] = C$ and if we define operators $\mathscr{A} = a_i^\dagger A_{ij} a_j$, and $\mathscr{B}$, $\mathscr{C}$ similarly, then

$$[A, B] = C \Leftrightarrow [\mathscr{A}, \mathscr{B}] = \mathscr{C} \tag{15.7}$$

In short, there is an isomorphism between matrix Lie algebras and operator Lie algebras based on the bilinear combinations $a_i^\dagger a_j$ of either boson or fermion operators.[1]

Remark 3. The operators $a_i^\dagger a_j$ are number conserving. If n_k particles exist in mode k and $N = \sum_{k=1}^{r} n_k$, then after the application of the operator $a_i^\dagger a_j$, mode i will have one additional particle, mode j will have one fewer, and the remaining modes will have unchanged numbers of particles. In short, the operators $a_i^\dagger a_j$ $(i \neq j)$ shift particles. Such operators are often called **shift operators**. All shift operators commute with the total number operator

$$\mathcal{N} = \sum_{i=1}^{r} n_i$$
$$n_i = a_i^\dagger a_i \tag{15.8}$$

Remark 4. If only one particle is present, then the operators $a_i^\dagger a_j$ can be replaced by projection operators:

$$a_i^\dagger a_j \Leftrightarrow |i\rangle\langle j| \tag{15.9}$$

Here $|i\rangle$ is the state with one particle in the ith mode (or level) and no particles in other levels. An alternative and equivalent interpretation that we will find useful is that a particle has r possible states $|1\rangle, |2\rangle, \ldots, |r\rangle$ and the operator $a_i^\dagger a_j$ shifts the particle from state j to state i.

2. The Classical Limit

A typical quantum system (a single atom, molecule, or nucleon) can exist in a very large number of states. Very often only a few of these states are important. When this is the case, we can number these states $|i\rangle$, $i = 1, 2, \ldots, r$. The most general operator in the finite-dimensional space $\mathbb{C}^r$ describing this "truncated" system has the form

$$\mathcal{O} = \sum_{i,j=1}^{r} |i\rangle A_{ij} \langle j| = \sum_{i,j=1}^{r} A_{ij} e_{ij} \tag{15.10}$$

Every operator in this space can be written as a linear superposition of the **single particle operators** e_{ij} which span the Lie algebra u(r) of the Lie group $U(r)$:

$$[e_{ij}, e_{rs}] = e_{is}\delta_{jr} - e_{rj}\delta_{si} \tag{15.11}$$

In many cases a quantum system contains many (N) identical subsystems, each with r internal states. The single particle operators $e_{ij}^{(\alpha)}$ describing the αth particle then commute with those describing the βth particle:

$$[e_{ij}^{(\alpha)}, e_{rs}^{(\beta)}] = 0 \qquad \alpha \neq \beta$$
$$= (15.11), \qquad \alpha = \beta \tag{15.12}$$

It is useful to define **collective operators** E_{ij} as follows:

$$E_{ij} = \sum_{\alpha=1}^{N} e_{ij}^{(\alpha)} \tag{15.13}$$

The collective operators have the same commutation relations as the single particle operators, for

$$
\begin{aligned}
[E_{ij}, E_{rs}] &= \sum_{\alpha=1}^{N} \sum_{\beta=1}^{N} [e_{ij}^{(\alpha)}, e_{rs}^{(\beta)}] \\
&= \sum_{\alpha=1}^{N} \sum_{\beta=1}^{N} [e_{ij}^{(\alpha)}, e_{rs}^{(\alpha)}]\delta^{\alpha\beta} \\
&= \sum_{\alpha=1}^{N} (e_{is}^{(\alpha)}\delta_{jr} - e_{rj}^{(\alpha)}\delta_{si}) \\
&= E_{is}\delta_{jr} - E_{rj}\delta_{si}
\end{aligned} \tag{15.14}
$$

is isomorphic with (15.11). Therefore if the internal dynamical group of a single particle is $U(r)$, then the internal dynamical group of N identical particles acting cooperatively is also $U(r)$.

When the number N of identical particles is large, it is convenient to define **intensive collective operators** by dividing E_{ij} by N.

$$
\frac{E_{ij}}{N} = \frac{1}{N} \sum_{\alpha=1}^{N} e_{ij}^{(\alpha)} \tag{15.15}
$$

The commutation properties of the intensive collective operators become very simple in the large-N limit:

$$
\left[\frac{E_{ij}}{N}, \frac{E_{rs}}{N}\right] = \frac{1}{N}\left(\frac{E_{is}}{N}\delta_{jr} - \frac{E_{rj}}{N}\delta_{si}\right) \tag{15.16}
$$

From (15.13), the matrix elements of the intensive collective operators E_{ij}/N are comparable with those of the single particle operators: of order 1. The matrix elements of the commutator of two intensive collective operators are of order $1/N$. In the limit $N \to \infty$, intensive collective operators commute. This means that they are simultaneously diagonalizable, so in some sense classical.

The derivation[2] of such a classical limit is beyond the scope of the present work. Instead, we present a heuristic argument for this limit. If a single particle wave function is

$$
|\psi\rangle = \mathrm{col}(z_1, z_2, \ldots, z_r) \tag{15.17}
$$

then the normalization condition $\langle\psi|\psi\rangle = 1$ requires

$$
\langle\psi|\psi\rangle = \sum_{j=1}^{r} z_j^* z_j = 1 \tag{15.18}
$$

The matrix element of a single particle shift operator e_{ij} is easily computed:

$$
\langle\psi|e_{ij}|\psi\rangle = z_i^* z_j \tag{15.19}
$$

Since $E_{ij}/N \sim e_{ij}$, we have for the classical limit of the intensive collective operators

$$\frac{E_{ij}}{N} \xrightarrow[N \to \infty]{\text{classical limit}} z_i^* z_j \tag{15.20}$$

These classical limit arguments can be pushed even further to obtain classical limits for intensive boson operators $a_i^\dagger a_j/N$, $a_i^\dagger/\sqrt{N}$, $a_j/\sqrt{N}$. These intensive operators commute in the limit $N \to \infty$, for

$$\begin{aligned} \left[\frac{a_i^\dagger a_j}{N}, \frac{a_k^\dagger}{\sqrt{N}}\right] &= \frac{1}{N}\left(\frac{a_i^\dagger}{\sqrt{N}}\right)\delta_{jk} \\ \left[\frac{a_i^\dagger}{\sqrt{N}}, \frac{a_j}{\sqrt{N}}\right] &= \frac{1}{N}(-I\delta_{ij}) \end{aligned} \tag{15.21}$$

Therefore it is not unreasonable to expect classical limits to exist. By inspection,

$$\begin{array}{ccc} E_{ij}/N & \sim & a_i^\dagger a_j/N \\ \downarrow & & \downarrow \\ z_i^* z_j & & (a_i^\dagger/\sqrt{N})(a_j/\sqrt{N}) \end{array} \tag{15.22}$$

we determine the classical limit $a_j/\sqrt{N} = z_j$, and so on.

Remark 1. The classical limit of the intensive collective operator E_{ij}/N is the expectation value of this operator with respect to a state $|\Psi\rangle = \prod_{\alpha=1}^{N} |\psi_\alpha\rangle$ which is a direct product of states $|\psi_\alpha\rangle$, each of which has the form (15.17). The classical limit of the intensive boson operator $a/\sqrt{N}$ is its expectation value with respect to the state

$$|\Psi_b\rangle = e^{-N|\mu|^2/2} \sum_{k=0}^{\infty} \frac{(\mu\sqrt{N})^k}{\sqrt{k!}} |k\rangle \tag{15.23}$$

where the factor $e^{-N|\mu|^2/2}$ is present for normalization purposes. In summary,

$$\frac{E_{ij}}{N} \to \langle\Psi|\frac{E_{ij}}{N}|\Psi\rangle = z_i^* z_j \tag{15.24a}$$

$$\frac{a}{\sqrt{N}} \to \langle\Psi_b|\frac{a}{\sqrt{N}}|\Psi_b\rangle = \mu \tag{15.24b}$$

The unitary condition $\langle\Psi|\Psi\rangle = 1$ requires the z_j to obey (15.18). No such condition is imposed on the complex parameter μ.

Remark 2. Since the expectation values of the operators E_{ij}/N, $a_k/\sqrt{N}$, and so on, describe the state of a system, the complex variables z_i, μ_k will be called state variables.

Important Remark. It is this replacement of intensive collective operators E_{ij}/N and intensive boson operators $a/\sqrt{N}$ by c numbers which is the key step in bringing the powerful machinery of Catastrophe Theory to bear on the difficult problems of quantum mechanics.

Caveat. The classical limit (15.24a) for intensive collective operators is true for the fully symmetric representations of the unitary groups, but for these representations this result is rigorously true. The classical limit (15.24b) for boson operators cannot be derived rigorously, but is valid for all cases studied in this chapter.

3. Collective Model Hamiltonians

Meshkov, Glick, and Lipkin (MGL)[3] proposed and studied a model nuclear Hamiltonian which was constructed from pseudospin operators $\mathbf{J}$ which have properties identical to the usual angular momentum operators. Their model has the form

$$\mathscr{H}_{MGL} = \varepsilon J_3 + \frac{V}{2N}(J_+^2 + J_-^2) + \frac{W}{2N}(J_+J_- + J_-J_+) \tag{15.25}$$

In generating this model it was assumed that each of N identical nucleons in a nucleus can be in one of two possible states separated by an energy ε. The parameters V and W represent quadrupole interaction strengths. The factor $1/N$ is present in the interaction terms for thermodynamic reasons.[4,5] All nucleons are assumed to undergo the same interaction. The operators J_3, $J_\pm$ are $(2J + 1) \times (2J + 1)$ angular momentum matrices.

Dicke proposed and studied a model Hamiltonian[6]

$$\mathscr{H}_D = \hbar\omega a^\dagger a + \sum_{\alpha=1}^{N} \varepsilon \tfrac{1}{2}\sigma^z_{(\alpha)} + \frac{\lambda}{\sqrt{N}} \sum_{\alpha=1}^{N} a^\dagger \sigma^-_{(\alpha)} + a\sigma^+_{(\alpha)} \tag{15.26}$$

describing the interaction between a single mode of the radiation field of energy $\hbar\omega$ and a system of N identical 2-level atoms. The energy difference between the atomic ground and excited state is ε, and the strength of the interaction, which is represented by λ, is essentially a dipole matrix element. The factor $N^{-1/2}$ is present for thermodynamic reasons. All atoms are assumed to undergo the same interaction. The operators σ^z, $\sigma^\pm$ are 2×2 Pauli spin matrices.

These two model Hamiltonians have more in common than is immediately apparent. For one thing, each is a prototype for a very extensive class of collective model Hamiltonians. For another, they exhibit very closely analogous critical properties. This analogy is by no means fortuitous. In fact, there is a physical reason for this connection which is described in Sec. 6 [cf. (15.47)]. In this section we simply delineate the two extensive classes of collective model Hamiltonians of which the MGL and Dicke models are archetypes.

A MODELS OF MGL TYPE. We suppose that each of N identical nucleons can be in any one of r possible states with energy $\varepsilon_1 < \varepsilon_2 < \cdots < \varepsilon_r$. In the absence of interactions the Hamiltonian can be written in terms of collective operators:

$$\mathscr{H} = \sum_{i=1}^{r} \sum_{\alpha=1}^{N} \varepsilon_i e_{ii}^{(\alpha)} = \sum_{i=1}^{r} \varepsilon_i E_{ii} = N \sum_{i=1}^{r} \varepsilon_i \left(\frac{E_{ii}}{N} \right) \tag{15.27i}$$

A typical interaction term in the Hamiltonian involves products of single particle scattering operators for two different nucleons. Such contributions can be written in the form

$$V^{(2)} = N^{-1} \sum_{\alpha \neq \beta} V'_{\alpha(ij), \beta(rs)} e_{ij}^{(\alpha)} e_{rs}^{(\beta)} \tag{15.28}$$

The factor N^{-1} is required by the following consideration. In any N-nucleon state $|\Psi\rangle$ the expectation value of any single particle operator $e_{ij}^{(\alpha)}$ has order of magnitude unity. If the scattering matrix elements V' have order of magnitude unity, then the contribution of terms to the right of the summation in (15.28) has order of magnitude N^2. As a result, the total Hamiltonian (15.27i) and (15.28) would not scale properly with N unless the factor N^{-1} were included in (15.28). The flat binding energy per nucleon curve[7] provides experimental confirmation that saturation and scaling occur in nuclei.

At this point it is useful to make what amounts to a mean field approximation. In this approximation we assume that the scattering matrix elements V' in (15.28) are independent of α and β. In other words, each particle "sees" the same interaction as any other particle. Thus the particle-particle interaction can be replaced by an average interaction as follows:

$$\frac{1}{N} \sum_{\alpha \neq \beta} \sum_{(ij)(rs)} V'_{\alpha(ij)\beta(rs)} e_{ij}^{(\alpha)} e_{rs}^{(\beta)} = \frac{1}{N} \sum_{(ij)(rs)} V_{(ij)(rs)} \sum_{\alpha} e_{ij}^{(\alpha)} \sum_{\beta} e_{rs}^{(\beta)}$$

$$= N \sum_{(ij)(rs)} V_{(ij)(rs)} \left(\frac{E_{ij}}{N} \right) \left(\frac{E_{rs}}{N} \right) \tag{15.27ii}$$

Here the sum has been extended to include self-interactions ($\alpha = \beta$) by rescaling the matrix elements: $(N - 1)V' = NV$.

The mean field approximation neglects nuclear range considerations and does violence to more realistic models of nuclear interactions. However, it simplifies computations considerably, replacing an intractable problem by a tractable one. Moreover, it may be a reasonably good approximation for systems containing large numbers of nucleons. Except for the most horrendous codes, most calculations on systems with more than a few nucleons involve some sort of mean field approximation at some stage.

For phenomenological reasons we may want to consider terms in a model Hamiltonian representing the simultaneous scattering of three or more nucleons. Under the mean field approximation and the scaling requirement, the 3-body interaction is represented by a term of the form

$$V^{(3)} = N \sum V_{(ij)(ij)'(ij)''} \left(\frac{E_{ij}}{N}\right)\left(\frac{E_{i'j'}}{N}\right)\left(\frac{E_{i''j''}}{N}\right) \tag{15.27iii}$$

Four nucleon terms and higher terms have a similar form. For physical reasons we expect the 2-nucleon interaction (15.27ii) to be more important than the 3-nucleon interaction (15.27iii), $V^{(3)}$ more important than $V^{(4)}$, and so on. The N-nucleon Hamiltonian is a sum of terms, in decreasing importance, of the form (15.27i), (15.27ii), (15.27iii), ... Such a Hamiltonian can be written

$$\frac{\mathscr{H}}{N} = h_Q\left(\frac{E}{N}\right) \tag{15.29}$$

For technical reasons we assume that h_Q is a polynomial of finite degree in the intensive collective operators E/N.

The MGL model Hamiltonian (15.25) is the prototype model Hamiltonian of the form (15.29). To obtain the MGL Hamiltonian as a special case of (15.29), we observe that, for $r = 2$,

$$\begin{aligned} E_{21} &= J_+ \\ E_{12} &= J_- \\ \tfrac{1}{2}(E_{22} - E_{11}) &= J_3 \end{aligned} \tag{15.30}$$

and set

$$V_{22} = \frac{\varepsilon}{2}, \qquad V_{(12)(12)} = V_{(21)(21)} = \frac{V}{2}$$

$$V_{11} = -\frac{\varepsilon}{2}, \qquad V_{(12)(21)} = V_{(21)(12)} = \frac{W}{2}$$

with all other scattering elements equal to zero.

When $r = 2$, the Hamiltonian (15.29) has the form $\mathscr{H} = Nh_Q(\mathbf{J}/N)$. If h_Q does not have the specific form (15.25) of the sum of a linear energy term (εJ_3) with quadratic interaction terms ($J_+^2 + J_-^2$, $\{J_+, J_-\}$) then $\mathscr{H}$ is said to be of MGL type.

When $r > 2$, the Hamiltonian $\mathscr{H}$ is said to be an extended MGL model if it is the sum of linear energy terms ($\varepsilon_i E_{ii}$) and quadratic interaction terms ($E_{ij}^2 + E_{ji}^2$, $\{E_{ij}, E_{ji}\}$). Otherwise it is said to be of extended MGL type. These definitions are summarized in Table 15.1.

Table 15.1 The MGL and Dicke Models are Prototypes for Larger Classes of Model Hamiltonians

	One Subsystem		Two Interacting Subsystems	
	Linear + Quadratic with Symmetry	Polynomial/ No Symmetry	Linear + Bilinear with Symmetry	Polynomial/ No Symmetry
$r = 2$ $SU(2)$	MGL	MGL type	Dicke	Dicke type
$r > 2$ $SU(r)$	Extended MGL	Extended MGL type	Extended Dicke	Extended Dicke type

B MODELS OF DICKE TYPE. We suppose now that each of N identical atoms or molecules can be in any one of r possible states with energy $\varepsilon_1 < \varepsilon_2 < \cdots < \varepsilon_r$. The subsystem consisting of the N identical atoms can interact with another subsystem consisting of the electromagnetic field. For simplicity we assume that only one field mode interacts with each level pair, and that the interaction is near resonant so that $\hbar\omega_{ji} \simeq \varepsilon_j - \varepsilon_i$. When the interactions are small, the principal contribution to the Hamiltonian can be written in terms of boson number operators and individual or collective atomic operators:

$$\mathscr{H} = \sum_{1 \le i < j}^{r} \hbar\omega_{ji} a_{ji}^{\dagger} a_{ji} + \sum_{i=1}^{r} \sum_{\alpha=1}^{N} \varepsilon_i e_{ii}^{(\alpha)} + V_{\text{Int}} \tag{15.31}$$

To model the interaction, we assume that the transition from state j to i ($j > i$) is associated with the emission (creation) of a photon in the near-resonant mode. If the dipole matrix element for this transition is λ_{ji}, then the interaction term may be represented

$$V_{\text{Int}} = \frac{1}{\sqrt{N}} \sum_{1 \le i < j}^{r} \sum_{\alpha=1}^{N} \lambda_{ji} a_{ji}^{\dagger} e_{ij}^{(\alpha)} + \lambda_{ji}^{*} a_{ji} e_{ji}^{(\alpha)} \tag{15.32}$$

The factor $N^{-1/2}$ is present for thermodynamic reasons.

The total Hamiltonian (15.31) + (15.32), representing an interaction between two subsystems, $\binom{r}{2}$ field modes, and N identical r-level atoms, can be written in the form

$$\frac{\mathscr{H}}{N} = h_Q\left(\frac{a^{\dagger}}{\sqrt{N}}, \frac{a}{\sqrt{N}}; \frac{E}{N}\right) \tag{15.33}$$

For technical reasons we assume that h_Q is a polynomial of finite degree in the intensive boson operators $a^{\dagger}/\sqrt{N}$, $a/\sqrt{N}$ and the intensive collective operators E/N.

The Dicke model Hamiltonian (15.26) is the prototype of model Hamiltonians of the form (15.33). To obtain the Dicke Hamiltonian as a special case of (15.33), we choose $r = 2$ and

$$\begin{aligned} a_{21}^{\dagger} &= a^{\dagger}, & e_{21}^{(\alpha)} &= \sigma_{\alpha}^{+} \\ a_{21} &= a, & e_{12}^{(\alpha)} &= \sigma_{\alpha}^{-} \\ & & e_{22}^{(\alpha)} - e_{11}^{(\alpha)} &= \sigma_{\alpha}^{z} \end{aligned}$$

and set

$$\hbar\omega_{21} = \omega, \qquad \begin{aligned} \varepsilon_2 &= +\frac{\varepsilon}{2}, \\ \varepsilon_1 &= -\frac{\varepsilon}{2}, \end{aligned} \qquad \lambda_{21} = \lambda \tag{15.34}$$

Model Hamiltonians of the form (15.33) are called Dicke models or extended Dicke models, depending on whether $r = 2$ or $r > 2$. Hamiltonians of the form (15.33) which include additional terms (e.g., E_{ij}^2, $E_{ij}E_{jk}a_{ik}$) besides the bilinear interaction terms of the form $a_{ji}^{\dagger}E_{ij}$ are called models of Dicke type ($r = 2$) or models of extended Dicke type ($r > 2$). This terminology is summarized in Table 15.1.

The classes of models of which the MGL and the Dicke models are prototypes differ in the following way. The latter models involve two interacting subsystems (atom and field), the former only one system (nucleons). A mean field approximation can be made for models of extended Dicke type, as was done to obtain models of extended MGL type. In this approximation, as far as the atoms are concerned, the field behaves classically, and as far as the field is concerned, the atoms look like a classical current.[8] In this mean field approximation the Hamiltonian h_Q is replaced by a semiclassical Hamiltonian which describes one subsystem quantum mechanically and the other classically:

$$h_Q\left(\frac{a^{\#}}{\sqrt{N}}, \frac{E}{N}\right) \rightarrow h_A\left(\left\langle\frac{a^{\#}}{\sqrt{N}}\right\rangle, \frac{E}{N}\right) \tag{15.35a}$$

$$h_Q\left(\frac{a^{\#}}{\sqrt{N}}, \frac{E}{N}\right) \rightarrow h_F\left(\frac{a^{\#}}{\sqrt{N}}, \left\langle\frac{E}{N}\right\rangle\right) \tag{15.35f}$$

Here $a^{\#} = a, a^{\dagger}$, and the expectation values are given in (15.24). The semiclassical atomic and field Hamiltonians h_A and h_F are complementary.

4. Algorithm for Ground State Energy Phase Transitions

The qualitative properties of the ground state of a quantum system may depend sensitively on the strengths of the interaction parameters which appear in the

Hamiltonian. These interaction parameters may be regarded as control parameters for the system. Then the ground state may undergo a qualitative transformation (phase transition) as the control parameters are changed.

A simple three-step algorithm exists for studying ground state energy phase transitions for the classes of models described in Sec. 3.[9] This algorithm is based on the classical limit results described in Sec. 2.

The three-step algorithm is

1. $\mathscr{H}/N = h_Q(a^{\#}/\sqrt{N}, E/N)$
2. $h_Q \to h_C = \langle h_Q \rangle = h_Q(\langle a^{\#}/\sqrt{N}\rangle, \langle E/N \rangle)$
3 $E_g/N = \min h_C$

A qualitative change in the ground state of the quantum operator $\mathscr{H} = Nh_Q$ occurs when a qualitative change in the minimum of the classical function h_C occurs. Four examples of the application of this algorithm are carried out in the following four sections.

5. Application 1: The MGL Model

For the MGL model (15.25) the first two steps in this simple algorithm are

$$\frac{\mathscr{H}}{N} = \tfrac{1}{2}\varepsilon \frac{(E_{22} - E_{11})}{N} + \tfrac{1}{2}V\left\{\left(\frac{E_{12}}{N}\right)^2 + \left(\frac{E_{21}}{N}\right)^2\right\} + \tfrac{1}{2}W\left\{\frac{E_{12}}{N}, \frac{E_{21}}{N}\right\} \tag{15.36i}$$

$$\begin{aligned} h_C = &-\tfrac{1}{2}\varepsilon(z_1^* z_1 - z_2^* z_2) + \tfrac{1}{2}V\{(z_1^* z_2)^2 + (z_2^* z_1)^2\} \\ &+ \tfrac{1}{2}W\{(z_1^* z_2)(z_2^* z_1) + (z_2^* z_1)(z_1^* z_2)\} \end{aligned} \tag{15.36ii}$$

The classical function h_C must be minimized with respect to z_1 and z_2, where $|z_1|^2 + |z_2|^2 = 1$. An overall phase of the wave function $|\Psi\rangle = \mathrm{col}(z_1, z_2)$ is arbitrary. We may choose this phase so that z_1 is real and nonnegative. Then

$$z_1 = \sqrt{1 - z_2^* z_2}$$

It is useful to define

$$z_2 = e^{-i\phi} \sin \tfrac{1}{2}\theta \tag{15.37}$$

Then

$$z_1 = \cos \tfrac{1}{2}\theta$$

Elementary trigonometric identities

$$|z_1|^2 - |z_2|^2 = \cos^2 \tfrac{1}{2}\theta - \sin^2 \tfrac{1}{2}\theta = \cos\theta$$

$$z_1^* z_2 = e^{-i\phi} \sin \tfrac{1}{2}\theta \cos \tfrac{1}{2}\theta = \tfrac{1}{2}e^{-i\phi} \sin\theta$$

can be used to transform the function h_C into the form

$$h_C = -\tfrac{1}{2}\varepsilon \cos\theta + \tfrac{1}{2}V(\tfrac{1}{2}\sin\theta)^2(e^{-2i\phi} + e^{+2i\phi}) + W(\tfrac{1}{2}\sin\theta)^2 \tag{15.38}$$

This function depends on the azimuthal angle ϕ unless $V = 0$. It may first be minimized over ϕ:

$$h'_C = -\tfrac{1}{2}\varepsilon \cos\theta + (W - |V|)(\tfrac{1}{2}\sin\theta)^2 \tag{15.38'}$$

A minimum occurs at the north pole ($\theta = 0$) when the quadrupole interaction parameters V, W are zero or small. The critical points of h'_C may be determined by the usual methods:

$$\frac{\partial h'_C}{\partial\theta} = \tfrac{1}{2}\sin\theta\{\varepsilon + (W - |V|)\cos\theta\} = 0 \tag{15.39}$$

This equation has solutions

$$\begin{aligned} \sin\theta &= 0 \\ \cos\theta &= \frac{\varepsilon}{|V| - W} \qquad \text{if } |V| - W > \varepsilon \end{aligned} \tag{15.40}$$

A second-order ground state energy phase transition occurs when $|V| - W = \varepsilon$. The ground state energy per nucleon is

$$\begin{aligned} \lim_{N\to\infty} \frac{E_g}{N} &= -\tfrac{1}{2}\varepsilon, & \varepsilon > |V| - W \\ &= -\tfrac{1}{2}\varepsilon \times \tfrac{1}{2}\left\{\frac{\varepsilon}{|V| - W} + \frac{|V| - W}{\varepsilon}\right\}, & \varepsilon < |V| - W \end{aligned} \tag{15.41}$$

This asymptotic value of E_g/N is compared in Fig. 15.1 with the value of E_g/N for finite values of N. These finite values were computed by numerical diagonalization of the Hamiltonian operator (15.25), which is a $(2J + 1) \times (2J + 1)$ matrix, where $J = N/2$.[10] Such a matrix diagonalization must be carried out separately each time the interaction parameters (V, W) are changed. It is clear from this figure that the numerical estimates for E_g/N converge to the analytically computed asymptotic value (15.41) as N increases.

The Hamiltonian (15.25) describes a competition between two processes. The term εJ_3 is minimum when all nucleons are in the ground state. The second term (V) is minimum when each nucleon is in a linear combination of ground and excited states. The third term (W) is minimum when all nucleons are in the ground state ($W > 0$) or else in a linear combination of ground and excited states ($W < 0$). The sum of these three contributions is minimized when

$$\begin{aligned} |\Psi\rangle &= \text{col}(1, 0), & \varepsilon > |V| - W \\ &= \text{col}\left(\left[\frac{|V| - W + \varepsilon}{2(|V| - W)}\right]^{1/2}, -\frac{V}{|V|}\left[\frac{|V| - W - \varepsilon}{2(|V| - W)}\right]^{1/2}\right), & \varepsilon < |V| - W \end{aligned} \tag{15.42}$$

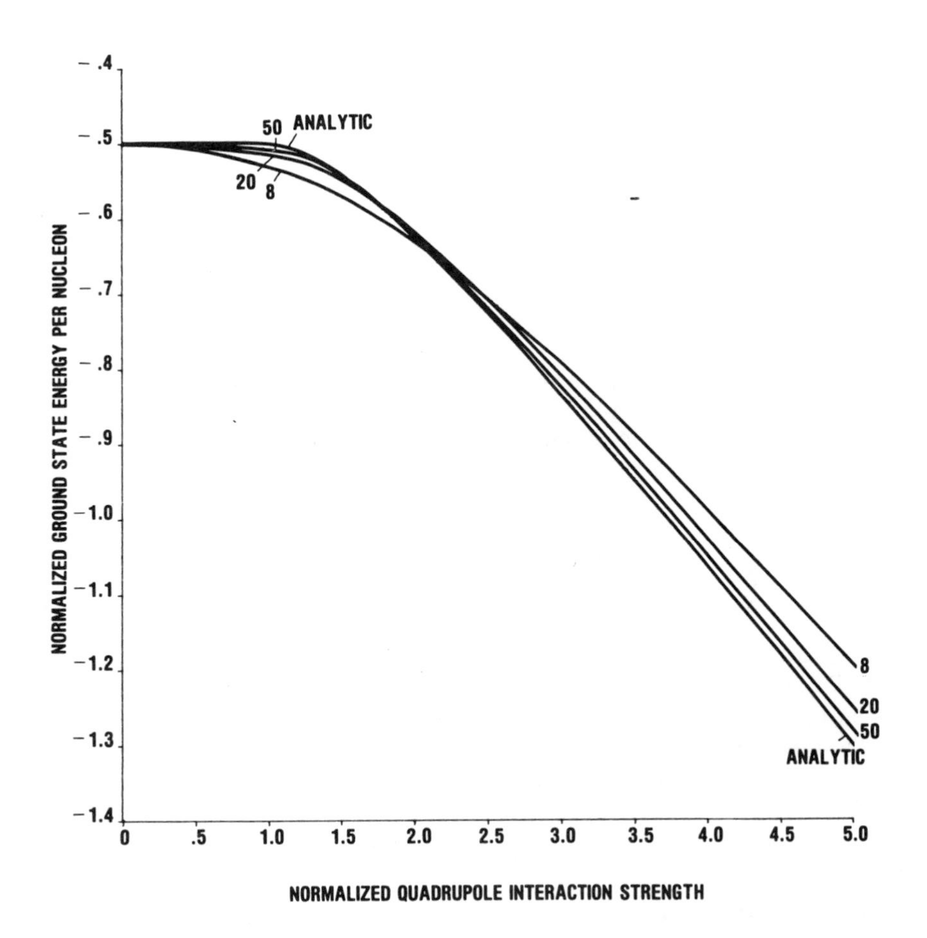

Figure 15.1 The ground state energy E_g/N for the MGL model [(15.36) with $W = 0$], computed by matrix diagonalization for $N = 8, 20, 50$ is compared with the analytic estimate of E_g/N given in (15.41) for $0 \leq |V|/\varepsilon \leq 5.0$.

When the interaction strength is large ($|V| - W > \varepsilon$), it pays energetically to excite each nucleon into a linear combination of ground and excited states. This excitation costs single nucleons energy (εJ_3), but there is a larger return on the interaction energy. It is for this reason that the nucleons prefer to be in an ordered state with $\sin\theta \neq 0$ rather than the ground state $\theta = 0$ in the strong interaction regime.

Remark. When $V = 0$, the minimum of h_C is independent of the azimuthal angle ϕ. We call this invariance a **gauge invariance**. The gauge invariance is broken when $V \neq 0$.

6. Application 2: The Dicke Model

For the Dicke model (15.26) the first step in the algorithm described in Sec. 4 leads to:

$$\frac{\mathscr{H}}{N} = \hbar\omega \frac{a^\dagger a}{N} + \tfrac{1}{2}\varepsilon \frac{(E_{22} - E_{11})}{N}$$
$$+ \lambda\left(\frac{a^\dagger}{\sqrt{N}}\right)\left(\frac{E_{12}}{N}\right) + \lambda^*\left(\frac{a}{\sqrt{N}}\right)\left(\frac{E_{21}}{N}\right) \tag{15.43i}$$

The classical limit h_C of this operator is obtained by the substitutions $a^\dagger/\sqrt{N} = \mu^*$, and so on. The resulting function can be written in terms of the angular variables (θ, ϕ) as described in the previous section. The result is

$$h_C = \hbar\omega\mu^*\mu - \tfrac{1}{2}\varepsilon \cos\theta + \lambda\mu^*(\tfrac{1}{2}e^{-i\phi}\sin\theta) + \lambda^*\mu(\tfrac{1}{2}e^{+i\phi}\sin\theta) \tag{15.43ii}$$

This function is invariant under the simultaneous transformations

$$\begin{aligned} e^{-i\phi} &\to e^{-i(\phi+\psi)} \\ \mu &\to \mu e^{-i\psi} \end{aligned} \tag{15.44}$$

The Dicke model possesses a gauge invariance, just as the MGL model does when $V = 0$.

The function h_C may be minimized most easily by eliminating μ, μ^* by means of the zero gradient condition:

$$\frac{\partial h_C}{\partial \mu} = \hbar\omega\mu^* + \lambda^*(\tfrac{1}{2}e^{i\phi}\sin\theta) = 0 \tag{15.45}$$

The resulting function of (θ, ϕ) is

$$h_C' = -\tfrac{1}{2}\varepsilon\cos\theta - \frac{\lambda^*\lambda}{\hbar\omega}(\tfrac{1}{2}\sin\theta)^2 \tag{15.45$'$}$$

This function is exactly the same as the function (15.38′) under the identification

$$\frac{|\lambda|^2}{\hbar\omega} = |V| - W \tag{15.46}$$

The Dicke model (15.26) always possesses the gauge invariance (15.44), while the MGL model (15.25) does only for $V = 0$. It would therefore appear that there is a very close relationship between the Dicke model and the MGL model with $V = 0$ and $W = -|\lambda|^2/\hbar\omega$.

The MGL and Dicke models undergo the same type of second-order ground state energy phase transitions since their respective potentials h_C' given by (15.38′) and (15.45′) are identical. However, the original Hamiltonians are not isomorphic, so the physical details of the two phase transitions are different. When

$|\lambda|^2 > \varepsilon\hbar\omega$, the energy cost in exciting each atom to a linear superposition of ground and excited states and having a nonzero number of photons in the field mode is more than offset by the energy returned through the polarization interaction. The atomic superposition $|\Psi_a\rangle$ and the field superposition $|\Psi_f\rangle$ which minimize the ground state energy, as well as the ground state energy per particle, are summarized in Table 15.2.

Remark. The close relationship between the critical properties of the MGL and Dicke models may appear surprising. This similarity may be made plausible by treating the operators which appear in the MGL model as if they were c numbers. Then the "equilibrium condition"

$$\frac{\partial h_Q}{\partial(a^\dagger/\sqrt{N})} = \hbar\omega\frac{a}{\sqrt{N}} + \lambda\left(\frac{E_{12}}{N}\right) = \left[\frac{a}{\sqrt{N}}, Nh_Q\right] \simeq 0 \qquad (15.47)$$

provides a relationship between the field and atomic operators. That is, the operator $a/\sqrt{N}$ "is proportional to" E_{12}/N. Eliminating the photon operators from the Hamiltonian (15.43i) by means of the constitutive relation (15.47) leads to the following effective Hamiltonian:

$$h_Q = \tfrac{1}{2}\varepsilon\frac{(E_{22} - E_{11})}{N} - \frac{\lambda^*\lambda}{\hbar\omega}\left(\frac{E_{12}}{N}\right)\left(\frac{E_{21}}{N}\right) \qquad (15.48)$$

It is therefore natural to expect very close similarities between the critical properties of MGL and Dicke Hamiltonians and, more generally, between extended MGL models and extended Dicke models. This argument is certainly not rigorous, but it does serve as a useful bridge connecting these two classes of models.

Table 15.2 Properties of the Dicke Model in the Limit $N \to \infty$

Normalized Coupling Constant	$\lvert\Psi_a\rangle = \text{col}(z_1, z_2)$	$\lvert\Psi_f\rangle$	$\frac{E_g}{N}$
$\frac{\lambda^2}{\varepsilon\hbar\omega} < 1$	$(1, 0)$	$\lvert 0\rangle$	$-\frac{\varepsilon}{2}$
$\frac{\lambda^2}{\varepsilon\hbar\omega} > 1$	$(\cos\tfrac{1}{2}\theta, e^{-i\phi}\sin\tfrac{1}{2}\theta)$	$e^{-N\lvert\mu\rvert^2/2}e^{\mu\sqrt{N}a^\dagger}\lvert 0\rangle$	$-\frac{\varepsilon}{2}\times\frac{1}{2}\left\lvert\cos\theta + \frac{1}{\cos\theta}\right\rvert$

$$\cos\theta = \frac{\varepsilon\hbar\omega}{\lambda^2}$$

$$\mu = \frac{-\lambda e^{-i\phi}\sin\tfrac{1}{2}\theta}{2\hbar\omega}$$

7. Application 3: Extended MGL Models

The extended MGL model Hamiltonians can be written as a sum of three terms:

$$\mathscr{H}_{MGL} = D + V + W \tag{15.49}$$

$$\frac{D}{N} = \sum_{i=1}^{r} \varepsilon_i \frac{E_{ii}}{N} \tag{15.50a}$$

$$\frac{V}{N} = \frac{1}{2} \sum_{1 \le i < j}^{r} V_{ij}\left(\frac{E_{ij}}{N}\right)^2 + V_{ji}\left(\frac{E_{ji}}{N}\right)^2 \tag{15.50b}$$

$$\frac{W}{N} = \frac{1}{2} \sum_{1 \le i < j}^{r} W_{ij}\left(\frac{E_{ij}}{N}\right)\left(\frac{E_{ji}}{N}\right) + W_{ji}\left(\frac{E_{ji}}{N}\right)\left(\frac{E_{ij}}{N}\right) \tag{15.50c}$$

Here E_{ii} are Hermitian and $E_{ij} = E_{ji}^{\dagger}$. The energies ε_i are real and $V_{ij} = V_{ji}^*$, $W_{ij} = W_{ji}^*$. The classical limit of $h_Q = \mathscr{H}/N$ is easily constructed from the rules of Sec. 2:

$$h_C = \sum_{i=1}^{r} \varepsilon_i z_i^* z_i + \frac{1}{2} \sum_{1 \le i < j}^{r} V_{ij}(z_i^* z_j)^2 + \text{cc} + \frac{1}{2} \sum_{1 \le i < j}^{r} W_{ij} z_i^* z_j z_j^* z_i + \text{cc} \tag{15.51}$$

where cc are complex conjugate terms.

The terms corresponding to D and W are invariant under the transformation $z_j \to z_j e^{-i\phi_j}$, while that corresponding to V is not. This corresponds to the gauge invariance of the operators D and W under $E_{ij} \to e^{i(\phi_i - \phi_j)} E_{ij}$. An extended MGL model with $V = 0$ has maximal gauge invariance. The more terms V_{ij} are nonzero, the more the gauge invariance is broken.

The phase of the coupling constants V_{ij} may be changed by changing the phases of the amplitudes z_j of the wave function. These r degrees of freedom can be exploited to make r of the $\binom{r}{2} = r(r-1)/2$ coupling constants V_{ij} positive or zero. For $r = 2$ or 3, all coupling constants V_{ij} may be made real, while for $r > 3$ this is no longer the case. For simplicity, we shall assume that all coupling constants V_{ij}, W_{ij} are real. When all V_{ij} are real, the classical limit of V has stationary values only when all z_j are real. It is therefore sufficient to minimize h_C over the sphere

$$x_1^2 + x_2^2 + \cdots + x_r^2 = 1 \tag{15.52}$$

where $z_j = x_j$ is real.

For the sake of simplicity we study extended MGL models for which only two levels i and j (with $\varepsilon_i < \varepsilon_j$) are connected by nonzero quadrupole interactions. With $z_j = x_j e^{-i\phi_j}$ we have

$$h_C = \sum_{k=1}^{r} \varepsilon_k x_k^2 + \tfrac{1}{2} V_{ij} x_i^2 x_j^2 (e^{2i(\phi_j - \phi_i)} + e^{-2i(\phi_j - \phi_i)}) + W_{ij} x_i^2 x_j^2 \tag{15.53}$$

When this is minimized over the azimuthal angles, we find

$$h_C = \sum_{k=1}^{r} \varepsilon_k x_k^2 + (W_{ij} - |V_{ij}|)x_i^2 x_j^2 \tag{15.54}$$

We take $W_{ij} = 0$ below for convenience without loss of generality.

The nature of the phase transition depends critically on whether the quadrupole interaction couples two excited states or the ground state with an excited state.[2,11]

1. $V_{j1} \neq 0$. In this case all order parameters except $x_1 = \cos \frac{1}{2}\theta$ and $x_j = \sin \frac{1}{2}\theta$ are zero at the minima of h_C, which becomes

$$h_C = \tfrac{1}{2}(\varepsilon_1 + \varepsilon_j) + \tfrac{1}{2}(\varepsilon_1 - \varepsilon_j)\cos\theta - |V_{j1}|(\tfrac{1}{2}\sin\theta)^2 \tag{15.55}$$

This function is identical in form with the function (15.38′). Therefore the system undergoes a second-order ground state energy phase transition as $|V_{j1}|$ becomes larger than $\varepsilon_j - \varepsilon_1$.

2. $V_{ji} \neq 0, j > i > 1$. In this case all order parameters except x_1, x_i, and x_j are zero at the minima of h_C. This becomes, upon the substitutions $x_i^2 + x_j^2 = \sin^2 \frac{1}{2}\theta \leq 1$, $x_j^2/(x_i^2 + x_j)^2 = y$,

$$h_C - \varepsilon_1 = \Delta_i(1 - y)\sin^2 \tfrac{1}{2}\theta + \Delta_j y \sin^2 \tfrac{1}{2}\theta - |V_{ji}|y(1 - y)\sin^4 \tfrac{1}{2}\theta \tag{15.56}$$

where $\Delta_j = \varepsilon_j - \varepsilon_1$. The equilibria are determined in the usual way:

$$\frac{\partial}{\partial y}(h_C - \varepsilon_1) = \{\ \}_1 \sin^2 \tfrac{1}{2}\theta = 0 \tag{15.57a}$$

$$\frac{\partial}{\partial \theta}(h_C - \varepsilon_1) = \{\ \}_2(\tfrac{1}{2}\sin\theta) = 0 \tag{15.57b}$$

These equations always have the solution $\theta = 0$. This state corresponds to having all nucleons in their ground state ($z_1 = 1$, $z_k = 0$, $k > 1$). This state is always a local minimum, and for this minimum $h_C = \varepsilon_1$.

Another solution can exist at $\theta = \pi$ provided that (15.57a) can be satisfied. This requires

$$\{\ \}_1 = \Delta_j - \Delta_i - |V_{ji}|(1 - 2y)\sin^2 \tfrac{1}{2}\theta = 0 \tag{15.58a}$$

This defines $y = \frac{1}{2} - ((\varepsilon_j - \varepsilon_i)/2|V_{ji}|)$. Since $0 \leq y \leq 1$, this critical point can exist only if $|V_{ji}| > \Delta_j - \Delta_i = \varepsilon_j - \varepsilon_i$. This critical point is also always a local minimum, and at this minimum

$$h_C - \varepsilon_1 = \tfrac{1}{2}(\Delta_j + \Delta_i) - \tfrac{1}{4}|V_{ji}| - \frac{1}{4|V_{ji}|}(\Delta_j - \Delta_i)^2 \tag{15.59a}$$

This local minimum has critical value ε_1 when $\sqrt{|V_{ji}|} = \sqrt{\Delta_j} + \sqrt{\Delta_i}$. A first-order ground state energy phase transition occurs as $|V_{ji}|$ increases through

$(\sqrt{\Delta_j} + \sqrt{\Delta_i})^2$. At this phase transition the order parameter x_1 jumps from $+1$ to 0 and the other order parameters change as follows:

$$x_i^2: \quad 0 \to \frac{\sqrt{\Delta_i}}{\sqrt{\Delta_j} + \sqrt{\Delta_i}}$$
$$x_j^2: \quad 0 \to \frac{\sqrt{\Delta_j}}{\sqrt{\Delta_j} + \sqrt{\Delta_i}} \tag{15.60}$$

A third solution can exist at $\theta \neq 0, \theta \neq \pi$, provided both (15.57a) and (15.57b) can be satisfied. In addition to (15.58a), this requires

$$\{\ \}_2 = \Delta_j y + \Delta_i(1 - y) - 2|V_{ji}|y(1 - y)\sin^2 \tfrac{1}{2}\theta = 0 \tag{15.58b}$$

Equations (15.58a) and (15.58b) are satisfied simultaneously by

$$y = \frac{\Delta_i}{\Delta_j + \Delta_i}$$
$$\sin^2 \tfrac{1}{2}\theta = \frac{\Delta_j - \Delta_i}{|V_{ji}|} \leq 1 \tag{15.59b}$$

This solution is the saddle separating the local minima at $\theta = 0$ and $\theta = \pi$. At this critical point $h_C - \varepsilon_1 = \Delta_j\Delta_i/|V_{ji}|$.

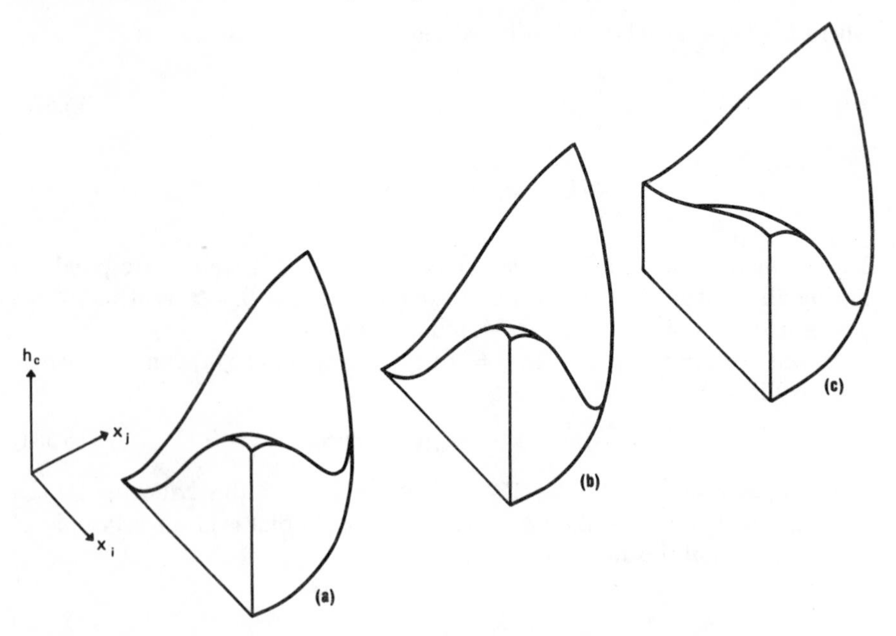

Figure 15.2 The shape of h_c (15.56) at $T = 0$ is shown for the multi-level extension of the MGL model Hamiltonian. Here x_i, x_j are the relevant order parameters, $r = (\varepsilon_i - \varepsilon_1)/(\varepsilon_j - \varepsilon_1) = 0.8$, and the quadrupole interaction V_{ij} couples two excited levels. The quadrant $0 \leq \theta \leq \pi$ is shown. A first-order phase transition occurs at $V_{ij} = 3.59$. (*a*) $V_{ij} = 2.5$; (*b*) $V_{ij} = 3.59$; (*c*) $V_{ij} = 6.0$.

The behavior of the system ground state as a function of increasing coupling constant $|V_{ji}|$ is as follows. For $|V_{ji}| < \varepsilon_j - \varepsilon_i$ there is only one minimum at $\theta = 0$. When $|V_{ji}| = \varepsilon_j - \varepsilon_i$, a degenerate critical point appears at $\theta = \pi$ in a catastrophe of fold type. As $|V_{ji}|$ increases, this degenerate critical point splits into a saddle at $\theta \neq \pi$ and a local minimum at $\theta = \pi$. This local minimum remains metastable until $|V_{ji}| = (\sqrt{\Delta_j} + \sqrt{\Delta_i})^2$. A first-order phase transition occurs as $|V_{ji}|$ becomes larger. The logarithm of the tunneling time from metastable to stable minimum is proportional to the height $\Delta_j\Delta_i/|V_{ji}|$ of the saddle barrier which separates them.

In Fig. 15.2 we show the function h_C for three values of the coupling constant V_{ji} for fixed values of the energy levels ε_k. In this sequence the minimum at $\theta = \pi$ is metastable, equal to, and stable with respect to the minimum at $\theta = 0$.

8. Application 4: Extended Dicke Models

The Hamiltonians h_Q for extended Dicke models, together with their classical limits, are

$$h_Q = \sum_{1 \le i < j}^{r} \hbar\omega_{ji} \frac{a_{ji}^\dagger a_{ji}}{N} + \sum_{1 \le i}^{r} \varepsilon_i \frac{E_{ii}}{N} + \sum_{1 \le i < j}^{r} \lambda_{ji} \left(\frac{a_{ji}^\dagger}{\sqrt{N}}\right)\left(\frac{E_{ij}}{N}\right) + \text{hc} \tag{15.61a}$$

$$h_C = \sum_{1 \le i < j}^{r} \hbar\omega_{ji} \mu_{ji}^* \mu_{ji} + \sum_{1 \le i}^{r} \varepsilon_i z_i^* z_i + \sum_{1 \le i < j}^{r} \lambda_{ji} \mu_{ji}^* z_i^* z_j + \text{cc} \tag{15.61b}$$

where hc means hermitian conjugate and cc means complex conjugate. The critical points of h_C are determined most easily by eliminating the photon order parameters μ, μ^* through the relation

$$\frac{\partial h_C}{\partial \mu_{ji}^*} = \hbar\omega_{ji}\mu_{ji} + \lambda_{ji} z_i^* z_j = 0 \tag{15.62}$$

The reduced function h_C' obtained by using the constitutive relation (15.62) to eliminate μ, μ^* from (15.61b) is

$$h_C' = \sum_{i=1}^{r} \varepsilon_i z_i^* z_i - \sum_{1 \le i < j}^{r} \frac{|\lambda_{ji}|^2}{\hbar\omega_{ji}} z_i^* z_i z_j^* z_j \tag{15.63}$$

This reduced function is identical to the classical function (15.51) (which describes the ground state critical properties of extended MGL Hamiltonians) under the identification

$$V_{ji} = 0, \qquad +W_{ji} = -\frac{|\lambda_{ji}|^2}{\hbar\omega_{ji}} \tag{15.64}$$

As a result, the gauge invariances and critical properties of these two models are isomorphic. There is a 1–1 correspondence between the polarization states of the N-nucleon system and the atomic subsystem of the interacting atom-field system. The quantum state of the field subsystem can be constructed as shown in Table 15.2.

That a relationship should exist between these two models is no longer surprising, in view of the discussion given on p. 383. In extended Dicke models the atoms interact with each other only through the field. An atomic polarization acts as a classical source driving the field subsystem. The field polarization state then carries this information to other atomic subsystems through the atom-field interaction. The field as an intermediary can be replaced effectively by atomic operators in the limit of large N. The following nonrigorous argument at least makes this plausible:

$$\frac{\partial h_Q}{\partial(a_{ji}^\dagger/\sqrt{N})} = \hbar\omega_{ji}\frac{a_{ji}}{\sqrt{N}} + \lambda_{ji}\left(\frac{E_{ij}}{N}\right) \simeq 0 \tag{15.65}$$

The $\simeq$ means that this operator, when acting on a direct product state $|\psi_a\rangle \otimes |\psi_f\rangle$, becomes arbitrarily small as this direct product approaches the exact ground state ($N \to \infty$). This relation between the field and atomic operators can be exploited to eliminate the field operators from h_Q (15.61a) to give

$$h'_Q = \sum_{i=1}^{r} \varepsilon_i \frac{E_{ii}}{N} - \sum_{1\leq i<j}^{r} \frac{|\lambda_{ji}|^2}{\hbar\omega_{ji}}\left(\frac{E_{ij}}{N}\right)\left(\frac{E_{ji}}{N}\right) \tag{15.66}$$

Since intensive collective operators commute [cf. (15.16)], their order in (15.66) is immaterial. Comparison of (15.66) with (15.50) reveals the reasons for the intimate connection between the critical properties of (extended) MGL and Dicke models.

9. Algorithm for Thermodynamic Phase Transitions

Our attention has been confined so far entirely to ground state energy phase transitions. These can occur at zero temperature as the coupling constants which characterize the system are changed. Normally a particular system has fixed coupling constants, so one must find clever methods to investigate such phase transitions. In solids this can be done by changing the isotropic pressure. In nuclei this can be done by investigating isotopic or isotonic sequences. Ground state energy phase transitions have been determined by such methods.

Thermodynamic phase transitions are generally simpler to study than ground state energy phase transitions under laboratory conditions. Here only one control parameter, the temperature T, is varied. In this section we describe a simple algorithm for studying thermodynamic phase transitions for the mean field class of model Hamiltonians described in Sec. 3.

At finite temperature a quantum mechanical system in thermodynamic equilibrium is not in a pure quantum state $|\psi_i\rangle$ but is rather a statistical mixture of pure states. The state $|\psi_i\rangle$ occurs with probability $P_i \simeq e^{-\beta E_i}$, where E_i

is the energy eigenvalue of $|\psi_i\rangle$ and $\beta = 1/kT$. The expectation value of any operator $\mathcal{O}$ is

$$\begin{aligned}\langle \mathcal{O} \rangle = \sum P_i \langle \psi_i | \mathcal{O} | \psi_i \rangle &= \mathrm{tr}(|\psi_i\rangle P_i \langle \psi_i |) \mathcal{O} \\ &= \mathrm{tr}\, \rho \mathcal{O}\end{aligned} \tag{15.67}$$

The operator ρ is called the **density operator**. In the zero-temperature limit only the state with lowest energy $|\psi_g\rangle$ contributes to this sum, and

$$\langle \mathcal{O} \rangle \xrightarrow{T \to 0} \langle \psi_g | \mathcal{O} | \psi_g \rangle \tag{15.68}$$

Only in this limit can the expectation value of an operator be constructed from a single pure state. For $T \neq 0$ the state of a quantum system is no longer determined by a minimum energy condition, but rather by a minimum free energy condition. The free energy F is determined from the partition function Z by

$$e^{-\beta F} = Z = \mathrm{tr}\, e^{-\beta \mathscr{H}} \tag{15.69}$$

For our purposes we are interested in the free energy per particle (nucleon, atom, molecule) in the limit of large N:

$$\lim_{N \to \infty} \frac{F}{N} = -\frac{1}{\beta N} \ln \mathrm{tr}\, e^{-\beta \mathscr{H}} \tag{15.70}$$

This can be computed for the class of Hamiltonians under consideration. This computation involves some technical group theoretical details, described below. The estimate leads to the algorithm which follows the technical discussion. Those uninterested in the technical details may skip over them directly to the algorithm.

Technical Discussion.[9] The estimate of F/N proceeds in a number of simple steps:

1. The Hilbert space on which the Hamiltonian $\mathscr{H}$ acts is broken up into a number of smaller subspaces. Each subspace is irreducible under the action of the dynamical group $U(r)$.[8,12]
2. Within each irreducible subspace upper and lower bounds are put on the sum $\mathrm{tr}\, e^{-\beta \mathscr{H}}$ using inequalities due to Bogoliubov and Lieb.[13]
3. These bounds are summed over all irreducible subspaces, including multiplicity, to give upper and lower bounds for the partition function Z.
4. The logarithms of these bounds are taken to construct bounds on the free energy F.
5. The logarithm of the (multiplicative) multiplicity factor becomes an additive term representing disorder, and eventually interpreted as entropy.
6. The difference between the upper and lower bounds on the intensive free energy F/N converges to zero in the limit $N \to \infty$.

This procedure therefore gives an estimate for F/N which becomes exact in the limit $N \to \infty$. This estimate is determined from the simple four-step algorithm:

1. $\mathscr{H}/N = h_Q(a^\#/\sqrt{N}, E/N)$
2. $h_Q \to h_C = \langle h_Q \rangle_T = h_Q(\langle a^\#/\sqrt{N}\rangle_T, \langle E/N \rangle_T)$
3. $\Phi = h_C - kT\mathscr{s}(\delta)$
4 $F/N = \min \Phi$

This algorithm is very similar to the algorithm for determining the intensive ground state energy E_g/N with two notable differences:

1. The expectation value $\langle h_Q \rangle_T$ is a finite-temperature expectation value, determined from (15.67). The classical limits $\langle \cdot \rangle_T$ are finite-temperature classical limits (i.e., taken in other than the representation space containing the ground state). For photon operators,

$$\begin{aligned} \left\langle \frac{a^\dagger}{\sqrt{N}} \right\rangle_T &= \left\langle \frac{a^\dagger}{\sqrt{N}} \right\rangle_{T=0} = \mu^* \\ \left\langle \frac{a}{\sqrt{N}} \right\rangle_T &= \left\langle \frac{a}{\sqrt{N}} \right\rangle_{T=0} = \mu \end{aligned} \tag{15.71}$$

For angular momentum operators,

$$\begin{aligned} \langle J_3 \rangle_T &= r \cos\theta \\ \langle J_\pm \rangle_T &= r \sin\theta\, e^{\pm i\phi} \end{aligned} \tag{15.72}$$

$$\begin{aligned} 0 \le r \le \tfrac{1}{2} \\ 0 \leftarrow \beta \rightarrow \infty \end{aligned} \tag{15.73}$$

For the groups $U(r)$ the classical limits $\langle E_{ij}/N \rangle_{T \neq 0}$ are known but not simple enough to justify their inclusion here.

2. The logarithm of the multiplicity factor is

$$\mathscr{s}(\delta) = -\sum_{i=1}^{r} \delta_i \ln \delta_i \tag{15.74a}$$

where

$$\begin{aligned} \delta_1 \ge \delta_2 \ge \cdots \ge \delta_r \ge 0 \\ \delta_1 + \delta_2 + \cdots + \delta_r = 1 \end{aligned} \tag{15.74b}$$

(The factors δ_i are ratios of Young partition lengths λ_i to the number N of particles: $\delta_i = \lambda_i/N$). For $SU(2)$, $\lambda_1 + \lambda_2 = N$, $\lambda_1 - \lambda_2 = 2J$, $r = J/N$, $\delta_1 = \frac{1}{2} + r$, $\delta_2 = \frac{1}{2} - r$, and

$$\mathscr{s}(r) = -\{(\tfrac{1}{2} + r)\ln(\tfrac{1}{2} + r) + (\tfrac{1}{2} - r)\ln(\tfrac{1}{2} - r)\} \tag{15.75}$$

where $0 \leq r = J/N \leq \frac{1}{2}$ appears in the classical limit of the angular momentum operators.

Example. For the case of N identical 2-level systems the algorithm proceeds as follows:[9]

1. $\mathcal{H}/N = h_Q(a/\sqrt{N}, a^\dagger/\sqrt{N}, J_3/N, J_+/N, J_-/N)$
2. $h_C(\mu, \mu^*, r, \theta, \phi) = h_Q(\mu, \mu^*, r\cos\theta, r\sin\theta\, e^{i\phi}, r\sin\theta\, e^{-i\phi})$
3. $\Phi(\beta) = h_C(\mu, \mu^*, r, \theta, \phi) - kT\mathscr{s}(r)$
4. $F(\beta)/N = \min_{\mu, \mu^*, r, \theta, \phi} \Phi(\beta)$

Remark. In the limit $T \to 0$, $F \to E_g$ and this algorithm reduces to the algorithm (Sec. 4) for determining E_g/N.

10. Application 1: The MGL and Dicke Models

The MGL Hamiltonian h_Q and the potential derived from it by this algorithm are

$$h_Q = \varepsilon \frac{J_3}{N} + \tfrac{1}{2}V\left[\left(\frac{J_+}{N}\right)^2 + \left(\frac{J_-}{N}\right)^2\right] + \tfrac{1}{2}W\frac{[J_+J_- + J_-J_+]}{N^2} \tag{15.76a}$$

$$\Phi_{MGL} = -\varepsilon r\cos\theta + \tfrac{1}{2}V(r\sin\theta)^2(e^{2i\phi} + e^{-2i\phi}) + W(r\sin\theta)^2 - kT\mathscr{s}(r) \tag{15.76b}$$

where $\mathscr{s}(r)$ is given in (15.75). Before discussing the thermodynamic properties of this model we construct the potential for the Dicke model also:

$$h_Q = \hbar\omega\frac{a^\dagger a}{N} + \varepsilon\frac{J_3}{N} + \lambda\left(\frac{a^\dagger}{\sqrt{N}}\right)\left(\frac{J_-}{N}\right) + \lambda^*\left(\frac{a}{\sqrt{N}}\right)\left(\frac{J_+}{N}\right) \tag{15.77a}$$

$$\Phi_D = \hbar\omega\mu^*\mu - \varepsilon r\cos\theta + \lambda\mu^* r\sin\theta\, e^{-i\phi} + \lambda^*\mu r\sin\theta\, e^{+i\phi} - kT\mathscr{s}(r) \tag{15.77b}$$

This potential may be studied by using the relation $\partial\Phi_D/\partial\mu^* = 0$ to eliminate μ, μ^* from (15.77b). Since the entropy term does not depend on the field order parameters, the details go through as in (Sec. 6), and the reduced thermodynamic potential for the Dicke model is

$$\Phi'_D = -\varepsilon r\cos\theta - \frac{\lambda^*\lambda}{\hbar\omega}(r\sin\theta)^2 - kT\mathscr{s}(r) \tag{15.78}$$

The critical properties of the Dicke model are isomorphic with the critical properties of the MGL model in the thermodynamic case [compare (15.78) with (15.76b)] as well as in the ground state case. The reasons are the same.

Before jumping into a detailed study of the function Φ_{MGL} we look first at the properties of the entropy term $-kT\mathscr{s}(r)$. The properties of $\mathscr{s}(r)$ are summarized in Fig. 15.3. We observe that while $\mathscr{s}(r)$ is bounded on the interval $[0, \frac{1}{2}]$, its derivative $\mathscr{s}'(r)$ is not. We note also that the function h_C is finite for all values of (r, θ, ϕ), $0 \leq r \leq \frac{1}{2}$. In the zero-temperature limit the most important contribution to the potential Φ is the energy term

$$\Phi \xrightarrow{T \to 0} h_C = \langle h_Q \rangle_{T=0} \tag{15.79}$$

In the high-temperature limit the most important contribution to the potential Φ is the entropy term

$$\Phi \xrightarrow{T \to \infty} kT \ln 2 + \text{small correction} \tag{15.80}$$

We are now prepared to discuss the thermodynamic critical properties of the MGL model. Without loss of generality we set $W = 0$. Minimizing with respect to ϕ leads to the reduced potential

$$\Phi'(\beta) = -\varepsilon r \cos\theta - |V|(r \sin\theta)^2 - kT\mathscr{s}(r) \tag{15.81}$$

In the zero-temperature limit only the energy term remains. This is extremized by choosing r as large as possible: $r = J/N = \frac{1}{2}$. With this choice h'_C reduces to (15.38′). The minimum of Φ' occurs at $\theta = 0$ if $\varepsilon > |V|$ and at $\cos\theta = +\varepsilon/|V|$ if $\varepsilon < |V|$.

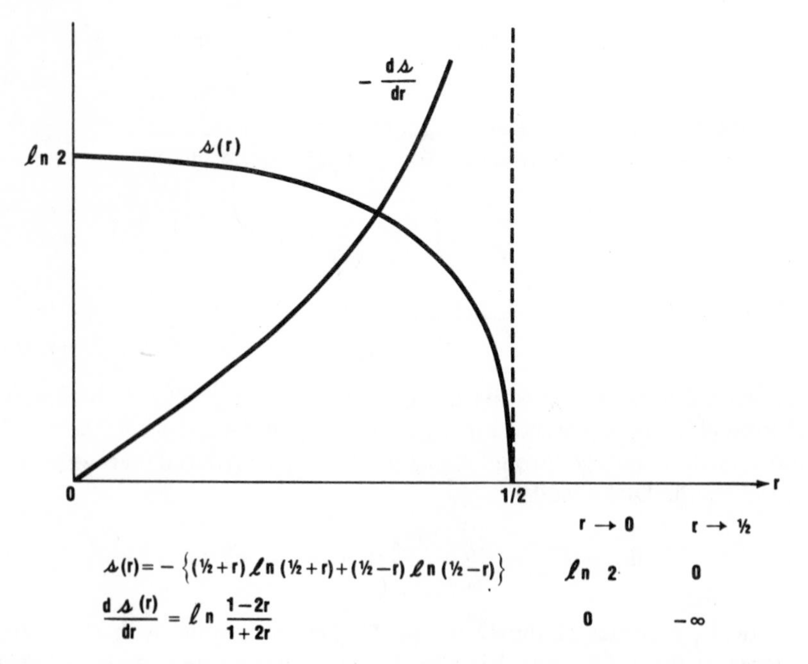

Figure 15.3 The intensive logarithm $\mathscr{s}(r)$ of the $SU(2)$ multiplicity factor and its derivative $\mathscr{s}'(r)$ depend on $r = J/N$ as shown.

In the high-temperature limit the entropy term dominates the energy term. The value of r at which the minimum occurs is determined from

$$\frac{\partial \Phi'}{\partial r} = -\varepsilon \cos\theta - 2r|V|\sin^2\theta - kT \ln\frac{1-2r}{1+2r} = 0 \qquad (15.82)$$

The term $\partial h_C/\partial r$ is bounded, so that the term $-kT\ \ln[(1-2r)/(1+2)]$ must be finite. This means that $r \to 0$ as $T \to \infty$. In this limit we may neglect the term $-2r|V|\sin^2\theta$ compared with the term $\varepsilon\cos\theta$. When the interaction terms in Φ' are neglected, the minimum over θ occurs at $\theta = 0$. Under this condition r is given approximately by

$$r(\beta) \simeq \tfrac{1}{2}\tanh\tfrac{1}{2}\beta\varepsilon \qquad (15.83)$$

Important Remark. As the temperature increases, the interaction terms become less important than the diagonal term εJ_3 because the interaction terms depend on r^2 while the diagonal term depends on r^1. In the high-temperature limit the interaction terms may be neglected when it comes to computing the values of the parameters (θ, ϕ) which minimize Φ. As far as these order parameters are concerned, the thermodynamic high-temperature behavior is the same as the ground state energy small interaction behavior. Increasing the temperature renormalizes the coupling constants toward zero.

We search for the thermodynamic phase transition by expanding Φ' about $\theta = 0$ and studying the coefficient of the quadratic term as a function of increasing r or decreasing temperature:

$$\Phi' = (-\varepsilon r - kT\mathscr{S}(r)) + (\tfrac{1}{2}\varepsilon r - |V|r^2)\theta^2 + \left(-\frac{\varepsilon r}{4!} + \frac{1}{3}|V|r^2\right)\theta^4 + \cdots \qquad (15.84)$$

The first term is minimized when

$$2r = \tanh\tfrac{1}{2}\beta\varepsilon \qquad (15.85)$$

[cf. (15.83)]. The coefficient of the quadratic term vanishes when

$$2r = \frac{\varepsilon}{|V|} \qquad (15.86)$$

When the quadratic term vanishes, the coefficient of the quartic term is positive. Therefore Φ' has a degenerate critical point of type A_{+3} when $\varepsilon/|V| < 1$ at

$$r_c = \frac{1}{2}\frac{\varepsilon}{|V|} \qquad \theta = 0 \qquad (15.87)$$

A second-order thermodynamic phase transition of the Ginzburg–Landau type occurs at the critical temperature defined by the **gap equation**

$$\frac{\varepsilon}{|V|} = \tanh\tfrac{1}{2}\beta_c\varepsilon \qquad (15.88)$$

11. Application 2: Extended MGL and Dicke Models

The thermodynamic critical properties of extended MGL models are equivalent to the thermodynamic critical properties of extended Dicke models. The argument demonstrating this equivalence differs in no way from the argument demonstrating the equivalence between the ground state critical properties of 2- and r-level MGL and Dicke models, and the argument used in the previous section showing the equivalence between the thermodynamic critical properties of 2-level MGL and Dicke models. As this is the case, the thermodynamic critical properties of either can be determined from those of the other.

We have previously exploited these equivalences to study the somewhat simpler MGL models in lieu of the Dicke models. In the present case the reverse procedure is simpler. This is so because the finite-temperature classical limits of the operators E_{ij}/N ($1 \le i, j \le r$, $r > 2$) are somewhat complicated. It is convenient to take the semiclassical limit (15.35) obtained by replacing all photon operators by their expectation values. Then the semiclassical Hamiltonian has the form:

$$\begin{aligned} \mathscr{H} = & \sum_{1 \le i < j}^{r} \hbar\omega_{ji} a_{ji}^{\dagger} a_{ji} + \sum_{1 \le i < j}^{r} \sum_{\alpha=1}^{N} \varepsilon_i e_{ii}^{(\alpha)} \\ & + \sum_{1 \le i < j}^{r} \sum_{\alpha=1}^{N} \lambda_{ji} \frac{a_{ji}^{\dagger}}{\sqrt{N}} e_{ij}^{(\alpha)} + \lambda_{ji}^{*} \frac{a_{ji}}{\sqrt{N}} e_{ji}^{(\alpha)} \end{aligned} \tag{15.89i}$$

$$\mathscr{H}_{SC} = N \sum_{1 \le i < j}^{r} \hbar\omega_{ji} \mu_{ji}^{*} \mu_{ji} + \sum_{\alpha=1}^{N} M^{(\alpha)} \tag{15.89ii}$$

$$M^{(\alpha)} = \sum_{i=1}^{r} \varepsilon_i e_{ii}^{(\alpha)} + \sum_{1 \le i < j}^{r} \lambda_{ji} \mu_{ji}^{*} e_{ij}^{(\alpha)} + \lambda_{ji}^{*} \mu_{ji} e_{ji}^{(\alpha)} \tag{15.89iii}$$

Each operator $M^{(\alpha)}$ is an $r \times r$ matrix representing the interaction of the αth particle with the classical external field. Under the mean field assumption, each r-level atom sees the same external field, so that $M^{(\alpha)} = M$ for all α. The free energy is

$$\begin{aligned} e^{-\beta F} &= \operatorname{tr} e^{-\beta \mathscr{H}} \\ &= e^{-\beta N \sum \hbar\omega_{ji}\mu_{ji}^{*}\mu_{ji}} \operatorname{tr} \exp\left[-\beta \sum_{\alpha=1}^{N} M \right] \end{aligned}$$

$$\operatorname{tr} \exp\left[-\beta \sum_{\alpha=1}^{N} M \right] = \operatorname{tr} \prod_{\alpha=1}^{N} e^{-\beta M} = \prod_{\alpha=1}^{N} \operatorname{tr} e^{-\beta M} = (\operatorname{tr} e^{-\beta M})^N \tag{15.90}$$

The free energy per nucleon is therefore

$$\begin{aligned} \frac{F}{N} &= \min \Phi \\ \Phi &= \sum_{1 \le i < j}^{r} \hbar\omega_{ji} \mu_{ji}^{*} \mu_{ji} - kT \ln \operatorname{tr}(e^{-\beta M}) \end{aligned} \tag{15.91}$$

The critical properties of the function Φ can be studied without too much difficulty when only one coupling constant λ_{ji} is nonzero.

The matrix M has only two off-diagonal matrix elements. The eigenvalues of M are easily computed:

$$\varepsilon_1, \varepsilon_2, \ldots, \hat{\varepsilon}_i, \ldots, \hat{\varepsilon}_j, \ldots, \varepsilon_r$$

and

$$\varepsilon_{\pm} = \left(\frac{\varepsilon_j + \varepsilon_i}{2}\right) \pm \left[\left(\frac{\varepsilon_j - \varepsilon_i}{2}\right)^2 + |\lambda_{ji}^* \mu_{ji}|^2\right]^{1/2} \tag{15.92}$$

As in the case of the ground state energy phase transition, the thermodynamic critical properties depend on whether the interaction couples a pair of excited states or an excited state with the ground state.

1. $\lambda_{j1} \neq 0$. In this case an interaction will force the ground state to a lower energy, so that a second-order phase transition is expected. This phase transition is determined by expanding the potential Φ in powers of $|\mu_{j1}|^2$. We find

$$\begin{aligned}\Phi &= \hbar\omega_{j1}|\mu_{j1}|^2 - kT \ln\left[z(\beta) + (e^{-\beta\varepsilon_1} - e^{-\beta\varepsilon_j}) \frac{|\lambda_{j1}^* \mu_{j1}|^2}{kT(\varepsilon_j - \varepsilon_1)}\right] \\ &= -kT \ln z(\beta) + \left\{\hbar\omega_{j1} + \frac{e^{-\beta\varepsilon_j} - e^{-\beta\varepsilon_1}}{z(\beta)} \frac{|\lambda_{j1}|^2}{\varepsilon_j - \varepsilon_1}\right\}|\mu_{j1}|^2 + \mathcal{O}(4)\end{aligned} \tag{15.93}$$

where

$$z(\beta) = \sum_{i=1}^{r} e^{-\beta\varepsilon_i} \tag{15.94}$$

The critical temperature, at which the coefficient of the quadratic term vanishes, is determined by

$$\frac{\hbar\omega_{j1}(\varepsilon_j - \varepsilon_1)}{|\lambda_{j1}|^2} = -\frac{e^{-\beta\varepsilon_j} - e^{-\beta\varepsilon_1}}{z(\beta)} \tag{15.95}$$

At this critical temperature the coefficient of the quartic term is positive. The multilevel system therefore undergoes a second-order thermodynamic phase transition of Ginzburg–Landau type (A_{+3}) at critical temperature T_c when $(\varepsilon_j - \varepsilon_1)\hbar\omega_{j1}/|\lambda_{j1}|^2 < 1$. Under this condition the system has undergone a similar ground state energy phase transition.

2. $\lambda_{ji} \neq 0, j > i > 1$. The thermodynamic critical behavior is much more interesting when the interaction couples a pair of excited states than when it

involves interaction between the ground state and an excited state. In this case

$$\Phi(\beta) = \hbar\omega_{ji}|\mu_{ji}|^2 - kT\ln\left[\textstyle\sum' e^{-\beta\varepsilon_k} + e^{-\beta(\varepsilon_j+\varepsilon_i)/2}(e^{\beta\theta} + e^{-\beta\theta})\right]$$
$$\theta^2 = \left(\frac{\varepsilon_j - \varepsilon_i}{2}\right)^2 + |\lambda_{ji}^* \mu_{ji}|^2 \tag{15.96}$$

Here the prime on the summation indicates exclusion of states i and j. For this case it is adequate to set λ_{ji} real and μ_{ji} positive or zero.

The critical behavior of $\Phi(\beta)$ can be studied by looking at the $T \to 0$ limit and by looking for bifurcations of ordered solutions from the critical point which exists at $\mu_{ji} = 0$ for all temperatures (by the symmetry under $\mu_{ji} \to -\mu_{ji}$). These special cases indicate that the order parameter has four particular values which characterize the thermodynamic critical properties of this model.[14]

In the zero-temperature limit only the lowest eigenvalue of the matrix M (15.92) contributes to the term containing the logarithm in (15.91). This minimum eigenvalue is either ε_1 or $\frac{1}{2}(\varepsilon_j + \varepsilon_i) - \theta$, depending on the value of μ_{ji}. It is useful to define $\Lambda_{ji}^2 = \lambda_{ji}^2/(\varepsilon_j - \varepsilon_i)\,\hbar\omega_{ji}$. Then it is a simple matter to check that $\Phi(\beta)$ has two stationary points with $\mu_{ji} > 0$ when $\Lambda^2 > \Lambda_1^2 = 1 + 2(\Delta_{i1}/\Delta_{ji})$, where $\Delta_{ji} = \varepsilon_j - \varepsilon_i$. The smaller of these nonzero solutions of $\partial\Phi/\partial\mu_{ji} = 0$ is always unstable. The larger is metastable with respect to the local equilibrium at $\mu_{ji} = 0$ for $\Lambda^2 < \Lambda_2^2$ and globally stable for $\Lambda^2 > \Lambda_2^2$, where $\Lambda_2 - \Lambda_2^{-1} = 2(\Delta_{i1}/\Delta_{ji})^{1/2}$.

Bifurcations from the critical point $\mu_{ji} = 0$ are determined by making a Taylor series expansion in powers of $|\mu_{ji}|^2$:

$$\Phi(\beta, \mu_{ji}) = -\beta^{-1}\ln z(\beta) + C_2(\beta)|\mu_{ji}|^2 + C_4(\beta)|\mu_{ji}|^4 + \cdots$$
$$z(\beta) = \sum_{i=1}^{r} e^{-\beta\varepsilon_i} \tag{15.97}$$
$$C_2(\beta) = \hbar\omega_{ji} - \frac{\lambda_{ji}^2}{\varepsilon_j - \varepsilon_i}\,\frac{e^{-\beta\varepsilon_i} - e^{-\beta\varepsilon_j}}{z(\beta)}$$

Nonzero solution branches bifurcate from the branch $\mu_{ji} = 0$ when $C_2(\beta) = 0$. The smallest value of λ_{ji} at which such bifurcation can occur is determined from

$$\max_{T\in[0,\infty)} \Lambda_3^2\,\frac{e^{-\beta\varepsilon_i} - e^{-\beta\varepsilon_j}}{z(\beta)} = 1 \tag{15.98}$$

The function $(e^{-\beta\varepsilon_i} - e^{-\beta\varepsilon_j})/z(\beta)$ goes asymptotically to zero as $T \to 0$ and as $T \to \infty$, with one maximum value within this range. For $\Lambda^2 > \Lambda_3^2$ there are two nonzero solutions which bifurcate from the branch $\mu_{ji} = 0$. The solution that bifurcates at the lower temperature is always unstable. The stability of the solution bifurcating at the higher temperature depends on the sign of $C_4(\beta)$. This is negative for $\Lambda^2 = \Lambda_3^2$ but becomes positive as T increases (when Λ^2 increases).

The thermodynamic critical properties of the extended MGL and Dicke models with one nonzero interaction between two excited states are determined by the value of the coupling constant λ_{ji} or the dimensionless coupling constant $\Lambda_{ji} = \lambda_{ji}/(\varepsilon_j - \varepsilon_i)$. There are four critical values of the coupling constant, as follows:

$$
\begin{aligned}
(\Lambda_{ji})_1^2 &= \frac{1 + \gamma_{ji}}{1 - \gamma_{ji}}, \qquad \gamma_{ji} = \frac{\varepsilon_i - \varepsilon_1}{\varepsilon_j - \varepsilon_1} < 1 \\
(\Lambda_{ji})_2^2 &= \frac{1 + (\gamma_{ji})^{1/2}}{1 - (\gamma_{ji})^{1/2}} \\
(\Lambda_{ji})_3^2 &= \min_{\beta > 0} \frac{z(\beta)}{e^{-\beta\varepsilon_i} - e^{-\beta\varepsilon_j}}
\end{aligned}
\tag{15.99}
$$

$(\Lambda_{ji})_4^2$ is determined by $C_2(\beta) = 0$ and $C_4(\beta) = 0$:

$$
\begin{aligned}
C_2(\beta) &\simeq 1 - (\Lambda_{ji})_4^2 \frac{e^{-\beta\varepsilon_i} - e^{-\beta\varepsilon_j}}{z(\beta)} \\
C_4(\beta) &\simeq \frac{e^{-\beta\varepsilon_i} - e^{-\beta\varepsilon_j}}{z(\beta)} + \frac{\beta(\varepsilon_j - \varepsilon_i)}{2} \left(\frac{e^{-\beta\varepsilon_i} - e^{-\beta\varepsilon_j}}{z(\beta)} \right)^2 \\
&\quad - \beta(\varepsilon_j - \varepsilon_i) \frac{e^{-\beta\varepsilon_i} - e^{-\beta\varepsilon_j}}{z(\beta)}
\end{aligned}
\tag{15.100}
$$

These two conditions fix simultaneously a critical temperature β_t and the coupling constant $(\Lambda_{ji})_4^2$.

The thermodynamic critical properties of this model are summarized as a function of increasing coupling constant Λ_{ji} in Fig. 15.4 and discussed below. The function $\Phi(\beta, \mu_{ji})$ has at most two critical points for $\mu_{ji} > 0$. The branch $\mu_{ji}^{>}(\beta)$ describes locally stable critical points of $\Phi(\beta, \mu_{ji})$, while $\mu_{ji}^{<}(\beta)$ describes locally unstable critical points. These may be connected (curves A, B, C, D, and one of the two curves E) or disconnected. The stability type of the connected branches $\mu_{ji}(\beta)$ changes at the point of vertical tangency. We now describe the critical behavior more fully, as a function of increasing Λ_{ji}.

1. $\Lambda < \Lambda_1$. $\mu_{ji} = 0$ at all temperatures.
2. $\Lambda_1 < \Lambda < \Lambda_2$ (curve A, $\Lambda = \Lambda_A$). At sufficiently low temperature a metastable ordered state exists. No thermodynamic phase transition occurs.
3. $\Lambda_2 < \Lambda < \Lambda_3$ (curves B, C, D, $\Lambda_B < \Lambda_C < \Lambda_D$). At $T = 0$ a stable ordered state exists. For $\Lambda = \Lambda_B$ this ordered state remains stable as T increases until $T = T_B$. Above this temperature the ordered state becomes metastable with respect to the disordered state $\mu_{ji} = 0$. If the Maxwell Convention is obeyed, a first-order phase transition occurs at $T = T_B$. If the Delay Convention is obeyed a zeroth-order phase transition occurs at the fold in curve B. Above this temperature there is only one critical point, the minimum at $\mu_{ji} = 0$. When the system is cooled from very high temperatures, a first-order phase transition

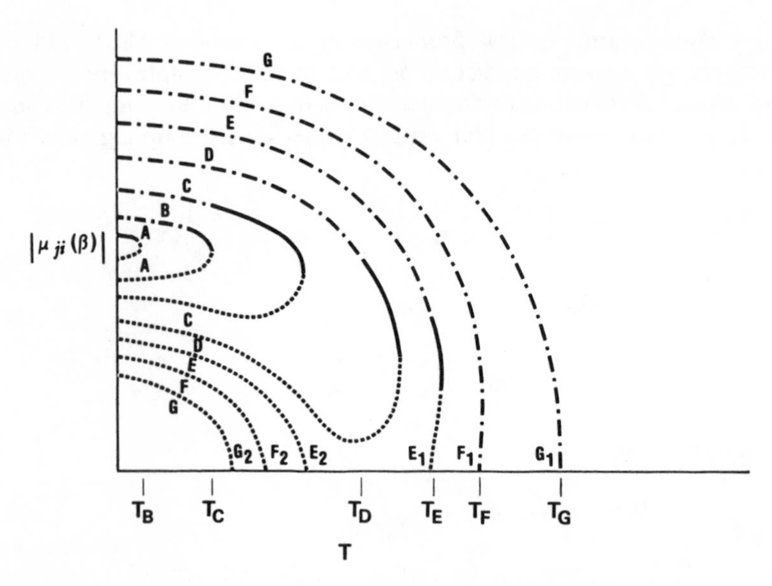

Figure 15.4 Values of the order parameter $\mu_{ji}(\beta)$ for which the potential $\Phi(\mu; \beta)$ is stationary are shown for seven values of the dimensionless scaled coupling constant Λ.

occurs at $T = T_B$ if the Maxwell Convention is obeyed. If the Delay Convention is obeyed, there is no phase transition at all. If, at very low temperatures, the system is kicked or otherwise molested (as by a graduate student, for example), it will then snap from the metastable disordered state into the ordered ground state in a zeroth-order phase transition. The ordered state of this system has a high-temperature spinodal point, while the disordered state has no low-temperature spinodal point. Increasing the coupling constant ($\Lambda \rightarrow \Lambda_C, \Lambda_D$) does not qualitatively change the thermodynamic critical properties of this system. The critical temperatures T_C, T_D at which the first-order phase transitions occur increase, as do the high-temperature spinodal points. As Λ increases, the distance between the spinodal point and the first-order phase transition temperature decreases.

4. $\Lambda_3 < \Lambda < \Lambda_4$ (curve E, $\Lambda = \Lambda_E$). Two branches bifurcate from the branch $\mu_{ji} = 0$. The low-temperature branch which bifurcates at E_2 represents local maxima, and is not of interest. The bifurcation of the high-temperature branch at E_1 is of type A_{-3}. This branch initially turns toward higher temperature, then turns around and decreases to zero. The branch E_1 is unstable between the points of vertical tangency. These two points define high- and low-temperature spinodal bounds for the first-order transition which occurs at T_E. The ordered state ceases to exist above the high-temperature spinodal point. The disordered state becomes unstable below the E_1 bifurcation point. It remains unstable (T decreasing) until the E_2 bifurcation point is reached. Below this temperature the disordered branch is metastable with respect to the ordered state. The first-order phase transition at T_E is surrounded by two spinodal points.

5. $\Lambda = \Lambda_4$ (curve F, $\Lambda = \Lambda_F = (\Lambda_{ji})_4$). The point $\mu_{ji} = 0$, $T = T_F = T_t$, $\Lambda_{ji} = (\Lambda_{ji})_4$ is a tricritical point. This is easily seen by the following argument. The function Φ is an even function, and the two control parameters β, Λ_{ji} can be chosen to make the two leading coefficients $C_2(\beta)$ and $C_4(\beta)$ vanish simultaneously. At this point the germ of $\Phi(\beta)$ is $\pm|\mu_{ji}|^6$. Global stability arguments indicate that the coefficient of this term must be positive. For $\Lambda = (\Lambda_{ji})_4$ fixed, a second-order phase transition occurs as T increases or decreases through T_F.

6. $\Lambda_4 < \Lambda$ (curve G, $\Lambda = \Lambda_G$). The branch bifurcating at G_1 is globally stable. The disordered branch is metastable for $0 \leq T < G_2$, unstable for $G_2 \leq T \leq G_1$, and stable for $G_1 = T_G \leq T < \infty$. A second-order thermodynamic phase transition occurs as T increases or decreases through T_G.

Remark. A first-order ground state energy phase transition occurs ($T = 0$) when $\Lambda = \Lambda_2$.

Remark. The effects of perturbations of the original Hamiltonian (15.89), due to the presence of a classical external field interacting with the atomic subsystem or a classical current interacting with the quantum-mechanical field, can now easily be determined. Symmetry-breaking perturbations will unhinge all the bifurcations, but will not significantly alter the number and types of critical points which occur for any combination of control parameters (Λ_{ji}, T) (*small* perturbations). Perturbations can easily be studied because the universal perturbation of the catastrophes A_2, $A_{\pm 3}$, A_5 which occur in this model are well known.

The interesting behavior just described occurs when just one of the coupling constants $\lambda_{ji} \neq 0$. What happens when many are nonzero?

We cannot present more than descriptive detail here. We assume that all $r(r-1)/2$ coupling constants λ_{ji} are real. Then each point $(\lambda_{21}, \lambda_{31}, \ldots, \lambda_{r,r-1}) \in \mathbb{R}^{r(r-1)/2}$ represents an r-level Dicke or MGL model. At sufficiently high temperature T all such models have a stable thermodynamic state with all order parameters $\mu_{ji} = 0$. An open set of points in $\mathbb{R}^{r(r-1)/2}$ describes models that have the following qualitative properties. As T decreases from ∞, a second-order phase transition occurs at T_1. Below this temperature there is one nonzero-order parameter until $T = T_2$, where a second second-order phase transition occurs. Below T_2 there are three nonzero-order parameters. This cascade of transitions continues as T decreases until $T = T_{r-1}$, below which all $r(r-1)/2$-order parameters μ_{ji} are nonzero. This sequence is summarized as follows:

Temperature	0 T_{r-1}	T_{r-2}	$\cdots$	T_3	T_2	T_1
Number of nonzero-order parameters	$\frac{r(r-1)}{2}$	$\frac{(r-1)(r-2)}{2}$	$\cdots 3+2+1$	$2+1$	1	0

(15.101)

The measure of this open set in $\mathbb{R}^{r(r-1)/2}$ is significant.

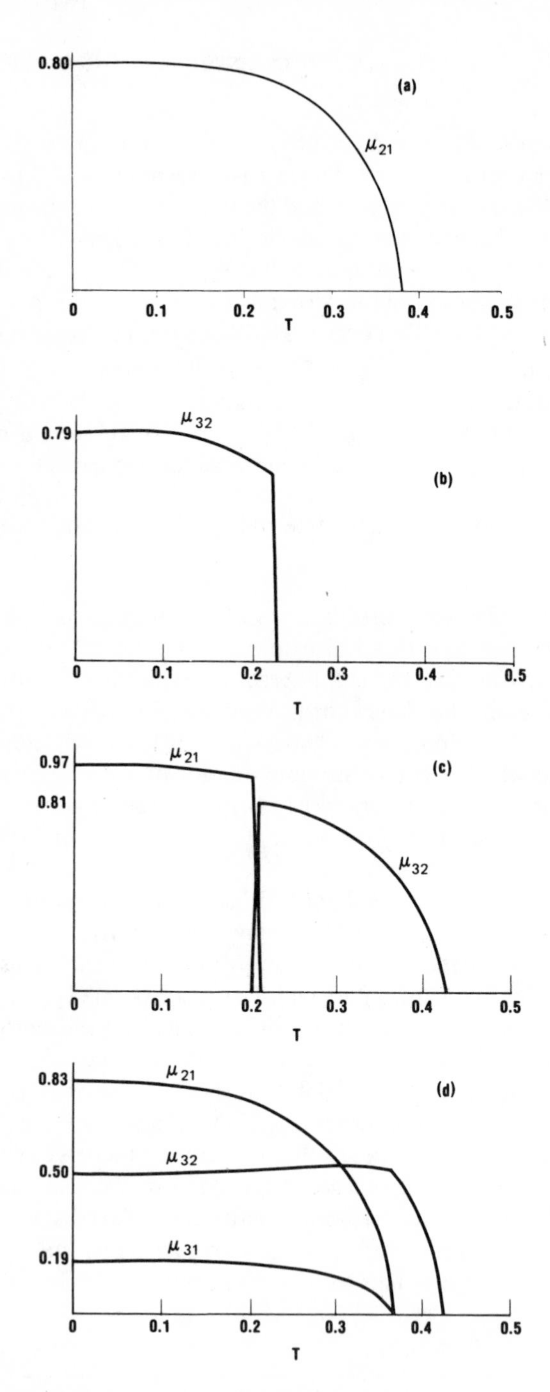

Figure 15.5 The thermodynamic critical properties of the 3-level Dicke model with resonant interaction $\hbar\omega_{ji} = \varepsilon_j - \varepsilon_i$ have been studied as a function of the dimensionless coupling constant $\Lambda_{ji} = \lambda_{ji}/(\varepsilon_j - \varepsilon_i)$. Here $\gamma_{32} = (\varepsilon_2 - \varepsilon_1)/(\varepsilon_3 - \varepsilon_1) = 0.2$ and $\mu_{21} = \langle E_{21}/N \rangle$, $\mu_{32} = \langle E_{32}/N \rangle$, $\mu_{31} = \langle E_{31}/N \rangle$. The following values of $(\Lambda_{12}, \Lambda_{23}, \Lambda_{13})$ have been used: (*a*) (2.0, 1.7, 0.8); (*b*) (0.0, 1.7, 0.0); (*c*) (2.0, 1.8, 0.7); (*d*) (2.0, 1.8, 0.8).

This open set contains separatrices on which two or more of the critical temperatures become equal (e.g., $T_2 = T_3$). The order of bifurcation changes on passing through such separatrices.

There are also open regions in $\mathbb{R}^{r(r-1)/2}$ describing models for which fewer than $(r-1)$ second-order phase transitions occur. Such open regions adjoin each other in a qualitatively obvious way.

Any of these second-order phase transitions may be replaced by a first-order phase transition. In addition, first-order phase transitions may occur between second-order phase transitions.

The simplest extended MGL and Dicke models exhibiting this rich variety of behavior are the 3-level models. The thermodynamic properties of these models were studied[15] under a resonant interaction assumption $\varepsilon_j - \varepsilon_i = \hbar\omega_{ji}$. The order parameters $\mu_{ji}(\beta)$ which minimize $\Phi(\beta)$ are shown in Fig. 15.5 for various values of the dimensionless coupling constants $\Lambda_{ji} = \lambda_{ji}/(\varepsilon_j - \varepsilon_i)$. Fig. 15.5*a* shows a single second-order phase transition. This is replaced by a single first-order phase transition in Fig. 15.5*b*. In Fig. 15.5*c* we see a high-temperature second-order phase transition, followed at lower temperature by a first-order phase transition. In these three cases at most one order parameter is nonzero at the global minimum of $\Phi(\beta)$. In the case of Fig. 15.5*c* there are two minima at $(\mu_{21} \neq 0, \mu_{31} = \mu_{32} = 0)$ and $(\mu_{21} = \mu_{31} = 0, \mu_{32} \neq 0)$. The critical values at these critical points become equal at the first-order phase transition. Finally Fig. 15.5*d* shows a cascade of two second-order phase transitions.

12. Crossover Theorem

In Secs. 5 and 6 we studied ground state energy phase transitions in MGL and Dicke models, and in Sec. 10 we studied their thermodynamic phase transitions. In synthesizing the results of these three sections, we conclude that if such a system undergoes a ground state energy phase transition (for fixed $T = 0$) as a function of increasing interaction parameters, then it also undergoes a thermodynamic phase transition for fixed interaction parameters as a function of increasing temperature. This relation is summarized by the gap equation

$$\frac{|V|}{\varepsilon} \tanh \tfrac{1}{2}\beta_c \varepsilon = 1 \tag{15.102}$$

which defines both the ground state energy critical properties ($\tanh \frac{1}{2}\beta_c\varepsilon \to 1$) and the thermodynamic critical properties. This equation is shown as a separatrix in the $|V/\varepsilon|$–T plane in Fig. 15.6*a*. Below the separatrix the order parameter $\sin\theta = 0$, while above this curve $\sin\theta \neq 0$ ($\cos\theta = +\varepsilon/|V|$).

At $T = 0$, as the order parameter $|V|$ is increased, the ground state at first undergoes no change (*a*, *b*). A second-order ground state energy phase transition occurs at *c*, above which the ground state is ordered (*d*, *e*). If the interaction strength $|V|$ is fixed and the temperature is increased, the thermodynamic

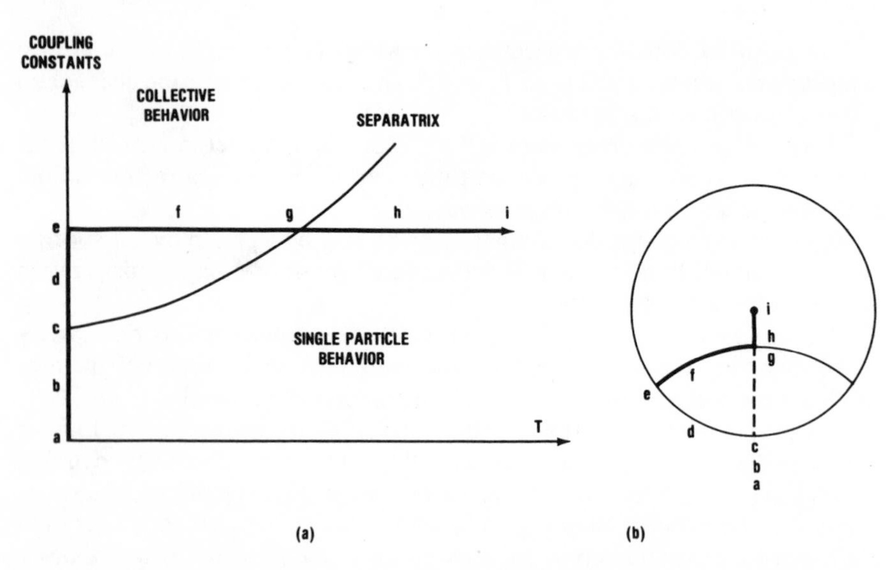

Figure 15.6 (*a*) The separatrix divides nuclear systems with nonzero order parameter (quadrupole moment) exhibiting cooperative behavior from those systems exhibiting single particle behavior. (*b*) The state of nuclear matter can be represented by a point in the sphere of radius $\frac{1}{2}$. In the zero-temperature limit this point lies on the sphere surface. In the absence of nuclear interactions this point lies at the south polar position from which the order parameter θ is measured. As the coupling constant is increased at $T = 0$, this point follows the positions $a \to e$, with a second-order ground state energy phase transition at c. As the temperature is increased, this point follows positions $e \to i$, with a second-order thermodynamic phase transition at g.

system state remains ordered, but the order parameter decreases (e, f) as T increases. Eventually a second-order thermodynamic phase transition occurs (g), and for higher temperature (h, i) the system remains in a disordered state. The values of the order parameters in the process $a \to i$ are shown in Fig. 15.6*b*.

The reasons for this behavior can be summarized easily for the MGL model. In the zero-temperature limit the ground state of h_C has $\theta = 0$ for $V = 0$ and for V small. The function

$$h_C = -\varepsilon r \cos\theta - |V|(r \sin\theta)^2 \tfrac{1}{2}(e^{2i\phi} + e^{-2i\phi}) \tag{15.103}$$

is an analytic function of the order parameters r, θ, ϕ and the control parameters ε, V, but the function

$$\frac{E_g}{N} = \min_{r=1/2,\theta,\phi} h_C \tag{15.104}$$

is not an analytic function of ε, V. It is analytic in the open intervals $|V|/\varepsilon > 1$ and $|V|/\varepsilon < 1$. The separatrix $|V|/\varepsilon = 1$ defines the zero-temperature cross-over of the minimum of h_C from one branch of $\nabla h_C = 0$ to another.

In the finite-temperature case the function Φ must be minimized with respect to the order parameters (r, θ, ϕ) at constant temperature. The function $\beta\Phi$ may just as well be minimized:

$$\beta\Phi = -\mathscr{s}(r) + \beta h_C \tag{15.105}$$

This is an analytic function of the order parameters (r, θ, ϕ) and the control parameters (ε, V, β), but the function

$$\frac{\beta F}{N} = \min_{r,\theta,\phi} \beta\Phi \tag{15.106}$$

is not analytic throughout the space $\mathbb{R}^2 = (|V|/\varepsilon, \beta)$. The minimum is analytic separately in the two open regions above and below the separatrix defined by the gap equation. High temperatures serve to renormalize the interaction strength $(V \to V' = 2rV)$. As a consequence, if $|V|/\varepsilon > 1$, then the expansion of min h_C in powers of $|V|/\varepsilon$ around $T = 0$, $|V|/\varepsilon = 0$, or of min Φ in powers of β around $|V|/\varepsilon > 1$, $\beta = 0$, will not converge to E_g/N or F/N. This requires a discontinuity in one of the derivatives of E_g/N with respect to $|V|/\varepsilon$ or of F/N with respect to β or kT. These discontinuities signal the ground state energy and the thermodynamic phase transitions, respectively.

This relation between ground state energy phase transitions and thermodynamic phase transitions is called the **Crossover Theorem.**[16,17] Its validity is not restricted simply to the MGL and Dicke model Hamiltonians. Its validity is much more widespread. Instead of presenting a proof, we simply indicate the main ingredients which enter into a proof of this theorem. They are all present, if incompletely stated, in the discussion above.

We assume that the Hamiltonian describing a system has the form

$$\mathscr{H}(\lambda) = \mathscr{H}_0 + \lambda\mathscr{H}_I \tag{15.107}$$

We assume further that $\mathscr{H}_0$ describes the energy level spectrum of the noninteracting system, and $\mathscr{H}_I$ describes the interaction. For extended models of MGL type we may take $\mathscr{H}_0 = \varepsilon_i E_{ii}/N$ and $\mathscr{H}_I$ as a polynomial in the generators E_{ij}/N. We further assume that no terms linear in E_{ij}/N occur in $\mathscr{H}_I$, and that $\langle\mathscr{H}_I\rangle$ is invariant under $z_i \to -z_i$. Under this assumption the state $z_i = 0$ $(i = 2, \ldots, r)$ is a global minimum of $\langle\mathscr{H}_0\rangle$ and remains a global minimum of $\langle\mathscr{H}(\lambda)\rangle$ for λ small, by symmetry. If $E_g(\lambda)/N$ depends nontrivially on λ for λ sufficiently large, then $E_g(\lambda)/N$ cannot be analytic on the entire line $\mathbb{R}^1$, for $E_g(\lambda)/N$ is independent of λ for λ in a small open interval around 0.

The existence of a thermodynamic phase transition is determined by similar analyticity arguments. It is necessary to minimize Φ and sufficient to minimize

$$\beta\Phi = -\mathscr{s}(\delta) + \beta[\langle\mathscr{H}_0\rangle + \lambda\langle\mathscr{H}_I\rangle] \tag{15.108}$$

In the infinite-temperature limit $\beta \to 0$, so that $-\mathscr{s}(\delta)$ (15.74a) must be minimized. This is minimized for $\delta_i = 1/r$ for an r-level system. In the high-temperature limit, β small, we may expand the terms $\mathscr{s}(\delta)$, $\langle\mathscr{H}_0\rangle$ and $\langle\mathscr{H}_I\rangle$ in powers of the small parameters $(\delta_i - 1/r)$. The three terms which appear in (15.108) have the following orders of magnitude:

$$-\mathscr{s}(\delta) \simeq -\ln r + \mathcal{O}\left[\left(\delta_i - \frac{1}{r}\right)^2\right]$$
$$\langle \mathscr{H}_0 \rangle \simeq \mathcal{O}\left[\left(\delta_i - \frac{1}{r}\right)^1\right] \tag{15.109}$$
$$\langle \mathscr{H}_I \rangle \simeq \mathcal{O}\left[\left(\delta_i - \frac{1}{r}\right)^2\right]$$

The last estimate for $\langle \mathscr{H}_I \rangle$ results from the assumption that linear terms in E_{ij}/N are absent from $\mathscr{H}_I$. In short, high temperatures renormalize the interaction term $\langle \mathscr{H}_I \rangle$ with respect to the unperturbed term $\langle \mathscr{H}_0 \rangle$ so that $\lambda \to \lambda' \sim \lambda\mathcal{O}[(\delta_i - 1/r)]$. To lowest order in these infinitesimals

$$\min \beta\Phi = -\ln r + \min\langle \mathscr{H}_0 \rangle \tag{15.110}$$

In short, the order parameters (other than δ_i) which determine the high-temperature behavior of Φ are identical to those that determine the low-interaction (λ small) behavior of $\langle \mathscr{H}(\lambda) \rangle$. The argument for a thermodynamic phase transition now follows by expanding Φ in a Laurent series expansion in β around $\beta = 0$. In short, if $E_g(\lambda_0)/N < E_g(0)/N$, a ground state energy phase transition must occur as λ decreases to zero with T fixed $(=0)$, and a thermodynamic phase transition must occur as T increases from zero with λ fixed $(=\lambda_0)$.

13. Structural Stability and Canonical Kernels

In the previous section we have demonstrated the connection between ground state energy phase transitions and thermodynamic phase transitions. We have said nothing about the type of phase transition encountered (first order, second order), the stability of the phase transitions under perturbation (structural stability), the terms which are important in producing phase transitions (canonical kernel), and the values of the interaction parameters and temperature at which phase transitions occur.

A great deal can be said about this spectrum of questions for thermodynamic phase transitions in models of Dicke type. We discuss thermodynamic phase transitions in particular because numerous interaction parameters may occur in this general class of models. To study ground state energy phase transitions it would first be necessary to delineate very specific models and the interaction (control) parameters which appear in them. For the study of thermodynamic phase transitions there is only one control parameter—temperature—so the arguments we present are valid for a very extensive class of model Hamiltonians.

Before describing this class of models, we first make some general considerations. The fold catastrophe A_2 is typically encountered in a 1-parameter family of functions. Such a catastrophe is associated with a zeroth-order (Delay Convention) or first-order (Maxwell Convention) phase transition. The cusp catastrophe A_{+3} is typically encountered in a 1-parameter family of functions only if an appropriate symmetry is imposed on the family. A symmetry-breaking perturbation will generally unhinge the bifurcation associated with the A_{+3} catastrophe and replace it by an A_2 catastrophe. The structural stability of thermodynamic phase transitions in Dicke models is therefore summarized as follows:

1. A zeroth- or first-order phase transition is structurally stable against all perturbations.
2. A second-order phase transition is structurally stable with respect to symmetry-preserving perturbations and unstable against symmetry-breaking perturbations. In the latter case it disappears altogether or returns at some distant point as a zeroth- or first-order phase transition.

Questions of structural stability can be answered without detailed knowledge of the models under consideration by exploiting only the symmetries of the model. To answer more detailed questions it is necessary to be somewhat more specific about the models under consideration. This is done most simply by introducing a pair of 3-component objects $\mathbf{u} = (u_3, u_+, u_-)$ and $\mathbf{v} = (v_3, v_+, v_-)$, together with three functions h, h_Q, h_C. The function $h = h(\mathbf{u}, \mathbf{v})$ is a function of the six arguments $\mathbf{u}, \mathbf{v}$. The quantum-mechanical operator $\mathscr{H}/N = h_Q$ is obtained from h through the operator substitutions indicated in Table 15.3, and subsequent symmetrization, if necessary. The classical limit $\langle \mathscr{H}/N \rangle = h_C$ is obtained from h by the c-number substitutions indicated in this table.

The following conditions are imposed on h, h_Q, h_C for the reasons indicated:[9]

1. h is a finite multinomial in all arguments, with finite coefficients. Reason: This is necessary for the rigorous proof of the classical limit relation

$$\left\langle h_Q\left(\frac{E_{ij}}{N}\right)\right\rangle = h_Q\left(\left\langle \frac{E_{ij}}{N}\right\rangle\right).$$

2. h_Q is Hermitian. Reason: This is a quantum-mechanical requirement.
3. h_C has a finite lower bound as a function of $\mu \in \mathbb{C}$ and for (r, θ, ϕ) in a sphere of radius $\frac{1}{2}$. Reason: This guarantees the existence of the limit $E_g/N = \min h_C$.
4. $h(u_3, u_+, u_-; v_3, v_+, v_-) = h(u_3, -u_+, -u_-; v_3, -v_+, -v_-)$. Reason: This symmetry suppresses the fold catastrophe A_2 and allows the cusp catastrophe A_{+3} to occur. This can be located by local differential methods.
5. h_C assumes a minimum value at $\mu = 0$ when $\mathbf{v} = 0$. Reason: In the high-temperature limit both the atomic subsystem and the field subsystem are disordered.
6. $\partial h/\partial v_3 < 0$ for $\mathbf{u} = 0, \mathbf{v} = 0$. Reason: The high-temperature state has $\theta = 0$.

Table 15.3 Operator Substitutions and c-Number Substitutions which Transform $h(u, v)$ to Operator h_Q and c-Number Function h_C (from Ref. 9)

	$h \to h_Q \equiv \dfrac{H(a, a^\dagger; J)}{N}$	$h \to h_C \equiv \dfrac{E(\mu, \mu^*; r, \theta, \phi)}{N}$	Parameter Properties
u_3	$\dfrac{n}{N} = \dfrac{a^\dagger a}{N}$	$\mu^*\mu = \dfrac{\alpha^*\alpha}{N}$	$\alpha \in \mathbb{C}, \mu \in \mathbb{C}$
u_+	$\dfrac{a^\dagger}{\sqrt{N}}$	$\mu^* = \dfrac{\alpha^*}{\sqrt{N}}$	$\alpha = \mu\sqrt{N}$
u_-	$\dfrac{a}{\sqrt{N}}$	$\mu = \dfrac{\alpha}{\sqrt{N}}$	
v_3	$\dfrac{J_3}{N} = \left(\sum_{j=1}^{N} \frac{1}{2}\sigma_j^3\right)/N$	$r \cos \theta$	$r \in [0, \frac{1}{2}]$
v_+	$\dfrac{J_+}{N} = \left(\sum_{j=1}^{N} \sigma_j^+\right)\Big/N$	$v^* = r \sin \theta e^{+i\phi}$	$\theta \in [0, \pi]$ $\phi \in [0, 2\pi]$
v_-	$\dfrac{J_-}{N} = \left(\sum_{j=1}^{N} \sigma_j^-\right)\Big/N$	$v = r \sin \theta e^{-i\phi}$	$(\theta, \phi) \in S^2$

Example. The Dicke model is obtained from

$$h(\mathbf{u}, \mathbf{v}) = -\hbar\omega u_3 + \varepsilon v_3 + \lambda(u_+ v_- + u_- v_+) \tag{15.111}$$

using the operator substitutions of Table 15.3

The high-temperature behavior for this class of models is determined by minimizing Φ or $\beta\Phi$ at constant temperature:

$$\frac{F}{N} = \min h_C - \beta^{-1}\mathscr{s}(r) \tag{15.112}$$

By assumptions 4–6 this has a minimum at $\mu = v = 0$ $(\theta = 0)$ at sufficiently high temperatures. This function must still be minimized with respect to r:

$$\frac{\partial}{\partial r}[h(0, 0, 0; r\cos\theta, 0, 0)_{\theta=0} - \beta^{-1}\mathscr{s}(r)] = \cos\theta\Bigg|_{\theta=0} \frac{\partial h}{\partial v_3}\Bigg|_{v_3=r}$$

$$+ \beta^{-1}\ln\left(\frac{1+2r}{1-2r}\right) = 0 \tag{15.113}$$

This provides a relation between r and T.

The branch $\mu = 0$, $\nu = 0$ is always a critical point of the function $\Phi(\beta)$. Bifurcations from this solution set ("thermal branch") are determined by looking for degenerate critical points along this branch. Such points are determined by isolating the terms of degree 2 in the small terms μ, μ^*, ν, ν^* in the Taylor series expansion of h_C around the thermal branch. Such terms can only arise from terms in h which are:

1. Linear in u_3
2. Degree 2 in $u_\pm$, $v_\pm$
3. Arbitrary degree in v_3

It is therefore sufficient to look at the germ of h around $\mathbf{u} = 0$, $\mathbf{v} = (v_3, 0, 0)$ of degree 2 in $u_\pm$, $v_\pm$ and of degree 1 in u_3. This germ has the form

$$h_{CK} = h + u_3 \frac{\partial h}{\partial u_3} + \tfrac{1}{2} M^\dagger A M + r M^\dagger B D + \tfrac{1}{2} r^2 D^\dagger C D \tag{15.114}$$

where

$$M = \begin{bmatrix} u_- \\ u_+ \end{bmatrix}, \quad D = \begin{bmatrix} v_- \\ v_+ \end{bmatrix}, \quad A = \begin{bmatrix} \dfrac{\partial^2 h}{\partial u_+ \, \partial u_-} & \dfrac{\partial^2 h}{\partial u_+ \, \partial u_+} \\ \dfrac{\partial^2 h}{\partial u_- \, \partial u_-} & \dfrac{\partial^2 h}{\partial u_- \, \partial u_+} \end{bmatrix} \tag{15.115}$$

All derivatives are evaluated at $\mathbf{u} = (0, 0, 0)$ and $\mathbf{v} = (v_3, 0, 0)$. From this function it is possible to obtain an operator $(h_{CK})_Q$ by the operator substitutions of Table 15.3. This operator may justifiably be called the **canonical kernel** of the original Hamiltonian operator h_Q, for it determines the bifurcations of ordered states from the thermal branch of h_Q.

The classical limit of this canonical kernel is constructed from the c-number substitutions given in Table 15.3. The germ of the classical limit is easily obtained by c-number substitutions, being careful only to note that

$$h_C\left(0, 0, 0; r - \frac{1}{2}\frac{\nu^*\nu}{r}, 0, 0\right) = h_C - \frac{1}{2}\left(\frac{\nu^*\nu}{r}\right)\frac{\partial h_C}{\partial v_3} \tag{15.116}$$

where again all derivatives are evaluated at $\mathbf{u} = (0, 0, 0)$ and $\mathbf{v} = (r, 0, 0)$. To second order, the function h_C can be written ($N = rD$)

$$h_C = h + \tfrac{1}{2} M^\dagger \left(\frac{\partial h}{\partial u_3} I_2 + A\right) M + M^\dagger B N + \tfrac{1}{2} N^\dagger \left(-\frac{1}{2r}\frac{\partial h}{\partial v_3} I_2 + C\right) N \tag{15.117}$$

This result can be expressed in terms of the single pair of order parameters μ, μ^* or v, v^* by standard manipulations:

$$h_C - h = \tfrac{1}{2}M^\dagger\left\{\frac{\partial h}{\partial u_3} I_2 + A - B^\dagger\left(-\frac{1}{2r}\frac{\partial h}{\partial v_3} I_2 + C\right)^{-1} B\right\}M \tag{15.118a}$$

$$= \tfrac{1}{2}N^\dagger\left\{-\frac{1}{2r}\frac{\partial h}{\partial v_3} I_2 + C - B^\dagger\left(\frac{\partial h}{\partial u_3} I_2 + A\right)^{-1} B\right\}N \tag{15.118b}$$

The condition for bifurcation is that an eigenvalue of either of these matrices vanishes. This condition determines a critical value r_c at which a bifurcation from the thermal branch occurs. The corresponding critical temperature is determined from the minimization condition (15.113):

$$-\left.\frac{\partial h}{\partial v_3}\right|_{r_c} = kT_c \ln \frac{1 + 2r_c}{1 - 2r_c} \tag{15.119}$$

Example. The function h and its canonical kernel

$$h = -\hbar\omega u_3 + \lambda(u_+ v_- + u_- v_+) + \kappa v_+ v_- + \gamma v_+^2 v_-^2 \tag{15.120}$$

$$h_{CK} = -\hbar\omega u_3 + \lambda(u_+ v_- + u_- v_+) + \kappa v_+ v_- \tag{15.121}$$

determine a second-order phase transition from (15.118) with

$$\frac{\partial h}{\partial u_3} = \hbar\omega, \qquad \frac{\partial h}{\partial v_3} = -\varepsilon \tag{15.122}$$

$$A = 0I_2, \qquad B = \lambda I_2, \qquad C = \kappa I_2 \tag{15.122}$$

The critical temperature is determined by (15.119), with r_c defined by

$$r_c = \frac{\varepsilon\hbar\omega}{\lambda^2 - \kappa\hbar\omega} \tag{15.123}$$

14. Dynamical Equations of Motion

All our efforts so far have been directed toward the study of static properties of the class of Hamiltonians introduced in Sec. 3. We turn our attention in this and the following section to the study of dynamical properties of these Hamiltonians. In this section we derive the dynamical equations of motion for models of MGL type. In the following section we investigate the steady-state properties of models of Dicke type.

We have studied static properties (E_g/N, $F(\beta)/N$) by means of a variational principle. The dynamical equations of motion can also be determined by a variational principle. This involves the "minimization" of the action integral

$$\mathcal{S} = \int_{t_1}^{t_2} \mathcal{L}(q, \dot{q}, t)dt \tag{15.124}$$

The Euler–Lagrange equations of motion are derived from this variational principle:

$$\frac{d}{dt}\left(\frac{\partial \mathscr{L}}{\partial \dot{q}_i}\right) - \frac{\partial \mathscr{L}}{\partial q_i} = 0 \tag{15.125}$$

A suitable Lagrangian for the purposes of obtaining a system of quantum-dynamical equations of motion is[18]

$$\mathscr{L}(q, \dot{q}, t) = \left\langle \Psi \left| i\frac{\partial}{\partial t} - \mathscr{H} \right| \Psi \right\rangle \tag{15.126}$$

The generalized coordinates which appear in $\mathscr{L}$ are the coordinates that parameterize the class of trial states $|\Psi\rangle$ used in this variational formulation.

The expectation value $\langle\Psi|\mathscr{H}|\Psi\rangle$ has already been estimated for the class of Hamiltonians under consideration (MGL). It therefore remains only to compute the expectation value of $i\partial/\partial t$. This is most simply done by observing that the N-particle state is the direct product of single particle wave functions:

$$|\Psi\rangle = \prod_{\alpha=1}^{N} |\psi_\alpha\rangle \tag{15.127}$$

The expectation value of $i\partial/\partial t$ for a single particle whose wave function is

$$|\psi_\alpha\rangle = \text{col}[z_1, z_2, \ldots, z_r] \tag{15.128}$$

is

$$\langle\psi_\alpha| i\frac{\partial}{\partial t} |\psi_\alpha\rangle = i\sum_{j=1}^{r} z_j^* \dot{z}_j = \frac{i}{2}\sum_{j=2}^{r} z_j^* \dot{z}_j - z_j \dot{z}_j^* \tag{15.129}$$

The last equality has been obtained by using the normalization condition

$$|z_1|^2 + |z_2|^2 + \cdots + |z_r|^2 = 1 \tag{15.130}$$

If each of the N-particle wave functions is identical, then the expectation value of $i\partial/\partial t$ in the N-particle state is just N times the single particle expectation value. The intensive Lagrangian is therefore

$$l(z, \dot{z}, t) = \frac{\mathscr{L}}{N} = \frac{i}{2}\sum_{j=2}^{r} z_j^* \dot{z}_j - z_j \dot{z}_j^* - h_C(z^*, z) \tag{15.131}$$

The equations of motion are determined from

$$\begin{aligned} \frac{1}{N}\frac{\partial \mathscr{L}}{\partial \dot{z}_j} &= \frac{i}{2} z_j^* \\ \frac{1}{N}\frac{\partial \mathscr{L}}{\partial z_j} &= -\frac{i}{2}\dot{z}_j^* - \frac{\partial h_C}{\partial z_j} \end{aligned} \tag{15.132}$$

They are

$$i\dot{z}_j^* = -\frac{\partial h_C}{\partial z_j}$$
$$-i\dot{z}_j = -\frac{\partial h_C}{\partial z_j^*} \tag{15.133}$$

What do these equations mean and how can they be interpreted? Both questions can be satisfactorily answered by transforming to "Cartesian coordinates" $z_j = (p_j - iq_j)/\sqrt{2}$. In this coordinate system the Euler–Lagrange equations of motion assume canonical Hamiltonian form, with h_C as the classical Hamiltonian:

$$\dot{p}_j = -\frac{\partial h_C}{\partial q_j}, \qquad 2 \leq j \leq r$$
$$\dot{q}_j = +\frac{\partial h_C}{\partial p_j}, \tag{15.134}$$

We have identified the quantum-dynamical equations of motion with the canonical equations of motion for a classical system. Therefore the entire body of classical mechanics theory is available for the description and interpretation of these quantum-mechanical equations. In particular, Hamiltonian motion is conservative. As a result, the equations of motion describe trajectories in the surface defined by $h_C = E$. For r-level systems, h_C is a function of $2(r-1)$ variables, so h_C is a $2(r-1)$-dimensional surface in $\mathbb{R}^{2(r-1)+1}$. The flat space E = constant in $\mathbb{R}^{2r-1}$ has dimension $2(r-1)$. The intersection (if there is one) of these two manifolds in $\mathbb{R}^{2r-1}$ has dimension $2r-3$. For 2-level systems this intersection is 1-dimensional, so the intersection completely characterizes the topological properties of the orbits.

We have described in detail in Chapter 9 the telltale signs that a catastrophe is present. These catastrophe fingerprints appear in the study of gradient dynamical systems. Hamiltonian systems are conservative, so it may be wondered if analogous catastrophe flags are encountered during a phase transition. The answer is "yes," as is now demonstrated in the context of the MGL model. We consider trajectories which lie a fixed energy δ above the ground state energy. The classical limit of h_Q is

$$\langle h_Q \rangle = \tfrac{1}{2}(\varepsilon + V)p^2 - \tfrac{1}{4}Vp^4 + \tfrac{1}{2}(\varepsilon - V)q^2 + \tfrac{1}{4}Vq^4 \tag{15.135}$$

The global minimum occurs at $(p, q) = (0, 0)$ when $|V| < \varepsilon$. The curves $\langle h_Q \rangle = \delta$ have the shape of ellipses which become increasingly elongated as $|V|$ approaches ε. When $|V|$ exceeds ε, the ground state no longer has $E_g/N = 0$ but rather $E_g/N = -(|V| - \varepsilon)^2/4|V|$. The curves determined by

$$\langle h_Q \rangle = -\frac{(|V| - \varepsilon)^2}{4|V|} + \delta \tag{15.136}$$

become increasingly pinched toward the origin as $|V|$ continues to increase. Eventually the energy difference between the saddle at $(p, q) = (0, 0)$ and the two global minima becomes equal to δ. The trajectory then assumes the form of a figure eight. For larger interaction strengths the orbits are disconnected, with one roughly circular closed orbit around each of the two minima. This change in the qualitative properties of a trajectory at a second-order phase transition is illustrated in Fig. 15.7.

This phenomenon of orbital elongation, pinching, and eventual fission which accompanies a second-order phase transition in a conservative dynamical system is called **critical elongation**. It is one of the analogs for conservative dynamical systems of the catastrophe flags which exist for dissipative gradient dynamical systems.

We close this section by indicating that the difference between conservative and dissipative dynamical systems "is a rotation through 90°." If $\mathscr{L} = \langle i(\partial/\partial t) - \mathscr{H}\rangle$ is the Lagrangian for a conservative system, then we may surmise that $\text{Ly} = \langle \pm(\partial/\partial t) - \mathscr{H}\rangle$ is the Lyapunov function for a corresponding

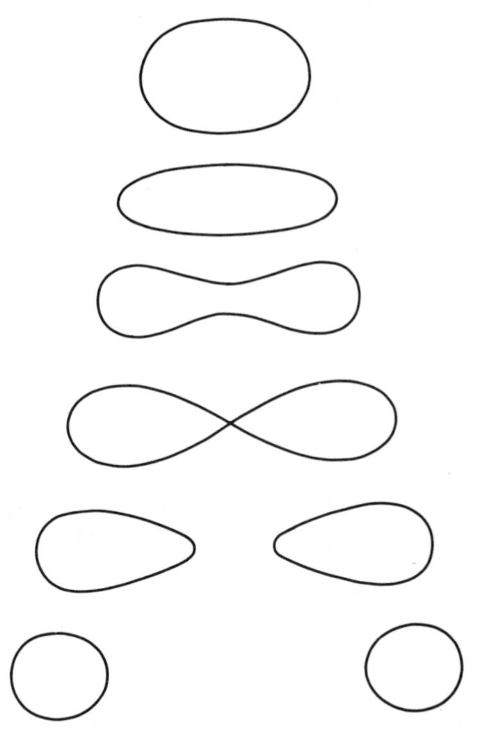

Figure 15.7 Trajectories of time-dependent Hartree-Fock orbits (in order parameter space) with an excitation energy δ above the ground state energy undergo a critical elongation, followed by a fission, as the interaction strength increases and a second-order ground state energy phase transition takes place.

dissipative system. Choosing the negative sign ($i(\partial/\partial t) \to e^{i\pi/2} i(\partial/\partial t)$) results in the following dynamical system equations of motion:

$$\begin{aligned} \dot{p}_j &= -\frac{\partial h_C}{\partial p_j} \\ \dot{q}_j &= -\frac{\partial h_C}{\partial q_j} \end{aligned} \tag{15.137}$$

For complicated model Hamiltonians $h_Q(E_{ij}/N)$, the classical function h_Q $(\langle E_{ij}/N \rangle)$ can be computed and its local minima determined by integrating the gradient system equations derived from the Lyapunov function. Once the locations of the local minima are known, the shape of the trajectories in the neighborhood of these minima can be determined by integrating the Hamiltonian equations of motion. The dissipative trajectories determined from the Lyapunov function exhibit all the usual catastrophe flags. They are also orthogonal to the conservative trajectories determined from the Lagrangian function. This orthogonality implies steepest descent. It also implies that for each catastrophe flag that occurs for trajectories of gradient dynamical systems there is a corresponding (dual) catastrophe flag for conservative systems.

15. Steady States Far from Equilibrium

Our first example of nonequilibrium phenomena involved conservative dynamical systems. Our second example will involve dissipative static systems. A system can be maintained in a steady state far from thermodynamic equilibrium when energy is fed into it and degraded as it passes through the system. In this process entropy is produced and carried off.

Variational procedures have been effective for computing the ground state energy per particle, the free energy per particle, and the dynamical equations of motion for the classes of Hamiltonians described in Sec. 3. At present there is no[19] general variational principle for determining the steady states of a system far from thermodynamic equilibrium, despite repeated claims to the contrary.[20–22]

The Lagrangian formulation described in the previous section is not applicable to the dissipative systems under consideration. The Lyapunov method can be used to determine the system steady state, provided some phenomenological assumptions are "put in by hand." The appropriate skullduggery will be specified below. In the meantime we will study the dynamical properties of systems through the Heisenberg equations of motion for the expectation values of operators.

If $\mathcal{O}$ is an operator and $\langle \mathcal{O} \rangle$ is its expectation value, then the time evolution of $\langle \mathcal{O} \rangle$ is governed by the Heisenberg equation of motion:

$$i\hbar \frac{d}{dt} \langle \mathcal{O} \rangle = \langle [\mathcal{O}, \mathcal{H}] \rangle + \left\langle i\hbar \frac{\partial \mathcal{O}}{\partial t} \right\rangle \tag{15.138}$$

If the operator $\mathcal{O}$ is not explicitly time dependent, the last term in (15.138) vanishes. All operators considered in this section are explicitly time independent, so $\partial\mathcal{O}/\partial t = 0$.

Under conditions of thermodynamic equilibrium the time derivative $d\langle \mathcal{O}\rangle/dt = 0$, so that $[\mathcal{O}, \mathscr{H}]$ has zero expectation value. If the operators $\mathcal{O}$ are allowed to range over all shift operators that occur in extended Dicke models (a_{ji}, E_{ji}, $j \neq i$), the set of expectation values $\langle[\mathcal{O}, \mathscr{H}]\rangle$ so obtained is identical to the system of equations obtained by performing the variation

$$\delta e^{-\beta F} = \delta \operatorname{tr} e^{-\beta\mathscr{H}} = 0 \tag{15.139}$$

over all the order parameters (μ_{ji}, $\langle E_{ji}/N\rangle, j \neq i$).

To obtain a phenomenological equation of motion for the expectation value $\langle \mathcal{O}\rangle$ in the presence of dissipation, we assume the following:

1. External constraints "clamp" the expectation value of $\mathcal{O}$ to some value $\langle \mathcal{O}\rangle_c$ which is at the control of the experimenter.
2. If the system is perturbed, the expectation value of $\mathcal{O}$ will relax back to its unperturbed value with a damping constant $\gamma_{\mathcal{O}}$, that is, the perturbation decays away like $\delta\langle \mathcal{O}\rangle e^{-\gamma_{\mathcal{O}} t}$.

The Heisenberg equation of motion for $\langle \mathcal{O}\rangle$, valid for a conservative system, can be modified to accommodate these phenomenological assumptions in a very straightforward way

$$i\hbar \frac{d}{dt}\langle \mathcal{O}\rangle \rightarrow i\hbar\left(\frac{d}{dt} + \gamma_{\mathcal{O}}\right)(\langle \mathcal{O}\rangle - \langle \mathcal{O}\rangle_c) \tag{15.140}$$

If $\mathcal{O}$ is not Hermitian, $\langle \mathcal{O}\rangle$ need not be real. In this case $\langle \mathcal{O}\rangle$ may be time dependent, even though its modulus is constant. It is useful to remove any fast time dependence which occurs in $\langle \mathcal{O}\rangle$ by exhibiting the time-dependent phase factor explicitly:

$$\langle \mathcal{O}\rangle = \langle \tilde{\mathcal{O}}\rangle e^{-i\omega t} \tag{15.141}$$

The value of the angular velocity ω can be determined by inspecting the Heisenberg equation of motion. If

$$\langle[\mathcal{O}, \mathscr{H}]\rangle = \hbar\omega\langle \mathcal{O}\rangle + \text{other terms} \tag{15.142}$$

then the coefficient $\hbar\omega$ of $\langle \mathcal{O}\rangle$ on the right-hand side of (15.142) may be identified with the angular velocity $\hbar\omega/\hbar$.

The steady-state behavior of the system is obtained by setting all time derivatives equal to zero. This is possible only after the rapid time dependence has been removed from the expectation values of non-Hermitian operators. The phenomenological equation for the operator expectation value $\langle \mathcal{O}\rangle$ (dropping tildes) is therefore

$$-i\hbar\gamma_{\mathcal{O}}\langle \mathcal{O}\rangle_c = -i\hbar\gamma_{\mathcal{O}}\langle \mathcal{O}\rangle + \text{other terms} \tag{15.143}$$

This equation can be obtained directly from the Heisenberg equation of motion for $\langle \mathcal{O} \rangle$ by a very simple algorithm:

$$\begin{array}{llll} 1. & i\hbar(d/dt)\langle \mathcal{O}\rangle = & \hbar\omega\langle \mathcal{O}\rangle & + \text{ other terms} \\ & \downarrow \text{phenomenological} & \downarrow \text{phenomenological} & \downarrow \text{identical} \\ 2. & -i\hbar\gamma_{\mathcal{O}}\langle \mathcal{O}\rangle_c = & -i\hbar\gamma_{\mathcal{O}}\langle \mathcal{O}\rangle & + \text{ other terms} \end{array} \tag{15.144}$$

Remark 1. The far-from-equilibrium state of many systems can be characterized by order parameters which are the expectation values of non-Hermitian operators. Under equilibrium conditions these expectation values are zero, and they remain zero under gentle displacements from equilibrium. Nonzero values for the expectation values of these shift operators are associated with phase transitions. Such transitions remove the system from the "thermodynamic equilibrium branch" to a different branch which is ordered.

Remark 2. The equations of state for a system in thermodynamic equilibrium are homogeneous while the equations for a driven dissipative system are inhomogeneous. The inhomogeneities are due entirely to the phenomenological terms of the form $\gamma_{\mathcal{O}}\langle \mathcal{O}\rangle_c$.

We are particularly interested in describing the steady state of a laser. A laser is an extremely useful far-from-equilibrium system which exhibits a phase transition when energy is pumped in faster than it can be dissipated by atomic relaxation processes alone. In the past, Dicke models have been a convenient tool for the study of lasers. We therefore study the far-from-equilibrium properties of extended Dicke models.

For these models

$$\begin{aligned} \mathcal{H} = & \sum_{1\le i<j}^{r} \hbar\omega_{ji} a_{ji}^{\dagger} a_{ji} + \sum_{i=1}^{r} \varepsilon_i E_{ii} \\ & + \frac{1}{\sqrt{N}} \sum_{1\le i<j}^{r} \lambda_{ji} a_{ji}^{\dagger} E_{ij} + \lambda_{ji}^{*} a_{ji} E_{ji} \end{aligned} \tag{15.145}$$

The commutator of $\mathcal{H}$ with a collective operator, an intensive collective operator, or a single particle operator is an operator of the same type. If we choose as system state variables the expectation values of single particle operators and intensive operators $\mu_{ji} = \langle a_{ji}/\sqrt{N}\rangle$, $\nu_{ij} = \langle E_{ij}/N\rangle = \langle e_{ij}^{(\alpha)}\rangle$, then the Heisenberg equations of motion for these operators are

$$\begin{aligned} i\hbar\dot{\mu}_{ji} &= -\hbar\omega_{ji}\mu_{ji} + \lambda_{ji}\nu_{ij} \\ i\hbar\dot{\nu}_{ij} &= (\varepsilon_j - \varepsilon_i)\nu_{ij} + \lambda_{ji}^{*}\mu_{ji}(\nu_{ii} - \nu_{jj}) \\ i\hbar\dot{\nu}_{ii} &= \lambda_{ji}\mu_{ji}^{*}\nu_{ij} - \lambda_{ji}^{*}\mu_{ji}\nu_{ij}^{*} \end{aligned} \tag{15.146}$$

The phenomenological equations of motion are obtained directly from the algorithm above. In obtaining these equations we make the following assumptions:

1. The damping rates associated with $\mu_{ji}, \nu_{ij}, \nu_{ii}$ are $\tilde{\gamma}_{ji}, \gamma_{ij}, \gamma_{ii}$.
2. The pump mechanism changes the occupation probabilities of the atomic states but preserves no phase information. Then the clamped expectation values of the shift operators are zero. In the case of either strong dissipation or weak coupling, $\nu_{ii} \simeq (\nu_{ii})_c$ is the occupation probability of the ith single particle state.

Under these assumptions the phenomenological equations take the form

$$0 = -i\hbar\tilde{\gamma}_{ji}\mu_{ji} + \lambda_{ji}\nu_{ij} \tag{15.147i}$$

$$0 = -i\hbar\gamma_{ij}\nu_{ij} + \lambda_{ji}^*\mu_{ji}(\nu_{ii} - \nu_{jj}) \tag{15.147ii}$$

$$\begin{aligned} -i\hbar\gamma_{ii}(\nu_{ii})_c = -i\hbar\gamma_{ii}\nu_{ii} &+ \left\{\sum_{k>i}\lambda_{ki}\mu_{ki}^*\nu_{ik} - \sum_{k<i}\lambda_{ik}\mu_{ik}^*\nu_{ki}\right\} \\ &- \left\{\sum_{k>i}\lambda_{ik}^*\mu_{ik}\nu_{ik} - \sum_{k<i}\lambda_{ki}^*\mu_{ki}\nu_{ki}\right\} \end{aligned} \tag{15.147iii}$$

This system of equations can be treated by the following steps:

1. Solve (15.147i) for μ_{ji} in terms of ν_{ij}.
2. Use this relation to eliminate μ_{ji} from (15.147ii) and (15.147iii).
3. Solve (15.147iii) for $(\nu_{ii} - \nu_{jj})$ in terms of the order parameters ν_{ij} and the pump parameters $(\nu_{ii})_c - (\nu_{jj})_c$.
4. The resulting systems of equations of state are then

$$\nu_{ij}\{\hbar^2\gamma_{ij}\tilde{\gamma}_{ji} + |\lambda_{ji}|^2(\nu_{ii} - \nu_{jj})\} = 0, \qquad i < j \tag{15.148i}$$

$$\begin{aligned} \nu_{ii} - \nu_{jj} = (\nu_{ii})_c - (\nu_{jj})_c &+ 2\sum_k \theta(k, i)\frac{|\lambda_{ik}\nu_{ki}|^2}{\hbar^2\tilde{\gamma}_{ki}\gamma_{ii}} \\ &- 2\sum_k \theta(k, j)\frac{|\lambda_{jk}\nu_{kj}|^2}{\hbar^2\tilde{\gamma}_{kj}\gamma_{jj}} \end{aligned} \tag{15.148ii}$$

The summation convention is not used in these expressions. All sums are explicitly shown. The function θ has the properties

$$\begin{aligned} \theta(k, i) &= -1, \qquad k < i \\ &= 0, \qquad k = i \\ &= +1, \qquad k > i \end{aligned} \tag{15.149}$$

Under conditions of thermodynamic equilibrium we assume that all shift operators have expectation value zero. If the system state is ordered in the absence of pumping, assumption 2 on p. 415 must be modified. The expectation values v_{ii} are then occupation probabilities given by

$$\begin{aligned} v_{ii} &= \frac{e^{-\beta\varepsilon_i}}{z(\beta)} \\ z(\beta) &= \sum_{i=1}^{r} e^{-\beta\varepsilon_i} \end{aligned} \tag{15.150}$$

The function $v_{ii} - v_{jj}$ is then positive. When the pump is turned on, the excited states become more high populated at the expense of lower lying levels. However, all order parameters v_{ij} remain zero until a population inversion is established between some level pair. Specifically, bifurcation of an ordered state $v_{ij} \neq 0$ can occur at a pump strength defined by

$$(v_{jj})_c - (v_{ii})_c = \frac{\hbar^2 \gamma_{ij} \tilde{\gamma}_{ji}}{|\lambda_{ji}|^2} = \left(\frac{\hbar\gamma_{ij}}{\lambda_{ij}}\right)\left(\frac{\hbar\tilde{\gamma}_{ji}}{\lambda_{ji}}\right) \tag{15.151}$$

If we put $r = 2$, we recover the phenomenological theory of a laser based on the Dicke model developed by Haken and others.

The r-level Dicke model predicts a rich variety of steady-state behavior. If only one coupling constant λ_{ji} is nonzero, a second-order phase transition can occur if the population inversion can be made sufficiently large. This can often be done by pumping a selected transition in a cooled system. The description of this phase transition differs in no way from that of the 2-level Dicke model, described in detail by Poston and Stewart.[23] In the event that more than one coupling constant is nonzero, first-order phase transitions are possible. Under the Delay Convention these transitions are located by the bifurcations of non-zero solutions from the solution set $[\mu_{ji}(c), v_{ij}(c)] = 0$ of (15.147) where the control parameters c are the population inversion $(v_{ii})_c$. Such phase transitions actually occur at spinodal points and are zeroth-order phase transitions. First-order phase transitions actually occur on the Maxwell set $\mathscr{S}_M$. This is well defined only if the equations (15.148) defining the steady state can be derived from a potential. If not, then with luck the first-order phase transition can be determined by a Maxwell Construction.

Once a stable solution set of (15.147) has been determined, the expectation values of all single particle operators $e_{ij}^{(\alpha)}$ are known. So also are the expectation values of all photon operators $a^\dagger a$, a, $a^\dagger$ for each mode. These expectation values can be used to compute the reduced density operators ρ_A, ρ_F for the atomic subsystem and for the field subsystem. The density operator for an r-level system is an $r \times r$ matrix. Since

$$\begin{aligned} v_{ij} &= \langle e_{ij} \rangle = \operatorname{tr} \rho_A e_{ij} \\ \rho_A &= \sum_{i,j} v_{ij} e_{ji} \end{aligned} \tag{15.152}$$

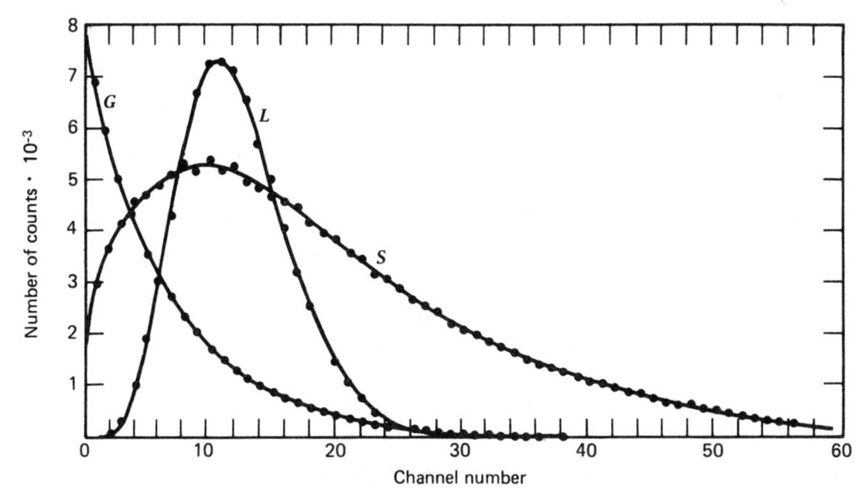

Figure 15.8 Photocount distributions from a laser below threshold (G) and far above threshold (L). (S) is a statistical mixture of light from lasers above and below threshold. Reprinted with permission from F. T. Arecchi, in: *Quantum Optics: Proceedings of the International School of Physics —Enrico Fermi* (*42*). R. J. Glauber (Ed.), N.Y.: Academic, 1969.

For a single field mode there are an infinite number of states. Nevertheless the reduced field density operator can be expressed as a function of the operator expectation values as follows:

$$\rho_F \simeq e^{M[a^\dagger a - \langle a^\dagger \rangle a - \langle a \rangle a^\dagger]} \tag{15.153}$$

where

$$\mathcal{N} = \frac{1}{e^{-M} - 1} = \langle a^\dagger a \rangle - \langle a^\dagger \rangle \langle a \rangle$$

defines the parameter M and the system noise $\mathcal{N}$ in terms of the correlation function $\langle a^\dagger a \rangle - \langle a^\dagger \rangle \langle a \rangle$.

This density operator has been used to predict a photocount distribution function for a laser operating at an arbitrary signal-to-noise ratio $\mathcal{S}/\mathcal{N}$, where $\mathcal{S} = \langle a^\dagger \rangle \langle a \rangle$. These predictions were carefully compared with photocount distributions observed for a laser operating below threshold and far above threshold. A comparison between the experimental and theoretical results which was made by Arecchi and De Giorgio[24] is shown in Fig. 15.8.

16. Multiple Stability

This chapter has been written to illustrate how Catastrophe Theory can play a useful role in the description of quantum-mechanical systems. Except for isolated remarks, we have actually not yet indicated how Catastrophe Theory can be useful. Almost the entire body of results developed so far could have been obtained by means of bifurcation theory. In this section we indicate how to go beyond bifurcation theory to obtain additional physical insights through the application of universal perturbation theory.

We have literally exploited the "Crowbar Principle" throughout this chapter. Most of the critical properties that we have studied depend on a single control parameter. Typical 1-parameter families of potentials possess at worst degenerate critical points of type A_2, and are of limited interest. We have throughout imposed symmetries on the Hamiltonian under consideration. These symmetries have been sufficient to suppress the fold catastrophe and allow the cusp catastrophes $A_{\pm 3}$ to appear naturally in the corresponding 1-parameter potentials. The consequent bifurcations were then studied as sections of these catastrophes. The remainder of the critical manifold can only be studied by introducing symmetry-breaking terms.

Exploitation of the "Crowbar Principle" for the purpose of reducing a nonlinear problem to a bifurcation problem introduces a number of computational conveniences. First, the bifurcation points are easily located. When a potential is available, these points are determined by the condition det $V_{ij} = 0$. If a potential is not available but a system of equations of motion $\dot{x}_i = F_i$ is available (previous section), bifurcation from a steady-state branch is determined from the condition det $F_{ij} = 0$. The symmetry initially present in the problem allows easier computation of all equilibrium or steady-state branches because the solutions of $\partial V/\partial x_i = 0$ or $F_i = 0$ occur in symmetry-related pairs. The stability along these branches can be determined by computing the eigenvalues of V_{ij} or F_{ij}. However, some elementary theorems obviate even this need: the stability along a bifurcating branch ($A_{\pm 3}$) is determined completely by the stability of the original branch and the type of catastrophe that occurs.

An arbitrary symmetry-breaking perturbation will often make it very difficult to determine the solution of $\nabla V = 0$ or $F = 0$. However, the qualitative properties of the perturbed system differ only adiabatically from those of the original system. Stability properties of a perturbed branch at a Morse critical point are identical to the stability properties of the unperturbed branch. Nor does the multiplicity of critical points, either equilibria or steady states, change under perturbation.* For these reasons the initial symmetry-restricted problem provides an adequate critical point skeleton which determines, by relatively simple calculations, the qualitative features of its perturbations. Second-order phase transitions which occur in symmetry-restricted models disappear under perturbation, but zeroth- and first-order phase transitions are structurally stable.

We can illustrate how catastrophe theory can enlarge our understanding of physical processes whose properties have been determined by a bifurcation analysis. This can be done in the context of the dynamical steady-state properties of the Dicke model far from thermodynamic equilibrium. If only isolated cusps occur in the equations of state (15.147), then a universal perturbation of the

* In the absence of a continuous gauge invariance.

initial quantum-mechanical model (15.145) is obtained by introducing one symmetry-breaking term for each bifurcation direction. As there are $\binom{r}{2} = r(r-1)/2$ independent-order parameters, an appropriate perturbation of this dimension will be universal. For example, we can introduce one classical external field with amplitude $\alpha_{ji}\sqrt{N}$ for each atomic resonance. Or we can introduce one classical current $j_{ji} = J_{ji}/N$ for each electromagnetic field mode. Introducing these perturbations changes the steady-state equations of motion derived from the extended Dicke model as follows:

$$\text{(15.147i):} \qquad \nu_{ij} \rightarrow \nu_{ij} + j_{ij} \tag{15.154i}$$

$$\text{(15.147ii):} \qquad \mu_{ji} \rightarrow \mu_{ji} + \alpha_{ji} \tag{15.154ii}$$

$$\text{(15.147iii):} \qquad \mu_{ji} \rightarrow \mu_{ji} + \alpha_{ji} \tag{15.154iii}$$

It is immaterial whether the symmetry breaking is introduced by means of classical fields or classical currents, or any independent combination of $r(r-1)/2$ classical fields and currents. These perturbations are equivalent on an abstract mathematical level and produce equivalent physical effects under the appropriate identification: $\alpha_{ji} = \alpha_{ji}\,(j_{ji})$.

A simplified set of equations of state can be obtained from (15.154i–iii) by following the steps leading from (15.147) to (15.148). We shall do this for perturbation $\alpha_{ji} = \alpha_{ij}^*$ arbitrary, $j_{ji} = j_{ij}^* = 0$. For diversion and more serious reasons (see below) we eliminate the order parameters ν_{ij} in favor of μ_{ji}. The equations of state are

$$\hbar^2\gamma_{ij}\tilde{\gamma}_{ji}\mu_{ji} + |\lambda_{ji}|^2(\mu_{ji} + \alpha_{ji})(\nu_{ii} - \nu_{jj}) = 0 \tag{15.155i}$$

$$\begin{aligned} \nu_{ii} - \nu_{jj} = (\nu_{ii})_c - (\nu_{jj})_c &+ 2\sum_k \theta(k,i)(\mu_{ki} + \alpha_{ki})^* \frac{\mu_{ki}\tilde{\gamma}_{ki}^*}{\gamma_{ii}} \\ &- 2\sum_k \theta(k,j)(\mu_{kj} + \alpha_{kj})^* \frac{\mu_{kj}}{\gamma_{jj}} \end{aligned} \tag{15.155ii}$$

A useful special case of these equations for r-level systems is obtained by setting $r = 2$. Then there is only one resonant mode, and the general model described here can be compared with a previously studied model.[25,26] To make this comparison, it is useful to define

$$\begin{aligned} \mu_{21} &= \mu \\ \alpha_{21} &= \alpha \\ \nu_{11} - \nu_{22} &= \sigma^z \\ \lambda_{21} &= \lambda \\ \gamma_{21} &= \gamma_\sigma \\ \gamma_{11} = \gamma_{22} &= \gamma_z \\ \tilde{\gamma}_{21} &= \gamma_a \end{aligned} \tag{15.156}$$

All these parameters may be taken real without loss of generality. Equations (15.155) reduce in this special case to

$$\hbar^2 \gamma_\sigma \gamma_a \mu + \lambda^2(\mu + \alpha)\left\{\langle\sigma^z\rangle_c + 4\mu(\mu + \alpha)\frac{\gamma_a}{\gamma_z}\right\} = 0 \tag{15.157}$$

This equation is a cubic in the order parameter μ. This is not surprising, since the initial symmetry-restricted system obeyed a cubic equation of state. This cubic can be transformed to canonical cusp catastrophe form:

(15.157): $$x^3 + Ax + B = 0 \tag{15.158i}$$

$$\begin{aligned} x &= \mu + \tfrac{2}{3}\alpha \\ A &= -\tfrac{1}{3}\alpha^2 + \frac{\gamma_z}{4\gamma_a}\left\{\langle\sigma^z\rangle_c + \frac{\hbar^2\gamma_\sigma\gamma_a}{\lambda^2}\right\} \\ B &= -\tfrac{2}{27}\alpha^3 - \frac{\alpha\gamma_z}{12\gamma_a}\left\{\langle\sigma^z\rangle_c + \frac{\hbar^2\gamma_\sigma\gamma_a}{\lambda^2}\right\} + \frac{\hbar^2\gamma_\sigma\gamma_a}{4\lambda^2}\alpha \end{aligned} \tag{15.158ii}$$

Equation (15.158) provides a universal perturbation of the equation of state for the symmetry-restricted 2-level laser. The physical control parameters are the population inversion $\langle\sigma^z\rangle_c$ and the external field α. The pump mechanism represented by the term $\langle\sigma^z\rangle$ destroys phase information, while the external field represented by the term α preserves phase information. These physical controls may therefore be interpreted as incoherent and coherent pump rates, respectively.

The equation of state (15.157) can be studied in terms of its canonical representation (15.158). The transformation between the physical control parameters $\langle\sigma^z\rangle_c$, α and the mathematical control parameters A, B is invertible, at least locally. Since the unique state in the single valued regime is stable, the upper and lower sheets in the multivalued regime are stable, and the central sheet is unstable. Any point on the stable part of the critical manifold can be reached by appropriate choice of the controls A, B, or $\langle\sigma^z\rangle_c$, α (and process, if necessary), at least mathematically. The physically accessible region is restricted by the constraint $-1 \leq \langle\sigma^z\rangle_c \leq +1$.

In a typical physical process, only one physical control parameter is varied at a time. For example, the external field may be absent ($\alpha = 0$). As the incoherent pump rate is increased, the system steady state remains disordered ($x = \mu = 0$) until a population inversion defined by

$$\langle\sigma^z\rangle_c = \langle e_{11}\rangle_c - \langle e_{22}\rangle_c = -\frac{\hbar^2\gamma_\sigma\gamma_\alpha}{\lambda^2} \tag{15.159}$$

Under these conditions the occupation probability of the excited state $|2\rangle$ exceeds that of the ground state $|1\rangle$ (recall $\varepsilon_2 > \varepsilon_1$) by $\hbar_2\gamma_\sigma\gamma_a/\lambda^2$, so a population inversion is required to initiate an ordered steady state (laser action). As the occupation probability difference cannot exceed 1, the condition for an order-

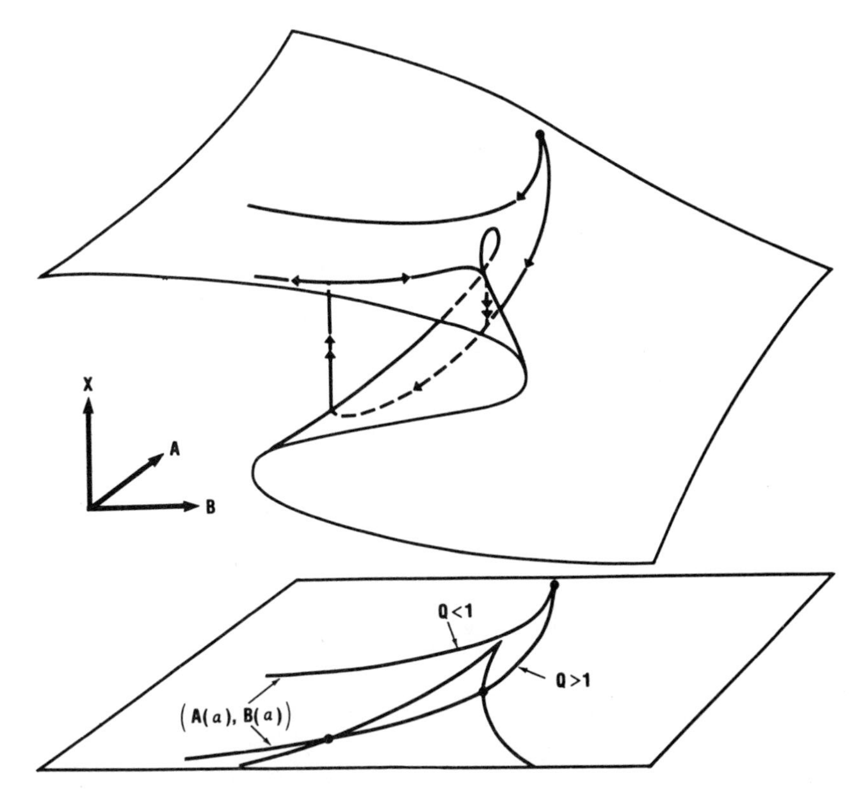

Figure 15.9 When the population inversion is held below threshold and the classical field α is increased, the control parameter trajectory follows a curved path. The state of the laser is determined by lifting this path to the cusp catastrophe manifold. A zeroth-order phase transition occurs if $Q > 1$ when the control parameter trajectory crosses the appropriate fold curve.

disorder phase transition to occur is $\lambda^2 > \hbar^2\gamma_\sigma\gamma_a$. No phase transition occurs if the coupling λ is too small or the dissipation $(\gamma_\sigma, \gamma_a)$ is too large.

If the population inversion $\langle\sigma^z\rangle_c$ is held constant and the external field is increased from zero, the trajectory $A(\langle\sigma^z\rangle_c, \alpha), B(\langle\sigma^z\rangle_c, \alpha)$ in control parameter space follows curved paths as shown in Fig. 15.9.

If the incoherent pump rate is held below threshold, then the trajectory may pass to the left of the cusp point or to its right, depending on the value of the ratio $Q \equiv \lambda^2\langle\sigma^z\rangle_c/8\hbar^2\gamma_\sigma\gamma_a$. If $Q < 1$, the trajectory passes to the left of the cusp and no phase transition takes place. If $Q > 1$, the trajectory hooks to the right around the cusp point and eventually crosses the left-hand fold line. These two qualitatively distinct types of trajectories have a separatrix defined by $Q = 1$. This trajectory goes through the cusp point.

In the case $Q > 1$, phase transitions occur when the incoherent pump rate $\langle\sigma^z\rangle_c$ is held fixed, or in particular is absent, and the coherent pump rate α is

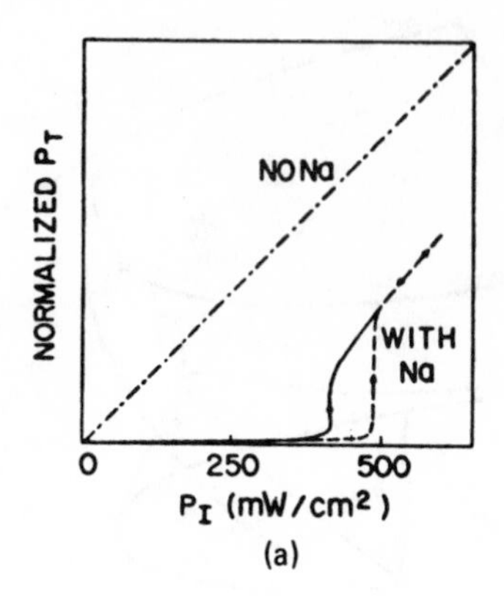

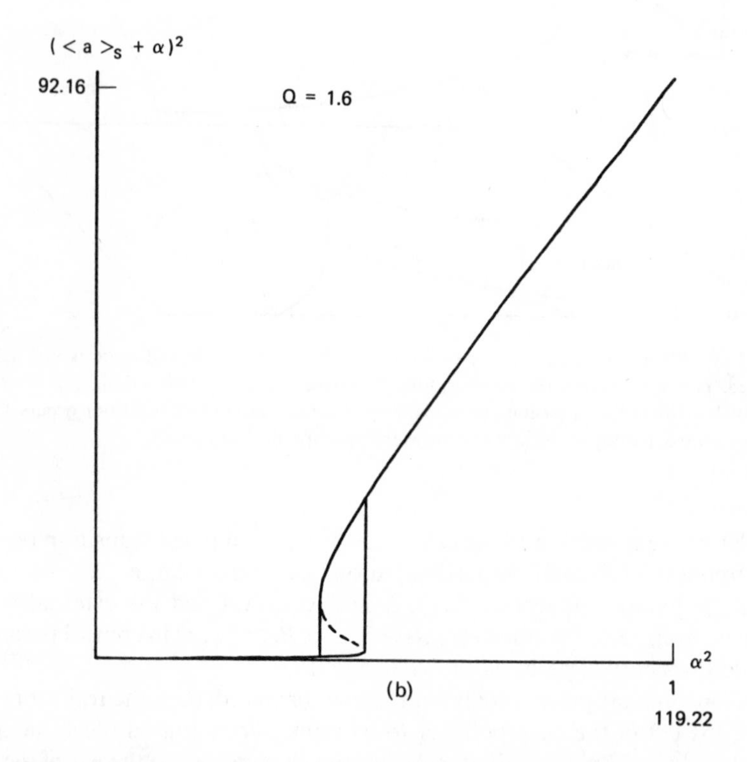

Figure 15.10 (*a*) Experimental data of Gibbs, McCall, and Venkatesan show a hysteresis cycle, Reprinted with permission from H. M. Gibbs, S. L. McCall, and T. N. C. Venkatesan, *Phys. Rev. Lett* **36** (1976), p. 1136. Copyright 1976 by American Institute of Physics. (*b*) Laser output intensity (ordinate) as a function of classical field input (abscissa) according to the model presented in (15.158), with $Q = 1.6$.

increased. When α is increased and the control point (A, B) crosses the left-hand fold curve, a zeroth-order phase transition occurs as the system state jumps from the lower sheet to the upper sheet. If α is then decreased, the system state must return to the lower sheet when the right-hand fold curve is crossed. If the Delay Convention is applicable, the system exhibits hysteresis.

Experiments of this nature have been performed by Gibbs, McCall, and Venkatesan.[27] Sodium vapor in a Fabry–Perot cavity tuned to one of the D lines was irradiated with a tunable dye laser which was slightly detuned from the resonance. The intensity of the signal emitted from the cavity was then measured as a function of the input intensity. This involves a comparison of $(\mu + \alpha)^2 = (x + \frac{1}{3}\alpha)^2$ with α^2. A comparison of the experimental curves with the predictions based on this model is presented in Fig. 15.10. This comparison is made for the ratio $Q = 1.6$.

The laser steady state is described by a point on the cusp catastrophe critical manifold under any combination of incoherent and coherent pumping. Every point in this manifold determines expectation values for the atomic operators and the field operators. These expectation values can be used to construct the reduced density operators for the atomic and field subsystems. Thus every point on the cusp catastrophe manifold defines a system density operator. These density operators can be used to determine fluctuation quantities, such as $\langle(\sigma^z - \langle\sigma^z\rangle)^2\rangle$. These variances become anomalously large near phase transitions (Chapter 9, Sec. 8).

The experimental data were taken with a sweep rate faster than the system equilibration time (tunneling time). In this way the spinodal lines (stability limits) surrounding the first-order phase transition are mapped out. A first-order phase transition occurs at $B = 0$. This will occur when the Maxwell Convention is applicable. This in turn will occur when the experiment cycle time is longer than the tunneling time. This time scale can be determined from the potential V associated with the equation of state (15.158) and the size of fluctuations around locally stable equilibria. The tunneling time scale is given in terms of these quantities by (8.31).

17. Summary

In this chapter we have exploited the results of Catastrophe Theory for the description of quantum-mechanical systems.

We have not treated quantum-mechanical systems in general. Rather, we have confined consideration to quantum-mechanical systems of a particular type. These systems contain a large number of identical subsystems (e.g., molecules, atoms, nucleons). While the description of these systems must be intrinsically quantum mechanical, the existence of large numbers of subsystems means that the associated intensive operators are almost commuting. To the extent that they commute, the treatment of these systems by classical techniques is a good approximation.

We have discussed two broad classes of model systems (Sec. 3). These include the MGL class of models in which only one type of system (nucleons) is present, and the Dicke class of models in which there are two interacting subsystems (atoms and field). The basic MGL and Dicke models serve as prototypes for the two general classes of models under consideration. These general models can involve multilevel systems as well as complicated polynomial interactions.

Under weak assumptions (Secs. 1, 2) the ground state energy for models of these classes can be determined by a simple recipe (Sec. 4). This involves replacing the operator Hamiltonian by its expectation value. The ground state energy can then be studied as a function of the coupling constants which appear in the Hamiltonian. In this way, ground state energy phase transitions can easily be studied (Secs. 5-8).

Thermodynamic phase transitions can be studied by a very similar method, involving the free energy instead of the ground state energy. The recipe for determining the thermodynamic system state (Sec. 9) was applied to models of both MGL and Dicke types (Secs. 10, 11) in order to discuss thermodynamic phase transitions. Systems which exhibit ground state energy phase transitions also exhibit thermodynamic phase transitions (Sec. 12). The conditions under which a second-order phase transition could be expected in any particular model of the classes studied were determined in Sec. 13.

The tools of Catastrophe Theory can be applied to nonequilibrium systems as well as equilibrium systems. The topological properties of the "orbits" of a quantum-mechanical system were determined in Sec. 14. The orbits determined by the time-dependent Hartree-Fock equations of motion exhibit critical elongation in the neighborhood of a degenerate critical point. This phenomenon is the dynamical system counterpart to critical slowing down for dissipative systems.

Dissipative systems were studied in Sec. 15. In particular, we studied the steady-state properties exhibited by Dicke models when they are pushed far from thermodynamic equilibrium. These models have often been used in the past to model laser systems. When the cusp critical point is unfolded in an appropriate way, new physical phenomena are to be expected (Sec. 16). One of these, optical bistability, was discussed.

Notes

The Dicke model was used by Haken and others to study the theoretical details of the laser phase transition.[28] These studies involved no external symmetry-breaking perturbations such as a classical external current or a satellite classical field, as introduced in Sec. 16. As a result, optical bistability was not encountered theoretically until 1976.[29] This long delay could have been avoided had the spirit and methods of Elementary Catastrophe Theory been familiar to physicists.

The thermodynamic phase transition in the Dicke model was first studied rigorously by Hepp and Lieb.[30] Their computations were simplified considerably by Wang and Hioe[31] who began the introduction of algorithms by using field coherent states. Their results, while correct, were not rigorous. An alternative algorithm, using atomic coherent states, was then introduced by Hepp and Lieb.[32] This algorithm was rigorous. Atomic and field coherent states were treated on an equal footing in the algorithm introduced by Gilmore[9] and presented in Sec. 9. The concept of canonical kernel was also introduced at that time. Once the phase transition could be studied by an algorithm involving potentials, the effect of an external symmetry-breaking perturbation could easily be studied.[33,34] These different studies of the Dicke model are summarized below:

Boundary Condition	External Symmetry-Breaking Perturbation Absent	Present
Equilibrium	1972 Ref. 30	1976 Ref. 33, 34
Nonequilibrium steady state	1959 Ref. 28	1976 Ref. 29

The close similarity in the behavior of the Dicke model subject to equilibrium and to nonequilibrium steady-state boundary conditions suggested that there might be a formal mathematical relationship ("analytic correspondence") between them. This relationship was exhibited by Gilmore and Narducci.[25,26]

This analytic correspondence has raised the following intriguing possibility. Under equilibrium boundary conditions both reduced density operators ρ_A and ρ_F have the structure

$$\rho = [\rho(\text{geometry})]^{M(\text{physics})}.$$

Here ρ(geometry) is an operator defined over the cusp catastrophe manifold. This operator has the form $\rho_{A \text{ or } F}(\text{geometry}) \simeq \exp - Nh_{A \text{ or } F}$, where the semiclassical Hamiltonians h_A, h_F are defined in (15.35). The number M(physics) depends on the system noise, and is in fact related to the system temperature ($M = \beta$). A similar Wigner–Eckart factorization exists for the density operator describing the nonequilibrium steady state of the Dicke model. The following questions remain unanswered:

1. In this case, what are the counterparts of the semiclassical Hamiltonians h_A and h_F?
2. What is the physical interpretation of M(physics)?
3. Can the nonequilibrium density operator be determined from a variational principle, as is the case under equilibrium boundary conditions?

References

1. R. Gilmore, *Lie Groups, Lie Algebras, and Some of Their Applications*, New York: Wiley, 1974.
2. R. Gilmore, The Classical Limit of Quantum Non-Spin Systems, *J. Math. Phys.* **20**, 891–893 (1979).
3. H. J. Lipkin, N. Meshkov, and A. J. Glick, Validity of Many-Body Approximation Methods for a Solvable Model: (I). Exact Solutions and Perturbation Theory, *Nucl. Phys.* **62**, 188–198 (1965).
4. D. Ruelle, *Statistical Mechanics*, New York: Benjamin, 1969.
5. R. Gilmore, Two Nonlinear Dicke Models, *Physica* **86A**, 137–146 (1977).
6. R. H. Dicke, Coherence in Spontaneous Radiation Processes, *Phys. Rev.* **93**, 99–110 (1954).
7. C. F. von Wei säcker, Zur Theorie der Kernmassen, *Z. Physik* **96**, 431–458 (1935).

8. F. T. Arecchi, E. Courtens, R. Gilmore, and H. Thomas, Atomic Coherent States in Quantum Optics, *Phys. Rev.* **A6**, 2211–2237 (1972).
9. R. Gilmore, Structural Stability of the Phase Transition in Dicke-like Models, *J. Math. Phys.* **18**, 17–22 (1977).
10. R. Gilmore and D. H. Feng, Studies of the Ground State Properties of the Lipkin-Meshkov-Glick Model via Atomic Coherent States, *Phys. Lett.* **76B**, 26–28 (1978).
11. R. Gilmore and D. H. Feng, Ground State Phase Transitions in Multilevel Extensions of the Lipkin-Meshkov-Glick Model, *Phys. Lett.* **85B**, 155–158 (1979).
12. R. Gilmore, Geometry of Symmetrized States, *Ann. Phys.* (*NY*) **74**, 391–463 (1972).
13. E. H. Lieb, The Classical Limit of Quantum Spin Systems, *Commun. Math. Phys.* **31**, 327–40 (1973).
14. R. Gilmore, Thermodynamic Phase Transition in the Dicke Model for Multilevel Systems, *J. Phys.* **A10**, L131–134 (1977).
15. R. Gilmore, S. R. Deans, and D. H. Feng, Phase Transitions and Geometric Properties of the Interacting Boson Model, *Phys. Rev.* **C** (March, 1981).
16. R. Gilmore and C. M. Bowden, Classical and Semi-classical Treatment of the Dicke Hamiltonian, in: *Cooperative Effects in Matter and Radiation* (C. M. Bowden, D. W. Howgate, and H. R. Robl, Eds.), New York: Plenum, 1977, pp. 335–355.
17. R. Gilmore and D. H. Feng, Ground State and Thermodynamic Phase Transitions of Nuclear Systems, in: *Proceedings of the International Meeting on Frontiers of Physics*, Republic of Singapore, New York: Plenum, 1980.
18. P. A. M. Dirac, Note on Exchange Phenomena in the Thomas Atom, *Proc. Camb. Phil. Soc.* **26**, 376–385 (1930).
19. M. J. Klein and P. H. E. Meijer, Principle of Minimum Entropy Production, *Phys. Rev.* **96**, 250–255 (1954).
20. I. Prigogine, *Etude Thermodynamique des Phénomènes Irreversibles*, Liege: Editions Desoer, 1947.
21. P. Glansdorff and I. Prigogine, *Thermodynamics of Structure, Stability, and Fluctuations*, New York: Wiley, 1971.
22. G. Nicolis and I. Prigogine, *Self-Organization in Nonequilibrium Systems*, New York: Wiley, 1977.
23. T. Poston and I. N. Stewart, *Catastrophe Theory and Its Applications*, London: Pitman, 1978, chap. 16.
24. F. T. Arecchi, Photocount Distributions and Field Statistics, in *Quantum Optics, Proceedings of the International Summer School of Physics "Enrico Fermi,"* Rendiconti 42, (R. J. Glauber, Ed.), New York: Academic Press, 1969, pp. 57–110.
25. R. Gilmore and L. M. Narducci, Relation Between the Equilibrium and Nonequilibrium Critical Properties of the Dicke Model, *Phys. Rev.* **A17**, 1747–1760 (1978).
26. R. Gilmore and L. M. Narducci, Laser as Catastrophe, in: *Coherence and Quantum Optics* (L. Mandel and E. Wolf, Eds.), New York: Plenum, 1977, pp. 81–91.
27. H. M. Gibbs, S. L. McCall, and T. N. C. Venkatesan, Differential Gain and Bistability Using a Sodium Filled Interferometer, *Phys. Rev. Lett.* **36**, 1135–1138 (1976).
28. H. Haken, Laser Theory, in: *Handbuch der Physik*, vol. XXV/2c, Berlin: Springer-Verlag, 1970.
29. R. Bonifacio and L. A. Lugiato, Cooperative Effects and Bistability for Resonance Fluorescence, *Optics Commun.* **19**, 1972–1976 (1976).
30. K. Hepp and E. H. Lieb, On the Superradiant Phase Transition for Molecules in a Quantized Radiation Field: The Dicke Maser Model, *Ann. Phys.* (*NY*) **76**, 360–404 (1973).

31. Y. K. Wang and F. T. Hioe, Phase Transitions in the Dicke Model of Superradiance, *Phys. Rev.* **A7**, 831–836 (1973).

32. K. Hepp and E. H. Lieb, Equilibrium Statistical Mechanics of Matter Interacting with the Quantized Radiation Field, *Phys. Rev.* **A8**, 2517–2525 (1973).

33. J. P. Provost, F. Rocca, G. Vallee, and M. Siruge, Lack of Phase Transition in the Dicke Model with External Fields, *Physica* **85A**, 202–206 (1976).

34. R. Gilmore, Persistence of the Phase Transition in the Dicke Model with External Fields and Counter-Rotating Terms, *Phys. Lett.* **55A**, 459–60 (1976).

CHAPTER 16

Climate

In this Chapter we indicate how the methods of Catastrophe Theory may be used to illuminate a profoundly important problem.

1. Prologue

The earth's surface records a history of many previous glacial epochs written in the language of geology.[1] During these times vast areas of the northern hemisphere were covered by year-round glaciers. Successive glacial times were separated by much warmer interglacial times. The glacial-interglacial alternation during the past million years has occurred with a periodicity of about 100,000 years (100 ky), with most of each period spent under glacial conditions. The interglacial times have had durations of 10 ky to 12.5 ky.

The advance and retreat of the ice sheets and the concurrent change in climate conditions have been intimately linked with large-scale migrations of both plant and animal species. In more recent times well-documented variations in climate (the greening of Iceland, the icing of Greenland) have forced the migrations of unfortunate peoples. The "Little Ice Age" which began in medieval times and ended in 1850 spread misery, starvation, warfare, plague, and death throughout Europe. The last cold spell of this period brought about the Irish Potato Famine. Climate-induced social and economic problems lead directly to large-scale political problems. Indeed, it is not farfetched to regard the social and political turbulence, the wars and invasions, the waxing and waning of kingdoms, and the shifting importance of geopolitical regions (as recorded by Plutarch or in the Old Testament, for example) to be the fallout of climate variations in more distant times and places that set in motion invasions from afar.

It is important to know if and when variations or changes in the climate will occur in the future. Climate variations comparable to those of biblical times or the "Little Ice Age" could create enormous political problems; climate changes associated with the transition from an interglacial to a glacial time would be devastating. The compression of humanity that could result from a new advance of the north polar icecap edge to 40° N.Lat. would have very unfortunate effects. Even before such a compression took place, the drastically reduced world food output attendant on the cooling global trends would seriously have affected the world population balance. These doomsday thoughts are not idle nightmares.[2] The interglacial time in which we live began about 10 ky ago. Further, there is evidence to suggest that the transition between glacial and interglacial climate conditions may occur very rapidly, within a human lifetime or two (about 100 years), or possibly even faster.

2. Pacemaker of the Ice Ages

A number of theories have been proposed to account for the ice ages which have occurred during the past half-million years. Some of these proposals are briefly described below.

A VARIATIONS IN THE SOLAR OUTPUT. This class of theories holds that the mean earth temperature changes directly in response to variations in the solar output. The sun is a nonequilibrium system which appears to be in a steady state. However, it oscillates with periods of 5 minutes, 27 days, 22 years, and undoubtably some longer term periods.[3,4] If one of the oscillation periods were commensurate with the ~ 100 ky periodicity of the ice ages, solar output variations could be the pacemaker for the ice ages if the solar output amplitude associated with this frequency component were "sufficiently large."

The sun offers its close observers rich rewards in the variety of surprises it holds.[4] For example, one recent solar surprise is the apparent shrinkage of the sun.[5] Analyses of observatory records for the past 140 years and other older historical data seemed to indicate that the sun is shrinking by about 0.1% per century, and that this shrinkage has been taking place during the past 400 years at least. This shrinkage indicates that some of the current solar output is derived from gravitational contraction. It is not yet clear how this is related to the failure to observe[7] the solar neutrino flux predicted by the standard solar model.[8] It is also not yet clear how this shrinkage will be reversed, and if there are important 100 ky components associated with these fluctuations.

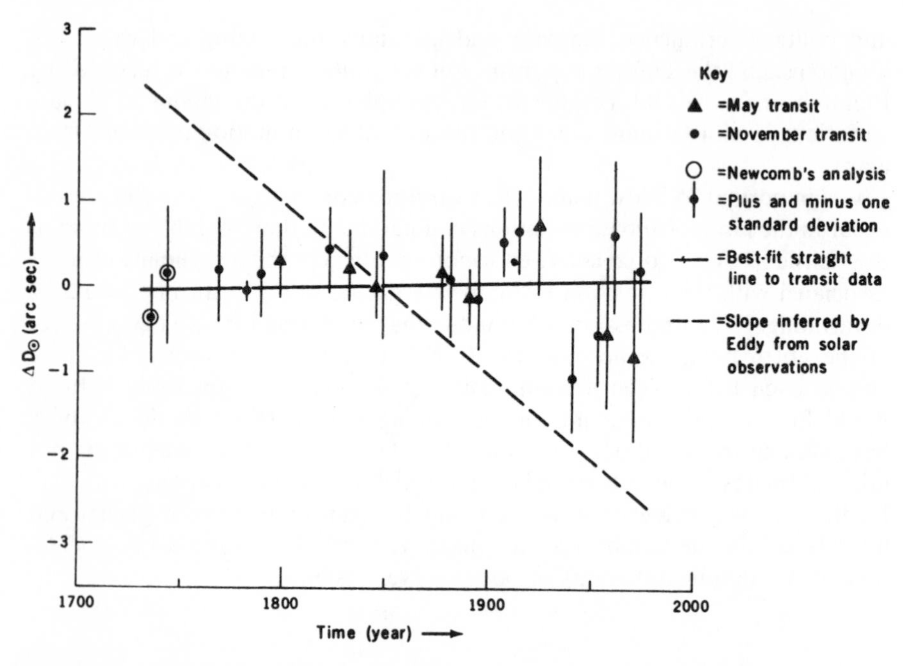

Figure 16.1 Residuals from Shapiro's analysis of observations of Mercury transits interpreted as due to possible changes $\Delta D_{\odot}$ in the diameter of the sun. The value $\Delta D_{\odot} = 0$ corresponds to $D_{\odot} =$ 1918.66 arc seconds; 1 arc second $\simeq$ 700 km. Reprinted with permission from I. I. Shapiro, *Science* **208** (1980), p. 52. Copyright 1980 by American Association for the Advancement of Science.

It is also not yet clear that this alleged shrinkage really exists. Analyses of transits of the planet Mercury in front of the sun in the period 1736–1973 by I. I. Shapiro[6a] indicate a shrinkage of less than 0.3 arc seconds per century with a 90 percent confidence limit (Fig. 16.1). This figure compares Shapiro's fit of the residuals (solid line) with the inferences of solar shrinkage presented in Ref. 5 (dashed line). An earlier analysis of solar transit data[6b] by Sofia et al. indicated a shrinkage of less than 0.6 arc seconds per century. An analysis of a more complete set of transit data by Morrison[6c] indicated a change of -0.14 ± 0.08 arc-seconds per century. At the present time there is no agreement[6d] on whether the sun is in fact shrinking and if so, by how much.

"The Sun is an obstinate reminder that while we may possess all the basic partial differential equations of classical and quantum physics, the rich variety of solutions ... extends far beyond present knowledge and imagination. The Sun, then, is our Virgil, acting as our guide through level after level of cooperative phenomena as we look more closely into the inferno."[4] In particular, sufficiently little is known about the sun that it is not yet possible to determine whether solar oscillations of the appropriate frequency, phase, and amplitude exist which could drive the ice ages.

B INTERSTELLAR DUST. As the solar system moves in its orbit through our galaxy, it may pass through interstellar dust. If it does, the solar output reaching the earth may be diminished.[9–11] If in the future it becomes possible to map out the density of interstellar dust in our neighborhood of the galaxy, then it may become possible to test for a correlation between earth ice ages and dust density fluctuations through which our solar system has passed. However, based on our insufficient knowledge, it is not yet possible to attempt such a correlation.

C VOLCANISM. Major volcanic eruptions (e.g., Krakatoa 1883) introduce enormous amounts of dust into the atmosphere. The increased dust content in the atmosphere would reduce the amount of sunlight incident on the earth's surface, initiating a cooling trend, increased cloud formations and reflection of sunlight, and greater precipitation.[12] Our knowledge of the mechanisms which may force periodicities on the earth's volcanic activity is at present too rudimentary even to attempt any correlation between volcanism and the periodic occurrence of ice ages.

D MAGNETIC FIELD REVERSALS. The earth's magnetic field has undergone a number of reversals in the past (Fig. 16.2).[13] The disappearance of the earth's magnetic field, even for brief periods, is expected to have significant effects on the chemistry of the upper atmosphere and the mutation/extinction rates of living organisms on the earth's surface. Although the magnetic field reversals are well documented, at present no theory adequately explains those that have occurred in the past nor predicts those that will occur in the future. Therefore, although correlations can be attempted between past glacial epochs and magnetic field reversals, no predictions for the future are possible.[14]

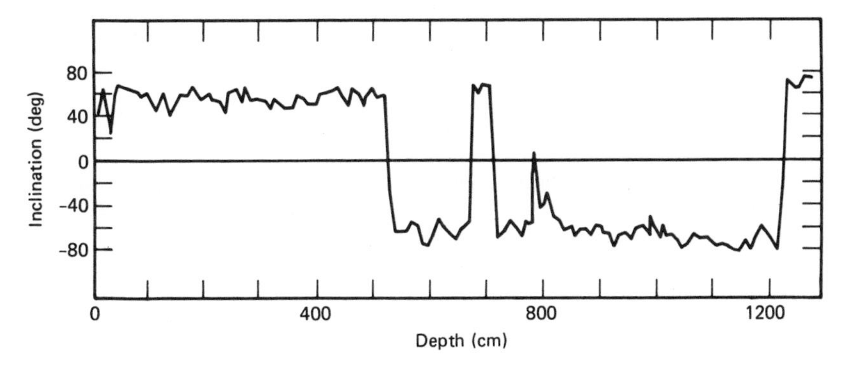

Figure 16.2 (*a*) Earth's magnetic field during the past 1.9 million years as determined from a north Pacific core. Reprinted with permission from D. Ninkovich, N. Opdyke, B. C. Heezen, and J. H. Foster, *Earth Planetary Sci. Lett.* **1**, (1966), p. 482. Copyright 1966 by North Holland Publishing Co.

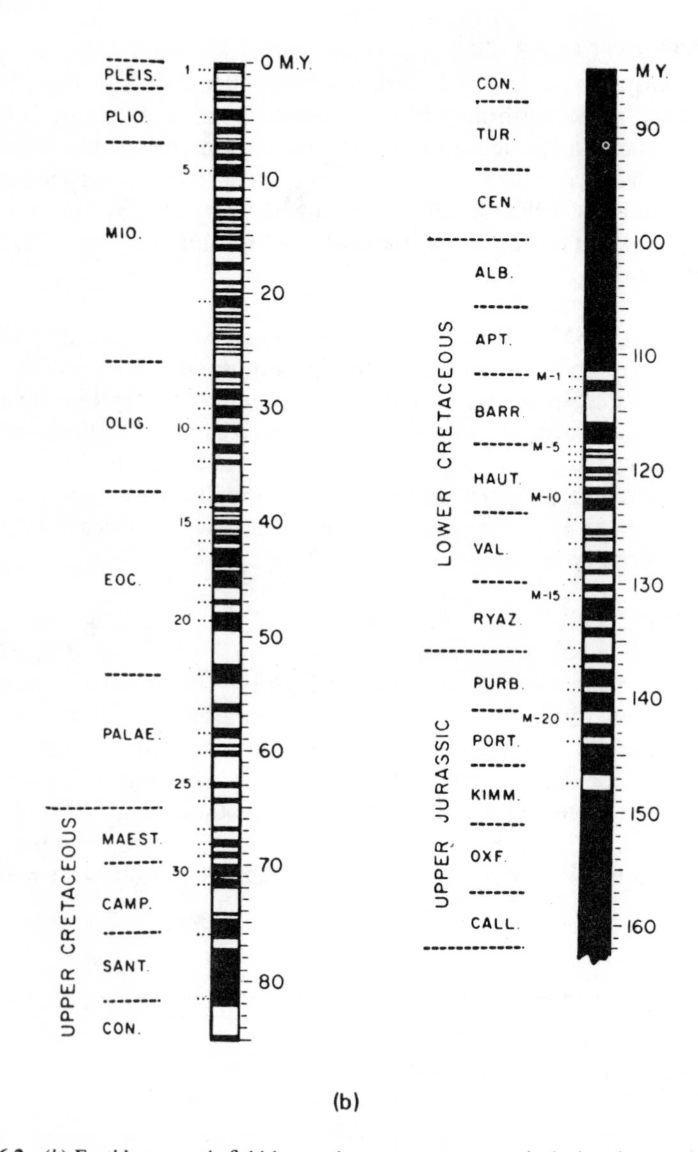

(b)

Figure 16.2 (*b*) Earth's magnetic field has undergone many reversals during the past 160 million years. Reprinted with permission from R. L. Larson and W. C. Pitman III, *Geol. Soc. Am. Bull.* **12**, 1972, p. 3641. Copyright 1972 by Geological Society of America.

E CLIMATE CHAOS. If it is ever possible to write down a system of equations which adequately describe the earth's climate, one thing is certain: they will be complicated (i.e., nonlinear). It is known that even simple nonlinear equations can exhibit very complicated and nonintuitive behavior. Such behavior will be described in Chapter 20 for a system of equations studied in detail by Lorenz. Based on the behavior of nonlinear time evolution equations, Lorenz[15] has proposed that no mechanism at all is necessary to explain the oscillating behavior of the earth's climate during the past half-million years. That is, this behavior is simply a manifestation of the nonlinear system of equations which governs the climate machine, even when no time-dependent forcing functions are present. This theory cannot yet be used to "explain" the previous ice ages for two reasons. Most obviously we do not know the system of equations (if one exists) that governs our climate. Even if we did, the solutions of such equations would depend very sensitively (divergence) on both the initial conditions and the input parameter values. Therefore observational data of arbitrarily great accuracy would be inadequate for determining the further evolution of the system.

F CHANGES IN THE EARTH'S ORBITAL GEOMETRY. The suggestion that variations in orbital geometry could be responsible for the earth's climate variations is not new.[3] This hypothesis was proposed in one form or another by Adhemer in 1824, Croll in 1876, and Koppen and Wegener in 1924.[16] However, the mathematical details were first worked out by Milankovitch.[17,18] This theory allows the correlations between the earth's climate history and the earth's orbital variations to be extended from the past to the future. If the historical correlation is good, this theory has predictive value.

Of the six classes of theories discussed above, not enough is presently known in five cases for any meaningful climate correlation to be attempted. This situation may change as we learn more about the sun, our galactic surroundings, the interior of the earth, magnetic fluctuations, and the behavior of nonlinear differential equations. However, enough is currently known about the earth's orbital properties for the Milankovitch Theory to be tested. For the time being we must content ourselves with the study of this theory and its implications.

3. The Milankovitch Theory

Milankovitch argued,[17] and then showed rigorously,[18–20] that changes in the earth's orbital parameters were responsible for changes in the amount of sunlight received at the top of the earth's atmosphere (insolation). The three orbital parameters which might drive climate variations are:

1. Eccentricity
2. Obliquity
3. Precession of the equinoxes

These are described more fully below.

A ECCENTRICITY. The motion of a planet in the presence of a central $1/r$ potential is an ellipse. The plane of the elliptic orbit is called the ecliptic. The shape of the orbit is characterized by the eccentricity e. Both the ecliptic and the eccentricity are invariants of the motion.

The orbital motion of the earth around the sun is perturbed by (among other things) the other planets. These noncentral planetary perturbations leave the plane of the elliptic orbit essentially unchanged. However, the eccentricity has varied during the past million years from 0.00 to about 0.06 (Fig. 16.3*a*). The variation is not harmonic (single frequency), but the frequency spectrum is dominated by components in the 90 ky to 105 ky range, with an average of about 93 ky.[21] Figure 16.3*b* shows the variation of the eccentricity during the past 500 ky. We are currently in a period of low orbital eccentricity.

The amount of energy received at the top of the earth's atmosphere varies by 0.1% between periods of roughly circular orbit and periods of maximum eccentricity. This is sufficient to change the mean earth temperature by a few degrees. Such a temperature variation is alone sufficient to produce climate extremes.[3]

When winter (e.g., northern hemisphere) occurs during perihelion (closest approach), the earth is on the average closer to the sun and the season is shorter. At the present time perihelion occurs during the northern hemisphere winter. Therefore northern winters are shorter and warmer than southern winters. During periods of very low eccentricity the northern and southern hemisphere winters would be comparable.

B OBLIQUITY. The spin angular momentum of the earth is not conserved because of the interaction of the sun and moon with the earth's quadrupole moment. As a result, the orientation of the earth's axis wanders among the "fixed stars." This wandering is conveniently visualized in terms of two processes: variation of the obliquity and precession of the equinoxes.

The obliquity ε is the angle between the plane of the earth's equator and the ecliptic, or alternatively, between the earth's axis and the pole of (perpendicular to) the ecliptic. The obliquity varies between 21.8° and 24.4°, with a period of approximately 41 ky (Fig. 16.4).

The earth's tilt has no effect on the total insolation received over the course of a year. Rather, it determines the seasonal temperature contrast. The contrast

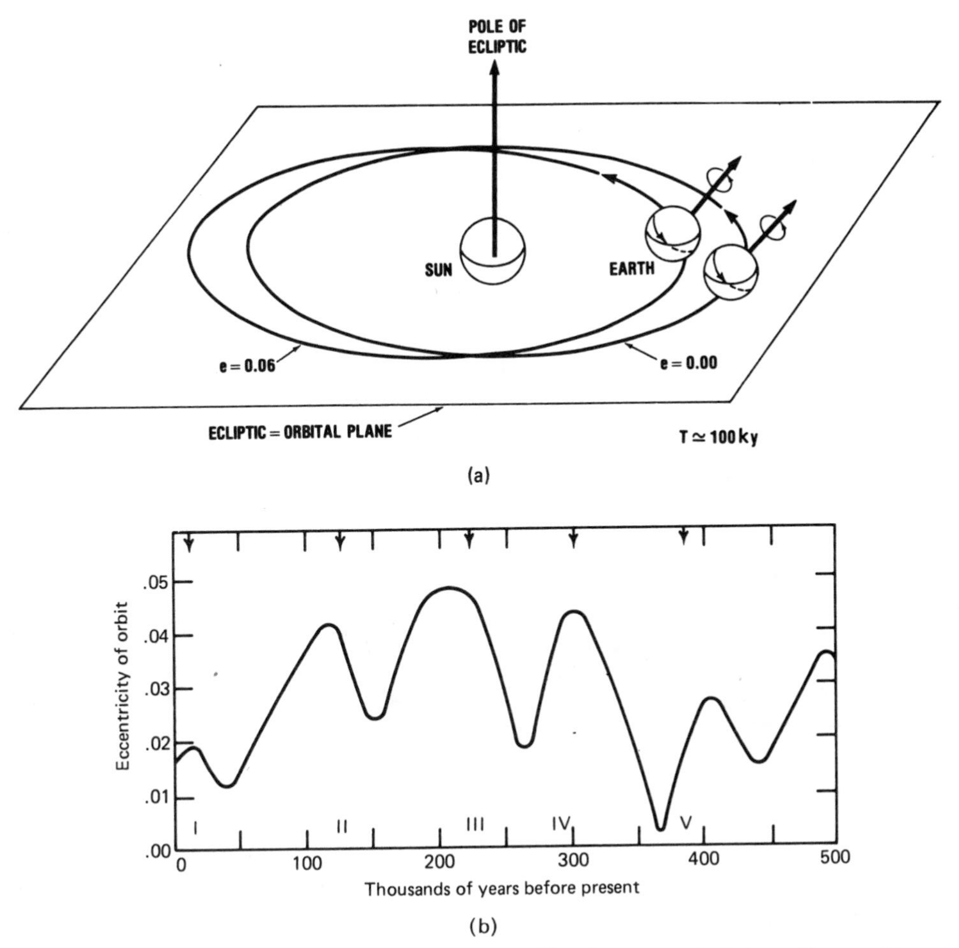

Figure 16.3 (*a*) Changes in the earth's orbit are caused by planetary perturbations. These perturbations leave the plane of the orbit unchanged, since the primary perturbations are due to coplanar planets. The eccentricity varies between 0.00 and 0.06. (*b*) The variation of the eccentricity of the earth's orbit is shown during the last 500 ky. Reprinted with permission from W. S. Broecker and J. van Donk, *Rev. Geophys. Space Phys.* **8** (1970), p. 188. Copyright 1970 by American Geophysical Union.

increases as the obliquity increases. This effect is synchronous in the two hemispheres. It is also latitude dependent, increasing with increasing latitude.

During periods of low contrast, the ice sheets are able to build up during the winter season but are not quickly melted during the warmer season, so that glaciation can occur. On the other hand, during periods of maximum contrast, ice cover grown during the winter season is eliminated during the summer season. Roughly speaking, there is a correlation between low and high insolation contrast periods and the glacial-interglacial alternations that have occurred.

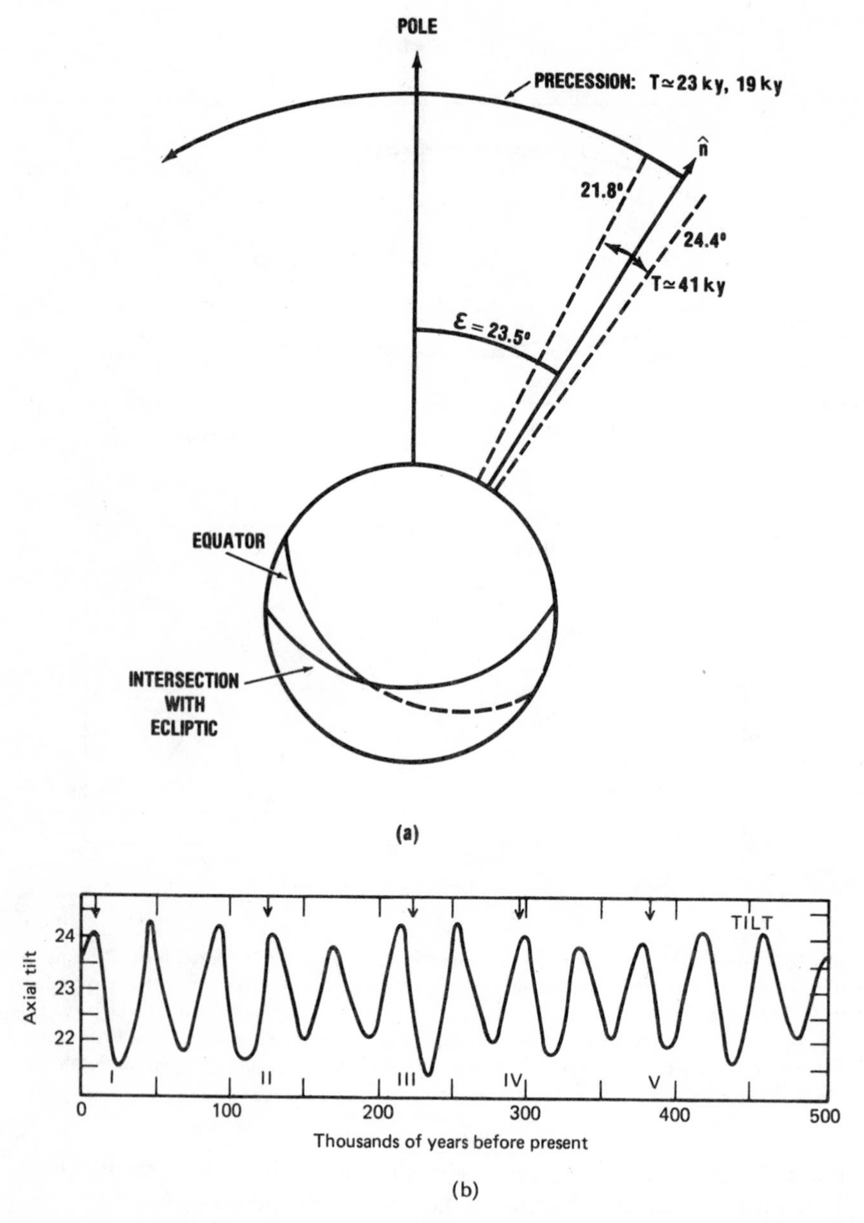

Figure 16.4 (*a*) The earth's axis precesses about the pole of the ecliptic with a period in the range of 19 to 23 ky. At the same time the angle between the pole and the axis varies between 21.8° and 24.4° with a period of about 41 ky. It is currently 23.5°. The precession is responsible for the precession of the equinoxes. The oscillation of the tilt angle is called the variation of the obliquity. (*b*) The variation of the earth's obliquity is shown for the last 500 ky. Reprinted with permission from W. S. Broecker and J. van Donk, *Rev. Geophys. Space Phys.* **8**, (1970), p. 188. Copyright 1970 by American Geophysical Union.

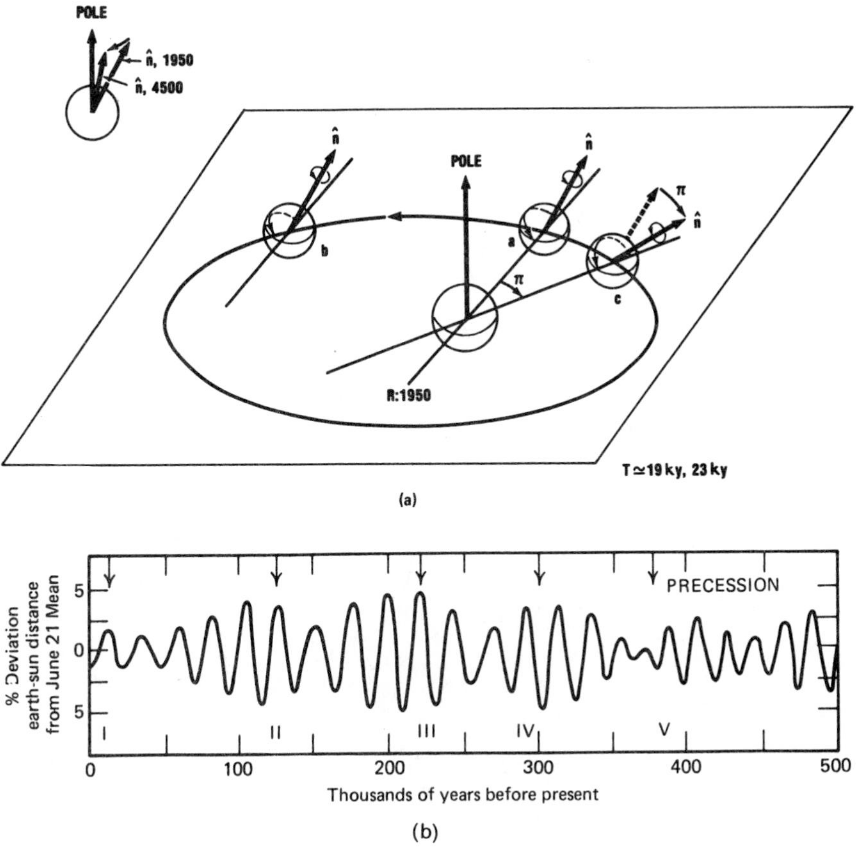

Figure 16.5 (*a*) The vernal equinox occurs when the line of intersection of the ecliptic (earth's orbital plane) and the earth's equatorial plane passes through the sun's center. This line rotates around the ecliptic because the earth's axis n precesses clockwise (inset) with components at $T \simeq 23$ ky and $T \simeq 19$ ky. a = vernal equinox, 1950, provides the reference line, line R, from which π is measured. b = spring day, later in 1950. c = vernal equinox when $\hat{n}$ has rotated through angle π. If $\pi \simeq 45°$, this orientation would occur in the year $1950 + (20 \text{ ky}/8) \simeq 4500$ AD. (*b*) The precession of the equinoxes has no direct effect on the earth-sun distance. However, it does have an effect on the earth-sun distance at a particular time in the earth's yearly or seasonal cycle. The plot of the variation (from mean value) of the earth-sun distance on June 21 shows the eccentricity modulated by the precession.[22] Reprinted with permission from W. S. Broecker and J. van Donk, *Rev. Geophys. Space Phys.* **8** (1970), p. 188. Copyright 1970 by American Geophysical Union.

C PRECESSION OF THE EQUINOXES. Not only does the tilt angle ε between the earth's axis $\hat{n}$ and the pole to the ecliptic vary, but the axis $\hat{n}$ also precesses around this pole. This precession is fairly complicated. Unlike the two other orbital parameter variations, this variation is not even approximately harmonic. Its frequency spectrum is split, with two component periodicities at approximately 23 ky and 19 ky.

This precession is responsible for the advance in the equinoxes. The plane of the ecliptic and the equatorial plane of the earth intersect in a line passing through the earth's center. This line moves with the earth as the earth moves in orbit around the sun. The vernal and autumnal equinoxes occur as this line passes through the sun's center. The orientation of this line changes as the vector $\hat{n}$ precesses around the pole of the ecliptic (Fig. 16.5). Since the line of necessity lies in the ecliptic, its orientation can be specified by measuring the angle π between it and the line for some reference year, chosen by convention to be 1950. Because of the precession of $\hat{n}$, the seasonal year is about 20 minutes shorter than the solar year.

Precession of the equinoxes does not result in any insolation change during the course of a year. Nor does it contribute to seasonal contrast when the earth is in a circular orbit. However, it does contribute to a seasonal contrast, the contrast increasing with eccentricity.

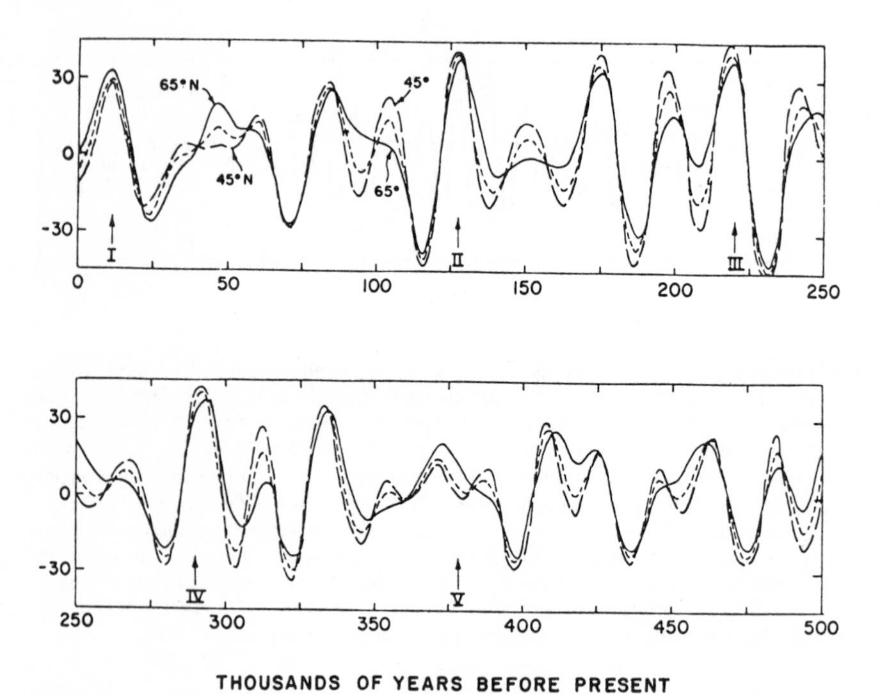

Figure 16.6 The three orbital effects can be superposed to determine the variation in the insolation at various latitudes during the past 500 ky. The vertical axis indicates difference from the mean caloric summer radiation at each latitude [900 at 45° N, 845 at 55° N, 775 at 65° N]. The maximum variations from mean are about $\pm 5\%$. Periods during which the obliquity and precession act asynchronously are apparent.[21,22] Reprinted with permission from W. S. Broecker and J. van Donk, *Rev. Geophys. Space Phys.* **8** (1970), p. 190. Copyright 1970 by American Geophysical Union.

Whereas changes in seasonal contrast in the northern and southern hemispheres due to changes in the obliquity are in phase and independent of orbital eccentricity, those due to precession of the equinoxes are out of phase and increase as orbital eccentricity increases. For example, the northern hemisphere now tilts away from the sun during perihelion. Therefore the northern hemisphere enjoys short warm winters and long cool summers. The southern hemisphere suffers short hot summers and long cold winters. In 8–10 ky the northern hemisphere will face the sun during perihelion. Then the northern hemisphere will suffer short hot summers and long cold winters, while the southern hemisphere enjoys short warm winters and long cool summers.

It is clear that the average yearly insolation received at the top of the atmosphere depends on all three orbital parameters. Changes from the long-term average summer insolation over the past 500 ky are shown in Fig. 16.6 for three different northern latitudes. Peaks in these curves correspond to maxima in seasonal contrast. It appears that the beginnings of the last four interglacial periods (indicated by arrows) correspond closely to (are initiated by ?) peaks in seasonal contrast.[22]

4. Test of the Milankovitch Theory

Among all the classes of theories proposed to explain the Pleistocene glaciations and discussed in Sec. 2, at present only the Milankovitch Theory is subject to refutation by comparison of its predictions for past behavior with experimentally determined geological evidence for the occurrence of past ice ages.

Hays, Imbrie, and Shackleton[23] set out to make just such a comparison. It was necessary to find appropriate geological samples which could be dated backward in time for the last 500 ky with an accuracy well below the shortest orbital period of 19 ky. It was further necessary to determine an appropriate set of measurements on these samples. Such measurements were required to give an accurate estimate of the temperature at one or more locations on the face of the earth.

The samples chosen were two deep sea cores. These were taken from a point in the South Pacific approximately centrally located between Africa, Antarctica, and Australia so as to minimize any possible perturbations from continental erosional runoff. The sediment accumulation rates provided the mechanism for making age estimates. The deeper the level in the core, the older the sediment. Certain globally occurring and reasonably well dated features, such as magnetic reversals and isotope anomalies, were used as date markers along the core lengths. Constant sedimentation rates were assumed between these markers. Sedimentation rates in these cores exceeded 3 cm/ky. Test samples (described below) were taken every 10 cm, so that the time interval between successive samples was less than 3 ky. In this way reasonably accurate time estimates were made going back 450 ky.

Three different types of analyses were performed on each of the small samples taken at 10 cm intervals. These analyses were performed on the skeletal remains of some particular species. When these small animals die, their skeletal remains settle to the bottom where they become part of the sediment. The relative abundance of these species and their isotopic content then provide information about temperature variations which is read backward in time by moving deeper into the core. The three analyses were as follows:

1. ^{18}O. The ^{18}O isotope content in skeletal remains is a measure of the abundance of this isotope in the world's oceans. The evaporation rate of water incorporating the heavier oxygen isotope is slightly less than the evaporation rate of normal water $H_2{}^{16}O$. During periods of glacial growth, the world's oceans will be depleted of water. The preferential evaporation of normal water then leads to an increase in the $^{18}O/^{16}O$ ratio in deposited skeletal remains. The variations in this ratio provide a record of climate variations in the northern hemisphere.
2. Ts. This involved a statistical analysis of the skeletal remains of several different types of radiolarians. The relative abundance of these different types is a measure of the water surface temperature above the core site.
3. *C. davisiana*. The relative abundance of a particular type of radiolarian (*C. davisiana*) to all other types used in the Ts analysis provides an estimate of the temperature and salinity gradients in Antarctic surface water.

These three tests made on the two subantarctic cores provide a record of temperature variations in the northern hemisphere, in subantarctic surface water, and in Antarctic surface water for the past 450 ky with a 3 ky resolution.

The detailed mechanisms involved in the transformation of orbit parameter variations into climate variations are not yet known. Nevertheless the Milankovitch Theory could be tested by making a "simplest possible" assumption: that the transformation is linear (Fig. 16.7*a*). Then frequencies in the system input (orbital variations) must appear in the system output (climate variations). The orbital input periods are 100 ky (eccentricity), 41 ky (obliquity), and 23 ky, 19 ky (precession of the equinoxes).

Periodicities in the climate variations were sought by Fourier analyzing the data from the three different types of analyses described above. The Fourier amplitudes were then plotted against frequency ($= 1/T$). In this form the data from the three types of analyses had periodic components in the ranges 87–119 ky, 37–47 ky, and 21–24 ky. The ratios of the areas under these peaks were typically 7:2:1. For the ^{18}O measurements the "peak" around 20 ky had structure and could be resolved into two components corresponding to periodicities 24 ky and 19.5 ky. The data also showed that glacial periods were associated with low eccentricity.

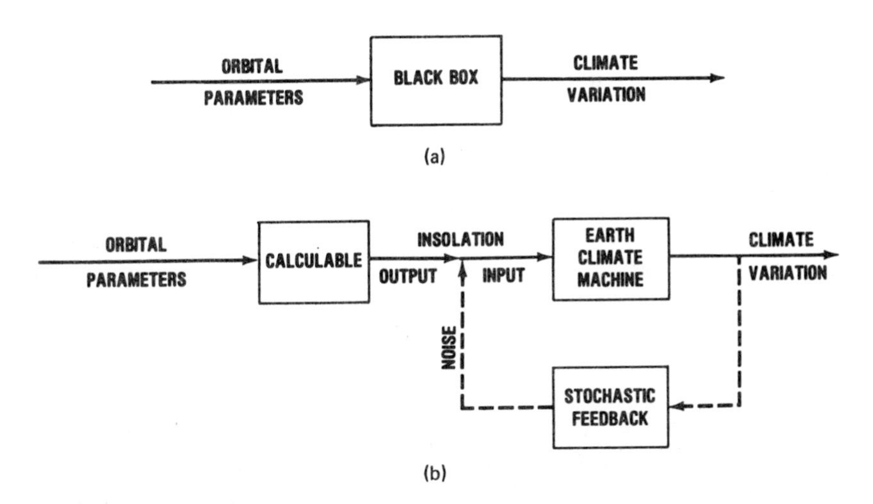

Figure 16.7 (*a*) The earth may be considered as a black box which converts variations in the orbital parameters into climate variations. If the "black box" can be represented by linear time-invariant equations, frequency components in the input must also appear in the output. (*b*) Variation in the insolation at the top of the atmosphere can be calculated from the known variation in the earth's orbital parameters. This insolation is the input to the earth's climate machine. Climate variations alter the earth's cloud cover and albedo, as do other events (volcanism) not directly connected with the earth's climate. A noisy feedback loop is included to represent these processes.

These comparisons of the astronomical data for the variations of the orbital parameters with the geological data for the variation of the Arctic, subantarctic, and Antarctic temperatures support the Milankovitch hypothesis that orbital variations are the driving force behind climate changes.[22, 23]

The Milankovitch Theory suggests that climate variations should occur on the same time scale as orbital variations, that is, slow compared to a human life. There is evidence that the glacial-interglacial transitions occur on a time scale of the order of a few hundred years. The occurrence of multiple time scales, as well as the presence of modality (glacial-interglacial) strongly suggests the occurrence of a catastrophe somewhere in the description of climate variation.

5. Connection with Catastrophe Theory

The Milankovitch theory offers no clues to the detailed mechanisms that transform orbital variations into climatic changes. It is, therefore, useful to decompose the transformation mechanism (Fig. 16.7*a*) into two linear systems operating in series (Fig. 16.7*b*). The first step converts the orbital data into energy fluxes received at the top of the atmosphere as a function of time and latitude (insolation data). This first step can be carried out accurately, making some reasonable astrophysical assumptions (constant solar output during the last million years, no major drifting dust clouds, no intergalactic dragons, etc.).

The problem comes in the second stage—determining the climate output in terms of the insolation input. The dichotomy made in Fig. 16.7*b* has now the advantage that the entire problem has been reduced to a problem in earth dynamics and thermodynamics alone, albeit complicated. In order to treat this problem, it is necessary to formulate models describing mass and energy transport in the world's oceans and atmosphere. This will involve a set of coupled nonlinear partial differential equations whose coefficients may not even be differentiable functions of some important state variables. (The albedo changes drastically at the water vapor condensation point.)

Since a general bifurcation theory for such complicated equations is nonexistent, a more modest approach might be more fruitful. Such an approach might proceed as follows. The most important components of the earth's climate engine are first isolated. These might include the major oceanic currents, trade winds, jet streams, and temperature, water vapor, and carbon dioxide density fields in each of these currents. A system of dynamical equations

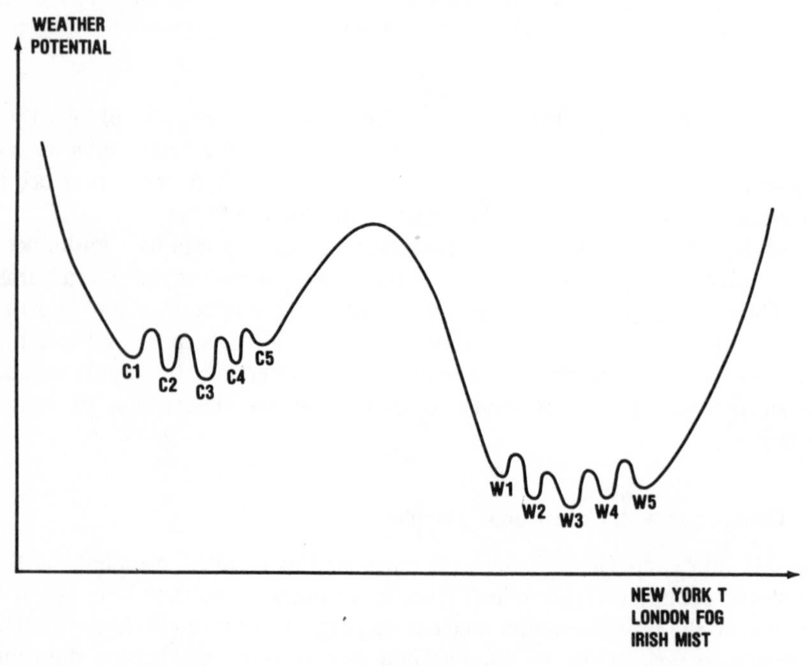

Figure 16.8 If it is possible to represent the state of the earth's climate by some sort of potential, then this potential might have two broad local minima, one at high temperature representing interglacial periods, the other at lower temperature representing glacial periods. Each broad minimum might have additional fine structure in the form of numerous local minima ($C1$–$C5$ and $W1$–$W5$).

describing mass and energy transfer might then be constructed for each of these currents. The source terms for these equations would include the periodic ($T = 1$ year) insolation inputs at the top of the atmosphere. The multiplicity of locally stable solutions could then be determined as a function of annual insolation input.

Remark. It is the hope of some geologists and climatologists that analysis of recent geological data (since 10,000 BC) will indicate at least a partial spectrum of locally stable oceanic and atmospheric current patterns during the current (and previous) interglacial period, and that analysis of data from the glacial epochs will indicate a spectrum of oceanic and atmospheric current patterns that are locally stable during those epochs.

To put these speculations into a somewhat more concrete form we could use a single variable to suggest the state of the earth's climate. This could be, for instance, the mean annual temperature in New York, the number of foggy days in London, or the density of mist in Ireland. All three measures probably have low values during glacial periods and high values during interglacial periods. Then the probability that the earth assumes some particular planetary

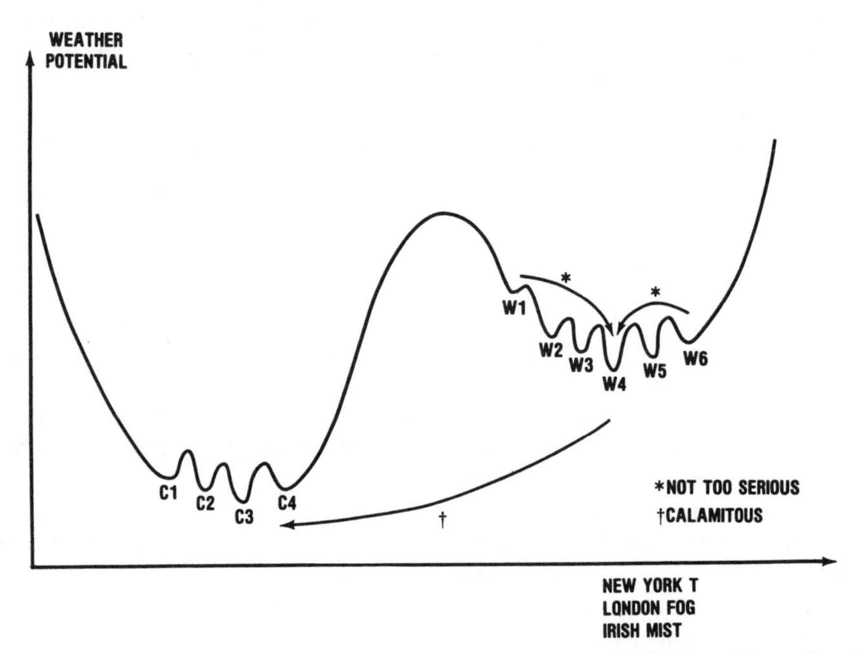

Figure 16.9 As the earth's orbital parameters slowly change, the relative heights of the two broad minima might alsc slowly change. If the current interglacial high-temperature minimum becomes metastable with respect to the lower temperature glacial minimum, transitions on a fast nongeological time scale may signal the onset of glaciation. Transitions $Wi \rightarrow Wj$ are less calamitous than transitions $Wi \rightarrow Cj$.

climatological configuration might be proportional to exp $-$ $[V(x)]$, where the function $V(x)$ is sketched in Fig. 16.8. This function suggests that it is possible to jump from one stable warm configuration to another (e.g., $W2 \rightarrow W3$), leading to only a "small" change in global climatic conditions. There is strong evidence to suggest that transitions of this type have occurred during the current interglacial period.

The overall shape of the function $V(x)$ will change as the orbital parameters change. During a glacial period $V(x)$ might have a shape as indicated in Fig. 16.9. Coming from a warm period, the transitions among the warm structures ($Wi \rightarrow Wj$) would be of much less importance than any transition from a "warm" configuration to a "cold" configuration ($Wi \rightarrow Cj$).

Remark. The transition from a system of coupled nonlinear partial differential equations with deterministic driving terms to a system of coupled nonlinear ordinary differential equations with stochastic driving terms described above is analogous to the reduction of a Fokker-Planck equation for a probability distribution to a system of Langevin equations for the important expectation values.

Remark. If the approach outlined in this section can actually be carried out, then it would be of interest to apply these techniques to an earth whose surface is extrapolated into the distant past and future. Such an extrapolation must take account of continental drift. The changing shape of the world's oceans will certainly be responsible for changing the (relative probabilities of the) locally stable ocean currents and, by coupling, the atmospheric currents as well.

6. Reduction to Cusp Form

If we are interested in the glacial-interglacial transition, then the fine structure shown in Figs. 16.8 and 16.9 is less important than the gross structure, and can be ignored. Then the existence of modality and the occurrence of fast transitions suggests the presence of a catastrophe underlying all the complicated equations. The presence of two major modes—cold and warm—suggests that the catastrophe which is present is the cusp catastrophe.

The identification of the cusp catastrophe tells us that there is only one important state variable x which is an unknown and probably complicated function of the climate machine state variables, also unknown. The latter state variables, and the dependence of the "bad" variable on them, must be determined by building models.

The cusp control parameters a, b depend on insolation inputs and therefore on time t through the Milankovitch connection. In this way the slow orbital variations could lead to fast transitions if the Delay Convention were approximately applicable.

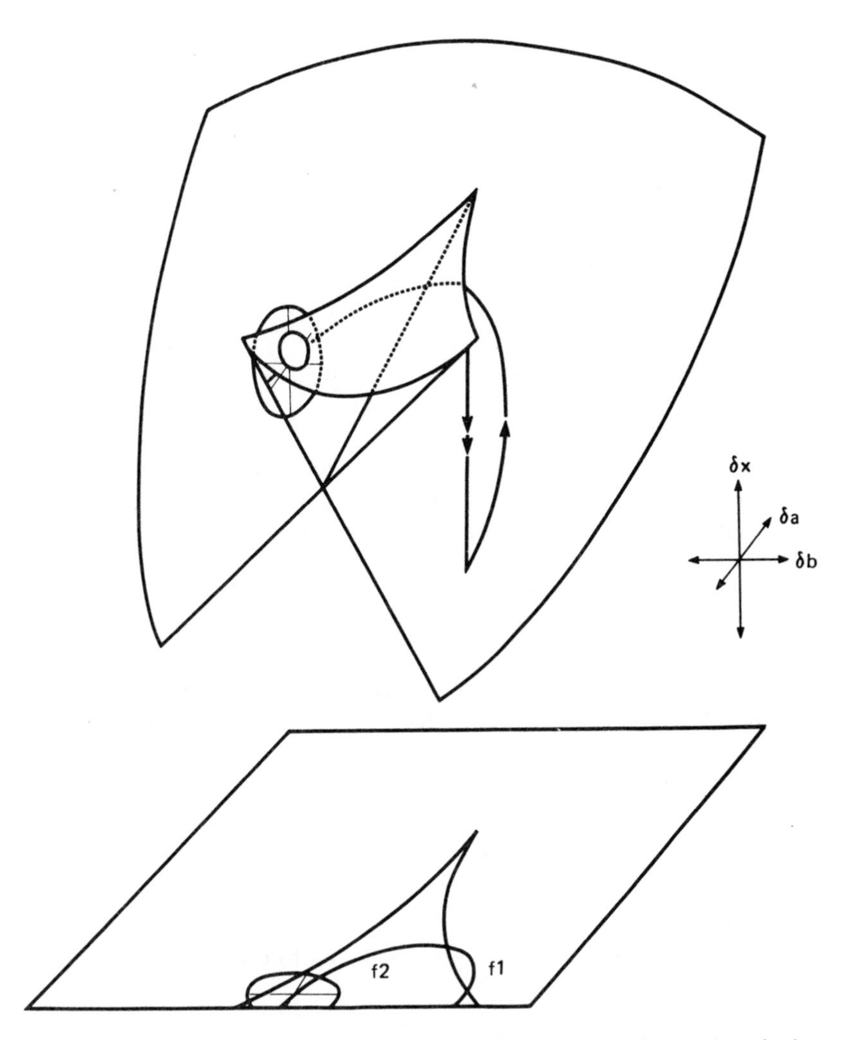

Figure 16.10 If the control parameters $a(t)$, $b(t)$ cross the fold line f1, nothing much can be done to prevent a sudden jump (double arrow). If the controls approach the fold line f2 so that the "noise ellipse" cuts the upper surface (top) or extends past the fold line (bottom), fluctuations which might cause a transition from the metastable minimum (middle sheet) to the stable minimum (bottom sheet) can be suppressed, leaving the system state in the warmer metastable configuration. Inset shows noise levels in the state variable and control parameter directions.

In any realistic model noise is an important consideration. Noise will occur both in the control parameters and in the state variables.

1. Noise in the control parameters. Although the insolation outputs at stage 1 (Fig. 16.7*b*) might be known precisely, only part of this energy is actually used to fuel the earth's climate machine. The remainder of the input radiation is

reflected by cloud cover and by the earth's surface. There are clearly short-term (daily and weekly) variations in the cloud cover that are probably unimportant. There are likely to be longer term variations in the cloud cover due to changing ocean and atmospheric currents. Volcanic activity also affects the fraction of incident radiant energy that actually drives the system. These and other random variations must be considered as noise on top of the "signals" $a(t), b(t)$.

2. Noise in the state variables. Each of the system state variables is a number obtained by averaging over all or part of a yearly cycle. Since the state variables are averaged quantities, their yearly values are subject to fluctuations. Such fluctuations must be considered as noise in the value of the important (Fig. 16.9) state variable x.

Noise in the control parameters $a(t)$, $b(t)$ may drive the climate from one mode to another (Fig. 16.10). Sufficiently long-lived fluctuations may also drive the system from a metastable to a stable state. Thus it is likely that sudden transitions between modalities may be triggered by noise in either the state variables or the control parameters. This could account for the two distinctly different time scales involved in the glacial-interglacial transitions.

Remark. Transitions in the configurations of the northpolar jet stream have been mimicked in simple laboratory experiments.[24] These experiments, which exhibit modality, sudden jumps, and hysteresis, have been interpreted in the context of multiple cusp catastrophes.

7. Toward a Useful Program

This Catastrophe Theory viewpoint of the Pleistocene climate catastrophes suggests some concrete steps that might be taken to combat the sudden onset of glaciation in the future:

1. Determine a set of state variables sufficiently large that their values completely define the state of the earth's climate.
2. Determine the dynamical equations describing the evolution of these state variables as a function of insolation.
3. Determine the potential (pseudopotential, Lyapunov function) which describes the equilibrium configuration of the quasistatic system.
4. Isolate the appropriate function of the state variables whose non-Morse behavior is responsible for the phase transitions.
5. Determine the trajectory of the control parameters $a(t)$, $b(t)$ in the cusp control space, and the mean noise levels associated with $a(t)$, $b(t)$, $x(t)$. If the trajectory crosses the fold lines, so that metastable equilibria disappear, then nothing much can be done to prevent neoglaciation.

6. If the trajectory approaches but does not cross the fold lines, so that it is only the noise fluctuations that drive the transition from metastability to stability, then something can be done to prevent neoglaciation.
7. Whenever a dangerous fluctuation in $a(t)$, $b(t)$ occurs, it can be counteracted in a way that suppresses the fluctuation and *steers* the system away from the nearby fold edge. The suppression mechanisms might involve changing the insolation inputs by varying the cloud cover or the earth's albedo at certain latitudes, or might involve tampering with the temperature, flow rate, or direction of one or more of the earth's ocean or atmospheric currents. Whenever a fluctuation in x occurs that could drive the system over the barrier separating the mestastable and the stable states (Fig. 16.10), it could be suppressed by similar methods.

8. Epilogue

The approach to the study of climate changes presented here can lead to the concise formulation of important questions. Although climate change appears to be forced by variations in the earth's orbital parameters, it is unfeasible to alter these parameters to avoid climate transformation. Were it feasible, it would be unwise. The program outlined above could indicate methods of "freezing" the earth's climate in the interglacial stage by applying sufficient forces only to specific parts of the earth's solar energy-climate condition machine.

Which parts? How much force? The prescription above must answer questions such as these. Each of the stages above will be difficult to carry out. Later stages would be more difficult; the last, the most difficult. Although the enormous technological capabilities required for the last stage may become available, the political capabilities will probably not.

References

1. R. F. Flint, *Glacial Geology and the Pleistocene Epoch*, New York: Wiley, 1947; *Glacial and Quaternary Geology*, New York: Wiley, 1971.
2. T. M. Laur, The World Food Problem and the Role of Climate, *EOS* **57**(4), 189–195 (1976).
3. J. R. Herman and R. A. Goldberg, *Sun, Weather, Climate*, Washington, D.C.: NASA, 1978.
4. E. N. Parker, Gearing up to Answer Questions Posed by the Sun, *Physics Today* **32**(9), 9–10 (1979).
5. J. A. Eddy and A. A. Boornazian, Secular Decreases in the Solar Diameter, 1863–1953, *Bull. Am. Astr. Soc.* **11**, 437 (1979), (01.19.03).

6a. I. I. Shapiro, Is the Sun Shrinking?, *Science* **208** (4), 51–53 (1980).

6b. S. Sofia, J. O'Keefe, J. R. Lesh, and A. S. Endol, Solar Constant: Constraints on Possible Variations Derived from Solar Diameter Measurements, *Science* **204**, 1306–1308 (1979).

6c. *Physics Today*, **33**(5), 21–22 (1980). See also Ref. 6a.

6d. *New York Times*, July 6, 1980, p. 21.

7. R. Davis, Solar Neutrinos II. Experimental, *Phys. Rev. Lett.* **12**, 303–305 (1964).

8. J. N. Bahcall, Solar Neutrinos. I. Theoretical, *Phys. Rev. Lett.* **12**, 300–302 (1964).

9. F. Hoyle and R. A. Lyttleton, The Effect of Interstellar Matter on Climate Variation, *Proc. Camb. Phil. Soc.* **35**, 405–415 (1939).

10. R. J. Talbot, D. M. Butler, and M. J. Newman, Climatic Effects During Passage of the Solar System Through Interstellar Clouds, *Nature* **262**, 561–563 (1976).

11. B. Dennison and V. N. Mansfield, Glaciations and Dense Interstellar Clouds, *Nature* **261**, 32–34 (1976).

12. J. P. Kennett and R. C. Thunell, Global Increase in Quaternary Explosive Volcanism, *Science* **187**, 497–503 (1975).

13. R. L. Larson and W. C. Pitman, World-wide Correlation of Mesozoic Magnetic Anomalies, and Its Implications, *Bull. Geol. Soc. Amer.* **83**, 3645–3662 (1972); D. Ninkovich, N. Opdyke, B. C. Heezen, and J. H. Foster, Paleomagnetic Stratigraphy, Ratios of Deposition and Tephrachronology in North Pacific Deep-Sea Sediments, *Earth Planet. Sci. Lett.* **1**, 476–492 (1966).

14. G. Wollin, D. B. Ericson, and W. B. F. Ryan, Variations in Magnetic Intensity and Climatic Changes, *Nature* **232**, 549–551 (1971).

15. E. N. Lorenz, Climate Determinism, in: *Causes of Climatic Change* (J. M. Mitchell, Ed.), *Meteorological Monogr.* **8**, Boston, Mass: Am. Meteorological Soc., 1968, pp. 1–3.

16. M. Koppen and A. Wegener, Die Klimate der geologischen Vorzeit, Berlin, 1924.

17. M. Milankovitch, Mathematische Klimalehre und astronomische Theorie der Klimaschwankungen, in: *Handbuch der Klimatologie* (W. Koppen and R. Geiger, Eds.), vol. I, pt. A. Berlin: Gebr. Bornträger, 1930, pp. 1–76.

18. M. Milankovitch, Die Chronologie des Pleistocans, *Bull. Acad. Sci. Math. Nat. Belgrade* **4**, 49 (1968).

19. Brouwer and van Woerkom, in: *Climate Change* (H. Shapley, Ed.), Cambridge, Mass: Harvard University Press, 1953.

20. A. D. Vernekar, Long Period Global Variations of Incoming Solar Variation, in: *Research on the Theory of Climate*, vol. 2, Rep. of the Travelers Research Center, Inc., Hartford, Conn., May 1968.

21. A. D. Vernekar, *Long Period Global Variations of Incoming Solar Radiation*, Meteorological Monogr. **12**, 34, Boston, Mass.: Am. Meterological Soc., 1972.

22. W. S. Broecker and J. van Donk, Insolation Changes, Ice Volumes, and the O^{18} Record in Deep Sea Cores, *Revs. Geophys. Space Sci.* **8**, 169–198 (1970).

23. J. D. Hays, J. Imbrie, and N. J. Shackleton, Variations in the earth's orbit: Pacemaker of the Ice-Ages, *Science* **194**, 1121–1132 (1976).

24. R. C. Lacher, R. McArthur, and G. Buzyna, Catastrophic Changes in Circulation Flow Patterns, *Am. Scientist* **65**, 614–621 (1977).

PART 3

Beyond Elementary Catastrophe Theory

In which we provide glimpses of the spectrum of mathematics which lies just beyond the fringes of Elementary Catastrophe Theory

CHAPTER 17

Beyond the Elementary Catastrophes

Elementary Catastrophe Theory is the study of the degeneracies in families of mappings F of the unbounded domain $\mathbb{R}^n$ into $\mathbb{R}^m$:

$$F: \quad \mathbb{R}^n \otimes \mathbb{R}^k \to \mathbb{R}^m \tag{17.1}$$

where $k \leq 5$, $n < \infty$, $m = 1$. To go beyond the elementary catastrophes, we must relax the conditions described here. In this chapter we discuss five ways of going beyond the elementary catastrophes:

1. $k > 5$
2. $n \to \infty$
3. $m > 1$
4. By looking at atypical (symmetry-restricted) functions
5. By restricting the domain of definition of F to a bounded closed subset $\mathcal{S} \subset \mathbb{R}^n$.

We relax separately each of these five restrictions in the first five sections.

1. $k > 5$

When the number of control parameters exceeds 5, a non-Morse germ can depend on parameters called moduli (see Chapter 3, Secs. 7–9). These first appear in the case of families of functions depending on $k = 6$ control parameters when these six degrees of freedom are used to annihilate all six quadratic coefficients at the critical point of a non-Morse function of $l = 3$ "bad" state variables. A nonsingular change of variables can then give canonical values (+1 or 0) to 9 of the 10 three-state-variable cubic coefficients, leaving one as a continuous parameter, or modulus. It is for this reason that Thom's list terminates at $k = 5$.

When more controls are present, more moduli may occur. There is a simple relationship between the number k of control parameters present, the degeneracy μ (called **Milnor number**) of the degenerate catastrophe germ, and the number of moduli left over:[1]

$$\mu = k + m + 1 \tag{17.2}$$

We have already encountered (17.2) in the case of simple ($m = 0$) germs, where the dimensionality k and degeneracy μ are related by $k = \mu - 1$.

Arnol'd has classified all non-Morse germs[1–3] with

1. $\mu \leq 16$
2. $k \leq 10$
3. $m \leq 2$

It is surprising that the most natural classification of non-Morse germs does not involve those with small μ or k, but those with small m.

1. $m = 0$. There are three types of simple germs. There is one infinite series A_μ ($\mu \geq 2$) depending on one state variable and a second infinite series D_μ ($\mu \geq 4$) depending on two state variables. In addition, there is one finite series E_μ ($\mu = 6, 7, 8$) depending on two state variables and containing three members. These germs are shown in Table 17.1.

   Each catastrophe has a $k = (\mu - 1)$-dimensional universal perturbation, and "throws off" 1-dimensional curves of catastrophe germs of one lower control dimension. These relations have already been discussed (Chapter 7), and are summarized in an "abutment diagram," which is given in Fig. 7.6 for all simple catastrophes.
2. $m = 1$.[4] There is one infinite series of unimodular germs:

$$T_{p,q,r} = x^p + y^q + z^r + axyz, \qquad a \neq 0, \quad \frac{1}{p} + \frac{1}{q} + \frac{1}{r} < 1 \tag{17.3}$$

depending on three state variables and one modulus a. Each function $T_{p,q,r}$ in this series is actually a 1-modal family of germs. The function $T_{p,q,r}$ has $\mu = p + q + r - 1, m = 1, k = p + q + r - 3$. It is useful also to consider the "boundary germs" for which $p^{-1} + q^{-1} + r^{-1} = 1$:

$$\begin{aligned} T_{3,3,3} &= x^3 + y^3 + z^3 + axyz, & \left(\frac{a}{3}\right)^3 &\neq -1 \\ T_{2,4,4} &= x^2 + y^4 + z^4 + ay^2z^2, & \left(\frac{a}{2}\right)^2 &\neq +1 \\ T_{2,3,6} &= x^2 + y^3 + z^6 + ay^2z^2, & \left(\frac{a}{3}\right)^3 &\neq -\left(-\frac{1}{2}\right)^2 \end{aligned} \tag{17.4}$$

Table 17.1 0-Modal Germs

	Infinite Series			Finite Series	
$l = 1$	A_μ	$x^{\mu+1}$	$\mu \geq 2$		
$l = 2$	D_μ	$x^2y + y^{\mu-1}$	$\mu \geq 4$	E_6	$x^3 + y^4$
				E_7	$x^3 + xy^3$
				E_8	$x^3 + y^5$

The first of these germs involves three state variables, but the latter two involve only two "bad" state variables.

In addition to the single infinite series (17.3) there are 14 exceptional 1-modal families which occur in six finite series containing one member (U), two members each (S, W), or three members each (Q, Z, E). Three unimodular families (E, Z, W) involve two state variables, while the remaining three (Q, S, U) involve three state variables. The 14 1-modal germs are given in Table 17.2.

As in the case of the 0-modal catastrophes, unimodal catastrophes throw off 1-dimensional curves of catastrophes of control dimension reduced by 1. For the infinite series $T_{p,q,r}$ we have the simple abutments

$$\begin{array}{ccc} & T_{p,q,r} & \\ \swarrow & \downarrow & \searrow \\ T_{p-1,q,r} & T_{p,q-1,r} & T_{p,q,r-1} \end{array} \tag{17.5}$$

Table 17.2 The Six Exceptional Series Containing 14 Exceptional Families

		$m = 1$		$m = 2, \mathbf{a} = a_0 + a_1y$
$l = 2$	Z_{11}	$x^3y + y^5 + axy^4$	Z_{17}	$x^3y + y^8 + \mathbf{a}xy^6$
	Z_{12}	$x^3y + xy^4 + ax^2y^3$	Z_{18}	$x^3y + xy^6 + \mathbf{a}y^9$
	Z_{13}	$x^3y + y^6 + axy^5$	Z_{19}	$x^3y + y^9 + \mathbf{a}xy^7$
	W_{12}	$x^4 + y^5 + ax^2y^3$	W_{17}	$x^4 + xy^5 + \mathbf{a}y^7$
	W_{13}	$x^4 + xy^4 + ay^6$	W_{18}	$x^4 + y^7 + \mathbf{a}x^2y^4$
	E_{12}	$x^3 + y^7 + axy^5$	E_{18}	$x^3 + y^{10} + \mathbf{a}xy^7$
	E_{13}	$x^3 + xy^5 + ay^8$	E_{19}	$x^3 + xy^7 + \mathbf{a}y^8$
	E_{14}	$x^3 + y^8 + axy^6$	E_{20}	$x^3 + y^8 + \mathbf{a}xy^8$
$l = 3$	Q_{10}	$x^3 + y^4 + yz^2 + axy^3$	Q_{16}	$x^3 + yz^2 + y^7 + \mathbf{a}xy^5$
	Q_{11}	$x^3 + y^2z + xz^3 + az^5$	Q_{17}	$x^3 + yz^2 + xy^5 + \mathbf{a}y^8$
	Q_{12}	$x^3 + y^5 + yz^2 + axy^4$	Q_{18}	$x^3 + yz^2 + y^8 + \mathbf{a}xy^6$
	S_{11}	$x^4 + y^2z + xz^2 + ax^3z$	S_{16}	$x^2z + yz^2 + xy^4 + \mathbf{a}y^6$
	S_{12}	$x^2y + y^2z + xz^3 + az^5$	S_{17}	$x^2z + yz^2 + y^6 + \mathbf{a}zy^4$
	U_{12}	$x^3 + y^3 + z^4 + axyz^2$	U_{16}	$x^3 + xz^2 + y^5 + \mathbf{a}x^2y^2$

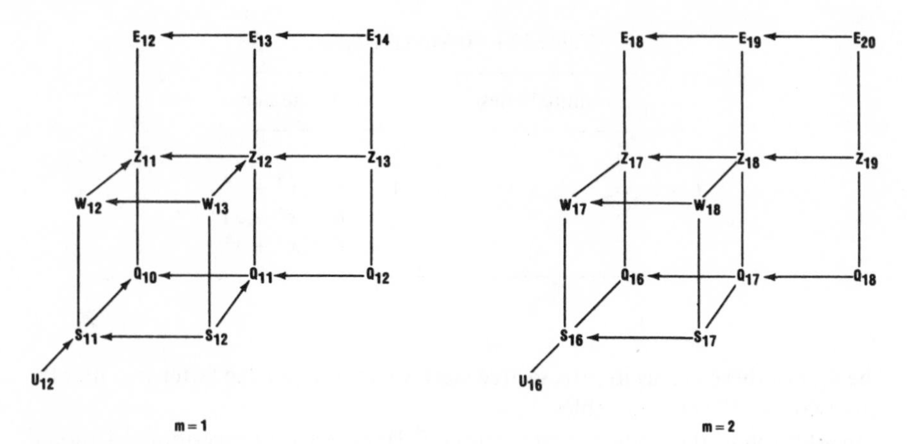

Figure 17.1 The 14 1-modal and 2-modal exceptional catastrophe germs abut in pyramids as indicated.

which may be built up into a 3-dimensional diagram, if desired. The end points of abutments of type (17.5) are the "boundary germs" given in (17.4). The abutment diagram for the six exceptional series containing 14 exceptional unimodal families is shown in Fig. 17.1.

The lowest member of each of the exceptional unimodal catastrophe series abuts onto a member of the infinite series $T_{p,q,r}$ as shown in Fig. 17.2. It is clear from this figure that the exceptional germs are near the boundary germs (17.4). Further, the unimodal boundary germs abut onto the zero modal exceptional germs as indicated in Fig. 7.6. The highest member of each of the exceptional unimodal catastrophe series is obtained from a catastrophe germ which we have not yet encountered. The germs feeding in from the right labeled J, Z, W, Q, S, U are bimodal, and N, V are 3-modal. The germ 0 is 5-modal.

3. $m = 2$.[5] In the bimodal case there are eight infinite series and six exceptional finite series containing a total of 14 families of germs. The germs of the exceptional families are given in Table 17.2, and their abutment diagram is shown in Fig. 17.1. Four of the infinite series (J, Z, W, $W^{\#}$) involve two state variables, the remaining four (Q, S, $S^{\#}$, U) involve three state variables.

These eight infinite series are defined by equations analogous to (17.3), with $a \to \mathbf{a} = (a_0 + a_1 y)$ and $a_0 \neq 0$. Each member in a series abuts on the next lower member of that family as well as members of some other families (like $D_{\mu+1} \to A_\mu$). Each of the eight infinite series has a boundary germ. The boundary germs for W and $W^{\#}$ are the same, as is also the case for S and $S^{\#}$. The six boundary germs are subject to constraints (on a_0) comparable to the constraints given in (17.4). The six bimodal boundary germs abut on the six highest families in the six finite unimodal exceptional series (Fig. 17.2).

Remark 1. It appears that the exceptional series of germs of modality m play the role of "buffers" between infinite series of germs of modality m and

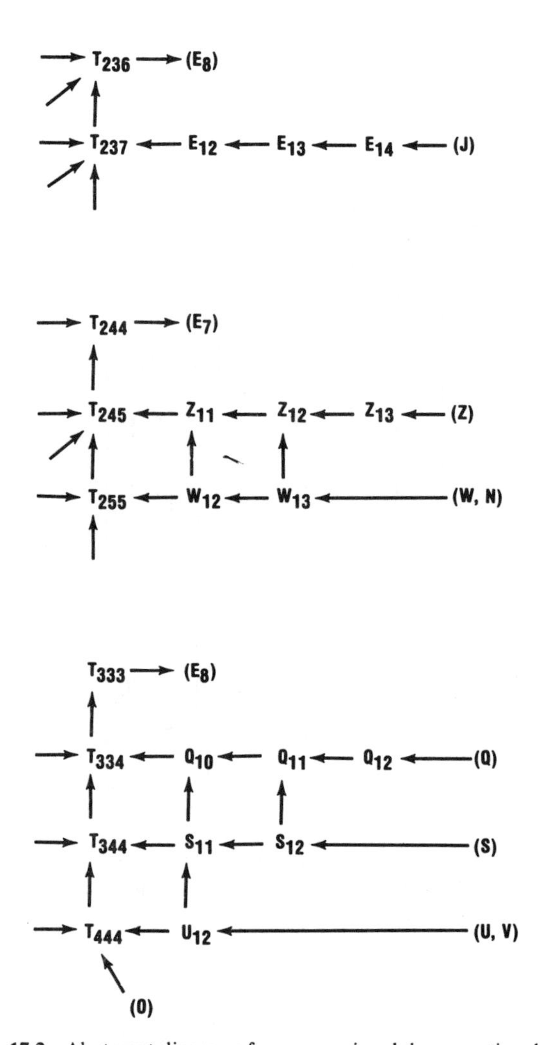

Figure 17.2 Abutment diagrams for some unimodular exceptional germs.

series of modality $m + 1$. The boundary germs of infinite series of modality $m + 1$ abut on highest members of finite (exceptional) families of modality m, and the bottom of the exceptional series abuts near, but not at, the boundary of an infinite series of the same modality. This is summarized schematically in Fig. 17.3.

Remark 2. Now that it is known that catastrophes with $k > 5$ can be classified and put into canonical form (modulo moduli), the program of studying these series and families of higher catastrophes seems of enormous abstract but little practical interest.

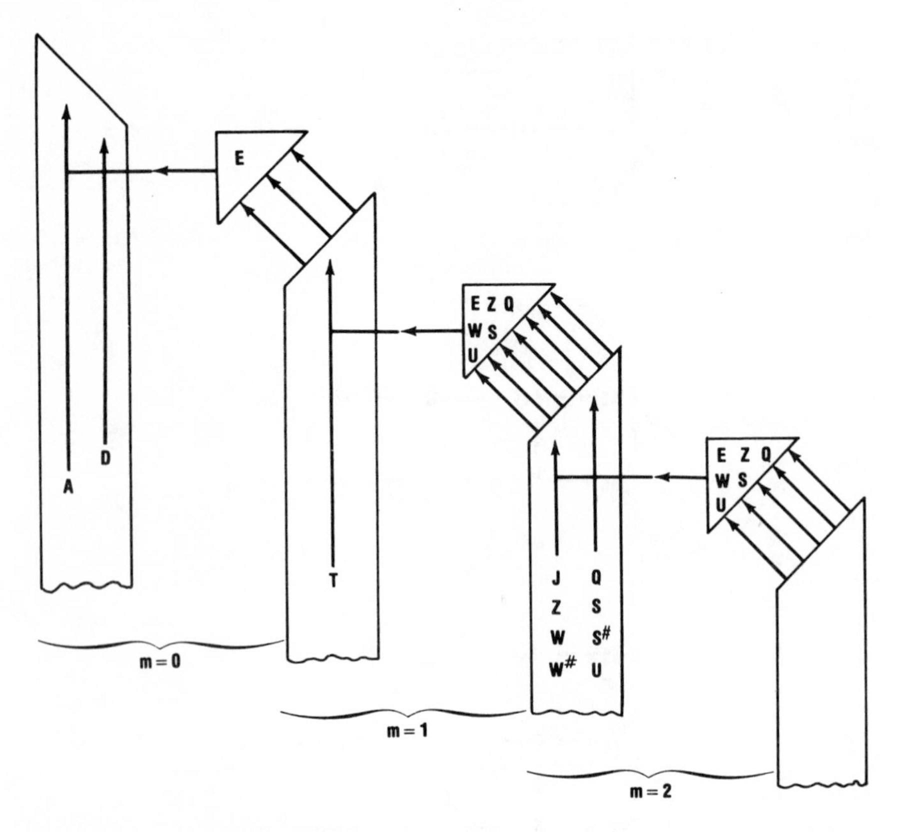

Figure 17.3 The exceptional catastrophe germs of modality m appear to serve as buffers between infinite families of catastrophe germs of modality $m + 1$ and those of modality m. These germs are all defined in Ref. 3.

2. $n \to \infty$

The results described in Chapter 2 are valid for typical families of functions defined on state spaces of dimension $n < \infty$. It is natural to ask if these results extend to the case $n \to \infty$.

To answer the question it is first necessary to reformulate our concept of the stability matrix as a symmetric operator on a Hilbert or a Banach space. The directions corresponding to the zero eigenvalues of the $n \times n$ stability matrix span the null space of the stability matrix in the finite-dimensional space. The generalization to the Hilbert or Banach space case is now clear. The "bad" directions belong to the null space of the symmetric operator. If the null space is finite dimensional and the nonzero eigenvalues of the symmetric operator are bounded away from zero, then the problem reduces to an essentially finite-dimensional problem, and there are no surprises.

In the case that the null space is not finite dimensional, then the dimension of the control space must also be infinite dimensional. It is not known yet what canonical forms (if any) the mapping assumes whose derivative is the symmetric stability operator.

In the finite state variable case the nonzero eigenvalues of the stability operator are necessarily bounded away from zero, because there are only a finite number of eigenvalues. In the infinite-dimensional case this is no longer necessarily true, so that we can expect new phenomena to occur and be associated with loss of stability. At the present time it remains to be seen what a complete list of such phenomena will either look like or correspond to physically.

3. $m > 1$

The extension of the program of Catastrophe Theory from functions ($m = 1$) to mappings ($m > 1$) will probably require a completely different approach from the one outlined in Part 1 of this book, and described in somewhat more detail in Part 4.

In Chapter 5 we saw that structurally stable functions were dense in the space of all functions mapping $\mathbb{R}^n \to \mathbb{R}^1$. As a result, any function could be approximated arbitrarily closely by structurally stable functions. A comparable statement would be desirable in the case of mappings $F: \mathbb{R}^n \to \mathbb{R}^m$. However, the

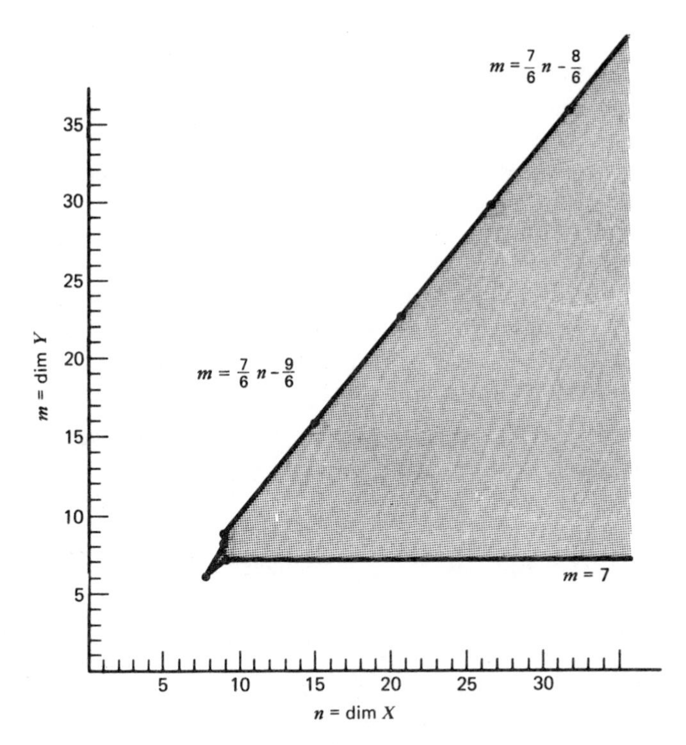

Figure 17.4 Stable mappings $F: \mathbb{R}^n \to \mathbb{R}^m$ are not dense for the dimension pairs (n, m) within the shaded region, including the boundary. Reprinted with permission from M. Golubitsky and V. Guillemin, *Stable Mappings and Their Singularities*, New York: Springer, 1973, p. 163. Copyright 1973 by Springer-Verlag, Inc.

comparable statement is not true. In fact, if we consider mappings F of a manifold X into a manifold Y:

$$X \xrightarrow{F} Y$$

then the set of structurally stable mappings $F: X \to Y$ is not dense in the set of all such mappings if the dimensions n, m of X, Y lie inside or on the boundary of the shaded region in Fig. 17.4.

In the case $m = n$, k-parameter families of mappings may be identified with autonomous dynamical systems. In Chapter 19 we will look at dynamical systems, study what happens under simple degeneracies, and search for universal perturbations in cases of small dimensionality.

Remark. It is not only natural, but necessary to ask questions of the type proposed in these three sections. Extensions of Catastrophe Theory in the three directions indicated might at last provide a sensible language in which to ask and possibly answer questions about the "collapse of the wavefunction" in quantum mechanics. Thus such an extension of Catastrophe Theory could provide a mechanism for treating questions which must now be regarded as philosophical.

4. Symmetry-Restricted Catastrophes

The functions and families of functions considered so far have been "typical" or "in general position." This means that there are no restrictions on or relations among the terms in the Taylor series expansion of the function about any point.

If some sort of symmetry constraint is imposed on a system, this constraint will appear through restrictions on the Taylor series expansion of the corresponding potential function about some point. It is then of interest to carry out the program of Elementary Catastrophe Theory for classes of symmetry-restricted functions. To illustrate how such a study might proceed, we consider the function $F(x, y; c)$ of two state variables (x, y) and k control parameters $c \in \mathbb{R}^k$. We assume that F has a critical point at the origin, and investigate the effect of four different symmetry restrictions on the Taylor series expansion of F.

Example 1. $F(x, y; c)$ is invariant under $x \to \pm x$. The Taylor series expansion of F can contain only even powers of x:

$$F(x, y; c) = F_{00} + F_{01}y + F_{20}x^2 + F_{02}y^2 + F_{21}x^2y + \cdots \tag{17.6i}$$

Since $\partial^{p+q}F/\partial x^p \partial y^q = F_{pq} = (-)^p F_{pq}$, all coefficients F_{pq} with p odd must vanish. Also F_{01} vanishes because the origin is a critical point.

Example 2. $F(x, y; c)$ is invariant under $x \to \pm x$ and $y \to \pm y$. Under this imposed symmetry restriction, $F_{pq} = (-)^p F_{pq} = (-)^q F_{pq} = (-)^{p+q} F_{pq}$ so that all coefficients F_{pq} with either p or q odd must vanish, and

$$F(x, y; c) = F_{00} + F_{20}x^2 + F_{02}y^2 + F_{40}x^4 + \cdots \tag{17.6ii}$$

Example 3. $F(x, y; c)$ is invariant under the 8-element group C_{4v} consisting of reflections through the lines $x = 0$, $y = 0$, $x = y$, $x = -y$ and rotations through $\pi/2, 2\pi/2, 3\pi/2, 4\pi/2 = 0$ radians. (The rotations are generated by pairs of reflections.) Invariance under reflections in the lines $x = 0$, $y = 0$ imposes the constraints discussed in Example 2 above. Invariance under reflections through the line $x = y$ imposes the additional constraint $F_{pq} = F_{qp}$, so that the Taylor series expansion of $F(x, y; c)$ is now

$$F(x, y; c) = F_{00} + F_{20}(x^2 + y^2) + F_{40}(x^4 + y^4) + F_{22}x^2y^2 + \cdots \quad (17.6\text{iii})$$

Example 4. $F(x, y; c)$ is invariant under the rotation group

$$F(x', y'; c) = F(x, y; c)$$

where

$$\begin{pmatrix} x' \\ y' \end{pmatrix} = \begin{pmatrix} \cos\theta & \sin\theta \\ -\sin\theta & \cos\theta \end{pmatrix} \begin{pmatrix} x \\ y \end{pmatrix}$$

Then F is a function only of the one rotation group invariant $r^2 = x^2 + y^2$, and

$$F(x, y; c) = F_0 + F_2 r^2 + F_4 r^4 + \cdots \quad (17.6\text{iv})$$

In each of the expansions presented above the Taylor series coefficients are functions of the k control parameters c, although this has not been explicitly indicated. The canonical forms and universal perturbations for the functions $F(x, y; c)$ can now be studied under each of the imposed symmetry restrictions using the methods of Chapters 3 and 4.

1. *Effect of the Control Parameters.* In general the k control parameters c can be chosen so that k of the leading terms in the Taylor series expansion are zero. If $k = 3$, the leading terms in the Taylor series expansions of $F(x, y; c)$ as given in (17.6i–iv) are, neglecting the constant term,

$$\begin{aligned} F(x, y; c) &\xrightarrow{\text{i}} (0) + F_{03}y^3 + F_{40}x^4 + \cdots \\ &\xrightarrow{\text{ii}} (0) + F_{40}x^4 + F_{22}x^2y^2 + F_{04}y^4 + \cdots \\ &\xrightarrow{\text{iii}} (0) + F_{60}(x^6 + y^6) + F_{42}x^2y^2(x^2 + y^2) + \cdots \\ &\xrightarrow{\text{iv}} (0) + F_8 r^8 + \cdots \end{aligned} \quad (17.7)$$

In case ii one of the three quartic coefficients can also be made to vanish. The results below depend on whether $F_{40} = 0$ or $F_{22} = 0$. Here the terms collected within parentheses and labeled 0 are those that may be annihilated by appropriate choice of the three control parameters.

2. *Effect of a Smooth Change of Variables.* It then remains to be determined which higher degree terms can be transformed away by a symmetry-preserving nonlinear transformation and which cannot. The terms that cannot be transformed away constitute the germ of the k-parameter family $F(x, y; c)$. These symmetry-restricted germs may depend on moduli. In the four cases considered

in Examples 1–4 a smooth symmetry-preserving transformation can be constructed which brings the functions in (17.7) to the following canonical forms:

$$F(x, y; c) \doteq F(x', y'; c) = y'^3 \pm x'^4 \tag{17.8i}$$

$$= \begin{cases} \pm x'^6 + ax'^2y'^2 \pm y'^4, & \text{if } F_{40} = 0 \\ \pm x'^4 \qquad\qquad \pm y'^4, & \text{if } F_{22} = 0 \end{cases} \tag{17.8ii}$$

$$= \pm(x'^6 + y'^6) + ax'^2y'^2(x'^2 + y'^2) \tag{17.8iii}$$

$$= \pm r'^8 \tag{17.8iv}$$

3. *Universal Perturbation.* After the residual functions (17.7) have been brought to canonical form by a smooth change of variables, it is desirable to determine the most general perturbation of the resulting germ. When the imposed symmetry group is sufficiently large, the universal perturbation consists precisely of those terms in (17.7) that have been annihilated by the appropriate choice of control parameters. If the symmetry group is so small that the symmetry is preserved by a change in the origin, linear terms may be present in the perturbation. Thus for the four cases considered in (17.6) and (17.7) the universal perturbations are

$$p(x, y) \to p_{01}y + p_{20}x^2 + p_{21}x^2y \tag{17.9i}$$

$$\to \begin{cases} p_{20}x^2 + p_{02}y^2 + \underline{p_{22}x^2y^2} + p_{40}x^4, & \text{if } F_{40} = 0 \\ p_{20}x^2 + p_{02}y^2 + p_{22}x^2y^2, & \text{if } F_{22} = 0 \end{cases} \tag{17.9ii}$$

$$\to p_{20}(x^2 + y^2) + p_{40}(x^4 + y^4) + p_{22}x^2y^2 + \underline{p_{42}x^2y^2(x^2 + y^2)} \tag{17.9iii}$$

$$\to p_2r^2 + p_4r^4 + p_6r^6 \tag{17.9iv}$$

The general case proceeds analogously. Suppose that F is a k-parameter family of functions of n state variables x and suppose that F is required to be invariant under some symmetry group G acting in $\mathbb{R}^n$. Let $\phi_1(x)$, $\phi_2(x), \ldots,$ $\phi_i(x), \ldots$ be the homogeneous irreducible invariant polynomials on $\mathbb{R}^n$ under G, and assume that $\phi_i(x)$ has degree d_i. The complete set of homogeneous invariant polynomials $\phi_i(x)$ may be determined by character analysis, as was done in Chapter 11, Secs. 8 and 9. Then $F(x; c)$ has an expansion

$$F(x; c) = \sum F_{p_1 \cdots p_i \cdots} \phi_1^{p_1}(x) \cdots \phi_i^{p_i}(x) \cdots \tag{17.10}$$

where the Taylor series coefficient $F_{p_1 \cdots p_i \cdots}$ is the coefficient of a term of degree $p_1d_1 + \cdots + p_id_i + \cdots$ which is invariant under G. We assume that G has no linear invariants, so that $0 \in \mathbb{R}^n$ is a critical point. Then the three stages in the study of Elementary Catastrophe Theory, applied to the symmetry-restricted function (17.10), are:

1. Arrange the coefficients $F_{p_1 \cdots p_i \cdots}$ in order of ascending degree. Use the k degrees of freedom of the control parameters to kill off the first k nonzero Taylor coefficients.

2. Find a smooth symmetry-preserving transformation which removes the "largest number" of higher degree terms. The function which remains is the germ of the function $F(x; c)$ invariant under the symmetry group G.

3. The universal perturbation consists of those terms initially annihilated by the appropriate choice of the control parameters, together with any terms in the germ multiplied by a modulus. Such terms are underlined in (17.9ii) and (17.9iii).

Remark 1. If $F(x;c)$ is assumed invariant under the symmetry group G (acting at the origin) and if $F(x; c)$ has minima but $x = 0$ is not a minimum, then there are several $x^0 \in \mathbb{R}^n$ at which $F(x; c)$ has a minimum. Moreover, the invariance group of $F(x; c)$ about x^0 is a subgroup of G. At x^0 the stability matrix of F may or may not be singular. To illustrate these ideas, we consider a function $F(x; c)$ of n real state variables $x_1, \ldots, x_n$ which is invariant under the real orthogonal group $SO(n)$. Then $F(x; c)$ has the Taylor series expansion about $0 \in \mathbb{R}^n$

$$F(x; c) = \text{constant} + F_2 r^2 + F_4 r^4 + F_6 r^6 + \cdots$$

where $r^2 = x_1^2 + \cdots + x_n^2$.

At $F_2 = 0$ the system undergoes a second-order phase transition if $F_4 > 0$ and a zeroth-order phase transition if $F_4 < 0$ but $F_6 > 0$. If $F_4 > 0$ and $F_6 = F_8 = \cdots = 0$, then F has a nonlocal minimum at

$$r^2 = -\frac{F_2}{2F_4} > 0 \qquad \text{for } F_2 < 0$$

$$F\left(r^2 = -\frac{F_2}{2F_4}; c\right) = -\frac{F_2^2}{4F_4} < 0$$

The minimum is nonlocal because F assumes the same value at all points at a distance $(-F_2/2F_4)^{1/2}$ from the origin. If we fix a point $x^0 = (0, \ldots, 0, r)$, then $F(x; c)$ is invariant under the rotation group $SO(n - 1)$ acting on the first $n - 1$ coordinates $x_1, \ldots, x_{n-1}$. The stability matrix of F, evaluated at x^0 (or any equivalent point) has rank 1, with one positive eigenvalue $\lambda = -4F_2 > 0$ and $n - 1$ eigenvalues zero. This reduction of symmetry on passing from a local minimum at $0 \in \mathbb{R}^n$ to be a nonlocal minimum at $r \neq 0$ is sometimes called "spontaneous symmetry breaking." This concept is illustrated in Fig. 17.5 for $n = 2$, where the group subgroup pair is $O(2)$ and $O(1)$ (including reflections).

Remark 2. It is regrettable that the powerful tool called Group Theory was shunned for so long ($\simeq$ 1929–1959) by the major part of the physics community (solid-state physicists excepted). When its importance was finally and fully recognized, it was adopted with a vengeance as if to make up for lost time. Now many physicists, particle in particular, tend to make Group Theory central to all endeavors. Thus it is that "spontaneous symmetry breaking" is a central topic about which much discussion revolves.

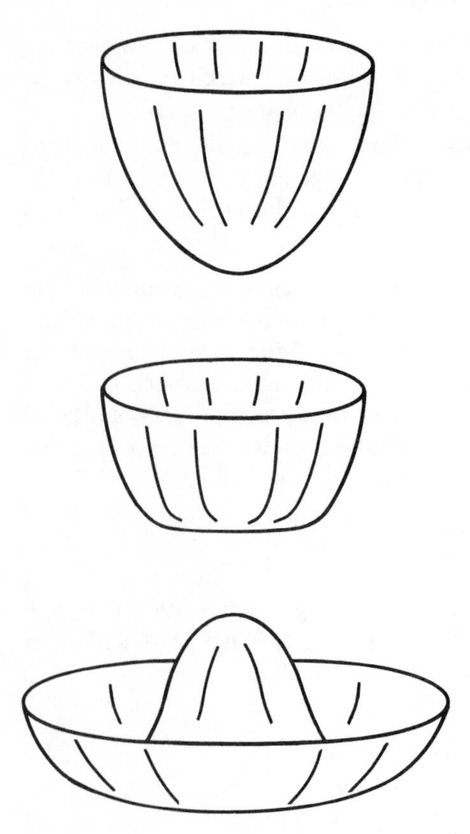

Figure 17.5 The potential $V = x^2 + y^2$ has an $O(2)$ symmetry when expanded about the minimum at $(x, y) = (0, 0)$. The potential $V = -(x^2 + y^2) + (x^2 + y^2)^2$ also has an $O(2)$ symmetry about the origin. However, about any minimum (e.g., $(x_0, y_0) = (1/\sqrt{2}, 0)$) the potential has only the $O(1)$ reflection symmetry $(1/\sqrt{2} + \delta x, \delta y) \to (1/\sqrt{2} + \delta x, -\delta y)$.

This is very unfortunate. There is nothing spontaneous about symmetry breaking. It is a consequence of some dynamics or some variational principle. The word "spontaneous" is a smokescreen behind which hides a vast ignorance of the nonlinear mathematics which ultimately provides a description for the physical processes at hand. It is this nonlinear mathematics which is of central importance. The presence of symmetries allows higher catastrophes to occur at the cost of fewer control parameters. Many different symmetries may play the same restrictive role for a given nonlinear dynamics. The nonlinear mathematics is central, symmetry peripheral. Symmetry exists to be broken. To regard "spontaneous symmetry breaking" as the driving force behind some physical process is to miss entirely that physical process.

5. Constraint Catastrophes

In much of our work so far we have sought to determine the state of a physical system by minimizing a potential. We have assumed implicitly that the potential was defined over all $\mathbb{R}^n$. When this is so, $\nabla V = 0$ is a necessary condition for the existence of a minimum. However, if V is defined only on a closed subset $S \subset \mathbb{R}^n$ (closed: $\bar{S} = S$), this is no longer true. The condition $\nabla V = 0$ is a necessary and sufficient condition for the existence of a local maximum, minimum, or other type of stationary point in the *interior* $S - \partial S$ of S, where ∂S is the boundary of S $[\partial S = \bar{S} \cap (\overline{\mathbb{R}^n - S})]$. But the condition $\nabla V = 0$ is neither necessary nor sufficient for the existence of a local maximum or minimum value on ∂S. In fact, the condition $\nabla V = 0$ at some point in ∂S is not even structurally stable and is destroyed by perturbations.

We illustrate these ideas first in the case $n = 1$ by considering the function $f(x) = x^2$, $-\infty < x < +\infty$. Under the perturbation εx,

$$F(x; \varepsilon) = f(x) + \varepsilon x = (x + \tfrac{1}{2}\varepsilon)^2 - (\tfrac{1}{2}\varepsilon)^2, \qquad -\infty < x < +\infty \tag{17.11}$$

so that the minimum moves from $x = 0$ to $x = -\varepsilon/2$. Therefore the perturbation has no qualitative effect on the critical properties of $f(x) = x^2$, a fact already determined in Chapter 4.

However, if we now consider the function $f(x) = x^2, 0 \leq x < \infty$, the perturbation εx has a pronounced qualitative effect. For $\varepsilon > 0$ there is a global minimum at $x = 0$, and at this minimum $dF/dx > 0$. For $\varepsilon < 0$ there is a global minimum in the interior of $[0, \infty)$ at $x = -\varepsilon/2 > 0$ at which $dF/dx = 0$, and a local maximum at $x = 0$ at which $dF/dx < 0$. The local minimum at $x = 0$ for $\varepsilon > 0$ and the local maximum at $x = 0$ for $\varepsilon < 0$ are present only because of the constraint $x \geq 0$ on the domain of definition of $F(x)$. The function $F(x; \varepsilon = 0)$ $= x^2$ is a separatrix between functions of two qualitatively different types: those with one global minimum and those with a local maximum as well as a global minimum. Equivalently the function $F(x; \varepsilon = 0)$ is the only member in the family of functions $F(x; \varepsilon)$ for which $dF/dx = 0$ *on* the boundary (∂S: $x = 0$) (Fig. 17.6).

The 1-dimensional catastrophe germs A_k: x^{k+1} subject to the constraint $x \geq 0$ may be treated similarly. Since the universal perturbation of restricted $A_1 = x^2$ is 1-dimensional, we expect the universal perturbations of A_k to have one more control parameter in the constrained case ($x \geq 0$) than in the unconstrained case. The reason for this is easy to see. In the unconstrained case the term x^k in the perturbation of $A_k = x^{k+1}$ could be transformed away by a simple shift of origin. In the constrained case the half-line is inhomogeneous (a translation makes it "look different"), so the position of the origin (boundary) is the additional required control parameter. The constrained catastrophe family A_k is given by

$$A_k\colon \quad x^{k+1} + \sum_{j=1}^{k} a_j x^j \tag{17.12}$$

The bifurcation set for the constrained A_2 catastrophe is shown in Fig. 17.7.

The argument proceeds analogously in the case $n > 1$ of many state variables. We assume for convenience that S is the closure of some nice simple connected open set, and that ∂S is a smooth manifold of dimension $n - 1$. The necessary

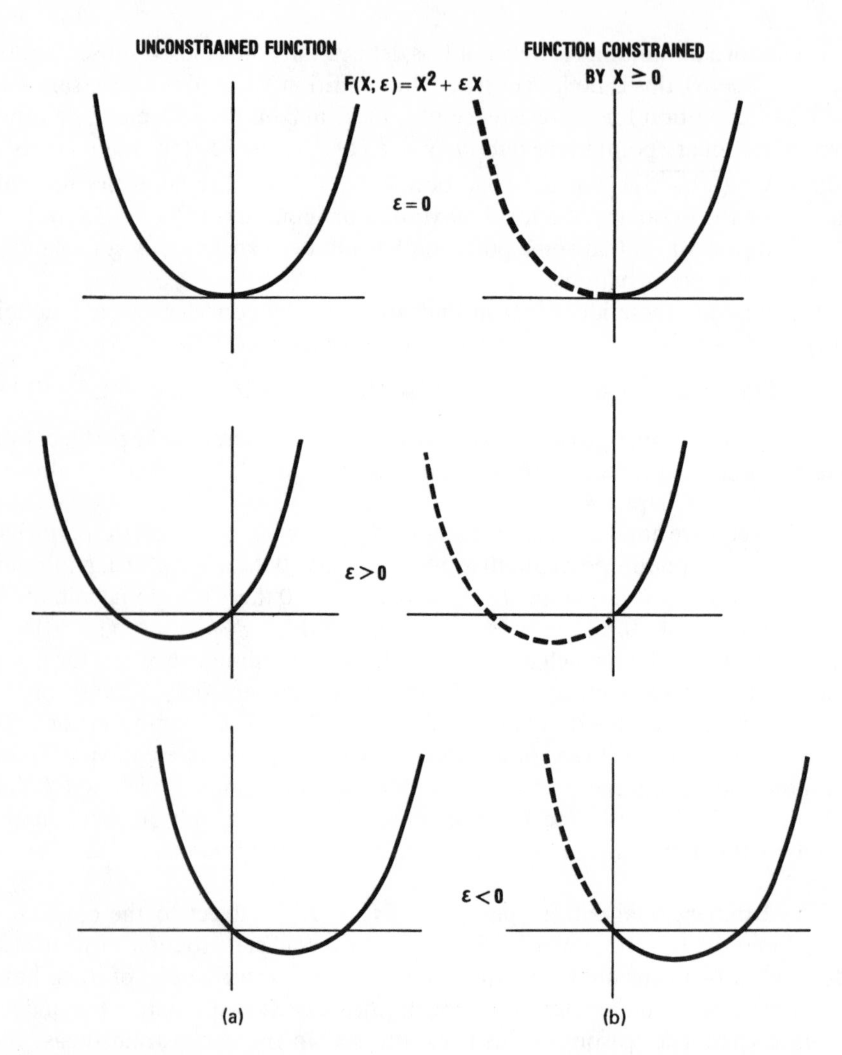

Figure 17.6 (*a*) A perturbation of the unconstrained function produces no qualitative changes in its properties. (*b*) A perturbation of the constrained and structurally unstable function ($\nabla F = 0$ for $x \in \partial S$) produces a qualitative change in its properties.

and sufficient condition for a minimum, maximum, or stationary point of F to occur in the interior of S is $\nabla F = 0$. In the interior of S there is nothing new, and in fact the catastrophes we have studied previously may also be called **interior catastrophes**. It is on the boundary of S that new features arise. Such catastrophes may by duality be called **exterior catastrophes**, or more simply **boundary catastrophes**. These structurally unstable functions have the property that $\nabla F = 0$ at some point in ∂S. Since we have assumed ∂S to be a smooth

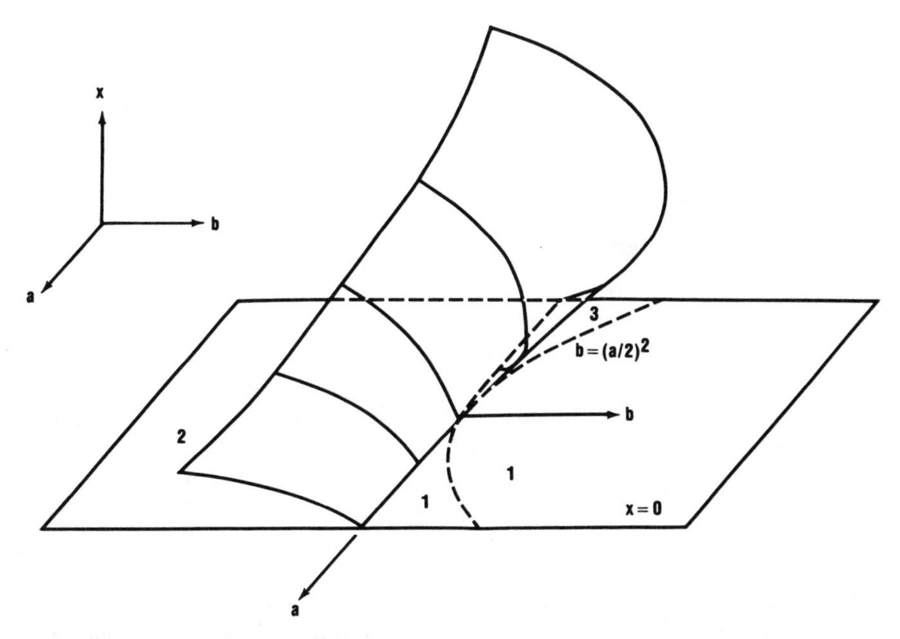

Figure 17.7 The local extrema and the stationary points of the constrained catastrophe $F(x; a, b) = \frac{1}{3}x^3 + \frac{1}{2}ax^2 + bx$ are shown in the x–a–b space $\mathbb{R}^1 \otimes \mathbb{R}^2$. The plane $x = 0$ represents local extrema for all values of the control parameters (a, b). The numbers in the control plane indicate the number of critical points, including local constrained maxima and minima, of the constrained function.

$(n-1)$-dimensional manifold, we may choose $x_2, \ldots, x_n$ as coordinates for ∂S, and use the coordinate x_1 as the constrained coordinate by imposing the condition $x_1 \geq 0$. This means $x_1 < 0$ lies outside S, $x_1 = 0$ is in ∂S, and $x_1 > 0$ is in $S - \partial S$. Pitt and Poston[7] have analyzed constraint catastrophes which have universal perturbations with small numbers of dimensions. Some of their results are presented in Table 17.3.[8]

Remark. It happens frequently that the state variables associated with some system are constrained to be nonnegative. Examples include tension (physics

Table 17.3 Constraint Catastrophes[7,8]

l	k	Germ	Perturbation
1	1	$\pm x^2$	ax
1	2	$\pm x^3$	$ax^2 + bx$
2	2	$xy \pm y^3$	$ax + by$
1	3	$\pm x^4$	$ax^3 + bx^2 + cx$
2	3	$\pm(xy + y^4)$	$ax + by + cy^2$
2	3	$\pm(x^2 + y^3)$	$ax + by + cxy$

and engineering), chemical concentrations (chemistry), population densities (ecology), and so on. When such constraints occur, changes in the qualitative nature of the system state must be determined through the use of the internal (Table 2.2) as well as the external (Table 17.3) catastrophes.

6. Summary

The results presented in Table 2.2 (Thom's list) are the response to a precisely formulated question: What are the simple non-Morse functions that will typically be encountered in k-parameter families of functions $F: \mathbb{R}^n \otimes \mathbb{R}^k \to \mathbb{R}^1$?

If it is desired to extend this table, then it is first necessary to extend this question. As is often the case, a well-posed question can be extended in a number of different ways. We have considered five different ways of enlarging the scope of this question; there are many other ways the question may be extended.

In Sec. 1 we dropped the term "simple" from the question and discussed series and families of catastrophe germs depending on 0, 1, or 2 moduli. These results, due primarily to Arnol'd, probably have little practical interest, although they have enormous intrinsic interest.

In the following section we dropped the restriction $n < \infty$ and thereby extended the domain of the function from the finite-dimensional space $\mathbb{R}^n$ to an infinite-dimensional space. Under certain conditions on the mapping F_{ij} (finite nullity, all nonzero eigenvalues bounded away from zero), the problem of classifying the non-Morse critical points reduces to a finite-dimensional (= nullity of F_{ij}) problem, and ordinary Catastrophe Theory methods and results are applicable. However, new mechanisms for the loss of stability are allowed by the transition from finite to infinite-dimensional spaces, and these are not yet clearly understood.

In Sec. 3 we dropped the restriction $m = 1$ and inquired about mappings $F: \mathbb{R}^n \to \mathbb{R}^m$. Although canonical forms exist for some singularities of such mappings, a major problem exists because stable mappings are not always dense in the space of all such mappings. As a result it is likely that new mathematical techniques will have to be developed in order to treat such cases.

A problem of great practical importance is to determine the types of qualitative changes which can occur in families of functions which are invariant under some symmetry group G. (We drop the word "typical" from the original question.) This problem can be treated by the methods already developed for studying the elementary catastrophes. We have illustrated these procedures by carrying out several examples in Sec. 4.

In Sec. 5 we discussed constraint catastrophes (unbounded domain → closed and bounded domain in the original question). Once again the procedures of Elementary Catastrophe Theory were applicable. Separatrices in families of functions occur when the point p at which $\nabla F = 0$ lies in the boundary ∂S of the domain of definition of the function F.

References

1. V. I. Arnol'd, Critical Points of Smooth Functions and Their Normal Forms, *Russian Math. Surveys* **30**(5), 1–75 (1975).
2. V. I. Arnol'd, Normal Forms of Functions in Neighborhoods of Degenerate Critical Points, *Russian Math. Surveys* **29**, 10–50 (1974).
3. V. I. Arnol'd, Local Normal Forms of Functions, *Invent. Math.* **35**, 87–109 (1976).
4. V. I. Arnol'd, Classification of Unimodal Critical Points of Functions, *Fun. Anal. Appl.* **7**, 230–231 (1973).
5. V. I. Arnol'd, Classification of Bimodal Critical Points of Functions, *Fun. Anal. Appl.* **9**, 43–44 (1975).
6. M. Golubitsky and V. Guillemin, *Stable Mappings and Their Singularities*, New York: Springer, 1973.
7. D. H. Pitt and T. Poston, Determinancy and Unfoldings in the Presence of a Boundary (to appear).
8. T. Poston and I. N. Stewart, *Catastrophe Theory and Its Applications*, London: Pitman, 1978.

CHAPTER 18

Gradient Dynamical Systems

The study of Elementary Catastrophe Theory is concerned with the equilibria of potentials and families of potentials, and how these equilibria change when external control parameters change. Many systems of physical interest are in equilibrium, although many more interesting physical systems are not in equilibrium. For this reason it is desirable to extend the methods and results of Elementary Catastrophe Theory to dynamic (not static) systems.

It is clearly necessary to go beyond Elementary Catastrophe Theory in order to describe the discontinuous phenomena—the sudden jumps—associated with the elementary catastrophes. This is because Elementary Catastrophe Theory cannot possibly describe the dynamical processes involved in the transition from one static equilibrium to another, as reference to Table 1.1 will make clear. In this chapter we make the simplest possible extension of Elementary Catastrophe Theory—to gradient dynamical systems. In the following chapter we treat autonomous dynamical systems.

In Sec. 1 we see the difficulties faced in extending the results of Elementary Catastrophe Theory, for gradient dynamical systems have no canonical forms akin to the list of elementary catastrophes. In Secs. 2–4 we review the methods of phase portraits, as applied to gradient systems. In Sec. 5 we consider how the results of Elementary Catastrophe Theory can be used to squeeze results easily from a problem which might otherwise require a great deal of time and effort to analyze. In Sec. 6 we indicate how to "march" along the critical points of a potential as the control parameters change their values. In Sec. 7 we describe the relation between Catastrophe Theory and Bifurcation Theory.

1. Noncanonical Form for Gradient Systems

Systems whose first-order equations of motion can be derived from a potential $V(x)$ by

$$\frac{dx_i}{dt} = -\frac{\partial V}{\partial x_i} \tag{18.1}$$

are called **gradient systems**. The equations $dx_i/dt = 0$ define the equilibria $\nabla V = 0$ of the system. The study of the equilibria of gradient systems, and how these equilibria move about and change character as the control parameters change, is called Elementary Catastrophe Theory.

The elementary catastrophes are canonical forms for the potential $V(x; c)$ in the neighborhood of non-Morse critical points. A typical potential $V(x; c)$ can be brought to the canonical form for one of the elementary catastrophes by a smooth change of variables when $k \leq 5$. Such transformations facilitate the enumeration and classification of the elementary catastrophes.

It might be hoped that gradient dynamical systems (18.1) could also be transformed to some canonical form by a smooth change of variables in the neighborhood of "bad points." This hope is defeated by the fact that (18.1) is already a canonical form because the left-hand side is a canonical form. Any smooth change of variables which might bring the right-hand side into the canonical form for one of the elementary catastrophes will undo the canonical structure of the left-hand side. This difficulty is not encountered in the study of Elementary Catastrophe Theory because the left-hand side of (18.1) (0) is canonical in all coordinate systems.

2. Phase Portraits: $\nabla V \neq 0$

In order to investigate the properties of the gradient system (18.1) in the neighborhood of some point x^0, we proceed in the canonical way. That is, we expand $V(x)$ in a Taylor series expansion about x^0 and keep only the leading terms. This method, called the method of phase portraits, has already been described briefly in Chapter 5, Sec. 1, but we will elaborate on it somewhat in this and the following two sections.

$$\begin{array}{lll} \nabla V \neq 0 & & \text{Sec. 2} \\ \nabla V = 0, & \det V_{ij} \neq 0 & \text{Sec. 3} \\ \nabla V = 0, & \det V_{ij} = 0 & \text{Sec. 4} \end{array} \tag{18.2}$$

These cases are treated separately as indicated.

If $\nabla V \neq 0$, then

$$\frac{dx_i}{dt} = -V_i + \mathcal{O}(1), \qquad V_i = \left.\frac{\partial V}{\partial x_i}\right|_{x^0} \tag{18.3}$$

Neglecting terms of order 1, (18.3) has solutions

$$(x(t) - x^0)_i = -V_i t$$

The behavior is locally linear, which is exactly the behavior expected at a point where the force $F = -\nabla V \neq 0$.

If the values of the control parameters are changed slightly from c^0 to $c^0 + \delta c^0$, the values of the first derivatives V_i will also change slightly. If $V_i(x^0; c^0) \neq 0$, then in general $V_i(x^0; c^0 + \delta c^0) \neq 0$ so that the qualitative nature of the phase portrait does not change in the neighborhood of a nonequilibrium point.

By previous arguments (Chapter 2, Sec. 1) in this approximation we can always choose a coordinate system so that $dx_1/dt = -1, dx_j/dt = 0, j = 2, \ldots, n$. In the neighborhood of x^0, the motion is rectilinear in the x_1 direction with constant velocity -1.

3. Phase Portraits: Equilibria

If $\nabla V = 0$ but $\det V_{ij} \neq 0$ at x^0, then the point x^0 is a nondegenerate equilibrium. In the neighborhood of x^0 the gradient equations of motion (18.1) become

$$\frac{d}{dt}\delta x_i = -V_{ij}\delta x_j + \mathcal{O}(2) \tag{18.4}$$

where $\delta x_i = x_i - x_i^0$. If we neglect terms of order 2, this is a local linear system. Such a system can be integrated immediately to give

$$\delta x_i(t) = (e^{-Vt})_{ij}\delta x_j(0) \tag{18.5}$$

where V is an $n \times n$ real symmetric matrix, as is $\exp(-Vt)$. However, it is more informative to perform a time-independent linear transformation on (18.4), first bringing it to canonical diagonal form

$$\frac{d}{dt} y_i = -\lambda_i y_i \qquad (\text{no sum}, \lambda_i \neq 0) \tag{18.6}$$

before solving. The solutions of (18.6) are trivial:

$$(y(t) - y^0)_i = \delta y_i(0)e^{-\lambda_i t} \tag{18.7}$$

For $\lambda_i > 0$, $y_i(t)$ approaches y_i^0 as $t \to +\infty$. For $\lambda_i < 0$, $y_i(t)$ moves away from y_i^0. The stability properties of a gradient dynamical system at a Morse equilibrium are characterized by the Morse i-saddle type of equilibrium.

If the values of the control parameters are changed slightly from c^0 to $c^0 + \delta c^0$, then the location x^0 of the equilibrium will change slightly [cf. (5.2)] and the values of the matrix elements V_{ij} at the equilibrium will also change slightly. The eigenvalues will change slightly, but the i-saddle type will remain unchanged, since no eigenvalue will undergo a sign change. As a result, a Morse critical point is structurally stable under perturbation.

However, the dynamical properties near the equilibrium are not necessarily structurally stable under perturbation when det $V_{ij} \neq 0$. Dynamically structurally unstable situations arise when two or more nonzero eigenvalues λ_i of V_{ij} become equal (dynamical degeneracy). The reason for this instability can be seen most easily in the case of a 2-dimensional dynamical system ($n = 2$). If the nondegenerate critical point occurs at $(x, y) = (0, 0)$ and the x and y axes are the principal axes, then the time evolution of the system from initial position $(\delta x, \delta y)$ $(\delta x \neq 0, \delta y \neq 0)$ is given by

$$\begin{aligned} x(t) &= \delta x e^{-\lambda_1 t} \\ y(t) &= \delta y e^{-\lambda_2 t} \end{aligned} \tag{18.8}$$

If both eigenvalues are positive, the system is (dynamically) stable.

We can ask *how* the system approaches the equilibrium. This is easily determined by computing the limit

$$\lim_{t \to \infty} \frac{y(t)}{x(t)} = \left(\frac{\delta y}{\delta x}\right) \lim_{t \to \infty} e^{(\lambda_1 - \lambda_2)t}$$

This limit is 0 for $\lambda_2 > \lambda_1$, $(\pm)\infty$ for $\lambda_1 > \lambda_2$. Thus the system approaches the equilibrium along the x axis if $\lambda_1 < \lambda_2$ and along the y axis if $\lambda_2 < \lambda_1$ (Fig. 18.1). The case $\lambda_1 = \lambda_2$ is (dynamically) structurally unstable, for the system then approaches equilibrium along a straight-line trajectory. Perturbation of this structurally unstable case $\lambda_1 = \lambda_2$ (Fig. 18.1*b*) results in one of the two dynamically structurally stable cases shown in Fig. 18.1*a* and *c*.

This behavior may be indicated schematically as shown in Fig. 18.1 *a*′–*c*′. Dynamical structural stability in n-dimensional gradient systems may be discussed by an extension of the method just described. If a stable nondegenerate equilibrium occurs at x^0, and principal axes $x_1, x_2, \ldots, x_n$ are chosen so that $\lambda_1 > \lambda_2 > \cdots > \lambda_n$ (>0), then the dynamical approach to equilibrium is as follows. The coordinate δx_1 approaches zero most quickly, followed by δx_2, etc. Equality of two or more eigenvalues represents a dynamically structurally unstable situation in which an arbitrarily small perturbation can radically change the trajectory along which the system approaches a locally stable equilibrium.

Remark. This type of structural instability for dynamical gradient systems is somewhat analogous to the Maxwell (equal depth) structural instability for static gradient systems.

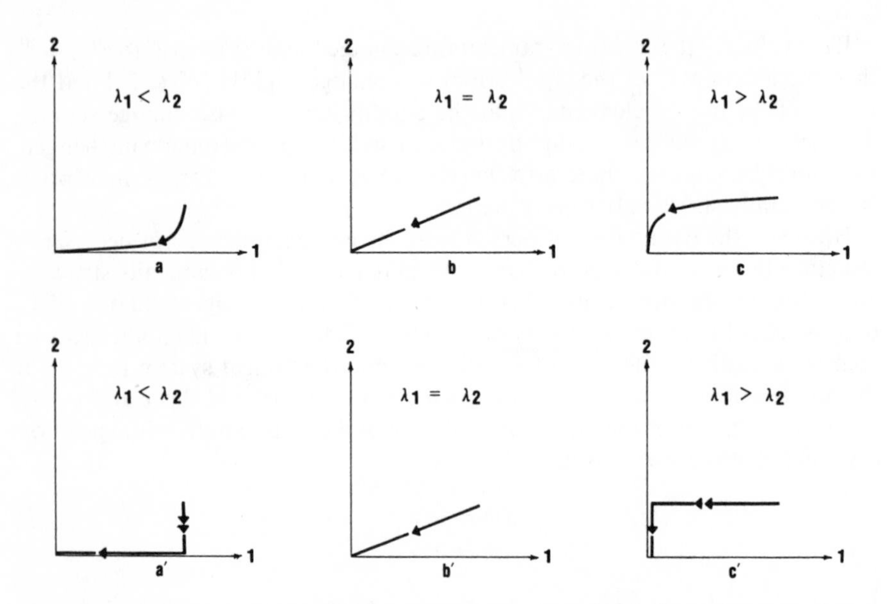

Figure 18.1 The trajectory which a system follows on approach to stable equilibrium is determined by the relative sizes of the positive eigenvalues of the stability matrix (a–c). These trajectories may be represented schematically using a double arrow to indicate faster motion than is indicated by a single arrow ($a' - c'$).

4. Phase Portraits: Degeneracies

If $\nabla V = 0$ and det $V_{ij} = 0$ at x^0, then the phase portrait in the neighborhood of x^0 may be determined by the methods described above. However, it is often easier to determine the phase portrait by:

1. Perturbing the potential
2. Indicating the phase portrait near each isolated critical point
3. Looking at the "far-field" behavior

We illustrate this technique with a series of examples.

Example 1. Determine the phase portrait associated with the potential $V(x, y) = x^3/3 + y^2/2$. The most general perturbation of this potential is

$$V'(x, y; a) = \frac{x^3}{3} + ax + \frac{y^2}{2} \tag{18.9}$$

The associated equations of motion are

$$\begin{aligned}\frac{dx}{dt} &= -\frac{\partial V'}{\partial x} = -x^2 - a \\ \frac{dy}{dt} &= -\frac{\partial V'}{\partial y} = -y\end{aligned} \tag{18.10}$$

There are two isolated critical points at $(x, y) = (+\sqrt{-a}, 0)$ and $(-\sqrt{-a}, 0)$ when $a < 0$. The stability matrix is

$$V_{ij} = \begin{bmatrix} 2x & 0 \\ 0 & 1 \end{bmatrix} \tag{18.11}$$

so that the x and y directions are the principal directions. The phase portraits for the isolated critical points are shown schematically in Fig. 18.2*a*. The phase portrait for the combined system is shown in Fig. 18.2*b* when $|a| < 1/4$. The phase portrait for the degenerate potential with $a = 0$ is shown in Fig. 18.2*c*. This can be obtained from Fig. 18.2*b* either by taking the limit $a \to 0^-$, or by keeping a fixed and extending the range of the x and y axes.

Example 2. The phase portrait associated with the potential $V(x, y) = x^4/4 + y^2/2$ may be determined in a completely analogous way. For simplicity we consider a symmetry-preserving perturbation of $V(x, y)$ of the form

$$V'(x, y; a) = \frac{x^4}{4} + \frac{ax^2}{2} + \frac{y^2}{2} \tag{18.12}$$

We need not consider the most general perturbation since we need only to "morsify" the degenerate critical point, and we will eventually take the limit of vanishing perturbation. The equations of motion are

$$\begin{aligned} \frac{dx}{dt} &= -\frac{\partial V'}{\partial x} = -x^3 - ax \\ \frac{dy}{dt} &= -\frac{\partial V'}{\partial y} = -y \end{aligned} \tag{18.13}$$

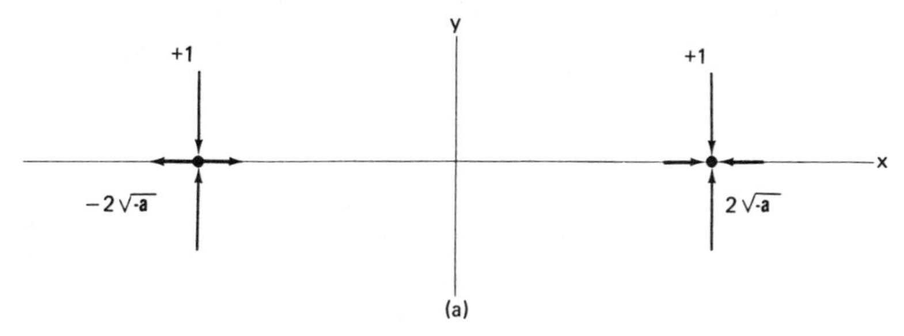

Figure 18.2 The properties of isolated critical points may be indicated by arrows. The arrows lie along the principal axes, pointing toward the critical point if the eigenvalue of V_{ij} is positive (stable), away if it is negative. The length of the arrows indicates the magnitude of the corresponding eigenvalue. (*a*) Properties of the critical points obtained from the non Morse critical point A_2 by morsification.

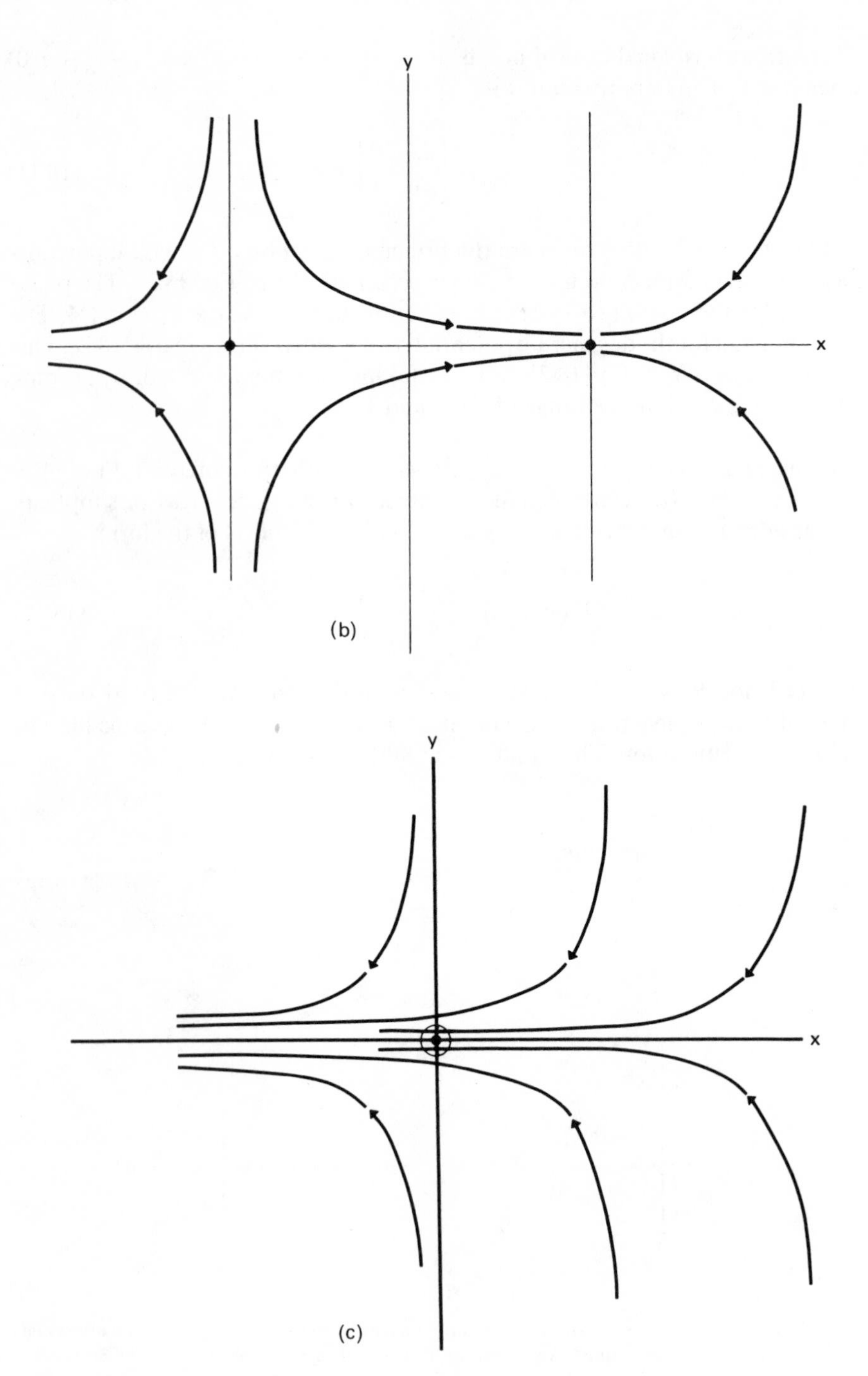

Figure 18.2 (*b*) Corresponding flows. (*c*) Flow near a degenerate critical point obtained by "demorsification" of (*b*).

The phase portraits around the nondegenerate critical points are indicated in Fig. 18.3 ($-\frac{1}{2} < a < 0$). The far-field limit, or $a \to 0^-$ limit, is shown in Fig. 18.3*c*.

Example 3. The phase portrait of the potential $V(x, y) = x^2y - y^3/3$ may be determined in a completely analogous way. The schematic phase portrait has been given already in Fig. 5.19. From this figure we can easily reconstruct the phase portrait for $V(x, y)$, as indicated in Fig. 18.4.

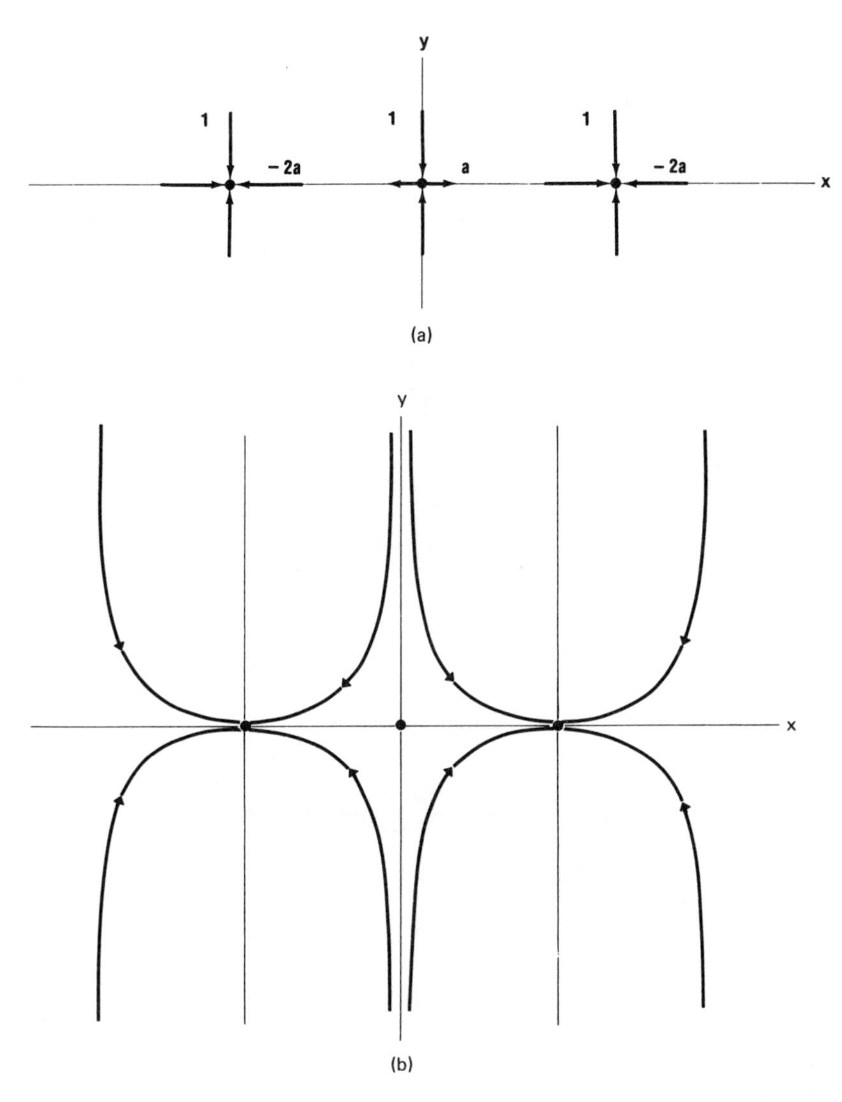

Figure 18.3 (*a*) Morsification of an A_{+3} critical point. (*b*) Corresponding flow.

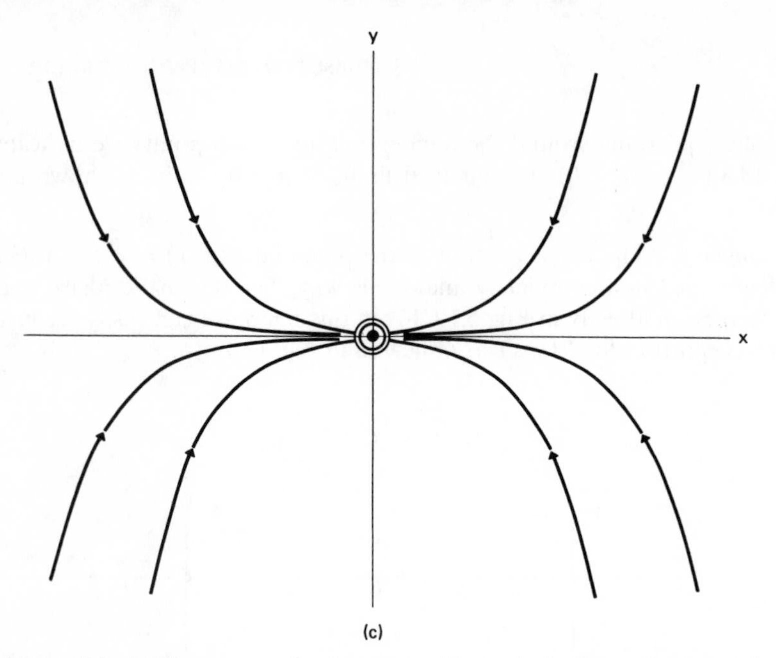

Figure 18.3 (c) Flow near an A_{+3} degenerate critical point by demorsification of (b).

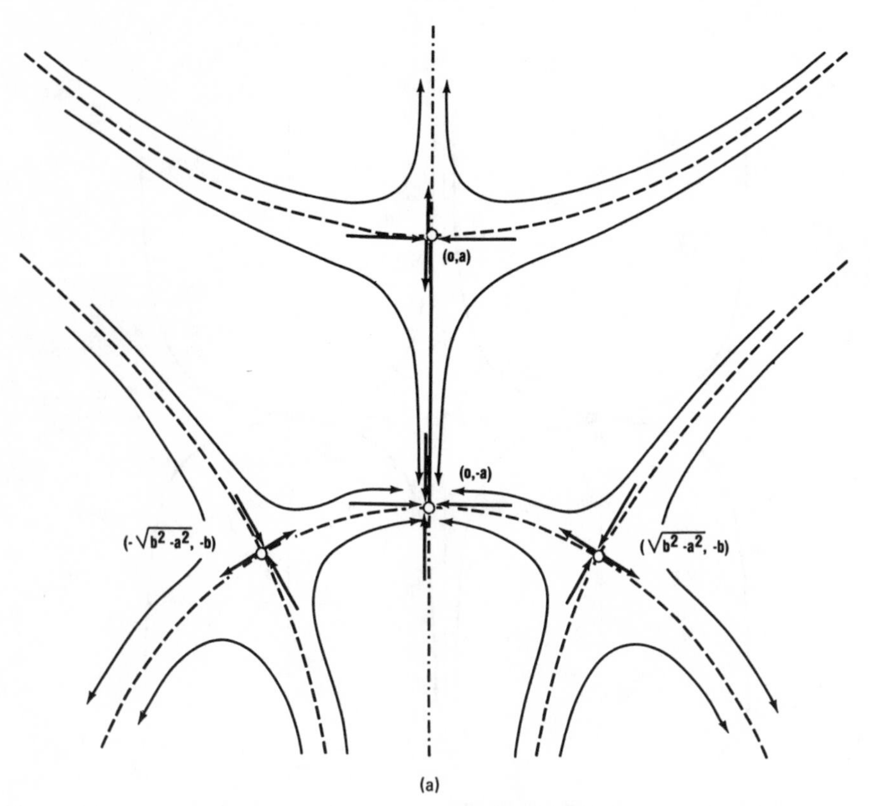

Figure 18.4 (a) Morsification and flows in the neighborhood of a germ $D_{-4} = x^2y - y^3/3$.

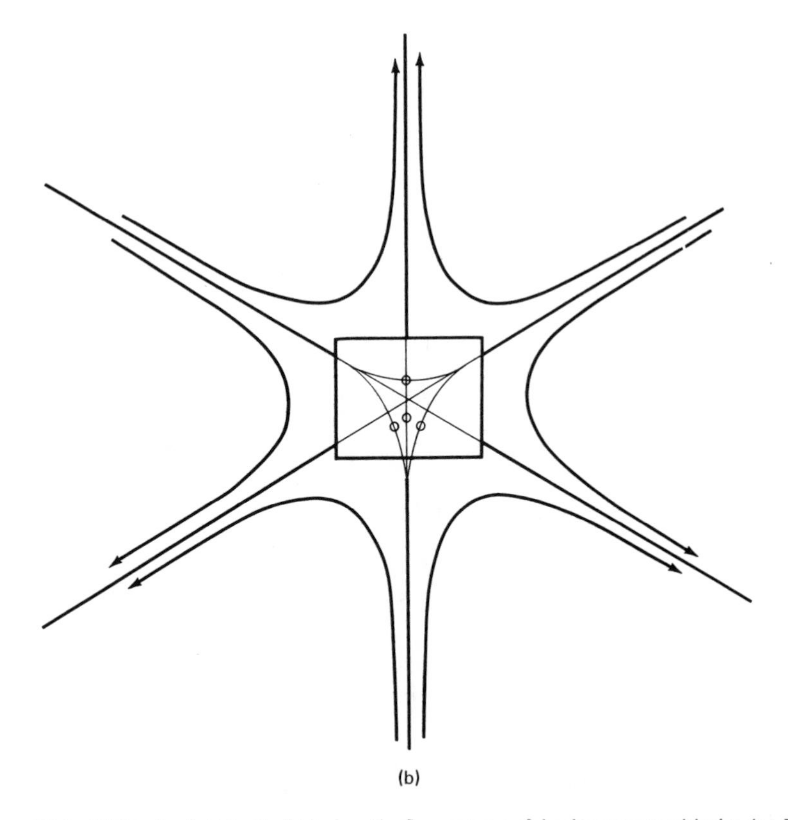

Figure 18.4 (*b*) The far-field limit of (*a*) gives the flow pattern of the degenerate critical point D_{-4}.

The phase portrait for the potential may easily be determined directly, without resorting to perturbation techniques. The equations of motion are

$$\begin{aligned} \frac{dx}{dt} &= -\frac{\partial V}{\partial x} = -2xy \\ \frac{dy}{dt} &= -\frac{\partial V}{\partial y} = -x^2 + y^2 \end{aligned} \tag{18.14}$$

The x component of velocity vanishes along the lines $x = 0$ and $y = 0$, while the y component vanishes along the lines $x = +y$ and $x = -y$ (Fig. 18.5). The structurally unstable trajectories along which the x and y components of velocity are proportional can be located by setting $x = \alpha y$ in (18.14) and solving for α

$$\begin{aligned} \alpha \frac{dy}{dt} &= -2\alpha y^2 \\ \frac{dy}{dt} &= (1 - \alpha^2)y^2 \end{aligned} \tag{18.15}$$

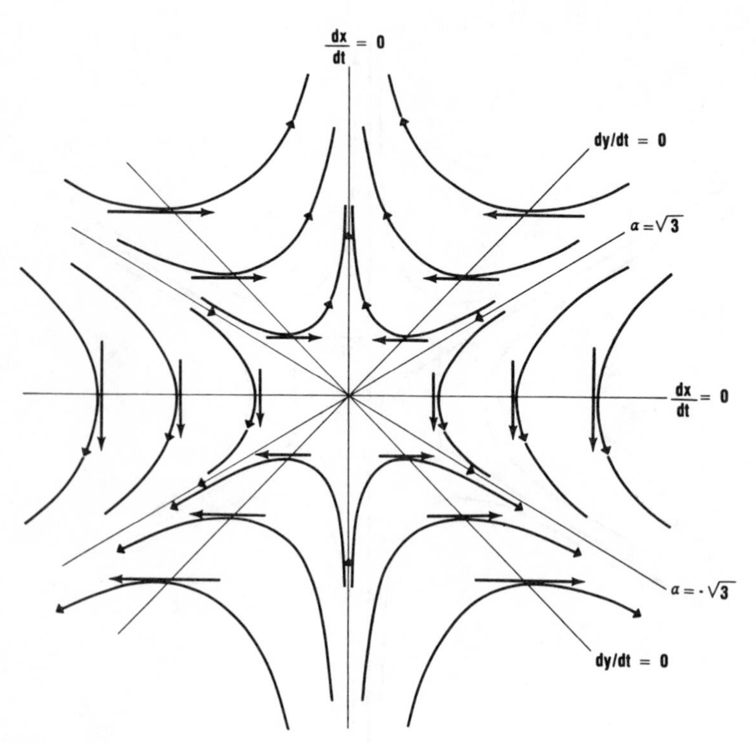

Figure 18.5 The flow in the neighborhood of a degenerate critical point can also be determined by elementary considerations. Here $\dot{x} = 0$ when $xy = 0$, $\dot{y} = 0$ when $x = \pm y$, and separatrices occur when $x = \alpha y$, $\alpha = 0, \pm\sqrt{3}$.

Then

$$\alpha(1 - \alpha^2)y^2 = -2\alpha y^2$$

For arbitrary y this has solutions:

$$\begin{aligned} \alpha &= 0 \quad (y \text{ axis}) \\ \alpha &= \pm\sqrt{3} \end{aligned} \tag{18.16}$$

These separatrices, along with the flow directions on these separatrices, are also shown in Fig. 18.5. From these two pieces of information (the two lines on which $dx/dt = 0$ and the two lines on which $dy/dt = 0$, as well as the three separatrices) it is a straightforward matter to determine the phase portrait.

It may be observed that Fig. 18.5 possesses some symmetry. This symmetry may be made manifest by performing a coordinate transformation and observing how the potential changes. If we rotate from coordinate system x, y to coordinate system x', y' (Fig. 18.6), the coordinate transformation is

$$\begin{bmatrix} x' \\ y' \end{bmatrix} = \begin{bmatrix} \cos\theta & \sin\theta \\ -\sin\theta & \cos\theta \end{bmatrix} \begin{bmatrix} x \\ y \end{bmatrix} \tag{18.17}$$

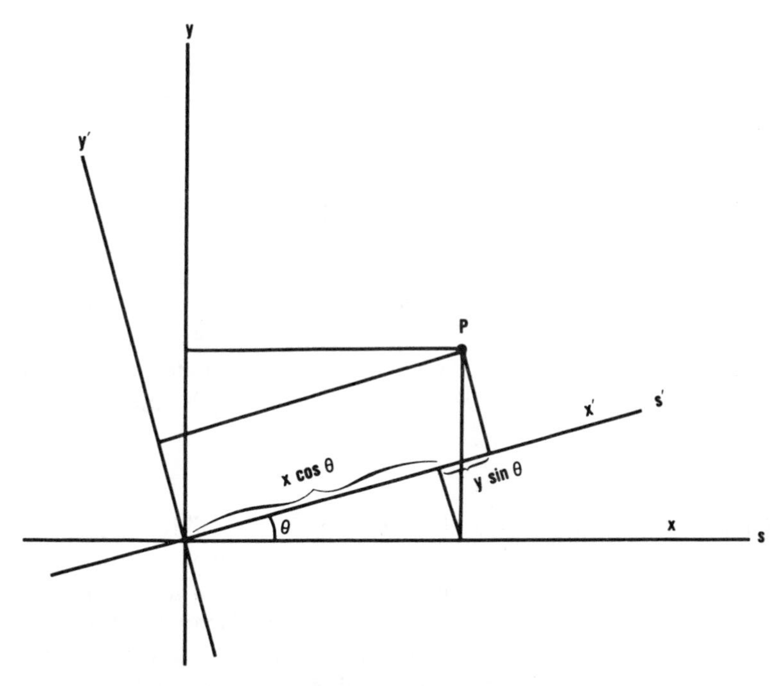

Figure 18.6 Simple rotation of coordinates. Here $x' = x\cos\theta + y\sin\theta$. Similarly, $y' = y\cos\theta - x\sin\theta$.

That is, if the coordinates of any point p are $x(p)$, $y(p)$ in the initial coordinate system S, and the coordinates of the same point p are $x'(p)$, $y'(p)$ in the new coordinate system S', then x, y and x', y' are related by (18.17) (Fig. 18.6). If we choose $\theta = 2\pi/6$ radians (cf. Fig. 18.5) and solve for x, y in terms of x', y', then

$$\begin{bmatrix} x \\ y \end{bmatrix} = \begin{bmatrix} \frac{1}{2} & -\frac{\sqrt{3}}{2} \\ +\frac{\sqrt{3}}{2} & \frac{1}{2} \end{bmatrix} \begin{bmatrix} x' \\ y' \end{bmatrix} \tag{18.18}$$

The function $V(x, y)$ may easily be expressed in the new coordinate system:

$$\begin{aligned} V(x, y) &= \qquad y \qquad \left(x + \frac{y}{\sqrt{3}}\right)\left(x - \frac{y}{\sqrt{3}}\right) \\ &\qquad\quad \downarrow \qquad\qquad \downarrow \qquad\qquad \downarrow \\ V(x', y') &= \left\{\frac{\sqrt{3}}{2}\left(x' + \frac{y'}{\sqrt{3}}\right)\right\}\left\{x' - \frac{y'}{\sqrt{3}}\right\}\left\{\frac{-2}{\sqrt{3}}y'\right\} \\ &= -y'\left(x' + \frac{y'}{\sqrt{3}}\right)\left(x' - \frac{y'}{\sqrt{3}}\right) \end{aligned} \tag{18.19}$$

Under a 60° rotation the sign of the potential is changed. The potential is left invariant under a reflection in the y axis: $(x, y) \to (-x, +y)$, and changes sign under reflection in the x axis: $(x, y) \to (x, -y)$. These symmetries of the potential account for the symmetries of the phase portrait which are apparent in Fig. 18.5.

Remark. For gradient systems associated with a potential $V(x; c)$ depending on k control parameters, it is important to know whether the set of points in $\mathbb{R}^n \otimes \mathbb{R}^k$ ($x \in \mathbb{R}^n$, $c \in \mathbb{R}^k$) at which $V(x; c)$ has degenerate critical points is closed with measure zero. If so, the complementary set, representing points at which V has nonzero gradient or nondegenerate equilibria, is open and dense in $\mathbb{R}^n \otimes \mathbb{R}^k$. The methods of analysis described above are then useful, because the phase portrait near a degenerate critical point can be approximated arbitrarily closely by the phase portraits of perturbed potentials with only isolated critical points.

5. The Branching Tree

We have seen that it is a reasonably simple matter to fill in the phase portrait of a gradient dynamical system once the location, principal axes, and eigenvalues have been determined. If the potential depends on one or more control parameters, then the location of the equilibria, the principal axes, and the eigenvalues will be functions of these controls. Degenerate critical points will occur naturally, and the phase portrait in the neighborhood of such catastrophes may be determined by the methods outlined in the preceding section.

In the case that only one or two state variables x, y are of interest and only one control parameter c is present, the location(s) of the critical points $(x(c), y(c))$ may be drawn as a function of the control parameter c. The number of isolated critical points will in general depend on the value of c, but this number changes only when a catastrophe occurs.

In this section we show how to study the phase portraits associated with a family of potentials $V(x, y; c)$. We shall choose the form of this family so that it is special enough to study easily yet general enough to be representative of a wide class of physically important potentials. We do this by imposing a symmetry restriction, that is, we demand that V be invariant under the transformations

$$\begin{aligned} x &\to \pm x \\ y &\to \pm y \end{aligned} \tag{18.20}$$

This means that $V(x, y; c)$ is an even function of x and y, or equivalently, is a function of x^2 and y^2:

$$V(x, y; c) = f(x^2, y^2; c) \tag{18.21}$$

The equations of motion are

$$\begin{aligned} \frac{dx}{dt} &= -\frac{\partial V}{\partial x} = -\frac{\partial f}{\partial x^2}\frac{d(x^2)}{dx} = -2x\frac{\partial f}{\partial x^2} \\ \frac{dy}{dt} &= -\frac{\partial V}{\partial y} = -\frac{\partial f}{\partial y^2}\frac{d(y^2)}{dy} = -2y\frac{\partial f}{\partial y^2} \end{aligned} \tag{18.22}$$

The critical points of V are determined in the usual way by setting $dx/dt = 0$, $dy/dt = 0$. Clearly the origin $(x, y) = (0, 0)$ is a critical point for all values of c. If $\partial f/\partial x^2 = 0$ at $(x, y) = (0, 0)$ for some value of c, then the origin is a degenerate critical point with x as the "bad" state variable. The stability matrix for V is

$$V_{ij} = 4\begin{bmatrix} \frac{1}{2}\frac{\partial f}{\partial x^2} + x^2 \frac{\partial^2 f}{\partial (x^2)^2} & xy \frac{\partial^2 f}{\partial x^2\, \partial y^2} \\ xy \frac{\partial^2 f}{\partial x^2\, \partial y^2} & \frac{1}{2}\frac{\partial f}{\partial y^2} + y^2 \frac{\partial^2 f}{\partial (y^2)^2} \end{bmatrix} \tag{18.23}$$

For the critical point at the origin, the x and y axes are the principal axes. Also, for all critical points in the x–c plane $y = 0$ and all critical points in the y–c plane $x = 0$, the x and y axes are principal directions. The x and y axes are generally not principal directions for critical points not in a symmetry plane.

From the discussion of Chapter 3 we know that only one eigenvalue generally vanishes at a non-Morse critical point when a single control parameter is present. This is also true in the present symmetry-restricted case. In a family of functions in general position depending on a single control parameter it is typical to encounter only the fold catastrophe A_2. However, the family $V(x, y; c)$ is not in general position, but rather in special position due to the symmetry restriction (18.20). As we have seen in Chapter 17, Sec. 4, it is then typical to encounter cusp catastrophes.

To see under what conditions the fold and the cusp can occur, we assume that (x_c, y_c) is a critical point, and we investigate the symmetries (if any) of $V(x, y; c)$ in the neighborhood of this critical point. This may be done by investigating the relations

$$V(x_c \pm \delta x, y_c \pm \delta y; c) \stackrel{?}{=} V(x_c + \delta x, y_c + \delta y; c) \tag{18.24}$$

This function is not generally invariant under $\delta x \to -\delta x$ if $x_c \neq 0$, but it is necessarily invariant if $x_c = 0$. Therefore if the "bad" state variable occurs in the x direction and $x_c \neq 0$, a fold catastrophe will occur, but if $x_c = 0$, a cusp catastrophe occurs instead. Similar considerations hold for the y direction. These considerations are summarized in Table 18.1.

Table 18.1 A Catastrophe Encountered in a 1-Parameter Family of Functions Invariant under $F(\pm x, \pm y; c) = F(x, y; c)$ is Typically a Fold or a Cusp Catastrophe According to the Symmetry at the Degenerate Critical Point

Bad State Variable	Location of Degenerate Critical Point			
	$x = 0, y = 0$	$x \neq 0, y = 0$	$x = 0, y \neq 0$	$x \neq 0, y \neq 0$
x	Cusp	Fold	Cusp	Fold
y	Cusp	Cusp	Fold	Fold

We are now in a position to see qualitatively what can happen to the symmetry-restricted potential $V(x, y; c)$ as the control parameter c is varied. Along the high-symmetry solution $x(c) = 0$, $y(c) = 0$, the eigenvalues associated with the x and y directions are functions of c. If the x eigenvalue goes through zero, then a cusp catastrophe occurs, and two new solutions bifurcate in the x-direction. We can compute the stability matrix along these new solutions. Since $x(c) \neq 0$, $y(c) = 0$ on these new solutions, the stability matrix is diagonal, and the eigenvalues are associated with the x and y directions. If the x eigenvalue vanishes along this new branch, a fold catastrophe occurs. If the y eigenvalue vanishes, then a cusp catastrophe occurs, and two additional solutions bifurcate in the y direction. Along these new solution sets $x(c) \neq 0$, $y(c) \neq 0$, and any catastrophes which occur must be folds. There are no new solutions bifurcating from critical branches with $x(c) \neq 0$, $y(c) \neq 0$.

The bifurcation diagram (Fig. 18.7) has the shape of a (Mondrian) tree. Therefore we call the solution sets $x(c)$, $y(c)$ for $\nabla V(x, y; c) = 0$ **branches**. The universal solution $x(c) = 0$, $y(c) = 0$, which is required by symmetry, is called the **trunk** or the **zeroth branch**. Primary branches bifurcate from the zeroth branch. Secondary branches bifurcate from the primary branch, and so on. In the present example there are no tertiary branches because there is no symmetry along secondary branches.

To make these considerations more concrete, we assume that $0 \leq c < \infty$ and

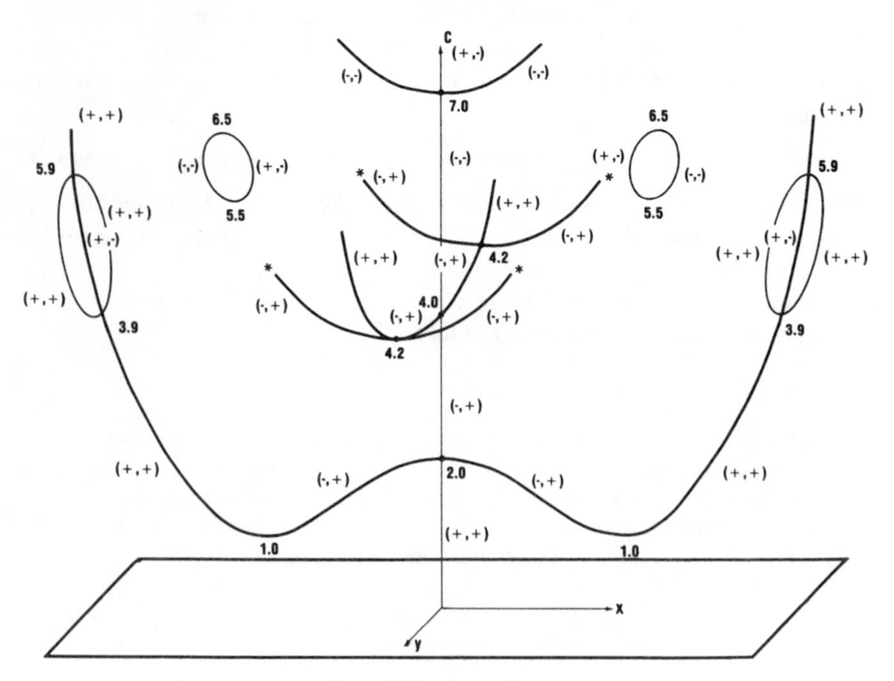

Figure 18.7 Branching tree.

that for $c = 0$, $V(x, y; c)$ has only one critical point at the origin. We also assume that along this zeroth branch, the eigenvalue associated with the x direction vanishes at $c = 2.0$ and $c = 7.0$, while the eigenvalue associated with the y direction vanishes at $c = 4.0$. A useful branch label (beside the pedigree "zeroth," "primary," . . .) is the inertia, or the sum of signs of the eigenvalues of the stability matrix. This is an invariant between bifurcations and **turnarounds** (described below), and changes along a branch every time an eigenvalue goes through zero and changes sign along that branch. The inertia along the zeroth branch may be determined without effort. At $c = 0$, $V(x, y; c)$ has a minimum, so the inertia is $(+, +)$. At $c = 2.0$, the x eigenvalue vanishes. Therefore the inertia is $(+, +)$ for $0.0 \leq c < 2.0$, and $(-, +)$ for $2.0 < c < 4.0$. In what follows, the first and second signs within parentheses refer to the x and y eigenvalues. At $c = 4.0$ the y eigenvalue vanishes, so that the inertia is $(-, -)$ for $4.0 < c < 7.0$. Finally at $c = 7.0$ the x eigenvalue vanishes again, so the inertia is $(+, -)$ for $7.0 < c$. These effortless considerations reveal that the solution $x(c) = 0$, $y(c) = 0$ is unstable for $c > 2.0$.

Let us now analyze the first of the primary branches. We assume that the catastrophe at $x = 0$, $y = 0$, $c = 2.0$ is of type A_{-3}. Since the bifurcation is in the x direction, in the neighborhood of the bifurcation point

$$V(x, y; c) \simeq \text{constant} + (2.0 - c)x^2 - ax^4 + by^2 \tag{18.25}$$

where $a > 0$, $b > 0$, and higher order terms and positive numerical factors are unimportant and have been suppressed. In the neighborhood of this non-Morse critical point, the primary branches have the form $x(c) \simeq \pm[(2.0 - c)/2]^{1/2}$, $y(c) = 0$. They "bend" with the canonical half-power law dependence toward smaller values of c. On the primary branch the x direction is unstable, but the y direction must still be stable by arguments involving continuity of the eigenvalues of $V_{ij}(x, y; c)$ in the neighborhood of $(x, y; c) = (0, 0; 2.0)$. Therefore the inertia on the primary branch near the bifurcation point is $(-, +)$.

Now this primary branch cannot continue indefinitely toward smaller values of c. This is because we have assumed that $V(x, y; c)$ has only one critical point at $c = 0$, that at $(x, y) = (0, 0)$. Nor can the primary branch simply disappear by ending. Further, in a 1-parameter family we can typically meet only the fold (no symmetry) or the cusp (symmetry). This is then uniquely what happens to the primary branch. For $0 < c < 2.0$ (say at $c = 1.0$), along the branch $x(c) \neq 0$, $y = 0$, the x eigenvalue vanishes a second time in a canonical fold catastrophe. We can say that the primary branch "turns around" and moves in the direction of increasing c. The inertia on the "upswing" is $(+, +)$ because the x eigenvalue has again gone through zero.

Remark 1. We may also say, more picturesquely, that the attractor and saddle "collide and annihilate" each other as a function of decreasing c at $c = 1.0$. Or for c increasing, the attractor and saddle are "spontaneously created" at $c = 1.0$, and the saddle is subsequently absorbed by the zeroth branch.

Along the primary branch $x(c) \neq 0$, $y(c) = 0$, $V(x(c) + \delta x, \delta y; c)$ is invariant under $\delta y \to -\delta y$, but not under $\delta x \to -\delta x$. Therefore vanishing eigenvalues for the x direction are associated with additional "turn-arounds," while vanishing eigenvalues for the y direction are associated with bifurcation from the primary to a secondary branch characterized by $y(c) \neq 0$. At these secondary bifurcation points the tangent to the secondary branch is perpendicular to the x–c plane. We shall assume that two secondary branches bifurcate from this primary branch at $c = 3.9$ and at $c = 5.9$, and that both bifurcations are of type A_{+3}. The potential in the neighborhood of $x(c) \neq 0$, $y = 0$, $c = 3.9$ has the form

$$V(x(c) + \delta x, y; c) \simeq a(\delta x)^2 + (3.9 - c)y^2 + by^4 \tag{18.26}$$

where $a > 0$, $b > 0$, and higher order terms have been truncated as before. There are two local solutions $y(c) \neq 0$ for $c > 3.9$, none for $c < 3.9$. Therefore this secondary branch bends upward. The secondary branch carries an inertia $(+, +)$, while the primary branch carries an inertia $(+, -)$ for $3.9 < c < 5.9$. At $c = 5.9$ the y eigenvalue again vanishes. The A_{+3} catastrophe at $x(c) \neq 0$, $y = 0$, $c = 5.9$ has the form

$$V(x(c) + \delta x, y; c) \simeq a(\delta x)^2 + (c - 5.9)y^2 + by^4 \tag{18.27}$$

$a > 0, b > 0$. This has two local solutions $y(c) \neq 0$ for $c < 5.9$, none for $c > 5.9$. Therefore the secondary branch bifurcating at this A_{+3} catastrophe initially turns down. This secondary branch also has inertia $(+, +)$. This branch may or may not be the same as the secondary branch bifurcating at $c = 3.9$. If it is, then the potential in the neighborhood of the primary branch has the approximate form

$$V(x(c) + \delta x, y; c) \simeq a(\delta x)^2 + \tfrac{1}{2}(c - 3.9)(c - 5.9)y^2 + by^4 \tag{18.28}$$

In this case we can say that the primary branch "throws off" stable secondary branches, at $c = 3.9$, and these branches are reabsorbed by the primary branch at $c = 5.9$. The other possibility is that the secondary branch bifurcating at $c = 5.9$ first turns down and later, for some value of $c > 0.0$, turns up again in a fold catastrophe for reasons previously indicated. The inertia along this branch on the upswing would be $(+, -)$. The first of these two possibilities is illustrated in Fig. 18.7.

We now turn our attention to the second primary branch, which bifurcates from the trunk at $c = 4.0$. We assume an A_{+3} catastrophe in the y direction. The primary branch turns upward and carries away an inertia $(-, +)$. This primary branch is characterized by $x = 0$, $y(c) \neq 0$. A secondary bifurcation in the x direction is still possible. We assume this occurs at $c = 4.2$ and has type A_{-3} so that

$$V(x, y(c) + \delta y; c) \simeq (c - 4.2)x^2 - ax^4 + b(\delta y)^2 \tag{18.29}$$

where $a > 0, b > 0$. There are two local solutions $x(c) \neq 0$ for $c > 4.2$, none for $c < 4.2$. Therefore the secondary branch bifurcating at this A_{-3} catastrophe turns upward. The inertia along the primary branch for $c > 4.2$ must be $(+, +)$ because the x eigenvalue has changed sign. On the secondary branch it must be

$(-, +)$ because this secondary branch $x(c) \neq 0$, $y(c) \neq 0$ is unstable in the x direction.

We consider now the third primary branch which bifurcates from the trunk at $c = 7.0$ in the x direction. The x eigenvalue goes from negative to positive as c increases through 7.0. If we assume an A_{-3} catastrophe at $(x, y; c) = (0, 0; 7.0)$, the primary branch must turn up and carry away an inertia $(-, -)$.

Remark 2. If a bifurcation from an n^{ary} to an $(n + 1)^{\text{ary}}$ branch occurs at $c = c^0$, then the inertia of the $(n + 1)^{\text{ary}}$ branch, which exists locally on only one side of c^0, is the same as the inertia of the n^{ary} branch on the opposite side of c^0. Equivalently, if for c increasing an $A_{\pm 3}$ catastrophe occurs with eigenvalue changing from $\pm$ to $\mp$, the bifurcating solution turns up or down according to

		Eigenvalue Change, c Increasing	
		$+ \to -$	$- \to +$
Catastrophe	A_{+3}	Up	Down
	A_{-3}	Down	Up

Remark 3. The tangent to the $(n + 1)^{\text{ary}}$ branch bifurcating from the n^{ary} branch at c^0 is parallel to the eigenvector of the stability matrix with vanishing eigenvalue, evaluated on the n^{ary} branch at c^0.

Remark 4. The solution set $x(c)$, $y(c)$ for $\nabla V(x, y; c)$ need not be connected. A saddle and a local extremum may "spontaneously form" as a function of increasing c. If neither is absorbed by a branch connected to the trunk, the total solution set is disconnected. These new solutions may undergo their own bifurcations and turnarounds. They may continue indefinitely toward increasing c, or they may collide and annihilate each other, forming isolated "bubbles"

Table 18.2 Location and Type of Critical Points for the Branching Tree Example

Location	Catastrophe Type	Bifurcation Direction	$m^{\text{ary}} \to n^{\text{ary}}$
2.0	A_{-3}	x	$0 \to 1$
4.0	A_{+3}	y	$0 \to 1$
7.0	A_{-3}	x	$0 \to 1$
1.0 $x \neq 0$	A_2	x	$1 \to 1$
3.9 $x \neq 0$	A_{+3}	y	$1 \to 2$
5.9 $x \neq 0$	A_{+3}	y	$1 \leftrightarrow 2$
4.2 $y \neq 0$	A_{-3}	x	$1 \to 2$
5.5 $x \neq 0$	A_2	x	Disconnected
6.5 $x \neq 0$	A_2	x	primary branch

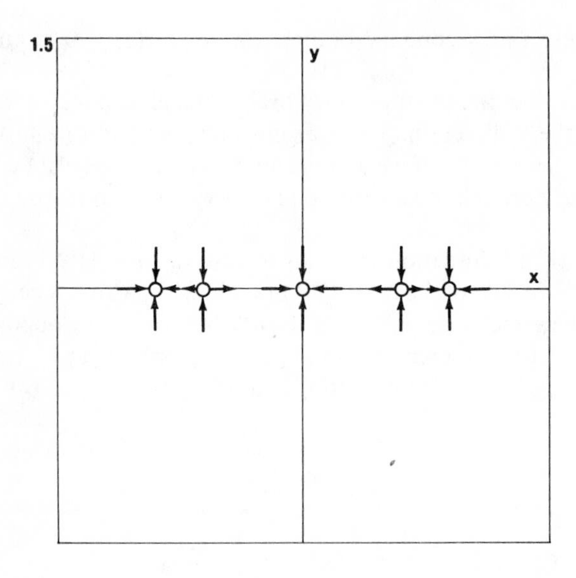

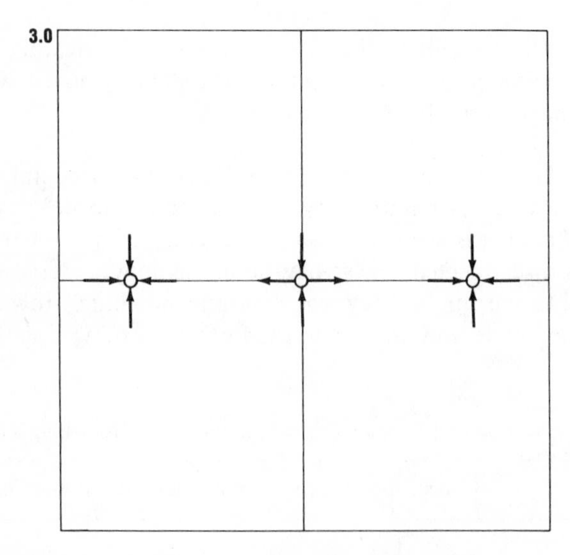

Figure 18.8 Intersections of the branching tree with various planes c = constant. Critical points within any plane are related in very precise ways with critical points in nearby planes. The skeleton of critical points within a plane can be used to reconstruct the flow pattern within that plane.

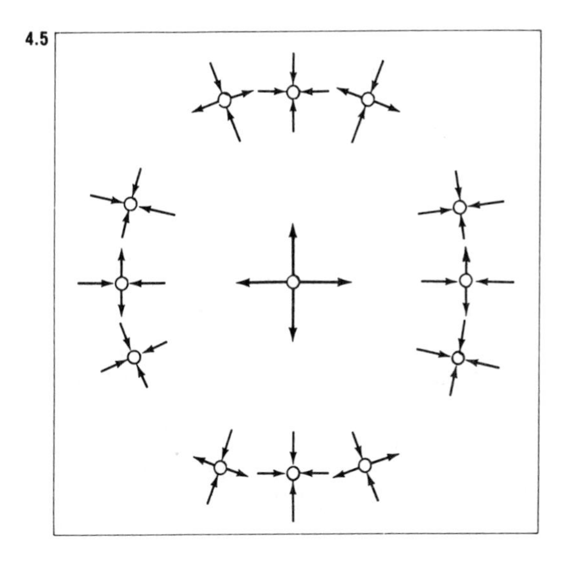

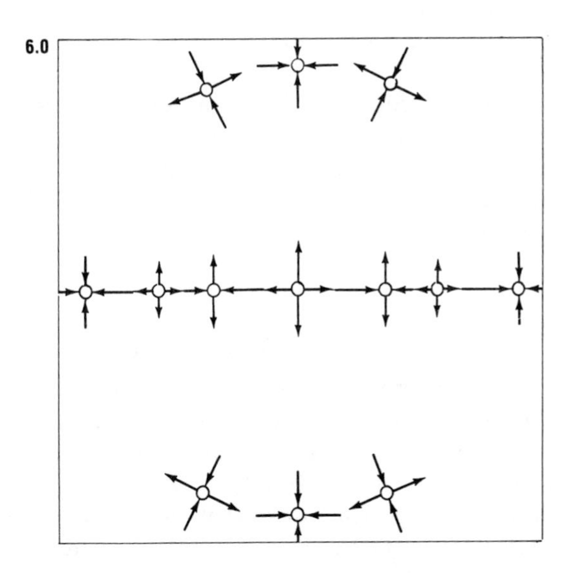

Figure 18.8 (*Continued*)

in state variable space. The inertia assignments on such bubbles are often easily guessed. Two disconnected "bubbles" are shown in Fig. 18.7 for $5.5 \leq c \leq 6.5$.

Remark 5. We have determined completely the qualitative properties of a gradient system simply from a knowledge of the location of the non-Morse critical points in $\mathbb{R}^2 \otimes \mathbb{R}^1$. The qualitative features of the "equilibrium tree" were determined by a knowledge of the location and type of the only catastrophes that could occur $(A_2, A_{\pm 3})$ in this 1-parameter example. Further, the inertia of the stability matrix on each solution branch was completely determined (without computation anywhere!) simply from the knowledge of the inertia at the one point $(x, y; c) = (0, 0; 0)$. We summarize in Table 18.2 the location, type, and implications of each non-Morse critical point assumed for this example.

Remark 6. The set of solutions $x(c)$, $y(c)$ to $\nabla V(x, y; c) = 0$ that is connected to the zeroth branch has a simple genaeology. Many primary branches can bifurcate from the trunk; they are labeled by the order in which they bifurcate. Likewise, many secondary branches can bifurcate from each primary branch, and they are also labeled accordingly. In the present example no tertiary branches can occur. For example, the first secondary branch from the second primary branch is labeled with asterisks (*) in Fig. 18.7.

In Fig. 18.8 we show all equilibria that occur for values $c = 1.5, 3.0, 4.5, 6.0$. These are simply intersections of the equilibrium tree with the four horizontal planes $c =$ constant. A phase portrait for the gradient system can easily be drawn in each plane by the methods indicated earlier in this chapter. Instead, we simply indicate the stability type of each equilibrium.

6. Marching

The qualitative (topological) methods developed in Secs. 2–4 and applied in Sec. 5 are useful because they provide a great deal of information about a gradient system (or a dynamical system) at the cost of relatively little labor. In fact, topology was originally invented as a branch of mathematics by Poincaré for precisely this reason.

We summarize here the principal steps that may be taken for the study of a gradient system that depends on external control parameters.

1. For a fixed value of the control parameters c, all equilibria $\nabla V = 0$ are determined. The stability type and principal axes of each equilibrium are determined. These can then be used as a skeleton on which the phase portrait may be "fleshed out" over the entire state variable space. Separatrices and basins of attraction can also be determined as described in Chapter 5, Secs. 1 and 2.
2. As the control parameters change, the locations of the equilibria also change.

In particular, the attractors, the basins of attraction, and the separatrices change. However, the qualitative features of the phase portrait remain unchanged as long as det $V_{ij} \neq 0$ along each equilibrium. The occurrence of a degenerate critical point forces a qualitative change in the system phase portrait, and conversely. In fact, if along one critical branch $x^i(c)$, det V_{ij} is zero at c^0, then there must be additional critical points in the neighborhood of $(x^0; c^0) \in \mathbb{R}^n \otimes \mathbb{R}^k$.

For gradient dynamical systems depending on a single control parameter s, the algorithm described above may be implemented by a procedure called **marching**. The relationship between the change in location of a critical point and the change in the control parameter s is given by (5.2):

$$V_{ij}\delta x^j + V_{is}\delta s = 0 \tag{18.30}$$

When the stability matrix V_{ij} is nonsingular, the changes δx^j are given by

$$\delta x^j = -(V^{-1})^{ji} V_{is} \delta s \tag{18.31}$$

Equation (18.31) is eminently practical for machine computations. By choosing the "marching step" size δs sufficiently small, it is a simple matter to march along a critical branch between degenerate critical points.

Of course, the interesting things happen precisely where V_{ij} becomes singular and the marching equation (18.31) fails. In Sec. 5 we have seen what kinds of interesting things can happen at a degenerate critical point:

1. The original critical branch turns around.
2. New branches bifurcate from the original branch.

If (x^0, s^0) is a degenerate critical point, the "bad" direction in state variable space is easily determined from the equation

$$V_{ij}\delta x^j = 0$$

That is, the eigenvector $(\delta x^1, \delta x^2, \ldots, \delta x^n)$ of V_{ij} with zero eigenvalue is tangent to the original branch at turnaround or to the bifurcating new branch.

It is a simple matter to determine whether a turnaround (A_2) or a bifurcation (A_3) is associated with a degenerate critical point. (We describe the situation for more complicated catastrophes later.) If $(\delta x)^0$ is the eigenvector of V_{ij} at (x^0, s^0) with zero eigenvalue and $(\delta x)^{-1}$ represents the change in the coordinates of the critical point during the march from $s^0 - \delta s$ to s^0, then

1. If $(\delta x)^{-1}$ is approximately $\parallel$ to $(\delta x)^0$, an A_2 catastrophe occurs at (x^0, s^0).
2. If $(\delta x)^{-1}$ is approximately $\perp$ to $(\delta x)^0$, an A_3 catastrophe occurs at (x^0, s^0).

The occurrence of a degenerate critical point is an obstruction to the integration of the "marching equation" (18.31). However, since the elementary catastrophes have canonical properties, if we know the type of catastrophe that occurs at a degenerate critical point, we can use its canonical properties to "integrate through" the degeneracy. We illustrate how this can be done for both the fold and the cusp catastrophes.

Fold. We assume that $V(x; s)$ has a degenerate critical point at $(x^0; s^0)$. The values of $(x^0; s^0)$ may be determined by extrapolation, as shown in Fig. 18.9. The critical curvature of the fold has a canonical square root dependence (cf. Fig. 18.9*a*), so that a plot of $(V'')^2$ versus s in the neighborhood of the degenerate

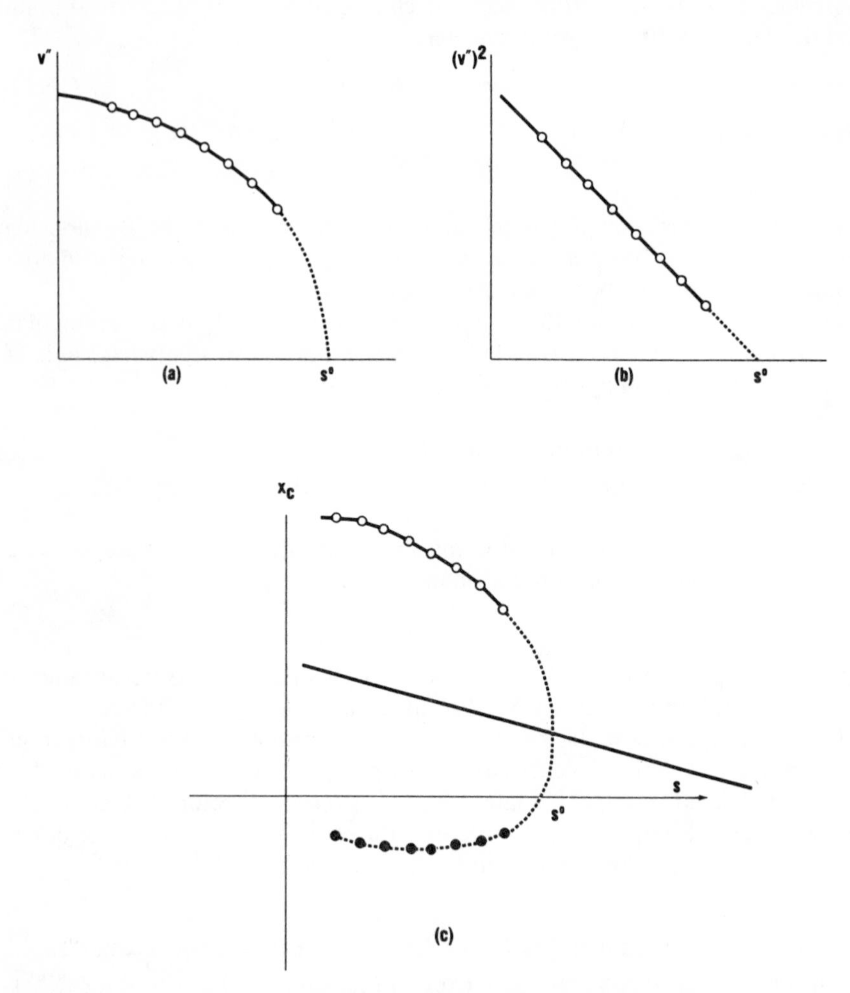

Figure 18.9 Marching along a critical point is feasible away from a catastrophe germ. When the stability matrix V'' becomes small (*a*), the canonical form for the appropriate catastrophe may be used to locate the germ by extrapolation (*b*). A canonical catastrophe cross section can then be used to step through the catastrophe germ (*c*).

critical point provides a good estimate of s^0 by a simple linear extrapolation (Fig. 18.9*b*). Once s^0 is known, a parabola tangent to the vertical hyperplane $s = s^0$ can be fitted through the critical points $x(s^0 - n\delta s)$, $n = 1, 2, \ldots$. The location of the branch after the turnaround at s^0 can be determined by reflection, as shown in Fig. 18.9*c*. Once we have "stepped through" the critical point in this manner, the marching equation can once again be used to determine the wanderings of the critical branch, taking care to change the sign of the marching step δs.

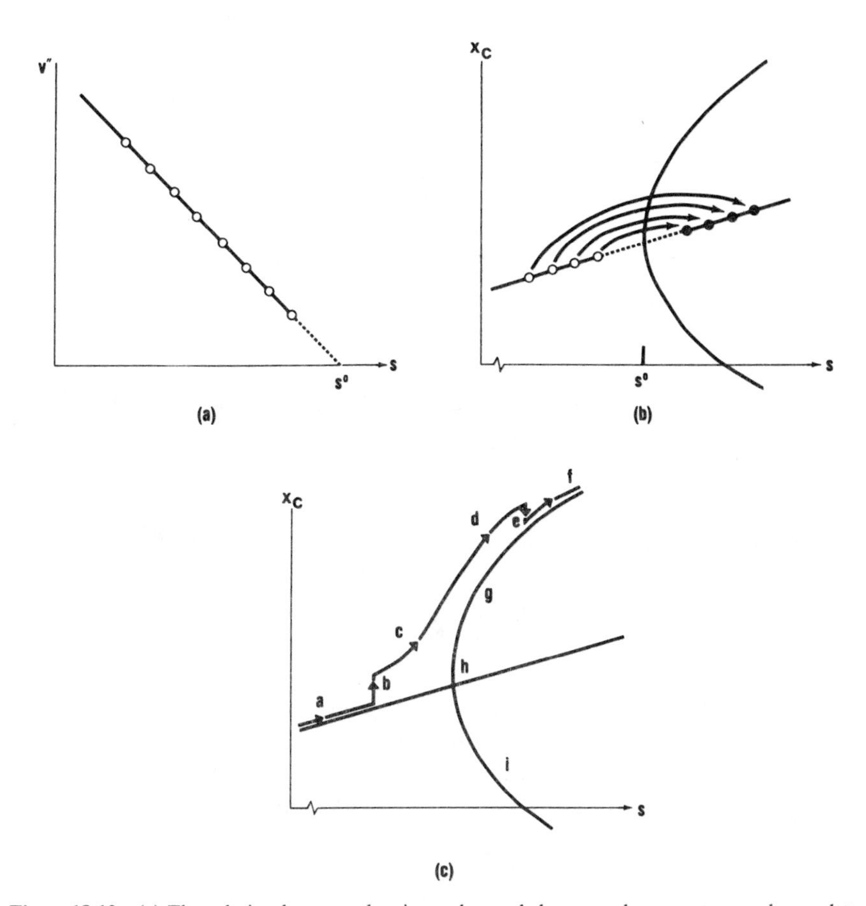

Figure 18.10 (*a*) The relation between the eigenvalue and the control parameter can be used to infer the type of catastrophe present as well as its location. (*b*) The initial branch of a cusp can be continued simply by extrapolating past the degenerate critical point. The extrapolations need not be linear. (*c*) Bifurcating branches may be located by "perturbation theory." Here we march past *a* on the initial branch, and at *b* we add a universal perturbation to the potential, taking care to remain at a critical point, now displaced. Marching resumes (*c*, *d*) past the degenerate critical point. At *e* the perturbation is removed, and the system state moves back to the bifurcated branch. Marching can resume in the same direction (*f*, . . .) or in the opposite direction (*g*, *h*, *i*, . . .) to map out the new branch. A canonical step-through procedure must be used to get past *h*.

Cusp. In this case the vanishing eigenvalue approaches zero linearly, so that the location of the cusp point may be determined by extrapolation, as in Fig. 18.10*a*. If the cusp point occurs at $(x^0 = x(s^0), s^0)$, then points on the initial branch on the far side of s^0 have coordinates given approximately by

$$x(s^0 + i\delta s) \simeq x^0 + (x^0 - x(s^0 - i\delta s)) \tag{18.32}$$

as shown in Fig. 18.10*b*. Points on the bifurcating branch may be determined as follows. Back up to $s - \Delta s$ (where $\Delta s \sim 10$ or more marching steps) and add a small perturbation to the potential which vanishes outside the interval $(s_0 - \Delta s, s_0 + \Delta s)$. A perturbation in general position will "unhinge" the cusp bifurcation. As a result, the marching equation can be integrated past the catastrophe without encountering obstructions and will lead to the bifurcating branch on the far side of $s^0 + \Delta s$.

The symmetrically placed bifurcating branch can now be determined by various methods:

1. Integrating from $s_0 + \Delta s$ along the unperturbed branch in the negative s direction and "stepping through" the cusp point
2. Returning to $s_0 - \Delta s$, using the negative of the original perturbation, and marching as before
3. Integrating the original perturbation in the negative s direction beginning on the central branch at $s_0 + \Delta s$
4. Symmetry

The behavior of the bifurcating solution set becomes more interesting when the nullity is 2. For example, this can occur for the potential

$$V(x, y; s) = k_1(s_1 - s)x^2 + k_2(s_2 - s)y^2 + x^4 + y^4, \qquad k_1 > 0, \quad k_2 > 0 \tag{18.33}$$

For $s_1 \neq s_2$ single bifurcations occur at four points, as indicated in Fig. 18.11*a*. As $s_1 \to s_0$, $s_2 \to s_0$, the bifurcating solution set at s_0 becomes 2-dimensional, with four distinct branches emerging from the "trunk" at s_0 (Fig. 18.11*b*). In the event that the excess nullity is a result of symmetry, as is the case for $k_1 = k_2 = k$,

$$V(x, y; s) = k(s_0 - s)(x^2 + y^2) + (x^2 + y^2)^2 \tag{18.33$'$}$$

the bifurcating solution set is 2-dimensional, corresponding to a parabola of revolution (Fig. 18.11*c*). On this solution set the stability matrix has one nonzero eigenvalue and one zero eigenvalue. The latter indicates that the stationary point (in this case, a minimum) corresponding to this "branch" is nonlocal in nature. The shape of the potential for various values of s is indicated in Fig. 18.11*d*. In general, when a function is restricted to be invariant under some continuous group, bifurcating critical sets are nonlocal in nature with one or more eigenvalues of the stability matrix being zero.

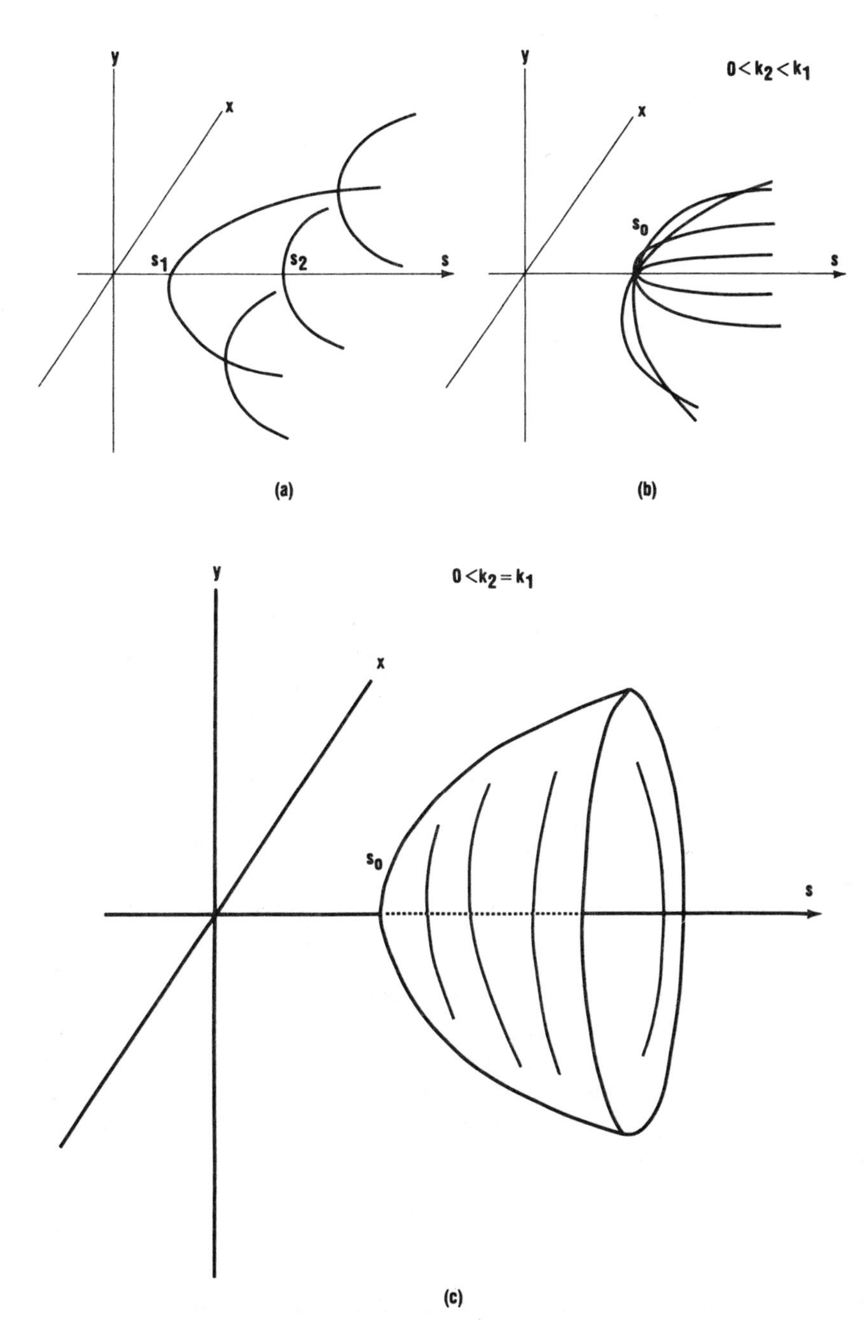

Figure 18.11 (*a*) The critical set of the potential (18.33) exhibits four catastrophes of type A_{+3}. (*b*) When the bifurcations from the trunk coincide ($s_1 = s_2$), all four branches bifurcate from the same point. (*c*) If the 2-dimensional space of bifurcating solutions is due to a symmetry (18.33′), the bifurcation set consists of a paraboloid.

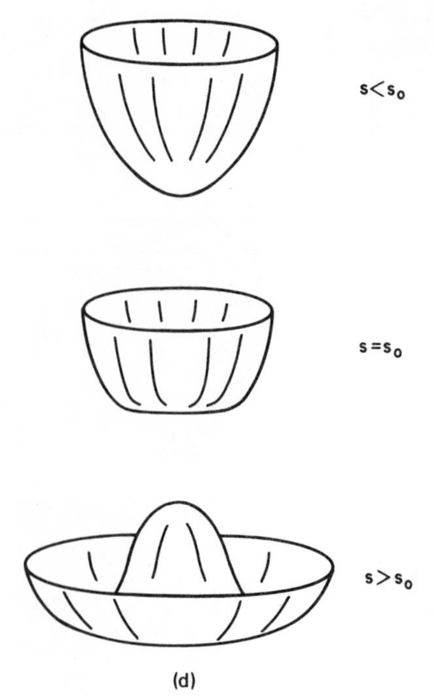

Figure 18.11 (*d*) The corresponding potential is invariant under the group $O(2)$, and is shown for $s < s_1$, $s = s_1$, and $s > s_1$.

7. Relation Between Catastrophe Theory and Bifurcation Theory

Bifurcation Theory is essentially the study of how new solutions to an equation or system of equations can "bifurcate" from some known solution as a parameter changes. An important class of bifurcation problems arises in the search for stationary values of a function. We can therefore expect a close relationship between Elementary Catastrophe Theory and this class of bifurcation problems. More generally, Bifurcation Theory, as a study of new and qualitatively different solutions to a set of equations, falls within the program of Catastrophe Theory.

To make our considerations concrete, we consider a potential function $V(x; s)$ of n state variables $x \in \mathbb{R}^n$ and a single control parameter $s \in \mathbb{R}^1$. We assume that $V(x; s)$ has a critical point at $x = 0 \in \mathbb{R}^n$ for all values of s. Then a Taylor series expansion around this critical point leads to

$$V(x; s) = V(s) + x^i V_i(s) + \frac{1}{2!} x^i x^j V_{ij}(s) + \cdots \tag{18.34}$$

As before, all derivatives are computed at the critical point, the constant term is unimportant, and $V_i(s) = 0$. The critical point remains isolated as long as $\det V_{ij}(s) \neq 0$. When the stability matrix becomes singular, another solution branch (of $\nabla V = 0$) is connected to the original branch. This branch may bifurcate from the initial branch, or the initial branch may "turn around" [in which case we cannot set the critical point at $x(s) = 0$ for all s]. Both possibilities have been discussed in the previous section.

The direction in which a new solution bifurcates is easily determined from (18.34). It is the direction in which a perturbation does not change the value of the potential in second order:

$$\delta x^i \delta x^j V_{ij} = \delta x^i (V_{ij} \delta x^j) = 0$$

Therefore the initial direction of a new bifurcating solution is given by the eigenvector(s) of the stability matrix with zero eigenvalue(s). In practice, the most frequently encountered bifurcation corresponds to the "trident," or symmetric cross section of the cusp catastrophe (Fig. 10.2). In fact, the example discussed in Section 5 consists of a number of isolated bifurcations of this type (plus "turnarounds").

As the parameter s changes, the number of solutions of the equation $\nabla V(x; s) = 0$ may change. If the new solutions are connected to the branch $x(s) = 0$, they may be tracked by the marching algorithm discussed in the previous section. However, if the total solution set consists of two or more disconnected components, the marching algorithm, starting from the trunk $x(s) = 0$, is inadequate for determining the disconnected solution sets.

However, we know that a change in the number of solutions of $\nabla V(x; s) = 0$ is connected with a catastrophe. It is therefore useful to consider the general bifurcation problem as a problem of determining the solution set of

$$\nabla V(x; c) = 0, \qquad x \in \mathbb{R}^n, \quad c \in \mathbb{R}^k \tag{18.35}$$

for some 1-dimensional curve

$$c(s) = (c_1(s), c_2(s), \ldots, c_k(s)) \in \mathbb{R}^k \tag{18.36}$$

Although there is a very large number of possible curves in control parameter space, the behavior of $V(x; c)$ at any point $c \in \mathbb{R}^k$ is canonical and known for the elementary catastrophes. For example, if the potential has a cusp critical point, then in the neighborhood of this point

$$V \doteq \frac{x^4}{4} + \frac{ax^2}{2} + bx \tag{18.37}$$

The general bifurcation problem then reduces to the problem of determining the solution set $\nabla V = 0$ for any path $a(s)$, $b(s)$ in control parameter space depending on the single control s (arc length). The solution set is simply the intersection of the cusp catastrophe manifold $\nabla V = 0$ with the 2-dimensional ruled surface $(x, a(s), b(s))$ whose points are parameterized by x, s. This gives a 1-dimensional solution set, as indicated in Fig. 18.12. Other paths give other solution sets (Fig. 18.13).

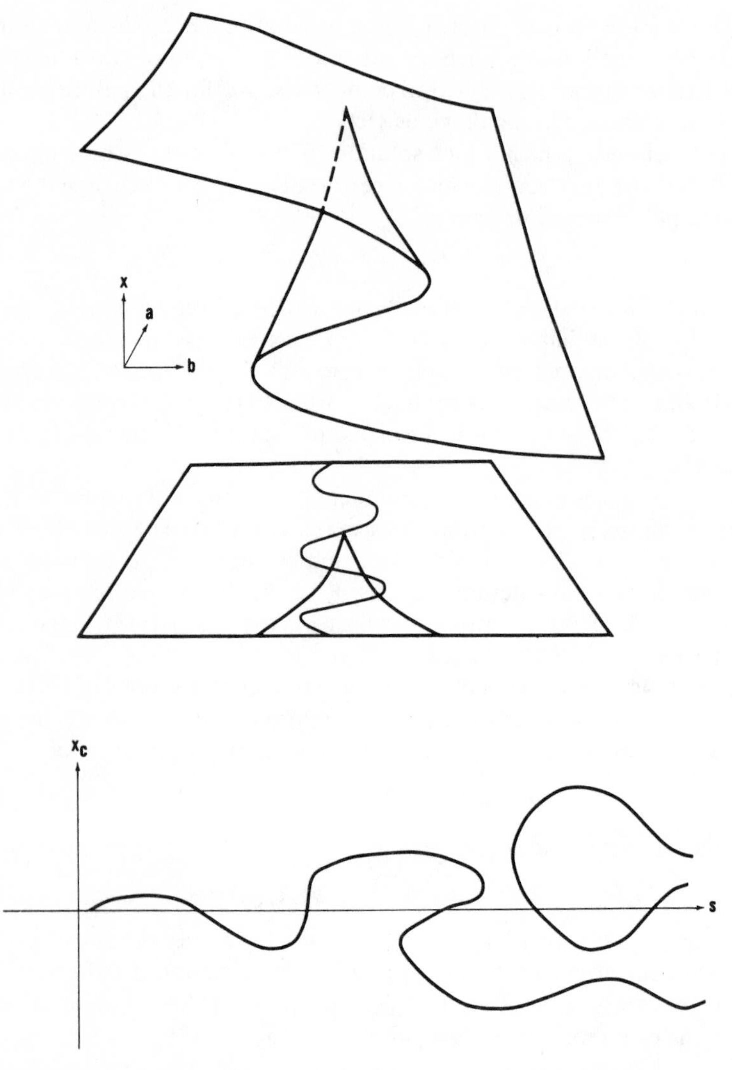

Figure 18.12 Simple curves in the cusp catastrophe control plane, when lifted to the critical manifold, can give rise to peculiar looking solution sets. Here s parameterizes distance along the curve in the control parameter plane.

Remark. The disconnected solution set shown in Fig. 18.14a can be associated with the catastrophe A_3, but the one shown in Fig. 18.14b cannot be associated with any catastrophe smaller than A_5.

In the event that the solution set is disconnected, the disconnected pieces may be determined using the marching algorithm in the following manner.

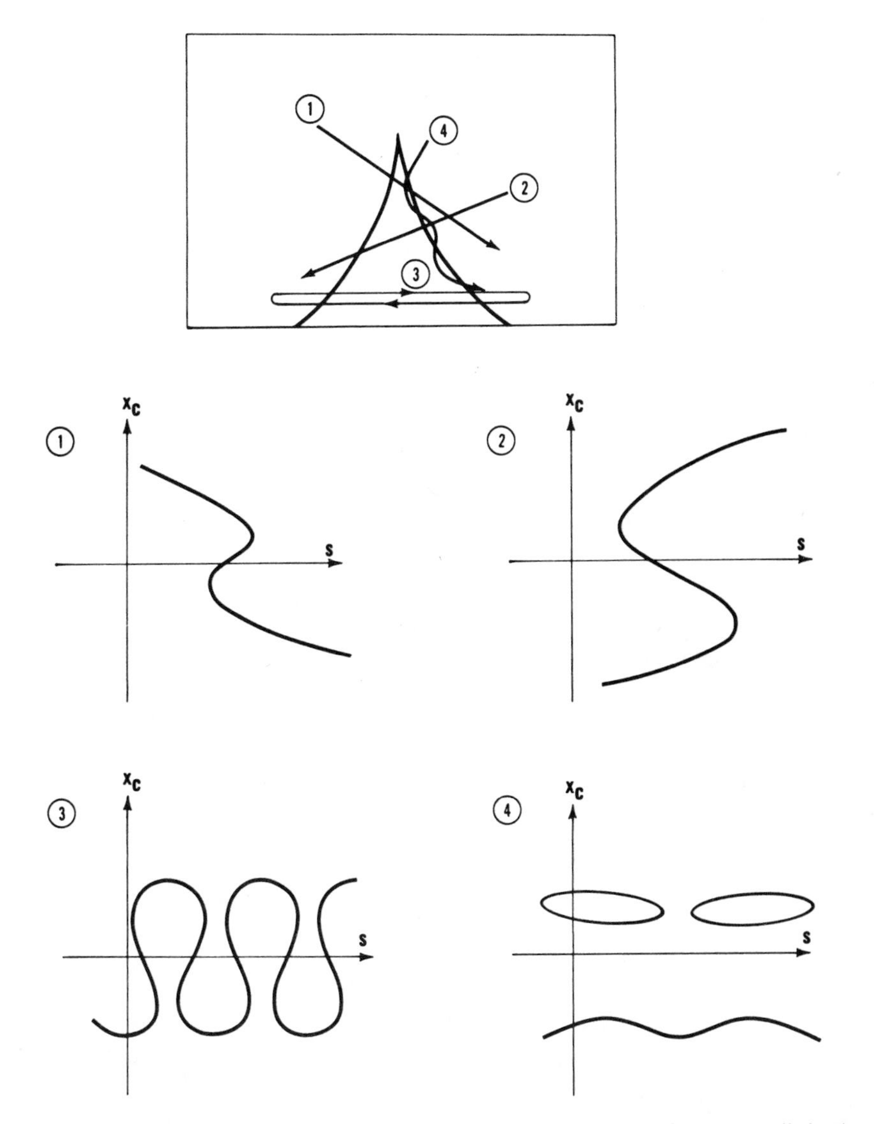

Figure 18.13 Different curves in the control parameter plane can give rise to very distinctive solution sets. Golubitsky and Schaeffer[1] have determined that there are at most 53 distinct irreducible solution sets that can be obtained from different curves in the cusp control plane.[1]

1. Add a perturbation to the function, or in particular change the values of the (control) parameters which appear in the function. This must be done "adiabatically," taking care to remain at an equilibrium while the perturbation is being carried out.
2. When the perturbation has been made, march along the solution set by varying the parameter s. If the topology of the new solution set differs from that

Figure 18.14 (*a*) This solution set can be obtained from the cusp. (*b*) This cannot. It requires a catastrophe possessing at least five critical points. The A_5 catastrophe is the most economical candidate.

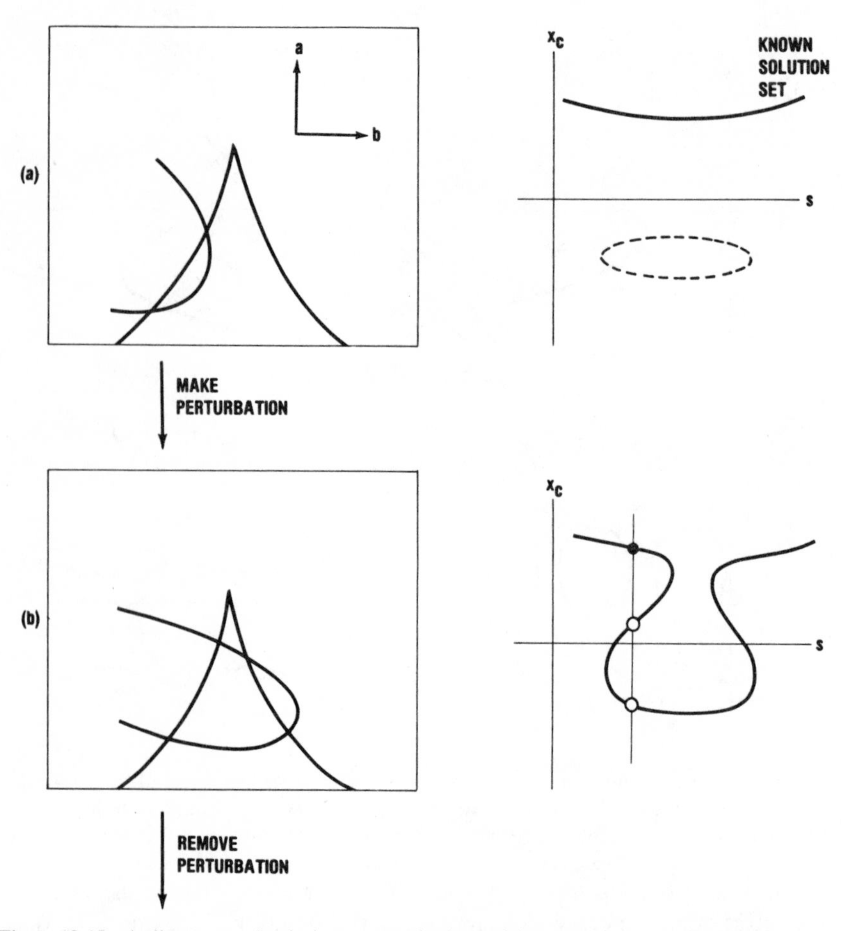

Figure 18.15 A disconnected solution set can be located by marching, together with the use of Catastrophe Theory methods. (*a*) The connected solution set is determined by marching. (*b*) For fixed value of the arc length *s*, a perturbation is adiabatically added until the total solution set is connected. As the perturbation is added, the single known critical point at fixed value of *s* (full circle) is followed. The kink in the solution set is located by marching. Standard step-through methods must be used. A new critical point at the same value of *s* is located.

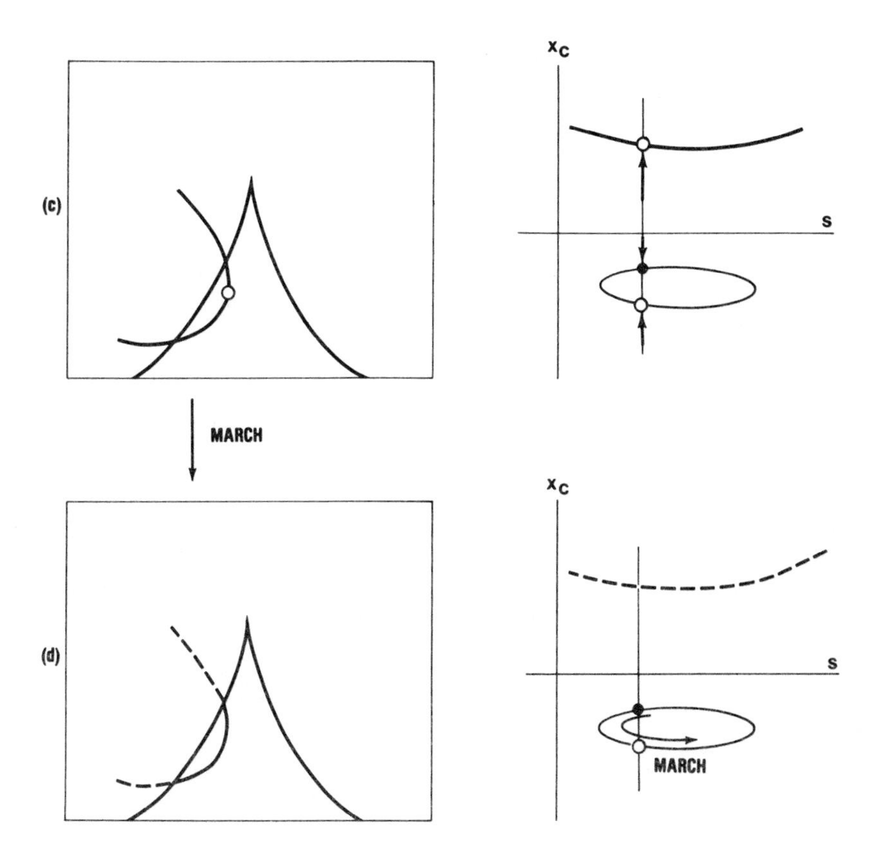

Figure 18.15 (*c*) This new critical point (full circle) is followed as the perturbation is adiabatically removed. (*d*) The disconnected solution set of the original unperturbed potential is mapped out by marching.

of the original solution set, disconnected solutions may have been "attached" to the original solution set by the perturbation.

3. Choose values of s where the numbers of solutions of $\nabla V = 0$ differ between the unperturbed and perturbed functions.
4. Adiabatically remove the perturbation, taking care to track each of the equilibria (fixed s) while the perturbation is being undone.
5. Marching determines some components of the disconnected solution set.

This process is illustrated in Fig. 18.15 for a simple example.

8. Summary

In this chapter we have studied both the statics and the dynamics of gradient dynamical systems. Although these dynamical systems cannot be transformed to canonical form except at equilibrium (Sec. 1), the methods and results of

Elementary Catastrophe Theory are useful for studying their dynamical properties. In the frequently encountered case $n = 2$, the dynamical properties are easily visualized using the method of phase portraits (Secs. 2–4). The phase portrait in the neighborhood of a degenerate critical point can be determined as the "far-field" limit of the phase portrait of a perturbed catastrophe germ.

In Sec. 5 we considered an example representative of a large class of physically interesting applications of Catastrophe Theory/Bifurcation Theory. It was remarkable that with so little effort we were able to infer so much about the approximate locations and stability types of the critical branches of some symmetry-restricted function of a single control parameter. In practice, examples of this type which are encountered in the sciences or in engineering problems are often simpler.

In Sec. 6 we saw how to determine the stationary values of a potential by the marching process. Catastrophes provide an obstruction to this process, but their known canonical forms provide a way through or around these obstructions. In Sec. 7 we indicated a relationship between Bifurcation Theory and Catastrophe Theory. Further, we saw how the methods of Catastrophe Theory (i.e., changing the control parameters) could lead to the discovery of disconnected solution sets and their subsequent determination by the marching process.

References

1. M. Golubitsky and D. Schaeffer, A Theory for Imperfect Bifurcation via Singularity Theory, *Commun. Pure Appl. Math.* **32**, 21–98 (1979).

CHAPTER 19

Autonomous Dynamical Systems

The terminology and methods developed for studying gradient dynamical systems can be used to study the properties of autonomous dynamical systems as well. In particular, terms, processes, and methods such as equilibrium, critical point, degenerate critical point, structurally (un)stable, morsification, general perturbation, and phase portrait, may be transferred from previous chapters to the present one. The transference of terminology and techniques is illustrated in Secs. 2 and 3.

In Sec. 4 we study the geometry of the linearized stability matrix. Such matrices are represented by points in $\mathbb{R}^{n^2}$. This space is divided into open regions describing qualitatively distinct structurally stable dynamical systems by a separatrix describing structurally unstable systems whose components are determined by the eigenvalues of the linearized stability matrix. The separatrix contains two types of components: those on which a change of dynamical stability occurs, and those on which it does not. The former components are analogous to the set determined by the condition det $V_{ij} = 0$ for a gradient system. Such components describe systems (dynamical and gradient) with degenerate critical points. Bifurcations are associated with these components. The latter components are analogous to the set in the control space for a gradient system for which two or more critical values are equal (Maxwell set). Bifurcations are not generally associated with this latter set for either gradient or dynamical systems. These ideas are illustrated in Sec. 4 for 2-dimensional dynamical systems.

The "Crowbar Principle" suggests that we look carefully at the various components of the separatrix in $\mathbb{R}^{n^2}$. We do this for $n = 2$ in Secs. 5, 7, and 8. In Sec. 5 we look at perturbations in the nonbifurcation part of the separatrix and see that the corresponding critical point is indeed isolated.

In Sec. 7 we study components of the bifurcation set on which one of the two eigenvalues goes through zero. The corresponding doubly degenerate critical point is called a saddle node. A 1-dimensional morsification of the saddle node is the dynamical system analog of the fold catastrophe, while a 2-dimensional morsification can lead to a "stability exchange."

In Sec. 8 we study components of the bifurcation set on which the real parts of complex conjugate eigenvalues are zero. The corresponding degenerate critical points are called vortices or centers. A 1-dimensional morsification of a vortex leads to a Hopf bifurcation. Hopf bifurcations depending on k control parameters are closely related to symmetry-restricted catastrophes of type $A_{\pm(2k+1)}$.

1. Reduction to Gradient Form

Many autonomous (= time-independent) systems of nonlinear differential equations can be written as gradient systems. Many more cannot. For this enlarged class of dynamical systems the equations of motion must be written directly in terms of the force components $F_i(x; c)$, since these cannot generally be derived from a potential

$$\frac{dx_i}{dt} = F_i(x; c) \tag{19.1}$$

Since this system of equations is more general than the corresponding gradient system (where $F = -\nabla V$), the types of behavior associated with such dynamical systems will be more varied ("richer") than for gradient systems.

Since gradient systems are intrinsically simpler to handle than dynamical systems, it is useful to give a criterion which allows to determine when a dynamical system is in fact a gradient system. For a gradient system

$$F_i = -\frac{\partial V}{\partial x_i} \tag{19.2}$$

so that

$$F_{ij} = \frac{\partial F_i}{\partial x_j} = -\frac{\partial^2 V}{\partial x_j\,\partial x_i} = -\frac{\partial^2 V}{\partial x_i\,\partial x_j} = \frac{\partial F_j}{\partial x_i} = F_{ji}$$

That is, the generalized curl of $F = -\nabla V$ is zero for a gradient system

$$(\text{curl } F)_{ij} = \frac{\partial F_j}{\partial x_i} - \frac{\partial F_i}{\partial x_j} \tag{19.3}$$

Conversely, if all $n(n-1)/2$ independent components of the curl of F for a dynamical system vanish, then the system is a gradient system.

Example 1. The dynamical system

$$\frac{dx}{dt} = 2xy$$
$$\frac{dy}{dt} = \mu x^2 - y^2 \tag{19.4}$$

is a gradient system if and only if $\mu = 1$, for

$$\begin{array}{ccc} \dfrac{\partial}{\partial y}(2xy) & \stackrel{?}{=} & \dfrac{\partial}{\partial x}(\mu x^2 - y^2) \\ \| & & \| \\ 2x & \stackrel{?}{=} & 2\mu x \end{array} \tag{19.5}$$

For $\mu = 1$ the corresponding potential function is the germ of the D_{-4} catastrophe: $-V(x, y) = x^2 y - \frac{1}{3}y^3$.

2. Phase Portraits: $F \neq 0$

The qualitative description of autonomous dynamical systems can be carried out in much the same way as for gradient systems. This involves the construction of phase portraits, which is particularly simple and convenient for 2-dimensional ($n = 2$) dynamical systems. For this reason we will confine our discussion largely to 2-dimensional systems in this chapter. The study proceeds exactly as in the previous chapter:

1. Describe phase portraits (Sec. 2).
2. Classify equilibrium types (Sec. 3).
3. Look at degenerate equilibria (Sec. 4).
4. Perturb the system in the neighborhood of degenerate equilibria (Secs. 5, 7, and 8).

The terminology as well as the methods of the previous chapter will be used in this chapter.

The phase portrait is constructed by indicating the direction of the force $F(x, y)$, and consequently the flow direction $(dx/dt, dy/dt)$, at every point in the x–y plane. The global flow may then be determined by connecting the arrows together "head to tail" in the obvious way.

Phase portraits for structurally stable dynamical systems are particularly informative. The system is structurally stable if a perturbation of the system

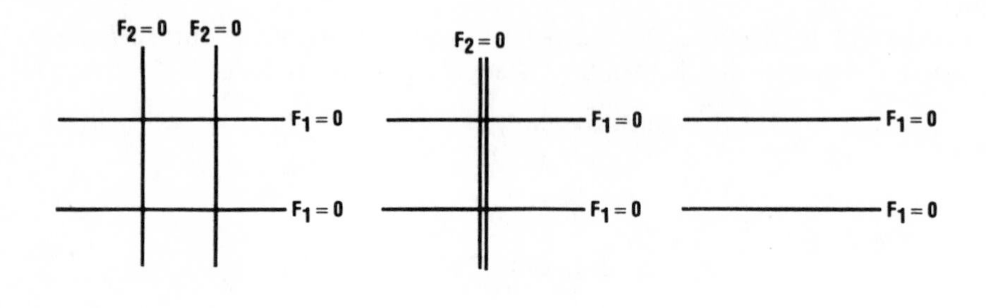

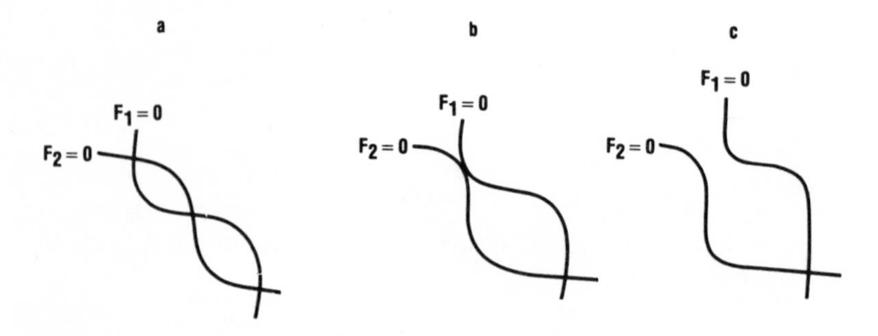

Figure 19.1 A structurally unstable dynamical system (*b*) is sandwiched between two structurally stable dynamical systems (*a*) and (*c*).

(19.1) has no qualitative effect on the number, location, and intersection properties of the curves $F_i = 0$. In Fig. 19.1 we show the curves $F_i = 0$ for two structurally unstable dynamical systems sandwiched between the curves $F_i = 0$ for the perturbed systems.

For a structurally stable dynamical system the curves $F_1 = 0$ and $F_2 = 0$ divide the plane into regions of alternating sign for dx/dt and dy/dt. This provides a useful skeleton from which the global behavior of the entire dynamical system may be inferred. This idea is illustrated in Fig. 19.2, which shows a part of the plane $\mathbb{R}^2$. In this region we assume that the equation $F_1(x, y) = 0$ has two solutions, and that $F_2(x, y) = 0$ also has two solutions. These four "hyperplanes" divide the part of the plane shown into nine open regions. If the signs of (F_1, F_2) are known in any one of these open regions, they may be determined in all other open regions. We have assumed that the signs of (F_1, F_2) are as shown in the upper left-hand region. It is then a simple matter to determine the flow directions in the remaining eight open regions and on the four curves $F_i = 0$. Once this has been done, the global behavior of the dynamical system is easily determined by "following the arrows." For the dynamical system whose phase portrait is shown in Fig. 19.2 there is one source, one sink, and two equilibria about which "spiral behavior" occurs.

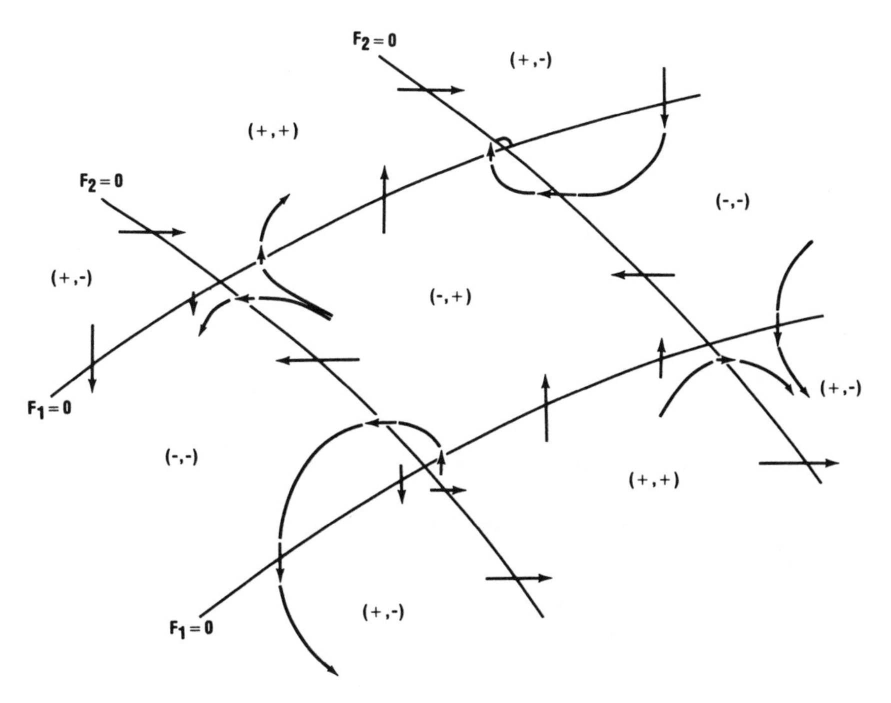

Figure 19.2 The qualitative properties and phase portrait of a 2-dimensional dynamical system are determined by the lines $F_i = 0$ and their intersections.

3. Phase Portraits: $F = 0$

As the next step in the global analysis of dynamical systems it is useful to search for the equilibrium points $F_1(x, y) = 0$, $F_2(x, y) = 0$. These occur at the intersections of the lines $F_1 = 0$ and $F_2 = 0$. The stability properties of a dynamical system at an equilibrium can be determined by linearizing the system about the equilibrium point. For an equilibrium at $x = x^0$ we set $\delta x = x - x^0$ and then

$$\frac{d}{dt}(x^0 + \delta x)_i = F_i(x^0 + \delta x) = F_i(x^0) + \frac{\partial F_i}{\partial x_j}\delta x_j + \mathcal{O}(2)$$
$$\frac{d}{dt}\delta x_i = F_{ij}\delta x_j + \mathcal{O}(2) \tag{19.6}$$

The local dynamical and structural stability properties are determined by the eigenvalues of the stability matrix $F_{ij} = \partial F_i/\partial x_j$ at the critical point x^0.

For a 2-dimensional dynamical system with an equilibrium taken at the origin (0, 0) without loss of generality, the linearized equations of motion have the form

$$\frac{d}{dt}\begin{pmatrix} x \\ y \end{pmatrix} = F\begin{pmatrix} x \\ y \end{pmatrix} \tag{19.7}$$

A very convenient representation for the real 2×2 matrix F is given in terms of the Pauli spin matrices as follows:

$$F = \begin{bmatrix} \lambda + r & s + \omega \\ s - \omega & \lambda - r \end{bmatrix} \tag{19.8}$$

The eigenvalues of F are

$$\lambda_{\pm} = \lambda \pm \sqrt{r^2 + s^2 - \omega^2} \tag{19.9}$$

There is a 1–1 correspondence between the points $(\lambda, \omega, r, s) \in \mathbb{R}^4$ and stability matrices F (19.8) for 2-dimensional dynamical systems. It is convenient to write $\mathbb{R}^4 = \mathbb{R}^1 \otimes \mathbb{R}^3$ with $\lambda \in \mathbb{R}^1$, $(\omega, r, s) \in \mathbb{R}^3$.

The cone $r^2 + s^2 - \omega^2 = 0$ in $\mathbb{R}^3$ acts as a separatrix between those stability matrices F for which both eigenvalues are real $(r^2 + s^2 - \omega^2 > 0)$ and those for which the eigenvalues form a complex conjugate pair $(r^2 + s^2 - \omega^2 < 0)$.

When both eigenvalues are real and unequal, the dynamical system is locally equivalent to a gradient system. In this case there is a change in dynamical stability when $\det F = \lambda^2 + \omega^2 - r^2 - s^2 = 0$. Thus for any point $(\omega, r, s) \in \mathbb{R}^3$

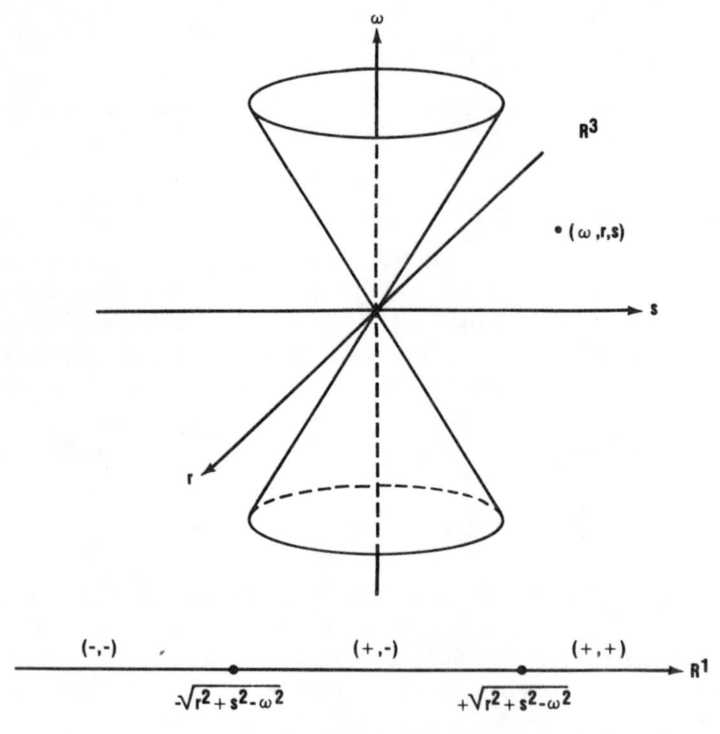

Figure 19.3 When (ω, r, s) lies outside the cone shown, both eigenvalues of the stability matrix are real. The dynamical system is then locally equivalent to a gradient system. The inertia of this system depends on λ and on $(r^2 + s^2 - \omega^2)^{1/2}$, as shown.

outside the cone $r^2 + s^2 - \omega^2 = 0$ the line $\mathbb{R}^1$ is divided into three open regions characterizing gradientlike systems with eigenvalues $(+, +), (+, -), (-, -)$, as shown in Fig. 19.3. In these three cases the critical point is called an **unstable node**, a **saddle**, and a **stable node**, respectively.

Points inside the cone $r^2 + s^2 - \omega^2 = 0$ characterize dynamical systems which are not locally equivalent to gradient systems. The qualitative properties

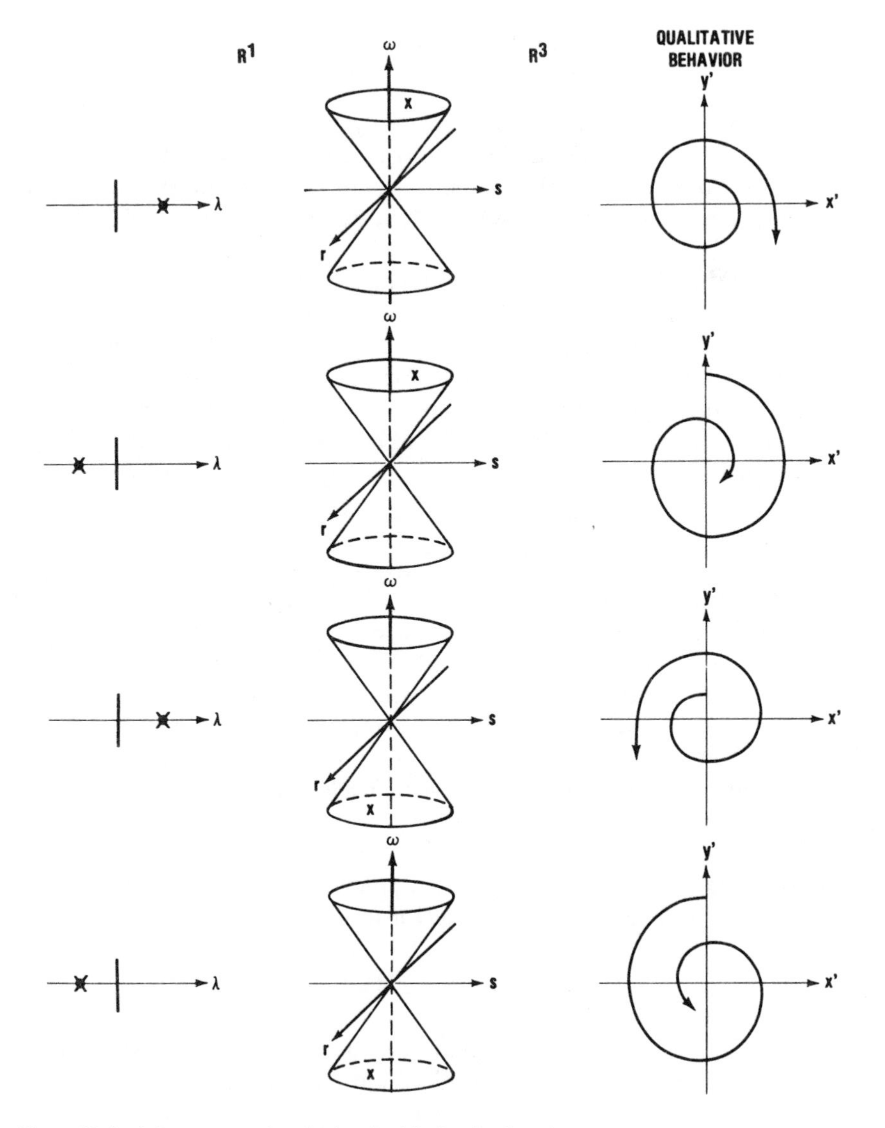

Figure 19.4 A focus around an isolated critical point is stable or unstable according to whether $\lambda < 0$ or $\lambda > 0$. The direction of circulation depends on the sign of ω.

of such systems are determined by the signs of the real and imaginary parts of the eigenvalues

$$\begin{aligned}\lambda_{\pm} &= \lambda \pm i\omega' \\ \omega' &= \text{sign}(\omega)\sqrt{\omega^2 - r^2 - s^2}\end{aligned} \tag{19.10}$$

After a similarity transformation the dynamical equations (19.7) can be brought to the form

$$\frac{d}{dt}\begin{pmatrix} x' \\ y' \end{pmatrix} = \begin{pmatrix} \lambda & \omega' \\ -\omega' & \lambda \end{pmatrix}\begin{pmatrix} x' \\ y' \end{pmatrix} \tag{19.11}$$

The system of equations (19.11) is dynamically stable if $\lambda < 0$ and unstable if $\lambda > 0$. The motion is clockwise if $\omega' > 0$ (i.e., $\omega > 0$) and counterclockwise if $\omega < 0$. These four qualitatively distinct types of behavior are summarized in Fig. 19.4. The critical point is called a **stable focus** if $\lambda < 0$, an **unstable focus** if $\lambda > 0$.

The space $\mathbb{R}^4$ representing 2×2 stability matrices F is partitioned into seven open regions representing seven qualitatively distinct types of dynamical behavior. These open regions are structurally stable. Three of these open regions represent dynamically stable systems. These open regions are separated by a separatrix containing components of dimensions 3, 2, 1, 0, and which therefore has measure zero. The separatrix in $\mathbb{R}^4$ will be discussed in more detail in the following section.

4. Geometry of the Stability Matrix: $n = 2$

At a critical point for an n-dimensional dynamical system the stability matrix F_{ij} is represented by a point in $\mathbb{R}^{n^2}$. This space is divided into a number of open regions parameterizing systems with qualitatively different dynamical stability properties as discussed in Sec. 3. These open regions are separated from each other by a separatrix in $\mathbb{R}^{n^2}$ which has components of dimensions $n^2 - 1$, $n^2 - 2, \ldots, 1, 0$, and hence measure zero. This separatrix parameterizes structurally unstable stability matrices. The structural instability may arise for two distinct reasons:

1. Two or more eigenvalues $\lambda_1, \lambda_2, \ldots, \lambda_n$ may be equal. This part $\mathscr{S}_a$ of the separatrix may be determined analytically from the expression

$$S_a \stackrel{\text{def}}{=} \prod_{j>i=1}^{n} (\lambda_j - \lambda_i) = 0 \tag{19.12a}$$

2. The real parts of one or more eigenvalues may be zero. This part $\mathscr{S}_b$ of the separatrix may be determined analytically from the expression

$$S_b \stackrel{\text{def}}{=} \prod_{i=1}^{n} (\text{Re}\,\lambda_i) = \prod_{i=1}^{n} \frac{(\lambda_i + \lambda_i^*)}{2} = 0 \tag{19.12b}$$

The structurally stable open regions in $\mathbb{R}^{n^2}$ are characterized by the condition $S_a \neq 0$, $S_b \neq 0$. The open regions describing dynamically stable systems are characterized by

$$\operatorname{Re} \lambda_i < 0, \qquad i = 1, 2, \ldots, n \tag{19.13}$$

No changes in dynamical stability are associated with points on the components of the separatrix $\mathscr{S}_a$, defined by $S_a = 0$, $S_b \neq 0$. These points in $\mathbb{R}^{n^2}$ define structurally unstable but nondegenerate dynamical systems. In the neighborhood of any point in this component of the separatrix are points describing isolated critical points of qualitatively different types but with the same inertia (cf. Fig. 18.1).

Changes in dynamical stability are associated with the components of the separatrix $\mathscr{S}_b$, defined by $S_b = 0$, $S_a \neq 0$. These points in $\mathbb{R}^{n^2}$ define structurally unstable degenerate dynamical systems. The occurrence of degeneracies of the critical points is associated with the bifurcation of new solutions from old solutions. As a result, the components of the separatrix determined by $S_b = 0$ (19.12b) are called the bifurcation set of F_{ij}.

We summarize in (19.14) the properties of points in $\mathbb{R}^{n^2}$ determined by the algebraic conditions (19.12):

	$S_b \neq 0$	$S_b = 0$
$S_a \neq 0$	$\mathbb{R}^{n^2} - (\mathscr{S}_a \cup \mathscr{S}_b)$: structurally stable	$\mathscr{S}_b$: degenerate critical points, bifurcation set
$S_a = 0$	$\mathscr{S}_a$: isolated critical points, equal eigenvalues, no bifurcation	$\mathscr{S}_a \cap \mathscr{S}_b$: degenerate critical points with equal eigenvalues, bifurcation occurs

(19.14)

Remark. The components of $\mathscr{S}_b$ of the bifurcation set for dynamical systems are analogous to the components of the bifurcation set for gradient systems. In both cases points in the bifurcation set describe systems with degenerate critical points. The components of $\mathscr{S}_a$ for dynamical systems are analogous to the Maxwell set, or the set of points in the control parameter space of gradient systems which represent structurally unstable potentials for which the critical values at two or more critical points are equal.

These general remarks can be used to simplify the description of the separatrix in $\mathbb{R}^4$ for the stability matrix of 2-dimensional dynamical systems. The set $\mathscr{S}_a$ describing isolated structurally unstable critical points is defined by

$$\begin{aligned} S_a = (\lambda_+ - \lambda_-) = 0 &\Rightarrow \omega^2 - r^2 - s^2 = 0 \\ S_b \neq 0 \qquad &\Rightarrow \qquad \lambda \neq 0 \end{aligned} \tag{19.15}$$

Points in $\mathscr{S}_a$ represent dynamical systems with two equal nonzero eigenvalues. Perturbation of such systems results in a dynamical system with real unequal eigenvalues or with complex conjugate eigenvalues. These components of the separatrix will be discussed in Sec. 5.

The bifurcation set $\mathscr{S}_b$ is determined from (19.12b). In the case that both eigenvalues are real ($r^2 + s^2 - \omega^2 > 0$), (19.12b) reduces to

$$\lambda_+ \lambda_- = 0 \Rightarrow \lambda^2 + \omega^2 = r^2 + s^2 \tag{19.16}$$

In the case that both eigenvalues are complex ($r^2 + s^2 - \omega^2 < 0$), (19.12b) reduces to

$$(\text{Re}\ \lambda_+)(\text{Re}\ \lambda_-) = \lambda^2 = 0 \Rightarrow \lambda = 0 \tag{19.17}$$

The components of the bifurcation set defined by (19.16) will be discussed in Sec. 7, and those determined by (19.17) will be discussed in Sec. 8.

Each of equations (19.15)–(19.17) involves one equation constraining the four real parameters (λ, ω, r, s). As a result, in a 1-control-parameter family of 2-dimensional dynamical systems $F_{ij}(c)$ it is possible to encounter the 3-dimensional components of the separatrix defined by (19.15)–(19.17) in a structurally stable way. However, it is not possible to encounter the 2-dimensional component $\mathscr{S}_a \cap \mathscr{S}_b$ defined by

$$\begin{aligned} \omega^2 &= r^2 + s^2, \qquad \omega \neq 0 \\ \lambda &= 0 \end{aligned} \tag{19.18}$$

in a structurally stable way unless $F_{ij}(c)$ depends on two or more control parameters.

In accordance with the "Crowbar Principle" it will be fruitful to study perturbations of those dynamical systems which are parameterized by points in the separatrix defined by (19.12). We turn our attention now to this problem.

5. Perturbations Around Equal Nonzero Eigenvalues

A dynamical system with two equal nonzero eigenvalues is structurally unstable against perturbations. A 2-dimensional dynamical system with this property is represented by a point in the cone $\omega^2 - r^2 - s^2 = 0$, $\lambda \neq 0$. The most general perturbation of such a dynamical system which leaves the isolated critical point fixed has the form

$$\begin{aligned} \frac{dx_i}{dt} &= F_i(x) + \delta F_i(x) \\ &= F_{ij}x_j + \delta F_{ij}x_j + \mathcal{O}(2) \end{aligned} \tag{19.19}$$

Since the eigenvalues of F_{ij} have nonzero real part, so also do the eigenvalues of the perturbed stability matrix $(F + \delta F)_{ij}$ for sufficiently small perturbations δF. As a result, the local properties of the perturbed system can be determined by studying the linearized form of (19.19):

$$\frac{dx_i}{dt} = (F + \delta F)_{ij}x_j \tag{19.20}$$

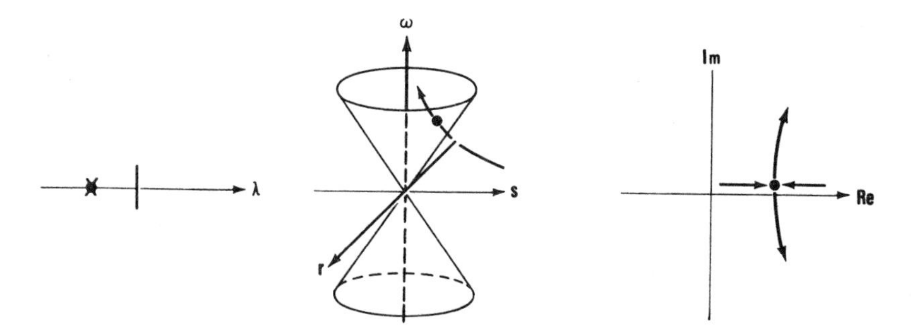

Figure 19.5 A 1-dimensional curve in $\mathbb{R}^4$ may encounter the 3-dimensional surface $\omega^2 = r^2 + s^2$ at isolated points in a structurally stable way. On crossing through this surface the two eigenvalues scatter off each other at right angles, as shown.

For the 2-dimensional dynamical system with F given by (19.8) we choose

$$\delta F = \begin{bmatrix} \delta\lambda + \delta r & \delta s + \delta\omega \\ \delta s - \delta\omega & \delta\lambda - \delta r \end{bmatrix} \tag{19.21}$$

The difference between the perturbed eigenvalues is

$$\lambda'_+ - \lambda'_- = 2\sqrt{r'^2 + s'^2 - \omega'^2} \tag{19.22}$$

where $r' = r + \delta r$, and so on. This difference is real if the perturbed point (ω', r', s') lies outside the cone, imaginary if it lies inside.

We now assume that the dynamical system (19.19) depends on k control parameters c. Then the coordinates (λ, ω, r, s) depend on c. In the case $k = 1$ the curve $(\lambda(c), \omega(c), r(c), s(c)) \in \mathbb{R}^4$ may encounter the cone $r^2 + s^2 - \omega^2 = 0$, $\omega \neq 0$, $\lambda \neq 0$ in a structurally stable way. This curve will pass from the outside of the cone to the inside, or vice versa. The eigenvalues of F will change from real and unequal to a complex conjugate pair in a "scattering process" as indicated in Fig. 19.5.

A path in $\mathbb{R}^4$ will not encounter the apex of the cone $\omega = 0$, $\lambda \neq 0$ unless $k \geq 3$. In the case $k \geq 3$ a curve through the cone apex may be approximated by a straight line segment. Such a line segment will either pass from the inside of the cone to the inside, or from the outside to the outside. In this case we have "head-on" collisions of the eigenvalues, as indicated in Fig. 19.6.

In the event that we are dealing with gradient systems rather than dynamical systems, $\omega = \delta\omega = 0$ and a path in the $\omega = 0$ subspace of $\mathbb{R}^4$ may encounter the apex of the cone $r = s = 0$, $\lambda \neq 0$ in a structurally stable way provided $k \geq 2$. In this case the eigenvalues collide in a head-on way as shown in Fig. 19.6*b*.

Instead of controlled paths in $\mathbb{R}^4$ we now consider perturbations δF of (19.19) which are "isotropic" in the sense that the probability distribution function for δF has the form

$$P(\delta\lambda, \delta\omega, \delta r, \delta s) = f(\delta\lambda, (\delta\omega)^2 + (\delta r)^2 + (\delta s)^2) \tag{19.23}$$

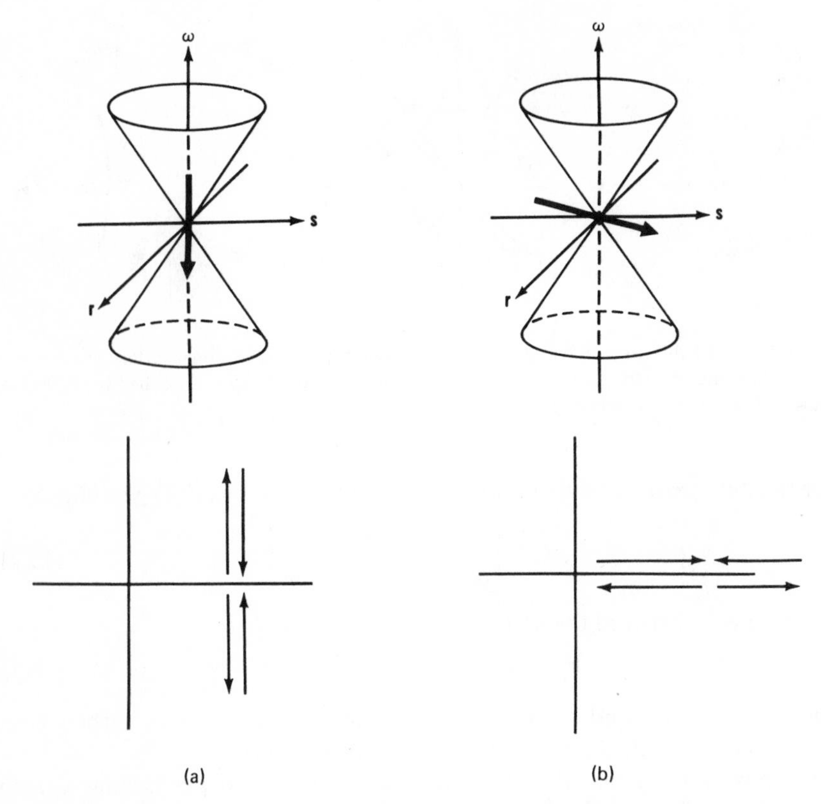

Figure 19.6 If the control parameter curve in $\mathbb{R}^4$ passes through the apex of the cone, the two eigenvalues will scatter off each other in the fashion of a "head-on" collision.

1. It is possible for a 1-parameter family of 2-dimensional dynamical systems to have isolated members with doubly degenerate eigenvalues. A perturbation of such a system, represented by a point on the cone $r^2 + s^2 - \omega^2 = 0$, $\omega \neq 0$, $\lambda \neq 0$, with a δF randomly distributed according to (19.23) will produce a dynamical system with real unequal eigenvalues with probability .5, and with complex conjugate eigenvalues with probability .5. Perturbations which leave the eigenvalues equal occur with measure zero.
2. Two equal nonzero eigenvalues with $\omega = 0$ are first stably encountered in 3-parameter families of 2-dimensional dynamical systems. In such a case a perturbation with a distribution of the form (19.23) will lead to a dynamical system with complex eigenvalues with probability $(1 - \cos \pi/4) = 1 - 1/\sqrt{2}$, and with real unequal eigenvalues with probability $\sqrt{2}/2$.
3. For a gradient system the equal eigenvalue degeneracy is first encountered in 2-parameter families. Perturbations of such a system will lead to a dynamical system with real unequal eigenvalues with probability 1.

6. Geometry of the Stability Matrix: $n > 2$

When $n > 2$ the geometry of the linearized stability matrix F_{ij} cannot easily be visualized by the methods used to construct Figs. 19.3–19.6. Alternative methods are called for to determine how the different subsets in $\mathbb{R}^{n^2}$ which parameterize the F_{ij} with various degeneracies fit together. One such technique, a close cousin of the diagrammatic technique presented in Chapter 7, Sec. 4, is discussed below. This technique can be used to determine how many connected open subsets in $\mathbb{R}^{n^2}$ exist which parameterize structurally stable linearized stability matrices. It can also be used, with the "method of contraction," to determine the spectrum of "abutments" or "adjacencies" of these open sets, that is, to determine which open sets are "near" which other open sets. ("Near" means share a boundary set.) When combined with the methods of Chapter 14, this method can also be used to determine the structure of the separatrix in $\mathbb{R}^{n^2}$.

We begin by observing that no matter how large n is, the n eigenvalues of the real $n \times n$ linearized stability matrix F_{ij} can be located in the complex plane $\mathbb{C}^1 = \mathbb{R}^2$. Each point in $\mathbb{R}^{n^2}$ uniquely determines an F_{ij} (and conversely), and each F_{ij} uniquely determines a distribution of eigenvalues in the complex plane (and not conversely). Now suppose that some point $p \in \mathbb{R}^{n^2}$ determines the $n \times n$ matrix $F_{ij}(p)$, and suppose also that $F_{ij}(p)$ has distinct eigenvalues, all of which have nonzero real part. Then all points sufficiently near p will determine $n \times n$ matrices $F_{ij}(p) + \delta F_{ij}$, and these perturbed matrices will also have distinct eigenvalues with nonzero real part (with no eigenvalues crossing the imaginary axis). Therefore the set of points in $\mathbb{R}^{n^2}$ parameterizing matrices F_{ij} with [cf. (19.12)]

$$S_a \neq 0$$

$$S_b \neq 0$$

is open (and dense) in $\mathbb{R}^{n^2}$, and the corresponding matrices are structurally stable (and dense) in the set of real $n \times n$ matrices. The open set in $\mathbb{R}^{n^2}$ is precisely $\mathbb{R}^{n^2} - (\mathscr{S}_a \cup \mathscr{S}_b)$.

There is a 1–1 correspondence between the connected open sets in $\mathbb{R}^{n^2} - (\mathscr{S}_a \cup \mathscr{S}_b)$ and the qualitatively distinct distributions of eigenvalues of an $n \times n$ matrix in the complex plane, provided the two possible flow directions around each focus are taken into account. The diagrammatic technique for determining the geometry of $\mathbb{R}^{n^2}$ is as follows:

1. List all possible qualitatively distinct distributions of n eigenvalues of a real $n \times n$ matrix in the complex plane. All eigenvalues must be distinct with non-zero real part.
2. To each distribution corresponds one connected open set in $\mathbb{R}^{n^2}$.
3. The union of these open sets is $\mathbb{R}^{n^2} - (\mathscr{S}_a \cup \mathscr{S}_b)$. This set is dense in $\mathbb{R}^{n^2}$.

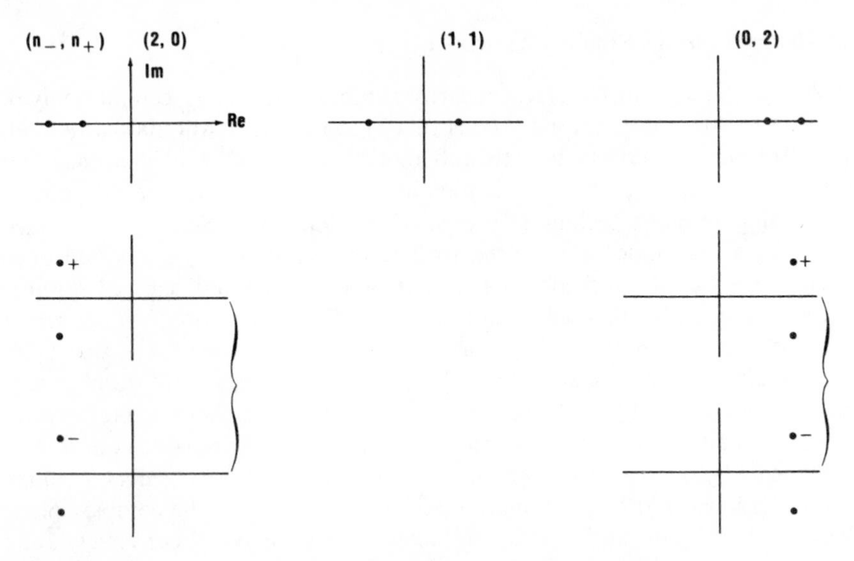

Figure 19.7 The two real roots may be distributed between the two half-planes in three ways, as shown in the top line. A complex conjugate pair may occur in either half-plane. Such a pair describes a focus which may rotate in either the positive or the negative direction.

We illustrate this diagrammatic method by taking a second look at an old problem ($n = 2$) and a first look at a new problem ($n = 3$) in the following two examples.

Example 1. Determine the geometry of the 2×2 stability matrices F_{ij}.

Solution. We shall speak of the geometry of $\mathbb{R}^{n^2}$ and the geometry of stability matrices interchangeably. A real 2×2 matrix has two eigenvalues which may both be real, or a complex conjugate pair. In the case that both eigenvalues are real, they may both be in the left half-plane, one in each half-plane, or both in the right half-plane. When the eigenvalues form a complex conjugate pair, this pair may be in either half-plane. Further, the rotation around the focus may be in the positive or negative direction. The five distinct distributions of eigenvalues, together with the finer resolution according to rotation direction, are shown in Fig. 19.7. To each diagram there corresponds a connected open set in $\mathbb{R}^4$: there are seven connected open sets in $\mathbb{R}^4$. Each of these open sets parameterizes a linearized dynamical system with qualitatively different behavior.

Example 2. Determine the geometry of 3×3 stability matrices.

Solution. All three eigenvalues may be real, or there may be one real eigenvalue and a complex conjugate pair. All qualitatively distinct eigenvalue distributions are shown in Fig. 19.8, but the existence of two possible flow directions

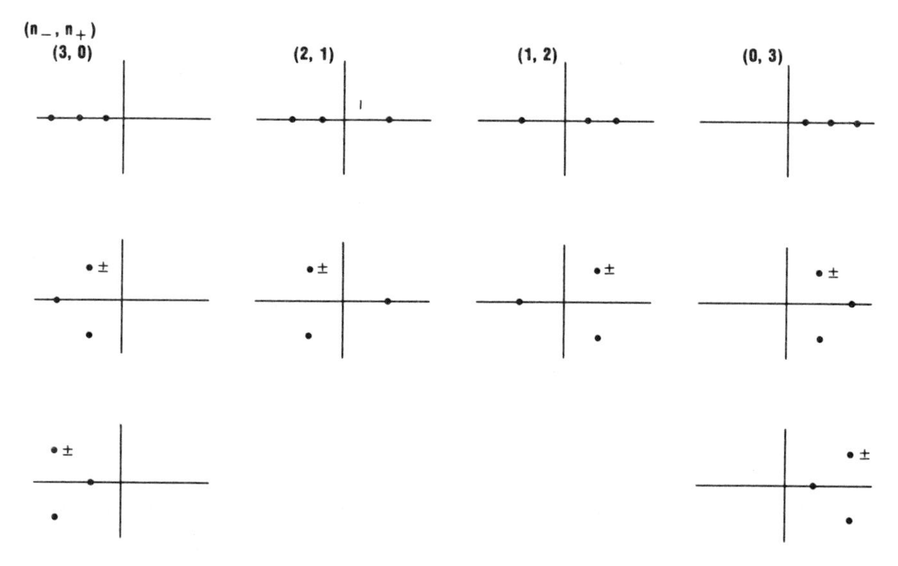

Figure 19.8 All ten qualitatively distinct distributions of the eigenvalues of a real 3×3 matrix are shown. Taking focus flow directions into account, a linearized 3-dimensional dynamical system can exhibit 16 qualitatively distinct types of behavior.

around each focus is only indicated. There is a total of 16 ($= 4 + 2 \times 6$) connected open sets, of which only five parameterize dynamically stable systems.

When combined with the method of contraction, the diagrammatic technique can be used to determine which open sets are contiguous and which are not. Briefly, two open sets are contiguous if they have a degenerate eigenvalue distribution in common.

Example 3. Determine which open sets in $\mathbb{R}^9$ describing stable systems are contiguous with the open set having three real negative eigenvalues.

Solution (Fig. 19.9). The boundary of this open set is defined by $S_a = 0$ (19.12a). All possible contractions of the three distinct negative eigenvalues to a degenerate case are performed. All other open sets with identical contractions are contiguous, and share the indicated component of the separatrix.

Example 4. Determine the open sets of 5×5 stability matrices which are contiguous through a component of the separatrix which parameterizes matrices with three degenerate negative eigenvalues and a complex conjugate pair in the right half-plane.

Solution. The five open sets are shown in Fig. 19.10. Also shown is an open set not near this component of the separatrix.

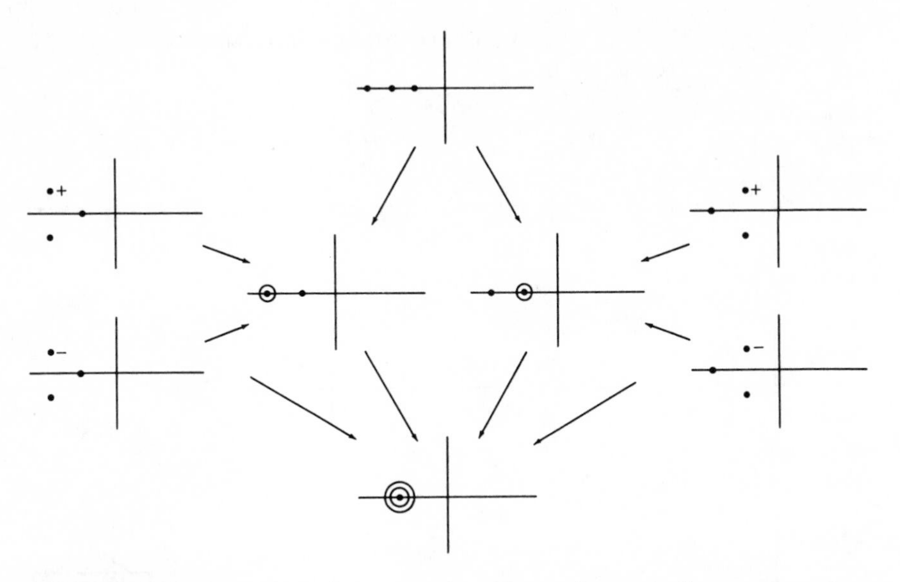

Figure 19.9 Diagrams which can be contracted to the same degenerate diagram represent contiguous open regions in $\mathbb{R}^{n^2}$. All five stable qualitatively distinct 3-dimensional systems are contiguous.

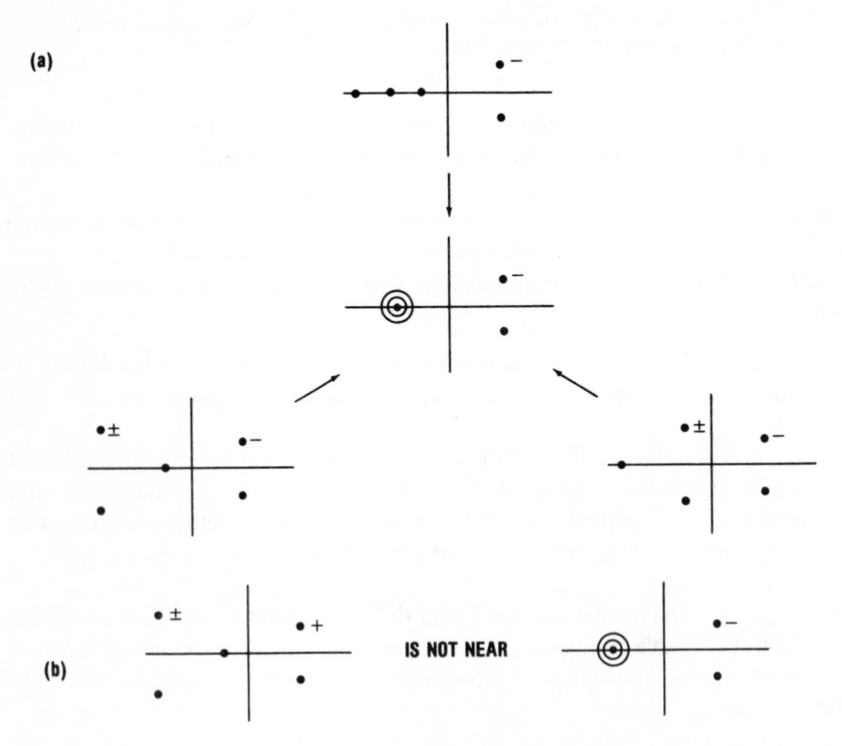

Figure 19.10 (*a*) The three structurally stable systems shown are contiguous through a component of the separatrix parameterizing the system with a threefold negative eigenvalue degeneracy. (*b*) Many open sets are not contiguous with the degenerate diagram. One is shown.

Eigenvalue distribution diagrams exhibiting degeneracies clearly describe points in a separatrix $\mathcal{S}_a$ in $\mathbb{R}^{n^2}$, but they do not uniquely characterize components of the separatrix. The dimensionalities of the various components of the separatrix are determined from the Jordan canonical form associated with a degenerate diagram, and through the methods described in Chapter 14. The following three examples illustrate the procedure.

Example 5. Determine the structure of the separatrix $\mathcal{S}_a$ in the space describing 2×2 stability matrices.

Solution (Fig. 19.11). The three open sets with two roots in the negative half-plane have a single component in common. This component parameterizes 2×2 matrices with a doubly degenerate negative eigenvalue. The stability matrix may have the form $\lambda\lambda$ (diagonal) or λ^2 (Jordan upper triangular). The diagonal matrix can only be encountered in 3-parameter families (cf. Table 14.1), so the corresponding component of the separatrix has dimension 1 $(= 4 - 3)$. The upper triangular matrix λ^2 can be encountered in 1-parameter families; it is parameterized by a $3(= 4 - 1)$-dimensional component of the

DEGENERATE DIAGRAM	JORDAN FORM	DIMENSION OF JORDAN FORM	DIMENSION OF SEPARATRIX COMPONENT
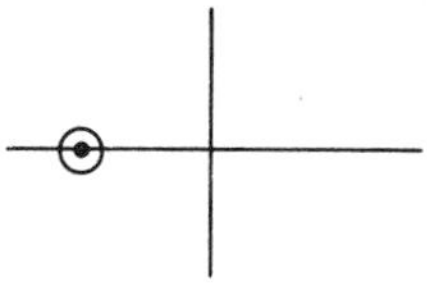	λ^2	1	3
	$\lambda\ \lambda$	3	1
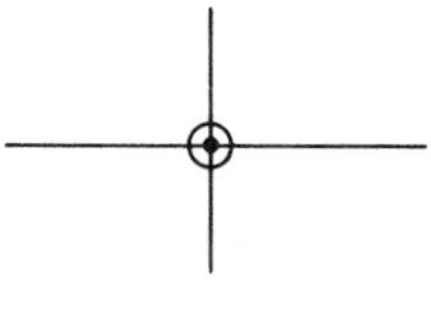	0^2	2	2
	00	4	0
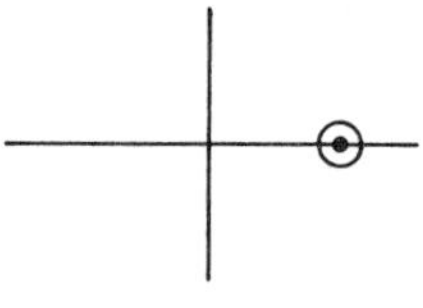	λ^2	1	3
	$\lambda\ \lambda$	3	1

Figure 19.11 The separatrix in $\mathbb{R}^4$ for 2×2 stability matrices has components whose dimensions may easily be computed. The components of dimension 0, 1, 2, 3 parameterize systems with degeneracies as shown.

separatrix. In terms of the parameterization given in (19.8), two eigenvalues can be equal only if $\omega^2 = r^2 + s^2$, that is, the corresponding point lies on the cone in $\mathbb{R}^3$. If the point lies at the apex $\omega = 0 (\Rightarrow r = s = 0)$, then λ is arbitrary, the corresponding matrix is diagonal, and the component of the separatrix has dimension 1. If $\omega \neq 0$, then F_{ij} is not diagonalizable. The four parameters (λ, ω, r, s) satisfy one constraint, and the corresponding component of the separatrix is 3-dimensional.

Similar considerations hold when the degenerate eigenvalues are in the right half-plane. When the degeneracy occurs at the origin, there is an additional constraint of the form $\lambda = 0$. This reduces the dimensionality of the components as follows:

$$\lambda^2: \ 3 \rightarrow 0^2: \ 2$$

$$\lambda\lambda: \ 1 \rightarrow 00: \ 0$$

Example 6. Determine the dimensions of the components of the separatrix $\mathscr{S}_a$ describing 3×3 matrices with three degenerate negative roots.

Solution. The associated Jordan forms may be λ^3, $\lambda^2\lambda$, $\lambda\lambda\lambda$. The associated Jordan-Arnol'd canonical forms have dimensions 2, 4, 8 (cf. Table 14.1). Therefore the three components of the separatrix parameterizing this degenerate diagram have dimensions $9 - 2, 9 - 4, 9 - 8$, or 7, 5, and 1.

Example 7. Determine the dimensions of the components of the separatrix parameterizing the degenerate distribution shown in Fig. 19.10.

Solution. The corresponding root structure is $(\lambda)^3\alpha\bar{\alpha}$. The three-fold degenerate root may have Jordan canonical forms as described in Example 6. The dimensions of the components of the separatrix $\mathscr{S}_a$ parameterizing this degenerate diagram are therefore $25 - 2 = 23$, $25 - 4 = 21$, $25 - 8 = 17$.

7. Perturbations of a Saddle Node

The bifurcation set (19.12b) for 2-dimensional dynamical systems contains components defined by (19.16) and (19.17). In the former case both eigenvalues are real and at least one eigenvalue is zero. We will confine ourselves to the case where only one eigenvalue is zero. Since the eigenvalues of F are unequal, we can find a coordinate system in which the stability matrix is diagonal. Then, up to nonvanishing terms of lowest order, the dynamical system equations may be written

$$\begin{aligned} \frac{dx}{dt} &= \lambda_1 x + \mathcal{O}(2), \qquad \lambda_1 \neq 0 \\ \frac{dy}{dt} &= Ax^2 + 2Bxy + Cy^2 + \mathcal{O}(3) \end{aligned} \tag{19.24}$$

For concreteness we choose $\lambda_1 < 0$ throughout this section.

This system may be treated by the method of phase portraits. First we keep only the lowest order nonvanishing terms which appear in the force expressions on the philosophy that we are looking only at local properties. By performing a scale change on y ($y \to y/C$ if $C \neq 0$), (19.24) can be brought to the form

$$\frac{dx}{dt} = \lambda_1 x$$
$$\frac{dy}{dt} = (y - m_1 x)(y - m_2 x) \tag{19.25}$$

where the coefficients m_i obey the equation

$$AC + 2Bm + m^2 = 0 \tag{19.26}$$

Two structurally stable cases arise: both slopes m_1, m_2 are real and unequal; or the m_i are complex conjugate. These cases are separated by the structurally unstable case $m_1 = m_2$.

For each case the root lines $dx_i/dt = 0$ and the corresponding phase portrait may easily be determined. In the case m_1 and m_2 are real and unequal (Fig. 19.12*a*), the resulting phase portrait resembles the far-field limit of the phase portraits for a saddle and a stable node, shown in Fig. 19.13. This is true also for the phase portrait in the degenerate case $m_1 = m_2$ (Fig. 19.12*b*). In the case that m_1 and m_2 are complex conjugates, the phase portrait still resembles that of a

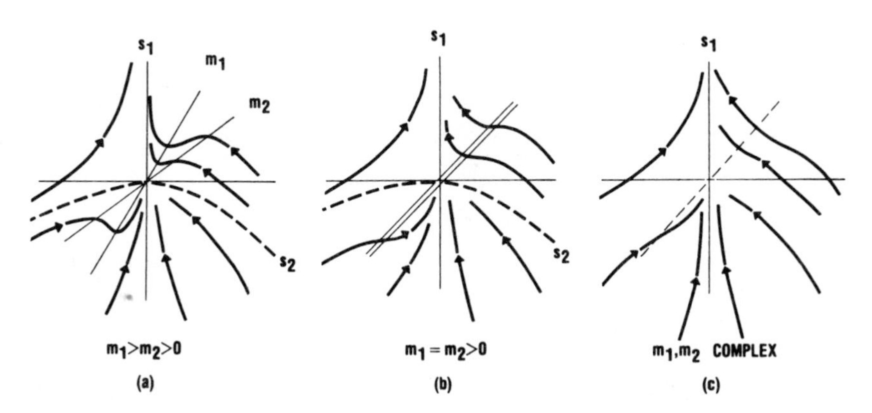

Figure 19.12 The phase portrait for the dynamical system (19.25) is shown for the three cases: (*a*) $m_1 > m_2 > 0$; (*b*) $m_1 = m_2 > 0$; (*c*) m_1, m_2 complex, Re $m_1 > 0$. In each case the far-field limit looks like the phase portrait of a saddle node. Here s_1, s_2 are separatrices.

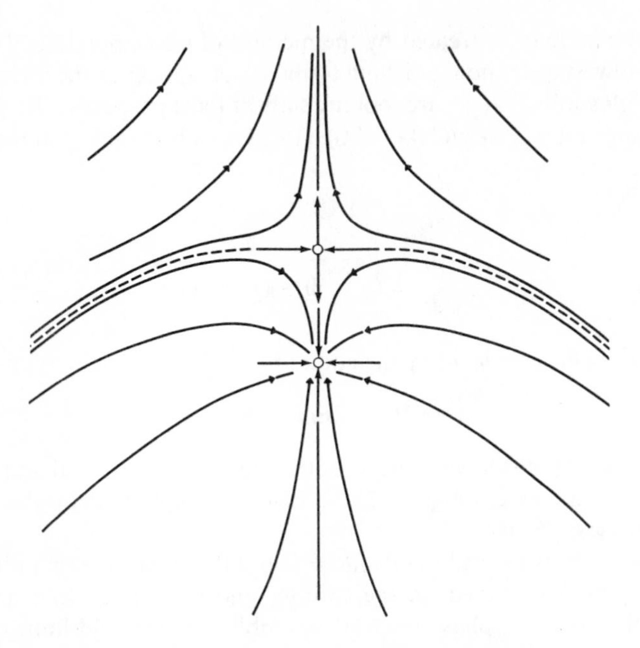

Figure 19.13 The phase portrait for a saddle node can be determined easily from the known phase portraits of the saddle and the node.

saddle node pair in the far-field limit, except that now there is no longer an equilibrium at the origin.

Based on our experience in Chapter 4, we expect that the most general perturbation of each equation for dx_i/dt would include terms whose degree was less than or equal to that of the lowest degree terms retained in the structurally unstable system. Thus we might take for our perturbation of the system (19.24)

$$\frac{dx}{dt} = a + (\lambda_1 + \delta F_{11})x + \delta F_{12}y$$

$$\frac{dy}{dt} = b + \delta F_{21}x + \delta F_{22}y + (A + \delta A)x^2 + 2(B + \delta B)xy + (C + \delta C)y^2 \tag{19.27}$$

where a, b, δF_{ij}, δA, δB, δC are all small. This perturbation is clearly too complicated to work conveniently with. We therefore simplify it according to the following considerations:

1. Since the original 2×2 matrix F_{ij} had two distinct eigenvalues $\lambda_1 \neq 0$ and $\lambda_2 = 0$, the perturbed matrix $F_{ij} + \delta F_{ij}$ will have two real distinct eigenvalues $\lambda'_1 \simeq \lambda_1$ and λ'_2 small and of first order.

2. The matrix $F_{ij} + \delta F_{ij}$ can be diagonalized by a linear transformation, since it has unequal eigenvalues.
3. This transformation induces only a small change in the quadratic coefficients $A + \delta A$, and so on. This in turn introduces only a small change $m_1 \to m_1'$, $m_2 \to m_2'$. This change is unimportant in the two structurally stable cases $m_1 \neq m_2$.
4. The origin of coordinates may be changed to remove a but not b.

Therefore the most general perturbation of the dynamical system (19.25) with $m_1 \neq m_2$ is

$$\begin{aligned}\frac{dx}{dt} &= \lambda_1 x \\ \frac{dy}{dt} &= b + 2cy + (y - m_1 x)(y - m_2 x)\end{aligned} \tag{19.28}$$

where b, c are small and may be considered as the dynamical system control parameters. The critical points for the perturbed dynamical system are

$$(x, y)_c\colon \quad \begin{matrix}(0, y_1) \\ (0, y_2)\end{matrix} \tag{19.29}$$

where the critical ordinates satisfy

$$y^2 + 2cy + b = 0 \tag{19.30}$$

The loci $y_c(c)$ are shown for three values of b ($b = 1, 0, -1$) in Fig. 19.14. The dynamical stability properties at each of these isolated critical points may easily be computed by performing a standard stability analysis. We do this explicitly in the case $b = 0$. In this case the two critical points have coordinates

$$\begin{aligned}&1.\quad (x_1 = 0, y_1 = 0) \\ &2.\quad (x_2 = 0, y_2 = -2c)\end{aligned} \tag{19.31}$$

The linearized equation of motion at the critical point 1 is

$$\frac{d}{dt}\begin{bmatrix}\delta x \\ \delta y\end{bmatrix} = \begin{bmatrix}\lambda_1 & 0 \\ 0 & 2c\end{bmatrix}\begin{bmatrix}\delta x \\ \delta y\end{bmatrix} \qquad \text{for } (x, y)_c = (0, 0) \tag{19.32}$$

This critical point has signature $(-, -)$ for $c < 0$ and $(-, +)$ for $c > 0$. Therefore this critical point is a stable node for $c < 0$ and a saddle for $c > 0$. The linearized equation of motion for the other isolated critical point is

$$\frac{d}{dt}\begin{bmatrix}\delta x \\ \delta y\end{bmatrix} = \begin{bmatrix}\lambda_1 & 0 \\ 2c(m_1 + m_2) & -2c\end{bmatrix}\begin{bmatrix}\delta x \\ \delta y\end{bmatrix} \qquad \text{for } (x, y)_c = (0, -2c) \tag{19.33}$$

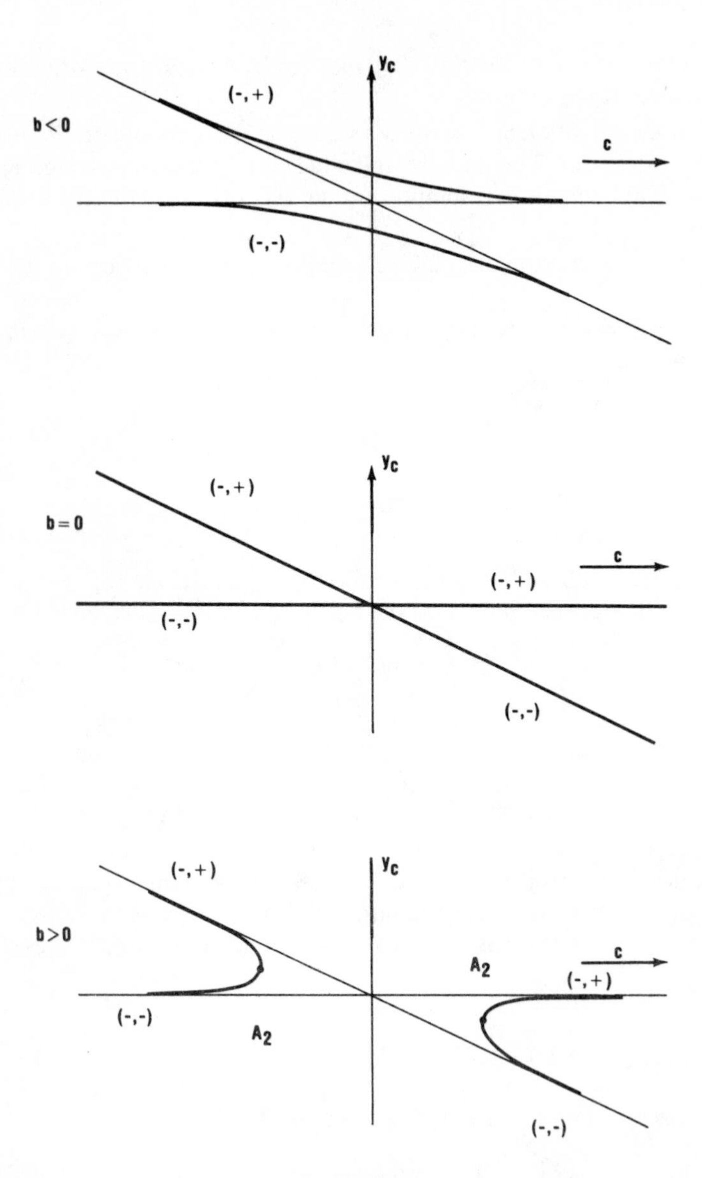

Figure 19.14 The equilibrium values of the ordinate y are shown as a function of the parameter c for three values of the parameter b. These parameters occur in the universal perturbation (19.28) of the degenerate dynamical system (19.25). The stability properties along each branch can be determined either by a local linear stability analysis [e.g., (19.32)] or by inspection of the appropriate equilibrium surface, shown in Fig. 19.15.

This isolated critical point is a stable node $(-, -)$ for $c > 0$ and a saddle $(-, +)$ for $c < 0$. These signatures are indicated in Fig. 19.14 ($b = 0$). This figure clearly shows that when $b = 0$ and the control parameter c passes through zero, there is an exchange of stability between the two critical points as they pass through each other.

The dynamical stability properties of the isolated critical points in the cases $b < 0$ and $b > 0$ may be determined in the same way. The signatures of the linearized system are shown along each of the curves $y(c)$ in Fig. 19.14. However, it is not necessary to go through a stability analysis for each of these cases. It is sufficient to realize that the three sets of curves shown in Fig. 19.14 are the b = constant cross sections of the surface (19.30). This 2-dimensional surface is shown in Fig. 19.15. There is a change of stability of critical points on this surface at the point of vertical tangency. It is therefore sufficient to determine the stability type for any one point on either the upper or the lower part of this folded surface. This has already been done in (19.32) and (19.33). In consequence, all critical points $(x_c, y_c) = (0, y_c)$ with y_c on the lower surface are stable nodes, while those with y_c on the upper surface are saddles.

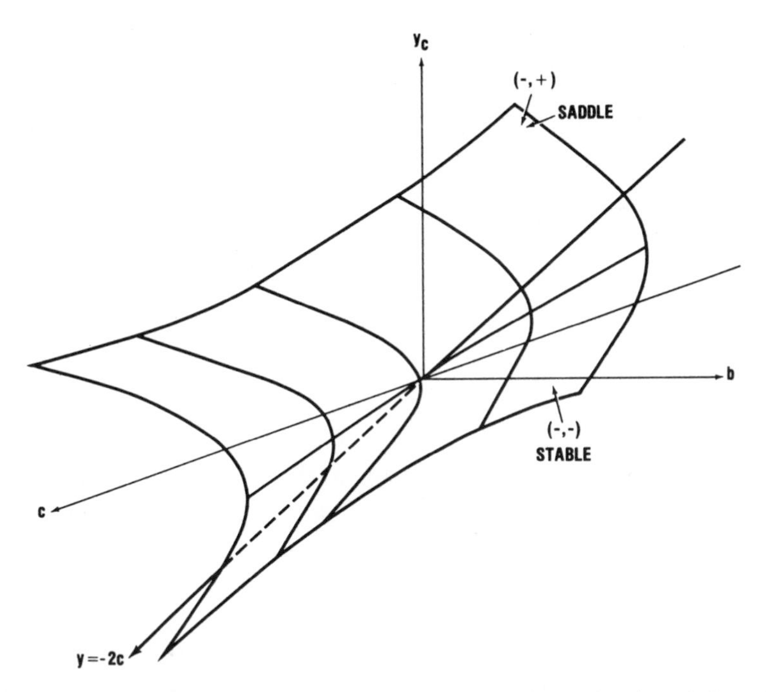

Figure 19.15 The equilibrium y coordinate for the system (19.28) is given by a 2-dimensional surface in the space $\mathbb{R}^1 \otimes \mathbb{R}^2 = (y; b, c)$, where b, c, are two control parameters which appear in the universal perturbation of (19.25). Points above the fold curve represent saddles, those below, stable nodes.

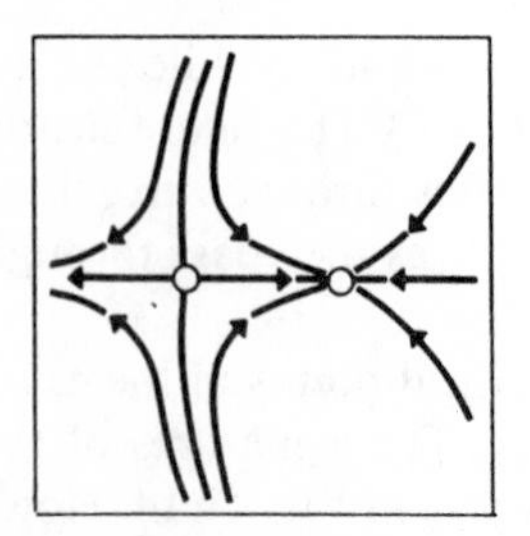
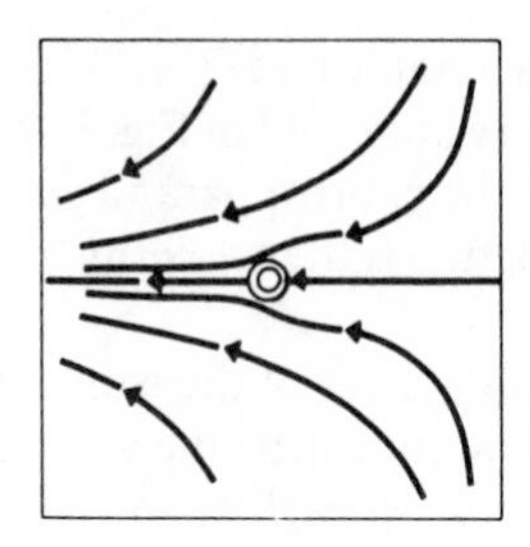
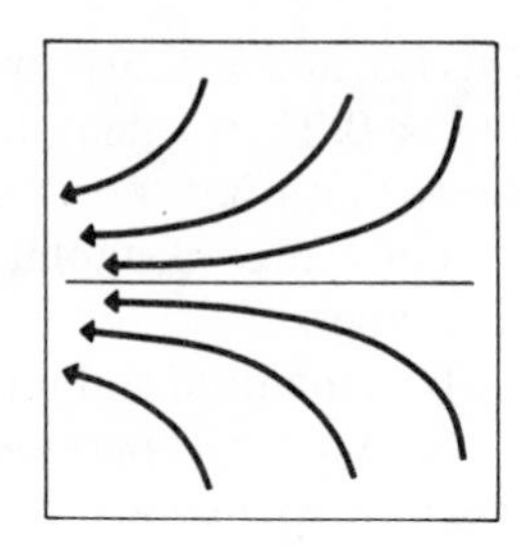

Figure 19.16 Morsification of a doubly degenerate critical point in a dynamical system (center) can result in either two isolated critical points (left) or no critical points at all (right). This system is analogous to the fold catastrophe for gradient systems.

Remark. Fig. 19.14 may be "interpreted" in terms of Feynman diagrams for particle scattering. In the case $b < 0$ there is a "soft" collision between the two "particles" as a function of increasing "time" c. The particles are the critical points, the collision is soft because the scattering is through a small angle. For $b = 0$ there is a "charge exchange" corresponding to a stability exchange. For $b > 0$ there is a hard scattering. This may be viewed as pair annihilation followed by pair creation.

In a k-parameter family of 2-dimensional dynamical systems a degenerate critical point of saddle node type may be encountered in a structurally stable way if $k \geq 1$. Morsification of such a doubly degenerate critical point will result in a dynamical system with either no critical points or two isolated critical points near the origin (Fig. 19.16). Such a morsification is analogous to a fold catastrophe in a gradient system. If $k \geq 2$ this degenerate critical point may be morsified by a path in b–c control parameter space with $b = 0$. Such a path, giving rise to a stability exchange, cannot occur in a structurally stable way in a dynamical system with $k < 2$.

8. The Hopf Bifurcation[1]

In the neighborhood of the bifurcation set in $\mathbb{R}^4$ defined by (19.17), the equations of motion can be written, in a suitable coordinate system, in the form

$$\frac{d}{dt}\begin{bmatrix} x \\ y \end{bmatrix} = \begin{bmatrix} \lambda & -\omega' \\ \omega' & \lambda \end{bmatrix}\begin{bmatrix} x \\ y \end{bmatrix} \tag{19.34}$$

On the bifurcation set itself, $\lambda = 0$ and $\omega' \neq 0$ if the number of control parameters does not exceed 1. When $\lambda = 0$, the phase portrait is as shown in Fig. 19.17. The phase portraits of these structurally and dynamically unstable systems are called **centers** or **vortices**.

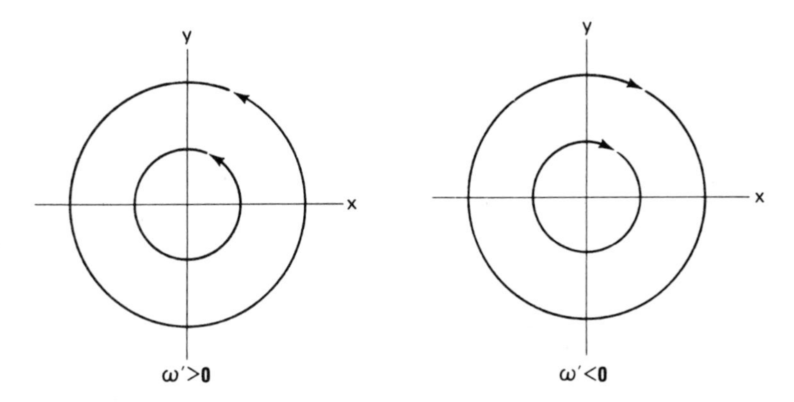

Figure 19.17 The structurally unstable flow pattern for a dynamical system with a pair of complex conjugate eigenvalues with zero real part has the form shown.

It is convenient to study the perturbations of vortices by transforming to polar coordinates. After such a transformation the dynamical equations are

$$\frac{dr}{dt} = \lambda r$$
$$\frac{d\theta}{dt} = \omega' \tag{19.35}$$

On the bifurcation set $\lambda = 0$, and as λ passes through zero, a change in dynamical stability occurs. As in previous cases, we expect a change in dynamical stability to be accompanied by the appearance of qualitatively new types of solutions. In this expectation we are not disappointed.

As usual, we can learn about new solutions by studying a perturbation of the degenerate dynamical system. To do this, we write

$$\frac{dr}{dt} = f_r(r, \theta) = \lambda r + \text{higher order terms}$$
$$\frac{d\theta}{dt} = f_\theta(r, \theta) = \omega' + \text{higher order terms} \tag{19.36}$$

On the bifurcation set the leading term in the Taylor series expansion of $f_\theta(r, \theta)$ is nonzero, so we expect the higher order terms in this series to be unimportant. Accordingly, we neglect them.

In a first approximation we may consider only rotationally invariant perturbations

$$f_r(r, \theta) = f_r(r) \tag{19.37}$$

In point of fact, sufficiently near the origin the coordinate system (r, θ) can always be chosen to bring an arbitrary smooth perturbation into this form.[2] Further, the radial function $f_r(r)$ may contain only odd powers of r. This is because the rotational invariance implies that under $x \to -x$ and $y \to -y$, $\dot{x} \to -\dot{x}$ and $\dot{y} \to -\dot{y}$. This symmetry would be destroyed by terms in $f_r(r)$ containing even powers of r.

A general perturbation of the dynamical system (19.35) thus has the form

$$\begin{aligned} \frac{dr}{dt} &= \lambda r + Ar^3 + Br^5 + \cdots \\ \frac{d\theta}{dt} &= \omega' \end{aligned} \tag{19.38}$$

We consider first the case of 1-parameter families of dynamical systems for which the real part of the eigenvalue goes through zero with nonzero speed: $\partial\lambda(c)/\partial c \neq 0$ when $\lambda(c) = 0$. In this case we may choose λ itself as the control parameter, by the implicit function theorem. When $\lambda = 0$, in general $A \neq 0$ (for a 1-parameter family), so that the Taylor series in (19.38) may be truncated beyond the term r^3. Further, the radial distance may be rescaled ($r \to |A|^{-1/2}r'$) to give the following canonical forms for the perturbed system:

$$\begin{aligned} \frac{dr}{dt} &= \lambda r \pm r^3 \\ \frac{d\theta}{dt} &= \omega' \end{aligned} \tag{19.39}$$

The $\theta(t)$ dependence of this perturbed dynamical system is trivially $\theta_0 + \omega' t$. It is the radial equation that gives rise to qualitatively new solutions.

We consider now the stationary values of r for the dynamical system with

$$\frac{dr}{dt} = \lambda r - r^3 \tag{19.40$-$}$$

There is always a stationary value at $r = 0$. This is an attractor (stable) for $\lambda < 0$ and a repellor (unstable) for $\lambda > 0$.) There are no other stationary values of r for $\lambda < 0$. For $\lambda > 0$ there is a **stable limit cycle** with a radius $r = \sqrt{\lambda}$. As the real part λ of the eigenvalues $\lambda_\pm$ increases through zero, the stable focus at the origin loses its stability, to become an unstable focus, and throws off a stable circular attractor whose radius increases canonically as $\sqrt{\lambda}$. This is called a **supercritical Hopf bifurcation**. The phase portrait for such a bifurcation is shown in Fig. 19.18.

For the dynamical system with

$$\frac{dr}{dt} = \lambda r + r^3 \tag{19.40$+$}$$

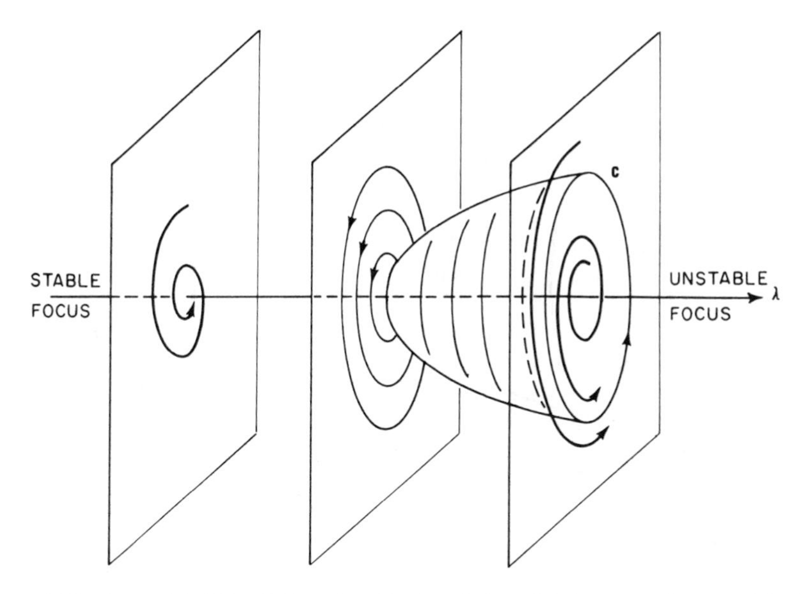

Figure 19.18 Supercritical Hopf bifurcation. For $\lambda < 0$, the origin is a stable focus, and the flow pattern spirals into the origin. As λ increases toward zero, the spiral motion becomes more and more circular. When $\lambda = 0$, the flow becomes a structurally unstable vortex. As λ increases through zero, the origin becomes an unstable focus, and the flow pattern spirals away from the origin. However, the far-field flow still spirals in toward a neighborhood of the origin. The separatrix between the stable and unstable spiral motions is a limit cycle, which is structurally stable and dynamically stable.

there is always an equilibrium at $r = 0$. Again, it is stable for $\lambda < 0$, unstable for $\lambda > 0$. However, for $\lambda < 0$ there is an **unstable limit cycle** with radius $r = \sqrt{-\lambda}$. As λ approaches zero from below, the repellor squeezes down on the stable focus at the origin, and finally squeezes it out of existence at $\lambda = 0$. This is called a **subcritical Hopf bifurcation**. Phase portraits for this bifurcation are shown in Fig. 19.19.

Remark. The dynamical system (19.36) cannot be derived from a potential because curl $F \neq 0$. However, if we ignore the constant angular dependence ($\dot{\theta} = \omega'$), the 1-dimensional radial equation can be written in the form of a gradient equation:

$$\frac{dr}{dt} = -\frac{d}{dr} V(r; \lambda)$$

$$V(r; \lambda) = -\tfrac{1}{2}\lambda r^2 \mp \tfrac{1}{4}r^4 \qquad \begin{pmatrix}\text{sub}\\ \text{super}\end{pmatrix}\text{critical} \tag{19.41}$$

In this sense, Hopf bifurcations are equivalent to symmetry-restricted Ginzburg–Landau ($A_{\pm 3}$) phase transitions "in the r direction."

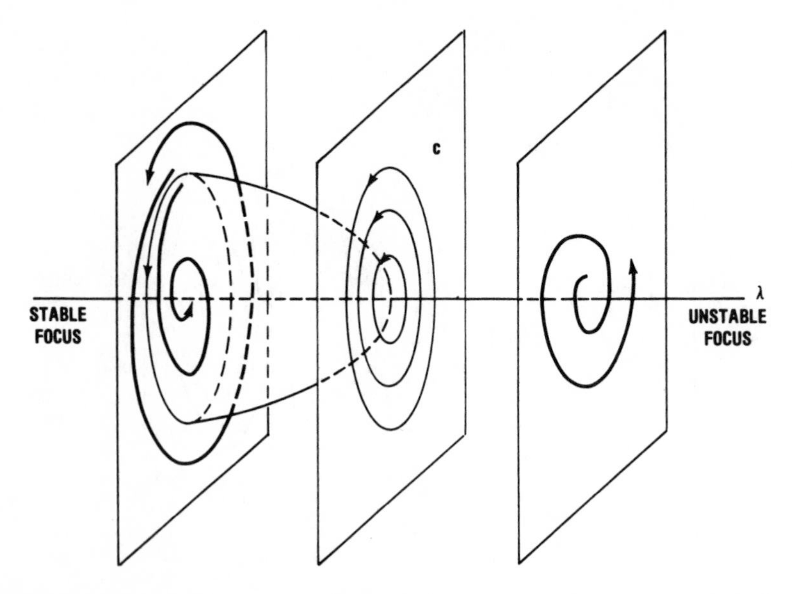

Figure 19.19 Subcritical Hopf bifurcation. For $\lambda < 0$, the origin is a stable focus. In the far–field limit the dynamical system behavior is unstable, with motion spiraling away to infinity. The stable region surrounding the origin and the unstable region containing "the point at infinity" share a separatrix, a limit cycle which is structurally stable and dynamically unstable. As λ increases toward zero, the limit cycle squeezes down on the attractor at the origin, and finally squeezes it out of existence at $\lambda = 0$. The flow pattern passes through the structurally unstable vortex phase when $\lambda = 0$, and for $\lambda > 0$ the origin becomes a repellor (unstable focus).

If the 2-dimensional dynamical system depends on $k > 1$ control parameters, the most general perturbation in the neighborhood of the bifurcation set $\lambda = 0$, $\omega' \neq 0$, has the form

$$\begin{aligned} \frac{dr}{dt} &= \sum_{j=1}^{k} a_j r^{2j-1} \pm r^{2k+1} \\ \frac{d\theta}{dt} &= \omega' \end{aligned} \tag{19.42}$$

The radial equation is clearly related to the symmetry-restricted catastrophe $A_{\pm(2k+1)}$.

Remark. The part $\mathscr{S}_a \subset \mathbb{R}^{n^2}$ of the separatrix of F_{ij} has components of dimension $n^2 - 1$ and lower. Therefore isolated members of 1-parameter families of n-dimensional dynamical systems may possess two equal nonzero eigenvalues in a structurally stable way. Similarly, the bifurcation set $\mathscr{S}_b \subset \mathbb{R}^{n^2}$ has components of dimension $n^2 - 1$ and lower. Therefore isolated members of 1-parameter families of n-dimensional dynamical systems may possess degenerate critical points of saddle node type or of vortex type in a structurally stable way. Further, it is possible to choose a coordinate system $x_1 = x, x_2 = y, x_3, \ldots, x_n$ for the n-dimensional system so that, in the neighborhood of the critical point, all the interesting behavior occurs in the x–y direction and is as described in Secs. 5, 7, and 8:

$$\frac{d}{dt}\begin{bmatrix} x \\ y \end{bmatrix} = \text{see Secs. 5, 7, 8}$$

$$\frac{dx_i}{dt} = \lambda_i x_i, \qquad i = 3, 4, \ldots, n$$

9. Summary

Autonomous dynamical systems are more complicated than gradient dynamical systems, and therefore have a richer structure. The former reduce to the latter when $\nabla \times F = \partial F_j/\partial x_i - \partial F_i/\partial x_j = F_{ji} - F_{ij} = 0$. A dynamical system is locally equivalent to a gradient system at a critical point if it has real unequal eigenvalues.

Autonomous dynamical systems can be studied using many of the same techniques and the terminology developed for the study of gradient systems. The critical points of an autonomous dynamical system are those points $x^0 \in \mathbb{R}^n$ at which the force $F = 0$. Degeneracies can occur at critical points. These degeneracies may be lifted by morsification and are associated with bifurcations. The methods of phase portraits provides a particularly convenient mechanism for the global description of 2-dimensional dynamical systems (Secs. 2, 3).

The dynamical and the structural stability properties of a dynamical system are determined by the eigenvalues of the stability matrix F_{ij} at a critical point. The matrices F_{ij} are parameterized by points in $\mathbb{R}^{n^2}$. The separatrix of points in $\mathbb{R}^{n^2}$ representing structurally unstable systems is determined from the condition $S_a S_b = 0$, where S_a and S_b are defined in terms of the eigenvalues λ_i of F_{ij} by (19.12). The components of the separatrix determined from $S_a = 0$, $S_b \neq 0$ describe structurally unstable systems with isolated critical points, while the components of the separatrix determined by $S_b = 0$ describe degenerate critical points and define the bifurcation set. The 3-dimensional components of the separatrix in $\mathbb{R}^4$ for 2-dimensional dynamical systems were studied in Sec. 4. The n-dimensional case was studied in Sec. 6.

Perturbations around the components $S_a = 0$, $S_b \neq 0$ of the separatrix were studied in Sec. 5. Such perturbations change the qualitative properties of a critical point but do not change the number of critical points.

The 3-dimensional components of the bifurcation set in $\mathbb{R}^4$ describe two distinct types of degenerate critical points. The doubly degenerate saddle node critical point and its perturbations were studied in Sec. 7.

In Sec. 8 we encountered a new type of dynamically unstable behavior, the vortex. Perturbations of vortices lead to descriptions of sub- and supercritical

Hopf bifurcations. Hopf bifurcations in 2-dimensional dynamical systems involving one or more control parameters are intimately related to symmetry-restricted catastrophes of type $A_{\pm(2k+1)}$.

In 1-parameter families of 2-dimensional dynamical systems the only degenerate critical points that can be encountered in a structurally stable way are saddle nodes and Hopf bifurcation points.

A general program for discussing the qualitative properties of autonomous dynamical systems, akin to the general program of Elementary Catastrophe Theory, may be formulated as follows:

1. Determine the bifurcation set in $\mathbb{R}^{n^2}$ associated with the stability matrix F_{ij} (19.12b).
2. Determine the most general perturbation for the dynamical systems described by points in each component of the bifurcation set determined in step 1.
3. Determine the geometry of each perturbation constructed in step 2.

This program is beyond the scope of the present work.

Notes

An English translation of Ref. 1 is to be found in Ref. 2. The qualitative (topological) theory of ordinary differential equations is presented in Ref. 3.

References

1. E. Hopf, Abzweigung einer periodischen Lösung von einer stationären Lösung eines Differentialsystems, *Ber. Math.-Phys. Kl. der Säch. Akad. der Wiss. zu Leipzig* **94** (19. Januar 1942).
2. J. E. Marsden and M. McCracken, *The Hopf Bifurcation and Its Applications*, New York: Springer, 1976.
3. V. I. Arnol'd *Ordinary Differential Equations*, Cambridge: MIT Press, 1973.

CHAPTER 20

Equations Exhibiting Catastrophes

In this chapter we continue our study of autonomous dynamical systems. Having gotten our toes wet in the previous chapter, we now take the plunge. Our aim is to list the spectrum of "irreducible flows" that can occur in an n-dimensional dynamical system. Once all such flows are listed, the behavior of an n-dimensional dynamical system can be modeled by placing various of these flows in different regions of state variable space. This is the analog of positioning Morse saddles M_i^n at various points in $\mathbb{R}^n$ to describe an arbitrary n-dimensional Morse function.

Methods for describing and constructing dynamical flows and invariant surfaces are described in Secs. 1, 4, and 9. In Sec. 1 we show how to construct 2-dimensional flows and limit cycles from 1-dimensional flows. Secs. 2 and 3 involve a more detailed study of the 2-dimensional flows called limit cycles.

In Sec. 4 we show how to construct 3-dimensional flows and invariant surfaces from the 2-dimensional structures. We place some of these basic building blocks together in simple ways to obtain strange dynamical system behavior in Secs. 5 and 6. The flow obtained in Sec. 6 was first studied in detail by Lorenz. This flow has been used to study nonlinear problems in hydrodynamics (Sec. 7) and electrodynamics (Sec. 8).

In Sec. 10 we introduce the Center Manifold Theorem. This is the dynamical system counterpart to the Thom Splitting Lemma (2.3). It plays the same simplifying role in dynamical systems theory as the splitting lemma does in its domain of validity. In particular, it makes the listing of the irreducible flows and their bifurcations a useful exercise.

We close this chapter (Sec. 11) with the Ruelle-Takens picture for the onset of turbulence in nonlinear systems of unspecified nature. This picture exploits the Center Manifold Theorem, the listing of invariant flows, the Hopf bifurcation, and the ideas of dynamical and structural stability and their loss.

1. Aufbau Principle. 1

A 1-dimensional dynamical system $\dot{x} = f(x)$ depending on no control parameters typically has only isolated critical points at which $f(x) = 0$. At such points typically $f'(x) \neq 0$. Thus a 1-dimensional dynamical system has typically only two kinds of critical points, attractors and repellors:

$$\frac{dx}{dt} = kx, \qquad k \neq 0 \tag{20.1}$$

At an attractor $k < 0$, while $k > 0$ for a repeller (Fig. 20.1).

What kind of critical points and critical behavior can we expect for 2-dimensional dynamical systems? We can get a handle on this question by building on the critical points of a 1-dimensional dynamical system. In Fig. 20.2 we begin with a Morse zero saddle $+x^2$ in $\mathbb{R}^1(f = -\nabla V)$ and "add" to it either a Morse zero saddle $+y^2$ or a Morse 1-saddle $-y^2$. The resulting isolated critical point is a Morse zero saddle M_0^2 or a Morse 1-saddle M_1^2, as shown in Fig. 20.2*a* and *b*. However, we can also add a flow as shown in Fig. 20.2*c*. In this case the critical point is no longer isolated. However, if we bend the flow back on itself (Fig. 20.2*c*), the resulting critical behavior is that of a stable "circular" motion. Such a critical trajectory (no longer a point) is called a stable limit cycle and denoted by $T^1 \times M_0^1$.

We can also build up 2-dimensional critical points and flows by starting from the unstable critical point M_1^1 and proceeding as in the previous paragraph. By adding a stable term $(+y^2)$, an unstable term $(-y^2)$, and a nonzero flow $(\sim y)$

Figure 20.1 In a 1-dimensional dynamical system an isolated critical point may be either an attractor or a repellor.

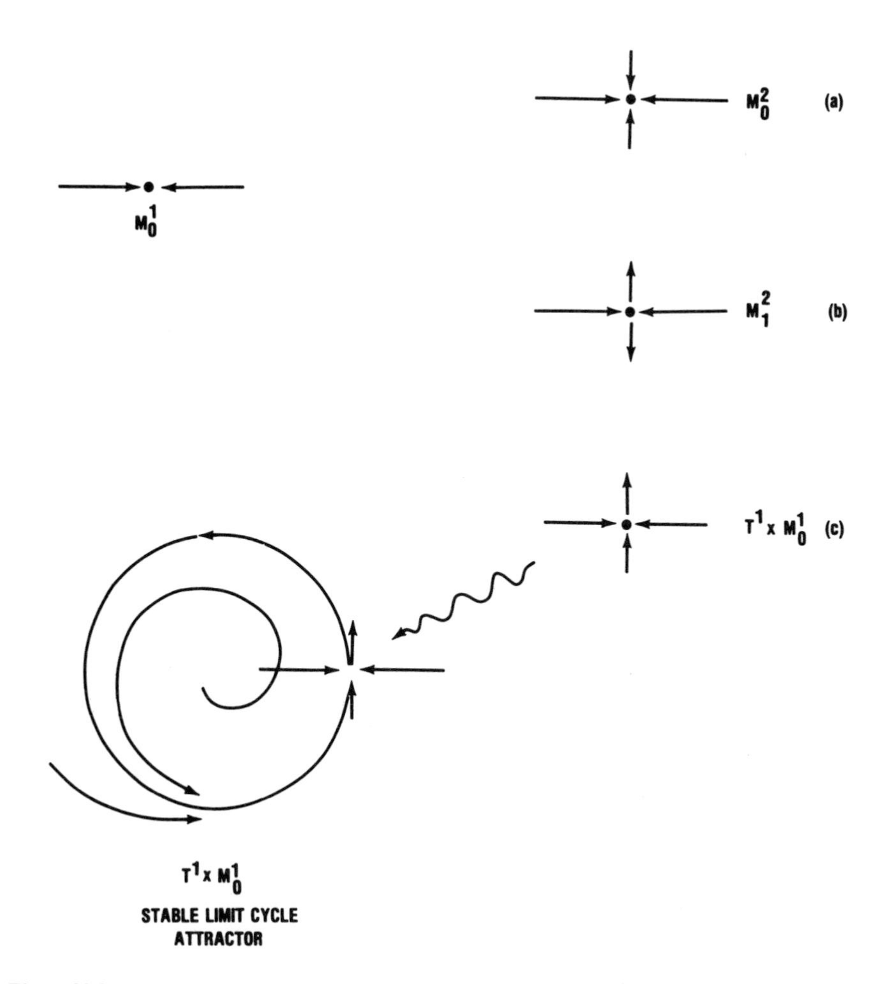

Figure 20.2 Starting from a stable 1-dimensional attractor ($V \simeq +x^2$), the addition of a flow term for the y direction of the form $(+y^2, -y^2, y)$ creates critical points of types M_0^2 and M_1^2 and a critical flow of type $T^1 \times M_0^1$ after the latter is bent back on itself.

we produce a Morse 1-saddle M_1^2 (Fig. 20.3*a*), a Morse 2-saddle M_2^2 (Fig. 20.3*b*), and an unstable limit cycle $T^1 \times M_1^1$ (Fig. 20.3*c*).

We may ask if all the structurally stable types of critical behavior possible for 2-dimensional dynamical systems have been constructed by this bootstrap procedure. The answer is no, we have missed the foci. One way to see that these must exist as a structurally stable type of critical point is to shrink the radius of the limit cycle down to zero (Fig. 20.4). In this $r \to 0$ limit the flow surrounding the stable limit cycle $T^1 \times M_0^1$ resembles that of a stable focus F_-^2; that surrounding the unstable limit cycle $T^1 \times M_1^1$ resembles that of an unstable focus F_+^2. This contraction procedure is equivalent to a Hopf bifurcation.

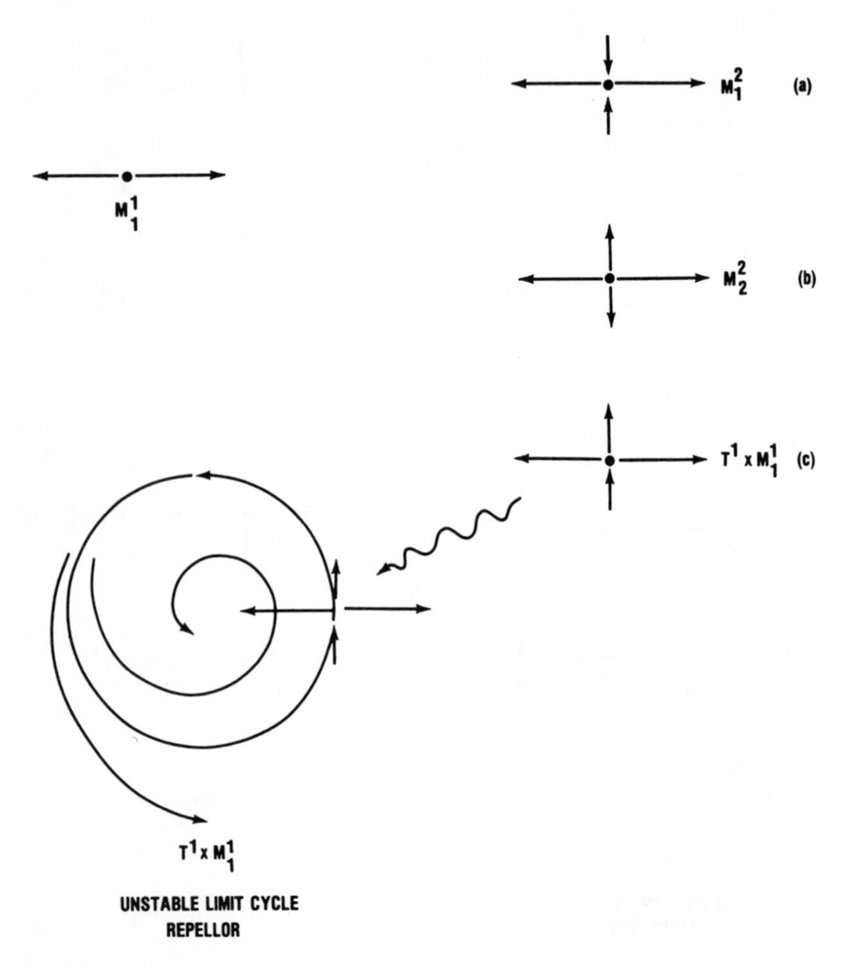

Figure 20.3 Starting from an unstable 1-dimensional critical point ($V \simeq -x^2$), the addition of a flow term for the y direction of the form $(+y^2, -y^2, y)$ creates critical points of types M_1^2 and M_2^2 and a critical flow of type $T^1 \times M_1^1$ after the latter is bent back on itself.

The isolated Morse saddles M_0^2, M_1^2, M_2^2 the foci F_-^2, F_+^2, and the limit cycles $T^1 \times M_0^1$, $T^1 \times M_1^1$ are the only kinds of structurally stable critical behavior possible for 2-dimensional dynamical systems.

Remark. We initially missed the foci $F_\pm^2$ for the following reason. We began with critical points in one dimension with real nonzero eigenvalue. In adding a flow in the second dimension, we have assumed a form $\dot{y} =$ constant or $\dot{y} = ky$, where k was a nonzero constant. In order to find complex conjugate eigenvalues, we would have to increase the dimension of the new flows by 2 instead of 1.

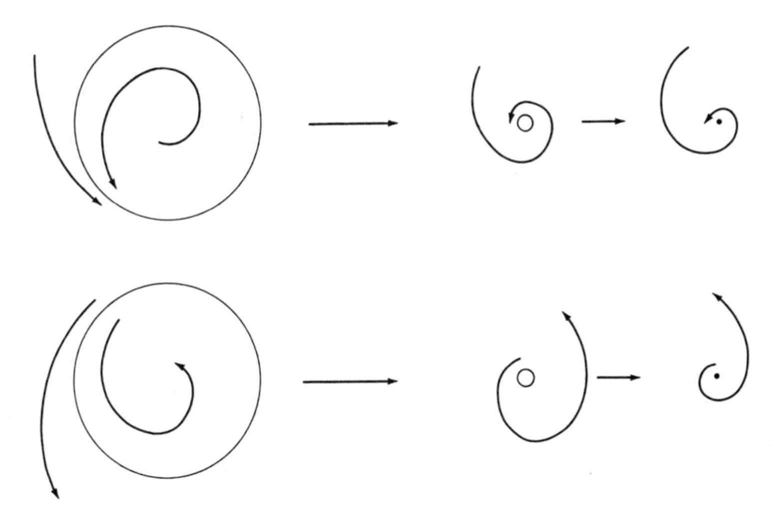

Figure 20.4 By shrinking the radius of a stable or unstable limit cycle, the interior focus (unstable or stable) is squeezed out of existence. The residual flow, as $r \to 0$, is that of a stable or unstable focus.

2. Van der Pol Oscillator. 1 Hopf Bifurcation

The simple harmonic oscillator and its perturbations exhibit many of the types of critical behavior described in Sec. 1. The simple harmonic oscillator

$$m\ddot{x} + kx = 0 \tag{20.2}$$

and its perturbations are the "hydrogen atom problem" of dynamical systems theory (simple enough to solve, complicated enough to be useful). The harmonic oscillator equation (20.2) can be written in standard dynamical system form as a pair of coupled first-order ordinary differential equations:

$$\begin{aligned} \frac{dx}{dt} &= p \qquad (m = 1) \\ \frac{dp}{dt} &= -x \qquad (k = 1) \end{aligned} \tag{20.3}$$

Solutions of (20.3) have the form $x = R \sin(\omega t + \phi)$, $p = R \cos(\omega t + \phi)$, $\omega = \sqrt{k/m}$. Since these trajectories are vortices (cf. Fig. 19.17), the system (20.3) is structurally unstable.

Remark. The harmonic oscillator equations (20.3) can be derived from a Hamiltonian. Every time-independent Hamiltonian is conservative and structurally unstable against perturbations of dissipative type. As a result, the ideas of structural stability can be applied in classical mechanics only after a great deal of care has been taken to specify the class of systems and types of perturbations under consideration.

The damped harmonic oscillator

$$\ddot{x} + \gamma\dot{x} + x = 0 \tag{20.4a}$$

is a structurally stable linear perturbation of the harmonic oscillator (20.2). This equation can be written in the standard dynamical system form as follows ($p \to y$):

$$\begin{aligned} \frac{dx}{dt} &= y - \gamma x \\ \frac{dy}{dt} &= -x \end{aligned} \tag{20.4b}$$

Equation (20.4) has solutions of the type $x \sim e^{\lambda t}$, where the eigenvalue λ obeys the equation

$$\lambda^2 + \gamma\lambda + 1 = 0 \tag{20.5}$$

If we consider γ as a control parameter, the 1-dimensional control space $\mathbb{R}^1$ is divided into four disjoint open regions representing structurally stable behavior by the two points $\gamma = \pm 2$ in the Maxwell set $\mathscr{S}_M$ and the point $\gamma = 0$ in the bifurcation set $\mathscr{S}_B$. This decomposition, together with representatives of the structurally stable and structurally unstable dynamical behavior, is shown in Fig. 20.5.

The harmonic oscillator (20.2) also has some physically interesting *nonlinear* perturbations which are structurally stable. One such class of perturbations has the form

$$\ddot{x} + \gamma f(x)\dot{x} + x = 0 \tag{20.6a}$$

or

$$\begin{aligned} \dot{x} &= y - \gamma F(x) \\ \dot{y} &= -x \end{aligned} \tag{20.6b}$$

We shall assume the following:

1. $(d/dt)F(x) = f(x)\dot{x}$.
2. $f(-x) = +f(x)$ and $F(-x) = -F(x)$.
3. $f(x)$ and $F(x)$ are smooth and differentiable.
4. $f(x)$ is finite on a finite interval of the x axis.
5. $\gamma \geq 0$.

The constant γ determines the strength of the damping term.

The van der Pol oscillator is obtained by setting $F(x) = x^3 + ax$; $f(x) = 3x^2 + a$ (in 20.6). This structurally stable perturbation of the harmonic oscillator can be studied using the methods of Catastrophe Theory when γ is either very small (this section) or very large (following section).

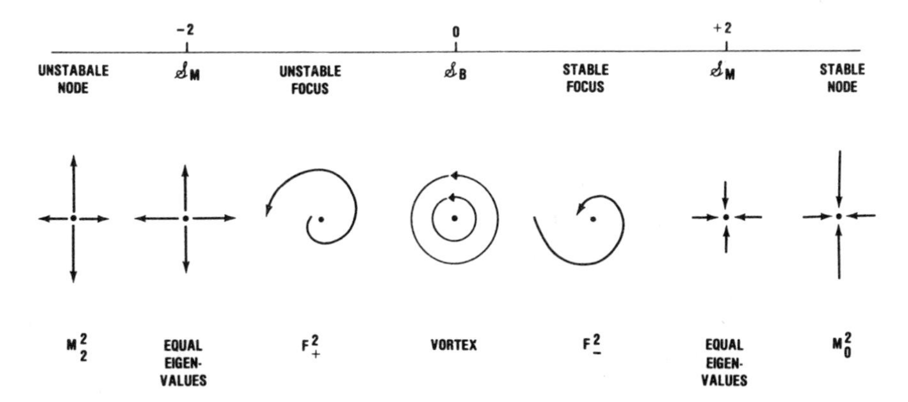

Figure 20.5 The control parameter space for the damped harmonic oscillator (20.4) is divided into four open regions by three points, two of which ($\gamma = \pm 2$) belong in the Maxwell set $\mathscr{S}_M$ and one of which belongs to the bifurcation set $\mathscr{S}_B$. The four open regions parameterize structurally stable harmonic oscillators.

First we observe that $F(0) = 0$ by assumptions 2 and 3. Therefore $(x, y) = (0, 0)$ is a solution of (20.6b) for all values of γ. The stability properties of this solution are determined by the value of $\gamma f(0)$, as shown in Fig. 20.5. If f depends on one or more control parameters, then when $\gamma f(0; a_1, \ldots) = \pm 2$ the corresponding system (20.6) is described by a point on the Maxwell set $\mathscr{S}_M$, but when $\gamma f(0; a_1, \ldots) = 0$ the local equilibrium at $(x, y) = (0, 0)$ is a center. We therefore expect a Hopf bifurcation to be associated with this structurally unstable behavior. We will determine the limit cycle behavior for γ small, and the Hopf bifurcation explicitly for $f(x) = 3x^2 + a$.

The properties of the dynamical system (20.6b) can be determined using the method of phase portraits (cf. Chapter 19). We are particularly interested in determining the "attractors" of (20.6). These may be isolated points or closed stable periodic orbits $T^1 \times M^1_0$. There is only one isolated critical point for (20.6), that at $(x, y) = (0, 0)$. If this point is a stable focus and if it is surrounded by a limit cycle, that limit cycle must be unstable. Similarly, if a stable limit cycle is present, the interior focus is unstable (Fig. 20.4). The shape of a stable limit cycle can be determined by integrating the dynamical system equations (20.6b) forward in time using arbitrary initial conditions $(x_0, y_0) \neq (0, 0)$. The shape of an unstable limit cycle can be determined by integrating backward in time, starting from a noncritical point. Once the contour of the limit cycle has been determined, the period of motion can be determined from the contour integrals

$$T = \oint dt = \oint -\frac{dy}{x} = \oint \frac{dx}{y - \gamma F(x)} \tag{20.7}$$

By eliminating dt from (20.6b) we find

$$x\,dx + y\,dy - \gamma F(x)\,dy = 0 \tag{20.8}$$

Since $x\,dx + y\,dy = \frac{1}{2}d(x^2 + y^2)$, the integral of (20.8) around a closed contour leads to the condition

$$\oint F(x)\,dy = 0 \tag{20.9}$$

This condition will be used below to determine the canonical dependence of the radius of a limit cycle under Hopf bifurcation

We now consider the case $|\gamma f(0; a)| \ll 1$. The focus at $(x, y) = (0, 0)$ is stable if $\gamma f(0; a) > 0$, unstable if $\gamma f(0; a) < 0$. Since the foci wind in or out slowly when $\gamma f(0; a)$ is small, the values of $x(t)$, $y(t)$ during the course of one revolution are given approximately by $x \simeq R \sin \omega t$, $y \simeq R \cos \omega t$ $(\omega = 1)$. In order to find the radius R of the approximately circular limit cycle (when it exists), we truncate $f(x; a) = a + 3x^2 + \mathcal{O}(4)$. Then

$$x \simeq R \sin \theta$$

$$y \simeq R \cos \theta$$

$$F(x; a) = x^3 + ax$$

$$\oint [R^3 \sin^3 \theta + aR \sin \theta](-R \sin \theta)\,d\theta = -\pi R^2(\tfrac{3}{4}R^2 + a) = 0 \tag{20.10}$$

For $a \neq 0$ there is an isolated critical point at $R = 0$, which is stable if $a > 0$, unstable if $a < 0$. If $a > 0$ there are no other nearby solutions, while if $a < 0$

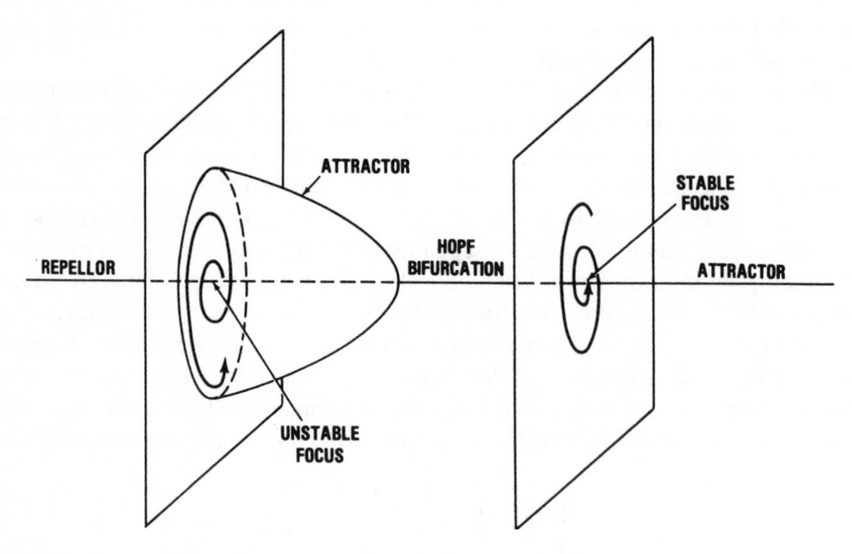

Figure 20.6 Subcritical Hopf bifurcation to a stable focus. The limit cycle has fixed period and its radius has a canonical $\frac{1}{2}$ power dependence on the parameter distance from the Hopf bifurcation point.

there is a stable limit circle of radius

$$R = \sqrt{\frac{-4a}{3}} \tag{20.11}$$

This subcritical Hopf bifurcation is illustrated in Fig. 20.6.[1] The period of this limit cycle is determined from (20.7) to be

$$T = \oint - \frac{d(R\cos\theta)}{R\sin\theta} = +2\pi \tag{20.12}$$

The positive sign indicates counterclockwise rotation. The period is related to the choice $\omega = \sqrt{k/m} = 1$.

3. Van der Pol Oscillator. 2 Relaxation Oscillations

In the case that γ (20.6) is very large, so also in general is y. Since it is not convenient to work with large quantities, we introduce a new phase space variable z according to

$$y = \gamma z, \qquad \gamma^{-1} \ll 1 \tag{20.13}$$

The dynamical system equations (20.6) then assume the form

$$\frac{dx}{dt} = \gamma(z - F(x)) \tag{20.14a}$$

$$\frac{dz}{dt} = -\gamma^{-1}x \tag{20.14b}$$

By eliminating t we can write an equation for the integral curves of (20.14) in the form

$$(z - F(x))\frac{dz}{dx} = -\frac{x}{\gamma^2}(\simeq 0) \tag{20.15}$$

For large values of γ, either

$$\frac{dz}{dx} \simeq 0$$

or

$$z - F(x) \simeq 0 \tag{20.16}$$

The motion of the dynamical system in the x–z plane is shown in Fig. 20.7 for $F(x) = x^3 + ax$ and $a < 0$. The two outside segments of the S-shaped curve are attractors, as can be seen by a linear stability analysis. The middle segment,

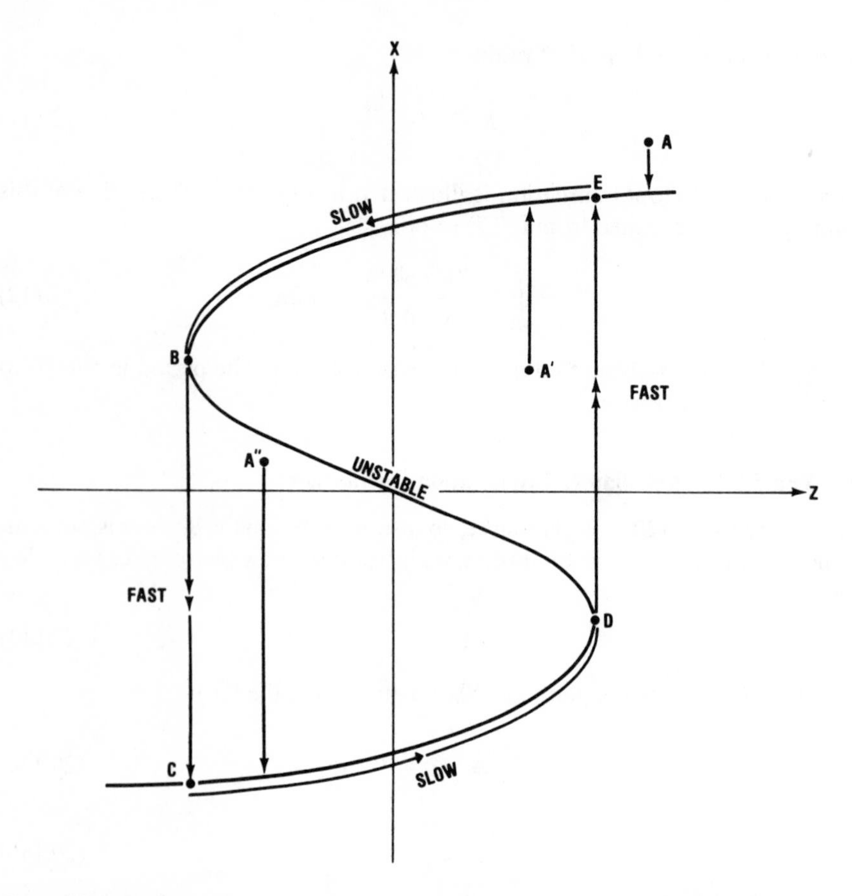

Figure 20.7 The curve $z - F(x) = 0$ is the familiar cusp cross section. The outer two branches are stable, the inner one is unstable. All flows approach this attractor, whether from A, A', A'', . . . , and then cycle around the path $EBCDE$ in a fast-slow-fast-slow relaxation oscillation.

between the two vertical tangents, is a repellor. For $x > 0$, $\dot{z} < 0$ and for $x < 0$, $\dot{z} > 0$. If the dynamical system starts from point A, it will quickly approach the curve $z = F(x)$ according to (20.14a), since γ is large. The state of the dynamical system will be represented by a point on this curve in phase space. This point moves slowly to the left because $\dot{z} = -\gamma^{-1}x$. When this point reaches the point B of vertical tangency, there will be a fast jump to point C on the lower sheet, followed by a slow motion along $z = F(x)$ to point D and a sudden jump to E.[1]

Other initial points A', A'' reach the attracting part of the cusp cross section $z = F(x)$ by fast motion in the z direction, as shown. Once the system reaches this curve, it cycles around the closed loop $BCDEB$ in an alternating fast-slow-fast-slow periodic orbit. Such an orbit is called a **relaxation oscillation**. The time scales involved are of order $1/\gamma$ for the fast jumps and of order γ for the slow traverses. As a result, the period is given approximately by

$$
\begin{aligned}
T &\simeq 2\int_E^B -\frac{\gamma\, dz}{x} = -2\gamma \int_E^B \frac{f(x)}{x}\, dx \\
&= 2\gamma \int_B^E \left(3x + \frac{a}{x}\right) dx \\
&= \gamma\left\{3(x_E^2 - x_B^2) + 2a \ln \frac{x_E}{x_B}\right\}
\end{aligned}
\tag{20.17}
$$

Relaxation oscillations occur for the van der Pol oscillator only in the multi-valued regime $a < 0$. For $a < 0$, $x_B = \sqrt{-a/3}$, $x_E = 2\sqrt{-a/3}$,

$$
T \simeq |\gamma a|(3 - 2 \ln 2) \simeq 1.6|\gamma a|,\ a < 0 \tag{20.18}
$$

Remark 1. For the structurally stable perturbation of the harmonic oscillator discussed in this and the preceding section, we have chosen $F(x; a) = x^3 + ax$ (van der Pol oscillator). A change of stability occurs at the critical point $(x, y) = (0, 0)$ as the parameter a decreases through zero. Associated with this change of stability is a Hopf bifurcation. A new circular (for a small) stable attractor bifurcates from the solution $(x, y) = (0, 0)$. As the parameter a continues to decrease, the limit cycle becomes squashed and distorted. Pushed to extremes ($|\gamma| \gg 1$) this limit cycle assumes the form of a relaxation oscillation.

Remark 2. Limit cycles are structurally stable under perturbation. If a system with an isolated limit cycle is perturbed, the new system will have an isolated limit cycle near the original isolated limit cycle. The only way to destroy a limit cycle in a 1-parameter family of 2-dimensional dynamical systems is to squeeze it down onto an interior focus (Fig. 20.8*a*) or to have it collide with and annihilate another limit cycle (Fig. 20.8*b*). This latter process may easily be identified with the fold catastrophe A_2 by taking a cross section of the flow transverse to the limit cycles. Limit cycles in $\mathbb{R}^n$, $n > 2$, are not structurally stable when they occur on tori T^k, $k > 1$.

Remark 3. We have identified the Hopf bifurcation and the relaxation oscillation of the van der Pol oscillator with the symmetry-restricted cusp catastrophe. Functions $F(x)$ associated with higher symmetry-restricted catastrophes A_k (20.6) can be used to construct dynamical systems with yet more interesting phenomena. The discussions given in these two sections go through for these more complicated systems.

Important Remark. Complicated dynamical systems become tractable when they involve multiple distinct and widely separated time scales.

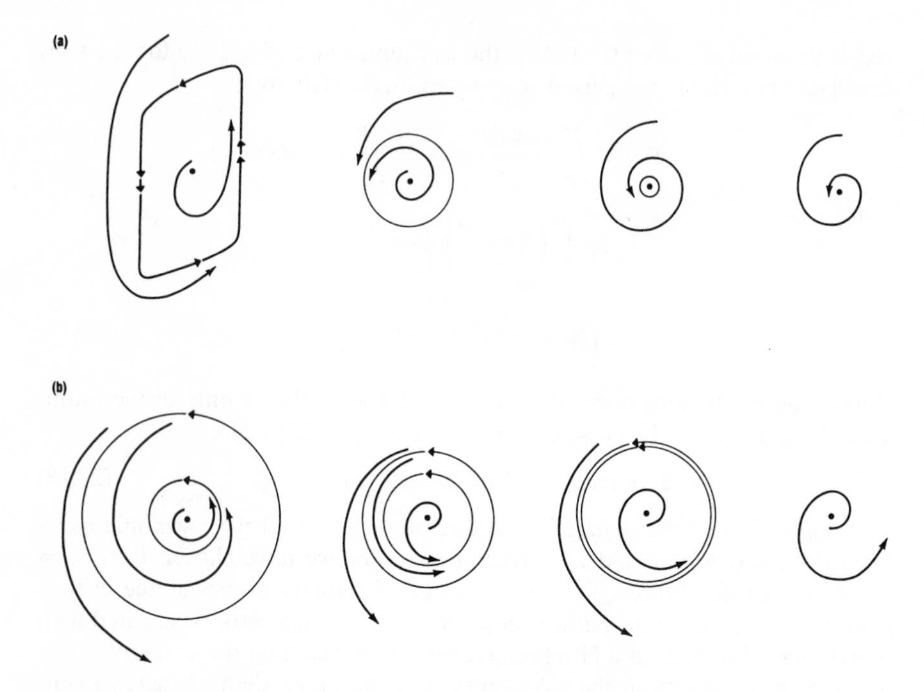

Figure 20.8 Structurally stable limit cycles can be destroyed in two ways. (*a*) The radius of the limit cycle may be shrunk to zero, in which case it disappears in a Hopf bifurcation. (*b*) It may collide with and annihilate another limit cycle of the opposite stability type and same flow direction. The two then disappear in what is essentially an A_2 fold catastrophe.

4. Aufbau Principle. 2

The Aufbau Principle was useful in constructing 2-dimensional critical points and flows from the simpler 1-dimensional critical points. This method can be extended to construct 3-dimensional critical behavior from the list of 2-dimensional critical structures.

This extension can be carried out as indicated in Fig. 20.2. In Fig. 20.9 we add terms of the form $+z^2$, $-z^2$, z to the stable isolated Morse critical point M_0^2. The resulting critical behavior is that of an isolated Morse minimum M_0^3, an isolated Morse 1-saddle M_1^3, and, bending the flow back on itself, of a stable limit cycle $T^1 \times M_0^2$. This construction is repeated in Figs. 20.10 and 20.11.

In Figs. 20.12 and 20.13 we again extend a structurally stable 2-dimensional flow by addition of the terms $+z^2$, $-z^2$, z. In Fig. 20.12 we begin with a stable 2-dimensional focus F_-^2. From this we construct respectively the stable focus $F_-^2 \times M_0^1$, the unstable focus $F_-^2 \times M_1^1$, and, bending the flow through the focus back on itself, the stable limit screw $\mathrm{Sc}_- = F_-^2 \times T^1$. In Fig. 20.13 we begin with an unstable 2-dimensional focus F_+^2 from which we construct $F_+^2 \times M_0^1, F_+^2 \times M_1^1, F_+^2 \times T^1$.

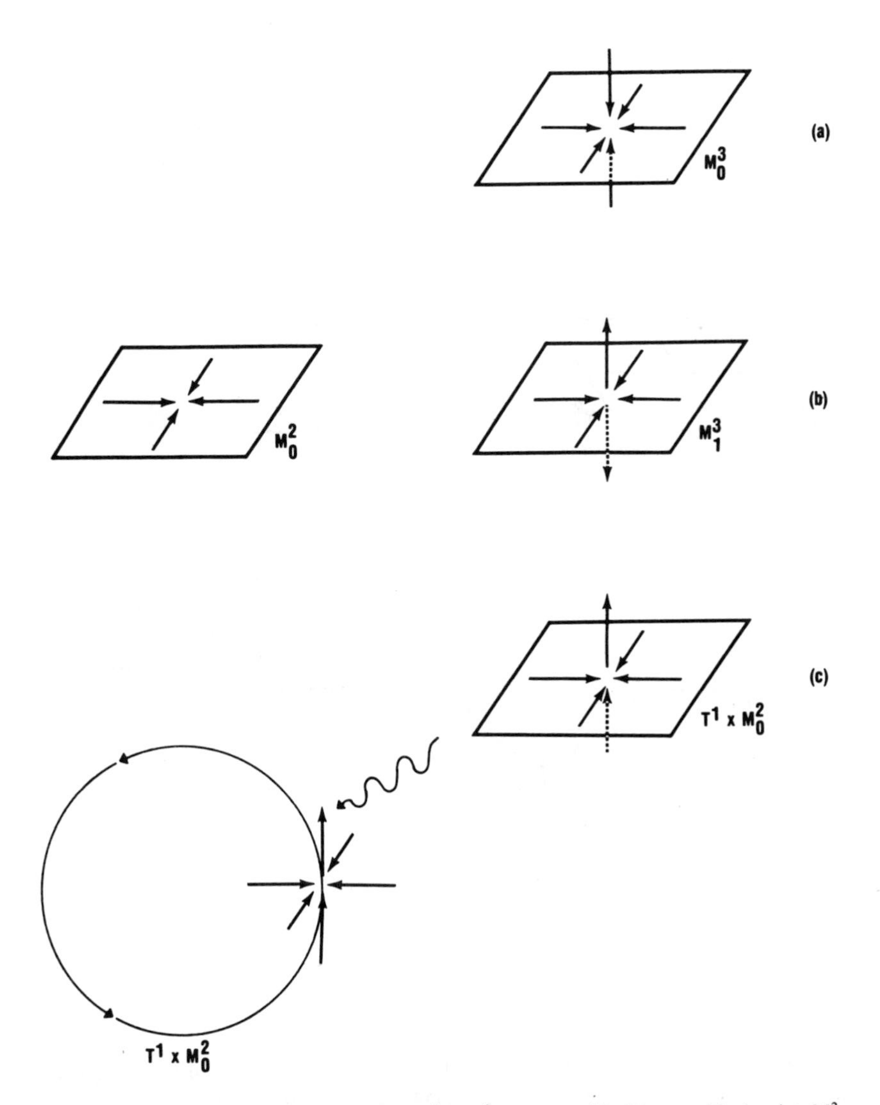

Figure 20.9 By adding a term of the form $+z^2$, $-z^2$, z to a stable Morse critical point M_0^2 we obtain isolated critical points M_0^3 (*a*), M_1^3 (*b*), and a stable limit cycle $T^1 \times M_0^2$ (*c*).

Stable and unstable limit cycles may be treated in the same way. The spectrum of structures obtained from a stable limit cycle $T^1 \times M_0^1$ is shown in Fig. 20.14, while those obtained from $T^1 \times M_1^1$ are shown in Fig. 20.15. The limit cycles $T^1 \times M_i^2 \in \mathbb{R}^3$, $i = 0, 1, 2$, have previously been obtained. The new structures are the stable torus $T^2 \times M_0^1$ and the unstable torus $T^2 \times M_1^1$.

The dynamical systems whose idealized flows are presented in Figs. 20.9–20.15 are particularly easy to describe when three distinct and widely separated

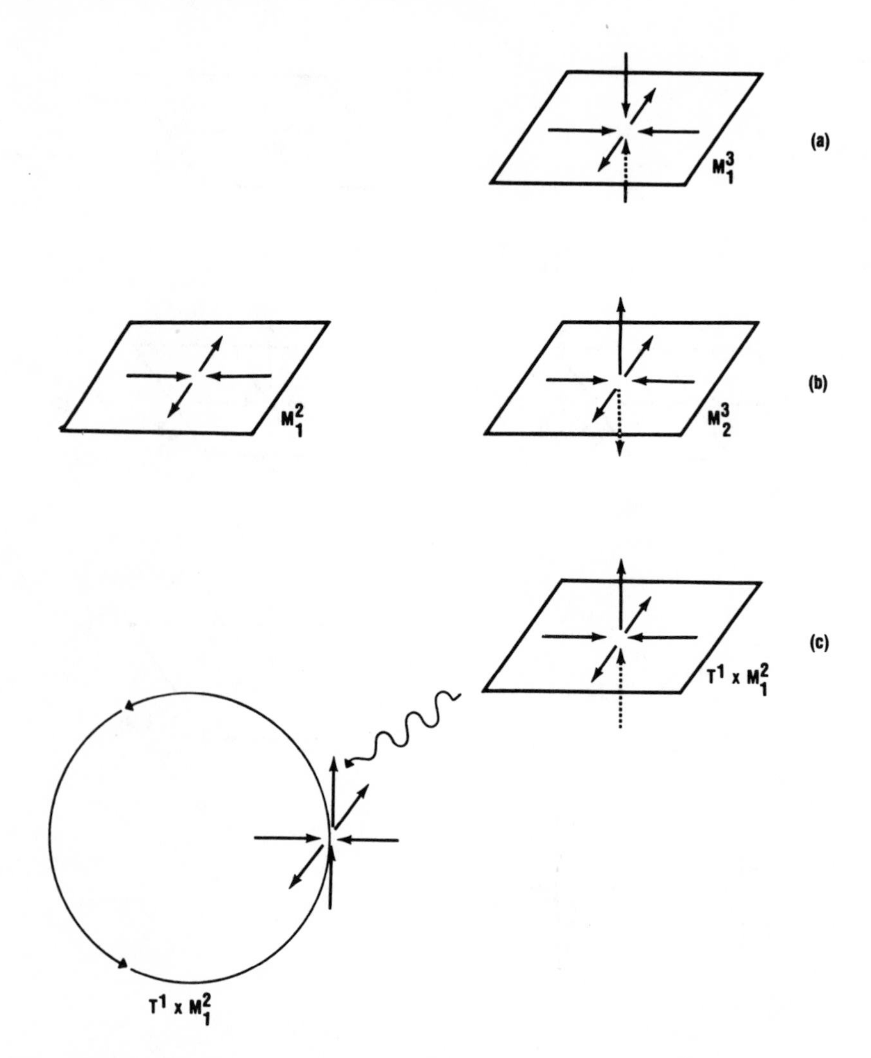

Figure 20.10 By starting with a Morse saddle M_1^2 and proceeding as in Fig. 20.9, we obtain isolated critical points M_1^3 (*a*), M_2^3 (*b*), and the unstable limit cycle $T^1 \times M_1^2$ (*c*).

time scales are present. For example, consider the flow $F_-^2 \times M_0^1$ for which the rotation rate ω around the z axis is much faster than either the radial relaxation rate γ_r or the z relaxation rate γ_z. If $\gamma_r \gg \gamma_z$, then the flow looks like a tornado (Fig. 20.16*a*) whereas if $\gamma_r \ll \gamma_z$, the flow looks like motion on a paraboloid (Fig. 20.16*b*). For a flow of type $F_-^2 \times M_1^1$ we assume that the rotational time scale is much faster than the radial time scale which is, in turn, much faster than the z time scale. Then the system state will execute rapid roughly circular rotation about the focus, the radius of the circular motion will decrease, while the motion remains near the plane $z = 0$. When the rotation radius becomes very

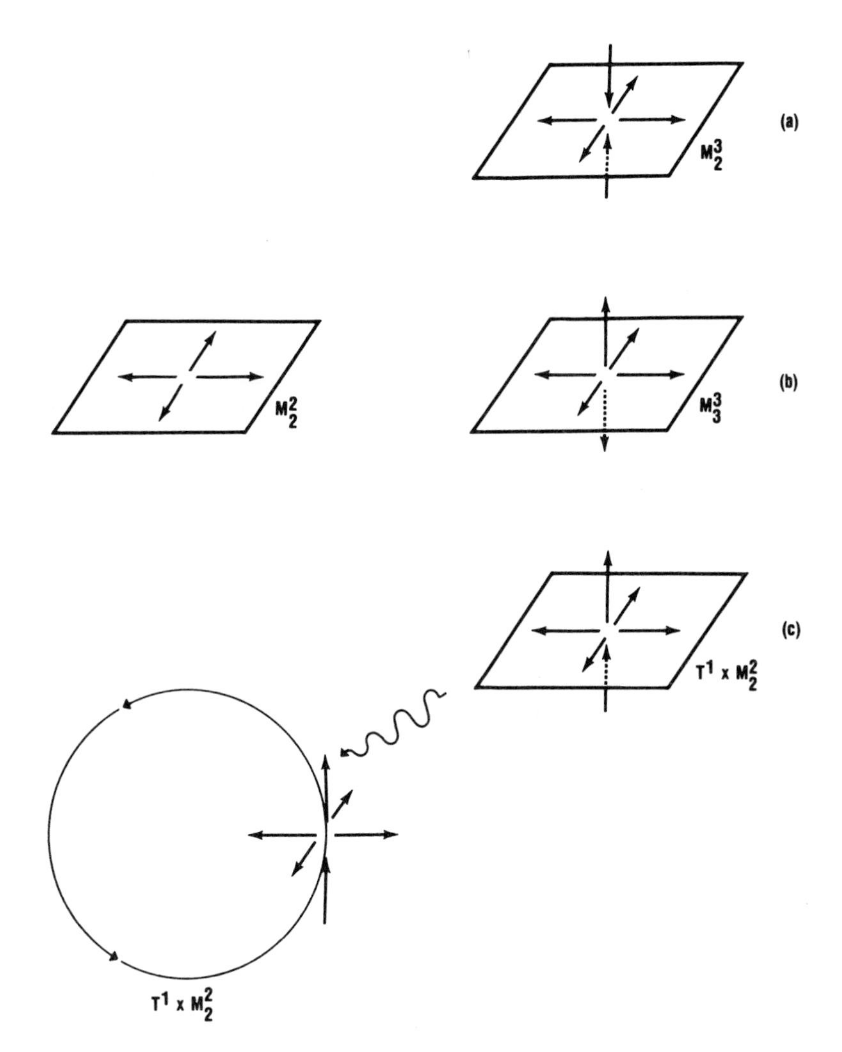

Figure 20.11 By starting with the unstable critical point M_2^2 and proceeding as before, we obtain isolated critical points of type M_2^3 (*a*), M_3^3 (*b*), and the unstable limit cycle $T^1 \times M_2^2$ (*c*).

small, the repulsion will send the system out roughly along the positive or negative z axis, depending on whether the motion was above or below the plane $z = 0$, which is unstable (Fig. 20.16*c*). The limit screw motion is also easily described when three distinct time scales are involved.

The attractor represented by the stable torus is somewhat different from the other attracting sets. The set of limit points associated with all the other stable flows form 0-dimensional or 1-dimensional sets. A system initially at an isolated critical point will remain at such a critical point. A system initially on a 1-dimensional limit set will travel around the limit set, visiting every point in that

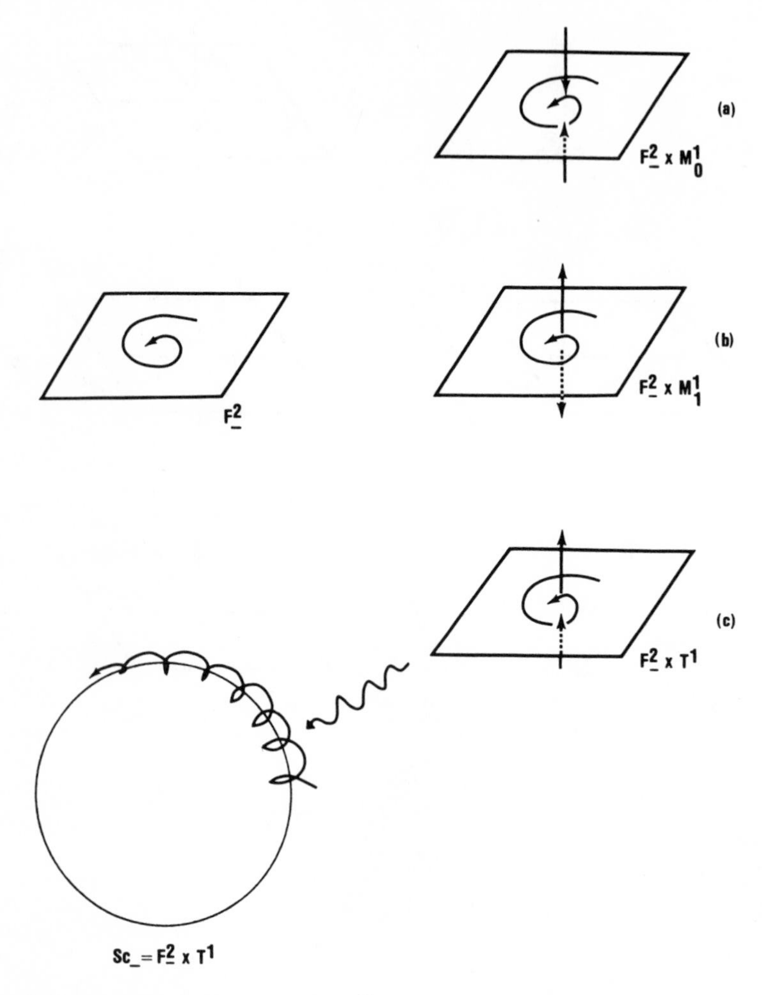

Figure 20.12 By starting with a stable focus F^2_- and proceeding in the usual way, we obtain stable and unstable flows of type $F^2_- \times M^1_0$ (*a*) $F^2_- \times M^1_1$ (*b*) at isolated critical points and the stable limit screw $F^2_- \times T^1 = \mathrm{Sc}_-$ (*c*).

set. This is no longer true for the limit tori. These sets are 2-dimensional. A dynamical system initially at a point in either a stable or an unstable torus will remain on the 2-dimensional torus, but will not visit all points on the torus. Roughly speaking, the dynamical system will traverse a set of measure zero or of measure one on the torus depending on whether the ratio of the rotation rates in the two directions is rational or irrational. It can now be seen why a limit cycle in $\mathbb{R}^3$ is unstable against perturbations. Suppose that the limit cycle wraps around a stable torus $T^2 \times M^1_0$ with rational frequencies (ω_1, ω_2) where, say, $\omega_1 = 1, \omega_2 = 1$. Under the perturbation $\omega_1 = 1, \omega_2 \to \omega_2' = \pi/3$, the perturbed

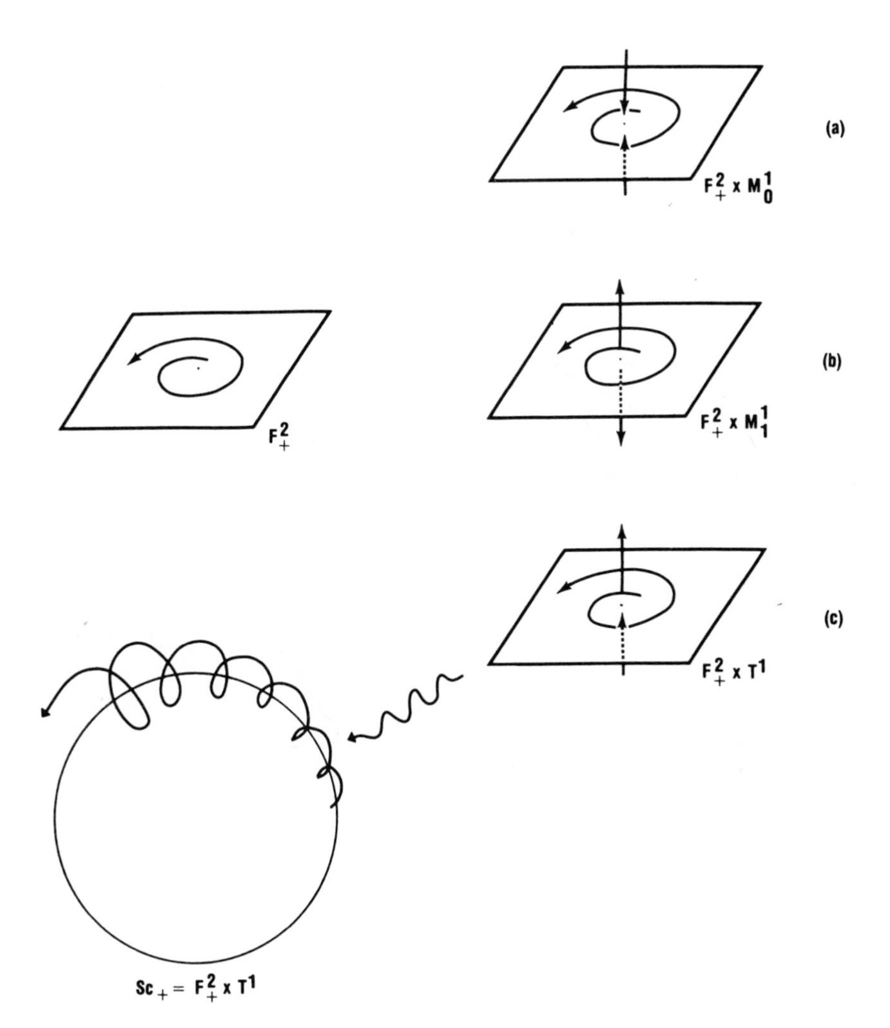

Figure 20.13 By starting with an unstable focus F^2_+ and proceeding in the usual way, we obtain the unstable flows $F^2_+ \times M^1_0$ (*a*), $F^2_+ \times M^1_1$ (*b*) about an isolated critical point and the unstable limit screw $F^2_+ \times T^1 = \mathrm{Sc}_+$ (c).

trajectory will never close on itself. Therefore limit cycles are structurally stable against perturbation in $\mathbb{R}^n$, $n \geq 2$ when they occur on an invariant torus T^k, $k = 1$, but they are structurally unstable when they occur on an invariant torus T^k with $1 < k \leq n - 1$.

Remark. We may ask whether the invariant surface T^2 is itself structurally stable. This surface is called invariant because any point in the surface will be mapped into another point in the surface under the dynamical system equations of motion. The result is that $T^2 \subset \mathbb{R}^3$ is structurally stable. Of course, higher

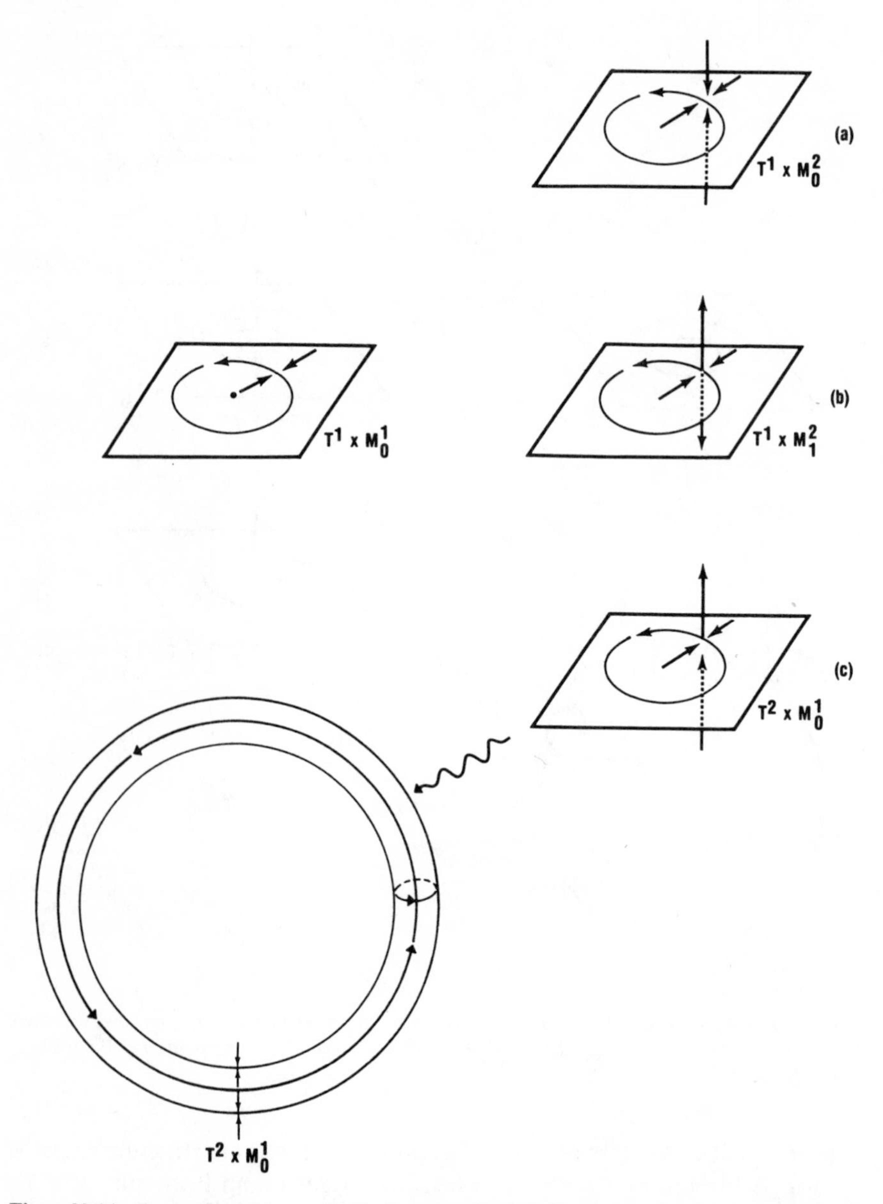

Figure 20.14 By starting with a stable limit cycle and proceeding in the usual way, we obtain the flow $T^1 \times M_0^2$ (stable) (*a*), $T^1 \times M_1^2$ (unstable) (*b*), and $T^2 \times M_0^1$ (stable) (*c*) None of these critical sets is a critical point.

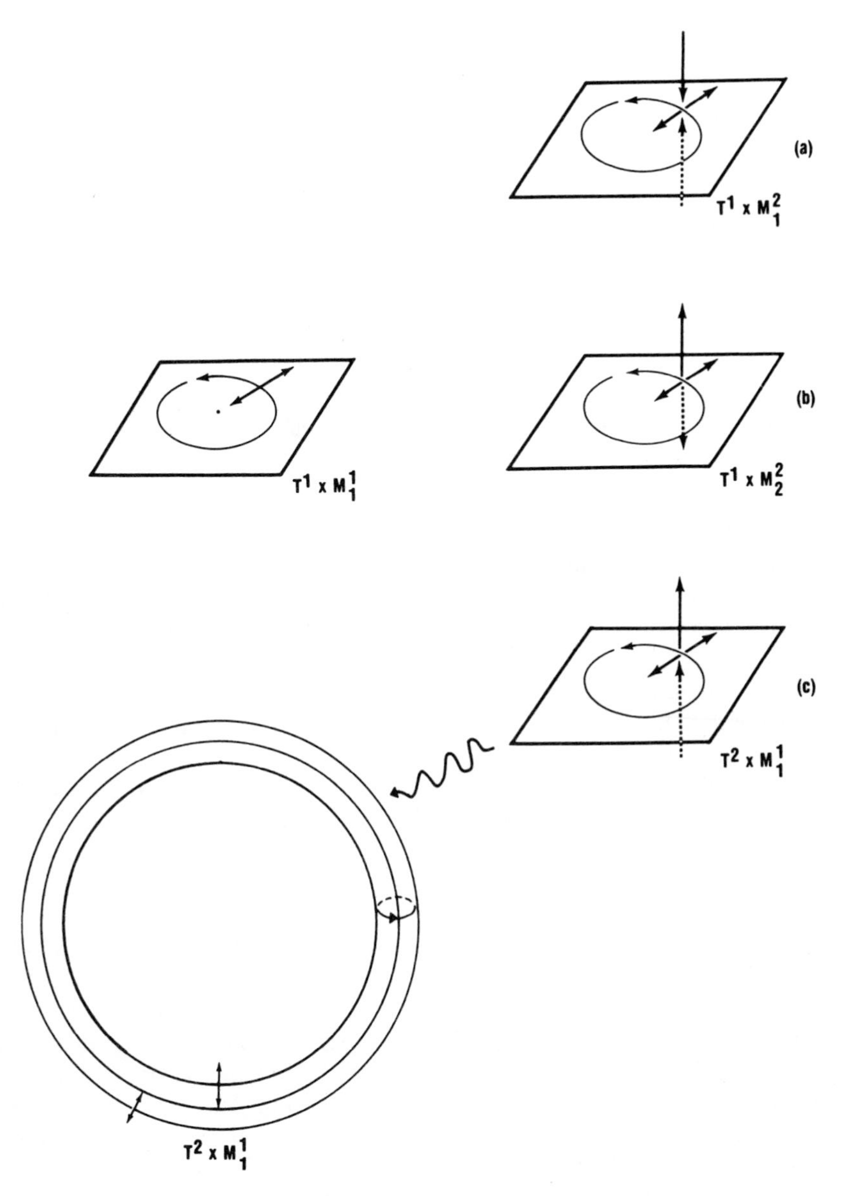

Figure 20.15 By starting with an unstable limit cycle and proceeding as before, we obtain nonlocal limit sets of the form $T^1 \times M_1^2$ (*a*), $T^1 \times M_2^2$ (*b*), and $T^2 \times M_1^1$ (*c*), none of which is stable.

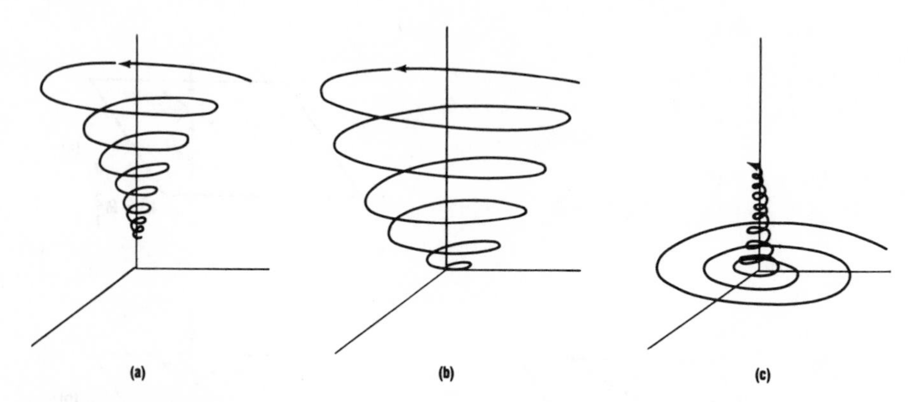

Figure 20.16 The qualitative properties of the flow in the neighborhood of a critical point of type $F^2_- \times M^1_0$ depend on the damping constants. (*a*) $\gamma_r \gg \gamma_z$. (*b*) $\gamma_r \ll \gamma_z$. For an unstable critical point of type $F^2_- \times M^1_1$ with $\omega \gg \gamma_r \gg \gamma_z$, the flow has the shape indicated in (*c*).

dimensional tori became less "rigid," so we should expect to reach some dimension n for which the invariant torus $T^{n-1} \subset \mathbb{R}^n$ is not structurally stable. This occurs for $n - 1 = 4$, about which more will be said later.

We have here encountered a new phenomenon which persists for some attractors of dynamical systems of all dimensions $n \geq 3$. The reason is simple to see. Limit sets of attractors are sets of dimensions $0, 1, 2, \ldots, n-1$ and measure zero in $\mathbb{R}^n$. Dynamical system trajectories are 1-dimensional. It is only when $\max(0, \ldots, n-1) \leq 1$ that a dynamical system trajectory can visit all points on an invariant surface. In higher dimensions invariant surfaces become more and more "porous" with respect to actual trajectories.

We can study 3-dimensional dynamical systems depending on parameters. Certain bifurcations and transformations can naturally occur in such families of dynamical systems. These bifurcations for 3-dimensional dynamical systems can often be inferred from the corresponding bifurcations in the underlying 2-dimensional system. For example, if two eigenvalues associated with an isolated point of type M^3_i or a 1-dimensional flow of type $T^1 \times M^2_i$ become equal, there is associated a transformation from "radial" to "spiral" behavior. In this sense the critical set M^3_i "stands near" the flow $F^2_- \times M^1_i$ $(i = 0, 1)$, M^3_i stands near $F^2_+ \times M^1_{i-2}$ $(i = 2, 3)$, and $T^1 \times M^2_0$ is near $\mathrm{Sc}_- = T^1 \times F^2_-$. The bifurcation of isolated critical points proceeds as for gradient systems. The stable screw Sc_- undergoes a Hopf bifurcation to an unstable screw and a stable invariant torus

$$\mathrm{Sc}_- \xrightarrow{\text{Hopf}} \mathrm{Sc}_+ + T^2 \times M^1_0 \tag{20.19}$$

The remaining bifurcations and concomitant flows may be obtained as easily:

$$M^3_1 \xrightarrow{A_3} M^3_2 + 2M^3_1 \tag{20.20}$$

5. Strange Building Blocks. 1 "Spiral Chaos"

In order to see what kinds of qualitative behavior may be associated with 3-dimensional dynamical systems, we consider one prototype in this section and another in the following section. Although these model dynamical systems may never have any contact with reality, their study is a first and brave attempt at opening a Pandora's box of strange behavior associated with dynamical systems in more than two dimensions.

In this section we consider the dynamical system of type $F_+^2 \times M_0^1$ (Fig. 20.13*a*). If we assume that the unstable focus lies in the attracting x–y plane $z = 0$, then the dynamical equations near the critical point are, in cylindrical coordinates,

$$
\begin{aligned}
\frac{dr}{dt} &= \gamma_1 r, \\
\frac{d\theta}{dt} &= \omega, \qquad \gamma_1, \gamma_2 > 0 \\
\frac{dz}{dt} &= -\gamma_2 z,
\end{aligned}
\tag{20.21}
$$

In order to get some interesting behavior for this system, we can add terms to (20.21) which have the effect of returning the flow from the neighborhood of (r large, $|z|$ small and $\neq 0$) to the neighborhood of (r small, $|z|$ large). The r–z cross section of such a return flow is indicated in Fig. 20.17*a*. More specifically, we can choose the cyclic flow in the r–z plane $\theta =$ constant to be of the relaxation-oscillation type as shown in Fig. 20.17*b*. The corresponding dynamical system equations then assume the form

$$
\begin{aligned}
\dot{z} &= \gamma(r - F(z)) \\
\dot{r} &= -\gamma^{-1}(z - z_0) \\
\dot{\theta} &= \omega
\end{aligned}
\tag{20.22}
$$

The description of this flow is reasonably simple for $\gamma \gg 1$, $\omega \gg \gamma^{-1}$. From an arbitrary initial condition A the system state rapidly approaches the surface of revolution $r = F(z)$. The system state then rapidly rotates around the z axis, while it slowly spirals to the drop-off point C. It quickly jumps to the bottom sheet of this surface of revolution at D, and then slowly spirals out to the edge E, where it jumps to the upper sheet at B and continues the process. The qualitative nature of the jumps $C \to D$ and $E \to B$ depends on the ratio γ/ω. When this ratio is large, the system state jumps from one sheet to a point approximately below or above it, without undergoing much of a rotation around the z axis (linear descent and ascent). When $\gamma/\omega \ll 1$, the system state rotates often about the z axis during the jump from one sheet to another (helical descent and ascent).

The dynamical system represented by (20.22) may be called a spiral chaotic attractor. Spiral motion clearly occurs on the sheets of revolution BC and DE.

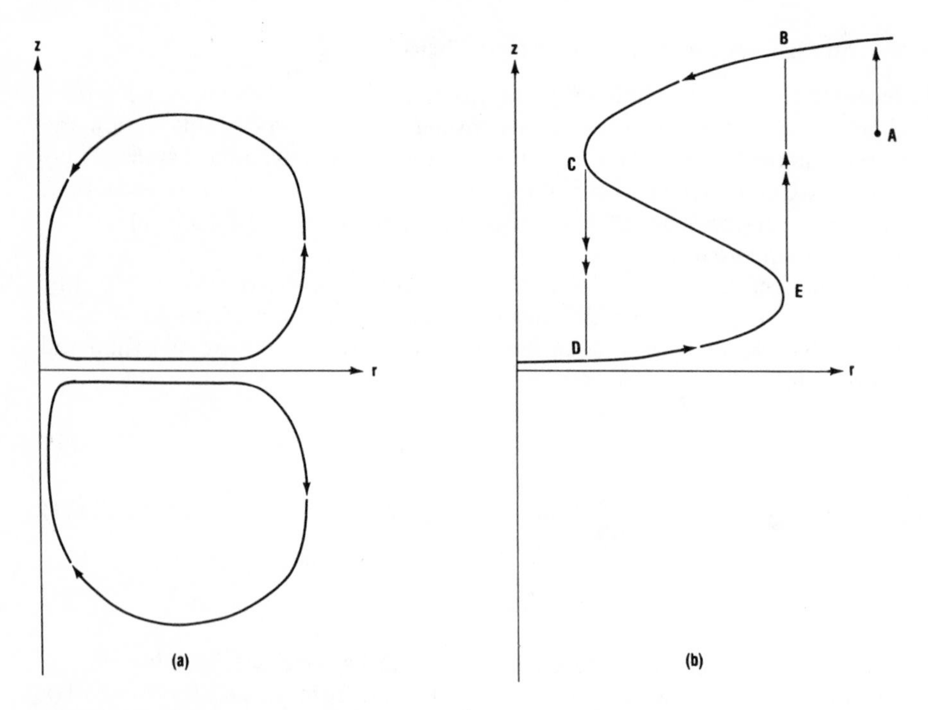

Figure 20.17 These 3-dimensional flows have $\dot{\theta} = \omega =$ constant and are limit cycles in the r–z plane. The flows exist on a stable limit torus and may or may not close on themselves. In (*a*) the limit cycle is smooth, while in (*b*) it has the form of a relaxation oscillation.

The angle θ (modulo 2π) at which sudden changes of z occur ($\gamma/\omega \gg 1$) may appear to be a random variable. In the limit $\gamma/\omega \ll 1$ the behavior may be called **spiral-helical chaos**.

The behavior is not chaotic. The angular increase during spiral motion and during linear/helical ascent/descent is invariant under rotation, so the "chaos" is associated with the sequence $n\Delta\theta \bmod 2\pi$, $n = 1, 2, \ldots$, where $\Delta\theta$ is very large. A better approximation to chaos may be obtained by deforming the surface $0 = r - F(z) \to f(x, y, z) = 0$ so that it is no longer a surface of revolution.

Rössler has studied a simple dynamical system $\dot{x} = F(x; c)$, $x \in \mathbb{R}^3$, $c \in \mathbb{R}^3$, which exhibits spiral chaos.[2] The dynamical system equations are

$$\begin{aligned} \dot{x} &= -y - z \\ \dot{y} &= x + ay \\ \dot{z} &= b + xz - cz \end{aligned} \tag{20.23}$$

This is probably the simplest dynamical system exhibiting a type of behavior not encountered in 2-dimensional dynamical systems, because $n = 3$ and because there is only one nonlinear term, and that of degree 2. This system exhibits a number of bifurcations as the control parameters $(a, b, c) \in \mathbb{R}^3$ are

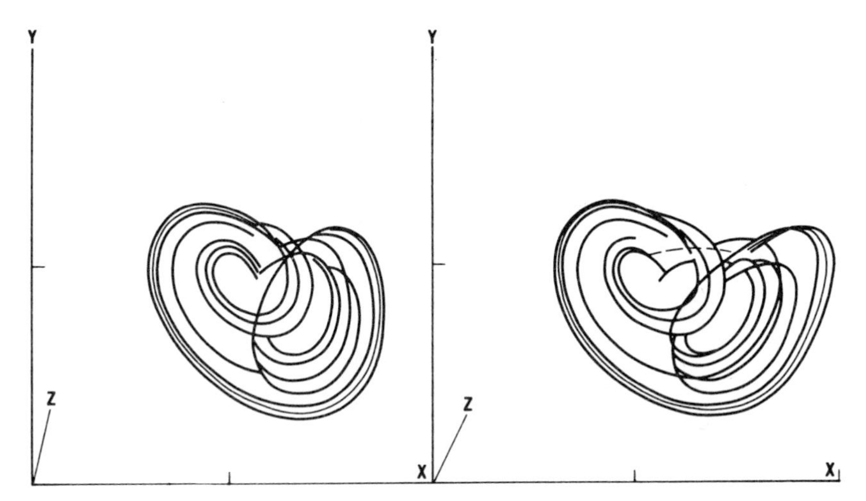

Figure 20.18 The flow associated with the dynamical system (20.23) was computed by Rössler for various values of the control parameters (a, b, c). One such flow pattern is presented here in stereo (cross your eyes until both figures fall on top of each other, then take an aspirin). Reprinted with permission from O. E. Rössler, *Z. Naturforsch.* **31a** (1976), p. 1668. Copyright 1976 by Verlag der Zeitschrift für Naturforschung.

changed. In Fig. 20.18 we reproduce a stereoscopic plot of the dynamical system trajectory in the (x, y, z) phase space. The outward spiral on the lower sheet and the inward spiral on the upper sheet are apparent. The linear descent is also clear, while the linear ascent may be inferred.

6. Strange Building Blocks. 2 Lorenz Attractor

If one local flow of type $F_+^2 \times M_0^1$ leads to interesting behavior, we may ask what bizarre kinds of behavior can occur when two flows of local type $F_+^2 \times M_0^1$ are present in $\mathbb{R}^3$. We present a schematic answer to this question in Fig. 20.19. Here we place one flow of type $F_+^2 \times M_0^1$ at the point $(x, y, z) = (x_0, y_0, 0)$. The x–y plane $z = 0$ is attracting, while the spiral motion in that plane is repelling. The other flow is placed at the point $(0, y_0, -x_0)$, the y–z plane $x = 0$ is locally attracting at this point, and the spiral motion in this plane is unwinding. We say nothing about the nonlocal behavior of the flows, away from these two critical points.

If we start the system at a point (x_0, y_0, ε), the system state will unwind in spiral motion while remaining near the plane $z = 0$. Eventually it will get sufficiently far from the first critical point and may become attracted to the $x = 0$ plane, which is attracting near the second critical point. When the system state gets near enough to this plane, it will begin to unwind in a spiral, until the system state gets far enough away from the second critical point to become attracted to the $z = 0$ locally attracting plane containing the first unstable focus. The process then repeats itself. The system state acts like a ping-pong ball being

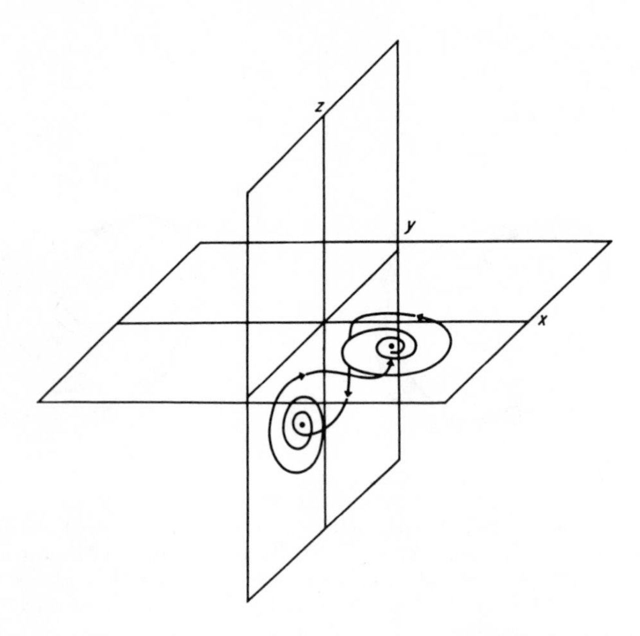

Figure 20.19 When a dynamical system has only two isolated critical points of type $F_+^2 \times M_0^1$ the flow can carry the system state back and forth between neighborhoods of the two critical points in a "chaotic" way.

tossed back and forth between the neighborhoods of two critical points of type $F_+^2 \times M_0^1$. This motion may appear to be chaotic.

A strange attractor of this type was analyzed in detail by Lorenz.[3] Lorenz was particularly interested in studying dynamical systems

$$\dot{x}_i = F_i(x) \tag{20.24a}$$

representing forced dissipative motion. Such systems can be studied by making a Taylor series expansion of the driving force $F_i(x)$ about any point:

$$F_i(x) = c_i - b_{ij}x_j + a_{ijk}x_j x_k + \mathcal{O}(3) \tag{20.24b}$$

Truncating each F_i beyond terms of order 2 may not capture the essence of all forced dissipative systems, but it does capture sufficient nonlinear complexity to keep mathematicians busy for years to come. It is also useful to assume that $a_{ijk}x_i x_j x_k$ vanishes identically and that $b_{ij}x_i x_j$ is positive definite.

A simple test suffices to determine when the dynamical system (20.24) has an attractor. Suppose that at $t = t_0$ we carve out a volume V in state space. Each point in this volume will move to a new point in a time interval $t - t_0$. We may ask how the volume changes. This may not be a good question for large time intervals, since the new "volume" may be very porous (Cantor set). For small time intervals, however, this is a well-posed question. Accordingly we let

$t - t_0 \to dt$. During this time interval each point in V will move only a short distance. Thus each point in the interior of V will move to some point in the interior of V during the infinitesimal time interval dt. It is only the points on the surface of V which are responsible for changing its shape and therefore volume during the interval dt. The volume change is therefore given by a surface integral:

$$V(t_0 + dt) - V(t_0) = \oint_{\partial V} dx_i \wedge dS_i \tag{20.25}$$

where dS_i is an oriented surface element. The time rate of change of the volume is therefore

$$\frac{dV}{dt} = \oint_{\partial V} \frac{dx_i}{dt} \wedge dS_i = \oint_{\partial V} F_i \wedge dS_i$$

where we have used the dynamical equation $\dot{x}_i = F_i$. This surface integral is related to the divergence of the vector field F, for

$$\lim_{V \to 0} \frac{1}{V} \frac{dV}{dt} = \lim_{V \to 0} \frac{\oint_{\partial V} F_i \wedge dS_i}{V} \stackrel{\text{def}}{=} \operatorname{div} F \tag{20.26}$$

In a locally Cartesian coordinate system $\operatorname{div} F = \sum \partial F_i/\partial x_i$. The result may be regarded as a generalization of Liouville's theorem. Nondissipative flows on classical phase space are divergence-free, so that phase space volume elements are conserved under Hamiltonian flows:

$$\dot{q}_i = \frac{\partial H}{\partial p_i}, \qquad \dot{p}_i = -\frac{\partial H}{\partial q_i}$$

(Liouville's theorem).

For the truncated class of nonlinear systems given in (20.24) we have $\operatorname{div} c_i = 0$ and $\operatorname{div} a_{ijk} x_j x_k = 0$, so that

$$\operatorname{div} F = \frac{\partial}{\partial x_i}(-b_{ij} x_j) = -b_{ii} = -\operatorname{tr} b < 0 \tag{20.27}$$

Since b_{ij} are constants, the measure of any volume decreases in time like $V(t) = V(0)e^{-t \operatorname{tr} b}$. The flows associated with (20.24b) shrink volumes to zero, so (20.24b) should have a measure zero attracting set.

The equations studied in detail by Lorenz arose in connection with the Bénard problem (Sec. 7). These equations are

$$\begin{aligned} \dot{x} &= -\sigma x + \sigma y \qquad\qquad\qquad, & \sigma > 0 \\ \dot{y} &= \quad rx - \; y \qquad - xz, & r > 0 \\ \dot{z} &= \qquad\qquad\quad - bz + xy, & b > 0 \end{aligned} \tag{20.28}$$

The two nonlinear terms clearly satisfy the condition $a_{ijk}x_i x_j x_k = 0$, and the linear terms satisfy the positive definiteness condition (b_{ij} may be made symmetric by the substitution $z \to z' = z - r + \sigma$). Since $\operatorname{tr} b = -(\sigma + 1 + b)$ is negative, the measure of any small volume approaches zero asymptotically as $t \to \infty$, even if (though) the volume becomes wildly distorted.

The dynamical system (20.28) possesses a critical point $(x, y, z) = (0, 0, 0)$ for all values of r. For $r > 1$ it possesses two additional critical points C, C' whose coordinates are

$$\begin{aligned} C&: \quad (x_0, x_0, z_0), \\ C'&: \quad (-x_0, -x_0, z_0), \end{aligned} \qquad z_0 = r - 1, \quad x_0 = \sqrt{bz_0} \tag{20.29}$$

There are no other critical points.

The stability properties of these three critical points can be determined by a standard linear stability analysis. Equations (20.28), linearized about any critical point, are

$$\frac{d}{dt}\begin{bmatrix} \delta x \\ \delta y \\ \delta z \end{bmatrix} = \begin{bmatrix} -\sigma & \sigma & 0 \\ r - z & -1 & -x \\ y & x & -b \end{bmatrix} \begin{bmatrix} \delta x \\ \delta y \\ \delta z \end{bmatrix} \tag{20.30}$$

At the critical point (0, 0, 0) the characteristic equation is

$$(\lambda + b)[\lambda^2 + (\sigma + 1)\lambda + \sigma(1 - r)] = 0 \tag{20.31}$$

This equation has eigenvalues

$$\begin{aligned} \lambda &= -\left(\frac{\sigma + 1}{2}\right) + \left[\left(\frac{\sigma + 1}{2}\right)^2 - \sigma(1 - r)\right]^{1/2} \\ \lambda &= -\left(\frac{\sigma + 1}{2}\right) - \left[\left(\frac{\sigma + 1}{2}\right)^2 - \sigma(1 - r)\right]^{1/2} \\ \lambda &= -b \end{aligned} \tag{20.32}$$

This critical point is of type $F^2_- \times M^1_0$ for $r < r_{-1} = 1 - \sigma^{-1}(\sigma + 1)^2/4$, of type M^3_0 for $r_{-1} \le r < 1$, and of type M^3_1 for $r > 1$.

The characteristic equation associated with the critical points C, C' is

$$\lambda^3 + (\sigma + b + 1)\lambda^2 + (r + \sigma)b\lambda + 2b\sigma(r - 1) = 0 \tag{20.33}$$

As $r \downarrow 1^+$, the three eigenvalues of (20.33) approach

$$\begin{aligned} &\lambda \to 0^- \\ &\lambda \to -(\sigma + 1) \\ &\lambda \to -b \end{aligned} \tag{20.34}$$

Thus as r increases through $+1$, the first of the three eigenvalues (20.32) at the critical point (0, 0, 0) increases through zero. There is a bifurcation of type A_{+3} at this point, with the two critical points of type M_0^3 leaving the origin with canonical square-root behavior [cf. (20.29)]. This bifurcation may be written

$$M_0^3 \xrightarrow{r \to r_1 = 1} M_1^3 + 2M_0^3 \tag{20.35}$$

Remark. In a 1-parameter family of equations exhibiting bifurcations related to elementary catastrophes, we would expect a fold "bifurcation" of type A_2. Instead we find a cusp bifurcation. This comes about because the symmetry intrinsic to the Lorenz equations (20.28) (invariant under $x \to -x$, $y \to -y$, $z \to +z$) suppresses the A_2 catastrophe, so that the only catastrophe which can typically occur is the A_3 catastrophe.

As r continues to increase above $+1$, no further qualitative changes occur in the critical point at the origin. Two qualitative changes occur for the critical points at C, C'. At some intermediate value r_f two of the three real negative eigenvalues (20.33) become equal. As r increases through r_f, the qualitative nature of C, C' changes from a stable Morse saddle to a stable focus:

$$M_0^3 \xrightarrow{r \to r_f} F_-^2 \times M_0^1 \tag{20.36}$$

Then, if $\sigma > b + 1$, at

$$r_4 = \frac{\sigma(\sigma + b + 3)}{\sigma - b - 1} \tag{20.37}$$

the real part of the complex conjugate pair of eigenvalues goes through zero. The critical points at C, C' change qualitative type from F_-^2 to F_+^2. A change in dynamical stability at a focus is associated with Hopf bifurcation. There is an inverted Hopf bifurcation as $r \to r_4$. Unstable limit cycles surrounding C, C' in the plane containing the spiral motion squeeze down on the stable foci as $r \to r_4$ with the canonical $\frac{1}{2}$ power dependence $(r_4 - r)^{1/2}$, and finally squeeze them out of existence. The bifurcation at r_4 may be summarized

$$F_-^2 \times M_0^1 + T^1 \times M_1^2 \xrightarrow{r \to r_4} F_+^2 \times M_0^1 \tag{20.38}$$

For $r > r_4$ there remain only three critical points, all unstable.

Study of the qualitative properties of the Lorenz equations as a function of increasing r has lead us into a no-mans-land for $r > r_4$. There is a time-honored method for studying problems in "no-mans-land." This is to approach the same area "from the other side." Accordingly, it is useful to study the qualitative properties of the Lorenz equations as a function of decreasing r. Such a study has been carried out by Robbins.[4] The results are as follows:

1. For $1/r = 0$ there is a stable symmetric periodic solution.
2. This stable symmetric periodic solution continues to exist for $1/r$ small, that is, it exists for r sufficiently large.
3. At $r = r_a$ the symmetric periodic solution loses its stability, and two stable asymmetric periodic solutions bifurcate from it.
4. At $r = r_b < r_a$ each asymmetric periodic solution loses its stability, and a pair of double-looped solutions with twice the period are created by bifurcation.
5. At $r = r_c < r_b$ these solutions lose their stability and a pair of four-looped solutions with four times the initial period bifurcate from each two-looped solution.
6. There appears to be a cascade of bifurcations of this type. The 2^n-looped solutions lose their stability at $r = r_n$ and two stable 2^{n+1}-looped solutions bifurcate from each. The daughter solutions have twice the period as the parents.
7. These bifurcation points appear to converge to an accumulation point r_∞ which is larger than r_4.
8. The no-man's-land of chaos exists for $r_4 < r < r_\infty$.

Equations (20.28) have been studied extensively, beginning with Lorenz who integrated them numerically using fixed control parameters $\sigma = 10$, $b = \frac{8}{3}$, and a single variable control parameter r. The bifurcation properties of the Lorenz model which we have just discussed are summarized in Fig. 20.20. This figure also summarizes a number of qualitative properties found through numerical integration. We now discuss the qualitative behavior of these equations as r increases from zero. For $0 < r < r_1 = 1$ there is only one critical point. This critical point is both a local and a global attractor. That is, any initial state will approach the origin as $t \to +\infty$. As r approaches 1, critical slowing down occurs, and as r increases through $+1$, the origin loses its stability while two Morse attractors M_0^3 split off. These are globally as well as locally stable. Except for a 1-dimensional set of points, every point in state space $\mathbb{R}^3$ will approach C or C' as $t \to \infty$. As r increases through $r_f = 1.345$, the qualitative change $M_0^3 \to F_-^2 \times M_0^1$ takes place. This affects neither the local nor the global stability of the attractors C, C'.

As r increases through $r_2 \simeq 13.926$, the two unstable trajectories from the origin return to the origin as $t \to +\infty$.[5] As r increases through r_2, the attractors C, C' lose the property of being global attractors. Instead, they are surrounded by neighborhoods N, N' in which they are local attractors. A point starting outside these neighborhoods may oscillate back and forth from the *neighborhood* of N to the neighborhood of N' (i.e., not *in* N or N') in an apparently random or chaotic way, until the trajectory enters either N or N'. Then the trajectory spirals down to C or C' with one Fourier frequency component. Such behavior has been called "mestastable chaos." In addition to these two types of behavior, there are infinitely many periodic orbits and infinitely many unstable turbulent orbits. These orbits have measure zero.

As r increases toward $r_3 \simeq 24.06$, the measure of N, N' decreases, the measure of the set of points from which metastable chaos can begin increases, and a kind of critical slowing down occurs. A point undergoing metastable chaos takes longer and longer to be captured by N or N' and be damped out as $r \to r_3$. At $r = r_3$ the two unstable trajectories from the origin approach an unstable periodic orbit as $t \to \infty$. As r increases through r_3, the chaotic repellor becomes a chaotic attractor. Of measure zero, its basin of attraction has positive measure.

As r increases toward $r_4 = 24.74$, the unstable limit cycles $T^1 \times M_1^2$ squeeze down on the foci $F_-^2 \times M_0^1$ so that the neighborhoods N, N' decrease in size and measure. The chaotic stable attractor increases its domain of attraction at the expense of N, N'. At $r = r_4$ the inverted Hopf bifurcation takes place. For $r > r_4$ there remains a "strange attractor" whose behavior we now discuss.

For $r > r_4$ the critical points C, C' are of type $F_+^2 \times M_0^1$. The planes containing the spiral motion of these foci are stable, although the unwinding spiral motion is itself unstable. A point in $\mathbb{R}^3$ will be attracted toward one of these two planes. Once near an attracting plane, the point will be caught in the whirlwind and spiral away from the focus. When, in its motion, the point crosses the plane $y = 0$, the point will be attracted toward the other plane in which the other unstable focus is suspended. The system point begins to fall toward the center of the other focus. As it approaches the attracting plane, it spirals away from the focus. Eventually the distance from the second focus must become sufficiently large that the point must cross the plane $y = 0$ again. The system point is then attracted back toward the center of the first unstable focus, and the motion repeats itself.

Lorenz has given a convenient representation for the chaotic trajectories which occur for $r > r_4$ (Fig. 20.21). The dynamical system trajectories lie near a thin (measure zero) attracting 2-dimensional surface. The projection of this surface onto the y–z plane is shown, together with the intersection of this surface with the planes $x =$ constant (value shown). For $z \lesssim 17$ the motion occurs on one sheet, but for $z \gtrsim 17$ there are two sheets, each one containing an unstable focus. The spiral unwinds (e.g., counterclockwise from C) on the upper sheet until it crosses the plane $y = 0$, at which point it begins unwinding from the other focus (C') in the opposite direction (clockwise). It continues to unwind from this focus on the lower sheet until it crosses the $y = 0$ plane, at which point it transfers its allegiance back to C.

This "strange attractor" is neither a 2-dimensional surface nor a soldered 2-dimensional manifold. In fact, it is a pathological topological object. Instead of two sheets joined by the unstable curves from the origin, there are infinitely

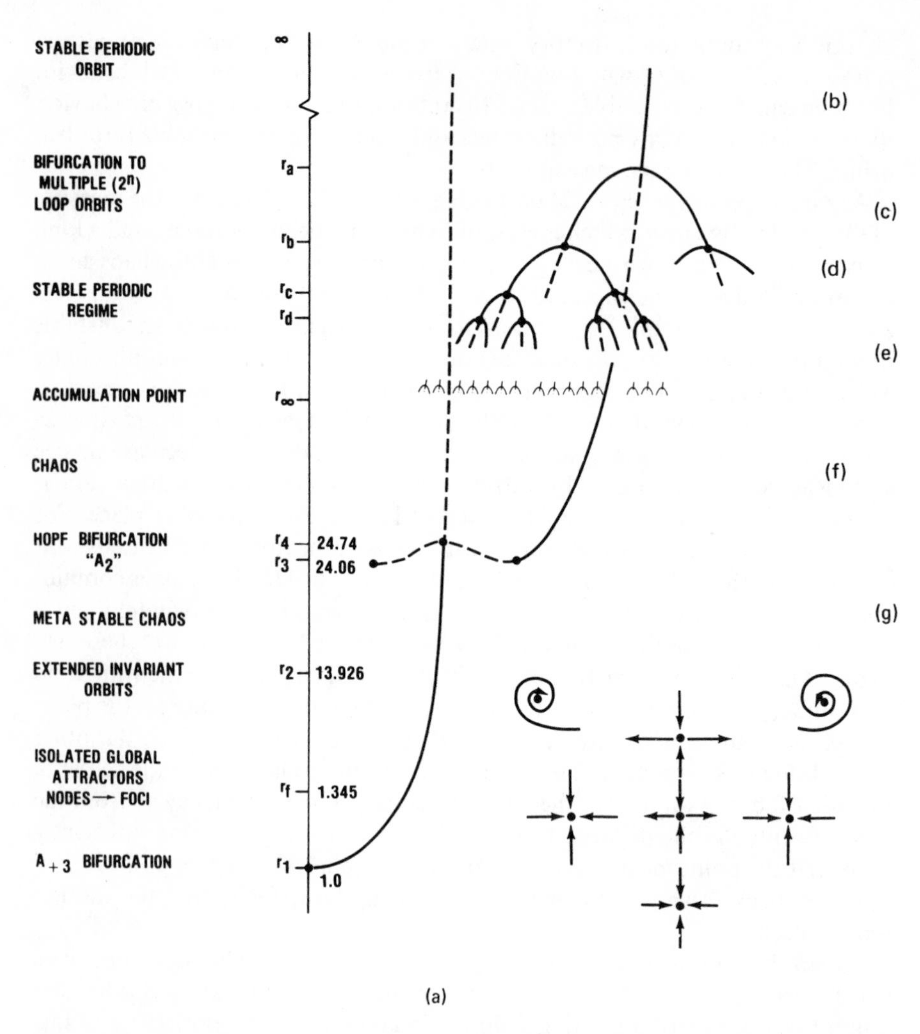

Figure 20.20 The qualitative properties of the stable attractor of the Lorenz equations (20.28) depend dramatically on the value of the control parameter r. Reprinted with permission from K. A. Robbins, *SIAM J. Appl. Math.* **36** (3), (1979), pp. 458, 466, 470. Copyright 1979 by SIAM.

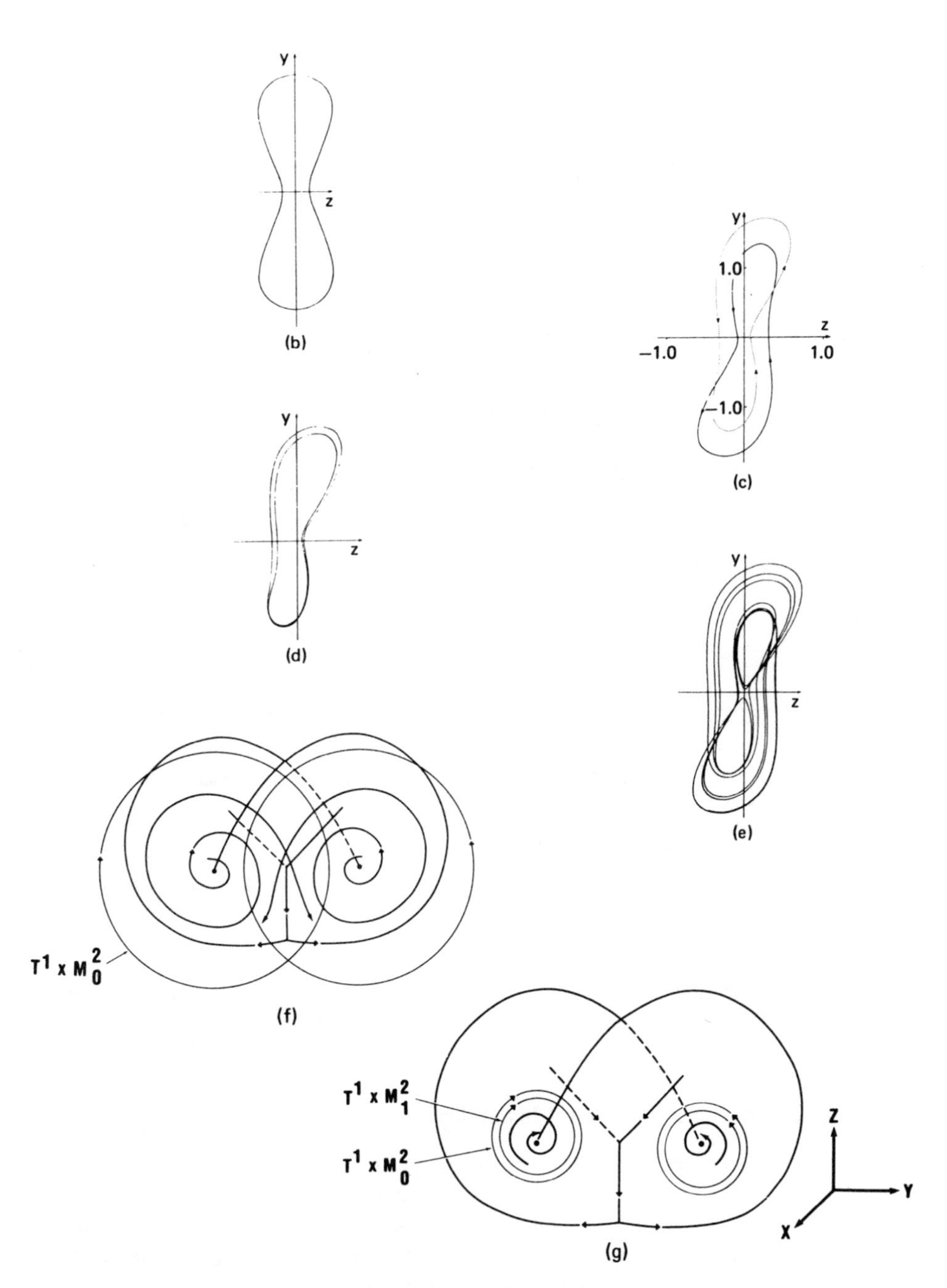

Figure 20.20 (*Continued*)

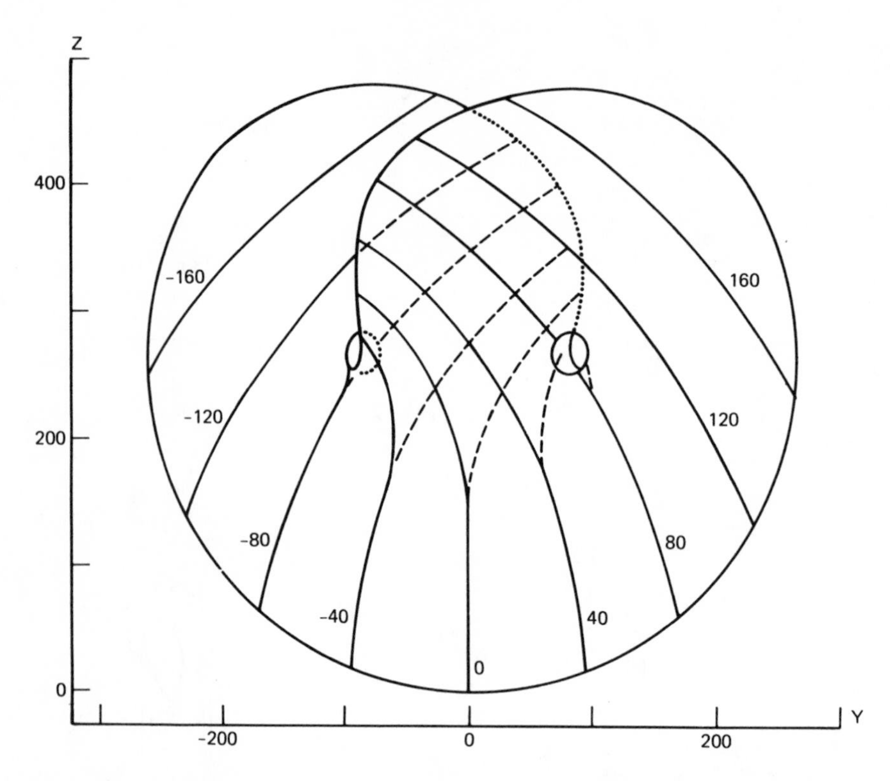

Figure 20.21 Lorenz has provided a useful heuristic model for the strange attractor. This involves a surface which is split into two sheets for z sufficiently large, as shown. The motion of the dynamical system alternates between flows of type $F_+^2 \times M_0^1$ on the upper and the lower sheet. In fact, this "surface" has a very peculiar topological structure. For this system, the control parameter $r = 28$ and the three state variables have been scaled up by a factor of 10. Reprinted with permission from E. N. Lorenz, *J. Atmos. Sci.* **20** (1963). Copyright 1963 by American Meteorological Society.

many. An arc transverse to the covers of this infinitely sheeted book intersects these leaves in a Cantor set.

7. Hydrodynamics. The Bénard Instability

When a fluid is heated gently from below (Fig. 20.22*a*), heat is transported from the bottom surface to the top surface by heat conduction. This heat flow is not accompanied by fluid motion. When the temperature gradient across the fluid is increased, heat cannot be transported fast enough by conduction alone, so the fluid helps out by transporting itself to the upper surface. Cooler liquid from the upper surface flows to the lower surface. A regular roll pattern is established (Fig. 20.22*b*). When the temperature gradient is made yet larger, chaos results (Fig. 20.22*c*).

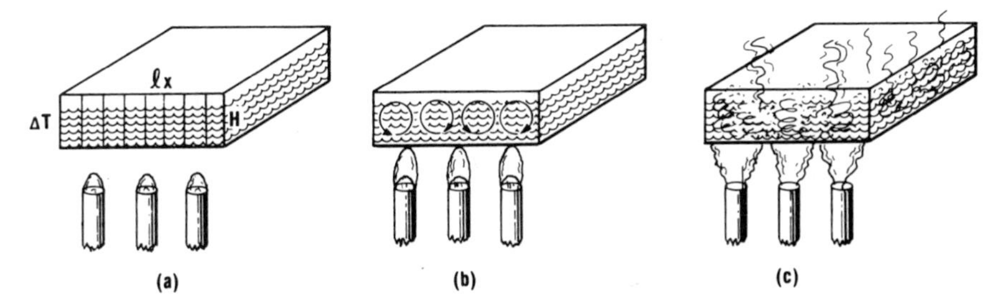

Figure 20.22 (*a*) At low heat input, heat is transported by conduction. (*b*) At moderate heat input, heat is transported by convection in orderly roll patterns. (*c*) At high heat input the fluid motion becomes chaotic.

The equations describing the heated fluid pictured in Fig. 20.22 are

$$\frac{\partial u_i}{\partial t} + u_j \frac{\partial u_i}{\partial x_j} = g\varepsilon\Delta T\delta_{i3} - \frac{1}{\rho}\frac{\partial P}{\partial x_i} + \nu\nabla^2 u_i$$

$$\frac{\partial T}{\partial t} + u_j \frac{\partial T}{\partial x_j} = \kappa\nabla^2 T \tag{20.39}$$

$$\frac{\partial u_i}{\partial x_i} = 0$$

Where

x_i	*i*th spatial coordinate
u_i	*i*th velocity field component
g	gravitational constant
ε	coefficient of thermal expansion
$\Delta T/H$	imposed temperature gradient
ρ	fluid density
T	fluid temperature field
P	fluid pressure field
ν	viscosity
κ	coefficient of thermal conduction
t	time

These complicated equations have been treated in the typical way. First, suitable physical assumptions and approximations are made to simplify them as much as possible. This includes scaling and introduction of dimensionless variables. Next the temperature and velocity fields are expanded in a Fourier series expansion once suitable boundary conditions (rectangular box) have been established. Then products of trigonometric functions are replaced by sums using standard trigonometric identities. Finally the linear independence of the trigonometric coefficients is used to relate time derivatives of the Fourier coefficients to nonlinear functions of these coefficients.

In short, an intractable problem is replaced by an intractable problem.

An alternative approach is to impose boundary conditions at the outset, constrain fluid motion to occur in a plane on the basis of physical intuition, and then lean heavily on these aids to effect a simplification in (20.39).

This results in a simpler set of equations:

$$\frac{\partial}{\partial t}\nabla^2\psi = -\frac{\partial(\psi, \nabla^2\psi)}{\partial(x, z)} + \nu\nabla^4\psi + g\varepsilon\frac{\partial\theta}{\partial x}$$
$$\frac{\partial}{\partial t}\theta = -\frac{\partial(\psi, \theta)}{\partial(x, z)} + \frac{\Delta T}{H}\frac{\partial\psi}{\partial x} + \kappa\nabla^2\theta \tag{20.40}$$

where ψ is the stream function ($u = \nabla\psi$) and $\theta = T(x, z, t) - T_{av}$, where T_{av} decreases linearly by ΔT between the bottom and top surfaces of the fluid. All motion is assumed to occur in the x–z plane y = constant.

The simplified equation (20.40) can be treated as per the recipe described above. Once again this results in the substitution of an intractable algebraic problem for an intractable differential equation. Numerical studies[6] of the coupled nonlinear equations resulting from (20.40) by this process indicated that, independent of initial conditions, all but three of the Fourier coefficients involved in the expansions of θ and ψ quickly approached zero. The three important Fourier coefficients did or did not, depending on the value of the dimensionless parameter

$$R_a = g\varepsilon H^4\frac{\Delta T/H}{\kappa\nu} \tag{20.41}$$

called the Rayleigh number, because Rayleigh first studied the bifurcation of time-dependent solutions from the time-independent solution of (20.39). This occurs at a critical value

$$R_c = \pi^4(1 + a^2)^3a^{-2}$$

where $a = H/l_x$, and l_x is the width of the container holding the fluid.

The mathematical processes of algebraic manipulation and truncation are generally noncommutative. However, in the case of (20.40) the numerical calculations indicated that they were commuting processes, at least for the range of parameters used. These results suggested to Lorenz that it might be more profitable to make the approximations before carrying out the algebra. Accordingly he introduced the following ansatz for the stream function and temperature difference:

$$\frac{a}{\kappa(1 + a^2)}\psi = \sqrt{2}X\sin\frac{\pi x}{l_x}\sin\frac{\pi z}{H}$$
$$\pi\frac{R_a}{R_c}\frac{\theta}{\Delta T} = \sqrt{2}Y\cos\frac{\pi x}{l_x}\sin\frac{\pi z}{H} - Z\sin\frac{2\pi z}{H} \tag{20.42}$$

Now only a minimum of algebra is required to determine the nonlinear equations of motion of the three Fourier coefficients $X(t)$, $Y(t)$, and $Z(t)$. The result is

$$\begin{aligned}
\frac{d}{d\tau}X &= -\sigma X + \sigma Y \\
\frac{d}{d\tau}Y &= rX - Y - XZ \\
\frac{d}{d\tau}Z &= -bZ + XY
\end{aligned} \tag{20.43}$$

where

$$\begin{aligned}
\tau &= \left(\frac{\pi}{H}\right)^2(1 + a^2)\kappa t \\
\sigma &= \frac{\nu}{\kappa} \text{ (Prandtl number)} \\
r &= \frac{R_a}{R_c} \\
b &= \frac{4}{(1 + a^2)} \leq 4
\end{aligned} \tag{20.44}$$

These are precisely the Lorenz equations, and this is precisely why they were studied and how they were derived in the first place.

There is a strong initial impulse to feel that such a crude ansatz as (20.42) can only lead to a "mutilation" of the nonlinear equation (20.40). That it is not so has been determined from numerical computations. The reason that this is not so is not yet known, although the Center Manifold Theorem (Sec. 10) suggests why this is the case.

8. Electrodynamics. The Laser-Spiking Instability

When a small electric potential is applied across a gas, only a very weak current flows through the gas tube (Fig. 20.23). When the potential gradient ($V/D \simeq \mathscr{E}$, electric field) becomes sufficiently large, the gas breaks down (ionizes) and sizable currents can flow through the tube. The I-V characteristics of a gas tube are highly nonlinear and exhibit hysteresis. The conduction process is accompanied by the emission of light. In fact, this is how neon lights work.

If the conduction tube is placed inside an appropriately tuned Fabry-Perot cavity, the feedback may become sufficiently great that laser action can occur. Below a certain pumping (current flow) threshold no laser action occurs; above, the laser is "on." The off-on laser transition is a bifurcation phenomenon. For moderate pumping rates the output of the laser cavity is stable in time and a smooth continuous function of the pump rate. However, beyond a second

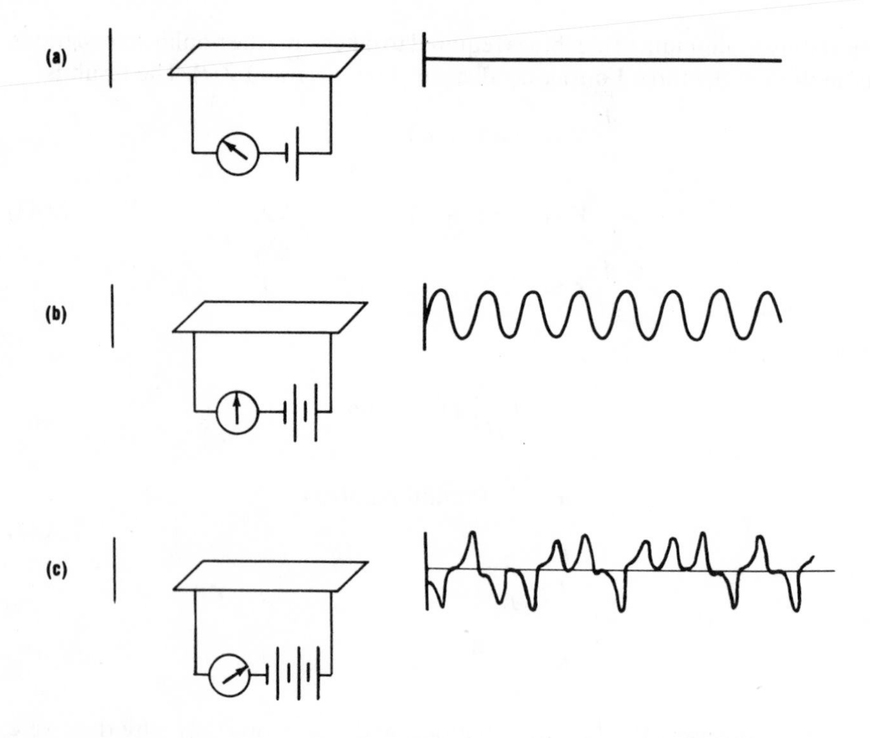

Figure 20.23 The current flow through a gas tube in a Fabry-Perot cavity determines the light output. (*a*) Low current in, no light out. (*b*) Intermediate current in, coherent light out. (*c*) High current in, chaotic light out.

critical threshold the laser begins to oscillate wildly in a chaotic way.[7] The stable attractors between the first and second thresholds are of type M_0^3 or $F_-^2 \times M_0^1$. The stable attractor above the second threshold is a strange chaotic attractor of the type studied by Lorenz.[8]

The bifurcation properties of the laser can be studied by looking at the combined Maxwell equations for the electromagnetic field and the equations of motion describing the matter in the laser cavity. In classical mechanics such coupled equations are called the Newton-Maxwell equations. The laser is a quantum-mechanical system. For this system the coupled equations of motion are called the Bloch-Maxwell equations. The important fields for the description of a laser are the electric field $\mathscr{E}(x, t)$, the polarization $\mathscr{P}(x, t)$ due to the non-vanishing of the electric dipole moment between various atomic or molecular levels, and the population inversion $\Delta(x, t)$ which is the difference between the number of atoms in the excited and ground states involved in the laser transition.

The Bloch-Maxwell equations are obtained from the Maxwell equations ($\partial^\mu F_{\mu\nu} = 4\pi j_\nu$; $\partial_\mu \tilde{F}^{\mu\nu} = 4\pi j^\nu$) by expressing the source charges and currents in terms of the microscopic quantities such as E1 matrix elements. They can be simplified by writing the electric field and the polarization field in the form

$$\begin{aligned} \mathscr{E} &= E(x, t)e^{i(\omega_0 t - kx)} + \text{complex conjugate} \\ \mathscr{P} &= P(x, t)e^{i(\omega_0 t - kx)} + \text{complex conjugate} \end{aligned} \tag{20.45}$$

where ω_0, k are the temporal and spatial periods, $e^{i(\omega_0 t - kx)}$ carries all the rapidly varying information, and $E(x, t)$, $P(x, t)$ are slowly varying functions. The coupled Bloch-Maxwell equations are

$$\begin{aligned} \left(\frac{\partial}{\partial t} + c\frac{\partial}{\partial x} + \kappa\right)E &= -2\pi i\omega_0 P \\ \left(\frac{\partial}{\partial t} + \gamma\right)P &= \frac{i}{3\hbar}|M_{ge}|^2 ED \\ \left(\frac{\partial}{\partial t} + \gamma_{\parallel}\right)(D - D_0) &= \frac{2i}{\hbar}(E^*P - EP^*) \end{aligned} \tag{20.46}$$

where

c	velocity of light
$\hbar$	Planck's constant divided by 2π
M_{ge}	E1 matrix element for the laser transition
D_0	population inversion produced by the external energy source
$\kappa, \gamma, \gamma_{\parallel}$	relaxation rates of the E, P, D fields, respectively

We begin by looking for space- and time-independent solutions of (20.46). This is done by setting all space and time derivatives equal to zero and looking for a solution of the resulting nonlinear equations. We find that

$$\begin{aligned} E &= 0 \\ P &= 0 \\ D - D_0 &= 0 \end{aligned} \tag{20.47}$$

is a solution for all values of D_0. For $D_0 \geq D_{0,1} = 3\hbar\kappa\gamma/(2\pi\omega_0|M_{ge}|^2)$ there are nonzero solutions with

$$\begin{aligned} D_{cw} &= D_{0,1} \\ |E|_{cw} &= \frac{\omega_0 h\gamma_{\parallel}}{4\kappa}|D_0 - D_{0,1}|^{1/2} \\ |P|_{cw} &= \frac{2\pi\omega_0}{\kappa}|E|_{cw} \end{aligned} \tag{20.48}$$

The "laser off" solutions (20.47) are stable below threshold (M_0^3) and unstable above threshold (M_1^3). The solutions (20.48) are initially stable above threshold (M_0^3).

The fields E, P, D may all be taken real. It is useful to study the stable solutions of (20.46) by scaling the fields in terms of their continuous wave solutions: $E/|E|_{cw}$, and so on. The coupled Bloch-Maxwell equations then assume the simpler form

$$\begin{aligned}
&\left(\frac{\partial}{\partial t} + \left[c\frac{\partial}{\partial x}\right] + \kappa\right)E' = \kappa P' \\
&\left(\frac{\partial}{\partial t} + \gamma\right)P' = \gamma E'D' \\
&\left(\frac{\partial}{\partial t} + \gamma_{\|}\right)D' = \gamma_{\|}\left(\frac{D_0}{D_{0,1}}\right) - \gamma_{\|}\left(\frac{D_0}{D_{0,1}} - 1\right)E'P'
\end{aligned} \tag{20.49}$$

The space derivative term in brackets may be removed by transforming to a moving coordinate system and can be dropped. The substitutions

$$\begin{aligned}
&E' = \alpha x, \qquad r = \frac{D_0}{D_{0,1}}, \qquad \alpha = \sqrt{b(r-1)} \\
&P' = \alpha y, \qquad \sigma = \frac{\kappa}{\gamma}, \qquad t \to t' = t\frac{\sigma}{\kappa} \\
&D' = z, \qquad b = \frac{\gamma_{\|}}{\gamma}
\end{aligned} \tag{20.50}$$

bring (20.49) into the form studied by Lorenz.[8]

9. Aufbau Principle. 3

It is clear from the foregoing that a knowledge of the type and properties of the critical points and flows that can typically by encountered in n-dimensional dynamical systems is a great asset in the understanding of the qualitative properties of such systems. The building up principle used in Secs. 1 and 4 was convenient for the purposes of constructing n-dimensional critical properties from those of dimension $n - 1$, when we went from $1 \to 2$ and from $2 \to 3$. However, this pictorial method runs out of usefulness at this dimension, for obvious reasons. If we want to extend this principle to higher dimensions, we must approach the problem from a new direction.

This new direction is suggested by Fig. 7.2 and Sec. 19.6.[9] Instead of describing the flows by arrows in $\mathbb{R}^n$, we look at the distribution of the roots of the stability matrix (in the complex plane) evaluated at a critical point or on a critical flow (invariant surface).

For $n = 1$ there is only one root at a critical point. The critical point is of type M_0^1 or M_1^1 (Fig. 20.24).

Figure 20.24 An isolated critical point in $\mathbb{R}^1$ is of type M_0^1 or M_1^1.

For $n = 2$ there are two roots at a critical point. They may both be real, in which case the critical point is of type M_i^2 ($i = 0, 1, 2$), or they may form complex conjugate pairs, in which case the critical point is of type $F_\pm^2$ (Fig. 20.25). Limit cycles are associated with a complex conjugate pair of eigenvalues crossing the imaginary axis. Perturbations of the structurally unstable case in which two nonzero roots occur on the imaginary axis are shown in Fig. 20.26. In Fig. 20.26*a* a perturbation moves both roots into the left half-plane, creating a stable focus. An unstable limit cycle may simultaneously be created (or a stable limit cycle destroyed). This is represented by a single point in the right half-plane. One of the two eigenvalues is removed because the critical set is no longer a point but a 1-dimensional flow ($\dot{\theta}$ = constant). The other small eigenvalue is that associated with the radial direction in a polar decomposition. In Fig. 20.26*b* we show how a different perturbation can lead to an unstable focus F_+^2 and a stable limit

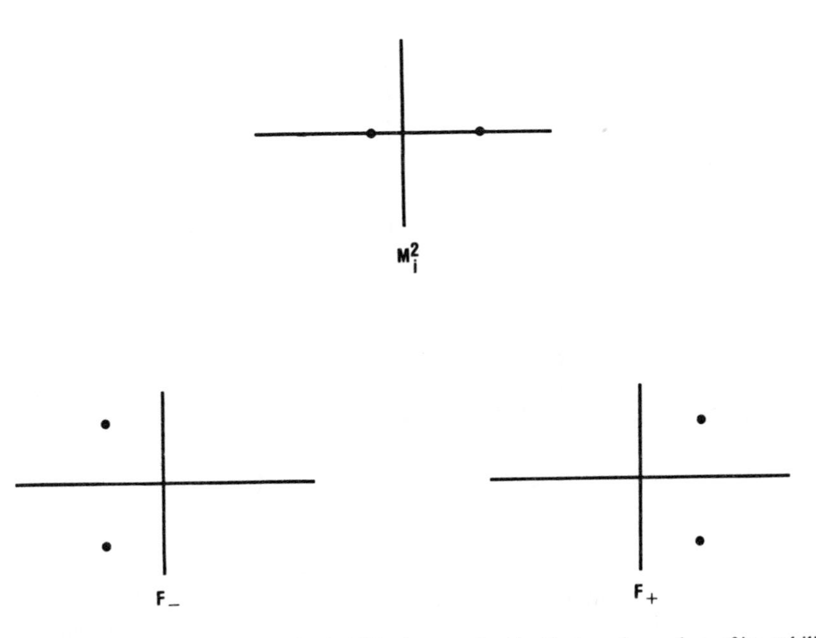

Figure 20.25 An isolated critical point in $\mathbb{R}^2$ is characterized by the two eigenvalues of its stability matrix. They may either be both real or they may be a complex conjugate pair.

cycle $T^1 \times M_0^1$. This latter bifurcation may be represented schematically by the equation

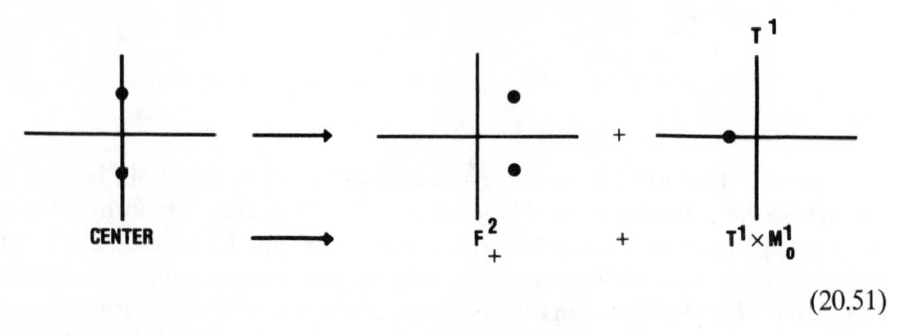

(20.51)

In the 3-dimensional case we may look at critical sets of dimensions 0 (critical points), 1 (limit cycles), and 2 (invariant tori). At a critical point the eigenvalues may all be real, or there may be a complex conjugate pair. The spectrum of isolated critical points contains those of type M_i^3 and $F_\pm^2 \times M_i^1$ (Fig. 20.27). On a 1-dimensional critical set there are two eigenvalues to consider. The critical flow spectrum contains flows of type $T^1 \times M_i^2$ and $T^1 \times F_\pm^2$. These latter are the unstable and stable screw fields $\mathrm{Sc}_\pm$ shown in Figs. 20.13 and 20.12. On a 2-dimensional critical set T^2 there is one leftover eigenvalue, so the

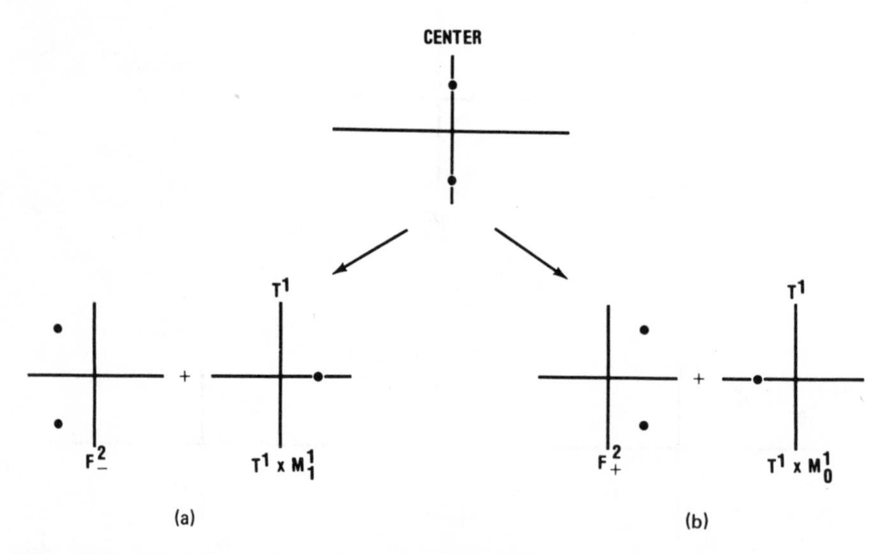

Figure 20.26 When the two eigenvalues at an isolated critical point are imaginary, the corresponding flow pattern is that of a vortex. This is a structurally unstable flow. Under a perturbation the two eigenvalues may move into the negative half-plane (stable focus). In this case an unstable limit cycle $T^1 \times M_1^1$ appears in a Hopf bifurcation. If the perturbation moves the complex eigenvalues into the right half-plane, a stable limit cycle is created.

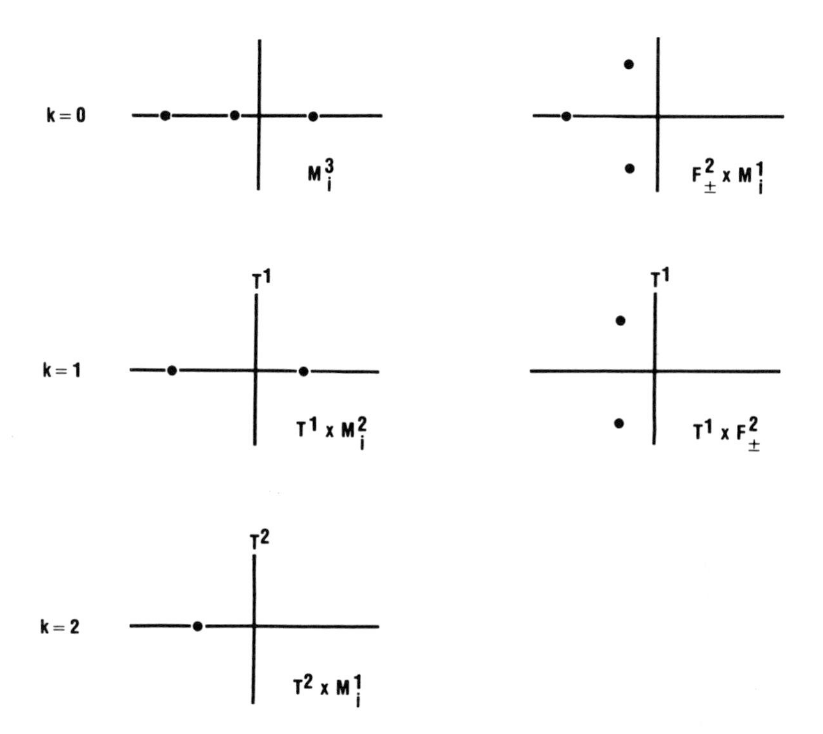

Figure 20.27 The critical set in $\mathbb{R}^3$ may have dimensionality $k = 0$, 1, or 2. The number of eigenvalues of the reduced stability matrix in the orthogonal directions is $3 - k$. These may be distributed in the complex plane as shown. The corresponding flows are of type $T^k \times M_i^{3-k}$ or $T^k \times F_+^2 \times M_i^{3-(2+k)}$.

corresponding flows are of type $T^2 \times M_i^1$. Fig. 20.27 contains all the information contained in Figs. 20.9–20.15. In addition, it is a simple matter to see which critical flows stand near other critical flows in the sense of bifurcations. For example

$$\begin{aligned} T^1 \times F_-^2 \xrightarrow{\text{Hopf}} & T^1 \times (F_+^2 + T^1 \times M_0^1) \\ &= \text{Sc}_+ + T^2 \times M_0^1 \end{aligned} \tag{20.52}$$

as previously noted.

Although we have gained nothing yet in the use of this method over that gained in Secs. 1 and 4, it greatly simplifies the analysis of critical surfaces for dynamical systems of dimensions $n \geq 4$. We treat the case $n = 4$ in Fig. 20.28 to illustrate. In the case of a k-dimensional invariant surface ($k < n$) there remain $n - k \geq 1$ eigenvalues to consider. For $k = 0$ (limit point) the stability types are M_i^4, $F_\pm^2 \times M_i^2$, $F_\pm^2 \times F_\pm^2$. In the case $k = 1$ the limit flow is a limit cycle T^1, and there remain three eigenvalues to consider. These have already been considered in Fig. 20.27. In the case $k = 2$ the limit flow exists on torus T^2, and the two

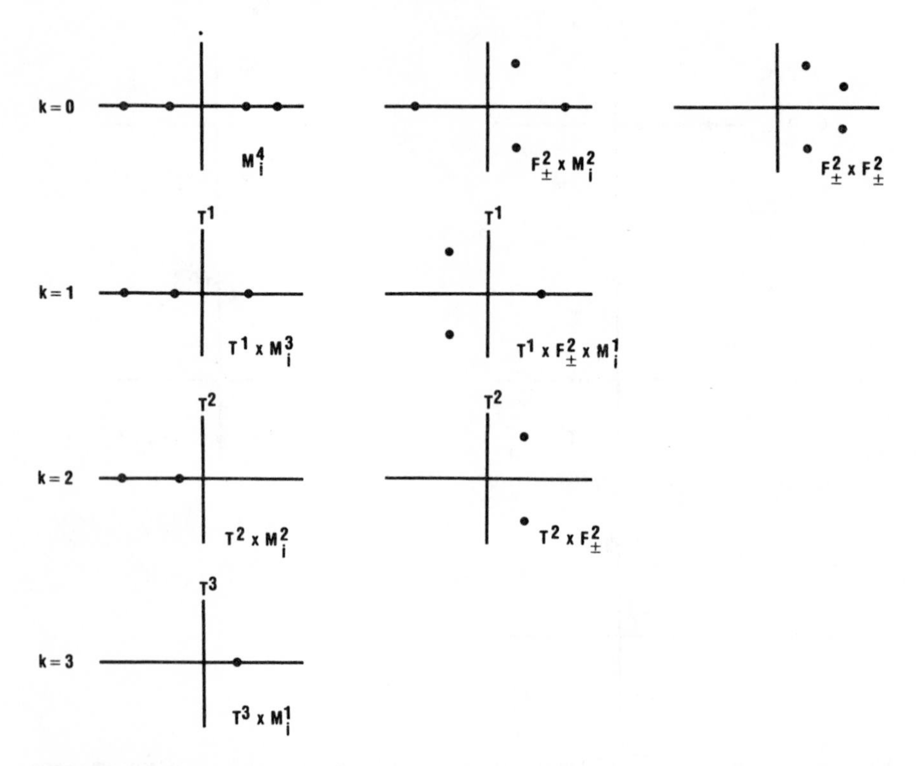

Figure 20.28 The critical set in $\mathbb{R}^4$ may have dimensionality $k = 0, 1, 2, 3$. The number of eigenvalues of the reduced stability matrix in the orthogonal directions is $4 - k$. These may be distributed in the complex plane as shown.

remaining eigenvalues may occur as shown in Fig. 20.25. For $k = 3$, Fig. 20.24 is apropos. The list of critical flows in $\mathbb{R}^4$ results:

$$
\begin{array}{llll}
k = 0, & M_i^4, & F_\pm^2 \times M_i^2, & F_\pm^2 \times F_\pm^2 \\
k = 1, & T^1 \times M_i^3, & T^1 \times F_\pm^2 \times M_i^1 & \\
k = 2, & T^2 \times M_i^2, & T^2 \times F_\pm^2 & \\
k = 3, & T^3 \times M_i^1 & &
\end{array}
$$

The bifurcation properties of these critical flows may easily be assessed. For example, since

$$
\begin{aligned}
T^1 \times F_-^2 \times M_0^1 \xrightarrow{\text{Hopf}} & T^1 \times (F_+^2 + T^1 \times M_0^1) \times M_0^1 \\
&= T^1 \times F_+^2 \times M_0^1 + T^2 \times M_0^2
\end{aligned}
$$

we see that $T^1 \times F_-^2 \times M_0^1$ stands near $T^2 \times M_0^2$ but not near $T^3 \times M_i^1$.

Critical flows in higher dimensions may be treated accordingly. There remains the question of whether all critical flows can be so obtained or, more to the point, whether the critical flows obtained by this building-up process are structurally stable under perturbation. This seems not to be the case for $n > 4$.[10]

10. Center Manifold Theorem

In this chapter so far we have looked at some specimens in the zoo of flows that may be encountered in dynamical systems. We have looked entirely at systems of small dimension, and we have not studied in any depth k-parameter families of dynamical systems. This approach would be adequate if there were some analog of the Splitting Lemma (2.4) valid for dynamical systems.

Such an analog exists. It is called the Center Manifold Theorem.[11] Suppose

$$\begin{gathered} \dot{x} = F(x; c), \qquad x \in \mathbb{R}^n, \quad c \in \mathbb{R}^k \\ F(x; c) \in \mathbb{R}^n \end{gathered} \tag{20.53}$$

is a k-parameter family of dynamical systems. A change of stability occurs at a critical point or flow whenever one or more of the eigenvalues of the associated stability matrix has vanishing real part. If the critical point occurs at x^0 for $c = c^0$, then linearization around the critical point gives the relation

$$\dot{x}_i = F_{ij}(x^0; c^0)\delta x_j + \text{higher order terms} \tag{20.54}$$

We can divide the linear vector space of displacements δx from x^0 up into three linear vector subspaces:

$$V = V_s + V_c + V_u \tag{20.55}$$

Here V_s is spanned by those eigenvectors of F_{ij} whose eigenvalues have negative real part. The unstable space V_u is spanned by the eigenvectors of F with positive real part. The central subspace V_c is spanned by those eigenvectors of F whose real part vanishes. It is this subspace which is critical because it is associated with the bifurcations of the system (20.53). This subspace can be locally extended to a manifold, called the **center manifold**.

If δx is an arbitrary displacement from x^0, then in the neighborhood of $(x^0; c^0)$ the dynamical system equations can be written in the simpler form

$$\begin{aligned} \delta x &= \delta v_s + \delta v_c + \delta v_u; \qquad \delta v_{\#} \in V_{\#} \\ \delta \dot{v}_s &= G_s \delta v_s \\ \delta \dot{v}_c &= G_c(\delta v_c; c) \\ \delta \dot{v}_u &= G_u \delta v_u \end{aligned} \tag{20.54'}$$

where G_s is an $s \times s$ matrix ($s = \dim V_s$), and G_u is a $u \times u$ matrix ($u = \dim V_u$). The operator $G(\delta v_c; c)$ is a nonlinear operator. Equation (20.54′) is the dynamical system analog of the Splitting Lemma (2.3). Away from the center manifold the dynamical system equations can be linearized. It is only on the central manifold of dimension $c \leq n$ that any higher order terms must be retained. All dynamical system bifurcations in the neighborhood of c^0 are determined from the operator $G_c(\delta v_c; c)$. This reduction in dimension greatly simplifies the study of bifurcations associated with dynamical systems (20.53).

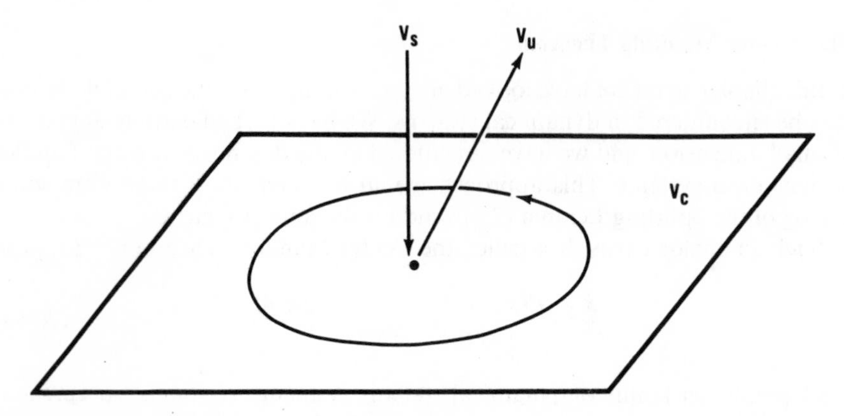

Figure 20.29 At a critical point in $\mathbb{R}^4$ with two imaginary eigenvalues and two real eigenvalues, one positive and one negative, a center manifold decomposition would have a 2-dimensional center manifold V_c and two 1-dimensional manifolds V_s, V_u.

In Fig. 20.29 we try to indicate a center manifold decomposition for a dynamical system in $\mathbb{R}^4$ at a critical point with two real eigenvalues, one positive, the other negative, and two conjugate imaginary eigenvalues. Any displacement δx with $\delta v_u \neq 0$ will eventually leave the neighborhood of the critical point. Since stable critical points are of primary interest in applications, we indicate in Fig. 20.30 the center stable decomposition for a dynamical system in $\mathbb{R}^3$ at a critical point with one real negative eigenvalue $-\lambda$ and two imaginary eigenvalues $\pm i\omega$. If $\lambda \gg \omega$, the system state will make a "linear descent" into the

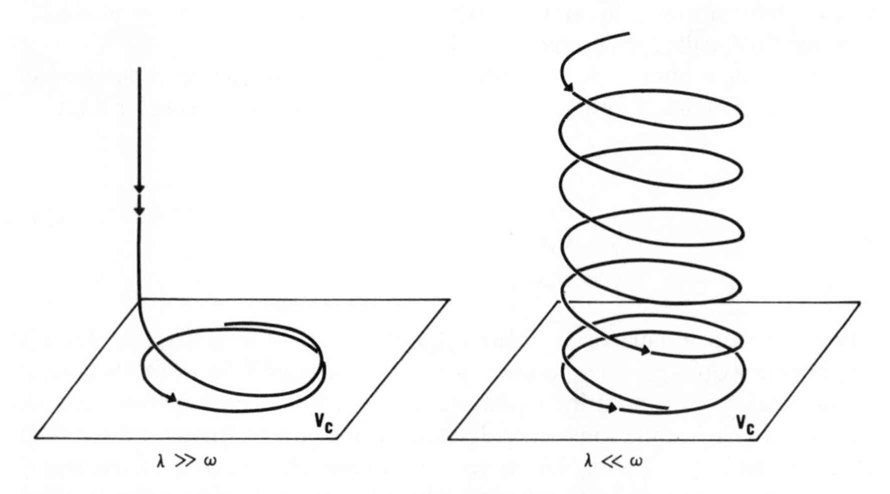

Figure 20.30 A critical point in $\mathbb{R}^3$ with two imaginary eigenvalues $\pm i\omega$ and one negative eigenvalue $-\lambda$ has a center stable manifold decomposition. When the eigenvalues are widely scaled, the dynamical system motion involves either linear descent ($\lambda \gg \omega$) or helical descent ($\omega \gg \lambda$).

center manifold. If $\lambda \ll \omega$, the descent will be helical. Associated with this center manifold is a Hopf bifurcation.

The types of bifurcations that can occur in k-parameter families of n-dimensional dynamical systems can be determined by:

1. Determining the components of the bifurcation set of $F_{ij}(x^0; c^0)$ and their various dimensionalities
2. Applying the Center Manifold Theorem at a typical point within each component of the bifurcation set
3. Applying the methods of the previous section to determine which dynamical flows stand near the particular structurally unstable flow parameterized by the point $c \in \mathscr{S}_B$ in the bifurcation set

It would be desirable to generate a dictionary of canonical mappings $G(\delta v_c; c)$ for dynamical systems in direct analogy with the list of canonical catastrophe functions now available for degenerate critical points.

11. Ruelle–Takens Picture of Turbulence

Numerous physical systems exhibit turbulent behavior when pushed sufficiently hard. How does this come about, and how can this be described mathematically?

Ruelle and Takens have proposed a mechanism by means of which turbulence can arise. This mechanism is elegant in its conceptual simplicity. It is applicable to nonlinear dynamical systems depending on only one control parameter c.

Here is the idea. Assume that in the neighborhood of $c = 0$ the n-dimensional dynamical system has a single stable state $x = 0$. As c is increased, this remains the unique solution until $c = c_1$, at which point a Hopf bifurcation takes place. For $c > c_1$ the solution at $x = 0$ is unstable, but a limit cycle T^1 (deformed circle) is nearby and is stable. The motion around this invariant set is periodic, with angular frequency ω_1. As c is further increased, this limit cycle remains stable until $c = c_2$, at which point the reduced $(n - 1) \times (n - 1)$ stability matrix (one direction is removed by the flow T^1) has a pair of complex conjugate eigenvalues with vanishing real part. A second Hopf bifurcation occurs, and for $c > c_2$ the original limit cycle T^1 is unstable but surrounded by a limit torus T^2. The flow on this torus is quasiperiodic with frequency components (ω_1, ω_2). The ω_i are generally functions of the control parameter c. As c increases, their ratio may vary rapidly between irrational (nonperiodic motion) and rational (periodic motion)

We now evaluate the reduced $(n - 2) \times (n - 2)$ stability matrix on the 2-dimensional invariant surface as we continue to crank up c. At c_3 a third pair of complex conjugate eigenvalues may have their real parts go through zero. Then T^2 becomes unstable for $c > c_3$, and is surrounded by a stable invariant torus T^3. This may also describe quasiperiodic motion with angular frequencies $(\omega_1, \omega_2, \omega_3)$.

At c_4 we have another Hopf bifurcation, signalled by two imaginary eigenvalues in the $(n-3) \times (n-3)$ reduced stability matrix $(n \geq 5)$. Beyond c_4, T^3 is unstable but is surrounded by a stable flow on $T^4 \subset \mathbb{R}^n$ $(n \geq 5)$. Ruelle and Takens[10] have shown that this last flow is nongeneric. Arbitrary perturbations (an open set in the space of perturbations of $F(x; c)$, $(c > c_4)$ of the quasiperiodic flow $(\omega_1, \omega_2, \omega_3, \omega_4)$ on T^4 possess strange attractors. Newhouse, Ruelle, and Takens[12] have subsequently shown that turbulent behavior may occur when three eigenvalues have crossed into the right half-plane.

This generic picture of the transition to turbulence is model independent. Nor need turbulence arise through a cascaded sequence of three or four local Hopf bifurcations. Fewer would suffice if nonlocal transitions occurred.

Remark. This picture is more economical of bifurcations than the Landau picture of the transition to turbulence. Landau's picture invokes an infinite cascade of bifurcations, as indicated schematically in Fig. 20.20 in the control-parameter range $r_a \geq r \geq r_\infty$. In the Landau picture, the system appears to behave chaotically only because of the large number of nonzero order-parameters.

Experimental and theoretical work seem to indicate that the Ruelle-Takens picture of the transition to turbulence appears to be at least on the right track. Gollub and Swinney[13] studied the Taylor instability (Fig. 20.31) using light-scattering techniques. Water was confined to an annular volume between a rotating steel cylinder of radius r_1 and a coaxial glass tube of inner radius r_2. The rotation period τ is related to the Reynolds number R through $R = 2\pi r_1 d/\nu\tau$, where ν is the kinematic viscosity. Light scattered from a small fluid volume carried information about the Fourier frequency components in the radial components of the fluid's velocity. This information was beaten out of the scattered light. Their plot of the reduced frequency $f^* = f\tau$ against the reduced Reynold's number $R^* = R/R_T$ is shown in Fig. 20.31. This is scaled so that the onset of turbulence occurs for $R^* = 1.0$. The onset of the various regimes occurs for the following values of R^*: $R_1^* = 0.064$, $R_2^* = 0.54 \pm 0.01$, $R_3^* = 0.78 \pm 0.03$, $R_4^* = 1.0$. The bifurcation to chaos at $R_4^* = 1.0$ was reproducible without hysteresis. The mode associated with f_2 dies out at $R^* = 0.78 \pm 0.04$. It appears that the disappearance of the mode associated with frequency f_2^* is directly related to the appearance of a new mode with frequency f_3^*. It also appears that a Hopf bifurcation at $R^* = 1.0$ leads to a quasiperiodic motion $(\omega_1, \omega_2, \omega_3)$ which is structurally unstable and which, under naturally occurring perturbations, then reduces to chaotic behavior. This behavior is characterized by a broad power spectrum. In this system the transition to turbulence involves a sequence of four bifurcations, one of which (at $R^* = 0.54$) may not be required for the occurrence of chaotic behavior.

Gollub and Benson[14] have used light-scattering techniques to study the Rayleigh–Bénard instability. Their findings are summarized in Fig. 20.32. Here R_c is the Reynolds number at which bifurcation to convective motion occurs.

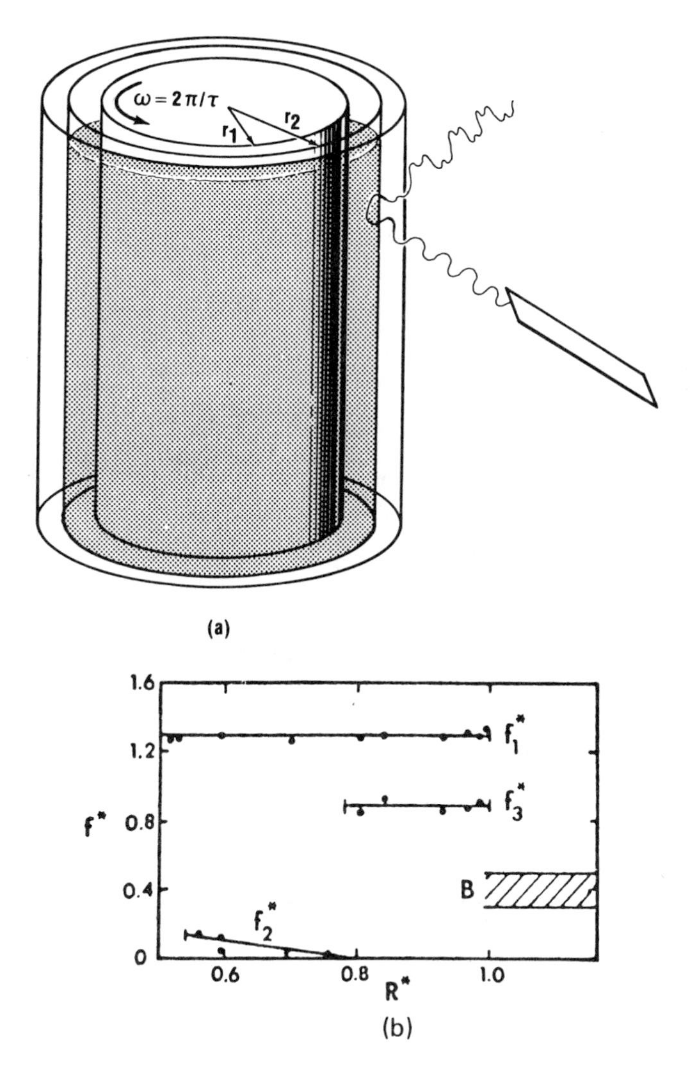

Figure 20.31 (*a*) Experimental arrangement for studying the Taylor instability by light scattering. (*b*) Experimental data showing the single periodic regime, two quasiperiodic regimes, and the turbulent regime. Reprinted with permission from J. P. Gollub and H. L. Swinney, *Phy. Rev. Lett.* **35** (1975), p. 929. Copyright 1975 by American Institute of Physics.

For $1 < R^* = R/R_c < 20$ the motion is convective, but with time-independent velocity field. In the range $20 < R^* \lesssim 50$ four distinct time-dependent velocity fields can exist in regions of overlapping Reynolds number. For $20 < R^* \lesssim 29$ there is a singly periodic regime. In this regime the power spectrum has Fourier components at the fundamental frequency and its harmonics. For $29 \lesssim R^* < 39.8$ this periodic motion develops a strong subharmonic in the power spectrum at half the fundamental frequency. This is analogous to the bifurcation to

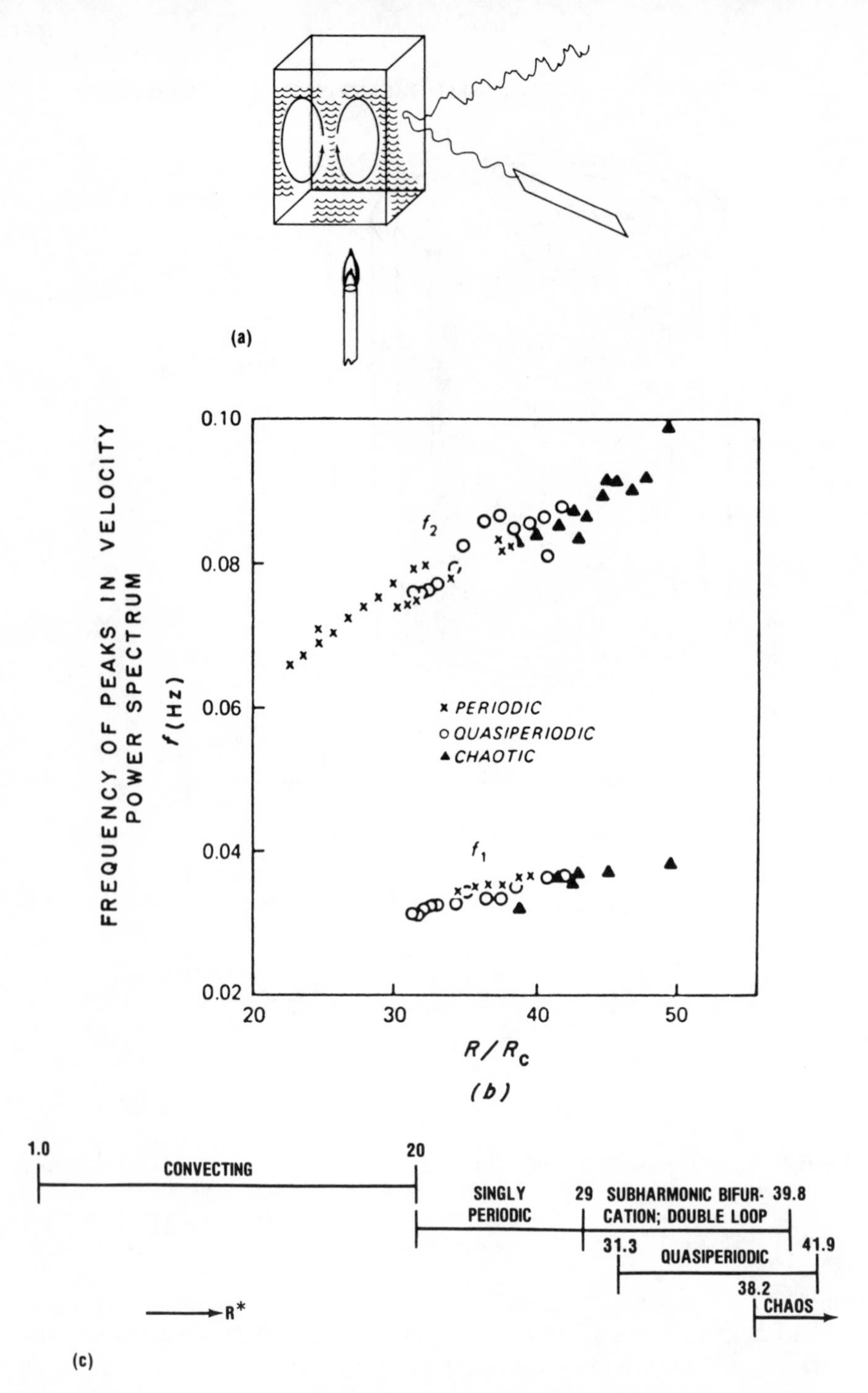

Figure 20.32 (*a*) Experimental arrangement for studying the Rayleigh-Bénard instability by light scattering. (*b*) Experimental data showing the periodic, quasiperiodic, and chaotic regimes. The compatibility of distinct regimes is shown more clearly in (*c*). Reprinted with permission from J. P. Gollub and S. V. Benson, *Phys. Rev. Lett.* **41** (1978), p. 948. Copyright 1978 by American Institute of Physics.

double-loop orbits at r_b in Fig. 20.20, except that this transition is surrounded by spinodal lines. In the range $31.3 < R^* < 41.9$ there is a quasiperiodic regime with incommensurate frequencies (ω_1, ω_2). A chaotic regime with a broad band in the power spectrum occurs for $38.2 < R^*$. The single frequency and subharmonic branch loses its stability at $R^* = 39.8$ and the quasiperiodic branch loses its stability at $R^* = 41.9$. For $41.9 < R^* < 50$ only the chaotic regime is observed.

The experimental findings have been qualitatively reproduced in a 14-state-variable dynamical system model studied by Curry.[15] This system of equations is obtained from the Navier-Stokes equation in the Boussinesq approximation. The procedure followed is precisely the one followed by Saltzman[6] and Lorenz,[3] except that the state variable space is expanded to 14 dimensions by retaining the 14 lowest even-parity Fourier coefficients. The qualitative agreement of the behavior of this model with the experimental findings of Gollub and Benson suggests that the model may be essentially correct, with errors introduced by truncation of the state variable space.

This theoretical model was studied by fixing the Prandtl number at 2.5 and varying the Rayleigh number. There appears to be only one stable orbital type for each value of reduced Rayleigh number R^*. The sequence of bifurcations is shown in Fig. 20.33. This model-dependent finding is compatible with the Ruelle-Takens (as modified with Newhouse) picture of the transition to turbulence. The experimental findings of Gollub and Swinney and Gollub and Benson suggest that the Ruelle-Takens picture is probably basically correct, but that Nature has found ways of inventing and concatenating the series of

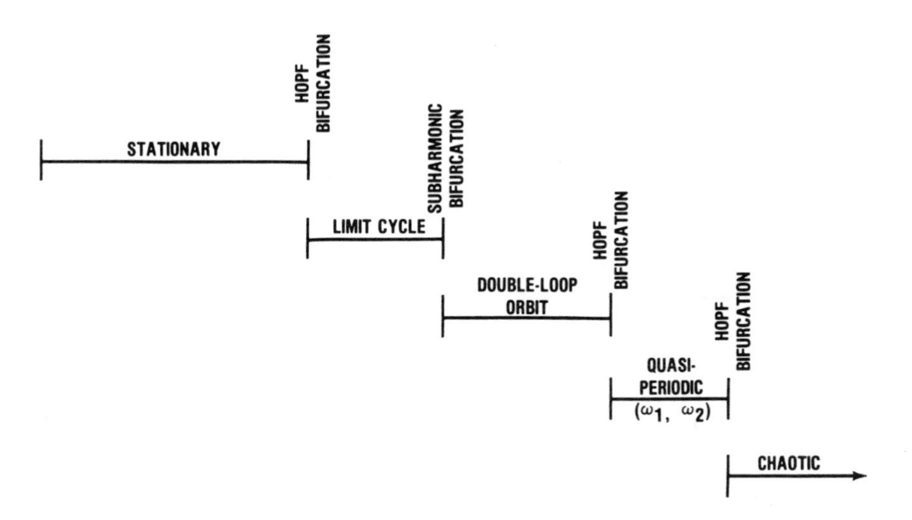

Figure 20.33 The critical behavior of the 14-state-variable model of the Navier-Stokes equation via the Boussinesq approximation shows only one type of stable behavior for each value of R. The sequence of orbital types clearly approximates those apparently observed for the Taylor and the Rayleigh-Bénard instabilities by light-scattering techniques. The results are from Ref. 15.

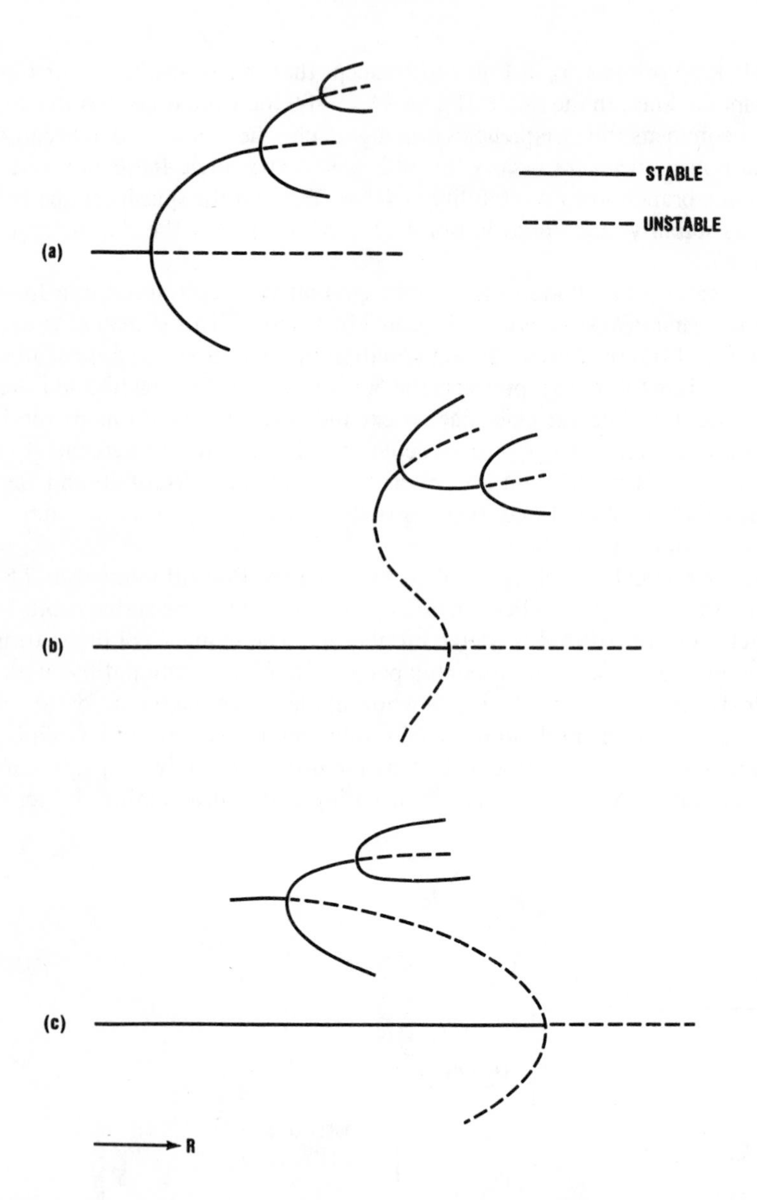

Figure 20.34 The Ruelle-Takens picture of the transition to turbulence involves a sequence of three Hopf bifurcations to a periodic orbit (ω_1), a quasiperiodic orbit (ω_1, ω_2), and another quasiperiodic orbit (ω_1, ω_2, ω_3) which can be perturbed to a chaotic orbit. (Other types of bifurcations can also occur.) The relative location of the three Hopf bifurcations can lead to systems in which all three stages are observed (*a*), two are observed (*b*), or only one is observed (*c*). Each bifurcation indicated above is a Hopf bifurcation.

three Hopf bifurcations so that the actual behavior of real physical systems is even more interesting than the sequential behavior proposed by Ruelle and Takens. In Fig. 20.34 we show schematically how 3, 2, or even 1 Hopf bifurcation could trigger the transition to turbulence.

12. Summary

Nonlinear systems are intrinsically more difficult to study than their linear counterparts. There is no vast body of theorems for the description of n-dimensional nonlinear dynamical systems as there is for linear dynamical systems. The best that we can do so far is to find interesting points and surfaces in $\mathbb{R}^n$ and then perform linear stability analyses about such invariant sets.

The aim of this chapter is to show how one can approach the problem of constructing these interesting invariant sets in a more or less organized way. This is done in a very graphic way in Secs. 1 and 4. The dimensionality of this pictorial technique depends on the dimension n of state space. The pictorial technique was reduced in Sec. 9 to a 2-dimensional graphical construction in the complex root plane of the (reduced) stability matrix. These methods allow a somewhat painless construction of dynamical flows and isolated nonlocal invariant surfaces, as well as an understanding of the bifurcations which exist among them.

We study some interesting nonlinear dynamical systems in the remaining sections. An interesting structurally stable flow is discussed in Secs. 2 and 3. In Sec. 2 we describe the creation of this flow, a limit cycle topologically equivalent to a circle T^1, through the Hopf bifurcation. As the nonlinearity in the structurally stable perturbation of the harmonic oscillator is increased, this flow becomes more and more squashed and deformed. In extremis, we find relaxation oscillations (Sec. 3).

In Secs. 5 and 6 we move onto flows in $\mathbb{R}^3$. We construct one interesting flow that feeds on itself from one of the basic 0-dimensional critical sets of $\mathbb{R}^3$ determined in Sec. 5 ($F_+^2 \times M_0^1$), together with the "feedback" mechanism of Sec. 4. This flow models the behavior of the simplest nonlinear dynamical system on $\mathbb{R}^3$. In Sec. 6 we place a pair of these 0-dimensional critical sets in $\mathbb{R}^3$. They can be arranged so that the state variable jumps back and forth in a chaotic way between the two foci. This is an extremely useful visualization for the behavior found in the Lorenz nonlinear equation. We have discussed the local critical properties (r_1, r_f, r_4) in some detail, although little is known about the nonlocal bifurcations (r_2, r_3). This system of nonlinear equations has been applied to the description of the Bénard instability (Sec. 7) and the laser-spiking instability (Sec. 8).

In Sec. 10 we decribe the Center Manifold Theorem. This is the analog for dynamical systems of the Thom Splitting Lemma for (the equilibria of) gradient systems. This theorem is important because it reduces the dimension of a complicated nonlinear problem from $n = \dim \mathbb{R}^n$ to $\dim V_c$. This latter number is just the number of eigenvalues of the stability matrix F_{ij} which have zero real part. Thus the study of bifurcations of the original dynamical system is reduced to a study of just those bifurcations that can occur on the center manifold. This justifies the methods described in Sec. 9 for determining the "irreducible flows" for n-dimensional dynamical systems.

In the final section we describe the Ruelle–Takens picture for the transition to turbulence through a sequence of four or three (or fewer) cascaded Hopf bifurcations. Two hydrodynamic experiments have been discussed which are compatible with this picture. The transition to turbulence of an optically bistable system[16] also appears to be compatible with this picture.

References

1. E. C. Zeeman, *Catastrophe Theory, Selected Papers 1972–1977*, Reading: Addison-Wesley, 1977.
2. O. E. Rössler, Different Types of Chaos in Two Simple Differential Equations, *Z. Naturforsch.* **31a**, 1664–1670 (1976).
3. E. N. Lorenz, Deterministic Nonperiodic Flow, *J. Atmos. Sci.* **20**, 130–141 (1963).
4. K. A. Robbins, Periodic Solutions and Bifurcation Structure at High r in the Lorenz Model, *SIAM J. Appl. Math.* **36**, 457–472 (1979).
5. Y. M. Treve, Theory of Chaotic Motion with Application to Controlled Fusion Research, in: *Topics in Nonlinear Dynamics*, A Tribute to Sir Edward Bullard (S. Jorna, Ed.), New York: American Institute of Physics, 1978, pp. 147–220.
6. B. Saltzman, Finite Amplitude Free Convection as an Initial Value Problem—I. *J. Atmos. Sci.* **19**, 329–341 (1962).
7. H. Risken and K. Nummedal, Self-Pulsing in Lasers, *J. Appl. Phys.* **39**, 4662–4672 (1968).
8. H. Haken, Analogy Between Higher Instabilities in Fluids and Lasers, *Phys. Lett.*, **53A**, 77–78 (1975).
9. V. I. Arnol'd, *Ordinary Differential Equations*, Cambridge: M.I.T. Press, 1973.
10. D. Ruelle and F. Takens, On the Nature of Turbulence, *Commun. Math. Phys.* **20**, 167–192 (1971).
11. J. E. Marsden and M. McCracken, *The Hopf Bifurcation and Its Applications*, New York: Springer, 1976.
12. S. Newhouse, D. Ruelle, and F. Takens, Occurrence of Strange Axiom A Attractors Near Quasi-Periodic Flows on T^m, $m \geq 3$, *Commun. Math. Phys.* **64**, 35–40 (1978).
13. J. P. Gollub and H. L. Swinney, Onset of Turbulence in a Rotating Fluid, *Phys. Rev. Lett.* **35**, 927–930 (1975).
14. J. P. Gollub and S. V. Benson, Chaotic Response to Periodic Perturbation of a Convecting Fluid, *Phys. Rev. Lett.* **41**, 948–951 (1978).
15. J. H. Curry, Chaotic Response to Periodic Modulation of a Convecting Fluid, *Phys. Rev. Lett.* **43**, 1013–1016 (1979).
16. H. M. Gibbs, F. A. Hopf, D. L. Kaplan, and R. L. Schoemaker, Observation of Chaos in Optical Bistability, *Phys. Rev. Lett.* **46**, 474–477 (1981).

PART 4

Catastrophe Theory for Joggers

In which we present in somewhat more detail the mathematics behind Elementary Catastrophe Theory.

CHAPTER 21

Thom's Theorem

In our presentation so far we have indicated how to use Thom's Theorem and discussed in detail the useful properties of catastrophe functions. However, we have avoided stating the theorem. In this chapter we establish the necessary mathematical details so that we can at least state the theorem (Sec. 5).

Thom's Theorem clearly involves perturbations of functions (cf. Table 2.2). In order to discuss perturbations of functions and of mappings, we must first establish a useful topology on the space of interest. This is done in Sec. 1.

Next Thom's Theorem is "qualitative" in nature, so we must establish some method for determining when two functions are qualitatively the same and when they are not. This is done in Sec. 2.

When we encounter a set of mathematical objects, we often want to study the properties shared by most elements in this set. A property is called generic if the subset for which the property is valid is open and dense in the original set. Genericity is treated in Sec. 3.

Thom's Theorem deals with manifolds and with the singularities of mappings of one manifold (the critical manifold) into another (control parameter space). Singularities can easily be discussed in terms of the Inverse Function Theorem: they occur where the theorem fails. This familiar theorem is discussed briefly in Sec. 4.

The additional concepts that are inputs to Thom's Theorem, such as transversality and unfolding, will be treated in subsequent chapters. Nor do we prove the theorem here. It is simply stated in Sec. 5.

1. Topology

In Part 1 we have been very casual about the meaning of the term "perturbation" or what we mean by saying that one function is "near" another. In order to formalize the concept of perturbation, we need to be able to construct a useful idea of "distance" between two functions, or equivalently, between one function and the zero function.

This means topology. There is one obvious topology that we can put on functions defined on $\mathbb{R}^n$. This is the "Taylor series topology." Basically we construct the Taylor series of $f(x)$ at $x^0 \in \mathbb{R}^n$

$$f(x) = f(x^0) + (x - x^0)_i f_i + \frac{1}{2!}(x - x^0)_i(x - x^0)_j f_{ij} + \cdots \tag{21.1}$$

All derivatives are evaluated at x^0. Then the function $f(x)$ is represented by a point in some Euclidean space. For example, if we look at the terms in (21.1) up to and including those of degree k, the Taylor coefficients $f, f_i, f_{ij}, \ldots, f_{ij\cdots k}$ belong to the space $\mathbb{R}^D$, where

$$D = \underset{\text{constant term}}{1} + \underset{\text{linear terms}}{n} + \underset{\text{quadratic terms}}{\frac{n(n+1)}{2!}} + \cdots + \underset{k\text{th-degree terms}}{\frac{(n+k-1)!}{k!(n-1)!}}$$

$$= \frac{(n+k)!}{n!k!} \tag{21.2}$$

The Taylor series of $f(x)$, evaluated at x^0 and truncated beyond terms of degree k, is called the **k-jet** of f at x^0 and denoted by the symbol $j^k f(x^0)$, or simply $j^k f$. For any smooth function $f(x)$, $j^k f(x^0) \in \mathbb{R}^D$.

Topologies on $\mathbb{R}^D$ are familiar and easy to work with. If f, g are two functions, the distance between f and g at x^0 may be defined as

$$\|f - g\|_{x^0} = |f(x^0) - g(x^0)| + \sum_i |f_i - g_i| + \sum_{ij} |f_{ij} - g_{ij}| + \cdots + \sum_{ij\cdots k} |f_{ij\cdots k} - g_{ij\cdots k}| \tag{21.3i}$$

For finite D another definition of distance may be given by

$$\|f - g\|_{x^0}^2 = |f(x^0) - g(x^0)|^2 + \sum_i |f_i - g_i|^2 + \sum_{ij} |f_{ij} - g_{ij}|^2 + \cdots + \sum_{ij\cdots k} |f_{ij\cdots k} - g_{ij\cdots k}|^2 \tag{21.3ii}$$

or by replacing

$$2 \to p, \qquad p \geq 1 \tag{21.3p}$$

Open sets, and therefore topologies, can be defined using the definition of distance given in (21.3i), (21.3ii), or (21.3p), that is, the ε neighborhood of f at x^0 contains all functions g whose distance from f is less than ε. For k (hence D) finite, the definitions (21.3i), (21.3ii), and (21.3p) give rise to equivalent topologies. In the limit $k \to \infty$, $D \to \infty$ and the topologies generated by the different distance definitions are distinct. The topology generated by the distance (21.3i) is then the most useful.

The C^k topology at x^0 is generated using the distance definition (21.3i). As $k \to \infty$, the corresponding toplogy is called the C^∞ topology at x^0. These topologies are denoted by $C^k(x^0; \mathbb{R}^1)$ and $C^\infty(x^0; \mathbb{R}^1)$.

The topologies described above allow us to define perturbations at a point, that is, $g(x)$ is a perturbation of $f(x)$ at x^0 if the distance between $g(x)$ and $f(x)$ *at* x^0 is small in the C^k (or C^∞) topology at x^0.

If the distance between $g(x)$ and $f(x)$ in the C^k (C^∞) toplogy at x^0 is less than ε *for all* $x^0 \in \mathbb{R}^n$, then $g(x)$ is said to belong to the ε neighborhood of $f(x)$ in the C^k (C^∞) topology. Two functions close together in the C^k topology have the property that the functions are almost equal everywhere on $\mathbb{R}^n$, and all corresponding derivatives up to degree k are also almost equal everywhere on $\mathbb{R}^n$. The corresponding topologies are denoted by $C^k(\mathbb{R}^n; \mathbb{R}^1)$ and $C^\infty(\mathbb{R}^n; \mathbb{R}^1)$.

Remark 1. These concepts can be extended from functions (f: $\mathbb{R}^n \to \mathbb{R}^1$) to mappings ($y$: $\mathbb{R}^n \to \mathbb{R}^m$). It is only necessary to replace the single Taylor expansion in (21.1) by m Taylor series expansions ($f \to y_r, r = 1, 2, \ldots, m$). There will then be one additional sum over r in each of the terms on the right-hand side of (21.3i). The corresponding topologies are denoted by $C^k(\mathbb{R}^n; \mathbb{R}^m)$ and $C^\infty(\mathbb{R}^n; \mathbb{R}^m)$.

Remark 2. All statements about perturbations in Part 1 (dense, open, etc.) are in terms of the C^∞ topology $C^\infty(\mathbb{R}^n; \mathbb{R}^1)$ or $C^\infty(\mathbb{R}^n; \mathbb{R}^m)$, as the case may be.

Remark 3. To construct a topology based on Taylor coefficients up to and including terms of degree k, the functions f must be at least k times differentiable. This is no problem for scientific or engineering applications, since the functions (potentials) encountered are usually smooth, so that derivatives of all orders exist. However, there is the implicit assumption that there is a 1–1 correspondence between functions and their Taylor series.

There is not. This is only true for analytic functions. A function $f(x)$ is analytic at x^0 if its Taylor series about x^0 converges to $f(x)$ in a neighborhood of x^0. To make the distinction between smooth and analytic functions concrete, mathematicians are fond of trotting out the following mathematical horror:

$$\begin{aligned} h(x) &= e^{-1/x^2}, \qquad & x \neq 0 \\ &= 0, & x = 0 \end{aligned} \tag{21.4}$$

All Taylor coefficients of $h(x)$ vanish at $x = 0$, so the Taylor series of the smooth function $h(x)$ converges to the function zero at $x = 0$. For analytic functions $f_1(x)$ and $g(x)$ we can define the function $f_2(x)$ by

$$f_2(x) = f_1(x) + g(x)h(x),$$

or

$$f_2(x) = f_1(x)[1 - h(x)]^k, \qquad k = 1, 2, \ldots$$

or

$$f_2(x) = \begin{cases} f_1(x)[1 - e^{-g(x)/x^2}], & x \neq 0, \\ f_1(0), & x = 0, \end{cases} \quad g(0) > 0 \tag{21.5}$$

The function $f_2(x)$ is smooth and has Taylor series identical to that of $f_1(x)$ at zero. But $f_2(x) \neq f_1(x)$, so that $f_2(x)$ is *not* analytic at zero. In short, for any analytic $f_1(x)$ there is a vast horde of smooth but nonanalytic functions $f_2(x)$ with the same Taylor series expansion.

In spite of this, the idea of mapping functions into Euclidean spaces $\mathbb{R}^D$ in order to generate a topology is useful and appealing. We use this approach as a heuristic tool. The results which we describe below hold for smooth as well as analytic functions.

2. Stable Functions

Roughly speaking, a function is stable, or structurally stable, if its qualitative properties are unaffected by an arbitrary perturbation. We have already seen in Chapter 4 that in an intuitive sense, potentials are stable at points where $\nabla V \neq 0$ and at Morse critical points, but not at degenerate critical points. We make the concept of (local) stability more precise as follows. Let $V(x_1, \ldots, x_n)$ be a potential, and let $\epsilon f(x_1, \ldots, x_n)$ be a perturbation. Then V is **stable** at $x^0 \in \mathbb{R}^n$ if there is a smooth change of coordinates $x'_j = x'_j(x_1, \ldots, x_n)$ so that the new (perturbed) function $V' = V + \epsilon f$ in the new coordinate system has the same structure as the old function V in the old coordinate system:

$$V'(x') \doteq V(x) \tag{21.6}$$

Example 1. At a Morse critical point we suppose

$$\begin{aligned} V(x) &= x_1^2 + x_2^2 + \cdots + x_n^2 \\ \epsilon f(x) &= \epsilon_1 x_1 + \epsilon_2 x_2 + \cdots + \epsilon_n x_n \end{aligned} \tag{21.7}$$

Then

$$V(x) + \epsilon f(x) = (x_1 + \tfrac{1}{2}\epsilon_1)^2 + \cdots + (x_n + \tfrac{1}{2}\epsilon_n)^2 - \tfrac{1}{4}(\epsilon_1^2 + \cdots + \epsilon_n^2)$$

Therefore if we choose

$$\begin{aligned} x_i' &= x_i + \tfrac{1}{2}\varepsilon_i \\ V'(x') &= x_1'^2 + \cdots + x_n'^2 - \tfrac{1}{4}(\varepsilon_1^2 + \cdots + \varepsilon_n^2) \end{aligned} \tag{21.8}$$

we have that

$$V'(x') = V(x) \tag{21.9}$$

As a result, functions are locally stable around Morse saddles.

Example 2. The function $f(x) = x^3$ is unstable at the degenerate critical point at $x = 0$ because a perturbation will either remove the critical points altogether (Fig. 6.1, $a_1 > 0$) or split them into nondegenerate critical points (Fig. 6.1, $a_1 < 0$).

Remark. Functions which are locally stable at all points where they are defined are called **stable**, **globally stable**, or **structurally stable**. In particular, Morse functions are globally stable. The function $f(x) = x^3$ is not globally stable because it is not locally stable at $x = 0$. The functions $F(x; a_1) = x^3 + a_1 x$ are Morse functions if $a_1 \neq 0$, so that these functions are globally stable.

Example 3. If we consider the *family* of functions A_3, then members of this family parameterized by points in the open regions, Region I or III (Fig. 21.1), are Morse functions. A perturbation $\varepsilon f = (\delta a/2)x^2 + (\delta b)x$ of such a function will produce a "nearby function" with properties qualitatively identical to those of the original function. The only unstable functions in this family are those functions parameterized by points on the cusp-shaped lines (Region II) or at the origin, because the perturbation $(\delta a, \delta b)$ moves those points into either Region I or Region III.

Remark. Sometimes another condition is placed on a function in order for it to be globally stable. This is that the function should assume different values at different critical points. Under such a condition, the family A_3 would not be stable on the half-line $a < 0, b = 0$ (dashed in Fig. 21.1) because the two minima at $x = \pm\sqrt{-a}$ have equal values (by symmetry). This is essentially a distinction between the local bifurcation set and the nonlocal bifurcation set (Maxwell set).

So far we have discussed the stability of functions and the stability of a particular function in a family of functions. But we have not discussed the stability of a *family* of functions itself. Suppose we have a 1-parameter family of functions with the property that some members of this family have two local minima and a local maximum, while other members of this family have a single minimum. To make these considerations concrete, we assume that this family has the form

$$V(x;\, s) = \tfrac{1}{4}x^4 + \tfrac{1}{2}a(s)x^2 + b(s)x, \qquad s \in \mathbb{R}^1 \tag{21.10}$$

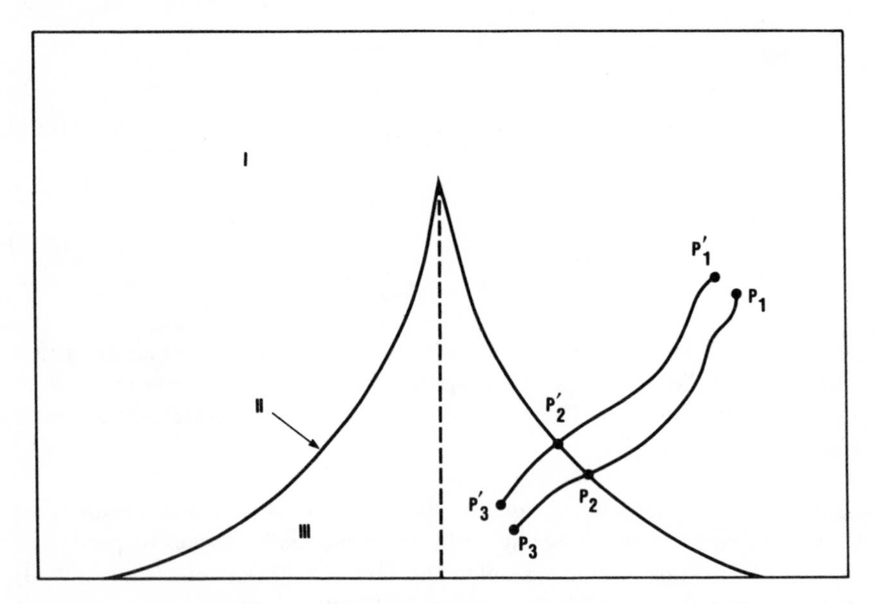

Figure 21.1 A curve connecting $p_3 \in$ Region III with $p_1 \in$ Region I must intersect the fold curve at some point p_2. Under a perturbation, the curve connecting p_3' with p_1' intersects the fold curve at a point p_2' near p_2. Thus degenerate critical points can stably be encountered in *families* of functions, even though they are structurally unstable when encountered in an isolated function.

and that the function $V(x;\, s_3)$ corresponds to the point p_3 in Fig. 21.1, while $V(x;\, s_1)$ corresponds to p_1. Any smooth 1-parameter family containing both $V(x; s_3)$ and $V(x; s_1)$ must contain all functions $V(x; s)$ on some line $(a(s), b(s))$ connecting p_3 with p_1. This means that a function $V(x;\, s_2)$ with a doubly degenerate critical point *must* be a member of this family.

Now suppose that we perturb (21.10) with the perturbation $\varepsilon f = \frac{1}{2}\delta a(s)x^2 + \delta b(s)x$. The perturbed family corresponds to the line connecting p_3' with p_1'. It is clear that this line must also cut the cusp-shaped line which acts as a separatrix between Regions I and III, so that the perturbed family contains a member $V'(x; s_2')$ with a doubly degenerate critical point. In fact, every curve connecting a point near p_3 with a point near p_1 must cut this separatrix. Thus we see that even though functions with degenerate critical points are not stable, they can occur stably in families of functions.

The concept of stability can be applied to functions ($\mathbb{R}^n \to \mathbb{R}^1$), families of functions ($\mathbb{R}^n \otimes \mathbb{R}^k \to \mathbb{R}^1$), mappings ($\mathbb{R}^n \to \mathbb{R}^m$), and families of mappings ($\mathbb{R}^n \otimes \mathbb{R}^k \to \mathbb{R}^m$). In the latter case the concept of stability works as follows:

1. Two families of mappings $f_1(x; a), \ldots, f_m(x; a)$ and $g_1(x; a), \ldots, g_m(x; a)$ are **equivalent** if there are smooth changes of coordinates:

$$x_i' = x_i'(x_1, \ldots, x_n; a_1, \ldots, a_k), \qquad 1 \le i \le n \tag{21.11a}$$

$$a_j' = a_j'(a_1, \ldots, a_k) \qquad 1 \le j \le k \tag{21.11b}$$

$$y_l' = y_l'(y_1, \ldots, y_m) \qquad 1 \le l \le m \tag{21.11c}$$

such that

$$y'_l[f_1(x;a),\ldots,f_m(x;a)] = g_l[x'_1(x;a),\ldots,x'_n(x;a);a'_1(a),\ldots,a'_k(a)] \quad (21.12)$$

In the notation of more modern mathematics (21.12) is a statement that the following diagram "commutes":

$$\begin{array}{ccccc} & \mathbb{R}^n \otimes \mathbb{R}^k & \xrightarrow{f} & \mathbb{R}^m & \\ & (x;a) & & (y) & \\ (21.11\text{a,b}) & \downarrow & & \downarrow & (21.11\text{c}) \\ & \mathbb{R}^n \otimes \mathbb{R}^k & \xrightarrow{g} & \mathbb{R}^m & \\ & (x';a') & & (y') & \end{array} \quad (21.13)$$

2. The family of mappings $f_1(x;a),\ldots,f_m(x;a)$ is stable if every perturbation $g_1(x;a),\ldots,g_m(x;a)$ is equivalent to it.

Remark. For the elementary catastrophes described in Table 2.2, the parameter space $\mathbb{R}^k$ is divided into open regions by separatrices of the type discussed in Chapter 5. Any functions parameterized by a point in one of these open regions is a Morse function, and therefore structurally stable. Functions parameterized by points on the separatrix are not structurally stable.

Remark. The families of functions (including their non-Morse functions) listed in Table 2.2 occur in a structurally stable way in typical k-parameter families of potentials $V(x_1, \ldots, x_n; c_1, \ldots, c_k)$ when these are reduced to the canonical form (2.4) near a degenerate critical point.

Remark. The concept of structural stability was first discussed by Andronov and Pontryagin[1] in terms of dynamical systems. A dynamical system was called structurally stable if the solutions of the equations of motion were not qualitatively altered by a perturbation of the dynamical system. A gradient system is a special case of a dynamical system, one which is characterized by a single function (its potential). As a consequence, the terminology "structurally stable" can be applied to gradient systems, and in particular to the functions that characterize such systems.

3. Generic Properties

Once a mathematical system depending on parameters has been defined, its properties may be discussed. Those properties possessed by an open dense subset of the system are called **generic properties**, or simply **generic**. The complementary subset in the system has measure zero.

Example 1 (Due to Sussman)[2]. We consider the mathematical system consisting of pairs of simultaneous linear equations in two unknowns x and y

$$\begin{aligned} a_{11}x + a_{12}y &= a_{13} \\ a_{21}x + a_{22}y &= a_{23} \end{aligned} \quad \text{or} \quad \begin{bmatrix} a_{11} & a_{12} \\ a_{21} & a_{22} \end{bmatrix} \begin{bmatrix} x \\ y \end{bmatrix} = \begin{bmatrix} a_{13} \\ a_{23} \end{bmatrix} \tag{21.14}$$

where the six coefficients a_{ij} are all real. The system (21.14) is parameterized by a point in $\mathbb{R}^6$. We can then ask the question: under what conditions does the system (21.14) possess a unique solution? Clearly, there is a unique solution if and only if $a_{11}a_{22} - a_{12}a_{21} \neq 0$. Now the 2×2 matrix in (21.14) can become singular for one of three reasons: (1) only one eigenvalue vanishes, or both eigenvalues are zero and (2) the matrix is diagonalizable or (3) it is not (i.e., it assumes Jordan form). It is easy to check that the set of points $(a_{11}, a_{12}, a_{21}, a_{22}) \in \mathbb{R}^4$ on which these possibilities occur have the following dimensions:

1. One eigenvalue $= 0$ 3 surface in $\mathbb{R}^4$
   Both eigenvalues $= 0$
2. Diagonalizable   origin of $\mathbb{R}^4$
3. Not diagonalizable   2 surface in $\mathbb{R}^4$

As a result, the set of points in $\mathbb{R}^6$ on which (21.14) does not have a unique solution breaks up into subsets of dimension 5 ($= 3 + 2$), 2, and 4. Since surfaces of dimension $< n$ in $\mathbb{R}^n$ have measure zero, the property of (21.14) having unique solutions is generic. If we are at a nongeneric point $(a_{11}, a_{12}, a_{13}, a_{21}, a_{22}, a_{23}) \in \mathbb{R}^6$, then a small perturbation $(\delta a_{11}, \delta a_{12}, \delta a_{13}, \delta a_{21}, \delta a_{22}, \delta a_{23}) \in \mathbb{R}^6$ (near the origin) will produce a system of equations with nonzero determinant and unique solution. So perturbation at a point where a generic property holds (stable point) will produce no qualitative changes in the generic property. Perturbation at an unstable point will produce drastic changes (in this case the number of solutions goes from ∞ to 1).

Example 2. In the family of functions A_3 the property of having isolated critical points is generic (i.e., Morse functions are generic).

Example 3. In 2-parameter families of functions the presence of a cusp is generic. The cusp point is associated with the degerate critical point of $\frac{1}{4}x^4$. If εf is any perturbation, from Table 2.2,

$$\tfrac{1}{4}x^4 + \varepsilon f \doteq V(x; a, b) \tag{21.15a}$$

Then for this family we find

$$V(x; -a, -b) + \varepsilon f = \tfrac{1}{4}x^4 \tag{21.15b}$$

Given the 2-parameter family $V(x; a, b)$ we can always find a perturbation which produces a function with a triply degenerate critical point. In short, (21.15b) can be used to "undo" what (21.15a) does.

"Generic" is a useful generalization of the mathematical concept of "dense." If a mathematical system has a generic property, then any element in that system can be approximated arbitrarily closely by elements with that property. For example, the non-Morse function x^3 can be approximated arbitrarily closely by the Morse functions $x^3 + a_1 x$ (let $a_1 \to 0$).

Remark. In Chapter 5 ("Crowbar Principle") we saw how the nongeneric points (at which $\nabla V = 0$) could be used to determine information about the generic points ($\nabla V \neq 0$) for a given function. Also, for a family of functions we saw how the nongeneric functions (with degenerate critical points) could be used to obtain information about the generic (Morse) functions.

Remark. The property of being a Morse function is generic in each of the families of catastrophes listed in Table 2.2.

Notation. Other terms sometimes used in place of "generic" are "**typical**" or "**in general position.**"[3]

4. Singularities of Mappings

An n-manifold is a smooth surface which is everywhere locally Euclidean of fixed dimension n. The sphere surface $x^2 + y^2 + z^2 = 1$ is a 2-dimensional manifold. A mapping f of one n-manifold $\mathscr{P}$ into another n-manifold $\mathscr{Q}$ is nonsingular at $p \in \mathscr{P}$ if it is locally invertible. The test for local invertibility is given explicitly by the Inverse Function Theorem. If $x_1, \ldots, x_n$ is a system of coordinates around $p \in \mathscr{P}$, and $f_1, \ldots, f_n$ a system around $f(p) \in \mathscr{Q}$, then f is locally invertible if and only if the Jacobian of the transformation is nonsingular:

$$\det \left| \frac{\partial f_i}{\partial x_j} \right|_p \neq 0, \qquad 1 \leq i, j \leq n \tag{21.16}$$

Speaking roughly but accurately, since $\mathscr{P}$ is an n-manifold, it looks locally like $\mathbb{R}^n$ at p. Similarly for $\mathscr{Q}$ at $f(p)$. Then the $n \times n$ matrix $\partial f_i / \partial x_j$ is locally the linear mapping of the (tangent space) $\mathbb{R}^n$ at p to the (tangent space) $\mathbb{R}^n$ at $f(p)$.

As an example we consider the set of critical points of (21.15):

$$\begin{aligned} \nabla(\tfrac{1}{4}x^4 + \tfrac{1}{2}ax^2 + bx) &= 0 \\ x^3 + ax + b &= 0 \end{aligned} \tag{21.17}$$

This is a 2-dimensional pleated manifold whose shape is shown in Fig. 21.2. The coordinates of any point on the manifold embedded in the 3-dimensional space $\mathbb{R}^3$ are

$$(x; a, b) = (\lambda_1; \lambda_2, -\lambda_1^3 - \lambda_1 \lambda_2) \tag{21.18}$$

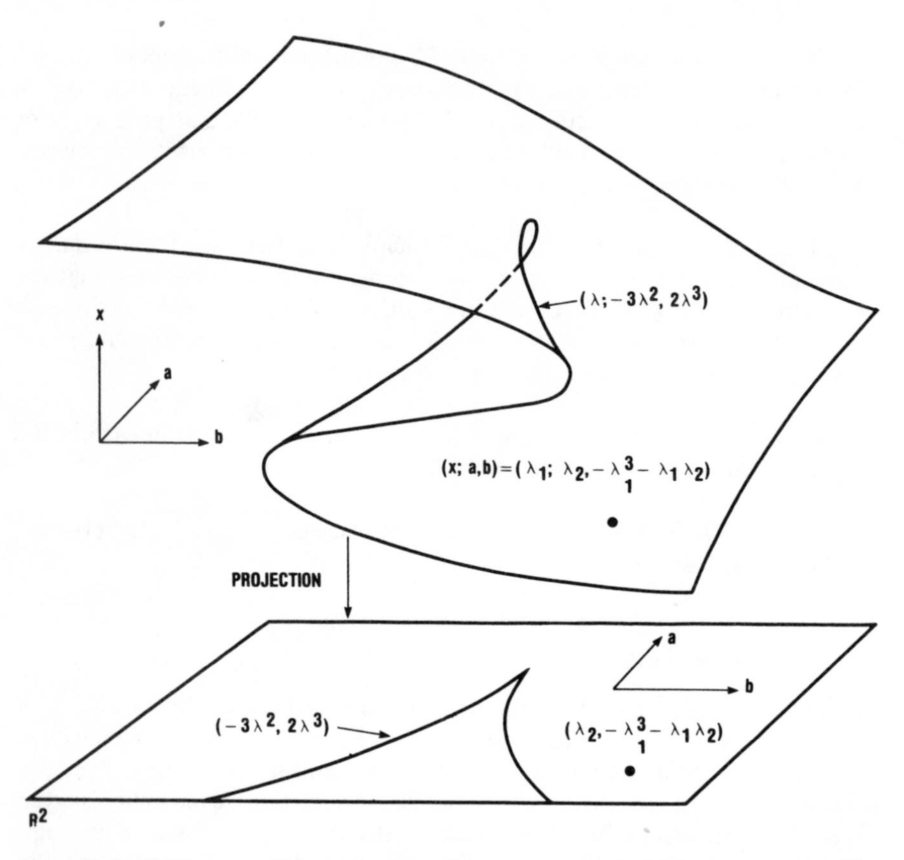

Figure 21.2 The equation $\nabla V(x;\ a,\ b) = 0$ represents a smooth 2-dimensional manifold. The semicubical parabola $(x; a, b) = (\lambda; -3\lambda^2, 2\lambda^3)$, which describes a point on the manifold at which the tangent plane is vertical, is also smooth. It is only in the projection mapping down into the control plane that singularities occur.

Now we consider the projection mapping of the 2-manifold (21.17) down onto the 2-dimensional control plane $\mathbb{R}^2$, defined by

$$\begin{aligned} &(x; a, b) \xrightarrow{\text{proj}} (a, b) \\ &a = \lambda_2 \\ &b = -\lambda_1^3 - \lambda_1\lambda_2 \end{aligned} \tag{21.19}$$

The Jacobian of this mapping is

$$\begin{vmatrix} \dfrac{\partial a}{\partial \lambda_1} & \dfrac{\partial a}{\partial \lambda_2} \\ \dfrac{\partial b}{\partial \lambda_1} & \dfrac{\partial b}{\partial \lambda_2} \end{vmatrix} = \begin{vmatrix} 0 & 1 \\ -3\lambda_1^2 - \lambda_2 & -\lambda_1 \end{vmatrix} = 3\lambda_1^2 + \lambda_2 \tag{21.20}$$

The Jacobian is nonzero, and the mapping invertible, except for $\lambda_2 = -3\lambda_1^2$. This defines a curve

$$(x; a, b) = (\lambda_1; -3\lambda_1^2, 2\lambda_1^3) \tag{21.21}$$

in $\mathbb{R}^3$. This is a smooth curve in $\mathbb{R}^3$ which indicates the set of points at which the tangent plane to the manifold (21.17) is "vertical." The projection of this curve down onto the control plane has a cusp-shaped kink:

$$(\lambda_1; -3\lambda_1^2, 2\lambda_1^3) \xrightarrow{\text{proj}} (a = -3\lambda_1^2, b = 2\lambda_1^3) \tag{21.22}$$

The equation of this cusp-shaped curve is

$$\left(-\frac{a}{3}\right)^{1/2} = \lambda_1 = \left(\frac{b}{2}\right)^{1/3}$$

or

$$\left(-\frac{a}{3}\right)^3 = \left(\frac{b}{2}\right)^2 \tag{21.23}$$

The curve (21.23) is the "shadow" that the fold in the manifold (21.17) would make if it were illuminated from faraway on the x axis.

We emphasize here that the manifold (21.17) is smooth, with no sharp corners or singularities. So also is the curve (21.21). This curve gives the location of all degenerate critical points in the family of functions $V(x; a, b)$. It is only the projection mapping (21.19) which contains singularities. The projection of (21.17) into $\mathbb{R}^2$ divides the control plane into open regions, Region I (V has one Morse critical point) and Region III (V has three Morse critical points). The fold lines, Region II, correspond to potentials having one nondegenerate and one doubly degenerate critical point, and the kink corresponds to the potential $x^4/4$. An arbitrary perturbation of (21.15a) may change the location and orientation of the separatrix, but it cannot change its shape. In short, the singularity in the projection mapping is stable against perturbations. This singularity is a generic property of projections of the critical manifold of 2-parameter families of functions onto the control plane.

5. Thom's Theorem

We are now in a position to state Thom's Theorem. If $f(x; c)$, $x \in \mathbb{R}^n$, $c \in \mathbb{R}^k$, is a smooth function mapping $\mathbb{R}^n \otimes \mathbb{R}^k \to \mathbb{R}^1$, we define the set of critical points $\mathscr{M}_f \subset \mathbb{R}^{n+k}$ as the set of points at which $\nabla_x f = 0$. We define the projection map $\chi_f : \mathscr{M}_f \to \mathbb{R}^k$ as in Sec. 4. We denote by F the set of all smooth functions mapping $\mathbb{R}^n \otimes \mathbb{R}^k \to \mathbb{R}^1$.

Theorem (*Thom*). If $k \leq 5$, there is an open dense set of functions $F_* \subset F$ with the following generic properties (for $f \in F_*$):

1. $\mathcal{M}_f$ is a manifold of dimension k.
2. Any singularity of χ_f is equivalent to an elementary catastrophe.
3. χ_f is locally stable under small perturbations of f.

In a form that is more immediately useful for purposes of applications, Thom's Theorem says that in a typical k-parameter family of functions ($k \leq 5$) of n variables, the only non-Morse critical points that can stably be encountered have the canonical forms listed in Table 2.2.

6. Summary

The essence of Elementary Catastrophe Theory for nonmathematicians is contained in Table 2.2. Nevertheless it would be inappropriate to discuss this subject without at least stating Thom's Theorem. The purpose of the present chapter is to make such a statement possible.

The statement of this theorem involves such concepts as perturbation, stability, genericity, and singularities. We have dealt to some extent with these concepts in previous chapters, but we treat them formally in the present chapter. These concepts are presented in Secs. 1–4, respectively. Once the notation is set, the definitions are made, and examples are given, it is possible at least to state this theorem. This is done in Sec. 5.

Notes

The mathematical concepts presented in this chapter (stability, genericity, and singularities) are well-presented in Refs. 3–7.

References

1. A. A. Andronov and L. S. Pontryagin, Coarse Systems, *Dokl. Akad. Nauk. USSR* **14**, 247–251 (1937).
2. H. J. Sussman, Catastrophe Theory, *Synthèse* **31**, 229–270 (1975).
3. V. I. Arnol'd, Singularities of Smooth Maps, *Russian Math. Surveys* **23**, 1–43 (1969).
4. M. Golubitsky and V. Guillemin, *Stable Mappings and Their Singularities*, New York: Springer, 1973.
5. T. Poston and I. N. Stewart, *Taylor Expansions and Catastrophes*, London: Pitman, 1976.
6. E. C. Zeeman, *Catastrophe Theory, Selected Papers 1972–1977*, Reading: Addison-Wesley, 1977.
7. T. Poston and I. N. Stewart, *Catastrophe Theory and Its Applications*, London: Pitman, 1978.

CHAPTER 22

Transversality

Transversality[1] is an extremely important idea for answering questions about stability and genericity. Two manifolds embedded in $\mathbb{R}^N$ intersect transversally at a point if the direct sum of their tangent spaces at the point has dimension N. Two manifolds intersect transversally if either their intersection is empty or they intersect transversally at all their points of intersection. Similar definitions hold for the intersection of one manifold $\mathscr{Q}$ in $\mathbb{R}^N$ with the image of a second manifold $\mathscr{P}$ which is mapped into $\mathbb{R}^N$ by a mapping $\mathscr{F}$. The power of transversality lies in the following two facts: (1) If $\mathscr{F}(\mathscr{P})$ intersects $\mathscr{Q}$ transversally in $\mathbb{R}^N$, so also does any perturbation of $\mathscr{F}$ (stability), and (2) the set of mappings $\mathscr{F}$ of $\mathscr{P}$ into $\mathbb{R}^N$ for which $\mathscr{F}(\mathscr{P})$ is transversal to $\mathscr{Q}$ is dense in the set of all mappings of $\mathscr{P}$ into $\mathbb{R}^N$ (genericity). We discuss transversality in Sec. 1, giving numerous examples to illustrate this concept.

The consequences of transversality bear important fruit when we identify $\mathbb{R}^N$ with the finite-dimensional linear vector space of p-jets of functions j^pf discussed in Chapter 21, Sec. 1.

In Sec. 2 we choose $\mathscr{Q}$ to be the linear vector subspace in $\mathbb{R}^N$ in which the coefficients of the linear terms in j^pf vanish. We also choose $\mathscr{P} = \mathbb{R}^n$ (state variable space) and $\mathscr{F}(\mathscr{P})$ to be the n-dimensional submanifold in $\mathbb{R}^N$ whose coordinates are parameterized by the point $x^0 \in \mathbb{R}^n$ about which the Taylor series expansion is made. The transversality theorem shows immediately that it is a generic property of functions to have only isolated critical points, and such functions are stable under perturbations.

In Sec. 3 and those that follow we choose $\mathscr{P} = \mathbb{R}^k$ (control parameter space) and $\mathscr{F}(\mathscr{P})$ as the k-dimensional manifold in $\mathbb{R}^N$ whose coordinates are the Taylor series coefficients of $j^p f$ evaluated at a critical point. In Sec. 3 itself we choose $\mathscr{Q} = V_l$, the set of points in $\mathbb{R}^N$ on which exactly l eigenvalues of $\partial^2 f/\partial x_i \partial x_j$ vanish. Then dim $V_l = N - l(l+1)/2$ and dim $\mathscr{F}(\mathscr{P}) = k$, so that non-Morse critical points with l "bad" state variables can only be encountered stably in k-parameter families of functions if $k \geq l(l+1)/2$.

In Secs. 4–7 we apply the same methods to study precisely which terms in the early part of a Taylor series expansion can stably be expected to vanish in a k-parameter family of functions at a non-Morse critical point with l "bad" state variables.

1. Transversality

Transverse means cross.

This notion can be made mathematically precise for curves or more generally for manifolds, or for mappings. Three important situations that we encounter occur when:

1. A manifold is transverse to another manifold.
2. A mapping is transverse to a manifold.
3. A mapping is transverse to another mapping.

We describe these situations now.

Case 1. Suppose that $\mathscr{P}$ is a manifold of dimension p in $\mathbb{R}^N$, and $\mathscr{Q}$ is a manifold of dimension q, also embedded in $\mathbb{R}^N$. These manifolds may or may not intersect. If they intersect at a point $x^0 \in \mathbb{R}^N$, we say they intersect **transversally at x^0**, or are **transverse at x^0**, if the tangent spaces to the two manifolds span the tangent space $\mathbb{R}^N$ at x^0.

Example 1 (Fig. 22.1). The two curves $\mathscr{P}$ and $\mathscr{Q}$ in $\mathbb{R}^2$, both of which are 1-dimensional manifolds, intersect transversally at points a and b, but not at point c (Fig. 22.1*a*). If either of the two manifolds $\mathscr{P}$ or $\mathscr{Q}$ is perturbed slightly, the point of double degeneracy c will either split into two isolated points of transverse contact c_1, c_2 (Fig. 22.1*b*) or else will fail to make contact (Fig. 22.1*c*).

Example 2 (Fig. 22.2). The curve $\mathscr{P}$ intersects the sphere $\mathscr{Q}$ in $\mathbb{R}^3$ transversally at points a and b but not at point c.

Example 3 (Fig. 22.3). In $\mathbb{R}^3$ the sphere $\mathscr{P}$ and the sphere $\mathscr{Q}$ intersect in a circle, which is a 1-dimensional manifold. The two spheres intersect transversally at every point on this circle. If the radii and centers of the two spheres are changed slightly, they will still intersect transversally in a circle, and further the new intersection circle will be near the original intersection circle.

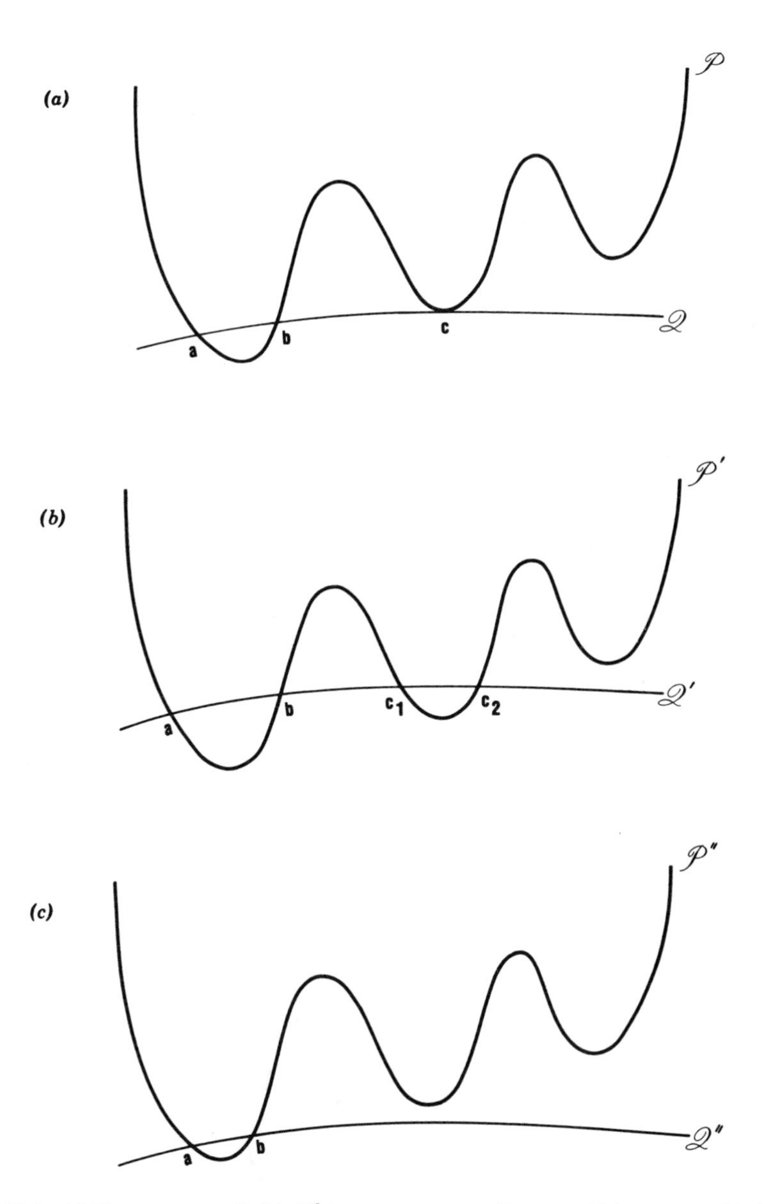

Figure 22.1 (a) The two curves $\mathscr{P}$, $\mathscr{Q}$ in $\mathbb{R}^2$ intersect transversally at a and b but not at c. Under an arbitrary perturbation, the perturbed curves $\mathscr{P}'$, $\mathscr{Q}'$ will intersect transversally at all points of contact with probability 1 (b), (c).

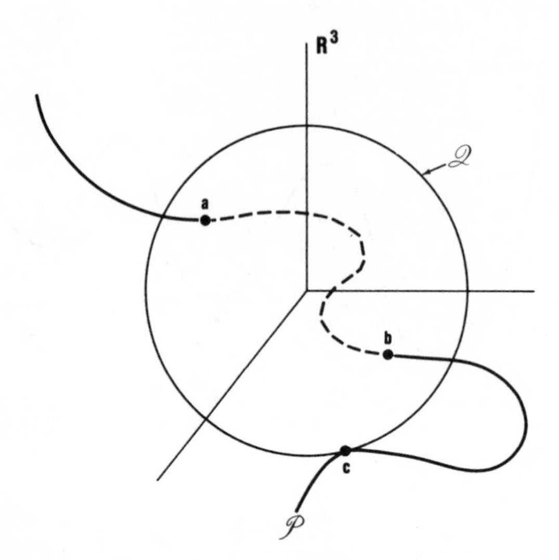

Figure 22.2 The points of contact (a, b) at which the curve $\mathscr{P}$ pierces the sphere surface $\mathscr{Q}$ in $\mathbb{R}^3$ are points of transversal intersection. These two manifolds do not intersect transversally at c.

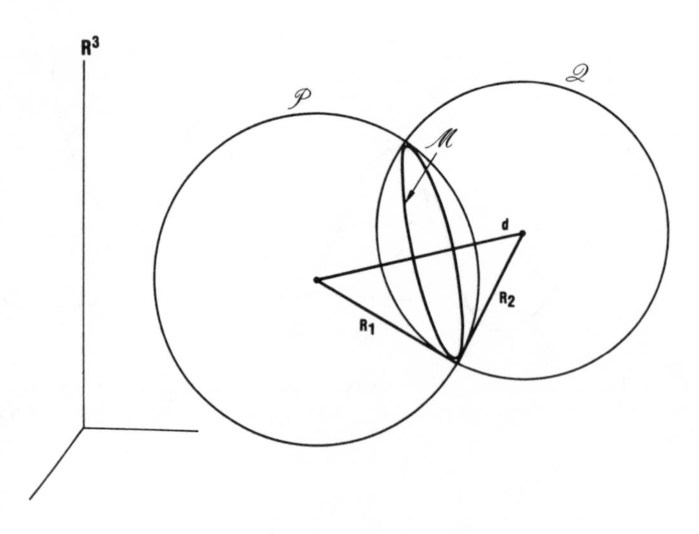

Figure 22.3 Two spheres $\mathscr{P}$ and $\mathscr{Q}$ in $\mathbb{R}^3$ intersect transversally in a circle $\mathscr{M}$ provided $R_1 + R_2 > d > |R_1 - R_2|$. The point of contact is not transverse if either inequality is replaced by an equality. No intersection occurs if either inequality is reversed.

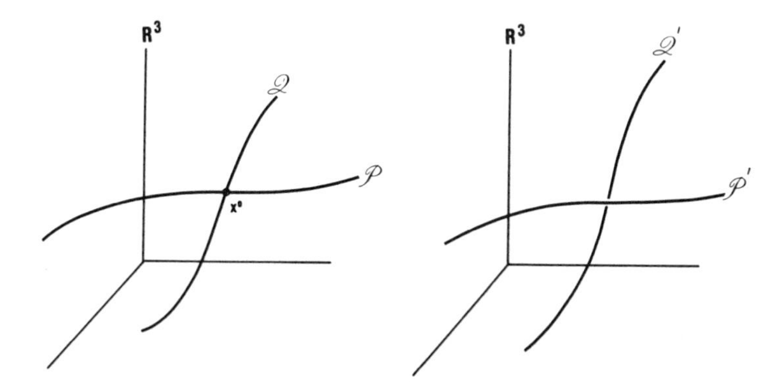

Figure 22.4 Two curves in $\mathbb{R}^3$ cannot intersect transversally.

Example 4 (*Fig. 22.4*). The 1-dimensional manifolds $\mathscr{P}$ and $\mathscr{Q}$ intersect at x^0 in $\mathbb{R}^3$. Since the tangent space to $\mathscr{P}$ at x^0 is 1-dimensional, as is the tangent space to $\mathscr{Q}$ at x^0, there is no way $\mathscr{P}$ and $\mathscr{Q}$ can intersect transversally in $\mathbb{R}^3$. If either $\mathscr{P}$ or $\mathscr{Q}$ is perturbed slightly, these manifolds will not intersect at all.

Example 5 (*Fig. 22.5*). The point $\mathscr{P}$ is a 0-dimensional manifold, while $\mathscr{Q}$ is a 2-dimensional manifold embedded in $\mathbb{R}^3$. $\mathscr{P}$ and $\mathscr{Q}$ cannot intersect transversally because the tangent spaces have dimensions 0 and 2. The tangent spaces cannot span $\mathbb{R}^3$, of dimension 3. If $\mathscr{Q}$ did intersect $\mathscr{P}$, a perturbation of either would result in a situation in which $\mathscr{P}$ and $\mathscr{Q}$ no longer intersected.

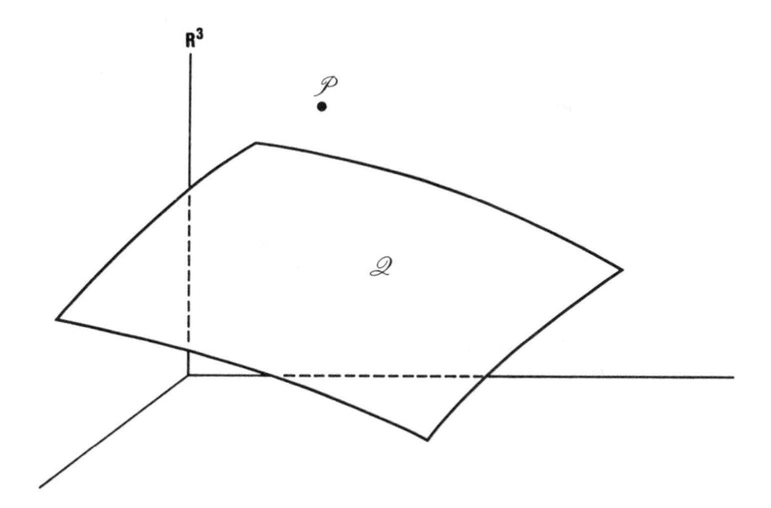

Figure 22.5 A point in $\mathbb{R}^n$ cannot be transverse to a manifold of dimension less than n in $\mathbb{R}^n$.

Definition. Two manifolds $\mathscr{P}$ and $\mathscr{Q}$ in $\mathbb{R}^N$ **intersect transversally**, or **are transverse**, if

1. They intersect transversally at all points of intersection, or
2. They do not intersect at all.

If $\dim \mathscr{P} + \dim \mathscr{Q} = p + q < N$, then the tangent spaces to $\mathscr{P}$ and $\mathscr{Q}$ at any point of intersection $x^0 \in \mathbb{R}^N$ cannot span $\mathbb{R}^N$. Therefore they can only be transverse if they do not intersect at all.

The important points about transversality of manifolds are as follows:

1. If $\mathscr{P}$ is transverse to $\mathscr{Q}$ in $\mathbb{R}^N$, then
   a. If $p + q < N$, $\mathscr{P}$ and $\mathscr{Q}$ do not intersect at all.
   b. If $p + q \geq N$, $\mathscr{P}$ and $\mathscr{Q}$ either do not intersect, or else intersect in a manifold of dimension $p + q - N \geq 0$.
2. Suppose that $\mathscr{P}$ and $\mathscr{Q}$ are perturbed slightly to manifolds $\mathscr{P}'$ and $\mathscr{Q}'$. Then if $\mathscr{P}$ is transverse to $\mathscr{Q}$, $\mathscr{P}'$ is transverse to $\mathscr{Q}'$. If $\mathscr{P}$ and $\mathscr{Q}$ do not intersect, $\mathscr{P}'$ and $\mathscr{Q}'$ do not intersect. If $\mathscr{P}$ and $\mathscr{Q}$ intersect transversally in a manifold $\mathscr{M}$, then $\mathscr{P}'$ and $\mathscr{Q}'$ intersect transversally in a manifold $\mathscr{M}'$ which is a perturbation of $\mathscr{M}$.
3. Suppose that $\mathscr{P}$ and $\mathscr{Q}$ are perturbed slightly to manifolds $\mathscr{P}'$ and $\mathscr{Q}'$. If $\mathscr{P}$ is not transverse to $\mathscr{Q}$, $\mathscr{P}'$ is transverse to $\mathscr{Q}'$.

Thus transversality is a generic property of manifolds. These concepts have been illustrated in Figs. 22.1–22.5.

Case 2. Suppose $\mathscr{Q}$ is a manifold of dimension q in $\mathbb{R}^N$ and $\mathscr{P}$ is a manifold of dimension p and *not* in $\mathbb{R}^N$. Suppose further that $\mathscr{F}$ maps $\mathscr{P}$ into $\mathbb{R}^N$. If $\mathscr{F}(\mathscr{P})$ and $\mathscr{Q}$ intersect at $x^0 \in \mathbb{R}^N$, the intersection at x^0 is transverse if the tangent spaces to $\mathscr{F}(\mathscr{P})$ and $\mathscr{Q}$ at x^0 span $\mathbb{R}^N$.

Definition. $\mathscr{F}$ is **transverse** to $\mathscr{Q}$ in $\mathbb{R}^N$ if[1]

1. $\mathscr{F}(\mathscr{P})$ is transverse to $\mathscr{Q}$ at all points of intersection, or
2. $\mathscr{F}(\mathscr{P})$ does not intersect $\mathscr{Q}$.

Example 6 (Fig. 22.6). We let $\mathscr{Q}$ be the 1-manifold $(1, 2, z)$ in $\mathbb{R}^3$, and $\mathscr{P}$ the 2-manifold $\mathbb{R}^2$. We map $\mathscr{P}$ into $\mathbb{R}^3$ using the mapping

$$(x, y) \in \mathbb{R}^2 \xrightarrow{\mathscr{F}} (2x, 3y, e^{-(x^2+y^2)}) \in \mathbb{R}^3$$

Then $\mathscr{F}(\mathscr{P})$ and $\mathscr{Q}$ intersect at the point $x^0 = (1, 2, e^{-0.69\dot{4}})$ in $\mathbb{R}^3$. The tangent

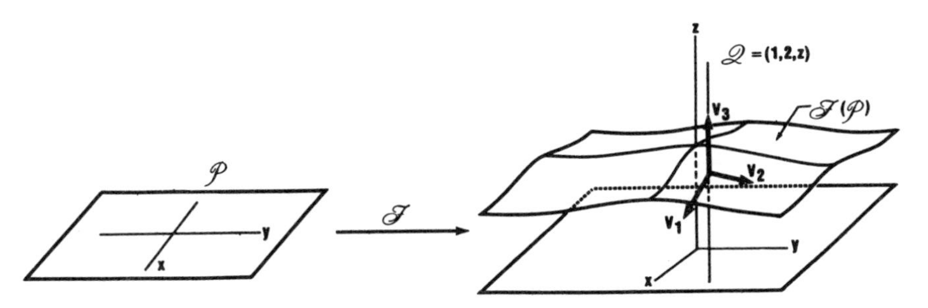

Figure 22.6 The vertical line $\mathscr{Q}$ intersects the surface $\mathscr{F}(\mathscr{P})$ transversally at a point.

space to $\mathscr{Q}$ at x^0 has basis vector $v_3 = (0, 0, 1)$. The tangent space to $\mathscr{F}(\mathscr{P})$ at x^0 has basis vectors

$$v_1 = \frac{\partial}{\partial x}(2x, 3y, e^{-(x^2+y^2)})|_{x^0} = (2, 0, -e^{-0.69\dot{4}})$$

$$v_2 = \frac{\partial}{\partial y}(2x, 3y, e^{-(x^2+y^2)})|_{x^0} = (0, 3, -\tfrac{4}{3}e^{-0.69\dot{4}})$$

Since

$$\det\begin{bmatrix} 2 & 0 & -e^{-0.69\dot{4}} \\ 0 & 3 & -\frac{4}{3}e^{-0.69\dot{4}} \\ 0 & 0 & 1 \end{bmatrix} = 6 \neq 0$$

the vectors v_1, v_2, v_3 span $\mathbb{R}^3$. Therefore $\mathscr{F}(\mathscr{P})$ is transverse to $\mathscr{Q}$ at their point of intersection. Under a perturbation the intersection remains transverse.

Example 7 (Fig. 22.7). We let $\mathscr{Q}$ be the x axis in $\mathbb{R}^2$ and map $\mathscr{P} = \mathbb{R}^1$ into $\mathbb{R}^2$ by the mapping

$$t \in \mathbb{R}^1 \overset{\mathscr{F}}{\to} (t, t) \in \mathbb{R}^2$$

The tangent vector to $\mathscr{Q}$ lies along the x axis. The tangent space to $\mathscr{F}(\mathbb{R}^1)$ at the point of intersection (0, 0) has basis vector (1, 1). These two vectors span $\mathbb{R}^2$, so the mapping $\mathscr{F}$ is transverse to the x axis.

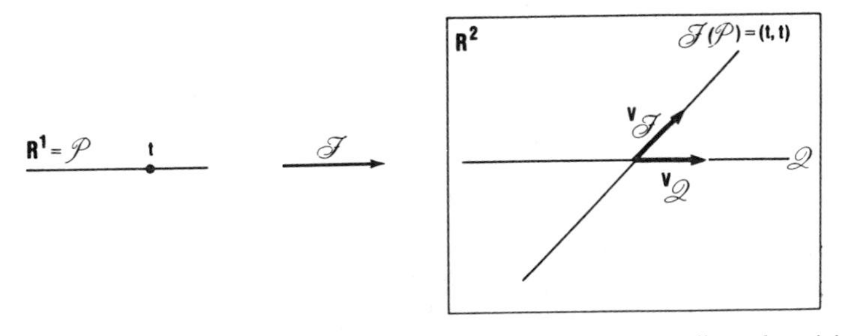

Figure 22.7 The curve $\mathscr{F}(\mathscr{P}) = (t, t)$ intersects the x axis transversally at the origin.

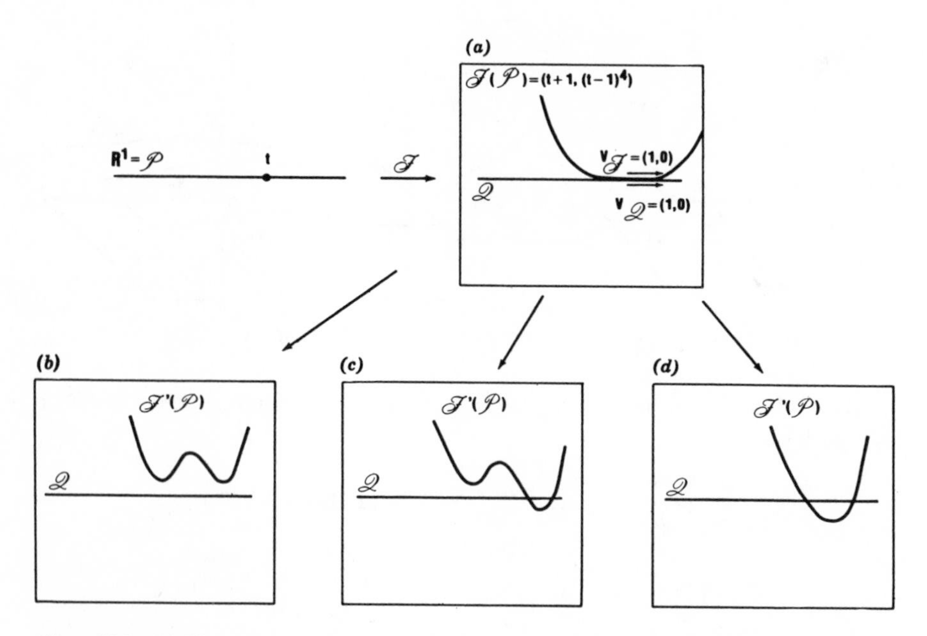

Figure 22.8 (*a*) The curve $\mathcal{F}(\mathcal{P})$ does not intersect the x axis transversally. There is a quartic osculation. The perturbed curve may intersect the x axis transversally at four points, at two points (*d*, *c*), or not at all (*b*).

Example 8 (*Fig. 22.8*). We let $\mathcal{Q}$ be the x axis in $\mathbb{R}^2$, let $\mathcal{P}$ be $\mathbb{R}^1$, and define $\mathcal{F}$ by

$$t \in \mathbb{R}^1 \rightarrow (\mathcal{F}_1(t), \mathcal{F}_2(t)) = (t+1, (t-1)^4) \in \mathbb{R}^2$$

Then there is one point of contact, at $x^0 = (2, 0)$. The tangent space to $\mathcal{Q}$ at $(2, 0)$ has basis vector $(1, 0)$. The tangent space to $\mathcal{F}(\mathbb{R}^1)$ at x^0 has tangent vector

$$\frac{d}{dt}(\mathcal{F}_1(t), \mathcal{F}_2(t))|_{t=1} = (1, 4(t-1)^3)|_{t=1} = (1, 0)$$

These two tangent vectors are parallel, so do not span $\mathbb{R}^2$ at $(2, 0)$. A slight perturbation of the mapping $\mathcal{F}$ will result in a transverse crossing as shown in Fig. 22.8 (b), (c), or (d).

Example 9 (*Fig. 22.9*). We again take $\mathcal{Q}$ as the x axis in $\mathbb{R}^2$, $\mathcal{P} = \mathbb{R}^1$, but this time choose $\mathcal{F}$ as follows:

$$t \in \mathbb{R}^1 \xrightarrow{\mathcal{F}} (\mathcal{F}_1(t), \mathcal{F}_2(t)) = (t^3, t^3) \in \mathbb{R}^2$$

The tangent vector of $\mathcal{F}(\mathbb{R}^1)$ at the single point of intersection is

$$\frac{d}{dt}(t^3, t^3)|_{t=0} = (3t^2, 3t^2)|_{t=0} = (0, 0)$$

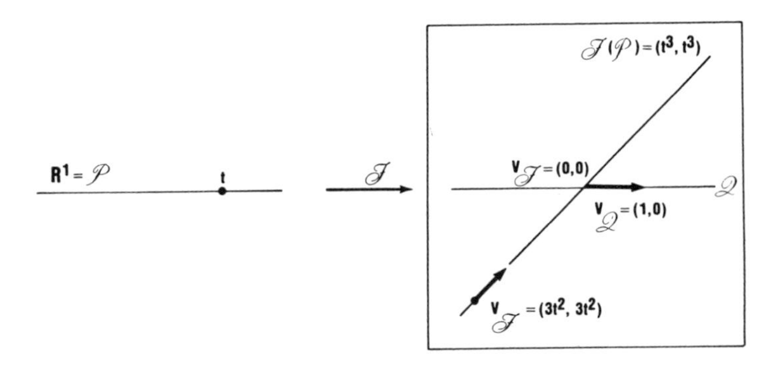

Figure 22.9 The curve $\mathscr{F}(\mathscr{P}) = (t^3, t^3)$ intersects the x axis at the origin, but the intersection is not transversal because the tangent vector (= velocity vector) $(3t^2, 3t^2)$ vanishes there.

As a result, $\mathscr{F}$ is not transverse to $\mathscr{Q}$.

Example 10. We let $\mathscr{Q}$ be a q-dimensional manifold in $\mathbb{R}^N$, and $\mathscr{P}$ be a p-dimensional manifold. Then the mapping $\mathscr{F}$ is given by

$$\mathscr{F}(\mathscr{P}) \to \mathbb{R}^N: \quad (\mathscr{F}_1(y_1, \ldots, y_p), \mathscr{F}_2(y_1, \ldots, y_p), \ldots, \mathscr{F}_N(y_1, \ldots, y_p))$$

If $\mathscr{F}(\mathscr{P})$ and $\mathscr{Q}$ intersect at $x^0 \in \mathbb{R}^N$ $[\mathscr{F}(y^0) = x^0]$, then the tangent space to $\mathscr{F}(\mathscr{P})$ at x^0 is spanned by the p vectors

$$\begin{aligned} v_1 &= \left(\frac{\partial \mathscr{F}_1}{\partial y_1}, \ldots, \frac{\partial \mathscr{F}_N}{\partial y_1}\right) \\ &\vdots \\ v_p &= \left(\frac{\partial \mathscr{F}_1}{\partial y_p}, \ldots, \frac{\partial \mathscr{F}_N}{\partial y_p}\right) \end{aligned} \tag{22.1v}$$

The q-dimensional tangent space to $\mathscr{Q}$ at x^0 is spanned by a q-tuple of vectors $u_1, \ldots, u_q$:

$$\begin{aligned} u_1 &= (u_{11}, u_{12}, \ldots, u_{1N}) \\ &\vdots \\ u_q &= (u_{q1}, u_{q2}, \ldots, u_{qN}) \end{aligned} \tag{22.1u}$$

The vectors $v_1, \ldots, v_p, u_1, \ldots, u_q$ span $\mathbb{R}^N$ at x^0 if the $(p + q) \times N$ matrix whose first p rows are given in (22.1v) and whose last q rows are given in (22.1u) contains a nonsingular $N \times N$ submatrix. This is clearly impossible if $p + q < N$.

The important points about the transversality of mappings to manifolds are as follows:[1,2]

1. If $\mathscr{F}(\mathscr{P})$ is transverse to $\mathscr{Q}$ in $\mathbb{R}^N$, then
   a. If $p + q < N$, $\mathscr{F}(\mathscr{P})$ and $\mathscr{Q}$ do not intersect at all.
   b. If $p + q \geq N$, $\mathscr{F}(\mathscr{P})$ and $\mathscr{Q}$ either do not intersect, or else intersect in a manifold of dimension $p + q - N \geq 0$.
2. If $\mathscr{F}(\mathscr{P})$ is transverse to $\mathscr{Q}$ and $\mathscr{F}$ is perturbed slightly to $\mathscr{F}'$, then $\mathscr{F}'(\mathscr{P})$ is transverse to $\mathscr{Q}$. If $\mathscr{F}(\mathscr{P})$ and $\mathscr{Q}$ do not intersect, $\mathscr{F}'(\mathscr{P})$ and $\mathscr{Q}$ do not intersect. If $\mathscr{F}(\mathscr{P})$ and $\mathscr{Q}$ intersect transversally in manifold $\mathscr{M}$, $\mathscr{F}'(\mathscr{P})$ and $\mathscr{Q}$ intersect transversally in a manifold $\mathscr{M}'$ which is a perturbation of $\mathscr{M}$.
3. The set of mappings $\mathscr{F}: \mathscr{P} \to \mathbb{R}^N$ which are transverse to $\mathscr{Q}$ is dense in the set of all mappings of $\mathscr{P}$ into $\mathbb{R}^N$. This means that any mapping $\mathscr{F}: \mathscr{P} \to \mathbb{R}^N$ (whether or not transverse to $\mathscr{Q}$) can be approximated arbitrarily closely by transverse mappings.

These concepts have been illustrated in Figs. 22.6–22.9.

Case 3. If $\mathscr{F}$ maps the p-manifold $\mathscr{P}$ into $\mathbb{R}^N$ and $\mathscr{G}$ maps the q-manifold $\mathscr{Q}$ into $\mathbb{R}^N$, we can discuss whether or not the mapping $\mathscr{F}$ is transverse to the mapping $\mathscr{G}$. This discussion generally follows the lines indicated in Case 2 and leads to similar conclusions. We will not carry out that discussion here.

Important Observation. The idea of transversality, particularly of mappings to manifolds (Case 2), leads to very important results when the Euclidean space $\mathbb{R}^N$ is the space of k-jets $j^k f \simeq \mathbb{R}^D$ discussed in Chapter 21. We apply the results discussed in this section to the structure established in Chapter 21 to derive results in the following two sections that closely parallel the material discussed in Chapter 2, Secs. 1 and 2.

2. Application 1: Linear Terms

For our first application we will consider the critical points of a smooth function f of n variables, and show that the critical points are isolated.

If we compute the p-jet of f according to (21.2), then the $D = (n + p)!/n!p!$ coordinates $f(x^0), f_i(x^0), f_{ij}(x^0), \ldots, f_{ij\cdots p}(x^0)$ are functions of the point $x^0 \in \mathbb{R}^n$ about which the Taylor series expansion has been made. Thus the D coordinates of n variables provide a parametric representation of an n-dimensional manifold in $\mathbb{R}^D$. An alternative and equivalent viewpoint is that the mapping $j^p f: \mathbb{R}^n \to \mathbb{R}^D$ given by $x^0 \in \mathbb{R}^n \to j^p f(x^0)$ is analogous to the mapping $\mathscr{F}(\mathscr{P})$ of Sec. 1, where now $\mathscr{P} = \mathbb{R}^n$.

The critical points of f occur where $\nabla f = 0$, that is, for $f_1(x^0) = \cdots = f_n(x^0) = 0$. So the critical points of f occur on the subspace in $\mathbb{R}^D$ in which all first derivatives vanish. This linear vector subspace of $\mathbb{R}^D$ has dimension $D - n$ and corresponds to the submanifold $\mathcal{Q}$. Since

$$\dim(\mathcal{P} = \mathbb{R}^n) + \dim(\mathcal{Q} = \mathbb{R}^{D-n}) \geq \dim(j^p f = \mathbb{R}^D)$$

$$n \qquad + \qquad D - n \qquad \geq \qquad D$$

the manifold $\mathcal{F}(\mathcal{P}) = j^p f(x^0)$ intersects the manifold $\mathcal{Q}$ in $\mathbb{R}^D$ transversely in a manifold of dimension $n + (D - n) - D = 0$, that is, in isolated points. This means that the typical critical points of $f(x)$ are isolated. Further, by the properties of transversal mappings (Sec. 1), if $f(x)$ is perturbed slightly, the critical points are moved slightly, but the intersection is still transverse. Thus, perturbation of a Morse function leads to a Morse function. Finally since transverse mappings are dense, any function $f: \mathbb{R}^n \to \mathbb{R}^1$ can be approximated arbitrarily closely by functions with isolated critical points.

3. Application 2: Quadratic Terms

For our second application we will consider the occurrence of non-Morse critical points in families of functions of n state variables and k control parameters. We will see that the number of eigenvalues of the stability matrix which can generically vanish at a point depends crucially on k.

The coordinates $f(x^0; c^0), f_i(x^0; c^0), f_{ij}(x^0; c^0), \ldots$ which appear in (21.1) are now functions of the $n + k$ variables $(x^0; c^0) \in \mathbb{R}^n \otimes \mathbb{R}^k$. We use up n of these degrees of freedom in locating critical points, as described in Sec. 2. That means that at a critical point, the nonzero coordinates in (21.1) are functions of the k control parameters c_α, $\alpha = 1, 2, \ldots, k$. In particular, at any critical point x^0 the matrix elements of the stability matrix $f_{ij}(c^0)$ are functions of k control parameters. Since there are $n(n + 1)/2$ independent [because $f_{ij} = f_{ji}$] matrix elements, the set

$$\mathcal{F}(k) = \{f_{11}(c^0), f_{12}(c^0), \ldots, f_{nn}(c^0)\} \tag{22.2}$$

is the parametric representation of a k-dimensional manifold in $\mathbb{R}^{n(n+1)/2}$. Alternatively, the quadratic terms (21.2) in the expansion (21.1) provide a mapping $\mathcal{F}$ of the control parameter space $\mathcal{P} = \mathbb{R}^k$ into $\mathbb{R}^N = \mathbb{R}^{n(n+1)/2}$.

Now we look at the subsets of points in $\mathbb{R}^{n(n+1)/2}$ on which $\det f_{ij} = 0$. Since f_{ij} is real and symmetric, $\det f_{ij} = 0$ only if one or more eigenvalues are zero. If we let V_l be the set of points in $\mathbb{R}^{n(n+1)/2}$ on which exactly l eigenvalues of the stability matrix are zero, then V_l is a manifold in $\mathbb{R}^{n(n+1)/2}$ and

$$\dim V_l = \frac{n(n + 1) - l(l + 1)}{2} \tag{22.3}$$

To see this last dimension statement clearly, we recall that a real symmetric matrix can be diagonalized by a real orthogonal transformation, so that

$$\mathcal{O}^{-1} f_{ij}\, \mathcal{O} = \left[\begin{array}{c|c} \begin{matrix}\lambda_1 & & \\ & \ddots & \\ & & \lambda_\ell\end{matrix} & 0 \\ \hline 0 & \begin{matrix}\lambda_{\ell+1} & & \\ & \ddots & \\ & & \lambda_n\end{matrix} \end{array}\right] \longrightarrow \left[\begin{array}{c|c} 0 & 0 \\ \hline 0 & \Delta \end{array}\right] \tag{22.4}$$

Here we have assumed that exactly l eigenvalues of f_{ij} are zero. The diagonal $(n - l) \times (n - l)$ submatrix Δ is determined by $n - l$ parameters. This equation can be inverted to give, using

$$\mathcal{O} = \begin{bmatrix} A & B \\ C & D \end{bmatrix}$$

$$f_{ij} = \begin{bmatrix} A & B \\ C & D \end{bmatrix}\begin{bmatrix} 0 & 0 \\ 0 & \Delta \end{bmatrix}\begin{bmatrix} A^t & C^t \\ B^t & D^t \end{bmatrix} = \begin{bmatrix} B\Delta B^t & B\Delta D^t \\ D\Delta B^t & D\Delta D^t \end{bmatrix} \tag{22.5}$$

Every real orthogonal $n \times n$ matrix is determined by $n(n - 1)/2$ real numbers [the group $O(n)$ has dimension $n(n - 1)/2$]. The independent parameters specifying $\mathcal{O}$ may be taken as those matrix elements above the main diagonal. In this case the submatrix B has dimension $l(n - l)$, the submatrices A and D have dimensions $l(l - 1)/2$ and $(n - l)(n - l - 1)/2$, respectively, and the submatrix C is fixed once A, B, and D have been specified. The $n \times n$ matrix f_{ij} in (22.5) is determined by the submatrices B, D, Δ, so

$$\begin{aligned} \dim f_{ij} &= \dim B + \dim D + \dim \Delta \\ &= l(n - l) + \frac{(n - l)(n - l - 1)}{2} + (n - l) \\ &= \frac{n(n + 1) - l(l + 1)}{2} \end{aligned} \tag{22.6}$$

We are now in a position to determine under what conditions it is typical (generic) for l eigenvalues of the stability matrix to vanish in a k-parameter family of functions. To do this, we use the transversality results of Sec. 1 for mappings, together with the identifications:

$\mathscr{P} = \mathbb{R}^k$	control parameter space
$\mathscr{F}(\mathscr{P})$	given in (22.2)
$\mathscr{Q} = V_l$	
$\mathbb{R}^N = \mathbb{R}^{n(n+1)/2}$	

Then exactly l eigenvalues of $f_{ij}(c)$ vanish if $\mathscr{F}(\mathscr{P})$ intersects V_l anywhere. This property is generic if the intersection is transversal. But $\mathscr{F}(\mathscr{P})$ and $\mathscr{Q}$ intersect transversely in $\mathbb{R}^N$ only if

$$k + \frac{n(n+1) - l(l+1)}{2} \geq \frac{n(n+1)}{2}$$
$$k \geq \frac{l(l+1)}{2} \tag{22.7}$$

We consider now several special cases.

Case 0: $k = 0$ (*No Control Parameters*). In this case locating a critical point fixes all matrix elements f_{ij}. So the stability matrix is represented by a point (0-dimension manifold) in $\mathbb{R}^{n(n+1)/2}$. We have seen already (Example 5 and Fig. 22.5) that there is no way a point can intersect transversely a manifold of dimension lower than $n(n+1)/2$ in $\mathbb{R}^{n(n+1)/2}$. From the properties of transverse mappings we conclude that the functions with Morse critical points at x^0 are dense in the set of all functions with critical points at x^0. Further, if f happens to have a non-Morse critical point at x^0, a slight perturbation will remove that degeneracy.

Case 1: $k = 1$ (*1 Control Parameter*). In this case $\mathscr{F}(\mathscr{P})$ is a 1-dimensional manifold in $\mathbb{R}^{n(n+1)/2}$. Such a manifold can intersect V_1 transversely in manifolds of dimension

$$k - \frac{l(l+1)}{2} = 1 - 1 = 0 \tag{22.8}$$

that is, at isolated points (Fig. 22.10 for $n = 2$). However, $\mathscr{F}(\mathscr{P})$ cannot intersect $V_2, V_3, \ldots$ transversally. If f is perturbed, the curve $\mathscr{F}(\mathscr{P})$ is also perturbed, but meets V_1 transversally near the original intersection points. We see that, in a 1-parameter family of functions, it is a generic property ($\mathscr{F}$ is dense) for a single eigenvalue to vanish at isolated points.

Case 2: $k = 2$ (*2 Control Parameters*). In this case $\mathscr{F}(\mathscr{P})$ is a 2-dimensional manifold in $\mathbb{R}^{n(n+1)/2}$. This manifold can intersect V_1 transversally in 1-dimensional submanifolds but cannot intersect V_2 for dimension reasons:

$$\begin{array}{llllll} & \dim V_l & + \dim \mathscr{F}(\mathscr{P}) & - \dim \mathbb{R}^{n(n+1)/2} & \\ l = 1: & \dfrac{n(n+1)}{2} - 1 + & 2 & - \dfrac{n(n+1)}{2} & = 1 \\ l = 2: & \dfrac{n(n+1)}{2} - 3 + & 2 & - \dfrac{n(n+1)}{2} & < 0 \end{array} \tag{22.9}$$

As a result, in a 2-parameter family of functions one eigenvalue can stably vanish, but not two.

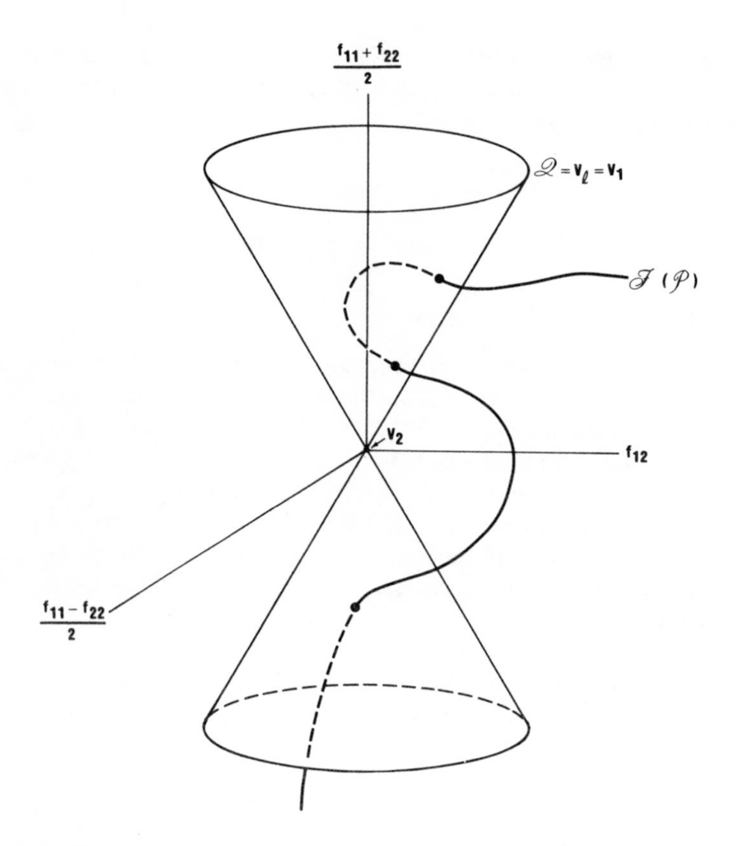

Figure 22.10 When only a single control parameter is present, $\mathscr{F}(\mathscr{P})$ is a 1-dimensional curve. For $n = 2$ the subset of points on which the stability matrix f_{ij} is singular has two components. These are V_1, on which only one eigenvalue is zero, and the apex of the cone V_2, where both eigenvalues vanish. $\mathscr{F}(\mathscr{P})$ can intersect V_1 transversally at isolated points because $1 + 2 - 3 = 0$, but $\mathscr{F}(\mathscr{P})$ cannot intersect V_2 transversally because $1 + 0 - 3 < 0$.

Case 3: $k = 3$ (*3 Control Parameters*). In this case $\mathscr{F}(\mathscr{P})$ is a 3-dimensional manifold which can intersect V_1 transversally in $\mathbb{R}^{n(n+1)/2}$ in 2-dimensional submanifolds, and which can intersect V_2 transversally at isolated points, but which cannot intersect V_3 transversally. As a result, in a 3-parameter family of functions two eigenvalues can stably vanish at isolated points.

Case 4: *k Control Parameters*. In this case $\mathscr{F}(\mathscr{P})$ has dimension k in $\mathbb{R}^{n(n+1)/2}$. Whether $\mathscr{F}(\mathscr{P})$ can intersect V_l transversally is determined by the difference

$$k - \frac{l(l+1)}{2} \tag{22.10}$$

If this number is greater than or equal to zero, $\mathscr{F}(\mathscr{P})$ and V_l can intersect transversally in a manifold of this dimension. If this number is less than zero, then these two manifolds do not intersect transversally. As a consequence, three eigenvalues cannot stably vanish until we reach 6-parameter families of functions, four eigenvalues until 10 parameter families are reached, and so on.

4. Application 3: Germs with $l = 1$

For our third application we will consider the qualitative properties of functions at non-Morse critical points. We will see that the number of terms in the Taylor series expansion which must be kept beyond the quadratic terms depends both on the dimension k of the control parameter space and on the number l of eigenvalues which vanish at the non-Morse critical point.

To begin, we assume that $f(x; c)$ of n state variables and k control parameters has a critical point at x^0 for $c = c^0$ at which exactly l eigenvalues of the stability matrix vanish. Then by the Splitting Lemma (2.3a) we can perform a smooth change of coordinates and consider only the interesting non-Morse function f_{NM} of l state variables $y_1, \ldots, y_l$ and k control parameters. The function f_{NM} may then be Taylor series expanded in powers of $(y - y^0)$, as in (21.1). Since y^0 is a critical point, all first derivatives are zero. When $c = c^0$, all second derivatives also vanish. Since the critical value of f_{NM} at $(y^0; c^0)$ is unimportant, we may set that equal to zero.

Therefore the qualitative nature of $f_{NM}(y; c)$ in the neighborhood of $(y^0; c^0) \in \mathbb{R}^l \otimes \mathbb{R}^k$ is determined by an expansion of type (21.1), beginning with the quadratic terms. We wish to determine how many of the early coefficients in the Taylor series expansion may stably vanish in a k-parameter family of functions. This is done by constructing the p-jet of f. This has dimension $D = (l + p)!/l!p! - l - 1$, since at the critical point $f_i(y^0) = 0$ (l constraints) and the critical value $f(y^0) = 0$ (1 constraint). Then the remaining coefficients $f_{ij}(c)$, $f_{ijk}(c)$, $\ldots$, $f_{ijk\cdots p}(c)$ provide a parametric representation for a k-dimensional manifold in $\mathbb{R}^D$.

In the case $l = 1$ only one eigenvalue vanishes, so there is only one "bad" variable y_1 which we relabel x for convenience. We also set $x^0 = 0$ for convenience. The analog of (21.1) is

$$j^p f_{NM}(x; c) = \frac{1}{2!} f_2(c) x^2 + \frac{1}{3!} f_3(c) x^3 + \cdots + \frac{1}{p!} f_p(c) x^p \tag{22.11}$$

The set of $p - 1$ coefficients

$$\mathscr{F}(\mathscr{P}) = \{f_2(c), f_3(c), \ldots, f_p(c)\} \tag{22.12}$$

is a parametric representation of a k-dimensional manifold in $\mathbb{R}^D = \mathbb{R}^{p-1}$. We define the linear vector subspaces T_j in $\mathbb{R}^{p-1}$ as follows:

$$\begin{aligned} T_2 &= \{0, f_3, f_4, \ldots, f_p\} \\ T_3 &= \{0, 0, f_4, \ldots, f_p\} \\ &\;\vdots \\ T_j &= \{0, 0, \ldots, 0, f_{j+1}, \ldots, f_p\} \end{aligned} \tag{22.13}$$

Then $\mathscr{F}(\mathscr{P})$ can intersect $\mathscr{Q} = T_j$ transversally in $\mathbb{R}^N = \mathbb{R}^{p-1}$ if

$$\begin{aligned} \dim \mathscr{F}(\mathscr{P}) + \dim T_j &\geq \dim \mathbb{R}^{p-1} \\ k + (p - j) &\geq p - 1 \end{aligned} \tag{22.14}$$

In short, the Taylor series coefficients $(f_0, f_1) f_2, \ldots, f_{j=k+1}$ can all vanish stably in a k-parameter family of functions, but the coefficients $f_2, \ldots, f_{j=k+2}$ cannot all vanish stably in a k-parameter family. In such a family the worst that f can behave at a non-Morse critical point is, neglecting factorial coefficients,

$$f(x) = f_{k+2} x^{k+2} + f_{k+3} x^{k+3} + \cdots \tag{22.15}$$

In a 1-parameter family of functions the leading nonvanishing term in the Taylor series expansion has degree ≤ 3; in a 2-parameter family it is x^4 (or x^3 or x^2 or x), and so on.

5. Application 4: Germs with $l = 2$

In this case two eigenvalues vanish. If we label the "bad" variables $y_1 = x$ and $y_2 = y$ and take the critical point at the origin for convenience, then the analog of (21.1) is, neglecting factorials,

$$f(x, y; c) = f_{20}(c)x^2 + f_{11}(c)xy + f_{02}(c)y^2 + f_{30}(c)x^3 + \cdots \tag{22.16}$$

Truncating beyond terms of degree p, there are $D = (p + 2)(p + 1)/2 - 3$ coefficients, starting with the quadratic terms. Thus the set

$$\mathscr{F}(\mathscr{P}) = \{f_{20}, f_{11}, f_{02}; f_{30}, f_{21}, \ldots\} \tag{22.17}$$

is a manifold of dimension k in $\mathbb{R}^D$. If $k \geq 3$, the three quadratic coefficients can vanish stably, so the leading terms are cubics. We have already seen in Chapter 3, Sec. 6, how the terms of degree 3 and higher can be put into a canonical form, provided that the coefficients of the cubic terms do not also all vanish.

6. Application 5: Germs with $l = 2, k \geq 7$

In this case it can stably happen that the coefficients of the three quadratic and the four cubic terms in the expansion (22.16) all vanish. Then the leading terms in the Taylor series expansion of $f(x, y; c)$ around the critical point are of degree 4. In this case the best we can do for a canonical form by means of a smooth change of variables is

$$f(x, y; c) \doteq \pm x'^4 + a x'^2 y'^2 \pm y'^4, \qquad a^2 - 4 \neq 0 \tag{22.18}$$

where a cannot be given a canonical value. We have seen already (Chapter 3, Sec. 7, and Table 3.1) that germs with $l = 2, k > 6$ depend on moduli.

7. Application 6: Germs with $l = 3$, $k \geq 6$

In the case $l = 3$ the Taylor series expansion of $f(x, y, z; c)$ about a critical point has six quadratic terms, ten cubic terms, and so on. By now-standard arguments, all six quadratic coefficients can stably vanish in families with $k \geq 6$. In the case $k = 6$ the ten remaining cubics can be transformed to a canonical form that depends on one modulus [cf. (3.49)].

8. Summary

The idea of transversality (Sec. 1), when applied to the space of truncated Taylor series expansions (Chapter 21, Sec. 1) leads to very important information about both functions and families of functions. The density, stability, and perturbation consequences of the transversality of manifolds and mappings provide a rigorous underpinning for the more intuitive arguments of Chapter 3, especially those surrounding (3.28).

When the concept of transversality was applied to the terms of degree 1 in the Taylor series expansion of a function (Sec. 2), we learned that having isolated critical points was a generic and stable property of functions.

When we discussed terms of degree 2 in the Taylor series expansion of a function about a critical point (Sec. 3), we proved that non-Morse functions with l "bad" variables could occur stably in k-parameter families of functions provided $k \geq l(l + 1)/2$.

The transversality arguments also give information about the structure of the germ of a non-Morse function at a degenerate critical point. For example, for $l = 1$ in a k-parameter family of functions it is possible for germs of the form x^j ($j \leq k + 2$) to occur stably, but the germ x^{k+3} cannot occur generically (Sec. 4). The case $l = 2$ cannot occur stably in a k-parameter family unless $k \geq 3$. In this case the germ depends on the cubic (or higher degree) terms. Further, a change of variables can bring these terms into some canonical form. The effects of a change of variables were discussed in Chapters 3 and 4, and will be discussed in a more systematic form in the following chapter. Suffice it to say that all the quadratic and cubic terms can stably vanish in $k \geq 7$-parameter families of functions. The remaining germ must then depend on at least one modulus (Sec. 6). The case $l = 3$ can stably occur only in $k \geq 6$-parameter families of functions, in which case the germ also depends on at least one modulus (Sec. 7). The other interesting cases studied in Chapter 3 (Sec. 9) can be studied rigorously by the methods described here.

Notes

Transversality is a particularly useful and important device for treating the stability and density of mappings and functions. In addition to the references cited within this chapter, transversality is discussed in Refs. 3–6.

References

1. R. Thom, Les singularités des applications differentiables, *Ann. Inst. Fourier* **6**, 43–87 (1955–1956).
2. M. Golubitsky and V. Guillemin, *Stable Mappings and Their Singularities*, New York: Springer, 1973.
3. R. Abraham and J. Robbin, *Transversal Mappings and Flows*, New York: Benjamin, 1967.
4. E. C. Zeeman, *Catastrophe Theory, Selected Papers 1972–1977*, Reading: Addison-Wesley, 1977.
5. T. Poston and I. N. Stewart, *Catastrophe Theory and Its Applications*, London: Pitman, 1978.
6. Y.-C. Lu, *Singularity Theory and an Introduction to Catastrophe Theory*, New York: Springer, 1976.

CHAPTER 23

Determinacy and Unfolding

The qualitative properties of a function in the neighborhood of a particular point in the space of state variables $\mathbb{R}^n$ can often be determined by studying the early part of its Taylor series expansion about that point

$$f(x) = f(0) + x_i f_i(0) + \frac{1}{2!} x_i x_j f_{ij}(0) + \cdots \tag{23.1}$$

We take the point to be the origin for convenience throughout this chapter.

To determine the local properties of $f(x)$ it is not necessary to consider the constant term in (23.1).

If $\nabla f \neq 0$, the qualitative properties of $f(x)$ are completely determined by the linear terms in (23.1). As a result, no information is lost by truncating (23.1) after the linear terms.

If $\nabla f = 0$ but $\det f_{ij} \neq 0$, the qualitative properties of $f(x)$ are completely determined by the quadratic terms in (23.1). As a result, no information is lost by truncating (23.1) beyond the quadratic terms.

If $\nabla f = 0$ and $\det f_{ij} = 0$, the higher terms are important, and one might hope that the qualitative properties of f are completely determined if (23.1) is truncated beyond terms of degree p, for some finite p. Determining whether a function $f(x)$ can in fact be truncated and, if so, for what value of p its Taylor series expansion can be truncated, is called the problem of determinacy. We study this in Sec. 2.

In fact, we do not simply truncate Taylor series expansions. We determine whether a smooth change of variables can transform away the tail of the Taylor series expansion. This was done in Chapter 3, but we simplify the procedure considerably in this chapter by considering only infinitesimal nonlinear transformations (Sec. 1).

The early terms in the Taylor series expansion of a function can naturally vanish when that function belongs to a family of functions. The problem of determining the most general family of functions of the smallest dimension which contains the original function is called the problem of unfolding. This is discussed in Sec. 3.

The examples given to illustrate the problems of determinacy and unfolding suggest natural algorithms for computing the determinacy and the unfolding of an arbitrary function. These two algorithms, given in Secs. 4 and 5, are two sides of the same coin. The coin itself, the canonical catastrophe germ, is represented by the terms not generated in either the determinacy or the unfolding algorithm. An algorithm for computing the germ of a function is presented in Sec. 6.

These algorithms are applied in Secs. 6–11 to view old problems in a new way, and to solve new problems.

In Sec. 12 we study multiple cusp catastrophes. These germs are of importance for studying the failure modes of optimized structures. The determinacy and the universal perturbations of these germs are computed. In the final section we use the rules of Secs. 4–6 to compute a complete list of simple germs and their universal perturbations.

1. Change of Variables

The most important tool used in Chapters 3 and 4 was the general nonlinear change of variables (3.1). With this tool we were able to make precise statements about the equivalence (qualitative similarity) of two functions, to supply constructive proofs of the Implicit Function Theorem, the Morse Lemma, the Splitting Lemma, and to determine both germs and perturbations of some non-Morse functions. However, the method used in those Chapters was unwieldy.

It is possible to drastically simplify the types of computations carried out in Chapters 3 and 4 by using infinitesimal rather than finite nonlinear coordinate transformations. The infinitesimal version of (3.1) is obtained by replacing all finite disposable parameters A in (3.2) by corresponding infinitesimal quantities:

$$\begin{aligned} x_i' &= x_i + \delta x_i \\ \delta x_i &= \delta A_i + \delta A_{i,j} x_j + \delta A_{i,jk} x_j x_k + \cdots \end{aligned} \tag{23.2}$$

The infinitesimal analogs of (3.3) are obtained through the obvious modifications.

The general finite nonlinear coordinate transformation (3.2) can be obtained by iterating the infinitesimal transformation (23.2) a large number of times. As a result, either the finite or the infinitesimal transformation can be used to approach the questions treated in Chapters 3 and 4. The infinitesimal approach leads directly and easily to algorithms for computing the determinacy of an arbitrary function and its unfolding.

Remark. The relation between the finite transformation (3.2) and the infinitesimal transformation (23.2) is much the same as the relation between a Lie group and a Lie algebra.[1]

2. Determinacy ("Truncating" the Taylor Series)

A function $f(x)$ is equivalent to, or has qualitatively the same properties as, a function f' if there is a smooth change of variables $x' = x'(x)$ such that

$$f'(x') = f(x) \tag{23.3}$$

If f is equivalent to f' and if the qualitative properties of f' are known, then f must have the same qualitative properties. It is therefore useful to find a polynomial of finite and small degree which is equivalent to f, if possible.

Definition. A function $f(x)$ is **p-determinate**[2,3] around zero if, for any other function f' with the same p-jet, there is a smooth change of variables such that

$$f(x) = f'[x'(x)] \tag{23.4}$$

Example 1. If $f(x) = x$ and $f'(x) = \sin x$, then $j^1 f = x$, $j^1 f' = x$, and $j^1 f = j^1 f'$. By choosing $x' = \sin^{-1} x (|x| < 1, |x'| < \pi/2)$, we find

$$f'[x'(x)] = \sin x' = \sin(\sin^{-1} x) = x = f(x) \tag{23.5}$$

Remark 1. Assume that $f(x), f'(x)$ are two functions with $f(0) = f'(0) = 0$ and $j^1 f \neq 0, j^1 f' \neq 0$. Then the Implicit Function Theorem guarantees $f(x) = f'[x'(x)]$, since both f and f' can be transformed to the same canonical form. In fact, a smooth and even linear change of coordinates $x' = x'(x)$ can always be chosen so that

$$j^1 f'[x'(x)] = j^1 f(x) \tag{23.6}$$

Remark 2. Assume that $f(x), f'(x)$ are two functions with $f(0) = f'(0) = 0$ and $j^1 f = j^1 f' = 0$, but for which $j^2 f$ and $j^2 f'$ are both nonzero and nonsingular. Then if $j^2 f$ and $j^2 f'$ have the same index or inertia (their quadratic forms have the same number of positive and negative eigenvalues), the Morse Lemma guarantees that f and f' can be transformed into each other by a smooth coordinate change, since both are equivalent to the same canonical form. In fact, a linear change of coordinates $x' = x'(x)$ can always be chosen so that

$$j^2 f'[x'(x)] = j^2 f(x) \tag{23.7}$$

Remark 3. Assume that $f(x)$ is p-determined. If

$$j^p f(x) = j^p f'(x) \tag{23.8a}$$

then

$$f(x) = f'[x'(x)] \tag{23.8b}$$

In particular, one function $f'(x)$ whose p-jet is $j^p f$ is the polynomial $j^p f$ of degree p. Therefore if we take $f' = j^p f$, (23.8b) says that $f(x)$ is equivalent to a polynomial of degree p. It is for this reason that statements about "truncating a Taylor series expansion" are equivalent to statements about "transforming away the Taylor tail."

Remark 4. In terms of determinacy, the problem of canonical forms outlined in Chapter 2 reduces to the problem of finding classes of functions which are determinate for

1. $p = 1$ (2.1)
2. $p = 2$ (2.2)
3. $p \geq 3$ (2.3) and (2.4).

Remark 5. Not every function is p-determinate for any finite p. Clearly, the function e^{-1/x^2} (21.4) cannot be determinate at 0, since its Taylor series coefficients are all zero. But simple analytic functions are not necessarily finitely determined. For example, the function $f(x, y) = x^2 y$ is not finitely determined. This is most easily seen by considering what qualitative effect a perturbation of the form $y^{\text{very high power}}$ has on $x^2 y$:[4]

$$\begin{aligned} f(x, y) &= x^2 y \\ g(x, y) &= x^2 y + y^{10^6} \\ h(x, y) &= x^2 y + y^{10^6 + 1} \end{aligned} \tag{23.9}$$

These three functions all have the same p-jet ($p < 10^6$), but their root structures are drastically different (Fig. 23.1). These three functions are clearly qualitatively different.

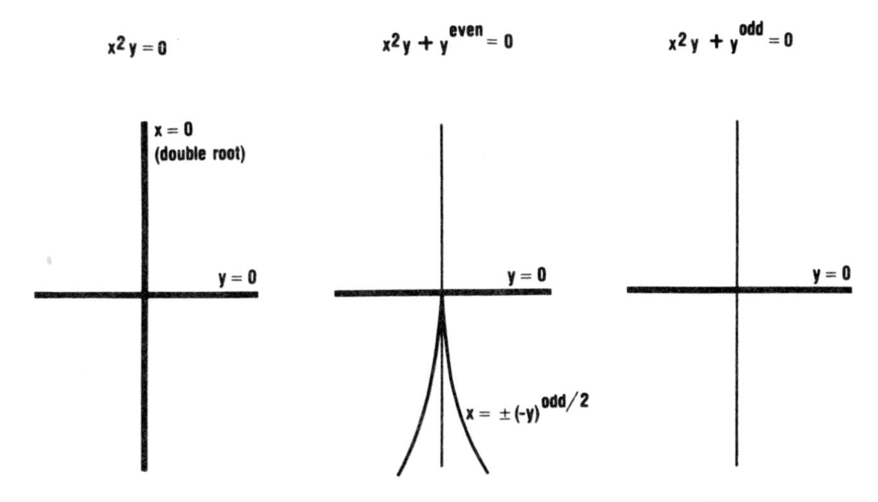

Figure 23.1 The function x^2y is not finitely determined. The root structure of (x^2y + perturbation), where the perturbation has the form $y^{\text{very high power}}$, depends dramatically on whether the power is even or odd.

We have seen already that instead of answering the question: "Is $f(x)$ equivalent to some polynomial $f'(x)$ of degree p?" it is much simpler to turn the question backwards and ask "What class of functions $f(x)$ is equivalent to the polynomial $f'(x) = j^p f(x)$?" The two questions are equivalent because any smooth nonsingular nonlinear transformation $x' = x'(x)$ can be inverted $x = x(x')$ according to the Inverse Function Theorem. Thus the two questions may be phrased mathematically:

$$f'(x') \stackrel{?}{=} f[x(x')] \tag{23.10a}$$

$$f(x) \stackrel{?}{=} f'[x'(x)] \tag{23.10b}$$

We show now by example that (23.10b) can be handled simply, particularly using the infinitesimal nonlinear transformation (23.2).

Example 2. What is the class of functions of two variables $f(x, y)$ which is equivalent to the germ of E_6: $f'(x, y) = x^3 + y^4$, given in the last line of Table 2.2?

Solution. We use the infinitesimal expressions $x' = x + \delta x$, $y' = y + \delta y$ given in (23.2) in (23.10b) with f' as given above:

$$\begin{aligned} f'[x', y'] &= (x + \delta x)^3 + (y + \delta y)^4 \\ &= (x^3 + y^4) + 3x^2\delta x + 4y^3\delta y + \mathcal{O}(2) \end{aligned} \tag{23.11}$$

The function $f(x, y)$ differs from the germ $(x^3 + y^4)$ by the additional terms which appear in (23.11). The terms of order 2 and higher in the infinitesimals

δx, δy may be neglected. Since we are interested in the qualitative properties of $f(x, y)$ *at a point*, we allow only homogeneous nonlinear infinitesimal transformations, with $\delta A_i = 0$ in (23.2). The first-order infinitesimal terms which appear as the correction to the germ $(x^3 + y^4)$ in (23.11) are then precisely those that are underlined in the diagram below:

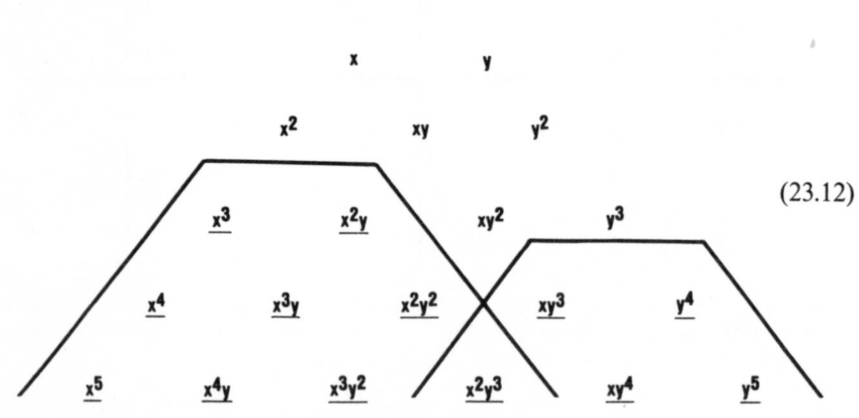

(23.12)

The terms in the shadow emanating from x^2 come from $3x^2\delta x$ in (23.11), with $\delta A_1 = 0$. Those emanating from y^3 come from $4y^3\delta y = (\partial f/\partial y)\delta y$, with $\delta A_2 = 0$.

The coefficients of the underlined monomials in (23.12) are all infinitesimals of first order. By iterating the homogeneous nonlinear transformation (23.2) sufficiently often, these coefficients can all be made finite and arbitrary.[1,3] Therefore the class of functions $f(x, y)$ which is equivalent to the catastrophe germ E_6: $f'(x, y) = x^3 + y^4$, is

$$f(x, y) = A_{30}x^3 + A_{21}x^2y + \sum_{p+q\geq 4} A_{pq}x^p y^q \tag{23.13}$$

with $A_{30} \neq 0$, $A_{04} \neq 0$.

3. Unfolding (Universal Perturbations)

Functions with non-Morse critical points may occur stably in families of functions depending on one or more control parameters. It is possible to study the effects of a perturbation on a function with a degenerate critical point by embedding the non-Morse function $f(x)$ in a larger family of functions $F(x; a)$ which contains $f(x)$ as a special case:

$$\begin{aligned} f &= f(x), \qquad x = (x_1, \ldots, x_l) \\ F &= F(x; a), \qquad a = (a_1, \ldots, a_r) \\ f(x) &= F(x; a)|_{a=0} \end{aligned} \tag{23.14}$$

Definition. The family of functions $F(x; a)$ is called an ***r*-dimensional unfolding** of $f(x)$.[3,4]

The larger the family, the more general the perturbation of $f(x)$ which it can describe. The hope is to find a family that is sufficiently large that it can describe all possible qualitatively distinct perturbations of $f(x)$, but which is sufficiently small that it is easy to work with.

Definition. The r-dimensional unfolding $F(x; a)$ is **versal** if any other unfolding $F'(x'; a')$ of $f(x)$ can be obtained from it by a smooth change of variables:

$$\begin{aligned} x_i' &= x_i'(x; a) \\ a_\alpha' &= a_\alpha'(a), \qquad \alpha = 1, 2, \ldots, r' \end{aligned} \tag{23.15}$$

where r' is not necessarily equal to r.

Definition. The unfolding $F(x; a)$ is called a **universal unfolding** of $f(x)$ if it is a versal unfolding of minimal dimension.

Example 1. For the germ $f(x) = x^3$ we have the following unfoldings:

$$\begin{aligned} F_1(x; a_1, a_2) &= x^3 + a_1 x + a_2 x^2 && \text{versal} \\ F_2(x; b) &= x^3 + \phantom{a_1 x +{}} bx^2 && \text{nonversal} \\ F_3(x; c) &= x^3 + cx && \text{universal} \end{aligned}$$

Since x^3 is 3-determinate, the most general perturbation is

$$p(x) = \varepsilon_0 + \varepsilon_1 x + \varepsilon_2 x^2 + \varepsilon_3 x^3 \tag{23.16}$$

with all ε_i small, $\varepsilon \in \mathbb{R}^D$, $D = (3 + 1)!/3!1! = 4$. The constant term may always be removed by adjusting the zero of the ordinate. It is unimportant for the discussion of local properties, and will be neglected here and below. If $\varepsilon_3 \neq 0$, it may be removed by rescaling the state variable x:

$$x \to x' = (1 + \varepsilon_3)^{1/3} x \tag{23.17}$$

Therefore ε_3 may be taken as zero. Then $F_1(x; a_1, a_2)$ is clearly a versal unfolding of $f(x) = x^3$, and the unfoldings $F_2(x; b) = F_1(x; 0, b)$ and $F_3(x; c) = F_1(x; c, 0)$ are both simply obtained from F_1. The mappings between F_2 and F_3 illustrate why F_2 is nonversal

$$\begin{aligned} F_2(x - \tfrac{1}{3}b; b) &= x^3 - \tfrac{1}{3}b^2 x = F_3(x; -\tfrac{1}{3}b^2) \\ F_3(x \pm \sqrt{-\tfrac{1}{3}c}; c) &= x^3 \pm \sqrt{-3c}\,x^2 = F_2(x; \pm\sqrt{-3c}) \end{aligned} \tag{23.18}$$

The smooth change of variables

$$\begin{aligned} x &\to x - \tfrac{1}{3}b \\ b &\to b \end{aligned} \tag{23.19}$$

maps F_2 into those functions F_3 with $c = -b^2/3 \leq 0$ only. On the other hand, F_3 is versal, since under the smooth change of variables

$$\begin{aligned} x &\to x + \tfrac{1}{3}a_1 \\ c &= a_2 - \tfrac{1}{3}(a_1)^2 \\ F_3(x; c) &= F_1(x; a_1, a_2) \end{aligned} \tag{23.20}$$

Put another way, the coefficients $\varepsilon_3, \varepsilon_2, \varepsilon_0$ may be eliminated from the perturbation (23.16) by rescaling, shift of x coordinate, and shift of y coordinate, respectively. Therefore the dimension of the smallest unfolding of $f(x) = x^3$ which can generate the family of functions $x^3 +$ (23.16) is 1, and $F_3(x; c)$ is a universal unfolding of $f(x) = x^3$.

Example 2. Find a universal unfolding of the non-Morse germ $E_6: f(x, y) = x^3 + y^4$.

Solution. The most general perturbation of a function of two variables is

$$p(x, y) = \sum_{i \geq 0, j \geq 0} p_{ij} x^i y^j \tag{23.21}$$

where we consider the coefficients p_{ij} as infinitesimals of order 1. Then

$$F(x, y; p) = x^3 + y^4 + \sum p_{ij} x^i y^j \tag{23.22}$$

Many of the perturbing terms can be transformed away by a smooth change of variables. To determine which perturbing terms can be transformed away and which cannot, we make an infinitesimal nonlinear coordinate transformation and proceed as in the previous section:

$$F(x', y'; p) = (x^3 + y^4) + (3x^2\delta x + 4y^3\delta y + \sum_{i,j} p_{ij} x^i y^j) + \mathcal{O}(2) \tag{23.23}$$

At the present time we are interested in embedding the germ $f(x, y)$ in a larger family $F(x, y; p)$, not in the qualitative properties of $f(x, y)$ at a fixed point. As a result, the infinitesimal nonlinear transformation in (23.23) above can be inhomogeneous. The terms that can arise from $3x^2\delta x$ and $4y^3\delta y$ are underlined in the diagram below:

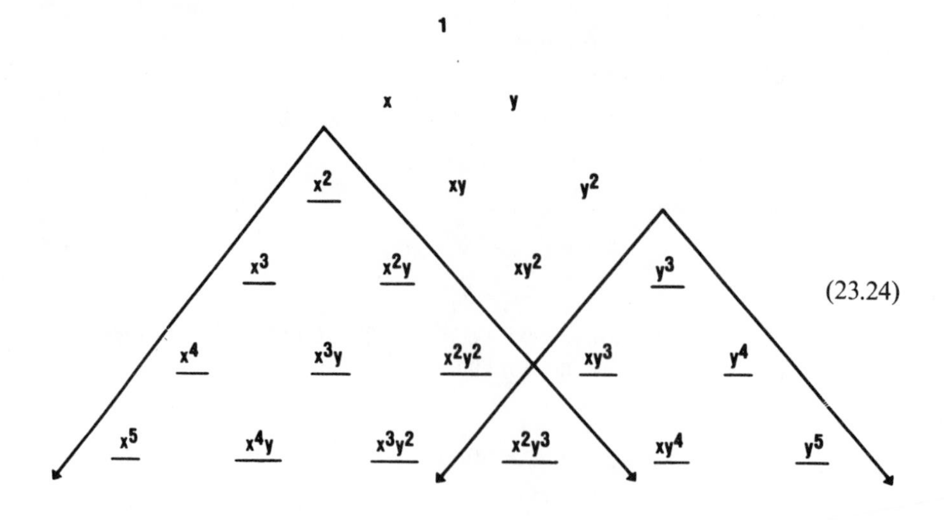

(23.24)

Therefore the nonlinear transformation may be chosen to transform away all terms $p_{ij}x^i y^j$ in the perturbation (23.23) which correspond to those underlined. Excluding the constant term, the universal unfolding of $f(x, y) = x^3 + y^4$ has dimension 5 and is

$$F(x, y; p) = x^3 + y^4 + p_{10}x + p_{01}y + p_{11}xy + p_{02}y^2 + p_{12}xy^2 \quad (23.25)$$

This may be compared with the canonical perturbation for E_6 given in Table 2.2.

4. Rules for Determinacy

A simple algorithm has been developed by Mather[5–10] for deciding if a function $f(x)$ is determinate and, if so, how much of its Taylor series must be retained to capture its qualitative properties. The motivation for, and proof of, this algorithm can be seen by carefully examining Example 2, Sec. 2.

The algorithm proceeds in a number of simple steps:

1. Assume that $f(x)$ is p-determinate.
2. Let $m(x)$ be the sequence of monomials in $x_1, x_2, \ldots, x_l$ of degree $1, 2, \ldots$:

$$m_j(x) = x_1, \ldots, x_l;\ x_1^2, x_1x_2, \ldots, x_l^2;\ x_1^3, \ldots \quad (23.26)$$

3. Compute the set of polynomials $R_{ij}(x)$ defined by

$$R_{ij}(x) = j^{p+1}\left\{\frac{\partial f}{\partial x_i} m_j(x)\right\} \quad (23.27)$$

4. Can all monomials of degree $p + 1$ be written as linear superpositions of the $R_{ij}(x)$ with constant coefficients?

If $f(x)$ is p-determinate, the answer to this question is "yes." Unfortunately the theorem underlying this algorithm is not an "if and only if" theorem, so the answer may be yes if $f(x)$ is not p-determinate.

An if and only if algorithm is obtained by replacing the infinitesimal homogeneous nonlinear transformation in Step 2 above by an infinitesimal homogeneous axis-preserving nonlinear transformation. Then the monomials $m_j(x)$ of Step 2 start at degree 2. If the answer to the question of Step 4 is "yes," $f(x)$ is p-determinate and might be $(p - 1)$-determinate.

Example 1. Compute the determinacy of

$$f(x, y) = \tfrac{1}{2}(x + y)^2 + \tfrac{1}{3}y^3 \quad (23.28)$$

Table 23.1 Determinacy Algorithm Applied to the Function
$f(x, y) = \frac{1}{2}(x + y)^2 + \frac{1}{3}y^3$

m_j		$j^4\left[\frac{\partial f}{\partial x_1} m_j\right]$	$j^4\left[\frac{\partial f}{\partial x_2} m_j\right]$
m_1	x	$x(x + y)$	$x(x + y) + xy^2$
m_2	y	$y(x + y)$	$y(x + y) + y^3$
m_3	x^2	$x^2(x + y)$	$x^2(x + y) + x^2y^2$
m_4	xy	$xy(x + y)$	$xy(x + y) + xy^3$
m_5	y^2	$y^2(x + y)$	$y^2(x + y) + y^4$
m_6	x^3	$x^3(x + y)$	$x^3(x + y)$
m_7	x^2y	$x^2y(x + y)$	$x^2y(x + y)$
m_8	xy^2	$xy^2(x + y)$	$xy^2(x + y)$
m_9	y^3	$y^3(x + y)$	$y^3(x + y)$
m_{10}	x^4	0	0
$\vdots$		$\downarrow$	$\downarrow$

Solution. Assume that $p = 3$. Then

$$\frac{\partial f}{\partial x} = x + y, \qquad \frac{\partial f}{\partial y} = (x + y) + y^2$$

$$m_1, m_2; m_3, \ldots = x, y; x^2, xy, y^2; x^3, \text{etc.} \tag{23.29}$$

The polynomials $R_{ij}(x)$ are given in Table 23.1. The complete set of monomials of degree $3 + 1 = 4$ can be expressed in terms of the $R_{ij}(x, y)$ as follows:

$$\begin{aligned} y^4 &= R_{25} - R_{15} \\ xy^3 &= R_{24} - R_{14} \\ x^2y^2 &= R_{23} - R_{13} \\ x^3y &= R_{27} - (R_{23} - R_{13}) \\ x^4 &= R_{26} - \{R_{27} - (R_{23} - R_{13})\} \end{aligned} \tag{23.30}$$

Since none of the R_{ij} appearing in (23.30) involve first-degree monomials $m_1 = x$, $m_2 = y$, the if and only if algorithm associated with axis-preserving transformations is applicable. Therefore $f(x, y)$ is either 3-determinate or 2-determinate. Since the stability matrix of $f(x, y)$ at zero has vanishing determinant, $f(x, y)$ cannot be 2-determinate.

Example 2. Find the determinacy of $f(x) = x^p$.

Solution. Assume that x^p/p is p-determinate. Then $\partial f/\partial x = x^{p-1}$ and $m_1, m_2, \ldots = x, x^2, \ldots$. As a result,

$$j^{p+1}\left\{\frac{\partial f}{\partial x} m_j\right\} = j^{p+1}\{x^{p-1} m_j\} = x^p, x^{p+1}, 0, \ldots \tag{23.31}$$

Since $x^{p+1} = R_{12}$, x^p is p-determinate.

Example 3. Compute the determinacy of $f(x, y) = x^2 y + y^p/p$.

Solution. Assume that $f(x, y)$ is p-determinate. Then

$$\begin{aligned} \frac{\partial f}{\partial x} &= 2xy \\ \frac{\partial f}{\partial y} &= x^2 + y^{p-1} \end{aligned} \tag{23.32}$$

Using the functions $m_j(x, y)$ as given in Example 1, it is easy to see that

$$\begin{aligned} x^{p+1} &= j^{p+1}\left\{\frac{\partial f}{\partial y} x^{p-1}\right\}, \qquad p - 1 > 2 \\ x^r y^s &= j^{p+1}\left\{\frac{1}{2}\frac{\partial f}{\partial x} x^{r-1} y^{s-1}\right\}, \qquad r + s = p + 1 \geq 3, \quad r \geq 1, \quad s \geq 1 \\ y^{p+1} &= j^{p+1}\left\{\frac{\partial f}{\partial y} y^2\right\} - j^{p+1}\left\{\frac{1}{2}\frac{\partial f}{\partial x} xy\right\} \end{aligned} \tag{23.33}$$

The function $f(x, y)$ is p-determinate.

This algorithm can be carried out in a systematic diagrammatic way for functions $f(x, y)$ of two variables when $\partial f/\partial x$ and $\partial f/\partial y$ are monomials.[11] The monomials 1; x, y; x^2, xy, and so on, are arranged in a triangular array à la Pascal. Then $\partial f/\partial x$ and $\partial f/\partial y$ correspond to points in this triangle. Multiplication by the monomials $m_j(x, y)$ corresponds to creating points in the "shadow" of $\partial f/\partial x$ and $\partial f/\partial y$. The shadow begins one or two rows below these derivatives, depending on whether or not the terms of degree 1 in x, y are included among the $m_j(x, y)$. Taking the $(p + 1)$-jet corresponds to ignoring all rows below the $(p + 1)$th row.

This method has already been used to compute the determinacy of

$$E_6\colon \quad f(x, y) = x^3 + y^4$$

in Sec. 2 (23.12).

5. Rules for Unfolding

Mather has also developed an algorithm to determine the universal unfolding of a function $f(x)$.[5–10] The rules for unfolding are literally the complement of the rules for determinacy.

The algorithm proceeds in a number of simple steps:

1. Find p, the determinacy of $f(x)$. It is sufficient to work with the polynomial $\bar{f}(x) = j^p f(x)$.
2. Let $n_j(x)$ be the sequence of monomials in $x_1, x_2, \ldots, x_l$ of degree $0, 1, 2, \ldots$:

$$n_j(x):\quad 1; x_1, x_2, \ldots, x_l; x_1^2, x_1 x_2, \ldots \tag{23.34}$$

3. Assume that $F(x; a)$ is an r-dimensional unfolding of $\bar{f}(x)$. Define

$$T_j(x) = \frac{\partial}{\partial a_j} j^{p+1} F(x; a)|_{a=0} \tag{23.35}$$

4. List all polynomials

$$S_{ij}(x) = j^p \left\{ \frac{\partial \bar{f}}{\partial x_i} n_j(x) \right\} \tag{23.36}$$

5. Can all monomials of degree $\leq p$ be expressed in the form

$$\text{any monomial of degree} \leq p = \sum s_{ij} S_{ij}(x) + \sum t_j T_j(x) \tag{23.37}$$

where s_{ij}, t_j are real numbers? If the answer is "yes," then $F(x; a)$ is a versal unfolding of $\bar{f}(x)$.

6. Is $T_j(x)$ a minimal set? If the answer is "yes," then $F(x; a)$ is a universal unfolding of $\bar{f}(x)$.

Remark 1. The polynomials $T_j(x)$ form a minimal set only if they are linearly independent.

Remark 2. The Implicit Function Theorem can be applied to the unfolding parameters a_j to give a canonical *linear* form for the universal unfolding of $\bar{f}(x)$:

$$F(x; a) = \bar{f}(x) + \sum_{j=1}^{r} a_j T_j(x) \tag{23.38}$$

Remark 3. The universal unfolding of a function $\bar{f}(x)$ is not necessarily unique (Examples 2 and 3 below), although it is unique modulo a smooth change of variables. Lists of the elementary catastrophes given by different authors may appear quite different, but they are equivalent under a smooth change of variables.

Example 1. Compute the universal unfolding of $f(x) = x^p \cos x$ around $x = 0$.

Solution. The function $x^p(1 - x^2/2! + \cdots)$ is p-determinate, so it is sufficient to work with $\bar{f}(x) = x^p$. The monomials $n_j(x)$ are

$$n_0, n_1, n_2, \text{etc} = 1, x, x^2, \text{etc.} \tag{23.39}$$

Then

$$j^p\left\{\frac{\partial \bar{f}}{\partial x} n_j\right\} = j^p\{x^{p-1} n_j\} = x^{p-1}, x^p, 0, \ldots \tag{23.40}$$

Every monomial of degree $\leq p$ can be expressed in the form (23.38), provided we choose

$$T_0, T_1, \ldots, T_{p-2} = x^0, x^1, \ldots, x^{p-2} \tag{23.41}$$

Since the constant term is unimportant, we have as a universal unfolding of $\bar{f}(x) = x^p$

$$F(x; a_1, a_2, \ldots, a_{p-2}) = x^p + \sum_{j=1}^{p-2} a_j x^j \tag{23.42}$$

Example 2. Compute the universal unfolding of $f(x, y) = x^2 y \pm y^3/3$.

Solution. This function is 3-determinate. The partial derivatives $\partial f/\partial x = 2xy$ and $\partial f/\partial y = x^2 \pm y^2$ as well as the monomials $n_j(x)$ and the 3-jets of their products $S_{ij}(x)$ are listed below:

$n_j(x, y)$	$S_{1j} = j^3\left(\frac{\partial f}{\partial x} n_j\right)$	$S_{2j} = j^3\left(\frac{\partial f}{\partial y} n_j\right)$	
$n_0 = 1$	xy	$x^2 \pm y^2$	
$n_1 = x$	$x^2 y$	$x(x^2 \pm y^2)$	(23.43)
$n_2 = y$	xy^2	$y(x^2 \pm y^2)$	
$n_3 = x^2$	0	0	
$\vdots$	$\vdots$	$\vdots$	

Neither of the monomials x, y of degree 1 can be expressed as a linear combination of the S_{ij}, so we must take $T_1(x, y) = x$ and $T_2(x, y) = y$. The set of polynomials of degree exactly 2 is a linear vector space of dimension 3. The functions $S_{11} = xy$ and $S_{21} = x^2 \pm y^2$ may be chosen as two of the basis vectors in this space.

We can choose as $T_3(x, y)$ any vector in this space which is linearly independent of S_{11} and S_{21}. This may be x^2, y^2, or $x^2 \mp y^2$. In Table 2.2 we have chosen $T_3 = y^2$, but in Chapter 5, Secs. 6 and 7, we have found the choices $T_3(x, y) = x^2 \mp y^2$ more convenient for computational purposes. All monomials x^3, x^2y, xy^2, y^3 of degree 3 can be expressed in terms of the S_{ij}, so that a universal unfolding of $f(x, y)$ is

$$F(x, y; a_1, a_2, a_3) = x^2y \pm \tfrac{1}{3}y^3 + a_1x + a_2y + a_3(x^2 \mp y^2) \tag{23.44}$$

Example 3. Compute an unfolding of $f(x, y) = x^2y + y^p/p$.

Solution. This function is p-determinate, with $\partial f/\partial x = 2xy$ and $\partial f/\partial y = x^2 + y^{p-1}$. The monomials $n_j(x, y)$ and the p-jets $S_{ij}(x, y)$ are listed below:

$$\begin{array}{ccc}
n_j(x, y) & S_{1j} = j^p\left(\dfrac{\partial f}{\partial x} n_j\right) & S_{2j} = j^p\left(\dfrac{\partial f}{\partial y} n_j\right) \\[2ex]
n_0 = 1 & xy & x^2 + y^{p-1} \\
n_1 = x & x^2y & x^3 + xy^{p-1} \\
n_2 = y & xy^2 & x^2y + y^p \\
n_3 = x^2 & x^3y & x^4 \\
n_4 = xy & x^2y^2 & x^3y \\
\vdots & \vdots & \vdots
\end{array} \tag{23.45}$$

All monomials x^ry^s, $r \geq 1$, $s \geq 1$, $r + s \leq p$, occur in the list S_{1j}, and the monomials x^r, $4 \leq r \leq p$, and y^p occur in the list S_{2j}. The term x^3 can be expressed as $x^3 = (x^3 + xy^{p-1}) - xy^{p-1}$, so that a minimal set of the $T_j(x, y)$ must include x, y, y^s, $1 < s < p - 1$, and either x^2 or y^{p-1}. A nonunique universal unfolding of $x^2y + y^p/p$ is

$$F(x, y; a_1, \ldots, a_p) = x^2y + \frac{y^p}{p} + \sum_{j=1}^{p-2} a_jy^j + \sum_{j=p-1}^{p} a_jx^{j-(p-2)} \tag{23.46}$$

Remark 4. Computation of the universal unfolding of a function $f(x, y)$ of two variables may be greatly simplified by a diagrammatic method when $\partial f/\partial x$ and $\partial f/\partial y$ are monomials.[2,11] In fact, this diagrammatic method was originally used in (23.24) to motivate Mather's algorithm for determinacy.

Example 4. Compute the unfolding of E_6: $x^3 + y^4$.

Solution. The diagrammatic method can be used to greatly facilitate this computation. The monomials x^py^q are arranged in Pascal's triangle. Then the "shadows" cast by multiplying $\partial f/\partial x = x^2$ and $\partial f/\partial y = y^3$ by the monomials n_j: $1; x, y; x^2, \ldots$ are outlined:

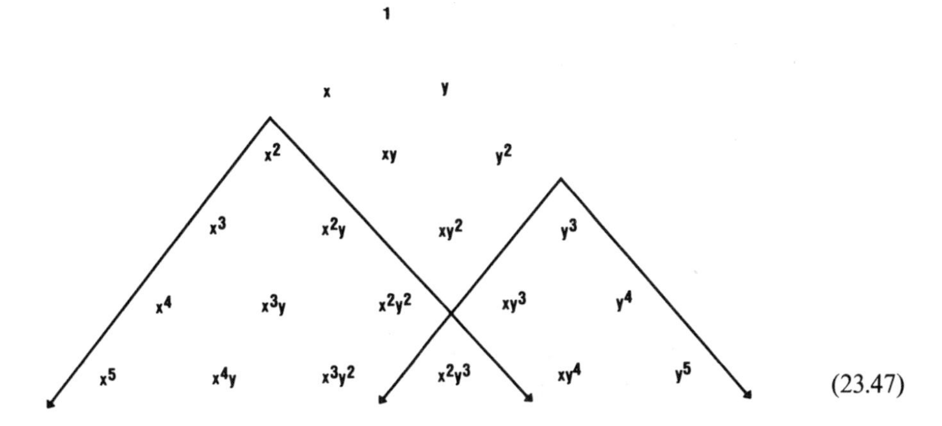

(23.47)

The remaining terms provide the universal perturbation of $E_6 = x^3 + y^4$. Neglecting the constant term, this universal perturbation is the 5-dimensional function

$$p(x, y) = a_1 x + a_2 y + a_3 xy + a_4 y^2 + a_5 xy^2 \tag{23.48}$$

6. Rules for Germs

By comparing Sec. 4 with Sec. 5 it is easy to see that the rules for determinacy and unfolding are complementary. In the diagram below the monomials $R_{ij}(x, y)$ generated in the determinacy algorithm for $f(x, y) = x^3 + y^4$ are underlined, and the unfolding terms $T_j(x, y)$ generated in the unfolding algorithm are overlined [cf. (23.12) and (23.47)].

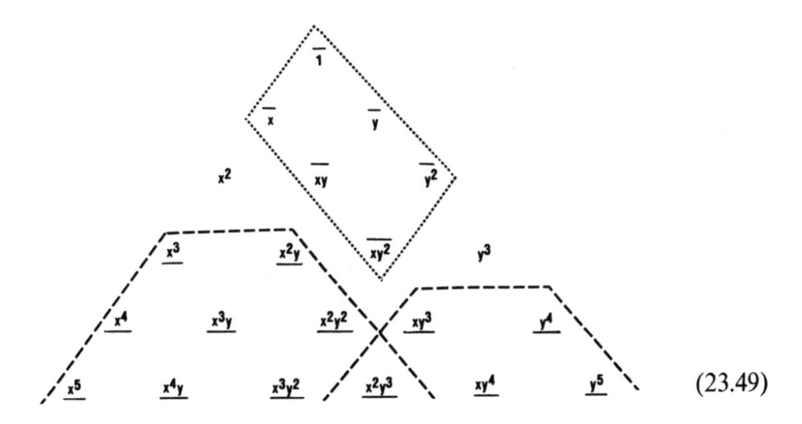

(23.49)

The sets of terms enclosed by dashed lines are in the "shadows" of x^2 and y^3. Only the terms $x^2 = \partial f/\partial x$ and $y^3 = \partial f/\partial y$ are neither underlined nor overlined. All underlined terms can be transformed away by smooth change of variables; the overlined terms are necessary for a universal unfolding, and the remaining terms are the derivatives of the germ of $f(x, y)$.

This "complementarity" immediately suggests an algorithm for determining the simplest possible germ to associate with any p-determinate function $f(x_1, \ldots, x_l)$.

This algorithm proceeds in a number of simple steps:

1. Find p, the determinacy of f. It is sufficient to work with the polynomial $\bar{f}(x) = j^p f(x)$.
2. Let V_p be the linear vector space spanned by all monomials in $x_1, \ldots, x_l$ of degree $\leq p$. Then dim $V_p = (p + l)!/p!l!$
3. Let V_D be the linear vector subspace of V_p spanned by all polynomials R_{ij} generated in the determinacy algorithm.
4. Let V_U be the linear vector subspace of V_p spanned by a minimal set of polynomials T_j generated in the unfolding algorithm.
5. Then $V_p - (V_D \oplus V_U) = V_p/(V_D \oplus V_U)$ is a linear vector space spanned by the first partial derivatives of the germ of f.

This algorithm "stands between" the two competing processes described in Chapters 3 and 4, or in Secs. 4 and 5 of the present chapter. On the one hand, the occurrence of control parameters in a family of functions can be used to eliminate some of the early terms in the Taylor series expansion about a critical point. On the other hand, a change of variables can be used to eliminate the later terms. These two competing processes "meet in the middle" at the catastrophe germ. The algorithm above provides the mechanism for constructing the germ.

Example 1. Find the canonical (simplest) germ f_{cg} associated with the function $f(x, y) = x^2y + y^3/3 + y^2/2$.

Solution. This interesting function (modulo a scale change) was encountered in (5.36).

STEP 1. The determinacy of $f(x, y)$ is computed by listing all monomials $1; x, y; x^2, \ldots,$ and all $(p + 1)$-jets R_{ij}. This is done in Table 23.2. This table lists all terms of degree ≤ 5 which arise in the determinacy algorithm; those of degree exactly 5 are enclosed in a dashed box. It is tempting to assume first that $f(x, y)$ is 3-determinate. In fact, all monomials of degree 4 occur in the table $(x^3y, x^2y^2, xy^3, y^4)$ or can be represented as a linear combination of functions in the table:

$$x^4 = R_{23} - R_{14} - R_{11}$$

Table 23.2 Algorithm for Determining the Germ of

$$f(x, y) = x^2y + \frac{y^3}{3} + \frac{y^2}{2}$$

				$j^5\left[\frac{\partial f}{\partial x} n_j\right]$		$j^5\left[\frac{\partial f}{\partial y} n_j\right]$
	n_0	1	S_{10}	xy	S_{20}	$x^2 + y^2 + y$
m_1	n_1	x	$R_{11} = S_{11}$	x^2y	$R_{21} = S_{21}$	$x^3 + xy^2 + xy$
m_2	n_2	y	$R_{12} = S_{12}$	xy^2	$R_{22} = S_{22}$	$x^2y + y^3 + y^2$
m_3	n_3	x^2	$R_{13} = S_{13}$	x^3y	$R_{23} = S_{23}$	$x^4 + x^2y^2 + x^2y$
m_4	n_4	xy	$R_{14} = S_{14}$	x^2y^2	$R_{24} = S_{24}$	$x^3y + xy^3 + xy^2$
m_5	n_5	y^2	$R_{15} = S_{15}$	xy^3	$R_{25} = S_{25}$	$x^2y^2 + y^4 + y^3$
m_6	n_6	x^3	$R_{16} = S_{16}$	x^4y	$R_{26} = S_{26}$	$x^5 + x^3y^2 + x^3y$
m_7	n_7	x^2y	$R_{17} = S_{17}$	x^3y^2	$R_{27} = S_{27}$	$x^4y + x^2y^3 + x^2y^2$
m_8	n_8	xy^2	$R_{18} = S_{18}$	x^2y^3	$R_{28} = S_{28}$	$x^3y^2 + xy^4 + xy^3$
m_9	n_9	y^3	$R_{19} = S_{19}$	xy^4	$R_{29} = S_{29}$	$x^2y^3 + y^5 + y^4$
m_{10}	n_{10}	x^4	$R_{1,10} = S_{1,10}$	0	$R_{2,10} = S_{2,10}$	x^4y
m_{11}	n_{11}	x^3y		↓	$R_{2,11} = S_{2,11}$	x^3y^2
m_{12}	n_{12}	x^2y^2			$R_{2,12} = S_{2,12}$	x^2y^3
m_{13}	n_{13}	xy^3			$R_{2,13} = S_{2,13}$	xy^4
m_{14}	n_{14}	y^4			$R_{2,14} = S_{2,14}$	y^5
m_{15}	n_{15}	x^5			$R_{2,15} = S_{2,15}$	0 ↓

However, we recall that the determinacy algorithm is not an if and only if algorithm. If f is 3-determinate, then all monomials of degree 4 can be expressed as linear combinations of the R_{ij}, but the fact that this is possible does not guarantee that f is 3-determinate because one of these terms (x^4) involves the term m_1 of degree 1. The if and only if statement can be used to prove that $f(x, y)$ is 4-determinate, because all terms of degree 5 either occur directly in the table ($x^4y, x^3y^2, x^2y^3, xy^4, y^5$) or can be expressed in terms of the $m_j(x, y)$ of degree ≥ 2:

$$x^5 = R_{26} - R_{17} - R_{13}$$

Remark. It is sufficient but not necessary to work with the polynomial $j^pf(x)$. In fact, it is sufficient to work with any polynomial truncation $j^{p'}f(x)$ of $f(x)$, $p' \geq p$, where p is the determinacy of $f(x)$. This is because the determinacy algorithm generates all monomials of degree $\geq p$, so that $V_p/(V_D \cap V_p) = V_{p'}/(V_D \cap V_{p'})$. It is simply desirable to work with the linear vector space $V_{p'}$ of smallest adequate dimension. This occurs for $p' = p =$ determinacy of $f(x)$.

Since we do not yet know whether $f(x, y) = x^2y + y^3/3 + y^2/2$ is 3-determinate or 4-determinate, we shall work in the space $V_{p'} = V_4$ spanned by monomials $x^r y^s$, $r + s \leq 4$. This space is 15-dimensional. If it should turn out that $f(x, y)$ is 3-determinate, then no harm and only a little extra work is involved in the computations. On the other hand, if $f(x, y)$ is 4-determinate and we were to work in V_3, an erroneous result would ensue.

STEP 2. The linear vector space V_p spanned by all monomials of degree ≤ 4 has dimension $(2 + 4)!/2!4! = 15$. The basis vector monomials are shown as dots in Fig. 23.2.

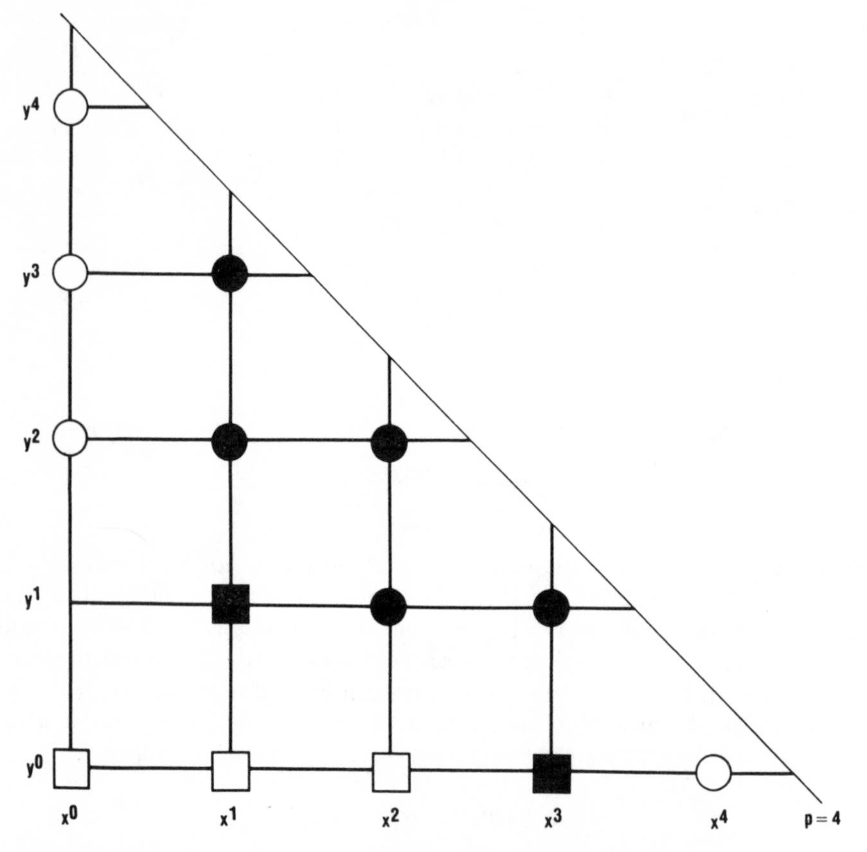

Figure 23.2 The space V_p used in the algorithm for computing the germ of a function $f(x, y)$ is spanned by the $(p + 1)(p + 2)/2$ monomials $x^i y^j$, $i + j \leq p$. For the function $f(x, y) = x^2y + y^3/3 + y^2/2$ various basis vectors occur in different parts of the algorithm, as indicated:

- ● occur directly in the determinacy algorithm;
- ○ occur indirectly in the determinacy algorithm;
- ■ terms not present in the determinacy algorithm but generated in the unfolding algorithm;
- □ basis vectors for V_U.

STEP 3. The linear vector space V_D is spanned by the polynomials R_{ij} appearing in Table 23.2. Those monomials that appear directly in Table 23.2 are shown as closed circles; those that can be expressed as linear combinations of the R_{ij} are shown as open circles. The linear vector space V_D is spanned by those monomials shown as open and closed circles in Fig. 23.2, together with the polynomial $R_{21} = x^3 + xy^2 + xy$. Dim $V_D = 10$.

STEP 4. The linear vector space spanned by the polynomials S_{ij} includes all basis vectors described above and the two additional basis vectors $S_{10} = xy$ and $S_{20} = x^2 + y^2 + y$. This space is spanned by the monomials shown as open and closed circles in Fig. 23.2, the two new monomials $xy = S_{10}$ and $x^3 = S_{21} - S_{12} - S_{10}$, shown as closed squares in Fig. 23.2, and the polynomial $S_{20} = x^2 + y^2 + y$. The dimension of this space is 12.

STEP 5. The complement of this space in V_p has dimension $15 - 12 = 3$ and is spanned by the $T_i(x, y)$. We may conveniently choose as basis vectors for this space V_U the three monomials $1, x, x^2$ (open squares). Then $V_D + V_U$ is spanned by the monomials shown in Fig. 23.2 as circles and open squares, together with the polynomial $S_{21} = x^3 + xy^2 + xy$. The space $V_p - (V_D \oplus V_U)$ has dimension $15 - (10 + 3) = 2$ and is spanned by two linearly independent combinations of the three monomials y, xy, x^3 which are also linearly independent of S_{21}. A convenient choice of basis vectors is the pair of monomials y, x^3.

STEP 6. The first partial derivatives of the germ of $f(x, y)$ span the 2-dimensional space constructed in Step 5:

$$\frac{\partial}{\partial x} f_{cg} \propto x^3$$

$$\frac{\partial}{\partial y} f_{cg} \propto y$$

Therefore $f(x, y) \doteq f_{cg} = \alpha x^4 + \beta y^2$, $\alpha \neq 0$, $\beta \neq 0$, and f is 4-determinate. Once the signs of α and β have been determined, the canonical germ may be taken as $\pm x^4 \pm y^2$ by a simple scale transformation. The stability matrix

$$f_{ij} = \begin{bmatrix} 0 & 0 \\ 0 & 1 \end{bmatrix},$$

so that the coefficient β of the y^2 term must be positive. The coefficient α of the x^4 term must be negative. This can be determined by solving the equation $f(x, y) = 0$. The root lines are shown in Fig. 23.3, together with the signs of the function in the three open regions into which these root lines partition $\mathbb{R}^2$. If α were positive, f_{cg} and therefore $f(x, y)$, would be positive definite. Therefore α must be negative and

$$f(x, y) = x^2 y + \frac{y^3}{3} + \frac{y^2}{2} \doteq -x^4 + y^2 = f_{cg}$$

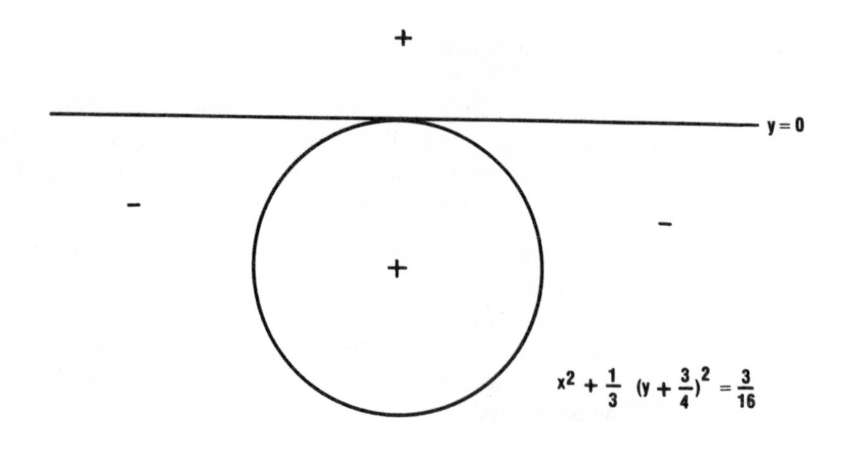

Figure 23.3 Roots of $f(x, y) = x^2y + y^3/3 + y^2/2 = 0$. The corresponding germ is not positive definite, so that $f \doteq -x^4 + y^2$.

7. Application 1: The Implicit Function Theorem

In the following sections we shall use the rules for determinacy, unfolding, and canonical germs to describe noncritical points, Morse critical points, and non-Morse critical points of various types. Throughout we investigate functions of l state variables whose value at the origin is zero.

Assume that $\nabla f \neq 0$. Then we can assume that f is 1-determinate and write

$$j^1 f = \sum_{i=1}^{l} a_i x_i \tag{23.50}$$

where at least one coefficient a_i is nonzero. For that coefficient (say a_1) $\partial \bar{f}/\partial x_1 = a_1 \neq 0$, and we can write any linear monomial x_j as

$$x_j = \frac{1}{a_1} j^1 \left(\frac{\partial f}{\partial x_1} x_j \right) \tag{23.51}$$

All monomials of degree exceeding one can be written similarly. Therefore f is 1-determinate. Further, all monomials of degree ≤ 1 can be expressed as a product of $\partial f/\partial x_1$ with a monomial $n_j(x)$ of degree ≤ 1. Therefore f needs no unfolding terms. Finally the 1-dimensional space V_p/V_D has basis vector 1, and the simplest canonical germ whose partial derivatives span this space is $f_{cg} = x_1$, so that

$$\nabla f \neq 0 \Rightarrow f \doteq x_1 \tag{23.52}$$

8. Application 2: The Morse Lemma

Assume that $\nabla f = 0$ but $\det \partial^2 f/\partial x_i\, \partial x_j = \det f_{ij} \neq 0$. It is tempting to assume that f is 2-determined. Then

$$\bar{f} = j^2 f = \tfrac{1}{2} x_i x_j f_{ij} \tag{23.53}$$

From this we determine $\partial\bar{f}/\partial x_i = f_{ij} x_j$, and since f_{ij} is a nonsingular matrix, each monomial x_j can be expressed as a linear combination of $\partial\bar{f}/\partial x_i$:

$$x_j = (f^{-1})_{ji} \frac{\partial \bar{f}}{\partial x_i} \tag{23.54}$$

The general monomial of degree 2 can be written as a linear combination of products of the form $\partial\bar{f}/\partial x_i$ and x_j for

$$x_i x_j = j^2\left(f_{ik}^{-1} \frac{\partial \bar{f}}{\partial x_k} x_j\right) = f_{ik}^{-1} j^2\left(\frac{\partial \bar{f}}{\partial x_k} x_j\right) \tag{23.55}$$

Here f^{-1} is the matrix inverse of f_{ij}. Since all monomials of degree ≤ 2 can be expressed in terms of the partial derivatives $\partial\bar{f}/\partial x_i$ multiplied by $n_0(x) = 1$ or $n_j(x) = x_j$, $1 \leq j \leq l$, the function $f(x)$ needs no unfolding terms. Finally the space V_p/V_D is spanned by the monomials $x_1, \ldots, x_l$, so the simplest function whose l partial derivatives span this space is simply the sum of l quadratic terms:

$$f \doteq \sum_{i=1}^{l} \lambda_i x_i^2, \qquad \lambda_i \neq 0 \tag{23.56}$$

9. Application 3: A_k

We consider now functions

$$f(x) = x^{k+1} g(x) \tag{23.57}$$

where $g(x)$ is smooth and $g(0) = 1$. This function is $(k + 1)$-determinate, as we have shown in Sec. 4, Example 2. The unfolding of $\bar{f}(x) = j^{k+1} f(x) = x^{k+1}$ has been determined in Sec. 5, Example 1. The space $V_p - (V_D \oplus V_U)$ is spanned by x^k. The canonical germ is the simplest function f_{cg} whose first derivative is proportional to $\partial\bar{f}/\partial x = x^k$. Therefore

$$f \doteq f_{cg} = x^{k+1} \tag{23.58}$$

and the most general perturbation is as given in Table 2.2.

10. Application 4: D_k

We consider now the functions

$$f(x, y) = x^2 y + y^{k-1} \tag{23.59}$$

This function is $(k - 1)$-determinate, as we have shown in Sec. 4, Example 3. Its universal unfolding was constructed in Sec. 5, Example 3. These results are summarized in Table 4.1.

11. Application 5: E_6, E_7, E_8

The germs E_6, E_7, E_8 are

$$\begin{aligned} E_6&: \quad f(x, y) = x^3 + y^4 \\ E_7&: \quad f(x, y) = x^3 + xy^3 \\ E_8&: \quad f(x, y) = x^3 + y^5 \end{aligned} \tag{23.60}$$

The rules for determinacy and unfolding are particularly easy to carry out for E_6 and E_8 because both $\partial f/\partial x$ and $\partial f/\partial y$ are monomials. These calculations are summarized diagrammatically in Fig. 23.4.

Carrying out the rules for determinacy and unfolding for E_7 is not quite so simple because $\partial f/\partial x = x^2 + y^3$ is not a monomial. Nevertheless it is not too difficult, and can even be done diagrammatically, as shown in Fig. 23.5. Since the monomials $b_2, d_2, e_2, \ldots$ occur in the shadow of $\partial f/\partial y = xy^2$, the monomials $b_1, d_1, e_1, \ldots$ can be expressed as linear combinations of the $S_{ij}(x)$. Similarly, since $f_1, \ldots$ occurs in this shadow, $f_2, \ldots$ can be expressed as linear combinations of the $S_{ij}(x)$. All monomials in the shadow of b_1 (including b_1), in the shadow of f_2 (including f_2), and in the shadow of xy^2 (excluding xy^2) can be expressed as

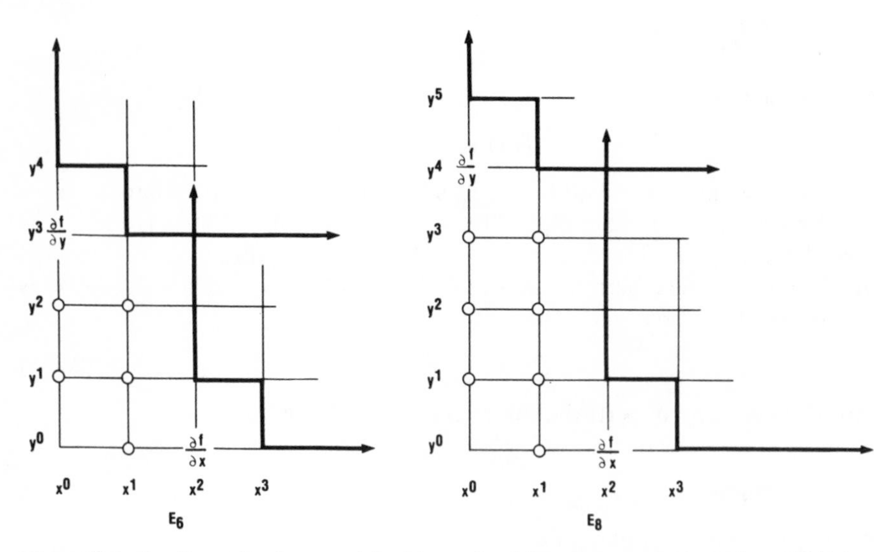

Figure 23.4 For E_6 and E_8 all monomials of degree 4 and 5 can be expressed in the form $(\partial f/\partial x_i)m_j$. The unfolding terms are represented by open circles. We exclude the constant term.

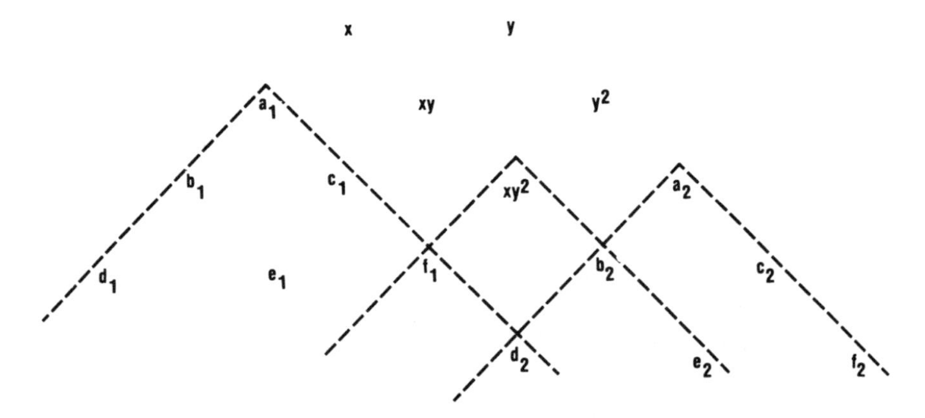

Figure 23.5 The diagrammatic technique can be used even when $\partial f/\partial x$ and $\partial f/\partial y$ are not monomials. Here it is applied to $E_7 = x^3 + xy^3$, where $\partial f/\partial y$ is a monomial but $\partial f/\partial x$ is not. Pairs of monomials arising from $(\partial f/\partial x)n_j = (x^2 + y^3)n_j$ in the unfolding algorithm are indicated by the same letters.

linear combinations of the $R_{ij}(x)$. All monomials of degree 5 can be so represented, so E_7 is 4-determinate.

The unfolding of E_7 must include the monomials x, y, xy, y^2 as well as some linear combination of a_1 and a_2 which is linearly independent of $a_1 + a_2$, and some linear combination of c_1 and c_2 which is linearly independent of $c_1 + c_2$. A (nonunique) universal unfolding of E_7 is given in Table 4.1.

12. Application 6: Multiple Cusps

The most common type of failure encountered in a single-element structural engineering problem is due to the occurrence of a cusp catastrophe. A complicated system may be composed of several structural elements, each of which can fail under some load in a catastrophe of cusp type. It is modern engineering practice (see Chapter 11) to design such a system so that once one failure mode occurs, there is not much residual strength left to prevent any of the other failure modes from occurring. In other words, it is modern engineering practice to design a system so that all failure modes occur under the same load value. If there are l failure modes, then at that load value, the related catastrophe germ is

$$f(x_1, \ldots, x_l) = \sum_{i=1}^{l} \lambda_i x_i^4, \qquad \lambda_i \neq 0 \tag{23.61}$$

It is important to know the imperfection sensitivity of multiple cusp catastrophe germs of this type. This means that it is important to know the most general perturbation of (23.61). Such a perturbation is easily constructed by following the rules for determinacy and unfolding.

The partial derivatives $\partial f/\partial x_i = x_i^3$ lead, under the determinacy algorithm, to monomials R_{ij} of the form

$$x_i^3(x_1^{p_1}x_2^{p_2}\cdots x_l^{p_l}), \qquad p_1 + p_2 + \cdots + p_l \geq 1 \tag{23.62}$$

That is, all monomials of the form

$$x_1^{q_1}x_2^{q_2}\cdots x_l^{q_l} \tag{23.63}$$

can be expressed in terms of the monomials $(\partial f/\partial x_i)m_j(x)$ provided that at least one of the q_i exceeds 2 and the sum of all q_i exceeds 3. The monomial of the form (23.63) of highest degree which cannot be expressed in terms of the $R_{ij}(x)$ has all exponents q_i equal to 2. Therefore the multiple cusp has determinacy $2l$, $l > 1$.

Remark. The double cusp is 4-determinate. This means that

$$f(x, y) = ax^4 + bx^3y + cx^2y^2 + dxy^3 + ey^4 + \text{higher degree terms} \tag{23.64}$$

is 4-determinate when $ae \neq 0$. In this case the higher degree terms can be "truncated." A linear transformation can be performed which brings $f(x, y)$ into canonical form:

$$f'(x', y') = \pm x'^4 + c'x'^2y'^2 \pm y'^4 \tag{23.65}$$

This is *not* a standard double cusp unless $c' = 0$. Thus even the simplest multiple cusp is not a simple catastrophe germ. In the case $l = 3$ the triple cusp is 6-determinate, so that all terms of degree 7 or higher can be "truncated." However, certain terms of degree 4, 5, 6 cannot be removed by any smooth transformation. Thus it is not sufficient for $f(x, y, z)$ to be a triple cusp if

$$j^4f(x, y, z) = x^4 + y^4 + z^4 \tag{23.66}$$

but it is sufficient if

$$j^6f(x, y, z) = x^4 + y^4 + z^4 \tag{23.66'}$$

The unfolding terms for the multiple cusps are just those monomials that are complementary to the S_{ij} which arise in the unfolding algorithm. These S_{ij} have the form (23.63), where now we require $q_1 + q_2 + \cdots + q_l \geq 0$. Thus the $T_j(x)$ consist of all monomials of the form (23.63) in which all exponents q_i are ≤ 2 (i.e., $q_i = 0, 1, 2$, each $i = 1, 2, \ldots, l$). There are 3^l such monomials, or $3^l - 1$ if the constant term is excluded. We conclude:

1. If, for $l > 1$,

$$j^{2l}f(x_1, \ldots, x_l) = \sum_{i=1}^{l} \lambda_i x_i^4, \qquad \lambda_i \neq 0 \tag{23.67}$$

then f is equivalent to a multiple cusp and

$$f(x_1, \ldots, x_l) \doteq \sum_{i=1}^{l} \lambda_i x_i'^4 \tag{23.67d}$$

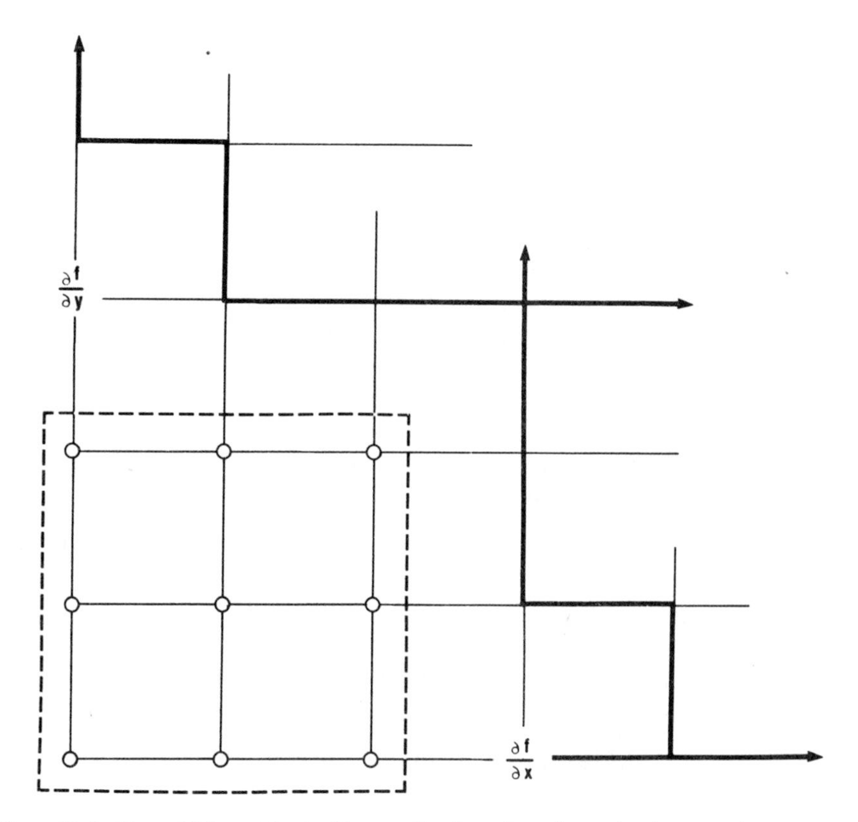

Figure 23.6 The unfolding of the double cusp $f(x, y) = x^4 + y^4$ contains the terms shown as open circles. These occur in the square bounded by the x and y axes and the horizontal and vertical lines emanating from the monomials $x^3 = \partial f/\partial x$ and $y^3 = \partial f/\partial y$.

2. The universal unfolding of f is

$$F(x; a) = \sum_{i=1}^{l} \lambda_i x_i^4 + \sum_{0 \leq q_i \leq 2} a_{1 \cdots l}^{q_1 \cdots q_l} x_1^{q_1} \cdots x_l^{q_l} \tag{23.67u}$$

The unfolding of the double cusp by the diagrammatic method is shown in Fig. 23.6. The unfolding terms are of the form $(x^0, x^1, x^2) \times (y^0, y^1, y^2)$ and are enclosed in a square by the dashed lines. In the triple cusp the unfolding terms are those that occur in the cube $(x^0, x^1, x^2) \times (y^0, y^1, y^2) \times (z^0, z^1, z^2)$. For the l-dimensional cusp the unfolding terms occur in the hypercube $(x_1^0, x_1^1, x_1^2) \times \cdots \times (x_l^0, x_l^1, x_l^2)$ of edge length 2.

13. Some Simple Germs

We close this chapter, part, and book with another view of Thom's remarkable result. We do this by answering a rather nebulously phrased question.

Question. Suppose one has a general family of potentials $V(x_1, \ldots, x_n; c_1, \ldots, c_k)$ depending on n state variables and k control parameters. Under what general conditions will only simple non-Morse germs occur, and what are they?

Solution. Consideration 1. The germ of V may depend on 1, 2,... "bad" state variables. In general l "bad" variables cannot be encountered in families depending on fewer than $k = l(l+1)/2$ control parameters. But we have seen that in a 6-parameter family, all 6 quadratic coefficients can stably vanish, and the canonical form for the cubic terms depends on a modulus. We conclude that simple germs will only be encountered in families of functions depending on fewer than 6 control parameters, in which case there can occur only one or two "bad" variables.

Consideration 2. The case in which there is only one "bad" state variable is easy to consider. If $f(x)$ is k-determinate, then

$$f(x) \doteq \pm x^k \tag{23.68d}$$

and the universal unfolding is

$$F(x; a_1, \ldots, a_{k-2}) = \pm x^k + \sum_{j=1}^{k-2} a_j x^j \rightarrow A_{k-1} \tag{23.68u}$$

Therefore the non-Morse germs that can stably occur in the case $k = 1, 2, 3, 4, 5$ are x^3, x^4, x^5, x^6, x^7.

Consideration 3. In the case in which there are two "bad" state variables x and y, we have only the cases $k = 3, 4, 5$ to consider. By the unfolding algorithm the monomials x and y will always occur among the unfolding terms. The germ that occurs depends on which additional unfolding terms are included. We consider the cases $k = 3, 4, 5$ separately, by listing the unfolding terms assumed and the canonical germ computed according to the algorithm of Sec. 6:

1. $k = 3$: x y xy — $x^3 + y^3$ D_4

   x y y^2 — $x^2y + y^3$ D_4

The former germ can be obtained from the latter, as discussed in Chapter 3, Sec. 6.

2. $k = 4$: x y y^2 y^3 — $x^2y + y^4$ D_5

The unfolding terms

$$\begin{array}{ccccc} & x & & y & \\ & & xy & & y^2 \end{array}$$

and

$$\begin{array}{ccc} x & & y \\ & xy & \\ & & xy^2 \end{array}$$

are not associated with any germ. This can be shown by proving some theorems regarding the shape of the perturbing terms in the Pascal triangle.

3. $k = 5$:

$$\begin{array}{cccccccc} & x & & y & & & & x^2y + y^5 \quad D_6 \\ x^2 & & & & y^2 & & & \\ & & & & & y^3 & & \\ & x & & y & & & & x^3 + y^4 \quad E_6 \\ & & xy & & y^2 & & & \\ & & & xy^2 & & & & \end{array}$$

This provides a list of all simple germs that occur stably and naturally in families of functions depending on $k(<6)$ control parameters, together with the universal unfoldings of these germs. Other allowed patterns of unfolding terms besides those suggested above may be used in the search for additional stable catastrophe germs, but those patterns lead to germs equivalent to those listed above.

14. Summary

A smooth nonlinear change of variables can be constructed by iterating an infinitesimal change of variables. This allows us to "linearize" the change-of-variables calculations performed with some effort in Chapters 3 and 4. This linearization technique is less useful for computational purposes, but more useful for proving theorems, than is the global transformation method described in Part 1.

With this infinitesimal approach it is a straightforward and simple process to determine which terms in a Taylor series expansion can be transformed away and which cannot. This determination, carried out for a few examples in Sec. 2, was formulated concisely into a set of rules for determinacy in Sec. 4. The terms that are in a sense complementary to those encountered in the determinacy algorithm are those that cannot be transformed away when a germ is perturbed. Thus these are precisely the terms required to provide the most general perturbation for a given germ. The rules for unfolding (a germ) are formulated concisely in Sec. 5 after a few examples have been worked out in Sec. 3.

The germ is the result of two competing processes: the control parameters can be used to kill off the early terms in the Taylor series expansion of a function, and smooth changes of variables can be used to transform away the later terms. The germ is the no-man's-land between these two processes. In fact, it falls between the two linear vector spaces constructed in the determinacy and the unfolding algorithms. More precisely, the quotient space $V_p/(V_D \oplus V_U)$ is spanned by the first partial derivatives of the canonical germ.

These three algorithms were then applied (Secs. 7 and 8) to study functions at noncritical and isolated critical points. The determinacy is 1 and 2, respectively, no unfolding terms are required, and the canonical germ may be taken as in (2.1) and (2.2).

These algorithms were next applied to the three families of simple germs A_k, D_k, E_k in Secs. 9–11 to compute their determinacy and discuss their unfoldings.

Multiple cusps, the type of catastrophe germs that are likely to become of increasing importance in the future as optimization algorithms become more important, were then discussed in Sec. 12. These provide examples of germs in which the terms "truncate beyond the degree of the germ" and "transform away the Taylor tail" are not synonymous. We also presented the general unfolding of this germ, of dimension $3^l - 1$.

We closed our work by showing in Sec. 13 a quick-and-dirty way of answering the question that Elementary Catastrophe Theory poses, and whose solution (presented in Table 2.2) required the development of the mathematical concepts treated (if somewhat less than rigorously) in Part 4 of this book.

References

1. R. Gilmore, *Lie Groups, Lie Algebras and Some of Their Applications*, New York: Wiley, 1974.
2. T. Poston and I. N. Stewart, *Taylor Series and Catastrophes*, London: Pitman, 1976.
3. T. Poston and I. N. Stewart, *Catastrophe Theory and Its Applications*, London: Pitman, 1978.
4. E. Ascher and T. Poston, Catastrophe Theory in Scientific Research, *Research Futures* **2**, 15–18 (1976), Battelle Memorial Institute, Ohio.
5. J. Mather, Stability of C^∞ Mappings I. The Division Theorem, *Ann. Math.*, **87**, 89–104 (1968).
6. J. Mather, Stability of C^∞ Mappings II. Infinitesimal Stability Implies Stability, *Ann. Math.* **89**, 254–291 (1969).
7. J. Mather, Stability of C^∞ Mappings III. Finitely Determined Map Germs, *Publ. Math. IHES* **35**, 127–156 (1968).
8. J. Mather, Stability of C^∞ Mappings IV. Classification of Stable Germs by R-Algebras, *Publ. Math. IHES* **37**, 223–248 (1969).
9. J. Mather, Stability of C^∞ Mappings V. Transversality, *Adv. Math.* **4**, 301–336 (1970).
10. J. Mather, Stability of C^∞ Mappings VI. The Nice Dimensions, in: *Proceedings of the Liverpool Singularities Symposium*, Lecture Notes in Mathematics **192** (C. T. C. Wall, Ed.), Berlin: Springer, 1971, pp. 207–253.
11. D. Siersma, Singularities of C^∞ Functions of Right-Codimension Smaller or Equal than Eight, *Indag. Math.* **25**, 31–37 (1973).

Epilogue

Thom initially introduced Catastrophe Theory in his book *Stabilité Structurelle et Morphogénèse* (1972). Catastrophe Theory was presented as a general theory of models by analogy. As such, the presentation was encrusted with a great deal of philosophical baggage. These accretions served to obscure the genealogy of Thom's Theorem, that is, as a descendent in a line of thought (qualitative or topological dynamics) originated by Poincaré. Further, his book was directed at biologists rather than mathematicians, physicists, chemists, or engineers. Thus, both in form and in substance, this work was incomprehensible and inaccessible to those who stood most directly in line to benefit from its contents.

Thom's profound insight initiated activity in a number of different areas ("bifurcation phenomenon"). Three lines of development deserve particular mention. These lines can be characterized by the names of Mather and Malgrange, Arnol'd, and Zeeman. Mather and Malgrange (and others) filled in the mathematical details require for the proofs of Thom's theorems. This school of activity laid the mathematical foundations for Catastrophe Theory and its future mathematical development.

Arnol'd first made the theory accessible to a wide audience by gathering the bits and pieces together and writing a series of review articles remarkable for their beauty, clarity, and conciseness. He then developed this subject in two directions. On the mathematical side he greatly extended the classification of singularities and the construction of their canonical forms. On the applied side he indicated how the occurrence of singularities in families of functions could be important in physics and the engineering disciplines.

Zeeman became the great popularizer of, and spokesman for, Catastrophe Theory. He saw clearly how Thom's theory of models could become a great asset and a useful working tool in those disciplines without a highly developed mathematical infrastructure. His early espousals of this point of view met with skepticism. Challenged, he produced a wide and fascinating spectrum of possible applications of Catastrophe Theory in the social and biological sciences.

For whatever reason, a reaction set in. This reaction, led by Sussman and Zahler, was directed, not at the insights of Thom, the developmental results of

Mather and Malgrange, or the classification and application results of Arnol'd, but only at the exploratory work of Zeeman. This reaction, though unfortunate, was no more than a tempest in a teapot. It was convincingly terminated in the pages of *Nature* (**270**, 381–384, Dec 1, 1977). The main result of this reaction was to delay by a few years the widespread acceptance of Catastrophe Theory as a useful working tool by scientists and engineers.

From time to time other arguments against the usefulness of Catastrophe Theory surface. These are of two general types.

1. Catastrophe Theory is topological, thus qualitative and therefore not quantitative, so not useful.
2. The elementary catastrophes of Thom belong to the study of bifurcations of the equilibria of gradient dynamical systems. Therefore they are inadequate for the description of realistic systems, which do depend on time.

Of these, the second is the more serious objection. In fact, I feel the first is a vapid objection. I present it here only because it is taken seriously in some quarters. The original intent of Poincaré's program was to develop a qualitative theory of dynamical systems. That the general theory is qualitative (or topological) does not at all prevent quantitative predictions for particular models that fall within the purview of the theory. Poincaré was not censured for developing a theory that was qualitative; neither were Andronov and Khaikin (*Theory of Oscillations*). To do so would be to miss entirely the point of the program, particularly when these theories could yield quantitative results when pressed. The same arguments apply to the elementary catastrophes. The cusp catastrophe manifold may be described qualitatively, but I certainly hope that the computer graphics in this book have left an indelible impression of their quantitative nature. The fitting of experimental data with curves derived from the elementary catastrophes (e.g., Fig. 15.8) should vouch also for the quantitative nature of the catastrophes.

The second objection, about the utility of the elementary catastrophes being circumscribed by their static nature, is somewhat more serious. While the observation is correct, the conclusion is sterile. The usefulness of Catastrophe Theory lies not so much in the canonical forms it provides as in the methods it employs. For nonmathematicians seeking to use Catastrophe Theory in a constructive way, I feel that there are six distinct levels on which Catastrophe Theory can be used:

1. *Organizational.* Known effects are put into a coherent framework, and seen to be related to each other. The canonical properties of catastrophes can then suggest new lines of development.
2. *Potential is known but complicated.* A gradient system may be described by a potential that is known but very complicated. The local structure of this

potential at a degenerate critical point can be uniquely determined by elementary considerations.

3. *Potential is unknown.* A gradient system may be described by a potential that is not known. Knowledge of modality, location of phase transitions, and susceptibility tensors may be adequate to determine degenerate critical points. The qualitative properties of the system follow from the type of degenerate critical point encountered.

4. *System is not a gradient system.* A dynamical system not derivable from the gradient of a potential may nevertheless have equilibria or steady states closely related to elementary catastrophes. This can occur if the nonlinear equations describing equilibria (equations of state) involve factors which are gradients of elementary catastrophes.

5. *Unknown equations govern a system.* If space variation or time evolution is slow, the existence of multiple locally stable modes may suggest that the system can be qualitatively or even semiquantitatively understood in terms of one of the catastrophe functions. Such understanding would involve non-negligible predictive ability, including the possibility of making educated guesses at the equations governing the system.

6. *Relative importance of variables is unknown.* In this case not only are the equations governing a system not known, but the variables which are important for describing the system may not be known. If limited modality exists, only a small number (1, 2, or 3) of functions of the state variables describing the system actually *drive* the system. These functions may be inferred in their linearized form at a degenerate critical point. When this is done, the qualitative behavior of the system in the neighborhood of such points is reasonably well determined.

The chapters in Part 2 of this book illustrate the applications of Catastrophe Theory at one or more of the levels of utility described above: thermodynamics (levels 1 and 5); elastic dynamics (1–3); aerodynamics (4); wave mechanics (1 and 4); quantum mechanics (1 and 4); and climate dynamics (4–6).

Of the levels of utility described above, level 6 seems closest in spirit to the observation of G. W. Hill and the program of Poincaré. Topological dynamics, or its scion Catastrophe Theory, seems an eminently suitable vehicle for attacking problems of such complexity that it is difficult even to known where to begin.

So where does that leave us? On the mathematics side, Catastrophe Theory has stimulated the development of several new areas of research. These will probably continue to flourish for a number of years. On the applications side the initial beneficiaries will undoubtedly be the fields of physics and mechanical engineering. The methods, or at least the results, of Catastrophe Theory will probably be extensively taught within five years. Extension into chemistry and other engineering disciplines may take somewhat longer. On a longer time-scale yet will be its entry into the biological and social sciences (about 15–25 years). In compensation for this delay will be the more important contributions it will provide. Physicists could live without Catastrophe Theory if necessary. The equations are there, their properties will eventually be discovered. This is not true for sociologists. Any method that is useful for suggesting the appropriate variable to study or even hinting at a useful set of equations must eventually have its important place in that field.

Is Catastrophe Theory useful and important? I am convinced that it is. The present work grew out of this conviction. Let those who will, turn their back on it and those who can, use it as an effective tool to further their own insights. I leave it to the reader to judge this question for himself or herself.

Author Index

Subject Index

References to definitions in **boldface.**

A

B

C

D

E

L

M

T

U

V

A CATALOG OF SELECTED

DOVER BOOKS

IN SCIENCE AND MATHEMATICS

RELATIVITY, THERMODYNAMICS AND COSMOLOGY, Richard C. Tolman. Landmark study extends thermodynamics to special, general relativity; also applications of relativistic mechanics, thermodynamics to cosmological models. 501pp. 5⅜ × 8½. 65383-8 Pa. $12.95

APPLIED ANALYSIS, Cornelius Lanczos. Classic work on analysis and design of finite processes for approximating solution of analytical problems. Algebraic equations, matrices, harmonic analysis, quadrature methods, much more. 559pp. 5⅜ × 8½. 65656-X Pa. $12.95

SPECIAL RELATIVITY FOR PHYSICISTS, G. Stephenson and C.W. Kilmister. Concise elegant account for nonspecialists. Lorentz transformation, optical and dynamical applications, more. Bibliography. 108pp. 5⅜ × 8½. 65519-9 Pa. $4.95

INTRODUCTION TO ANALYSIS, Maxwell Rosenlicht. Unusually clear, accessible coverage of set theory, real number system, metric spaces, continuous functions, Riemann integration, multiple integrals, more. Wide range of problems. Undergraduate level. Bibliography. 254pp. 5⅜ × 8½. 65038-3 Pa. $7.95

INTRODUCTION TO QUANTUM MECHANICS With Applications to Chemistry, Linus Pauling & E. Bright Wilson, Jr. Classic undergraduate text by Nobel Prize winner applies quantum mechanics to chemical and physical problems. Numerous tables and figures enhance the text. Chapter bibliographies. Appendices. Index. 468pp. 5⅜ × 8½. 64871-0 Pa. $11.95

ASYMPTOTIC EXPANSIONS OF INTEGRALS, Norman Bleistein & Richard A. Handelsman. Best introduction to important field with applications in a variety of scientific disciplines. New preface. Problems. Diagrams. Tables. Bibliography. Index. 448pp. 5⅜ × 8½. 65082-0 Pa. $11.95

MATHEMATICS APPLIED TO CONTINUUM MECHANICS, Lee A. Segel. Analyzes models of fluid flow and solid deformation. For upper-level math, science and engineering students. 608pp. 5⅜ × 8½. 65369-2 Pa. $13.95

ELEMENTS OF REAL ANALYSIS, David A. Sprecher. Classic text covers fundamental concepts, real number system, point sets, functions of a real variable, Fourier series, much more. Over 500 exercises. 352pp. 5⅜ × 8½. 65385-4 Pa. $9.95

PHYSICAL PRINCIPLES OF THE QUANTUM THEORY, Werner Heisenberg. Nobel Laureate discusses quantum theory, uncertainty, wave mechanics, work of Dirac, Schroedinger, Compton, Wilson, Einstein, etc. 184pp. 5⅜ × 8½. 60113-7 Pa. $4.95

INTRODUCTORY REAL ANALYSIS, A.N. Kolmogorov, S.V. Fomin. Translated by Richard A. Silverman. Self-contained, evenly paced introduction to real and functional analysis. Some 350 problems. 403pp. 5⅜ × 8½. 61226-0 Pa. $9.95

PROBLEMS AND SOLUTIONS IN QUANTUM CHEMISTRY AND PHYSICS, Charles S. Johnson, Jr. and Lee G. Pedersen. Unusually varied problems, detailed solutions in coverage of quantum mechanics, wave mechanics, angular momentum, molecular spectroscopy, scattering theory, more. 280 problems plus 139 supplementary exercises. 430pp. 6½ × 9¼. 65236-X Pa. $11.95

CATALOG OF DOVER BOOKS

NUMERICAL METHODS FOR SCIENTISTS AND ENGINEERS, Richard Hamming. Classic text stresses frequency approach in coverage of algorithms, polynomial approximation, Fourier approximation, exponential approximation, other topics. Revised and enlarged 2nd edition. 721pp. 5⅜ × 8½.
65241-6 Pa. $14.95

THEORETICAL SOLID STATE PHYSICS, Vol. I: Perfect Lattices in Equilibrium; Vol. II: Non-Equilibrium and Disorder, William Jones and Norman H. March. Monumental reference work covers fundamental theory of equilibrium properties of perfect crystalline solids, non-equilibrium properties, defects and disordered systems. Appendices. Problems. Preface. Diagrams. Index. Bibliography. Total of 1,301pp. 5⅜ × 8½. Two volumes. Vol. I 65015-4 Pa. $12.95
Vol. II 65016-2 Pa. $12.95

OPTIMIZATION THEORY WITH APPLICATIONS, Donald A. Pierre. Broad-spectrum approach to important topic. Classical theory of minima and maxima, calculus of variations, simplex technique and linear programming, more. Many problems, examples. 640pp. 5⅜ × 8½. 65205-X Pa. $13.95

THE MODERN THEORY OF SOLIDS, Frederick Seitz. First inexpensive edition of classic work on theory of ionic crystals, free-electron theory of metals and semiconductors, molecular binding, much more. 736pp. 5⅜ × 8½.
65482-6 Pa. $15.95

ESSAYS ON THE THEORY OF NUMBERS, Richard Dedekind. Two classic essays by great German mathematician: on the theory of irrational numbers; and on transfinite numbers and properties of natural numbers. 115pp. 5⅜ × 8½.
21010-3 Pa. $4.95

THE FUNCTIONS OF MATHEMATICAL PHYSICS, Harry Hochstadt. Comprehensive treatment of orthogonal polynomials, hypergeometric functions, Hill's equation, much more. Bibliography. Index. 322pp. 5⅜ × 8½. 65214-9 Pa. $9.95

NUMBER THEORY AND ITS HISTORY, Oystein Ore. Unusually clear, accessible introduction covers counting, properties of numbers, prime numbers, much more. Bibliography. 380pp. 5⅜ × 8½. 65620-9 Pa. $8.95

THE VARIATIONAL PRINCIPLES OF MECHANICS, Cornelius Lanczos. Graduate level coverage of calculus of variations, equations of motion, relativistic mechanics, more. First inexpensive paperbound edition of classic treatise. Index. Bibliography. 418pp. 5⅜ × 8½. 65067-7 Pa. $10.95

MATHEMATICAL TABLES AND FORMULAS, Robert D. Carmichael and Edwin R. Smith. Logarithms, sines, tangents, trig functions, powers, roots, reciprocals, exponential and hyperbolic functions, formulas and theorems. 269pp. 5⅜ × 8½. 60111-0 Pa. $5.95

THEORETICAL PHYSICS, Georg Joos, with Ira M. Freeman. Classic overview covers essential math, mechanics, electromagnetic theory, thermodynamics, quantum mechanics, nuclear physics, other topics. First paperback edition. xxiii + 885pp. 5⅜ × 8½. 65227-0 Pa. $18.95

SPECIAL FUNCTIONS, N.N. Lebedev. Translated by Richard Silverman. Famous Russian work treating more important special functions, with applications to specific problems of physics and engineering. 38 figures. 308pp. 5⅜ × 8½.
60624-4 Pa. $7.95

OBSERVATIONAL ASTRONOMY FOR AMATEURS, J.B. Sidgwick. Mine of useful data for observation of sun, moon, planets, asteroids, aurorae, meteors, comets, variables, binaries, etc. 39 illustrations. 384pp. 5⅜ × 8¼. (Available in U.S. only)
24033-9 Pa. $8.95

INTEGRAL EQUATIONS, F.G. Tricomi. Authoritative, well-written treatment of extremely useful mathematical tool with wide applications. Volterra Equations, Fredholm Equations, much more. Advanced undergraduate to graduate level. Exercises. Bibliography. 238pp. 5⅜ × 8½. 64828-1 Pa. $6.95

CELESTIAL OBJECTS FOR COMMON TELESCOPES, T.W. Webb. Inestimable aid for locating and identifying nearly 4,000 celestial objects. 77 illustrations. 645pp. 5⅜ × 8½. 20917-2, 20918-0 Pa., Two-vol. set $12.00

MODERN NONLINEAR EQUATIONS, Thomas L. Saaty. Emphasizes practical solution of problems; covers seven types of equations. ". . . a welcome contribution to the existing literature. . . ."—*Math Reviews*. 490pp. 5⅜ × 8½. 64232-1 Pa. $9.95

FUNDAMENTALS OF ASTRODYNAMICS, Roger Bate et al. Modern approach developed by U.S. Air Force Academy. Designed as a first course. Problems, exercises. Numerous illustrations. 455pp. 5⅜ × 8½. 60061-0 Pa. $8.95

INTRODUCTION TO LINEAR ALGEBRA AND DIFFERENTIAL EQUATIONS, John W. Dettman. Excellent text covers complex numbers, determinants, orthonormal bases, Laplace transforms, much more. Exercises with solutions. Undergraduate level. 416pp. 5⅜ × 8½. 65191-6 Pa. $9.95

INCOMPRESSIBLE AERODYNAMICS, edited by Bryan Thwaites. Covers theoretical and experimental treatment of the uniform flow of air and viscous fluids past two-dimensional aerofoils and three-dimensional wings; many other topics. 654pp. 5⅜ × 8½.
65465-6 Pa. $16.95

INTRODUCTION TO DIFFERENCE EQUATIONS, Samuel Goldberg. Exceptionally clear exposition of important discipline with applications to sociology, psychology, economics. Many illustrative examples; over 250 problems. 260pp. 5⅜ × 8½.
65084-7 Pa. $7.95

LAMINAR BOUNDARY LAYERS, edited by L. Rosenhead. Engineering classic covers steady boundary layers in two- and three-dimensional flow, unsteady boundary layers, stability, observational techniques, much more. 708pp. 5⅜ × 8½.
65646-2 Pa. $15.95

LECTURES ON CLASSICAL DIFFERENTIAL GEOMETRY, Second Edition, Dirk J. Struik. Excellent brief introduction covers curves, theory of surfaces, fundamental equations, geometry on a surface, conformal mapping, other topics. Problems. 240pp. 5⅜ × 8½. 65609-8 Pa. $6.95

CATALOG OF DOVER BOOKS